Lecture Notes in Networks and Systems

Volume 145

The series "Lecture Notes in Networks and Systems" publishes the latest developments in Networks and Systems—quickly, informally and with high quality. Original research reported in proceedings and post-proceedings represents the core of LNNS.

Volumes published in LNNS embrace all aspects and subfields of, as well as new challenges in, Networks and Systems.

The series contains proceedings and edited volumes in systems and networks, spanning the areas of Cyber-Physical Systems, Autonomous Systems, Sensor Networks, Control Systems, Energy Systems, Automotive Systems, Biological Systems, Vehicular Networking and Connected Vehicles, Aerospace Systems, Automation, Manufacturing, Smart Grids, Nonlinear Systems, Power Systems, Robotics, Social Systems, Economic Systems and other. Of particular value to both the contributors and the readership are the short publication timeframe and the world-wide distribution and exposure which enable both a wide and rapid dissemination of research output.

The series covers the theory, applications, and perspectives on the state of the art and future developments relevant to systems and networks, decision making, control, complex processes and related areas, as embedded in the fields of interdisciplinary and applied sciences, engineering, computer science, physics, economics, social, and life sciences, as well as the paradigms and methodologies behind them.

**** Indexing: The books of this series are submitted to ISI Proceedings, SCOPUS, Google Scholar and Springerlink ****

More information about this series at http://www.springer.com/series/15179

G. Ranganathan · Joy Chen · Álvaro Rocha
Editors

Inventive Communication and Computational Technologies

Proceedings of ICICCT 2020

Springer

Editors
G. Ranganathan
Department of Electronics
and Communication Engineering
Gnanamani College of Technology
Namakkal, Tamil Nadu, India

Joy Chen
Department of Electrical Engineering
Dayeh University
Changhua, Taiwan

Álvaro Rocha
Department of Informatics Engineering
University of Coimbra
Coimbra, Portugal

ISSN 2367-3370 ISSN 2367-3389 (electronic)
Lecture Notes in Networks and Systems
ISBN 978-981-15-7344-6 ISBN 978-981-15-7345-3 (eBook)
https://doi.org/10.1007/978-981-15-7345-3

This Springer imprint is published by the registered company Springer Nature Singapore Pte Ltd.
The registered company address is: 152 Beach Road, #21-01/04 Gateway East, Singapore 189721, Singapore

We dedicate this book to all the participants of the conference, ICICCT 2020

Preface

2020 International Conference on Inventive Communication and Computational Technologies (ICICCT 2020) was held on Gnanamani College of Technology, Namakkal, India, during May 28 and 29, 2020. ICICCT 2020 aims to cover the recent advancement and trends in the area of communication and computational technologies to facilitate knowledge sharing and networking interactions on emerging trends and new challenges.

ICICCT 2020 tends to collect the latest research results and applications on data communication and computer networking, software engineering, wireless communication, VLSI design and automation, networking, Internet of things, cloud and big data. It includes a selection of 84 papers from 264 papers submitted to the conference from universities and industries all over the world. All of the accepted papers were subjected to strict peer-reviewing by two–four expert referees. The papers have been selected for this volume because of quality and relevance to the conference.

We would like to thank the guest editors Dr. Joy Chen, Professor, Department of Electrical, Engineering, Dayeh University, Taiwan, and Dr. Álvaro Rocha, Professor, Department of Informatics Engineering, University of Coimbra, Portugal, for their valuable guidance and technical support for the article selection process.

ICICCT 2020 would like to express our sincere appreciation to all authors for their contributions to this book. We would like to extend our thanks to all the referees for their constructive comments on all papers; especially, we would like to thank the organizing committee for their hard work. Finally, we would like to thank the Springer publications for producing this volume.

Namakkal, India

Dr. G. Ranganathan
Conference Chair
ICICCT 2020

About the Conference

We welcome you to the 2020 International Conference on Inventive Communication and Computational Technologies (ICICCT 2020) was held on May 28 and 29, 2020, at Gnanamani College of Technology, Namakkal, Tamil Nadu. ICICCT 2020 provides a highly competitive forum for reporting the latest developments in the research and application of communication and computational technologies. We are pleased to present the proceedings of the conference as its published record.

The conference particularly encouraged the interaction of research students and developing academics with the more established academic community in an informal setting to present and to discuss new and current work. Their contributions helped to make the conference as outstanding as it has been. The papers contributed the most recent scientific knowledge known in the field of modern communication systems which includes informatics, data communication and computer networking, wireless communication, electronics, software engineering, machine learning and optimization, VLSI design and automation, networking, computing systems, social networks, Internet of things, cloud and big data, etc.

We hope that this program will further stimulate research in advanced electronics and communication technologies, artificial intelligence and capsule networks, data communication and computer networking and communicating things networks. We feel honored and privileged to serve the best recent developments in all the areas of communication technologies to you through this exciting program.

We thank all authors and participants for their contributions.

Dr. G. Ranganathan
Conference Chair
ICICCT 2020

Contents

Editors and Contributors

About the Editors

Dr. G. Ranganathan, Principal, Ranganathan Engineering College, Coimbatore, India. He has done his PhD in the Faculty of Information and Communication Engineering from Anna University, Chennai in the year 2013. His research thesis was in the area of Bio Medical Signal Processing. He has total of 29+ years of experience both in industry, teaching and research. He has guided several project works for many UG and PG Students in the areas of Bio Medical Signal Processing. He has published more than 35 research papers in International and National Journals and Conferences. He has also co-authored many books in electrical and electronics subjects. He has served as Referee for many reputed International Journals published by Elsevier, Springer, Taylor and Francis, etc. He has membership in various professional bodies like ISTE, IAENG etc., and has actively involved himself in organizing various international and national level conferences, symposiums, seminars etc.

Dr. Joy Chen is currently a full professor of Department of Electrical Engineering Dayeh University at Changhua Taiwan. Prior to joining the Dayeh University, he worked at the Control Data Company (Taiwan) as a technical manager since Sep. 1985 to Sep. 1996. His research interests include wireless communications, spread spectrum technical, OFDM systems, and wireless sensor networks. He has published a large number of SCI Journal papers in the issues addressed physical layer for wireless communication systems. Moreover, he also majors in developing some applications of the IOT (Internet of Thing) techniques and Dr. Joy I.-Z. Chen owned some patents authorized by the Taiwan Intellectual Property Office (TIPO).

Álvaro Rocha holds the title of Honorary Professor (2019), Information Science Aggregation (2011), PhD in Information Technology and Systems (2001), Master in Management Informatics (1995) and Degree in Applied Mathematics (1990). He is currently Professor at the University of Coimbra, Researcher at CISUC - Center for Informatics and Systems at the University of Coimbra, Collaborating Researcher at LIACC - Laboratory for Artificial Intelligence and Computer Science, Collaborative Researcher at CINTESIS - Research Center for Information Technology and Systems. He is also President of AISTI - Iberian Association of Information Systems and Technologies, President of the Portuguese Chapter of the IEEE SMC Society - Systems, Man, and Cybernetics, Editor of RISTI - Iberian Journal of Information Systems and Technologies, and Editor of the Journal of Information Systems Engineering & Management. He has also served as Vice President of Experts at Horizon 2020 of the European Commission, as an Expert in the Ministry of Education, Universities and Research of the Italian Government, and as an Expert in the Ministry of Finance of the Latvian Government.

Contributors

B. Aakash Sastra Deemed to be University, Thanjavur, India

N. V. Abiramy Computer Science and Engineering, Dr. N.G.P Institute of Technology, Coimbatore, India

T. S. Abishek Department of CSE, SSN College of Engineering, Chennai, India

Jayashree Agarkhed Computer Science and Engineering Department, Poojya Doddappa Appa College of Engineering, Kalaburagi, Karnataka, India

Harsh Agarwal Department of Computer Engineering and Applications, GLA University, Mathura, India

Osho Agyeya School of Computing Science and Engineering, VIT, Chennai, India

K. Aishwarya Lakshmi Department of Computer Science, PES University, Bangalore, India

A. Sherly Alphonse Department of Information Technology, Ponjesly College of Engineering, Nagercoil, India

Aishwarya Alva Department of Information Science, Dayananda Sagar Academy of Technology and Management, Bangalore, India

Ganesh Anam Department of Electronics and Communication Engineering, Amrita School of Engineering, Coimbatore, Amrita Vishwa Vidyapeetham, Coimbatore, India

N. Ani Brown Mary Sarah Tucker College, Tirunelveli, Tamil Nadu, India

Ani Harish Rajiv Gandhi Institute of Technology, Kottayam, Kerala, India

A. Anitha Ph.D Research Scholar, Department of Computer Science, Kongunadu Arts and Science College, Coimbatore, Tamil Nadu, India

Nazneen Ansari St. Francis Institute of Technology, Mumbai, India

K. S. Anusha Department of Electronics and Communication Engineering, Amrita School of Engineering, Amrita Vishwa Vidyapeetham, Coimbatore, India

Tanvir Anzum Computer Science and Engineering Department, Daffodil International University, Dhaka, Bangladesh

C. Aravind Department of Computer Science and Engineering, Dr.N.G.P. Institute of Technology (Dr.N.G.P. IT), Coimbatore, India

M. Aravinda Kumar Department of Electronics and Communication Engineering, Amrita School of Engineering, Amrita Vishwa Vidyapeetham, Coimbatore, India

Nikhil Arora Delhi Technological University, Delhi, India

P. S. G. Aruna Sri Department of Electronics and Computer Engineering, KLEF, Vaddeswaram, Guntur, India

Gokulalakshmi Arunachalam Department of Bio Medical Engineering, SNS College of Technology, Coimbatore, India

B. N. Arunakumari BMS Institute of Technology and Management, Bengaluru, India

R. Ashwin Kumar Department of Electronics and Communication, SASTRA Deemed University, Thanjavur, Tamil Nadu, India

Rachna Asthana Dr. Ambedkar Institute of Technology for Handicapped, Kanpur, India

Suganya Athisayamani School of Computing, Sastra Deemed to be University, Thanjavur, India

T. Avinash Department of Electronics and Communication, SASTRA Deemed University, Thanjavur, Tamil Nadu, India

S. Ayyasamy Department of Computer Science and Engineering, Dr.N.G.P. Institute of Technology (Dr.N.G.P. IT), Coimbatore, India

V. R. Balaji Department of Electronics and Communication Engineering, Sri Krishna College of Engineering and Technology, Coimbatore, Tamil Nadu, India

D. Balasubramaniam Vel Tech Rangarajan Dr. Sagunthala R&D Institute of Science and Technology, Chennai, India

Malti Bansal Department of Electronics and Communication Engineering, Delhi Technological University (DTU), Delhi, India

Prateek Bansal Delhi Technological University, Delhi, India

Namratha Bhat St. Francis Institute of Technology, Mumbai, India

N. P. G. Bhavani Department of EEE, Meenakshi College of Engineering, Chennai, India

A. A. Bini Computer Science and Engineering, Indian Institute of Information Technology, Kottayam, Kerala, India

Volodymyr Busygin Oles Honchar Dnipro National University, Dnipro, Ukraine

Su-Qun Cao Faculty of Electronic Information Engineering, Huaiyin Institute of Technology, Huaian City, Jiangsu Prov., China

S. Chandrasekaran Department of Computer Science and Engineering, Department of Tamil, Bharathiyar University, Chennai, Tamil Nadu, India

Saurav Singh Chauhan Department of CSE, Delhi Technological University, Delhi, India

Amit Chaurasia Department of Computer Engineering and Applications, GLA University, Mathura, India

Kalagotla Chenchireddy Department of EEE, Karunya Institute of Technology and Sciences, Coimbatore, Tamil Nadu, India

Nithya Chidambaram Department of Electronics and Communication, SASTRA Deemed University, Thanjavur, Tamil Nadu, India

P. Chinnasamy Assistant Professor, Department of Information Technology, Sri Shakthi Institute of Engineering and Technology, Coimbatore, India

Deepika Dasari Department of Electronics and Computer Engineering, KLEF, Vaddeswaram, Guntur, India

Saurabh Daware Department of Information Technology, Terna Engineering College Nerul, Navi Mumbai, India

Deepali Dhaka Department of Computer Science, Jamia Millia Islamia, New Delhi, India

D. Dhanush Department of Electronics and Communication Engineering, Amrita School of Engineering, Amrita Vishwa Vidyapeetham, Coimbatore, India

N. M. Dhanya Department of Computer Science and Engineering, Amrita School of Engineering, Coimbatore, Amrita Vishwa Vidyapeetham, India

A. Dinesh Kumar Computer Science and Engineering, KL Deemed to be University, Vaddeswaram, Andhra Pradesh, India

J. R. Dinesh Kumar Department of Electronics and Communication Engineering, Sri Krishna College of Engineering and Technology, Coimbatore, Tamil Nadu, India

M. N. Doja Jamia Millia Islamia, New Delhi, India

Vj Duy Du Delhi Technological University, Delhi, India

Ashish Dwivedi Department of Computer Engineering and Applications, GLA University, Mathura, India

Endalkachew Emare College of Informatics, Bule Hora University, Bule Hora, Ethiopia

Shaista Farheen Dayananda Sagar College of Engineering, Bengaluru, Karnataka, India

V. S. Felix Enigo Department of CSE, SSN College of Engineering, Chennai, India

Dhanashree Gaikwad School of Electrical Engineering, MIT Academy of Engineering Alandi, Pune, India

A. Ganesan Department of EEE, RRASE Engineering College, Chennai, India

A. Ganesh Department of Electronics and Communication Engineering, Amrita Vishwa Vidyapeetham, Kollam, India

Vaishnavi Ganesh Priyadarshini Indira Gandhi College of Engineering, Nagpur, India

C. Ganesh Babu Department of Electronics and Instrumentation Engineering, Bannari Amman Institute of Technology, Sathyamangalam, Tamil Nadu, India

Durgaprasad Gangodkar Graphic Era Deemed to be University, Dehradun, India

Veeranna Gatate Computer Science and Engineering Department, Poojya Doddappa Appa College of Engineering, Kalaburagi, Karnataka, India

S. Gayathri Department of Electronics and Communication Engineering, Amrita Vishwa Vidyapeetham, Kollam, India

K. Geetha SASTRA Deemed to be University, Thanjavur, India

P. Geetha Department of Computer Science and Engineering, Dhanalakshmi College of Engineering, Chennai, Tamil Nadu, India

V. Geetha Lekshmy Department of Computer Science and Applications, Amrita Vishwa Vidyapeetham, Amritapuri, India

Priyam Goel Delhi Technological University, Delhi, India

J. Gokul Prasath Department of Electronics and Communication Engineering, Amrita School of Engineering, Amrita Vishwa Vidyapeetham, Coimbatore, India

Shreya S. Gowda Masters in Networking and Internet Engineering, Electronics and Communication Engineering Department, JSS Science and Technology University (SJCE), Mysuru, India

Paras Gupta Department of Electronics and Communication Engineering, Amrita School of Engineering, Amrita Vishwa Vidyapeetham, Coimbatore, India

Saloni Gupta Delhi Technological University, Delhi, India

Vidhi Gupta Harcourt Butler Technical University, Kanpur, India

Vishal Gupta School of Electrical Engineering, MIT Academy of Engineering Alandi, Pune, India

Deivanai Gurusamy College of Informatics, Bule Hora University, Bule Hora, Ethiopia

Maddukuri Harika Department of Electronics and Computer Engineering, Koneru Lakshmaiah Education Foundation, Vaddesswaram, AP, India

S. Harini School of Computer Science and Engineering, VIT, Chennai, India

Nazmul Hasan Computer Science and Engineering Department, Daffodil International University, Dhaka, Bangladesh

Prasad B. Honnavali Department of Computer Science, PES University, Bangalore, India

Akshay Ingle School of Electrical Engineering, MIT Academy of Engineering Alandi, Pune, India

Nilesh Kumar Jadav Department of Computer Engineering, Marwadi University, Rajkot, Gujarat, India

C. S. Jagan Sai Kumar Department of Electronics and Communication Engineering, Amrita School of Engineering, Amrita Vishwa Vidyapeetham, Coimbatore, India

Nusrat Jahan Computer Science and Engineering Department, Daffodil International University, Dhaka, Bangladesh

Anmol Jain Department of Information Technology, Delhi Technological University (DTU), Delhi, India

Vinod Jain Department of Computer Engineering and Applications, GLA University Mathura, Mathura, India

D. A. Janeera Department of Electronics and Communication Engineering, Sri Krishna College of Engineering and Technology, Kuniyamuthur, India

M. V. Jayan Government Engineering College, Thrissur, Kerala, India

S. Jayanthy Department of Electronics and Communication Engineering, Sri Ramakrishna Engineering College, Coimbatore, India

V. Jegathesan Department of EEE, Karunya Institute of Technology and Sciences, Coimbatore, Tamil Nadu, India

M. J. Jeyasheela Rakkini SASTRA Deemed to be University, Thanjavur, India

Daniel Felips Jhosiah Department of Computer Science and Engineering, Dr.N.G.P. Institute of Technology (Dr.N.G.P. IT), Coimbatore, India

D. John Aravindhar Hindustan Institute of Technology and Science, Chennai, India

J. Judeson Antony Kovilpillai Department of Electronics and Communication Engineering, Sri Ramakrishna Engineering College, Coimbatore, India

Surbhi Kakar Department of Computer Science, Jamia Millia Islamia, New Delhi, India

N. S. Kalyan Chakravarthy QIS College of Engineering and Technology, Ongole, Andhra Pradesh, India

Venkatesh Kamath Department of Information Technology, Terna Engineering College Nerul, Navi Mumbai, India

M. Kameswara Rao Department of Electronics and Computer Engineering, KLEF, Vaddeswaram, Guntur, India

Jinesh M. Kannimoola Department of Computer Science and Applications, Amrita Vishwa Vidyapeetham, Amritapuri, India

S. P. Karthi Department of Electronics and Communication Engineering, Sri Krishna College of Engineering and Technology, Coimbatore, Tamil Nadu, India

P. V. S. M. S. Kartik Department of Computer Science and Engineering, Amrita School of Engineering, Coimbatore, Amrita Vishwa Vidyapeetham, Coimbatore, India

Navjeet Kaur Chitkara University Institute of Engineering and Technology, Chitkara University, Patiala, Punjab, India

Addepalli Kavya Department of Electronics and Computer Engineering, Koneru Lakshmaiah Education Foundation, Vaddesswaram, AP, India

Tucha Kedir College of Informatics, Bule Hora University, Bule Hora, Ethiopia

Vaishali Khairnar Department of Information Technology, Terna Engineering College Nerul, Navi Mumbai, India

Roshan Bapurao Kharke Department of CSE, Delhi Technological University, Delhi, India

Uday Pandit Khot Department of Electronics and Telecommunication, St. Francis Institute of Technology, Borivali, Mumbai, India

Maksym Khylko National Academy of Sciences of Ukraine, Kiev, Ukraine

Olena Khylko Kiev National Taras Shevchenko University, Kiev, Ukraine

S. Koushik Department of Electronics and Communication Engineering, Amrita Vishwa Vidyapeetham, Kollam, India

Shridevi S. Krishnakumar School of Computing Science and Engineering, VIT, Chennai, India

Aswathi Krishnan Department of Electronics and Communication Engineering, Amrita Vishwa Vidyapeetham, Kollam, India

Aishwary Kumar Department of Information Technology, Delhi Technological University (DTU), Delhi, India

Anuj Kumar Department of Computer Engineering and Applications, GLA University Mathura, Mathura, India

Ashok Kumar Chitkara University Institute of Engineering and Technology, Chitkara University, Patiala, Punjab, India

Nihal Kumar Delhi Technological University, Delhi, India

Rahul Vinod Kumar Department of Computer Science and Engineering, Amrita School of Engineering, Amrita Vishwa Vidyapeetham, Coimbatore, India

Rajesh Kumar Department of Computer Science and Engineering, Thapar Institute of Engineering and Technology, Patiala, India

Sumit Kumar Electronics & Telecommunication Department, Symbiosis Institute of Technology, Symbiosis International (Deemed University), Pune, India

N. Lakshitha Karthik Masters in Information Technology, Information Science and Engineering Department, Rashtreeya Vidyalaya College of Engineering (RVCE), Bengaluru, India

Anjaneyulu Lokam Electronics and Communications Engineering, National Institute of Technology, Warangal, India

Omkar Lubal Department of Information Technology, Terna Engineering College Nerul, Navi Mumbai, India

Rajeshwari Malekar Electronics & Telecommunication Department, Symbiosis Institute of Technology, Symbiosis International (Deemed University), Pune, India

Sandhya Mallavarapu Electronics and Communications Engineering, National Institute of Technology, Warangal, India

Yogita Deepak Mane Department of Computer Engineering, St. Francis Institute of Technology, Borivali, Mumbai, India

Veenu Mangat UIET, Panjab University, Chandigarh, India

K. Manjunathachari GITAM deemed to be University, Hyderabad, Telangana, India

Siddhi Martal St. Francis Institute of Technology, Mumbai, India

Priyanka Meel Delhi Technological University, Delhi, India

Monica Mehrotra Department of Computer Science, Jamia Millia Islamia, New Delhi, India

B. M. Mehtre Centre of Excellence in Cyber Security, Institute for Development & Research in Banking Technology (IDRBT), Hyderabad, India

Varsha Mittal Graphic Era Deemed to be University, Dehradun, India

Navya Mohan Department of Electronics and Communication Engineering, Amrita School of Engineering, Amrita Vishwa Vidyapeetham, Coimbatore, India

Monika UIET, Panjab University, Chandigarh, India;
Department of CS, Shaheed Rajguru College of Applied Sciences for Women University of Delhi, Delhi, India

Boris Moroz University of Technology, Dnipro, Ukraine

S. Mythili Associate Professor and Head, Department of Information Technology, Kongunadu Arts and Science College, Coimbatore, Tamil Nadu, India

Nikhitha Namburu Department of Electronics and Computer Engineering, KLEF, Vaddeswaram, Guntur, India

Yalamanchili Kavya Nandini Department of Electronics and Computer Engineering, KLEF, Guntur, India

Gayathri Narayanan Department of Electronics and Communication Engineering, Amrita Vishwa Vidyapeetham, Kollam, India

Daniya Nausheer Department of Information Science, Dayananda Sagar Academy of Technology and Management, Bangalore, India

Purav Nayak Delhi Technological University, Delhi, India

B. N. Neethu Department of Electronics and Communication Engineering, Sri Ramakrishna Engineering College, Coimbatore, India

R. Neethu Department of Computer Science, University Institute of Technology, Mukhathala, Kollam District, Kerala, India

R. Nidhya Computer Science and Engineering, Madanapalle Institute of Technology & Science, Madanapalle, Andra Pradesh, India

Venkatram Nidumolu Department of Electronics and Computer Engineering, Koneru Lakshmaiah Education Foundation, Vaddesswaram, AP, India

Umesh Kumar Nimesh Delhi Technological University, Delhi, India

S. Padmavathi Assistant Professor, Department of Information Technology, Sri Shakthi Institute of Engineering and Technology, Coimbatore, India;
Department of Computer Science and Engineering, Amrita School of Engineering, Amrita Vishwa Vidyapeetham, Coimbatore, India

K. Padmaveni Hindustan Institute of Technology and Science, Chennai, India

Kalyani Palutla GITAM deemed to be University, Hyderabad, Telangana, India

Shruti Ajithkumar Panicker Department of Computer Science and Engineering, Amrita School of Engineering, Amrita Vishwa Vidyapeetham, Coimbatore, India

Suja Sreejith Panicker School of Computer Engineering & Technology, MIT World Peace University, Pune, Maharashtra, India

Bhaskar Pant Graphic Era Deemed to be University, Dehradun, India

Anoop Kr. Patel Department of Computer Engineering, National Institute of Technology, Kurukshetra, India

N. D. Patel Centre of Excellence in Cyber Security, Institute for Development & Research in Banking Technology (IDRBT), Hyderabad, India;
School of Computer & Information Sciences (SCIS), University of Hyderabad (UoH), Hyderabad, India

Sohan Pawar St. Francis Institute of Technology, Mumbai, India

Sai Surya Kiran Pokala Computer Science and Engineering, Indian Institute of Information Technology, Kottayam, Kerala, India

E. Prabhu Department of Electronics and Communication Engineering, Amrita School of Engineering, Coimbatore, Amrita Vishwa Vidyapeetham, Coimbatore, India

P. Prakash Department of Computer Science and Engineering, Amrita School of Engineering, Coimbatore, Amrita Vishwa Vidyapeetham, Coimbatore, India

Baddepaka Prasad Osmania University, Hyderabad, India

Shivam Prasad School of Computing Science and Engineering, VIT, Chennai, India

D. Praveen Kumar Dayananda Sagar College of Engineering, Bengaluru, Karnataka, India

K. Prince Wesly Department of Computer Science and Engineering, Dr.N.G.P. Institute of Technology (Dr.N.G.P. IT), Coimbatore, India

Priya Department of Electronics and Communication Engineering, Delhi Technological University, Delhi, India

A. Priya B. S. Abdur Rahman Crescent Institute of Science and Technology, Chennai, India

K. Priyadharsini Department of Electronics and Communication Engineering, Sri Krishna College of Engineering and Technology, Coimbatore, Tamil Nadu, India

Akash Kumar Pujari Department of CSE, SSN College of Engineering, Chennai, India

V. Purushothaman Vel Tech Rangarajan Dr. Sagunthala R&D Institute of Science and Technology, Chennai, India

Anisha Radhakrishnan Department of Computer Science and Engineering, Amrita School of Engineering, Amrita Vishwa Vidyapeetham, Coimbatore, India

J. E. Raghavendra Prasad Department of Electronics and Communication Engineering, Amrita School of Engineering, Amrita Vishwa Vidyapeetham, Coimbatore, India

Rahul Department of CSE, Delhi Technological University, Delhi, India

Akshit Rai Department of Computer Engineering and Applications, GLA University, Mathura, Uttar Pradesh, India

S. Rajashree Department of Computer Science, PES University, Bangalore, India

S. Rakesh Assistant Professor, Department of Information Technology, Sri Shakthi Institute of Engineering and Technology, Coimbatore, India

V. N. V. Sri Ram Department of Computer Science and Engineering, Amrita School of Engineering, Coimbatore, Amrita Vishwa Vidyapeetham, Coimbatore, India

S. Ramachandram Osmania University, Hyderabad, India

Aishwarya Ramachandran Department of Computer Science and Engineering, Amrita School of Engineering, Amrita Vishwa Vidyapeetham, Coimbatore, India

K. Ranjani Department of Electronics and Instrumentation Engineering, SNS College of Technology, Coimbatore, India

M. Ranjithkumar PG & Research Department of Computer Science, Government Arts College (Autonomous), Coimbatore, Tamil Nadu, India

M. Kameswara Rao Department of Electronics and Computer Engineering, KLEF, Vaddeswaram, Guntur, India

Hema D. Raut Electronics & Telecommunication Department, Symbiosis Institute of Technology, Symbiosis International (Deemed University), Pune, India

Aswathy Ravikumar School of Computer Science and Engineering, VIT, Chennai, India

J. V. R. Ravindra Centre for Advanced Computing and Research Laboratory (C-ACRL), Department of Electronics and Communication Engineering, Vardhaman College of Engineering, Shamshabad, Telangana, India

Pranav Ray Department of CSE, Delhi Technological University, Delhi, India

C. R. Rene Robin Department of Computer Science and Engineering, Jerusalem College of Engineering, Chennai, Tamil Nadu, India

Amirtharajan Rengarajan Department of Electronics and Communication, SASTRA Deemed University, Thanjavur, Tamil Nadu, India

E. Rishi Kiran Centre for Advanced Computing and Research Laboratory (C-ACRL), Department of Electronics and Communication Engineering, Vardhaman College of Engineering, Shamshabad, Telangana, India

L. Robert PG & Research Department of Computer Science, Government Arts College (Autonomous), Coimbatore, Tamil Nadu, India

A. Robert Singh School of Computing, Kalasalingam Academy of Research and Education, Krishnankoil, India

S. B. RudraSwamy Electronics and Communication Engineering Department, JSS Science and Technology University (SJCE), Mysuru, India

D. Ruth Anita Shirley Department of Electronics and Communication Engineering, Sri Krishna College of Engineering and Technology, Kuniyamuthur, India

Paridhi Sachdeva Delhi Technological University, Delhi, India

M. Sachithanantham Department of Computer Science and Engineering, Amrita School of Engineering, Amrita Vishwa Vidyapeetham, Coimbatore, India

Jessica Sadavarte Mukesh Patel School of Technology, Management and Engineering, Mumbai, India

B. M. Sagar Information Science and Engineering Department, Rashtreeya Vidyalaya College of Engineering (RVCE), Bengaluru, India

Ishita Sagar Department of Electronics and Communication Engineering, Delhi Technological University (DTU), Delhi, India

C. H. Sai Hemantha Department of Electronics and Computer Engineering, KLEF, Vaddeswaram, Guntur, India

N. Sai Pooja Reddy Centre for Advanced Computing and Research Laboratory (C-ACRL), Department of Electronics and Communication Engineering, Vardhaman College of Engineering, Shamshabad, Telangana, India

T. Saikumar Department of Computer Science and Engineering, Amrita School of Engineering, Amrita Vishwa Vidyapeetham, Coimbatore, India

Vipul Salunke School of Computer Engineering & Technology, MIT World Peace University, Pune, Maharashtra, India

S. C. Sandeep Dayananda Sagar College of Engineering, Bengaluru, Karnataka, India

M. Saravanan Vel Tech Rangarajan Dr. Sagunthala R&D Institute of Science and Technology, Chennai, India

Aditya Sarin Mukesh Patel School of Technology, Management and Engineering, Mumbai, India

Hanna Sashchyk Kiev National Taras Shevchenko University, Kiev, Ukraine

J. Sathiya Priya Department of Electronics & Communication Engineering, Sri Krishna College of Engineering & Technology, Coimbatore, Tamil Nadu, India

A. S. Selvakumar Department of Computer Science and Engineering, Amrita School of Engineering, Amrita Vishwa Vidyapeetham, Coimbatore, India

Deepanshi Sengar Department of Computer Engineering, National Institute of Technology, Kurukshetra, India

M. Senthil Department of Electronics and Communication Engineering, Amrita School of Engineering, Amrita Vishwa Vidyapeetham, Coimbatore, India

Tazeen Shaikh Mukesh Patel School of Technology, Management and Engineering, Mumbai, India

Prashant Giridhar Shambharkar Delhi Technological University, Delhi, India

Abhilasha Sharma Delhi Technological University, Delhi, India

Arpit Sharma Department of Computer Engineering and Applications, GLA University, Mathura, India

Manmohan Sharma Lovely Professional University, Phagwara, India

Laxmikant Shevada Electronics & Telecommunication Department, Symbiosis Institute of Technology, Symbiosis International (Deemed University), Pune, India

Gagan Deep Shivamadhu Department of Information Science, Dayananda Sagar Academy of Technology and Management, Bangalore, India

M. Shrenik Dayananda Sagar College of Engineering, Bengaluru, Karnataka, India

S. Shrinidhi Department of Electronics and Communication Engineering, Amrita School of Engineering, Coimbatore, Amrita Vishwa Vidyapeetham, Coimbatore, India

Gennady Shvachych National Metallurgical Academy of Ukraine, Dnipro, Ukraine

M. R. Sindhu Department of Electrical and Electronics Engineering, Amrita School of Engineering, Coimbatore, Amrita Vishwa Vidyapeetham, Coimbatore, India

Prateek Singh School of Computing Science and Engineering, VIT, Chennai, India

A. Sivanesh Kumar Department of Computer Science and Engineering, Saveetha School of Engineering, Saveetha Institute of Medical and Technical Sciences, Chennai, India

G. Smilarubavathy Computer Science and Engineering, Dr. N.G.P Institute of Technology, Coimbatore, India

Sona Solani Department of Computer Engineering, Marwadi University, Rajkot, Gujarat, India

P. Sreenivasa Reddy Dayananda Sagar College of Engineering, Bengaluru, Karnataka, India

A. Srilakshmi Sastra Deemed to be University, Thanjavur, India

Swati Srivastava Department of Computer Engineering and Applications, GLA University, Mathura, Uttar Pradesh, India

V. Srividhya Department of EEE, Meenakshi College of Engineering, Chennai, India

A. Subahar Vel Tech Rangarajan Dr. Sagunthala R&D Institute of Science and Technology, Chennai, India

B. Subramanian Department of Computer Science and Engineering, Amrita School of Engineering, Amrita Vishwa Vidyapeetham, Coimbatore, India

K. Sujatha Department of EEE, Dr. MGR Educational & Research Institute, Chennai, India

A. Sujith Department of Computer Science, Research Centre, University of Kerala, Thiruvananthapuram, Kerala, India

Konjeti B. V. N. S. Sumanth Department of Computer Science and Engineering, Amrita School of Engineering, Coimbatore, Amrita Vishwa Vidyapeetham, Coimbatore, India

K. A. Sumithra Devi Department of Information Science, Dayananda Sagar Academy of Technology and Management, Bangalore, India

Seba Susan Department of Information Technology, Delhi Technological University (DTU), Delhi, India

J. N. Swaminathan QIS College of Engineering and Technology, Ongole, Andhra Pradesh, India

R. Swathy Assistant Professor, Department of Information Technology, Sri Shakthi Institute of Engineering and Technology, Coimbatore, India

V. Swetha Department of Computer Science and Engineering, Dr.N.G.P. Institute of Technology (Dr.N.G.P. IT), Coimbatore, India

Deveshi Thanawala Mukesh Patel School of Technology, Management and Engineering, Mumbai, India

Jonnakuti Lakshmi Thirusha Department of Electronics and Computer Engineering, KLEF, Guntur, India

S. Udhayakumar Department of Computer Science and Engineering, Saveetha School of Engineering, Saveetha Institute of Medical and Technical Sciences, Chennai, Tamil Nadu, India

D. Uma Nandhini Department of Computer Science and Engineering, School of Computing, Vel Tech Rangarajan Dr. Sagunthala R&D Institute of Science and Technology, Chennai, Tamil Nadu, India

M. Vaishnavi Reddy Centre for Advanced Computing and Research Laboratory (C-ACRL), Department of Electronics and Communication Engineering, Vardhaman College of Engineering, Shamshabad, Telangana, India

Swathi Vangala Centre for Advanced Computing and Research Laboratory (C-ACRL), Department of Electronics and Communication Engineering, Vardhaman College of Engineering, Shamshabad, Telangana, India

M. Varsha Nair Department of Electronics and Communication Engineering, Amrita Vishwa Vidyapeetham, Kollam, India

Mahima Varshney Department of Computer Engineering and Applications, GLA University, Mathura, Uttar Pradesh, India

S. Venkata Sai Santosh Department of Electronics and Computer Engineering, KLEF, Vaddeswaram, Guntur, India

Aditya Venkatesh Department of Information Science, Dayananda Sagar Academy of Technology and Management, Bangalore, India

Mahesh Vibhute School of Electrical Engineering, MIT Academy of Engineering Alandi, Pune, India

O. Vignesh QIS College of Engineering and Technology, Ongole, Andhra Pradesh, India

D. Vijai Srinivas VTEC Engineering, Coimbatore, India

S. Vinuja Department of Electronics and Communication Engineering, Amrita School of Engineering, Coimbatore, Amrita Vishwa Vidyapeetham, Coimbatore, India

Ankur Vishwakarma Department of Computer Engineering and Applications, GLA University, Mathura, India

Adithya Viswanathan Department of CSE, SSN College of Engineering, Chennai, India

Nunna Vivek Department of Electronics and Computer Engineering, Koneru Lakshmaiah Education Foundation, Vaddesswaram, AP, India

Rajeev Wankar School of Computer & Information Sciences (SCIS), University of Hyderabad (UoH), Hyderabad, India

Akhil Yadav Department of Electronics and Communication Engineering, Amrita School of Engineering, Amrita Vishwa Vidyapeetham, Coimbatore, India

Anupam Yadav Department of Computer Engineering and Applications, GLA University Mathura, Mathura, India

Nagaraju Yesho CDAC, Hyderabad, Telangana, India

Mathematical Modeling of the Data Processing Problems of Heat Experiments Based on Multiprocessor Computing Complexes

Gennady Shvachych, Boris Moroz, Maksym Khylko, Hanna Sashchyk, Olena Khylko, and Volodymyr Busygin

Abstract The research aims at solving problems associated with identifying parallel structures of algorithms and programs, and their reflection in computer architecture. This paper highlights the modeling processes of multidimensional non-stationary problems based on multiprocessor computing systems. An algorithm is developed for solving the coefficient inverse heat conduction problem. In the above studies, the inverse heat conduction problem is interpreted as the optimal control problem. The problem of choosing a regularization parameter is discussed. The regularization parameter is selected so that the residual of the approximate solution could be comparable in magnitude with the accuracy degree of the problem's initial data. Such a choice of the regularization parameter is easily realized when modeling computations by a multiprocessor computing system. It is manifested in iterative methods, which allows obtaining results close in accuracy to optimal. It is noted that the parallel algorithm basis of numerical minimization is the procedure for establishing the minimum function of many variables. The test modeling results of problems using multiprocessor computing systems are presented.

G. Shvachych
National Metallurgical Academy of Ukraine, Dnipro, Ukraine
e-mail: busygin2009@gmail.com

B. Moroz
University of Technology, Dnipro, Ukraine
e-mail: fau.inter.dept@gmail.com

M. Khylko
National Academy of Sciences of Ukraine, Kiev, Ukraine
e-mail: theeuroway@gmail.com

H. Sashchyk · O. Khylko
Kiev National Taras Shevchenko University, Kiev, Ukraine
e-mail: the.busygin@gmail.com

V. Busygin (✉)
Oles Honchar Dnipro National University, Dnipro, Ukraine
e-mail: busygin2009@gmail.com

G. Ranganathan et al. (eds.), *Inventive Communication and Computational Technologies*, Lecture Notes in Networks and Systems 145,
https://doi.org/10.1007/978-981-15-7345-3_1

Keywords Multiprocessor system · Mathematical model · Extreme problems · Thermo physics · Parallel structures · Residual principle · Coefficient problems · Inverse heat conduction problems

1 Introduction

In metallurgical production, many diverse and interconnected processes include heat transfer and mass exchange, hydrodynamic processes in melts, as well as a change in the substance aggregation state, deformation phenomena under power and thermal loads, etc. Part of the issues to improve the metallurgical production efficiency can be solved by ergonomic methods [1], or technological innovations [2], or organizational and economic [3]. Yet most of these processes, considered for their improvement, can be described based on differential equations of continuum mechanics, which reflect objective laws of conservation of mass, momentum, and energy. In mathematical terms, these systems are of multidimensional nonlinear differential equations that, like laws of chemistry and thermodynamics, describe interconnected processes, and their interaction.

Moreover, the practice of recent years shows that neither the intensification of metallurgical production processes nor the constructive improvement of metallurgical equipment variety is possible without studying and analyzing the heat transfer phenomena by mathematical modeling methods [4, 5]. A theoretical study of the heat and mass transfer process is mainly based on their numerical simulation using computer technology. Also, fundamental problems in a potentially endless increase in peak computer performance disappear with cluster computing systems development. Parallel computing systems rapidly develop, and with the advent of computing clusters [6, 7], parallel computing has become available to many. As a rule, mass processors, standard network technologies, and freely distributed software are used to build clusters. These circumstances made so-called big problems of metallurgical thermophysics [8–10] possible to solve.

At the same time, problems arising in the development of parallel computing systems that meet unique features are, as a rule, paramount and require in-depth study and research [11, 12]. Indeed, distributed computer modeling covers the whole spectrum of modern computer technology: supercomputers, cluster computing systems, local and global networks. Besides, distributed modeling allows solving problems requiring huge processing time, integrate mathematical models that are processed on various (including geographically distant) computing systems.

This paper highlights the issues of mathematical modeling of multidimensional non-stationary problems of metallurgical thermophysics.

2 Research Topic Relevance and Research Purpose

In many cases, the mathematical support of non-stationary thermal experiments is based on methods for solving the inverse heat conduction problem (IHCP), which include boundary thermal conditions determination, identification of heat and mass transfer processes, restoration of external and internal temperature fields, etc. However, at present, the main field of the IHCP application remains the processing and interpretation of the results of the thermal experiments. It was here where the most considerable theoretical and applied successes were achieved in methods' effectiveness and the breadth of their practical use. The formulation of inverse heat conduction problems in metallurgical thermophysics is more easily formulated by the concept of the cause-effect relation [13, 14]. At the same time, the boundary conditions and their parameters, initial conditions, thermophysical properties, etc., should be attributed to standard features of the heat exchange process. The main goal of the direct heat transfer problems is in this formulation of determining the cause-effect relations.

Furthermore, on the contrary, if, according to certain experimental information on the temperature field, it is necessary to restore causal features, then one or another IHCP formulation is applied. Today, stable methods of solving (regularization methods) of various ill-posed problems have been developed. In the proposed study, special attention is paid to one of the most promising areas in solving IHCP—reducing them to extreme settings [15, 16]. On this basis, the IHCP solutions are interpreted as the optimal control problem. This paper proposes an approach that reduces the functional minimizing problem via direct problem implementation to minimizing the many variables' functions. At the same time, studies of the accuracy and noise immunity of the method for solving IHCP are coming to the fore, and the problem of choosing the regularization parameter (the moment of computational termination) is discussed as well. It is shown that here the residual principle takes the central place: the regularization parameter is selected so that the residual of the approximate solution is comparable in magnitude with the accuracy degree of the problem's initial data. Such a regularization parameter option is easily realized when modeling computations by a multiprocessor computing system. It is especially clearly manifested in iterative methods, and considering the circumstances described above, this is precisely what gives results close in accuracy to optimal ones. So, if there is a computing system with the N number of processors, then it is possible to compute in parallel N values of functions that implement the minimum functional separation by solving controlled mathematical models. It is noted that the basis of the parallel algorithm of numerical minimization is the procedure for establishing the minimum of many variables' function.

So the primary research purpose aims at solving problems associated with identifying parallel structures of algorithms and programs and their reflection in the computers' architecture in solving a wide range of applied problems.

3 Mathematical Modeling Features of Thermal Problems

The effectiveness of the proposed approach is illustrated by several features of modeling inverse problems of studying the materials' thermophysical properties. Let us consider the modeling features of the indicated class of problems. Thus, the thermal problems of metallurgical thermophysics are considered. Their statement is formulated in terms of cause-effect relationships. Under the accepted mathematical model, the boundary conditions and their parameters, initial conditions, thermophysical properties, etc. are related to causal features of the heat exchange process.

It should be noted that solving heat conduction problems (HCP) allows determining various causal features of heat transfer processes in a solid-body-environment system from given temperature fields known from a thermal or numerical experiment. Causal features usually mean the coefficients of equations, initial fields, boundary conditions, features of the integration domain, etc. The IHCP is incorrectly posed; therefore, methods for their solution are more complicated than the corresponding direct problems. The article formulates a general approach to solving a large class of IHCP. An algorithm is developed, and the features of solving the coefficient IHCP are revealed.

4 Mathematical Formulation of Research Problem

When solving IHCP, it is first of all necessary to illuminate the controllability conditions of mathematical models (MM) that allow, by methods of mathematical modeling, bringing the system into a given thermal state using control actions. The MM is known and includes several causal features, which denoted by the R-input parameters' vector. Let a discrete analog of MM and a computational algorithm be developed. The MM sensitivity to vector R variations, i.e., it is shown that the desired solution to a specific MM problem is determined not only by the functions of spatial coordinates and time but also by the R-input parameters' function. Thus, to evaluate the reliability of the obtained MM solution, it is necessary to study its behavior with variations in the input data. When studying the MM sensitivity, variations in input parameters are assumed to be given. Equally important are the formulation and methods of solving inverse problems, which essence is to evaluate the vector R-input parameters from the actual information about the simulated system known from the experiment. The mathematical modeling process of this class of problems involves several stages.

First, the development of an algorithm and a computational program for solving the direct MM problem, the computational algorithm for solving that implements the transformation.

$$T = (x, t, R) \tag{1}$$

determining the MM temperature state vector as a function of independent variables and input parameters of the vector *R*.

Second, when solving IHCP as a quality criterion for identifying the parameters *R* on MM solutions, it is necessary to introduce into the algorithm a particular functional characterizing the model as a whole or the deviations between the measured $T_e(t_j)$ and the computed values of the state vector $T_p(t_j)$. Let us choose the standard mean square residual as deviation the measure.

$$J\left(T_p, T_e\right) = \left(T_p - T_e\right)^2, \tag{2}$$

wherein T_p is the thermal state vector value computed by the MM model. The components of the $T_e(t_j)$ vector can be determined on a discrete set of points of a given domain of temperature function definition. As a rule, in practice, several such criteria are used for the model's quality assessment.

Thus, the IHCP corresponding to their extreme formulation is solved using well-known numerical methods of optimization theory. Considering that the computation of the vector *gradJ*() inherent in these methods is a serious mathematical problem, how this can be avoided. Note that if MM implements transformation (1), then at each step of such transformations, it is possible to compute the values of functional (2). It allows, by repeated variation of the vector *R*-input parameters, constructing a sequence of changes in functional (2) that would include the point of its minimum. Thus, if this is feasible, then the IHCP solution is reduced to minimizing the function of many variables. This algorithm most looks merely for one variable of the vector *R*. The solving IHCP algorithm, in this case, includes separation of the interval containing the minimum functional point and the procedure for its refining. The update of the minimum coordinate can be implemented as follows. Suppose that the functional *J*(*R*) (2) has a sufficient analyticity margin concerning the vector *R*-input parameters. Let us represent its value by a segment of the Taylor series near a minimum.

$$J_{p+\varepsilon_k,1}(R) = J_{p,1} + \varepsilon_R J_{p,2} + \varepsilon_R^2 J_{p,3} + \cdots, \tag{3}$$

wherein

$$\begin{aligned}
\varepsilon_R &= \frac{R - R_p}{R_{p+1} - R_p} \in [-1, +1], \\
J_{p,2} &= \left(J_{p+1,1} - J_{p-1,1}\right)/2, \\
J_{p,2} &= \left(J_{p+1,1} + J_{p-1,1} - 2J_{p,1}\right)/2
\end{aligned} \tag{4}$$

are known Taylor components, and $p = 1, 2, 3, \ldots$ are the numbers of grid nodes.

Having saved three terms in (3), differentiating concerning ε_R, and equating to zero, constructed the interpolation formula

$$R = R_p - \frac{(R_{p+1} - R_p)}{2} \cdot \frac{J_{p,2}}{J_{p,3}}, \tag{5}$$

wherein all the notations correspond to those adopted above. Thus, the IHCP solution from this class of problems reduces to separating the minimum and its refinement by iterations according to Formula (5). This algorithm is tested below on solutions of coefficient IHCP.

5 Construction Features of the Controlled MM for the Coefficient IHCP Problem

Let us consider the one-dimensional problem of unsteady heat conduction described by the quasilinear equation.

$$C_{UV}(T)\frac{\partial T}{\partial \tau} = \frac{1}{x^K}\frac{\partial}{\partial x}\left[x^K\left(\lambda(T)\frac{\partial T}{\partial x}\right)\right], x \in [0, 1], \tau \in [0, \infty), \tag{6}$$

wherein

$$\begin{aligned} \tau &= a_0 t/L^2, a_0 = \lambda_0/C_{v0}, \\ x &= x/L, C_v = C_v/C_{v0}, \\ \lambda &= \lambda/\lambda_0, k = \overline{0, 1, 2}, \end{aligned} \tag{7}$$

are dimensionless input data, k is a parameter of the sample (plate, cylinder, and ball) shape.

Equation (6), after time discretization, is transformed at grid domain nodes ($p = 1, 2, \ldots, m_x - 1$) into an ordinary differential equations system of a two-point type.

$$T''_{p+\varepsilon_x,1}(\varepsilon_x) + 2A_p T'_{p+\varepsilon_x,1}(\varepsilon_x) - B_p T_{p+\varepsilon_x,1}(\varepsilon_x) = B_p T O_{p+\varepsilon_x,1}(\varepsilon_x), \tag{8}$$

wherein

$$\left.\begin{aligned} A_p &= \tfrac{1}{2}\left(\tfrac{\lambda_{p,2}}{\lambda_{p,1}} + k\tfrac{Dx1}{x_p}\right) \\ B_p &= \tfrac{Dx1}{Dt1}\tfrac{CV_{p,1}}{\lambda_{p,1}} \end{aligned}\right\}. \tag{9}$$

The functions $T_{p+\varepsilon_{x},1}(\varepsilon_x), TO_{p+\varepsilon x,1}(\varepsilon_x)$ in (9) are assigned to the current and previous time layers, respectively.

The solution of the investigated differential equation according to the direct method is presented in an analytical form by nodes

$$T_{p+\varepsilon_x,1}(\varepsilon_x) = T^*_{p+\varepsilon_x,1}(\varepsilon_x) + C_p e^{\beta_1 \varepsilon_x} + D_p e^{-\beta_2 \varepsilon_x}, \tag{10}$$

wherein

$$\beta_1 = \Omega_p - A_p, \quad \beta_2 = \Omega_p + A_p, \quad \Omega_p = \sqrt{A_p^2 + B_p}, \tag{11}$$

are the roots of the characteristic equation;

C_p, D_p are the integration constants;

$T^*_{p+\varepsilon x1}(\varepsilon x)$ is a particular solution to the inhomogeneous Eq. (8).

The final sub-node solution to this problem takes the following form

$$T_{p+\varepsilon_x,1}(\varepsilon_x) = T^*_{p+\varepsilon_x,1}(\varepsilon_x) + F1\frac{e^{-\beta_1(1-\varepsilon_x)}}{\text{Det}}\left(1 - e^{-2\Omega_p(1+\varepsilon_x)}\right) + F2\frac{e^{-\beta_1(1+\varepsilon_x)}}{\text{Det}}\left(1 - e^{-2\Omega_p(1-\varepsilon_x)}\right), \tag{12}$$

wherein

$$F1 = \left(T_{p+1,1} - T^*_{p+1,1}\right), F2 = \left(T_{p-1,1} - T^*_{p-1,1}\right), \text{Det} = \left(1 - e^{-4\Omega_p}\right) \tag{13}$$

are the grid complexes.

Setting in (12) $\mathcal{E}_x = 0$, obtained MM algebraic analog in the form of a system of linear differential equations of a tridiagonal structure

$$C_p T_{p+1,1} - T_{p,1} + D_p T_{p-1,1} = f_{p,1}, \quad p = \overline{1, 2m-1}, \tag{14}$$

wherein

$$\left.\begin{array}{c} C_p = \frac{e^{-\beta_1}}{1+e^{-2\Omega_p}} \\ D_p = \frac{e^{-\beta_2}}{1+e^{-2\Omega_p}} \\ f_{p,1} = \left(C_p T^*_{p+1,1} - T^*_{p,1} + D_p T^*_{p-1,1}\right) \end{array}\right\}. \tag{15}$$

The quadratic dependence of the argument $TO_{p+\varepsilon x1}(\varepsilon x)$ concretizes the form of the initial function $\mathcal{E}_x$

$$TO_{p+\varepsilon_x,1}(\varepsilon_x) = TO_{p,1} + \varepsilon_x TO_{p,2} + \varepsilon_x^2 TO_{p,3}, \tag{16}$$

wherein

$$\left.\begin{array}{c} TO_{p,2} = \frac{1}{2}\left(TO_{p+1,1} - TO_{p-1,1}\right) \\ TO_{p,3} = \frac{1}{2}\left(TO_{p+1,1} + TO_{p-1,1} - 2TO_{p,1}\right) \end{array}\right\}. \tag{17}$$

Considering these dependencies, particular solution of inhomogeneous Eq. (8), which is included in the MM (14) in an implicit form, takes the following form

$$TO^{*}_{p+\varepsilon_x,1}(\varepsilon_x) = M_0 + \varepsilon_x M_1 + \varepsilon_x^2 M_2, \tag{18}$$

wherein

$$\left.\begin{array}{c} M_0 = TO_{p,1} + b_1 TO_{p,2} + 2b_2 TO_{p,3} \\ M_1 = TO_{p,2} + b_1 TO_{p,3} \\ M_2 = TO_{p,3} \\ b_1 = 2\frac{A_p}{B_p},\ b_2 = 4\frac{A_p^2}{B_p^2} + \frac{1}{B_p} \end{array}\right\}. \tag{19}$$

The solution of linear algebraic equations system (14) for given input data is entirely simply implemented by the sweep method. Thus, developed the first MM being necessary to solve the coefficient IHCP in the above (extreme) statement. Let us designate it as Model 1 and should also add its gradient analog to the (temperature) MM (14). By differentiating function (12) $\mathcal{E}_x$ concerning and setting $\mathcal{E}_x = 0$, constructed the gradient Model 2

$$T_{P,2}(\varepsilon_x)_{\varepsilon_x=0} = \left\{T^{*}_{P,2}(0) + F1\frac{\ell^{-\beta_1}}{\text{Det}}\left[\beta_1 + \beta_2\ell^{-2\Omega p}\right]\right. \\ \left. - F2\frac{\ell^{-\beta_2}}{\text{Det}}\left[\beta_2 + \beta_1\ell^{-2\Omega p}\right]\right\}, \tag{20}$$

wherein the function $T^{*}_{p+\varepsilon 2}(\varepsilon X)$ is computed by the formula

$$T^{*}_{p+\varepsilon_x,2}(\varepsilon X) = M_1 + 2\,\varepsilon X M_2 \tag{21}$$

Thus, the identification MM, which includes Model 1 (14) and Model 2 (20), allows formulating a solution to the coefficient IHCP in an extreme setting according to the scheme described above.

The proposed approach is implemented as a package of application programs.

6 Brief Illustration of the Application Package

This section of the proposed researches considers an application package designed to process the results of a thermophysical experiment. Note that the primary purpose of creating such a software product was to provide substantial assistance to the researcher at all stages of a thermophysical experiment. The package is developed considering the requirements of object-oriented programming. The modeling

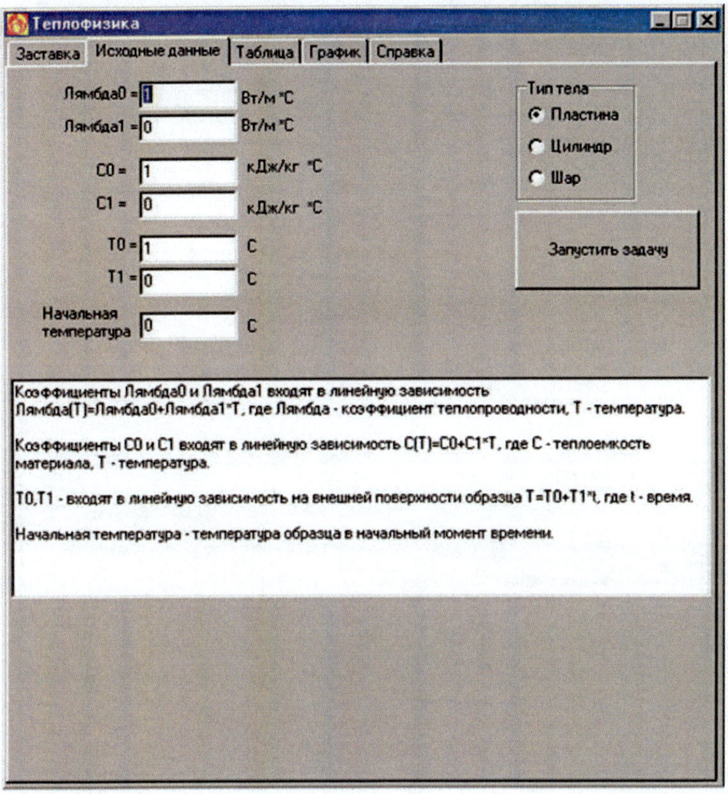

Fig. 1 Initial data input

was implemented by a multiprocessor computing system, which features to solve problems of this class are described in [11, 15–18].

The operation is represented in Figs. 1, 2, 3, 4 and 5. As shown in Fig. 1, the initial data input is carried out interactively. The problem solution results are displayed both as tables (Fig. 2) and depicted by the required graphical dependencies (Figs. 3 and 4)

7 The Solution of the Test Coefficient IHCP Using Mathematical Modeling

As a test problem let us consider a cylindrical sample made of a material with thermophysical properties [19] (coke from gas coal p. 41, Table 42-molded coke):

$$\lambda = 0.161 + 0.024 \times 10^{-2} \cdot T$$

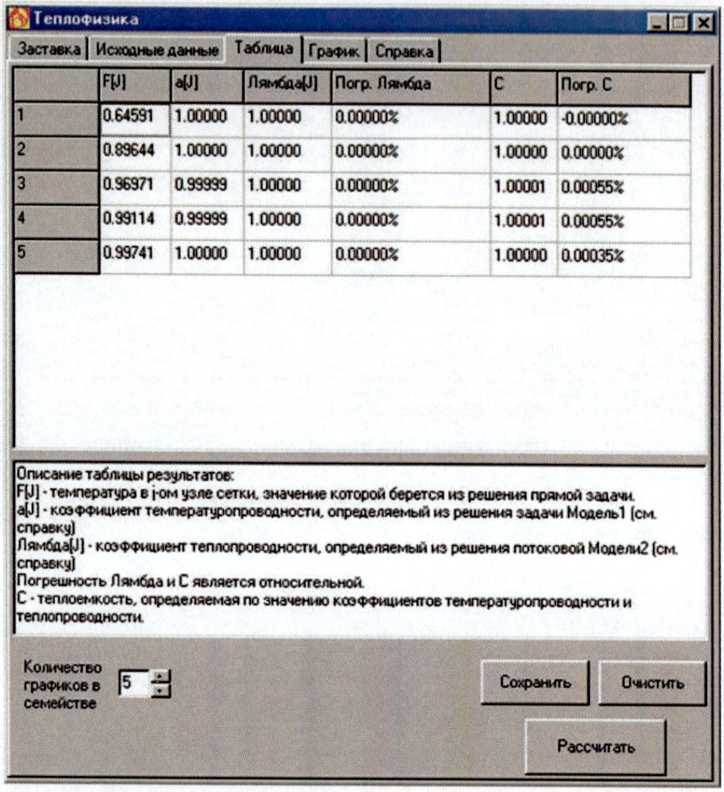

Fig. 2 IHCP solution results (as a table)

$$C = 1.281 + 0.208 \times 10^{-2} \cdot T, \tag{22}$$

The density of coke from gas coal $p = 1912$, kg/m^3. With such thermophysical properties, simulated the temperature field of a cylindrical shape sample ($k = 1$). For given time-linear temperature change at sample boundary ($TL = 20 + 100 \cdot \tau$), the temperature field for a particular time moment $\tau = a_0 t / R^2 = 0.5$, wherein $a_0 = \lambda_0 / p\ c_0 (\lambda_0 = 1, c_0 = 1)$, shown in Figs. 5, 6 and 7.

The coefficient exact values of thermal conductivity and thermal diffusivity are, respectively, equal $\lambda(f_2) = 0.166$, $a\,(f_2) = 0.054$, wherein f_2 is the temperature change in the second node along the cross-section of the sample. The minimum residuals presented in Figs. 6 and 7 exactly correspond to these values.

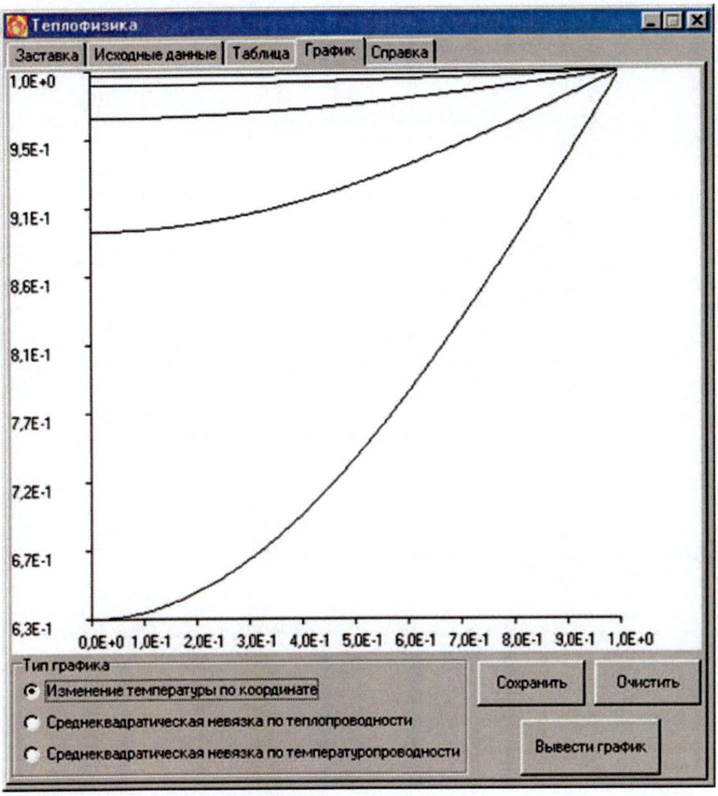

Fig. 3 The computation results of the temperature distribution

8 Conclusion

The following points could be emphasized as:

1. Supercomputers are currently inaccessible due to the enormous cost and service price. In this regard, a real alternative is cluster-type computing systems by which the simulation results are covered in this paper.
2. Being a relatively new technology, cluster-type parallel computing systems are useful in solving a large class of non-stationary multidimensional problems, while allowing to increase the productivity and quality of computations.
3. The software developed in this paper can be used to plan and process the results of a thermophysical experiment. The algorithms developed in the application program package are simply reconstructed to solve other coefficient and boundary problems of thermal conductivity.

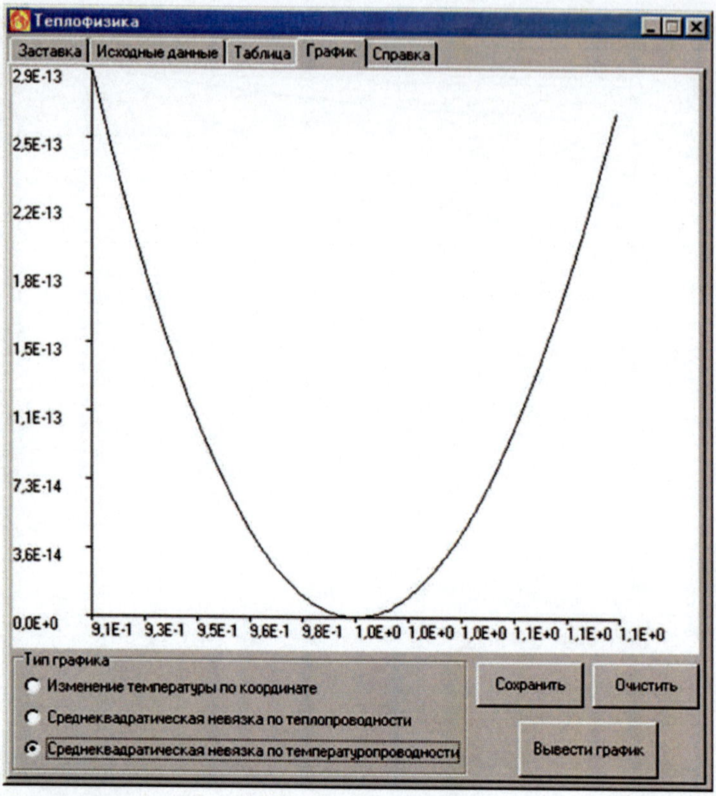

Fig. 4 The computation results of the temperature distribution

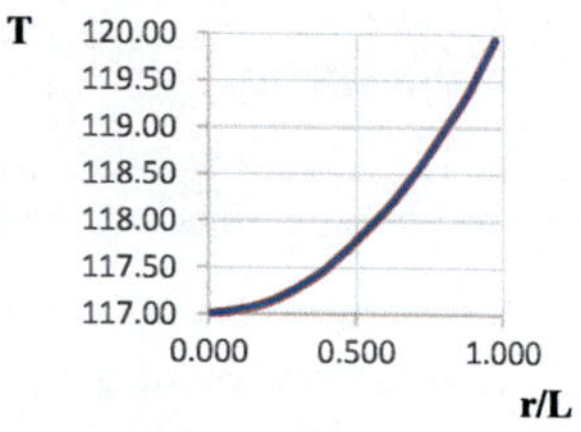

Fig. 5 The change in temperature over the cross-section of the sample at time moment $\tau = \tau_1 = 0.5$

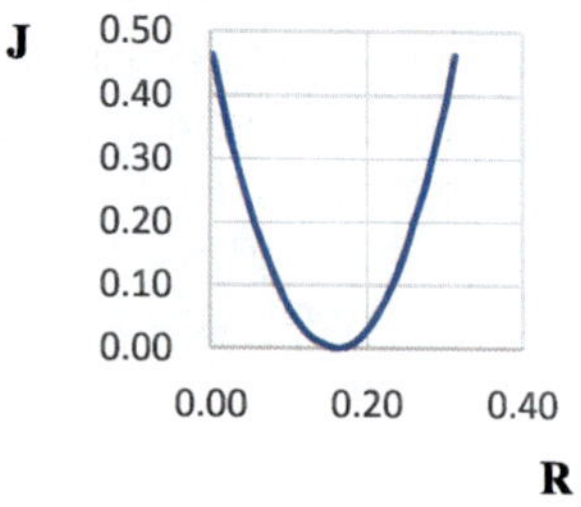

Fig. 6 The solution of the coefficient IHCP with $R = \lambda$

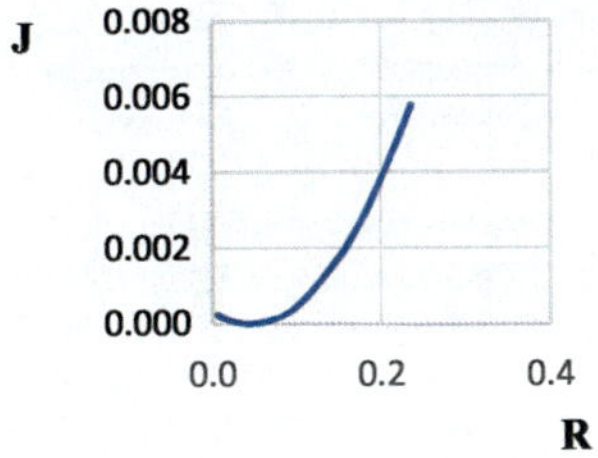

Fig. 7 The solution of the coefficient IHCP with $R = a$ with control relative to the coefficient of thermal diffusivity

4. The developed algorithms for solving thermophysical problems are highly accurate and efficient: the test solution for IHCP with accurate input data coincides with the thermophysical features of the sample material.
5. The developed software for processing the results of a thermophysical experiment is self-regulating. Moreover, it is quite merely tuned to the solution of others and, in particular, of boundary IHCP.

References

1. Shevyakov O, Krupskyi OP, Slavska YA (2017) Ergonomic provision of modernizing management processes of metallurgical production in Ukraine and China. Naukovyi Visnyk Natsionalnoho Hirnychoho Universytetu 1:134–143
2. Guo X, Zhou Y, Zha G, Jiang W, Yang B, Ma W (2020) A novel method for extracting metal Ag and Cu from high value-added secondary resources by vacuum distillation. Sep Purif Technol 242:116787. https://doi.org/10.1016/j.seppur.2020.116787
3. Dzhandieri GV (2020) Diagnostics of efficiency and optimization of the organizational and economic system of ferrous metals recycling. Chernye Metally 2020(1):56–62
4. Ivaschenko VP, Shvachych GG, Sobolenko OV (2015) The system of automated control of temperature regimes of thermal treatment of steel products. Metall Min Ind 1:142–147
5. Ivaschenko VP, Shvachych GG, Shmukin AA (2008) Parallel computations and applied problems of metallurgical thermophysics. Syst Technol Reg Interuniv Collect Sci Works 3(56):123–138
6. Bashkov EA, Ivaschenko VP, Shvachych GG (2011) Prospects of application of modern communication technologies and research of their influence on the efficiency of multiprocessor computing systems. In: Proceedings. Donetsk National Technical University. Informatics, cybernetics and computing series, vol 14, issue 188. DonNTU, Donetsk, pp 100–112
7. Ivaschenko VP, Bashkov EA, Shvachych GG et al (2011) Information systems, and technologies: monograph. Scientific-Innovation Center, Krasnoyarsk
8. Rouche PJ (1980) Computational fluid dynamics, Moscow
9. Tikhonov A et al (1966) Equations of mathematical physics. Science, Moscow
10. Kozdoba LA (1992) Computational thermophysics. Scientific Opinion, Kiev
11. Shvachych G, Ivaschenko O, Busygin V, Fedorov Y (2018) Parallel computational algorithms in thermal processes in metallurgy and mining. Naukovyi Visnyk Natsionalnoho Hirnychoho Universytetu, pp 129–137
12. Shvachych G, Busygin V (2018) Effective algorithms for solving coefficient problems of high accuracy order. In: System technologies. Proceedings, vol 4, issue 117, pp 86–94, Ukraine
13. Alifanov OM (1979) The heat transfer processes identification of flying objects. Mechanical Engineering, Moscow (in Russian)

14. Kovalenko ND, Shmukin AA, Guzhva MI, Makhin VV (1988) Non-stationary thermal processes in power plants of aircraft. Naukova Dumka, Kiev (in Russian)
15. Ivaschenko V, Shvachych G, Kholod Y (2014) Extreme algorithms for solving problems with higher-order accuracy. Applied and fundamental research. Publishing House Science and Innovation Center, Ltd. St. Louis, Missouri, USA, pp 157–170
16. Shvachych GG (2005) Determination of the thermophysical properties of materials based on solutions of coefficient IHCP in an extreme setting. Theory Pract Metall 1–2:104–108 (in Russian)
17. Shvachych GG (2009) High-performance computing system of cluster type MPP-architecture of problems. In: Proceedings of the international conference. Scientific basis for the introduction of new technologies in the era of the New Renaissance, Ashgabat, pp 342–343 (in Russian)
18. Shvachych GG (2004) Design features of parallel computational algorithms for PCs in heat and mass transfer problems. East Eur J Adv Technol 3:42–47 (in Russian)
19. Tereschenko VS (1999) Thermophysical properties of industrial materials. Dnipro

Analysis of Image Processing Techniques to Segment the Target Animal in Non-uniformly Illuminated and Occluded Images

Shruti Ajithkumar Panicker, Rahul Vinod Kumar, Aishwarya Ramachandran, and S. Padmavathi

Abstract Non-uniformly illuminated images are a class of images that, from a subjective perspective, are difficult to analyze. The excess noise and the lack of properly defined boundaries all contribute to making these images a difficult dataset for any form of analysis or segmentation. This calls for proper feature extraction and specific enhancement to make these images ready for efficient information gathering. This paper aims to visualize the features that can be enhanced using image enhancement techniques to identify the target animal in a non-uniformly illuminated and occluded image, thereby enhancing the recognition power of the proposed system. This paper uses a method to approximately detect and locate the position of the animal in an image. Segmentation Using Region Adjacency Graphs, Interactive Foreground Extraction using GrabCut Algorithm and DeepLab model for semantic image segmentation have also been analyzed.

Keywords Image enhancement · Low-light images · Non-uniform illumination · Feature extraction · Image segmentation

S. A. Panicker · R. V. Kumar (✉) · A. Ramachandran · S. Padmavathi
Department of Computer Science and Engineering, Amrita School of Engineering, Amrita Vishwa Vidyapeetham, Coimbatore, India
e-mail: cb.en.u4cse16143@cb.students.amrita.edu

S. A. Panicker
e-mail: cb.en.u4cse16151@cb.students.amrita.edu

A. Ramachandran
e-mail: cb.en.u4cse16102@cb.students.amrita.edu

S. Padmavathi
e-mail: s_padmavathi@cb.amrita.edu

G. Ranganathan et al. (eds.), *Inventive Communication and Computational Technologies*, Lecture Notes in Networks and Systems 145,
https://doi.org/10.1007/978-981-15-7345-3_2

1 Introduction

Images captured in environments with non-uniform illumination, such as those with a single point light source tend to have low contrast and a lot of noise which drastically reduces the information available for processing. A lot of images tend to be non-uniformly illuminated, but in this paper, we have utilized CCTV images obtained under conditions with poor lighting or captured at night-time. Images taken at low-light conditions tend to have minimal visibility. Hence, they aren't instrumental to be utilized in computer vision algorithms. These images tend to affect the performance of most vision-based algorithms, so it is highly recommended to enhance the quality of these images by using various image enhancement techniques. Enhanced images provide better results for automated image processing techniques and they help human viewers to improve their perception of the features in the image. The properties of the image are modified and the choice of attributes and the way they are changed are specific to a given task. After successfully enhancing the image using various techniques discussed, the process of segmentation is applied to the image by extracting the ROI (Region Of Interest) for image analysis. In this paper, we apply several mechanisms to enhance the low light image and segment the target image from its respective background.

2 Related Work

Paper [1] uses a PCA framework for improving non-uniformly illuminated images with decomposed luminosity-chrominance components. In paper [2], image enhancement is achieved by the use of spatial domain methods that involve a direct transformation of the image pixel. An auto encoder-based methodology is utilized in paper [3] to detect signal features from non-uniformly illuminated images and brighten images without saturating the lighter parts in the images. Papers [4, 5] use LIME methodology, wherein the intensity is estimated pixel-wise by finding the maximum value in the R, G, and B channels, by use of an illumination map. The non-uniformly illuminated images are enhanced in papers [6, 7] by a set of functions that provide details about the significant attributes of the image. Direct restoration of a normal image from a low light image is proposed in paper [8], by a ConvLSTM. In paper [9], existing digital image enhancement methods are reviewed. Paper [10] utilizes a modified version of the CLAHE algorithm to enhance vein images obtained using IR radiation. In paper [11], a dual channel-based method is proposed for enhancing images that have been non-uniformly illuminated. In paper [12], a histogram modification based algorithm is used to attain a wide and dynamic range that aids in contrast enhancement. Papers [13–15] utilize the CLAHE algorithm for enhancing low-light images by altering the local contrast of the image. In paper [16], the image is divided into several homogeneous regions by utilizing the MST algorithm, after which graphs(RAG) are extracted from the image and a graph matching process is performed to segment the target image. Papers [17, 18] utilize techniques like morphological processing and NCST for feature extraction of images and then employ the GrabCut method to obtain the segmentation [19–21].

3 Proposed Work

3.1 *Feature Detection Using Value Thresholding of Image*

To properly classify the different images according to the level of threat, it is important to detect the animal present in the image. The form of the animal has to be detected and labeled. But we know that in a low-contrast image this is difficult due to the inherent noise present in the image. Before we discuss the method used to achieve effective feature detection, we need to properly specify how we are going to measure this effectiveness. The shape schematic of an image is embedded in its edges. First, we detect the edges in an image and by using these filters. Then, by improving those parts of the image which have these edges, the sharpness of the image will visibly increase and the image will be suitable for segmentation.

Step 1: First, we enhance the image by using a simple enhancer that inverts the low-light image, applies the haze removal algorithm to the inverted low-light image, and then finally inverts the resulting image to obtain an enhanced image.
Step 2: Now we compare the edges for the original image and enhanced image.

3.2 *Orientation of the Subject Concerning the Camera*

The FAST features algorithm checks if a corner is present by determining a circular area around the potential pivot of a corner. The test detects a corner if a contiguous section of pixels is either brighter compared to the center plus a threshold or darker compared to the center minus a threshold.

Step 1: Detect the strongest two points from the set of FAST features. This gives us the two brightest points from the image. In a low-light CCTV image, these points will give us the eyes of the animal due to the retina reflecting a major portion of the light coming in. But, in some cases, the brightest point may just be something other than an eye.
Step 2: Find the Euclidean distance between these two points. This is done to check if the points are close enough to qualify as eyes.
Step 3: Check if the points are within the predetermined radius. If yes, then we can conclude that both eyes have been detected and that the animal is front-facing, i.e., facing towards the camera. If no, then we can conclude that only one eye has been detected and that the animal is sideways facing.

3.3 Extracting the ROI (Region of Interest)

In the above operation, we realized that the image enhancement in itself did not show a great improvement over the original image. This calls for a more specific and targeted enhancement to ensure we get a proper feature definition and extraction.

Step 1: Define a Region of Interest to further get an accurate outline of the subject, in this case, an animal. A method to obtain the ROI is by finding the eye of the animal and then taking a predefined area around it.
Step 2: The eye is chosen since it will be the brightest spot in the image and hence will be the easiest to find. The dimensions of the area are predefined by trial-and-error over a few images. The detection of eyes is done by getting the set of points of FAST features and extracting the strongest points alone. Since the eye is the brightest part of the image due to the retina reflecting the major portion of the light.
Step 3: Now, the predefined dimensions are used to crop the image with the position of eye as the reference point. Then, we follow the usual methodology to get EDGE features and compare them.

This yields a specific form of the animal being highlighted and gives a better and definitive understanding of the animal, and hence an efficient approach has been discovered.

3.4 Segmentation Using Region Adjacency Graphs

A Region Adjacency Graph (RAG) is a technique used for image segmentation, where the image is divided into numerous regions or segments.

Step 1: First of all, the input image is split into several segments, where each segment represents a node.
Step 2: A function is defined to obtain the edge weight as the difference between the average color values in the two adjacent nodes and to represent the edges as green lines and the centroids as yellow lines which would divide the RAG into disconnected regions, thus representing the target animal as a single connected component.
Step 3: These connected nodes are merged to form one connected component by the process of hierarchical merging. The results obtained after merging the components vary based on two important factors, the number of segments into which the image is divided and the threshold value, which determines the extent to which the target image is disconnected from the rest of the background. The edge weight between the two regions is set as the threshold value.
Step 4: Initially, the threshold value is kept constant and the number of segments is increased. We observe that as the number of segments increases, the target image becomes easier to perceive, but at some point, it starts to get distorted again.

Step 5: Later, the number of segments is kept constant and the threshold value is increased. As seen previously, when the threshold value was increased, it became easier to interpret the target image, but at some point, it got completely distorted again.

Hence, we observe that there is no ideal number of segments or a threshold value to obtain the most comprehensible result. The values differ for every image.

3.5 Interactive Foreground Extraction Using GrabCut Algorithm and DeepLab Model for Semantic Image Segmentation

An image segmentation technique known as the GrabCut method is based on graph cuts. The object to be segmented is bounded by the user-specified bounding boxes, and then estimates the color distribution of the target object and background using the gaussian mixture model. Then, the algorithm iteratively segments the region to get the best possible results.

Step 1: The user inputs the rectangle. All the content outside this rectangle is taken as background and everything inside is unknown.
Step 2: An initial foreground and background pixel labeling are done by the computer-based on the data we give.
Step 3: The foreground and background is modeled using a Gaussian Mixture Model (GMM).
Step 4: New pixel distribution is created by the GMM based on the data given. Possible front and background pixels are labeled from unknown pixels based on their color statistics relative to other pixels specified by the user (if the foreground and background are specified by the user, it is considered hard labeling, which means that it is unchanged).
Step 5: A graph is constructed from the pixel distribution. The nodes present in the graphs are called pixels. All the front pixels are connected to the source node and all the background pixels are connected to the sync node.
Step 6: The edge information or pixel comparison is called the weight between the pixels in terms of the pixel's foreground/background probability. Margins get less weight if there is a wide color difference in pixel color.
Step 7: The graph into two separating nodes with a minimum cost function for both source and sink node using an algorithm known as the min-cut algorithm (minimum weight). The aggregate of all weights of the edges that are cut is known as cost function.

The front connects all pixels to the source node and the background to the sinking node. This process is repeated until the classification is combined.

Fig. 1 **a** This shows the original image before enhancement [1], **b** shows the original image after enhancement

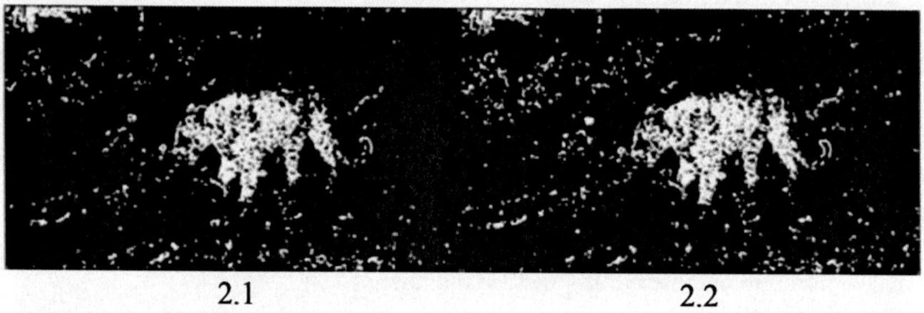

Fig. 2 **a** This shows the edge features of the original image, **b** shows the edge features of the enhanced image

4 Result Analysis

4.1 Dataset

The dataset consists of color and nighttime CCTV images of tigers, cheetahs, leopards, bears, deer, and elephants in non-uniformly illuminated conditions.

4.2 Feature Detection Using Value Thresholding of Image

Figures 1 and 2 show the output of steps 1 and 2 in Sect. 3.1.

4.3 Orientation of the Subject Concerning the Camera

See Figs. 3 and 4.

Fig. 3 This shows that the sample image with strongest features used to detect that subject is sideways facing. The two bright points are not within the radius to be classified as eyes

Fig. 4 This shows that the sample image with strongest features used to detect that subject is front facing. The two bright points are within the radius to be classified as eyes

4.4 Extracting the ROI (Region of Interest)

Figure 5 shows the output of step 2 in Sect. 3.3.

Figure 6 shows the output of step 3 in Sect. 3.3.

Fig. 5 Detecting eyes by getting the set of points of FAST features and extracting only the strongest points

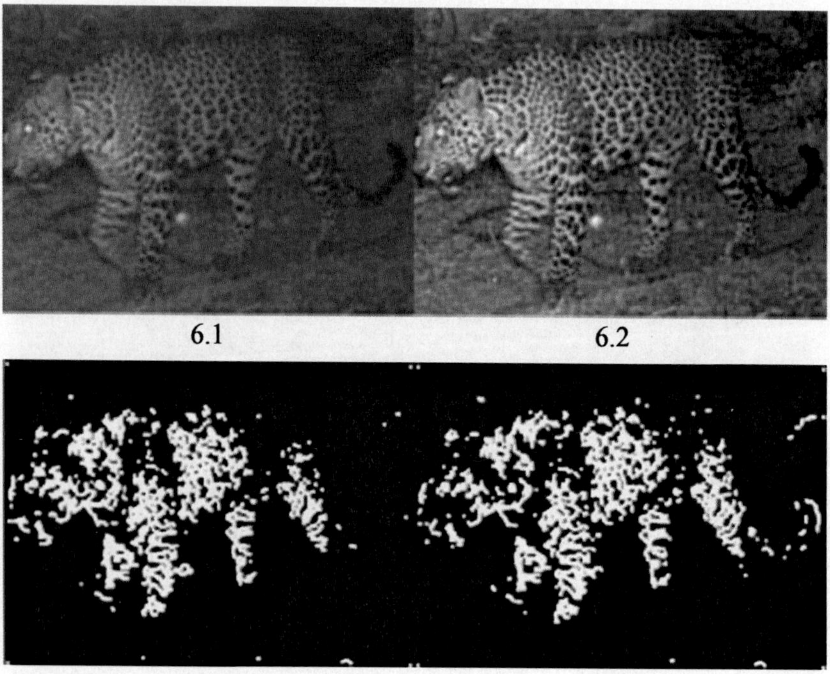

6.1 6.2

6.3 6.4

Fig. 6 **a** This shows the original image. **b** Shows the enhanced image. **c** Shows the EDGE features of the original image. **d** Shows the EDGE features of the enhanced image. All the images are cropped precisely to zero in on the ROI

4.5 Segmentation Using Region Adjacency Graphs

Figure 7 shows the output of step 2 in Sect. 3.4.

Figure 8 shows the output of step 4 in Sect. 3.4.

Figure 9 shows the output of step 5 in Sect. 3.4.

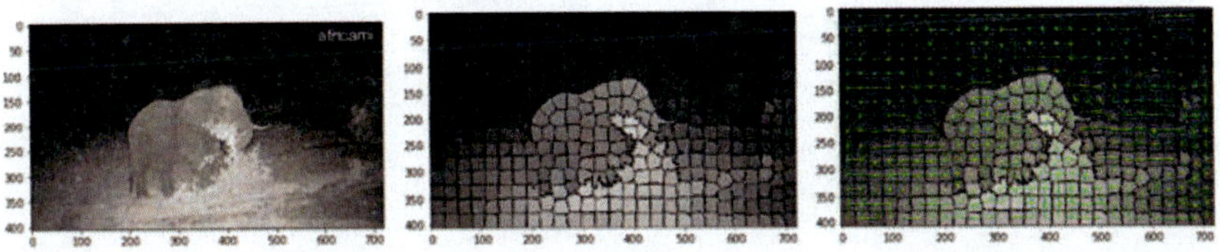

7.1 7.2 7.3

Fig. 7 **a** Shows the target image. **b** Shows the image after the initial segmentation. In **c**, the target image is represented as a single connected component, separated from the background

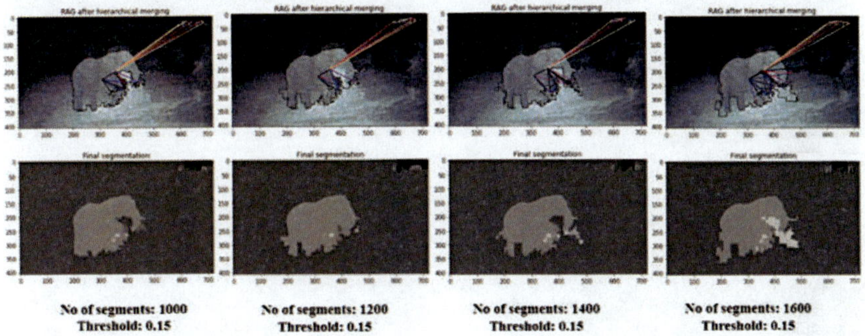

Fig. 8 This figure shows how the segmented image differs after hierarchical merging when the number of segments is increased while keeping the threshold value constant

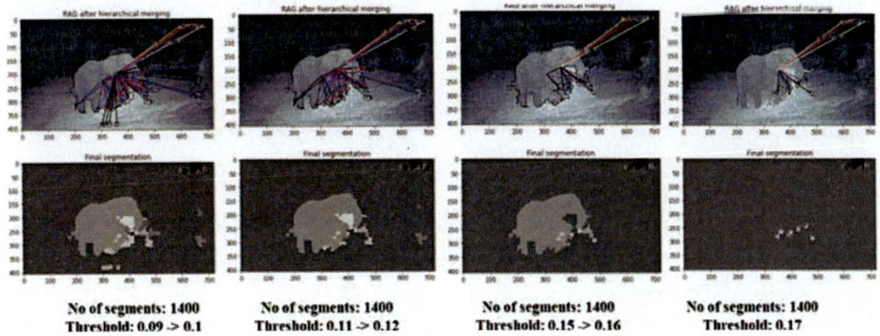

Fig. 9 This figure shows how the segmented image differs after hierarchical merging when the threshold value is increased while keeping the number of segments constant

10.1 10.2

Fig. 10 **a** Shows the animal image without occlusion. **b** Shows the result after segmentation

11.1 11.2

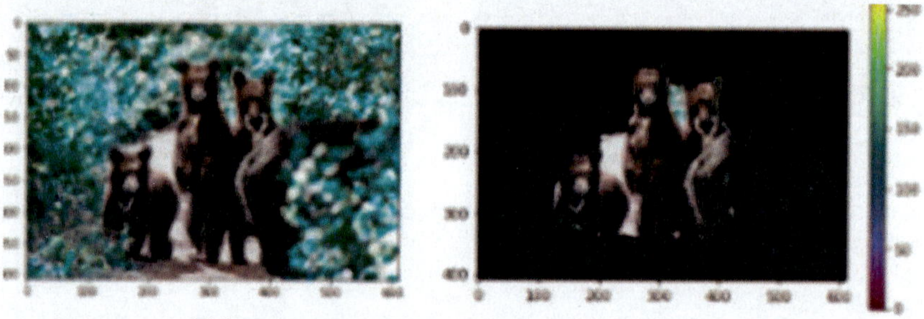

11.3 11.4

Fig. 11 **a** Shows the animal image with occlusion of less than 90%. **b** Shows the result after segmentation. **c** Shows the animal image with occlusion greater than 90%. **d** Shows the result after segmentation

4.6 Interactive Foreground Extraction Using GrabCut Algorithm and DeepLab Model for Semantic Image Segmentation

Figures 10 and 11 show the output of step 7 in Sect. 3.5. Figure 10 involves images without occlusion.

Figure 11 involves images with occlusion.

5 Conclusion

EDGE features play a major role in the subject detection. First, they are used to find the position of the subject in the frame which can help in isolating the animal out in the later part of the algorithm. The image enhancement helps in getting a clean image without the implicit noise. Thereby, effective segmentation is obtained that can help in the detection. Segmentation using **Region Adjacency Graphs** provides optimal segmentation results for night time images. It helps in detecting the target animal in the image, but the ideal value of the number of segments that the image must be divided into, and the threshold value differs depending on the dataset and is not constant. If the dataset consists of images with fairly similar shades, it needs to undergo contrast enhancement before being segmented, to receive optimal results. Segmentation using the **GrabCut algorithm** provides the best results for low light color images. Results show that the target animal was successfully segmented from the background even when it was partially occluded or the background color was similar to that of the target animal. The main disadvantage of this algorithm is that, if the segmentation initialized by user interaction is improper, it may provide us with unsatisfactory results. Furthermore, the segmentation process will be successful only if the occlusion of the target animal in the image is less than 90%.

The future scope of this paper involves automating the fast feature detection and extraction process, obtaining the best values for the number of segments and threshold value automatically, computerized construction of bounding box for efficient implementation of the GrabCut algorithm, and enhancing the algorithm to work effectively on night-time images. Once this has been done, it can find applications in security systems to detect the presence of animals in public places during the night-time, when the level of threat is high.

References

1. Priyanka SA, Wang Y-K, Huang S-Y (2019) Low-light image enhancement by principal component analysis
2. Rahman S, Rahman MM, Hussain K, Khaled SM, Shoyaib M (2014) Image enhancement in spatial domain: a comprehensive study. IEEE
3. Lore KG, Akintayo A, Sarkar S (2016) LLNet: a deep autoencoder approach to natural low-light image enhancement
4. Guo X, Li Y, Ling H (2017) LIME: low-light image enhancement via illumination map estimation. IEEE Trans Image Process 26(2):982–993
5. Gu Z, Chen C, Zhang D (2018) A low-light image enhancement method based on image degradation model and pure pixel ratio prior
6. Loh YP, Liang X, Chan CS (2019) Low-light image enhancement using Gaussian process for features retrieval
7. Ma S, Ma H, Xu Y, Li S, Lv C, Zhu M (2018) A low-light sensor image enhancement algorithm based on HSI color model
8. Xiang Y, Fu Y, Zhang L, Huang H (2019) An effective network with ConvLSTM for low-light image enhancement. In: Lin Z et al (eds) Pattern recognition and computer vision

9. Iwasokun GB, Akinyokun OC (2014) Image enhancement methods: a review. Br J Math Comput Sci
10. Bandara AMRR, Kulathilake KASH, Giragama PWGRMPB (2017) Super-efficient spatially adaptive contrast enhancement algorithm for superficial vein imaging. Int J Comput Appl (0975-8887)
11. Shi Z, mei Zhu M, Guo B, Zhao M, Zhang C (2018) Nighttime low illumination image enhancement with single image using bright/dark channel prior. EURASIP J Image Video Process 13
12. Sun C-C, Ruan S-J, Shie M-C, Pai T-W (2005) Dynamic contrast enhancement based on histogram specification. IEEE Trans Consum Electron 51(4):1300–1305
13. Dai D, Van Gool L (2018) Dark model adaptation: semantic image segmentation from daytime to nighttime
14. Daniel Nesa Kumar C, Aruna R (2018) Contrast limited adaptive histogram equalization (Clahe) based color contrast and fusion for enhancement of underwater images. IOSR Journal of Engineering (IOSRJEN)
15. Singh BB, Patel S (2017) Efficient medical image enhancement using CLAHE enhancement and wavelet fusion. Int J Comput Appl (0975-8887)
16. Akmal RM, Santoso J (2019) Image graph matching based on region adjacency graph. In: 5th international conference on science in information technology (ICSITech), Yogyakarta, Indonesia, pp 176–181
17. Kang F, Wang C, Li J (2018) A multiobjective piglet image segmentation method based on an improved non interactive GrabCut algorithm (2018)
18. Zhang Y, Yuan J, Liu H, Li Q (2017) GrabCut image segmentation algorithm based on structure tensor. J China Univ Posts Telecommun
19. Thirumurthy B, Parameswaran L, Vaiapury K (2018) Visual-based change detection in scene regions using statistical-based approaches. J Electron Imaging (2018)
20. Hrudya P, Nair LS, Adithya SM, Unni R, Vishnu Priya H, Poornachandran P (2013) Digital image forgery detection on artificially blurred images, In: International conference on emerging trends in communication, control, signal processing & computing applications (C2SPCA)
21. Sathya S, Parameswaran L, Karthika R (2018) Query by example—retrieval of images using object segmentation and distance measure. In: Lecture notes in computational vision and biomechanics

Enhanced Speeded Up Robust Feature with Bag of Grapheme (ESURF-BoG) for Tamil Palm Leaf Character Recognition

A. Robert Singh, Suganya Athisayamani, and A. Sherly Alphonse

Abstract Palm leaf was one of the vital writing tools in ancient times. Coarse, and contrasting colored surface and different colored letters are the hallmarks of this manuscript. In this paper, feature recognition using Enhanced Speeded Up Robust Feature with Bag of Grapheme (ESURF-BoG) is used to recognize the characters in Tamil palm leaf manuscripts. This method aims to detect the strongest critical points from the input character with different orientations. These key point features were created for training image as a model called Bag of Grapheme (BoG) with code word creation. Hence, unsupervised key point features were extracted, and pattern matching is performed for the testing image. The proposed architecture is used to recognize the characters from a vast data set of optical palm leaf manuscripts. This method is compared with the classic feature detectors SURF, min-Eigen feature detector and HOG feature detectors in terms of accuracy of recognition. The proposed method outperformed the classic models in terms of the recognition rate.

Keywords Character recognition · Pattern matching · SURF · BoG

1 Introduction

Hand-written character recognition is one of the emerging problems for both online and offline [1]. The live hand writing can be recognized using online recognition and the characters already written can be recognized using offline. Pattern recognition is an important component of any recognition system used for data processing and corresponding decision making. It is the discipline which develops the machine to

A. Robert Singh
School of Computing, Kalasalingam Academy of Research and Education, Krishnankoil, India

S. Athisayamani (✉)
School of Computing, Sastra Deemed to be University, Thanjavur, India
e-mail: suganyarobert@gmail.com

A. S. Alphonse
Department of Information Technology, Ponjesly College of Engineering, Nagercoil, India

G. Ranganathan et al. (eds.), *Inventive Communication and Computational Technologies*, Lecture Notes in Networks and Systems 145,
https://doi.org/10.1007/978-981-15-7345-3_3

perform perceptual tasks as like humans. There is a practical need to find efficient way of recognizing character from hand-written image and real images. Pattern recognition algorithms make available answer for all inputs and tells most matching patterns of the input image, by considering their statistical variation. This is different to pattern matching algorithms, which concentrates on exact matches of input with pre-existing patterns. Pattern matching is the process of categorizing set of pixels for the presence of some pattern from the raw data.

In this paper, the real-time camera captured palm leaf manuscript image is taken as an input. The difficulty faced in such input image is complex variations among the character shapes and also deformity in character due to deterioration. As far as compared to document image handling, palm leaf image is difficult because there is some discontinuity present in the character. The need of recognizing such inscription is to preserve and interpret the data written on palm leaf in a digital form. As compared to the normal feature extraction methods called structural features do not recognize the character because of the presence of discontinuity, it misclassifies the character. Hence, pattern matching guesses those discontinuity and recognizes which character it is from the templates already stored in the database.

2 Related Works

There are various feature extraction and pattern matching methods for various applications. Fujisawa et al. [2] used directional pattern matching technique for recognizing kanji characters in both machine-printed and hand-written data sets. Based on the local stroke components in the character, the directional planes were generated. For direction feature extraction, the frequency response was analyzed using gradient mask. Based on the similarity properties like continuity, stroke direction and thickness, stroke angles and direction position, the performance of character recognition was improved.

Prasad et al. [3] proposed a novel solution for recognizing hand-written Gujarati characters. Here, feed-forward neural network is used with the back propagation learning algorithm. The images are subdivided into templates and linear cross correlation is performed to find similar patterns between the already trained and input testing images. The results were validated using series of statistical tests called template classification, cross correlation coefficient analysis. The recognition rate is improved even for the similar characters. Schmid et al. [4] introduced assessment criteria for interest point detectors called repeatability rate which assesses the geometric stability under various transformations and information content deals with individuality of features. Based on the above criteria, various interest point detectors were compared like contour-based, intensity-based and parametric model-based methods.

Liu et al. [5] proposed structural feature matching for recognizing hand-written Chinese characters. Structural matching helps predicting the stroke for structural interpretation. Structural matching is performed by developing a model for each reference characters through attributed relational graph. By matching with the reference character, the strokes and the inner strokes were identified. Stroke and consistent matching recognize the character in KAIST image database and achieve promising results. Leutenegger et al. [6] proposed a novel method called Binary Robust Invariant Scalable Key point (BRISK) has three stages called key point detection, description and pattern matching. Key point detection is performed on scale space by constructing n octaves followed by sampling and rotation estimation. Descriptor matching among the selected key points was computed by finding hamming distance. Based on the brightness present in the image, the circular patterns were formed which detect the major key features help faster computation and limited computation power.

Fischer et al. [7] developed word spotting system for the scanned hand-written document image based on hidden Markov model (HMM). The input to this system is the text lines and keyword strings. The images were normalized to adopt for different writing styles and the local features were extracted using sliding window. Then, a score is assigned to the text line based on the keyword. Based on the threshold value, the keyword can be spotted. The system is trained using HMM which is mapped with the keyword text line which identifies the keywords present in the scripts.

Mohammad et al. [8] implemented optical character recognition (OCR) algorithm for hand-written documents to convert into an interpretable form. The system works on three stages: (i) Preprocessing by grayscale conversion (ii) Feature extraction by scanning the text lines from upper left corner to the lower right corner. (iii) Rrecognition by pattern matching by generating the binary format of the template image. The algorithm is trained for the similar characters; hence, it does not meet out font variations and bad handwriting.

In this paper, an enhanced SURF feature detector along with a group of feature codes is used to match the key points.

3 Overview of ESURF-BoG for Character Recognition in Palm Leaf Manuscript

The architecture of the proposed ESURF-BoG is shown in Fig. 1.

3.1 Preprocessing

The preprocessing is the procedure to prepare the raw input images ready to apply the specific function. In a camera captured palm leaf manuscript image, there are a number of factors affects the character recognition. These factors include: non-uniform light illuminations, complex palm leaf backgrounds, discontinuity in the character, deteriorated palm leafs, text alignment and linguistic complexities. In normal images, there will be a clean text on white background hence processing such

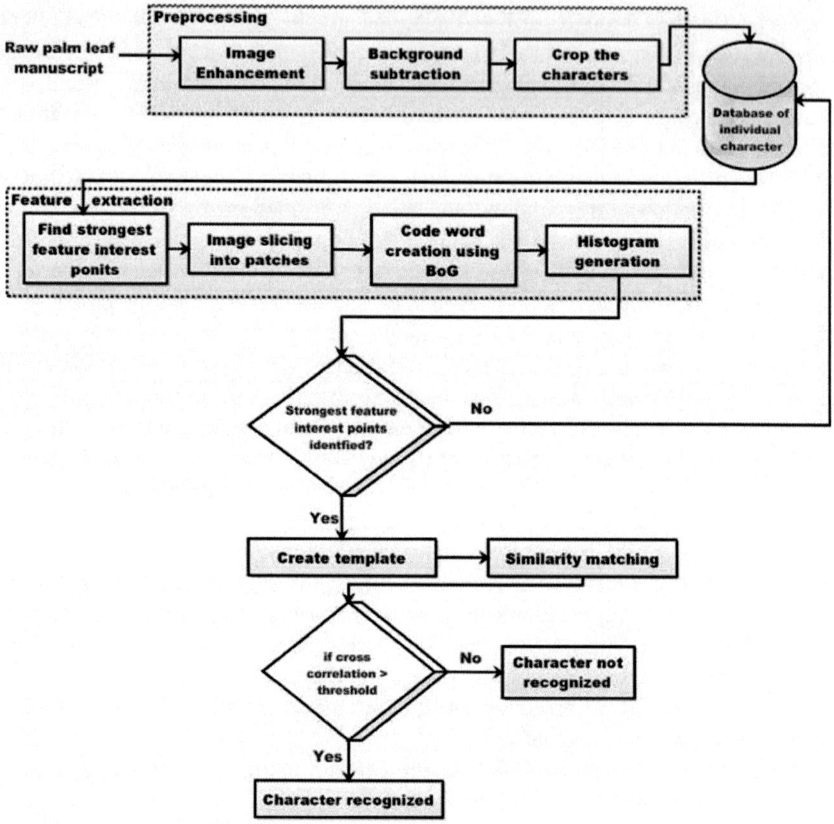

Fig. 1 Architecture of ESURF-BoG

image is easier compared to our application. Therefore, an effective image enhancement should be done to increase the perception of pixel information present in the image. In poor illumination condition, the enhancement technique helps to highlight the useful character pixel by suppressing the background pixels. The captured image is contrast stretched by stretching the pixel intensity values in dynamic range from 0 to $L - 1$. Next, the image is applied to homomorphic filter which increases the contrast by normalizing the brightness of the image. The filter helps reducing multiplicative noise caused by illumination and reflectance. The noise can be reduced by working on frequency domain and taking a logarithm function for the intensity values. The background subtraction is an adaptive thresholding method using the HSV color model. The significance of palm leaf manuscript is the bright, rigid color backgrounds. The third component value (V) is used for thresholding where the other two components H and S are nullified for any pixel that satisfies the thresholding. For an image I, the foreground image F is calculated as given in Eq. (1).

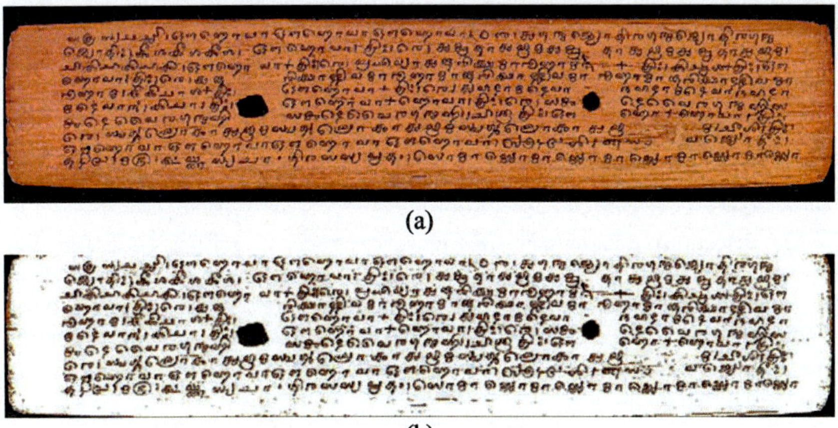

Fig. 2 a Original image. b Background subtracted image

$$F(x, y) = \begin{cases} 1 \text{ if } & V_I(x, y) > \text{th} \\ 0 & \text{else} \end{cases} \quad (1)$$

The output of background subtraction is shown in Fig. 2. From the enhanced image, each character is cropped and stored in the database.

3.2 Image-Based Feature Extraction

Feature extraction is the method of finding the abstraction of image pixel information and making a local decision at which pixel point has feature information. It starts deriving feature attributes from the initial set of input data by subsequent learning of character features by generality steps which will be better than human interpretation. Character recognition using Speeded-Up Robust Feature (SURF) algorithm developed by Bay et al. [9] undergoes three main stages called (i) Feature point detection (FPD), (ii) Confined region description (CRD) and (iii) Feature matching (FM). FPD is the process of locating the strongest points called interest points on the character edge pixels by eliminating the background pixel. SURF constructs the scale space as a level of octaves and spaced uniformly. The integral image is constructed and on which second derivative filter is applied. Need of constructing integral image is to up sample the filter instead of down sampling. Then local maxima sample space is found on which quadratic interpolation is applied. The integral image is given by Eq. (2)

$$I(X) = \sum_{i,j=0}^{i \le x, j \le y} I_{\text{in}}(x, y) \quad (2)$$

where $I_{in}(i, j)$ is the input image and $I(X)$ is obtained integral image where each point at X is $(x, y)^T$. The integral image is calculated based on the response of the rectangular area say P, Q, R, S as shown in Eq. (3). The rect area property performs fast convolutions with different size of box filters.

$$\text{rect area} = P - Q - R + S. \tag{3}$$

The blob detection is done based on Hessian matrix by deriving scalar-valued function by partial derivations. In general, the matrix is given by Eq. (4). The determinant value helps finding the maxima and minima values by second-order derivative. If the determinant value is negative, the point is not local extremum, and if it is positive, the point is extremum.

$$H = \begin{bmatrix} \frac{\partial^2 h}{\partial z_1^2} & \frac{\partial^2 h}{\partial z_1 \partial z_2} & \cdots & \frac{\partial^2 h}{\partial z_1 \partial z_n} \\ \cdots & & \ddots & \vdots \\ \frac{\partial^2 h}{\partial z_n \partial z_1} & \frac{\partial^2 h}{\partial z_n \partial z_2} & \cdots & \frac{\partial^2 h}{\partial z_n^2} \end{bmatrix} \tag{4}$$

The derivatives are calculated by convolution with kernel. The kernels were constructed for the Gaussian matrix x, y and four entries of the Hessian matrix are constructed. The use of Gaussian helps smoothing during convolution. Now let us construct Hessian matrix $H(q, \rho)$, let $q = x, y$ be the space point, ρ be the scale, $G_{xx}(q, \rho)$ be the convolution of the second-order Gaussian derivative given by Eq. (5). Hence, the interest points can be detected by finding Hessian matrix at every pixel.

$$H(q, p) = \begin{pmatrix} G_{xx}(q, p) & G_{xy}(q, p) \\ G_{yx}(q, p) & G_{yy}(q, p) \end{pmatrix} \tag{5}$$

Box filter of 9×9 matrix is considered for the approximation of gaussian function. The strongest points can be found from the Gaussian scale value and here, it is set to be 1.2, because the SURF can handle only lowest scale space. The strongest 128 points were created for each character and stored in the database. For testing an image, the strongest points will be identified and compared with the image stored in the database. Once the features are matched with the data base, it labels the output with the translated version of the Tamil alphabets. Hence, here, it does feature-to-feature comparison The SURF algorithm for training and testing phase is given in Algorithm 1. By using SURF algorithm, the features are extracted for each character and various orientations of the character. Hence,we need n number of comparison by searching the whole database for recognizing which character it is. Even though SURF algorithm able to detect the features and recognize the character, the time complexity is high. Because it searches the whole database, it compares all characters and their variations. Hence, to reduce the time complexity, SURF algorithm is upgraded with the creation of code word model for each character called BoG. Code word model is created for 12 vowels, 18 consonants characters, each has 25 variations. These 25 images are sliced into 500 patches called BoG and stored as a histogram form. The same is repeated for all the 30 characters. The proposed algorithm steps are shown in Algorithm 2.

Algorithm 1 SURF algorithm

Feature extraction in training phase

```
Read the file;
    ScalingFactor = 0.5;
    imageRGB = imresize(readimage(character,ext),scalingFactor);
    totalimages = numel(character);
    for
      n = 1:totalimages do
      boxchar1 = imread(input file name);
      boxchar = rgb2gray(boxchar1);
      strongest keypoints = SURFFeatures detect (boxchar);
      plot(selectStrongest(points,100));
      [Features, Points]=extractfeatures(boxchar, Points);
    endfor
```

Feature extraction in testing phase

```
char= dir(full folder (imagecharfolder,.jpg);
      totalimages = numel(char file);
      char image1 = imread (char file);
      charimage = rgb2gray(charimage1);
      strongestpoints = detectSURF features(char image);
      plot(select strongest(strongestkeypoints,300));
   [char features, strongest key points]=extract features(char image, strongest key points);
```

Matching Features

```
probabox Sets = match features(box features, char features);
    if
      isempty(box sets)
      then display (Nomatch);
    else
      matchedstrongestkeypointsbox = strongestkeypointsbox(boxsets(: 1), :);
      matchedcharpoints = charpoints(boxsets(: 2), :);
      figure;
      show;
      matched features(boxchar, char image,matched strongest points box);
    endif
```

Algorithm 2 Proposed BoG Algorithm

Step1 :**Begin**
Step2 :**Extract** and **Store** the different variations of single characters 'D'
Step3 : **Fix** the proposition of training and testing module (0.7, 0.3)
Step4 : **Slice** the image based on the **SRUF** feature
Step5 :**Construct** the Histogram of images
    Identify SURF points in each image in the folder
    Group SURF points to construct small set of patches
    Count the number of patches
    Plot histogram bar
Step6 : **Select** top 500 histogram to construct common visual code for group characters
Step7 : **Construct** SURF points for test image
Step8 : **Construct** code word for test image
Step9 : **Match** common features with the test image
Step10 :Label the character 'Z' with the modern Tamil alphabets
Step 11 : **End**

4 Result Analysis and Discussion

4.1 Data Set Collection

The proposed method is tested on a huge set of camera captured palm leaf manuscript provided by Agamaa Academy [10]. The method is implemented using MATLAB 2020. The characters are cropped and used for training and testing. In this paper, 5736 Tamil palm script pictures from 24 bundles of Noolaham online archive [11] and the palm leaf manuscripts from the Agama academy digital library [10] are used to generate the training set, and the testing images are obtained from [10] and the Tamil Heritage Web archive [12].

4.2 System Setup

The proposed feature detection and matching method was implemented using MATLAB 2020 platform. For ESURF method, 128 strongest keypoints are detected and the same is used for the recognition of all characters. First, the SURF features are extracted from training palm leaf manuscript image, which predicts 128 strongest key points from the image. The same 128 points are extracted for all the character from the palm leaf manuscripts and stored in the database. Once the testing images are given, the SURF features are extracted from the image and compared with database which already contains the trained image features. The drawback found in implementing SURF algorithm for recognizing single character, it searches whole database and compares with huge volume of data set which is trained. Hence, this is much time consuming.

To resolve the drawbacks of time consumption, modified SURF-BoG feature algorithm is proposed. Totally, at present, there are 12 vowels and 18 consonants. But in ancient characters, seven vowels and 17 consonants, totally 24 characters were used. Here, 25 samples with different orientations from each 24 character say 600 characters were cropped for training the system. For a single character, 25 variations of samples were taken and those characters are sliced into 500 patches. Each character with 500 patches compared all together to get a unique finalized similar patch called BoG is stored in histogram form in the database and this process is called as code-word creation. Such histograms are created for all the 24 characters and stored in the database. Once testing character is given, again it slices the images into 500 patches and generates histogram. Hence, the testing histogram searches for the matching histogram in the database. By correlation template matching between the character in the database and the testing character, the strongest interest points are generated as shown in Fig. 3. Table 1 compares the recognition rate obtained by the proposed method and the classic feature detection methods like SURF [9], min-Eigen [13] and histogram of oriented gradients (HOG) [14]. The histogram form stored in the data base is also shown. The results show that the proposed method outperforms all the classic methods. The least recognition rate of SURF is due to the similarity between the characters as shown in Table 2.

The least accuracy of SURF is misclassification of characters with similar feature points as shown in Table 2.

The experiments were performed on camera captured Tamil palm leaf inscription image. The recognition rate R can be calculated as given in Eqs. (6) and (7)

$$\text{NM} = N - \text{SM} \tag{6}$$

$$R = \frac{\text{SM}}{N} \times 100 \tag{7}$$

where SM is the similar match, NM is non-similar match, N is the number of test samples. The computation results for some of the sample images were shown in Table 3. The result shows that the ESURF-BoG achieves better recognition rate than

(a) (b) (c)

Fig. 3 **a** Image in data base, **b** testing image and **c** strongest interest points identified

Table 1 Comparison of recognition rate uing SURF, HOG, min-Eigen and ESURF-BoG feature detectors along with histogram plot

Test Character	Histogram Image	SURF	HOG	min-Eigen	ESURF-BoG
	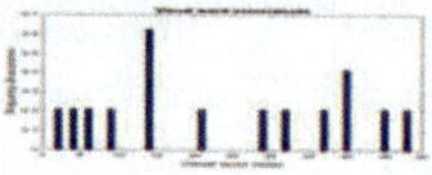	0.74	0.78	0.77	0.80
	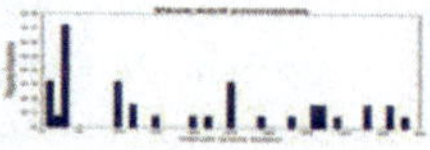	0.77	0.76	0.78	0.84
		0.90	0.91	0.89	0.95
		0.86	0.88	0.90	0.92
		0.76	0.77	0.78	0.80
	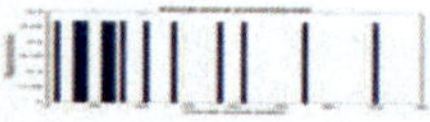	0.74	0.74	0.75	0.76

Table 2 Misclassified characters by SURF due to similarity between the characters

Processed character	Wrongly recognized character
 The vowel *ae*	ெ co character for *ae*
 The consonent *na*	ை co character for *ai*
 The consonent *zhu*	மு The consonet *mu*
 The consonent *Nna*	⊠ The consonent *tha*
 The consonent *pu*	⊠ The consoent *pa*

Table 3 Comparison on recognition rate in SURF and SURF-BoG

S.No	Sample test characters	Modern Tamil Character	SURF N	SURF SM	SURF NM	SURF R%	ESURF-BoG N	ESURF-BoG SM	ESURF-BoG NM	ESURF-BoG R%
1		அ	25	15	10	60	25	20	5	80
2		ஆ	25	13	12	52	25	21	4	84
3		இ	25	16	9	64	25	23	2	92
4		உ	25	11	14	44	25	19	6	76
5		எ	25	16	9	64	25	20	5	80
6		ஐ	25	14	11	56	25	23	2	92
7		ஒ	25	13	12	52	25	20	5	80
8		க	25	16	9	64	25	21	4	84
9		ங	25	12	13	48	25	19	6	76
10	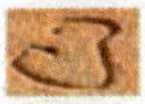	ச	25	14	11	56	25	17	8	68
11		ஞ	25	10	15	40	25	16	9	64
12		ட	25	16	9	64	25	23	2	92
13		ண	25	14	11	56	25	21	4	84
14		த	25	17	8	68	25	24	1	96
15		ந	25	20	5	80	25	20	5	80
16		ப	25	21	4	84	25	24	1	96
17		ம	25	22	3	88	25	24	1	96
18		ய	25	18	7	72	25	19	6	76
19		ர	25	20	5	80	25	22	3	88
20		ல	25	21	4	84	25	23	2	92
21		வ	25	19	6	76	25	21	4	84
22		ழ	25	15	10	60	25	17	8	68
23		ள	25	17	8	68	25	19	6	76
24		ன	25	20	5	80	25	23	2	92

SURF. The characters with complex structures like 'Sa', 'Gna' and 'zha' are recognized with less recognition rate. To improve the recognition rate of these characters, some complex features can be used.

5 Conclusion

The huge collection of palm leaf manuscripts is not yet transcripted. In digital era, capturing the palm leaf images and processing those images using image processing techniques give higher accuracy in developing automated character recognition system. Usual feature extraction methods like structural features cannot identify all features in all character because of the presence of deteriorated character. Therefore, the recognition rate will be reduced. Hence, the proposed system handles pattern matching concept to identify the strong interest points. Proposed ESURF-BoG algorithm handles all type of character and achieves higher recognition rate and reduces time consumption by construction of automated character recognition system. In future, some additional complex features can be used to improve the recognition rate of the characters with complicated structures.

References

1. Renuka R, Suganya V, Arun Kumar B (2014) Online hand written character recognition using digital pen for static authentication. In: International conference on computer communication and informatics (ICCCI), pp 1–5
2. Fujisawa H, Liu CL (2003) Directional pattern matching for character recognition revisited. In: Proceedings of the seventh international conference on document analysis and recognition, vol 2. IEEE Computer Society
3. Prasad JR, Kulkarni UV, Prasad RS (2009) Offline handwritten character recognition of Gujrati script using pattern matching. In: 3rd International conference on anti-counterfeiting, security, and identification in communication, pp 611–615
4. Schmid C, Mohr R, Bauckhage C (2000) Evaluation of interest point detectors. Int J Comput Vis 37(2):151–172
5. Liu CL, Kim IJ, Kim JH (2001) Model-based stroke extraction and matching for handwritten Chinese character recognition. Pattern Recogn 34(12):2339–2352
6. Leutenegger S, Chli M, Siegwart RY (2011) BRISK: binary robust invariant scalable key points. In: IEEE international conference on computer vision (ICCV), pp 2548–2555
7. Fischer A, Keller A, Frinken V, Bunke H (2012) Lexicon-free handwritten word spotting using character HMMs. Pattern Recogn Lett 33(7):934–942
8. Mohammad F, Anarase J, Shingote M, Ghanwat P (2014) Optical character recognition implementation using pattern matching. Int J Comput Sci Inf Technol 5(2):2088–2090
9. Bay H, Ess A, Tuytelaars T, Van Gool L (2008) Speeded-up robust features (SURF). Comput Vis Image Underst 110(3):346–359
10. Shivachariar RV (2019) Agama academy. Online; Accessed 4 July 2019
11. Noolaham Foundation (2015) Noolaham. http://www.noolaham.org/wiki/index.php/. Online; accessed 4 July 2019
12. Subashini (2017) Tamil heritage. http://thfcms.tamilheritage.org/category/palm-leaf/. Online; accessed 4 July 2019

13. Shi J, Tomasi C (1994) Good features to track. In: Proceedings of the IEEE conference on computer vision and pattern recognition, pp 593–600
14. Dalal N, Triggs B (2005) Histograms of oriented gradients for human detection. In: IEEE computer society conference on computer vision and pattern recognition, vol 1, pp 886–893

PlagoBot: A Confluence of Plagiarism and RPA

Venkatesh Kamath, Omkar Lubal, Saurabh Daware, and Vaishali Khairnar

Abstract Plagiarism in research is the unethical practice of using words or ideas (either planned or accidental) of others, without proper acknowledgement. Plagiarism, not only affects the credibility of the author but also hinders the quality of research and inventions. Organizations like Universities use the plagiarism detection software, to manually check the articles and send the results. The whole process becomes pretty mundane and repetitive to carry out in large number of papers received daily. In this paper, Robotic Process Automation (RPA) is proposed to detect plagiarism without any human intervention. This automated process leads to better human utilization, and a more efficient and streamline plagiarism detection process.

Keywords Plagiarism · Plagiarism detection · Robotic process automation

1 Introduction

Robotic process automation (or RPA) is a type of business process automation technology based on robotic software (bots) or artificial intelligence (AI) employees. In typical process automation tools, the software developer generates a list of actions to automate a back-end device function and interface using an internal application programming interface (API) or a specialized scripting language. In comparison, RPA systems create an action list by watching the user perform the task in the Graphic

V. Kamath (✉) · O. Lubal · S. Daware · V. Khairnar
Department of Information Technology, Terna Engineering College Nerul, Navi Mumbai, India
e-mail: kamath797@gmail.com

O. Lubal
e-mail: omkarlubal@gmail.com

S. Daware
e-mail: saurabhdaware99@gmail.com

V. Khairnar
e-mail: vaishalikhairnar@ternaengg.ac.in

G. Ranganathan et al. (eds.), *Inventive Communication and Computational Technologies*, Lecture Notes in Networks and Systems 145,
https://doi.org/10.1007/978-981-15-7345-3_4

User Interface (GUI) framework and then execute the automation by repeating the higher achievement in the GUI [1].

RPA helps companies to simplify at a fraction of the expense and time previously experienced. RPA is also non-intrusive in nature and leverages existing infrastructure without compromising underlying structures, which would be difficult and expensive to upgrade. With RPA, cost management and enforcement is no longer an ongoing expense, but a by-product of automation [2].

2 Why RPA Over Traditional Automation Methods

Traditional automation applications are written in programming languages using various APIs and libraries, thus increases the complexity of the application to debug. Whereas in the case of RPA, it mimics actions of the user on the UI, and these actions can be easily programmed into the application through ways like dragging and dropping the required activities to carry out a particular task. The users need comparatively less training than software programmers in order to start creating, tailoring, and utilizing the application according to their own needs.

3 Methodology

Robotic Process Automation is the core feature in our project, as the bot will utilize the features of RPA and will be responsible to carry out the complete workflow of plagiarism detection.

1. The bot is given access to the incoming mailbox it keeps scanning for any new emails that arrive at a fixed interval of time.
2. As soon as the bot detects that a new mail has arrived it reads the mail and downloads the attachment that came with it.
3. Now the bot extracts the content from the document and stores it in a variable.
4. After extraction of the contents of the document, this content is inputted to a plagiarism detection software to give the percentage of plagiarism in the document.
5. Based on this percentage obtained, it is compared with a threshold value and then necessary actions are carried out by the bot.
6. If the percentage of plagiarism is found to be greater than the threshold value then a "paper rejected" mail is sent to the candidate with the amount of percentage detected.
7. If the percentage of plagiarism is found to be less than or equal to the threshold value then a "paper accepted" mail is sent to the candidate or a notification will be sent to the necessary authority if required (Fig. 1).

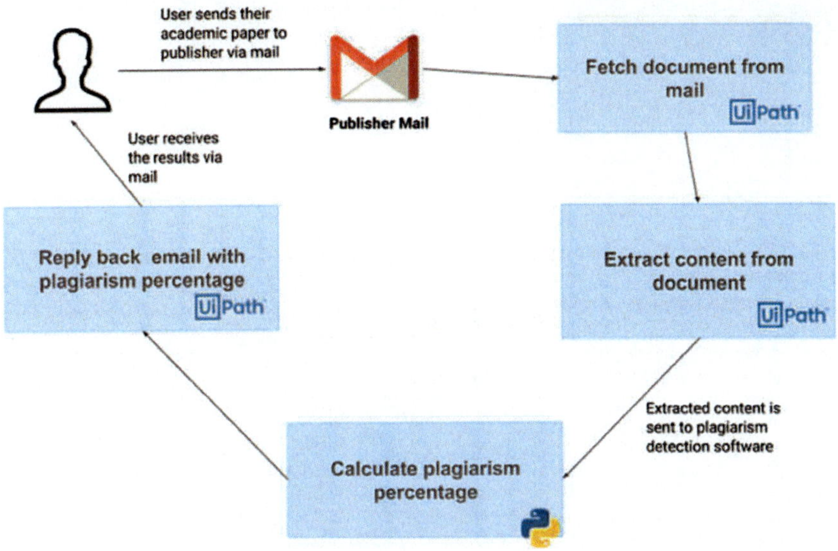

Fig. 1 High-level flow diagram

4 System Design

In Fig. 2, Flowchart activity is depicted. The most important aspect of flowcharts is that, unlike sequences, they present multiple branching logical operators, and enable them to create complex business processes and connect activities in multiple ways [3].

Figure 3 shows the variables that have defined for MainFlowChart activity and its type.

The rectangular start button is the Start Node that marks the point of entry into the workflow.

Get IMAP Mail Messages

This activity is responsible for fetching the IMAP (Internet Message Access Protocol) emails from the specified server and login information. This activity belongs to UiPath.Mail.IMAP.Activities.GetIMAPMailMessages package

As seen in Fig. 4,

(1) DisplayName contains the name that you want to give to the activity.
(2) 2)TimoutMS specifies the amount of time you have to wait for network streaming operations before throwing an error.
(3) MailFolder field contains the name of the mail folder to retrieve the messages.
(4) Port specifies the incoming mail (IMAP) server's port number, Server specifies the name of the email server host.

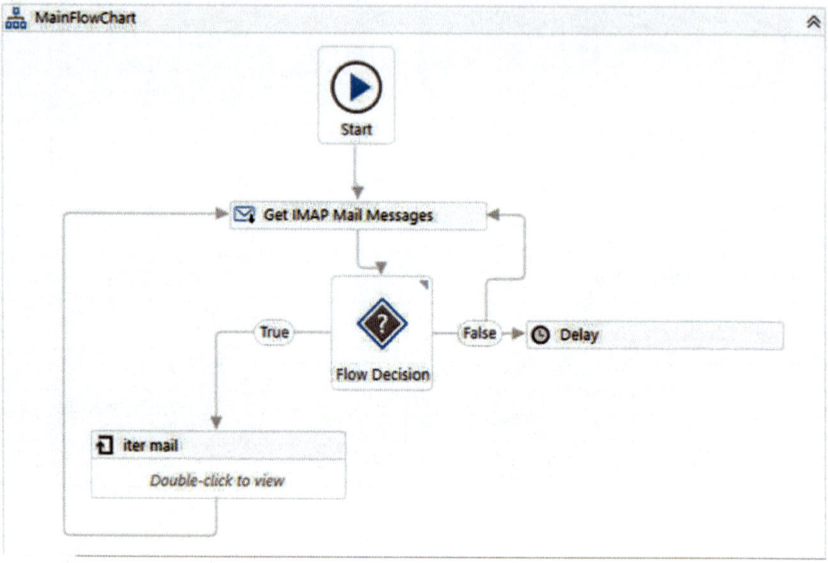

Fig. 2 Main flow chart

Name	Variable type	Scope	Default
RetrivedMails	List<MailMessage>	MainFlowChart	Enter a VB expression
Create Variable			

Fig. 3 Variables defined for the *MainFlowChart* activity

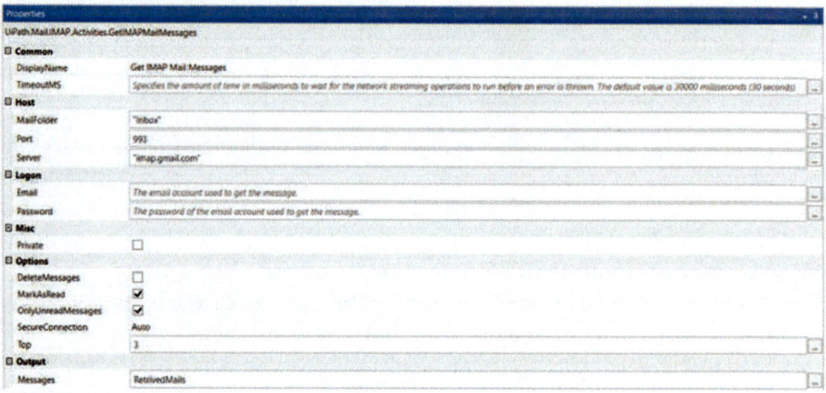

Fig. 4 Properties of *Get IMAP Mail messages*

Fig. 5 Properties of *Flow Decision*

(5) The email specifies the name email address of the account that our bot is going to use.
(6) Password specifies the password of the account that our bot is going to use.
(7) Checking MarkAsRead asks the bot to mark the messages it has processed as read so it won't process them again.
(8) Checking OnlyUnreadMessages specifies to read-only unread messages.
(9) Top specifies the number of emails to fetch at a time from the mail inbox.
(10) Messages specify the name of the collection of MailMessage objects that our bot is going to fetch.

A. **Flow Decision**

This activity is a part of the System.Activities.Statements package. This activity provides conditional switching between two branches based on a given condition. It works like an if-else condition. As in our case if the condition provided turns out to be true then the control will be passed to the next activity, i.e. iter mail or if it turns out to be False then the control is given to the Delay activity.

As seen in Fig. 5, the condition is RetrivedMails.Count > 0, this RetrivedMails variable contains the collections of the mail messages that got from the previous activities, and count function of C# gives us the count of emails in the variable. So if the count is greater than 0 then return TrueLabel, i.e. True else return FalseLabel, i.e. False.

B. **Delay**

When the Flow Decision activity shows False, as the name suggests it delays the execution of the next activity in its workflow for a specified amount of time (Fig. 6).

This delay is added in order to let new incoming mails come to the mailbox.

After the delay of 1 s, the control will be passed again to Get IMAP Mail messages activity to fetch any new incoming mails.

C. **iter mail**

This activity is an iterative type of activity that is going to use to process the mail objects stored in the RetrivedMails collection variable. This belongs to the UiPath.Core.Activities package and is a For Each type of activity that allows us to perform various activities on each element of the iterable variable that is provided.

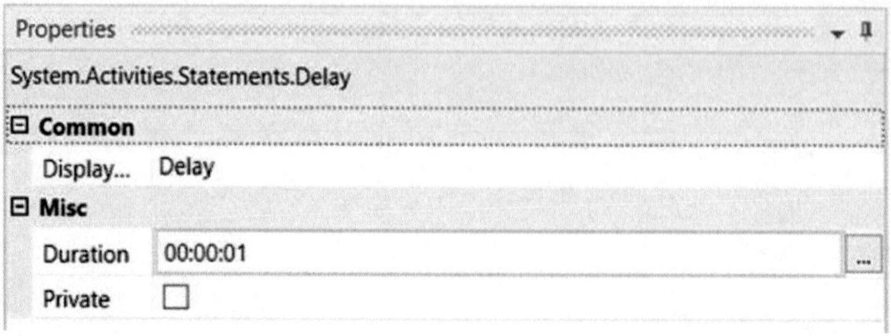

Fig. 6 Properties of the *Delay* activity

As seen in Fig. 7, the single mail object is stored in the item variable, and then passing it on to the next activity.

Now, let us understand the activities inside the iter mail activity.

(1) *Save Attachments*

As the name suggests this activity is responsible for storing the attachments of the MailMessage object.

(2) *If*

This activity gives us an If-Then-Else model, such that if the given condition is true then execute the Then block else execute the Else Block.

(3) *Sequence*

This activity is responsible for executing the set of activities it is provided in a given sequence and is a part of the System.Activities.Statements package.

ExtractContent

This activity is responsible for extracting text content from the attachments and passing it on to the next activity for further processing. It consists of a sequence activity containing Open Browser with an inbuilt sequence activity for content extraction activity.

Also, define two arguments as shown in Fig. 8 .fileName of type string that will contain the file path of the saved attachment and will be passed to this activity and pdfcontent of type string that will contain the content that will extract from the document and will be the output of this activity (Fig. 9).

An Open Browser activity is used, a part of UiPath.Core.Activities package, to open the document, as browsers are able to handle any type of documents like word or pdf, this is in order to avoid using any proprietary software and avoid any incompatibility issues.

This sequence activity contains the following activities.

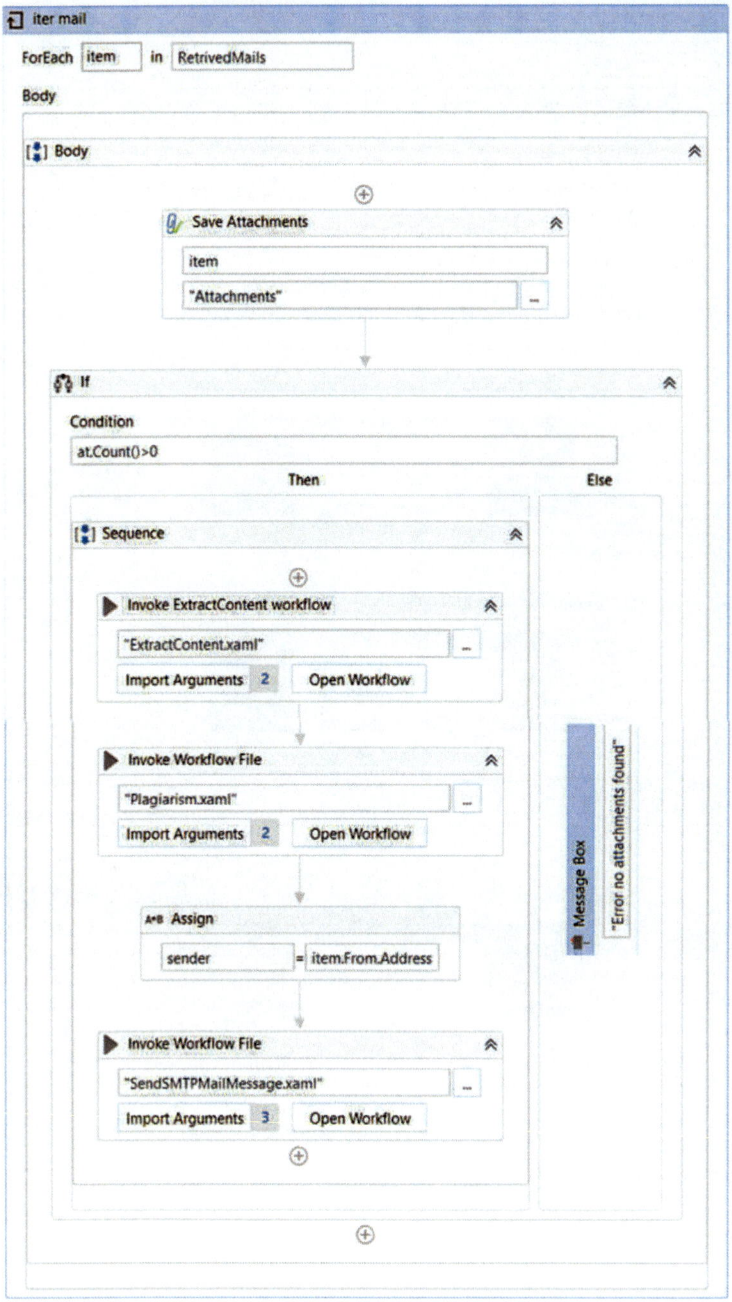

Fig. 7 *iter mail* activity

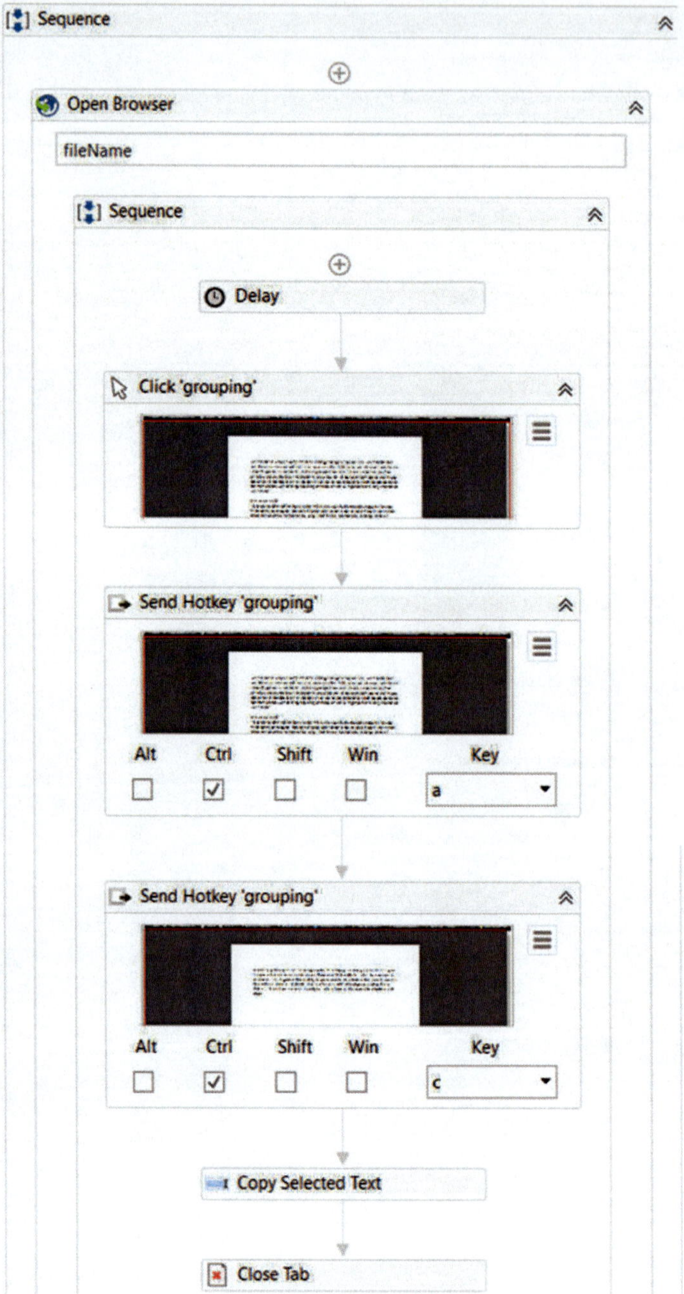

Fig. 8 *ExtractContent* complete workflow

Name	Direction	Argument type	Default value
fileName	In	String	*Enter a VB expression*
pdfcontent	Out	String	*Default value not supported*

Fig. 9 Arguments of *ExtractContent* activity

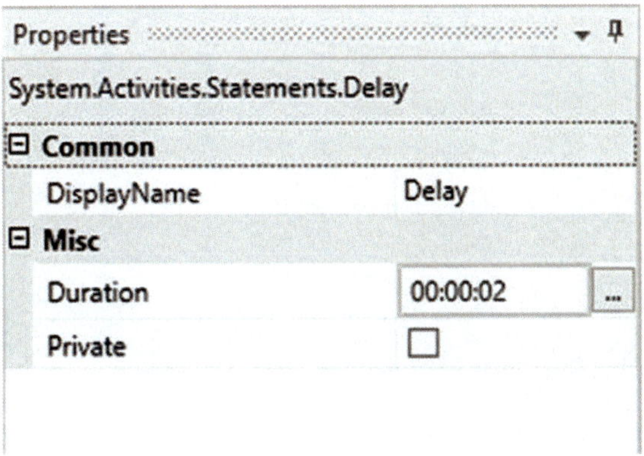

Fig. 10 Arguments of *Delay* activity

(1) **Delay**

As mentioned before this activity provides the delay functionality, and this is added in order to provide some time for the Open Browser activity to open the document as sometimes opening a document might take some time depending on the system's performance (Fig. 10).

(2) **Click Activity**

This activity is used to simulate a click on the current browser window, in order to shift focus to the current browser window, to perform copy-paste operations.

(3) **Send Hotkey (For Select All)**

This activity is used to simulate keyboard shortcuts and is going to use this activity to send Ctrl + A which is equivalent to Select all content on the current screen (in Windows).

(4) **Send Hotkey (Copy Selected Content)**

This activity is used to send Ctrl + C to the UI element and copy the selected content to the clipboard.

(5) **Copy Selected Text**

This activity is used to copy the text from the clipboard onto a variable and specify the name of the variable in the result field of the properties as shown in Fig. 11.

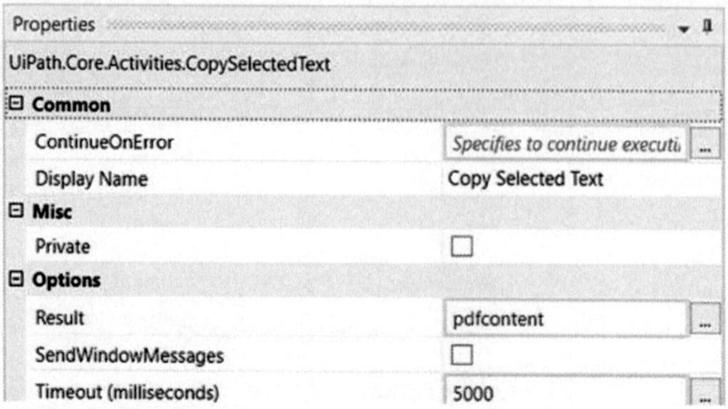

Fig. 11 Properties of *Copy Selected Text*

(6) Close Tab

This activity is used to close the browser tab.

D. Plagiarism

This activity contains a sequence of activities that perform plagiarism detection on the content.

In order to detect plagiarism, a python script is created that performs plagiarism detection using Google Search API and the Levenshtein Distance method (Fig. 12).

The sequence in this activity contains a Python Scope that accommodates all python activities and objects, also this activity is responsible for initializing the python environment that is specified in its properties as shown in Figs. 13 and 14.

The Python Scope activity contains the following activities.

(1) Message Box

This activity displays a message to the user, to indicate that the plagiarism detection process has started.

(2) Load Python Script

This activity is a part of UiPath.Python.Activities package and is used to load and execute python scripts (Fig. 15).

Name	Variable type	Scope	Default
loadOutput	PythonObject	Do	*Enter a VB expression*
invokeOutput	PythonObject	Do	*Enter a VB expression*
pythonOutput	Object	Do	*Enter a VB expression*

Fig. 12 Variables defined in *Plagiarism* workflow

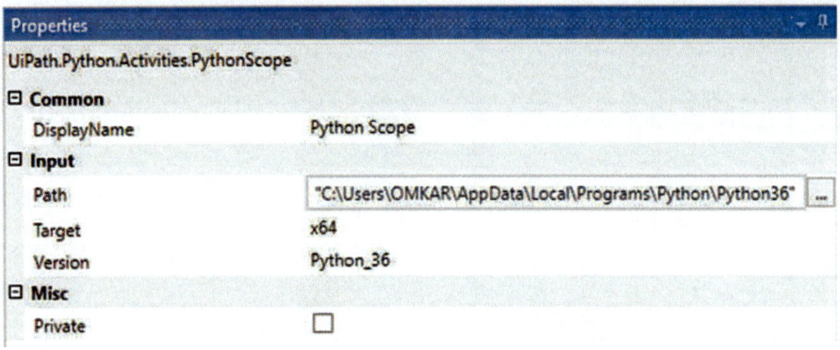

Fig. 13 Properties of *Python Scope*

(3) **Invoke Python Method**

This activity provides the functionality of invoking or calling a python method from the script (Fig. 16).

(4) **Get Python Object**

This activity converts the python object into a .NET data type. In the properties tab, python Object field is used to specify the variable's name that is currently storing the output of the invoked method (Fig. 17).

(5) **Assign**

This activity is used to set the values of various workflow variables. The pythonOutput variable that is storing results from the previous activity will be assigned to an argument named percent, and only its string data will be assigned as a percent is a string type of argument. This argument will then be passed to another workflow for further use.

(6) *Message Box*

Another message box will be shown to the user indicating the plagiarism detection process has ended.

E. **Assign**

After the content extraction and plagiarism detection process has ended, now send the results to the candidate, and for this, require their mail id, and this id is already stored in the item variable and can be accessed using.From.Address. This obtained email id is then assigned to the sender argument which will be passed on to the next workflow as mailTo argument, responsible for sending result mail to the candidate (Fig. 18).

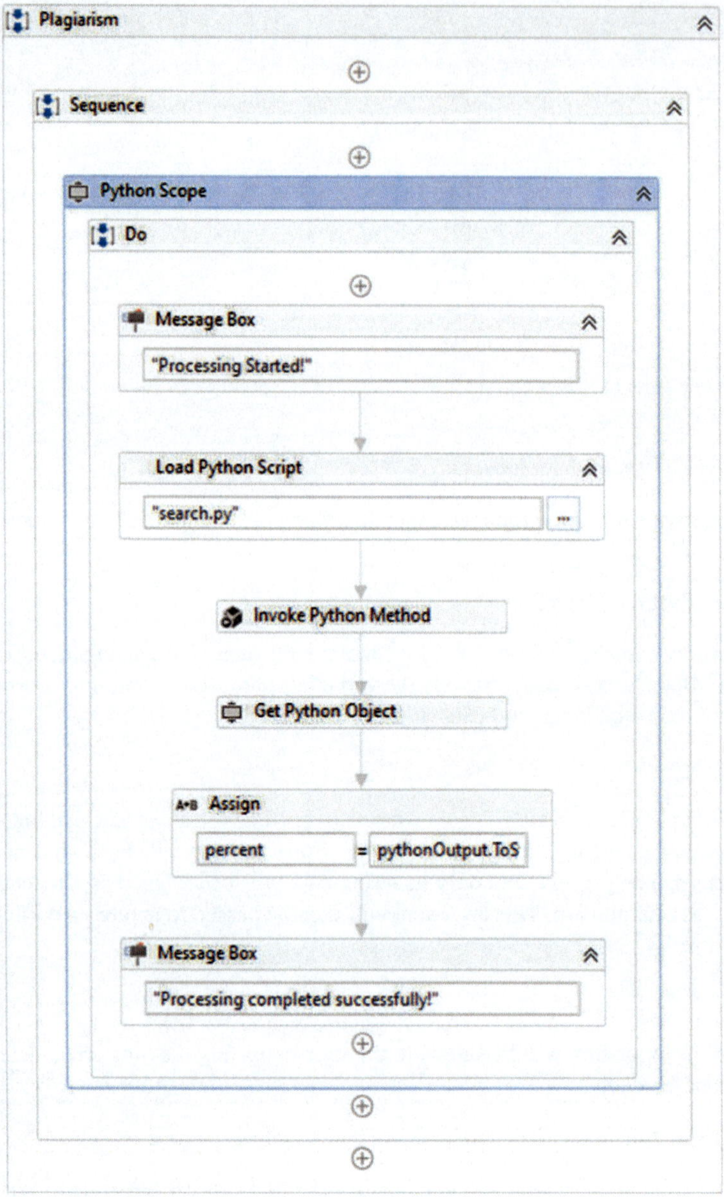

Fig. 14 *Plagiarism* complete workflow

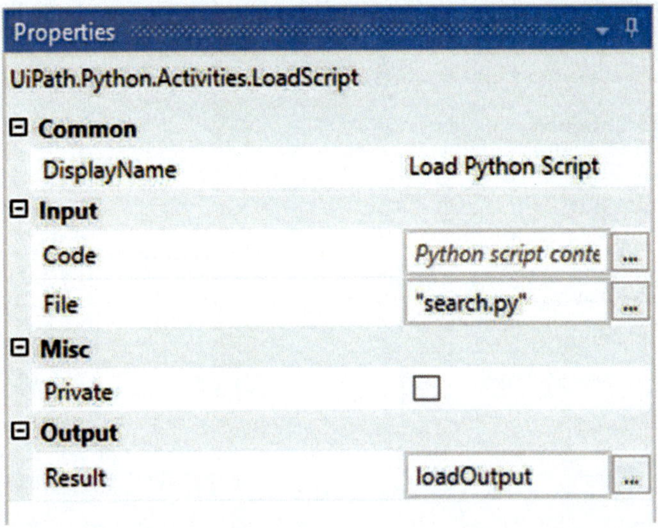

Fig. 15 Properties of *Load Python Script*

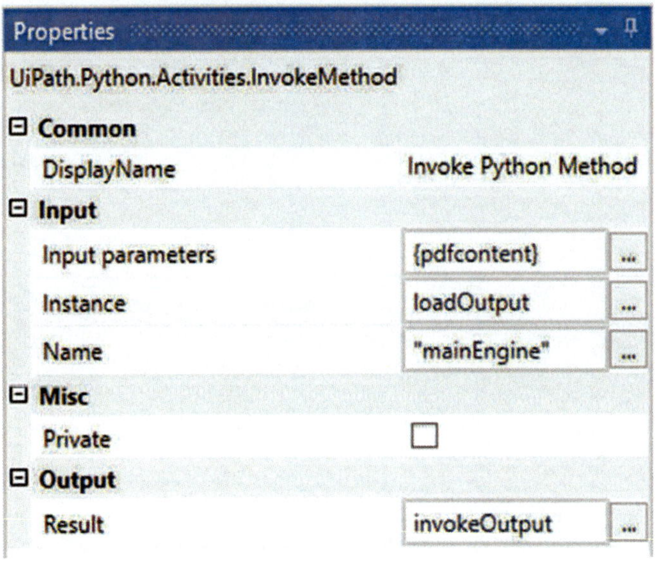

Fig. 16 Properties of *Invoke Method*

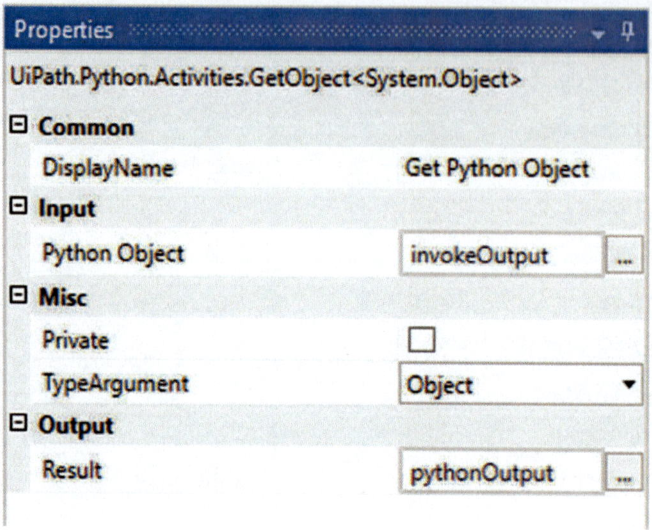

Fig. 17 Properties of *Get Python Object*

Fig. 18 Properties of *Assign*

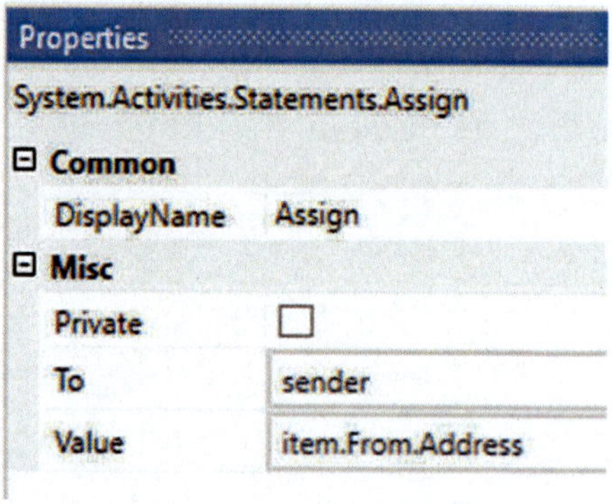

F. **SendSMTPMailMessage**

This activity is used to send mail messages using the SMTP (Simple Mail Transfer Protocol). This activity belongs to UiPath.Mail.SMTP.Activities package (Fig. 19).

Also, have to set certain fields in the properties tab of this activity like:

(1) Port number to 465 to support SMTP.
(2) The host specifies the host server address that is to be used.
(3) Email and password fields require you to enter your valid email credentials from which you have to send the mail.
(4) The name field will specify the name of the sender.

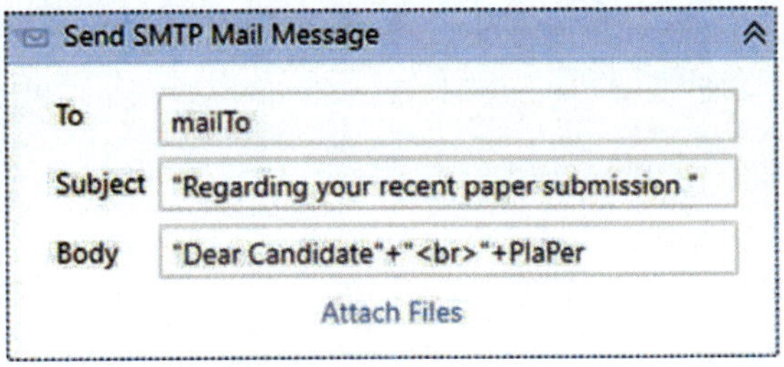

Fig. 19 SendSMTPMailMessage activity

So at the end, an email will be sent to the candidate that will contain the result of their submission and this concludes the processing of single document, after this the next document will be provided for processing by the ForEach activity if it is available, else the control will be given back to the Get IMAP Mail messages to fetch new incoming mails.

5 Plagiarism Detection Process

In order to perform plagiarism detection, a plagiarism detection script is created using python and passes the content to the script and it provides us with the percentage of plagiarism, highlighting certain areas of the content where it is found as well as the sources from where the content is copied.

Google Search API is used to search the content via Google's Search Engine and pass the content to the API and it provides us with relevant links that contain data similar to ours.

(1) Visit the link, and convert the data on the link to string using Web Scraping techniques.
(2) Apply the Levenshtein Distance algorithm to this fetched data.
(3) This algorithm gives us the difference between two strings, in the form of the number of single-character edits(insertions, deletions, substitutions) required to convert one string to another. It gives us the difference percentage between the strings, and convert this into the similarity percentage by subtracting it with 100%. Thus, now have the plagiarism percentage of the content.
(4) Repeat this process for all other links and generate a final plagiarism report of the complete document (Fig. 20).

In the end, our script will give us a final report consisting of highlighted content representing that it is plagiarised and also the source. This report is then sent to the RPA bot for further execution.

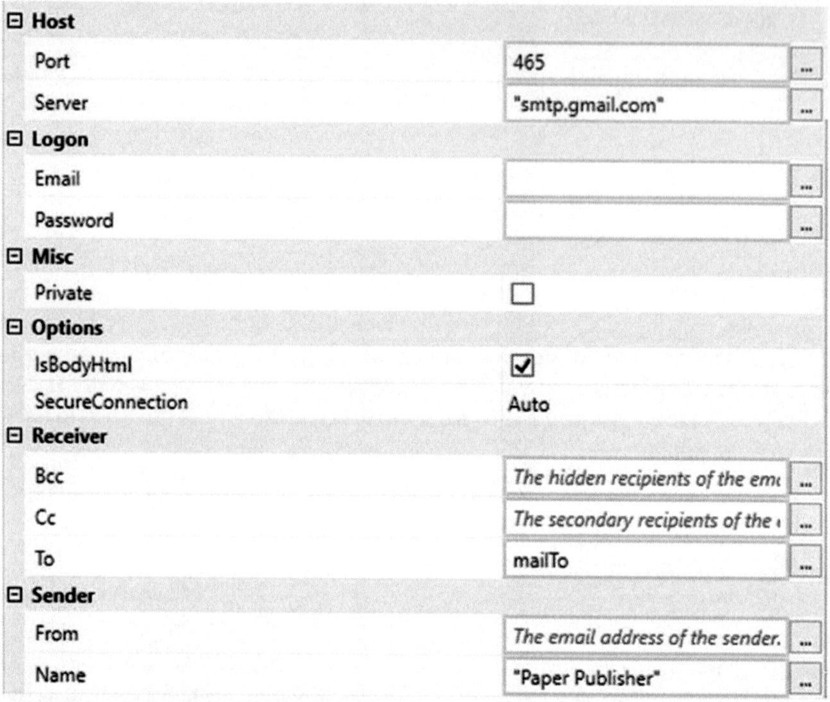

Fig. 20 Properties of *Send SMTP Mail Message*

6 Features of the System

This bot can be easily deployed on any daily use office computer or a PC with an internet connection, thus changes to the system on which our bot is being deployed will be minimal.

The main advantage of using an RPA bot is that they can be kept running 24 × 7 with minimal expense and supervision. The bot can accept submissions at any time of the day, and produce insightful results for the candidate as well as for its stakeholders like Paper Publishers or Educational Institutes.

Another important feature of our system is its modularity which makes it easy to make changes and attach new functionalities to the bot like integrating various other plagiarism detection methods or third-party software and thus increasing the plagiarism detection accuracy.

As shown in Fig. 21, the plagiarism report that will be generated and sent to the candidate.

Person Tracking using Facial Recognition

Abstract: Currently the process of finding a person is by manually looking at the surveillance and track accordingly which gets difficult for the person or the officials to find a person and may take up to days to track a person also there are many missing complaints which make a burden on the officials. A facial recognition system is a technology capable of identifying or verifying a person from a digital image or a video frame from a video source. There are multiple methods in which facial recognition systems work, but in general, they work by comparing selected facial features from given image with faces within a database. It is also described as a Biometric Artificial Intelligence based application that can uniquely identify a person by analyzing patterns based on the person's facial textures and shape.[1][2][3] 2

Create a person tracking system using Facial Recognition. Track a person using face recognition Stop unauthorized person from entering prohibited areas. Find lost person and find criminals. Facial recognition is a way of recognizing a human face through technology. A facial recognition system uses biometrics to map facial features from a photograph or video. It compares the information with a database of known faces to find a match. Facial recognition can help verify personal identity, but it also raises privacy issues. 1 We will develop a system which will use Machine learning and Image classification to detect a person and track it with less time by just providing the photos of the person.

1: https://us.norton.com/internetsecurity-iot-how-facial-recognition-software-works.html **(41% copied)**
2: https://en.wikipedia.org/wiki/Facial_recognition_system **(32% copied)**

Plagiarism Percentage: **73%**
Sorry but your paper is rejected due to high plagiarism. Please try again later.

Fig. 21 Plagiarism report generated

7 Conclusion

This paper shows the development of an RPA bot that performs the complete plagiarism detection process end to end without any human intervention.

This bot will help curb the prevalent problem of plagiarism more efficiently and also lead to better human resource utilization by helping it concentrate on more important matters rather than on repetitive and mundane processes.

References

1. Wikipedia Contributors (2020) Robotic process automation. In: Wikipedia, The Free Encyclopedia. Retrieved from https://en.wikipedia.org/w/index.php?title=Robotic_process_automation&oldid=942667705 (26 Feb)
2. Robotic Process Automation (RPA) (2020). Retrieved from https://www.computersi.com/solutions/robotic-process-automation-rpa/ (2 Mar)
3. Flowcharts (2020). Retrieved from https://docs.uipath.com/studio/docs/flowcharts (20 Mar)

MAGE: An Efficient Deployment of Python Flask Web Application to App Engine Flexible Using Google Cloud Platform

B. Aakash and A. Srilakshmi

Abstract In this digitalized era, most of the data is available as images. Text extraction plays an important role in finding vital and valuable information from images and is useful in processing, retrieving, editing, documenting, etc. In this paper, a cloud-based solution is proposed for efficiently extracting text using API. Large datasets with images can be uploaded to cloud storage using cloud vision API and Google text-to-speech API would extract text with flexibility. This can be is accomplished by Google cloud platform (GCP), using python flask. and can be deployed to Google cloud using python flask.

Keywords Application programming interface · Cloud vision · Google cloud platform

1 Introduction

Extracting text from different languages or formats is a challenging issue. The image text can be of any form such as scanned images, newspaper clippings, magazines, posters, etc. If the exact text is extracted successfully then interpreting the image can also be done. In the proposed work, Google cloud plays a major role in text extraction using API. The cloud vision API can perform well with even PDF and TIFF (Tagged image file format). The only limitation is that the conversion for images up to 2000 is done accurately. In our proposed work, the text extraction is deployed in one of the service called App Engine of Google cloud platform. The vision API has two annotated features such as TEXT_DETECTION and DOCUMENT_TEXT_DETECTION. The extracted text is converted to speech using text-to-speech and the audio is downloadable. Different fonts and line breaking texts in

B. Aakash · A. Srilakshmi (✉)
Sastra Deemed to be University, Thanjavur, India
e-mail: srilakshmi@cse.sastra.edu

B. Aakash
e-mail: aakashbabu.2000@gmail.com

G. Ranganathan et al. (eds.), *Inventive Communication and Computational Technologies*, Lecture Notes in Networks and Systems 145,
https://doi.org/10.1007/978-981-15-7345-3_5

bounding boxes are retrieved accurately. The image is uploaded to Google cloud storage which is capable of handling large image databases. The output of the PDF or TIFF format is then written to JSON file and the same is uploaded to cloud storage bucket. The image stored in bucket can be restricted with access permissions so that other users cannot access the image. The other users apart from the owner of the image are given only reader rights. The proposed work extracts text efficiently and accurately with security.

2 Related Work

Many research techniques have been proposed from image to text conversion which basically uses OCR techniques. Many ideas have been proposed with recent deep learning techniques and worked out with different types of images. Researchers used a modified deep learning approach for a large number of speaker identification. Using Mel Frequency Cepstral Coefficient feature vectors are represented. Using the bidirectional LSTM technique they achieved 99% accuracy [1]. Using the web scraping technique the text is extracted. Each word is ranked in terms of frequencies using rank size procedure. It works with different images also [2]. Generative Adversarial Network plays the main role from image to text conversion. U-Net architecture performs well than any other and gives impressive results. The three combined techniques GAN loss, L1 loss, and L2 loss are combined and tested to give accurate results [3]. The text is extracted from natural scenery images by using combined algorithms. It is implemented using three stages of algorithms such as detection and localization, and image preprocessing such as segmentation and the last stage as optical character recognition(OCR) [4]. In this paper, HAAR DWT is the technique used that separates the texted and non-texted regions by segmenting the images. After performing encoding and decoding process the morphological operations are done to achieve better extraction [5]. It uses a seam carving method for grayscale and binary images. The proposed work is language independent and implemented two algorithms for medial seams and separating seams. It gives better results when compared with other algorithms [6]. Three methods have been proposed to perform better extraction such as text classification, text detection, and character identification. The results are compared with benchmark datasets and achieve better results [7]. It performs a hybrid approach by merging two methods namely texture method and connected component method. An automatically constructed MLP text classifier is used which increases the accuracy parameter to high range [8]. Feature-based image classification is done using color autocorrelation histogram method and another method namely scale-invariant feature transform method. It is compared with other techniques but this hybrid approach yields better results [9]. Using CNN with modified parameters and aggregating the feature maps local features are proposed. Low-level features are concentrated more. They also proposed a probabilistic hash retrieval method to learn the semantic features of the image [10]. The Flickr dataset is taken and multimodal sentiment analysis has been proposed and the mid-level features are more concentrated [11]. Images captured from CCTV cameras at different angles and different heights are considered. The proposed method extracts features from corner points, area, and cavity of the CCTV captured images using Delaunay triangulation method. The proposed method outperforms existing method [12]. Mining and visualization techniques are applied to extract dense and abstract text from the image. To extract the key information term frequency-inverse document frequency algorithm is used. The proposed work outperforms and reduces the time and workload to extract the information from the images [13]. It is a Chinese text recognition system to identify the Chinese text. A new temporal convolutional recurrent network is created as the proposed work. Compared to the existing methods the proposed one outperforms and gives faster prediction [14]. The proposed work focuses on helping visually challenged people by extracting text from image and then converting to audio. Several preprocessing has been done to achieve high accuracy. Scanned images are processed using LattePanda Alpha system [15].

3 Implementation

In the proposed work, as discussed the Google cloud services are used. This new approach uses a web application to perform text extraction from image. Then the extracted text is converted to speech using API. The same is done for image dataset containing images of different font, blurred text, different language, various bounding boxes, newline, etc. a new approach uses web application to perform the text extraction from the image and convert the text into audio file. Then, the complete web application is deployed in the App Engine, a serverless deployment in Google Cloud Platform (GCP). For example, the application allows a user to upload a photo of a handwritten image and gives the text as the output along with the audio file as output where he can listen to it or download the audio file. The proposed architecture is worked out as specified in Fig. 1.

The above flow of the entire process is performed by using the following Google cloud platform (GCP) Resources, service, and API's offered by the provider.

- **Google Service Accounts**: The Service Account is used to access the Google Cloud APIs when testing locally. Once the service account is created and activated the appropriate permissions are given to it and finally, service Account key is created.
- **Google App Engine**: With App Engine, there are no servers to maintain. It is enough to upload the application only. Hence the entire web application is uploaded into the App Engine.
- **Virtualenv**: Though it is not a Google Resources a virtual environment is needed to be created for running the python-3. For a flask web framework it is necessary to use a virtual environment.

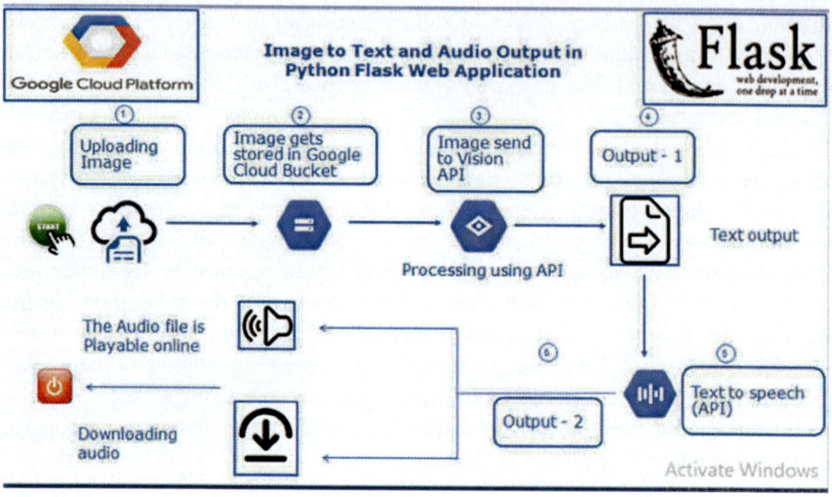

Fig. 1 Proposed methodology

- **Cloud Storage Bucket**: It is necessary to create a storage bucket to store the image dataset for processing one by one. Even after the conversion, the audio version of the text is stored to bucket again.
- **Cloud Monitoring**: This resource helps us to monitor the https traffic from the end-user who uses our web application from the host website also shows the CPU load.
- **Cloud Vision API**: It is the most important API that extracts the text from the image. This API is free to use and thereby helps us extract the text easily.
- **Cloud Text-to-Speech API:** This API is used to convert the text to audio. A request is made by passing the text as a parameter and the output is provided by the API as an audio file.

The services can be used on-demand. Since all the APIs are not enabled by Google cloud it is necessary to enable the text-to-speech API. Once logged into the cloud platform with the necessary credentials it is mandatory to set the current workspace to point the generated project ID using

>gcloud config set project [project_id]

Next in the Cloud Shell command-line, clone the github repository which has the current work's web application code. This cloning can be done by running the following command in the Cloud Shell

>git clone https://github.com/itsmeaakash77/gcp-project.git

The cloned repository consists of necessary python, .yaml and requirements file needed to run the application successfully and for further deployment to app engine service. It is now required to change the directory to GCP project using

>cd project_id

Now in order to authenticate the API requests and to set an environment variable for the project the following command has to be executed

>export YOUR_PROJECT_ID = [YOUR_PROJECT_ID]

As shown in the flow diagram Fig. 1 a service account has to be created to access the google cloud API. This is particularly used when the app has to be tested locally by the end-user. Once created the service account appropriate permissions can be set by using the below two-step command

>gcloud iam service-accounts create qwiklab \ –display-name "My Qwiklab Service Account"

>gcloud projects add-iam-policy-binding ${PROJECT_ID}\–member ServiceAccount:qwiklab@{PROJECT_ID}.iam.gserviceaccount.com\ –role roles/owner

Once the above command is executed without errors then service account key is generated and stored in JSON file by the name key.json in the home directory. Using this key the API required for this work can be accessed easily. The command is as follows

>gcloud iam service-accounts keys create ~/key.json \–iam-account qwiklab@${PROJECT_ID}.iam.gserviceaccount.com

Using the absolute path of the generated key an environment variable can be set for the service account key that is generated and to test the cloned web application locally a virtual environment is created. This is done as a two-step process.

>export GOOGLE_APPLICATION_CREDENTIALS="/home/${USER}/key.json"

>virtualenv -p python3 env

The python scripts which are written runs in local host of the virtual environment and it needs to be activated. From the cloned app github repository the dependencies from the requirements file are installed which is required by the application. The requirements .txt file consist of a list of package dependencies that is needed for the project. It also downloads all the package required for virtual environment.

>source env/bin/activate

.>pip install -r requirements.txt

>gcloud app create

A storage bucket is created using the cloud storage service where an image dataset can be uploaded. In order to set the environment variable CLOUD_STORAGE_BUCKET equal to the name of the project id, the below command is executed.

>export CLOUD_STORAGE_BUCKET=${PROJECT_ID}

>gsutil mb gs://${PROJECT_ID}

Finally, start the application using,

>python main.py

Once the application starts, preview the app using the web preview option as shown in Fig. 2 in the Cloud Shell toolbar and choose "Preview on port 8080."

A tab is opened in a browser and it gets connected to the server that just started. The screen will appear like this as shown in Fig. 3.

And finally, the application can be deployed to App Engine once tested successfully. App Engine Flexible uses a file called *app.yaml* to describe an application's

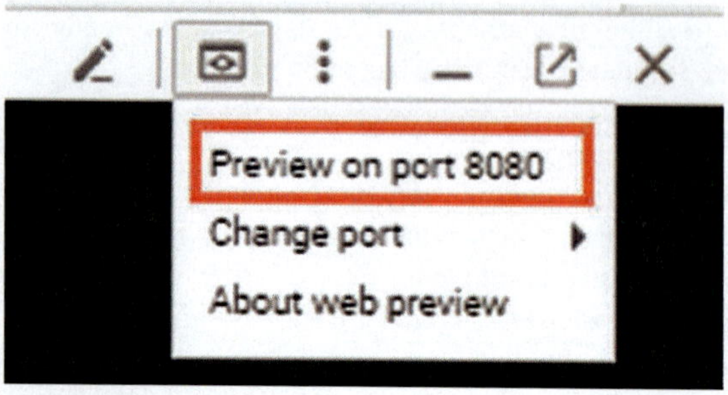

Fig. 2 Application hosted in port 8080

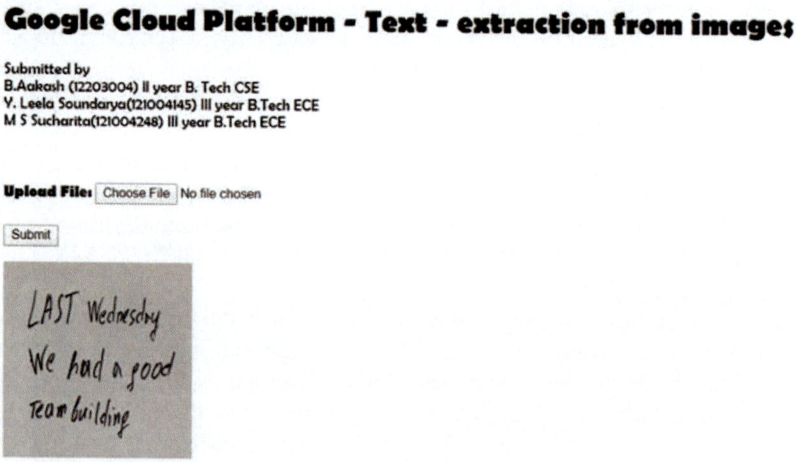

Fig. 3 Text extraction from images

deployment configuration. If this file is not present, App Engine will try to make a guess of the deployment configuration. However, it is a good idea to provide this file.

Next, to make necessary changes to the app.yaml file regarding any configuration information such an update can be done in here using nano editor.

>nano app.yaml

After making necessary changes to the yaml file the deployment of app is done. The next important step is to deploy the app using the command as shown in Fig. 4. This is done using the following command

Fig. 4 gcloud editor

>gcloud app deploy

After hitting the enter key an url(https://<PROJECT_ID>.appspot.com) is generated as shown here. This is the final and the last step of the work where the image extraction, conversion and downloading has been done.

4 Results and Discussions

This is how the website looks when the link is clicked. This works well for any kind of image. In future, the noise is removed from message and performs some image preprocessing techniques to get perfect output. The video images can also be tested in a similar way. Figure 5 shows the sample output of end-user handwritten images. Then the remaining Figs. 6 and 7 are the output for the images in blur form and different languages. Hence the image is independent of language, font, size, angle, appearance, etc. The extracted text is downloaded as audio and downloaded in mp3 format. From the below sample outputs, it is clear that the API's used works well at any instant and produces better results. The testing was done with nearly 50 images and received 100% accuracy.

Fig. 5 Audio enabled text

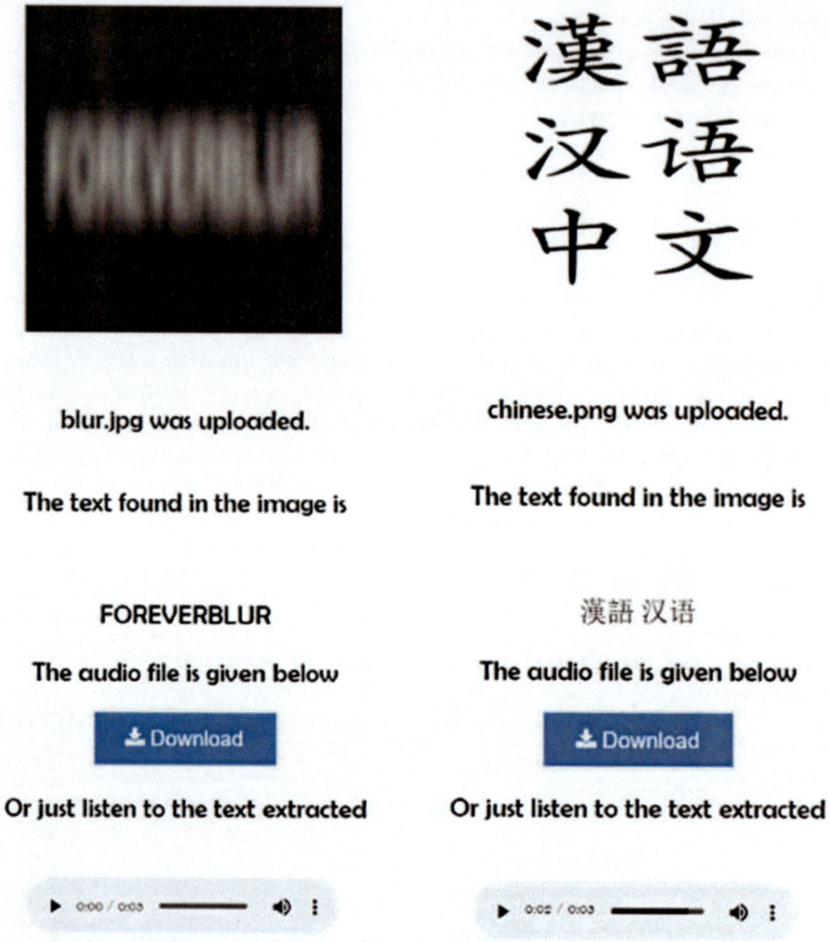

Fig. 6 Blur and Chinese text conversion

In Fig. 5, you can see that any file format from the local disk can be selected for conversion. Once submitted the conversion from image to text takes place. The results are also stored in json format which retrieves the image stored in bucket. Also using the API, the bounding box of the text direction (*X* and *Y* axis) are also stored in json file. If the text is broken to next line another bounding box is created and its XY values are stored in json for further processing along with the corresponding text name.

The above-shown results are done in Google cloud platform and random images are stored in cloud bucket and tested individually.

Fig. 7 Tamil text extraction

5 Conclusion and Future Work

By making use of Google cloud and its services the successful deployment of app has been done using App Engine. This is helpful for visually challenged people in greater way. Further, in the future, try to incorporate online images and perform preprocessing. Mathematical equations extraction and evaluation also have to be tried in the future. Also, if the image is with different languages and can try to extract it and then add a feature named Google translate to convert to common language. In the combination of images, also try out a technique to evaluate with stroked text. Mirrored text images also need to be considered and evaluated in the future.

References

1. Nammous MK, Saeed K, Kobojek P (2020) Using a small amount of text-independent speech data for a BiLSTM large-scale speaker identification approach. J King Saud Univ Comput Inf Sci
2. Ficcadenti V, Cerqueti R, Ausloos M (2019) A joint text mining-rank size investigation of the rhetoric structures of the US Presidents' speeches. Expert Syst Appl 123:127–142
3. Kundu S, Paul S, Bera SK, Abraham A, Sarkar R (2020) Text-line extraction from handwritten document images using GAN. Expert Syst Appl 140:112916

4. Zhang H, Zhao K, Song YZ, Guo J (2013) Text extraction from natural scene image: a survey. Neurocomputing 122:310–323
5. Kumar A, Rastogi P, Srivastava P (2015) Design and FPGA implementation of DWT, image text extraction technique. Procedia Comput Sci 57:1015–1025
6. Saabni R, Asi A, El-Sana J (2014) Text line extraction for historical document images. Pattern Recogn Lett 35:23–33
7. Yi C, Tian Y (2013) Text extraction from scene images by character appearance and structure modeling. Comput Vis Image Underst 117(2):182–194
8. Jung K, Han J (2004) Hybrid approach to efficient text extraction in complex color images. Pattern Recogn Lett 25(6):679–699
9. Sriman B, Schomaker L (2019) Multi-script text versus non-text classification of regions in scene images. J Vis Commun Image Represent 62:23–42
10. Peng X, Zhang X, Li Y, Liu B (2019) Research on image feature extraction and retrieval algorithms based on convolutional neural network. J Vis Commun Image Represent 102705
11. Zhao Z, Zhu H, Xue Z, Liu Z, Tian J, Chua MCH, Liu M (2019) An image-text consistency driven multimodal sentiment analysis approach for social media. Inf Process Manage 56(6):102097
12. Roy S, Shivakumara P, Pal U, Lu T, Kumar GH (2020) Delaunay triangulation based text detection from multi-view images of natural scene. Pattern Recogn Lett 129:92–100
13. Sun J, Lei K, Cao L, Zhong B, Wei Y, Li J, Yang Z (2020) Text visualization for construction document information management. Autom Constr 111:103048
14. Gan J, Wang W, Lu K (2020) In-air handwritten Chinese text recognition with temporal convolutional recurrent network. Pattern Recogn 97:107025
15. Manoharan S (2019) A smart image processing algorithm for text recognition, information extraction and vocalization for the visually challenged. J Innov Image Process (JIIP) 1(01):31–38

Traffic Sign Recognition System (TSRS): SVM and Convolutional Neural Network

Nazmul Hasan, Tanvir Anzum, and Nusrat Jahan

Abstract TSRS (Traffic Sign Recognition System) may play a significant role in the self-driving car, artificial driver assistance, traffic surveillance as well as traffic safety. Traffic sign recognition is necessary to overcome the traffic-related difficulties. The traffic sign recognition system consists of two parts—localization and recognition. In the localization part, traffic sign region is located and identified by creating a rectangular area. After that, in recognition part, the rectangular box provided the result for which traffic sign is located in that particular region. In this paper, we describe an approach toward the traffic signs recognition system. Here, we worked on 12 selected signs for traffic sign detection and recognition purpose. In this intention, we used a support vector machine (SVM) and convolutional neural network (CNN) individually to detect and recognize the traffic signs. We obtained 98.33% accuracy for SVM with an 80:20 train and test data ratio. On the other hand, the test result was 96.40% accurate for CNN.

Keywords Image processing · SVM · CNN · Traffic sign

1 Introduction

With the rapid growth of technological development, vehicles have become an essential part of our routine lives. It makes road traffic more and more intricate, which leads to more traffic accidents every year. In recent times, road accidents are happening regularly in an increasing manner across the world. The leading reason for most

N. Hasan · T. Anzum · N. Jahan (✉)
Computer Science and Engineering Department, Daffodil International University, Dhaka, Bangladesh
e-mail: nusratjahan.cse@diu.edu.bd

N. Hasan
e-mail: nazmul15-7914@diu.edu.bd

T. Anzum
e-mail: tanvir15-7890@diu.edu.bd

G. Ranganathan et al. (eds.), *Inventive Communication and Computational Technologies*, Lecture Notes in Networks and Systems 145,
https://doi.org/10.1007/978-981-15-7345-3_6

road accidents is the ignorance or unawareness of the traffic sign. The meaning of traffic sign is any object, device, or mark on the road that object is to carry to road users, or any specified class of road users, restrictions, prohibitions, warnings, or information of any explanation. Therefore, it includes not only marks on posts but also road markings, delineators, road studs, traffic signals, and other traffic control devices to provide smooth car driving.

Traffic sign detection and recognition system (TSRS) is an important issue to reduce traffic and increase the sustainability of self-driving cars without any incidence. TSRS plays a crucial role in an autonomous vehicle, smart driving, and smart traffic system. The traffic sign is used in Bangladesh since the 1930s and those are inadequate in these current traffic situations. So, we need to underpin TSRS to provide an autonomous vehicle to reduce traffic as well as a road accident.

An approach assembles with a video camera and an active computer with the vehicle is a simple driver assistance system based on the frame by frame analysis of the motion frames that can be developed and thereby generated the alert signals accordingly. So, the driver would be able to take the decision effortlessly. Road sign analysis is one of the significant aspects of an automated driver support system. Real-time qualitative road sign analysis is the root of any updated transport system. In this paper, we proposed this system to recognize a traffic sign with supervised classification algorithms. The outcome of this study can be used for recognizing traffic signs and directions also to slow down or direct drivers to another safe route.

A simple block diagram to present the overall prospective has depicted in Fig. 1. With the help of this diagram, we demonstrated the proposed system towards an automated car. In our proposed approach images are captured from the vehicle using a video camera. After the pre-processing input data classification algorithm was established to detect and recognize the traffic signs. Finally, we introduce a driver alert system to minimize the error of car driving approach.

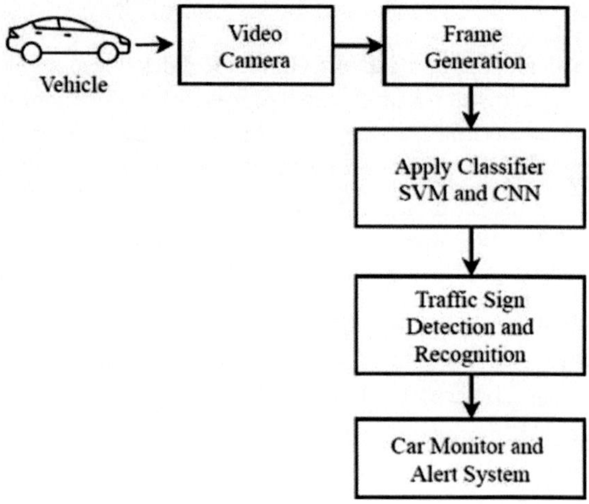

Fig. 1 Basic block diagram for TSRS

2 Related Work

This section highlights some recent related research works. An autonomous car is a recent research topic. We found a lot of researches on this field and the working progress is too high.

In 2019, Wei-Jong Yang et al. proposed an approach to recognize traffic signs. They worked with shaped based detection algorithms and for classification purposes, they choose a convolutional neural network. After simulation, they got 97% sign recognition accuracy [1]. In this paper, [2] author proposed SVM-based classification algorithm to recognize traffic signs. Here, they considered eight types of road signs. For training purposes, they used 600 different images for each sign, and for test purposes, 120 images were considered. In this paper, they tested individual signs with real data and their accuracy level was 66.6% to 100%. Prashengit Dhar et al. in 2017, proposed a Traffic Sign Recognition (TSR) system. Here, they were used the HSV color model and deep CNN for automatic feature extraction as a classifier. After this study, they achieved a 97% accuracy [3]. In 2019 [4] Aashrith et al. used Convolutional Neural Networks (CNN) to recognize traffic signs. They found 99.18% accuracy on using Belgium Data and German Traffic Sign Benchmark (GTSDB), whereas they got 99.50% accuracy almost 0.32% improvement at accuracy. In 2016 at paper [5], Di Zang et al. classified their dataset using Support Vector Machine then detection part is done by CNN. Their accuracy was almost 96.50%, and they used the GTSDB dataset. In 2017, at paper [6], Ardianto et al. used SVM to classified objects and got 91.5% accuracy in detection purposes and they also used the GTSDB dataset. They improved it to add a feature called Histogram of Oriented Gradients (HOG) that help to increase its accuracy by up to 98%. In 2017 Shi et al., applied SVM to detect which region of the image contains a traffic sign [7].

Pavly Salah Zaki and et al. worked on traffic sign detection multi-object detection systems. Faster recurrent convolutional neural networks and single-shot multi-box detector with several feature extractors they used in 2019 to detect traffic signs. However, they underpin F-RCNN to get the best results. Here, they used the GTSDB dataset. The GTSDB holds complete 900 images, were 800 for training, and 100 for testing [8]. In 2020, Yanmei Jin and et al. also worked with the GTSRB dataset to propose a Single Shot Detector algorithm combine with multi-feature fusion and they called it MF-SSD. However, in this time they divided the total 900 images into 600 training and 300 as test data to detect traffic signs [9]. In 2017, Yassmina Saadna and Ali Behloul discussed an approach to detect and recognize traffic signs. Their main goal was to find detection methods for localizing the regions of interest that contain traffic signs. They divided the methods into three steps—color-based, shape-based, and finally learning-based methods [10]. In 2016, they proposed an approach to detect and classify the traffic signs. It has two main steps: road sign detection, and classification with recognition. To classify the traffic signs they used neural network and four types of traffic signs: Stop Sign, No Entry Sign, Give Way Sign, and Speed Limit Sign is used. It considered a total of 300 sets of images, and they got 90 and

88% accuracy for detection and recognition purposes [11]. So, traffic sign detection and recognition is necessary to build an autonomous car driving system.

3 Proposed Methodology

To recognition traffic signs we focused on two machine learning algorithms. Much recent research works on traffic signs used SVM and CNN. For detecting and recognizing traffic sign we used these two algorithms. In the next step we also evaluated the proposed approach with CNN and SVM.

3.1 Data Collection and Preparation

To complete this study a dataset was built from the cropping video frame. We also collect some random video and crop traffic sign area to build a real dataset. Then, we categorized its classes and split the whole data into training and validation dataset. We have a total of 1200 images to propose the SVM and CNN models.

We have considered 12 different classes and each class completed with 100 images. Split the whole dataset as training and validation purposes. To generate a model with SVM, 80% data for training, and 20% data for testing purposes is considered. Meanwhile, for CNN classification we have divided the dataset into two classes train and validation. For training purpose, data volume was 1080 images and 120 images for testing purpose, where a total of 12 different traffic signs was contained. Table 1 represents the selected 12 traffic signs. Here, we illustrated every traffic sign with sample an image.

In the next section, we are attending to elaborate on the work procedure of our proposed approach to detect and recognize traffic signs.

3.2 Convolutional Neural Network

The convolutional neural network is a class of deep learning neural networks. A convolutional neural network represents a huge breakthrough in image recognition. In this recent era, CNN is one of the most popular machine learning algorithms. Many research fields focused on CNN to achieve the highest accurate result. A CNN model contracts with the following four layers:

- Convolutional layers
- Relu function
- Polling layers
- Fully connected layer.

Table 1 12 traffic signs with description

Description	Traffic sign	Description	Traffic sign
Turn left		Danger	
Stop		40 km/h	
Only left		30 km/h	
Only right		Turn right	
Road merges ahead		Pedestrian	
Speed breaker		Bike	

Here, Convolutional operates on two images in 2D format. One as an input image, and others as a filter of the input image, to produce an output image. Convolution of two functions f and g is defined with following the equation

$$(f * g)(i) = \sum_{j=i}^{m} g(j) * f(i - j + m/2) \quad (1)$$

The rectified linear unit is another step to implement the convolutional layer. In this layer, applying an activation function "relu" onto feature maps to increase non-linearity in the network. On the other hand, Polling progressively decreases the size of the input representation. It makes it possible to detect objects in an image. It helps to decrease the number of parameters and the amount of computation required. Polling also helps to control overfitting. There are many types of polling. But in this paper, we used max polling.

Fully connected layer, in this step an artificial neural network work as a CNN. This step combines our features into more attributes. These will predict the class level with better accuracy. The error is computed and then occurred backpropagation. The weights and feature detectors are adjusted to help optimize the performance of the model. This process continues layer by layer. A sample flowchart for our proposed model on CNN and SVM is given in Fig. 2.

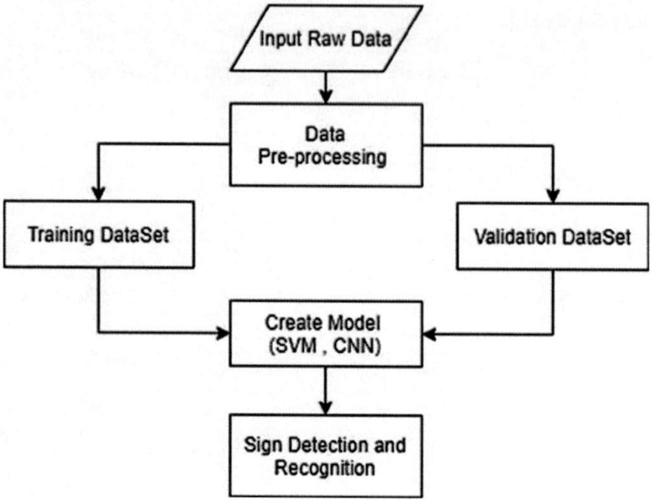

Fig. 2 Flowchart of our model

3.3 Support Vector Machine (SVM)

SVM is a discriminative classifier that is defined by a separating hyperplane. Mainly, SVM is a familiar supervised machine learning algorithm. It is usually used in classification problems. In the SVM model, we can plot each item as a point in n-dimensional space with the value of each feature being the value of a particular coordinate (where *n* is several features). It will be clearly understood with an example of two classes. We can understand the process of the SVM model with the help of Fig. 3.

To understand the work process of SVM we needed to know the ROI module. This region follows three steps. The first one is color transformation, which converts the

Fig. 3 Split data into two classes (SVM)

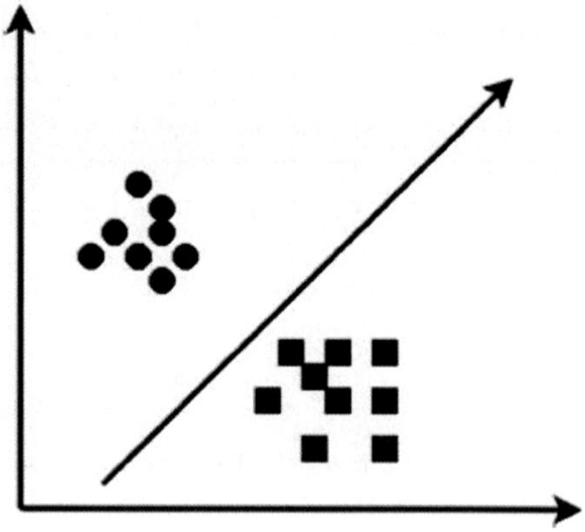

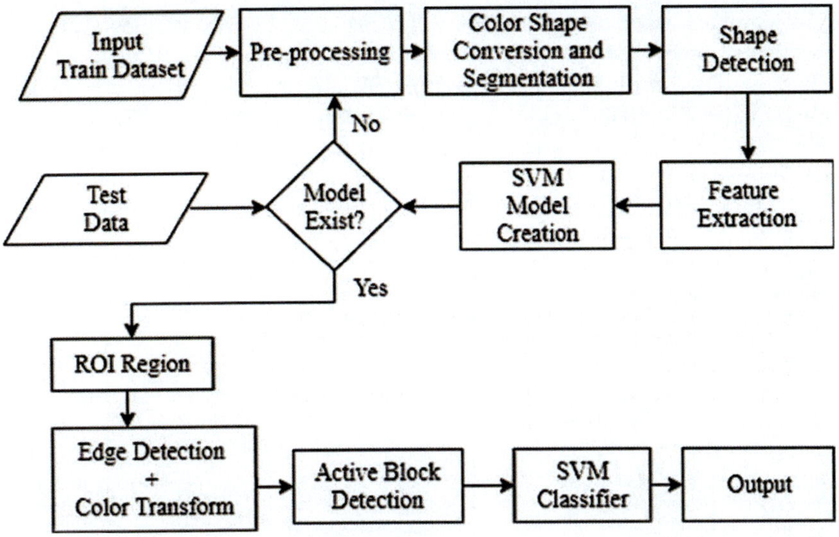

Fig. 4 SVM for traffic sign detection and recognition

RGB value of each pixel to gray pixel. The second step is to control shape matching over the gray images to find possible sign locations. Finally, it refines the ROI. This module exploits the regularity of traffic signs with their color and shapes with high competence.

In this paper, to describe the methodology, the size of the cells and blocks are varied to get different image sizes. We used 9-bin histograms for this methodology. To train the model we rescale the training images to 32 × 32 pixel images. HOG descriptors to calculate and train the SVM classifier. The n resulting support vectors are merged to a single one by multiplying every vector by its weight "*a*" and adding it with a global vector "*v*". Equation 2 for computing the support vector [12].

$$v = \sum_{i}^{n} ai \, . \, vi \tag{2}$$

Now, we illustrated the whole work process of SVM in Fig. 4. Here, we depict all the steps of SVM to recognize the traffic signs for TSRS.

4 Result Analysis

In this section, we are going to discuss the result which was obtained from SVM and CNN. SVM classification for Traffic sign detection and recognition provided 98.33% accuracy (80:20 data split ration), while in the CNN method we achieved

99.56% training accuracy and 96.40% validation accuracy. The training accuracy, validation accuracy, and loss of CNN model are visualized in Figs. 5 and 6.

For the SVM method, we were getting 98.33% accuracy when we splitted the total dataset into 80:20 ration for training and testing purposes. However, we achieved 99.17% accuracy when considered 90:10 ration. Figure 7 for presenting the performance of SVM.

After that, we developed a system using SVM to evaluate the results with real-time video. The detection part uses image processing techniques that create contour on

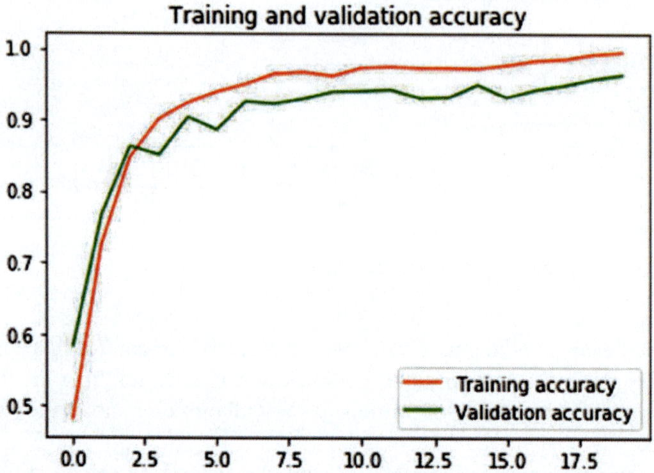

Fig. 5 Accuracy of training and validation

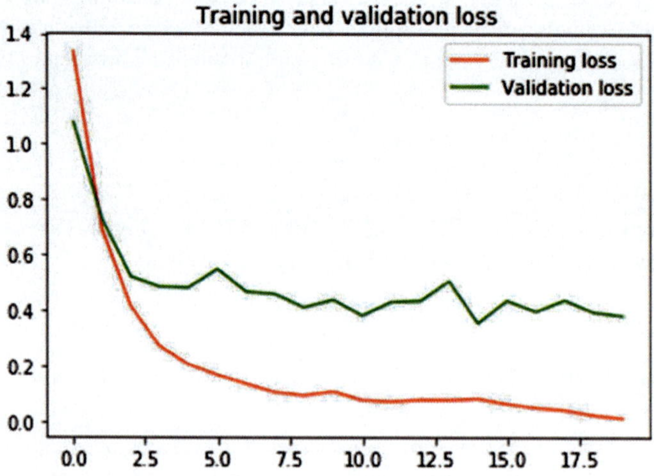

Fig. 6 Line chart for training and validation loss

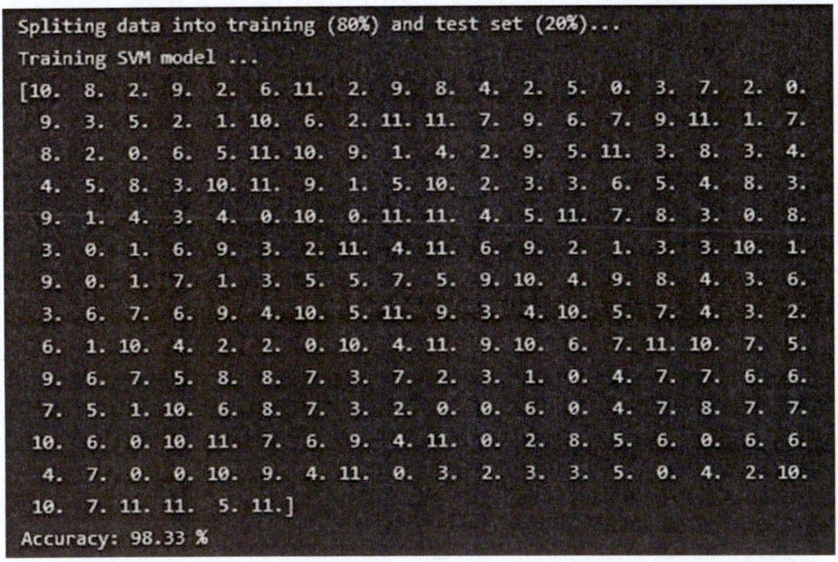

```
Spliting data into training (80%) and test set (20%)...
Training SVM model ...
[10.  8.  2.  9.  2.  6. 11.  2.  9.  8.  4.  2.  5.  0.  3.  7.  2.  0.
  9.  3.  5.  2.  1. 10.  6.  2. 11. 11.  7.  9.  6.  7.  9. 11.  1.  7.
  8.  2.  0.  6.  5. 11. 10.  9.  1.  4.  2.  9.  5. 11.  3.  8.  3.  4.
  4.  5.  8.  3. 10. 11.  9.  1.  5. 10.  2.  3.  3.  6.  5.  4.  8.  3.
  9.  1.  4.  3.  4.  0. 10.  0. 11. 11.  4.  5. 11.  7.  8.  3.  0.  8.
  3.  0.  1.  6.  9.  3.  2. 11.  4. 11.  6.  9.  2.  1.  3.  3. 10.  1.
  9.  0.  1.  7.  1.  3.  5.  5.  7.  5.  9. 10.  4.  9.  8.  4.  3.  6.
  3.  6.  7.  6.  9.  4. 10.  5. 11.  9.  3.  4. 10.  5.  7.  4.  3.  2.
  6.  1. 10.  4.  2.  2.  0. 10.  4. 11.  9. 10.  6.  7. 11. 10.  7.  5.
  9.  6.  7.  5.  8.  8.  7.  3.  7.  2.  3.  1.  0.  4.  7.  7.  6.  6.
  7.  5.  1. 10.  6.  8.  7.  3.  2.  0.  0.  6.  0.  4.  7.  8.  7.  7.
 10.  6.  0. 10. 11.  7.  6.  9.  4. 11.  0.  2.  8.  5.  6.  0.  6.  6.
  4.  7.  0.  0. 10.  9.  4. 11.  0.  3.  2.  3.  3.  5.  0.  4.  2. 10.
 10.  7. 11. 11.  5. 11.]
Accuracy: 98.33 %
```

Fig. 7 Result of SVM

each frame and find all ellipse or circles among those contours. Then, the detection part marked as a categorized traffic sign. Some sample results are shown in the following figures.

Figures 8, 9, and 10 to understand the performance of SVM with the help of some real-time data. If we implement the full system with CNN then it will be work well then SVM as we have mentioned their accuracy level. After complete our study we obtained that TSRS can be an established and sustainable system. Many kinds of research are going on with traffic issues and self-care driving purposes. We needed an accurate model to implement this system to provide a better solution in every case.

Fig. 8 Recognize "Danger" sign

Fig. 9 Recognize drive "30 km/h" sign

Fig. 10 Recognize "Turn Right" sign

5 Conclusion

This paper study was to represent an original effective traffic sign detection and recognition approach towards the design of TSRS. As a recent research topic TSRS is getting popular day by day. In this study, it is done using SVM and CNN classification algorithms to decline extensive traffic difficulties. In our experiment, we obtained the highest training accuracy from CNN 99.56%, while the test accuracy was 96.40%. We showed the real-time evaluation results of SVM, where the system performed 98.33% accurately. Many research focused on SVM and CNN to solve this specific problem.

In the future, we aim to increase the number of traffic sign classes with a large amount of quality data. As in machine learning research, maintaining data volume and data quality is the most important and time-consuming part. To provide a complete system to overcome the traffic issues, our ambition is to implement a system with distance calculation from car to a traffic sign.

References

1. Yang WJ, Luo CC, Chung PC, Yang JF (2020) Simplified neural networks with smart detection for road traffic sign recognition. In: Arai K, Bhatia R (eds) Advances in information and communication. FICC 2019. Lecture notes in networks and systems, vol 69. Springer, Cham
2. Dubey AR, Shukla N, Kumar D (2020) Detection and classification of road signs using HOG-SVM method. In: Elçi A, Sa P, Modi C, Olague G, Sahoo M, Bakshi S (eds) Smart computing paradigms: new progresses and challenges. Advances in intelligent systems and computing, vol 766. Springer, Singapore
3. Dhar P, Abedin MZ, Biswas T, Datta A (2017) Traffic sign detection—a new approach and recognition using a convolution neural network. In: 2017 IEEE region 10 humanitarian technology conference (R10-HTC), Dhaka, pp 416–419
4. Aashrith V, Smriti S (2019) Traffic sign detection and recognition using a CNN ensemble. In: IEEE international conference on consumer electronics (ICCE), 7 Mar 2019
5. Di Z, Junqi Z, Dongdong Z (2016) Traffic sign detection based on cascaded convolutional neural networks. In: Conference on software engineering, artificial intelligence, networking and parallel/distributed computing (SNPD)
6. Sandy A, Chih-Jung C, Hsueh-Ming H (2017) Real-time traffic sign recognition using color segmentation and SVM. In: Conference on systems, signals and image processing (IWSSIP)
7. Jian-He S, Huei-Yung L (2017) A vision system for traffic sign detection and recognition. In: IEEE 26th symposium on industrial electronics (ISIE)
8. Zaki PS et al (2019) Traffic signs detection and recognition system using deep learning. In: Conference on intelligent computing and information systems (ICICIS)
9. Jin Y, Fu Y, Wang W, Guo J, Ren C, Xiang X (2020) Multi-feature fusion and enhancement single shot detector for traffic sign recognition. IEEE Access 8:38931–38940
10. Saadna Y, Behloul A (2017) An overview of traffic sign detection and classification methods. J Multimed Inf Retrieval 6:193–210. https://doi.org/10.1007/s13735-017-0129-8
11. Sheikh MAA, Kole A, Maity T (2016) Traffic sign detection and classification using colour feature and neural network. In: Conference on intelligent control power and instrumentation (ICICPI), Kolkata, pp 307–311
12. Fatin Z, Bogdan S (2011) Warning traffic sign recognition using a HOG-based K-d tree. In: IEEE intelligent vehicles symposium (IV)

A Comparative Study of Machine Learning Algorithms for Gas Leak Detection

J. E. Raghavendra Prasad, M. Senthil, Akhil Yadav, Paras Gupta, and K. S. Anusha

Abstract A gas leak detection system considers various factors for detecting leaks. Sensors are placed around the leak-prone areas, and the presence of a leak is determined based on the concentration values of the sensors. The models produce a variety of results depending on the type of algorithm used to determine the leak. An error in leak detection may cause harmful consequences if the gas is explosive or corrosive in nature. In this paper, we take the concentration values for consideration and applying 4 machine learning techniques namely decision tree, random forest, ACF, and Naïve Bayes to a concentration data of a 20-sensor network, and then the results have been compared. The experimental results show that the random forest has the best performance when compared to the other algorithms.

Keywords ACF · Random forest · Decision tree · Naïve Bayes · Gas leak

J. E. Raghavendra Prasad (✉) · M. Senthil (✉) · A. Yadav · P. Gupta · K. S. Anusha
Department of Electronics and Communication Engineering, Amrita School of Engineering, Amrita Vishwa Vidyapeetham, Coimbatore, India
e-mail: cb.en.u4eie16037@cb.students.amrita.edu

M. Senthil
e-mail: cb.en.u4eie16049@cb.students.amrita.edu

A. Yadav
e-mail: cb.en.u4eie16001@cb.students.amrita.edu

P. Gupta
e-mail: cb.en.u4eie16035@cb.students.amrita.edu

K. S. Anusha
e-mail: ks_anusha@cb.amritanet.edu

G. Ranganathan et al. (eds.), *Inventive Communication and Computational Technologies*, Lecture Notes in Networks and Systems 145,
https://doi.org/10.1007/978-981-15-7345-3_7

1 Introduction

Gas leaks occurring in industries are very hazardous because exposure to these gases can be very harmful and in extreme cases fatal. Most of the leaks that occur in the industry are unidentified, and these gasses emit to the environment for a prolonged period causing pollution to the environment. There have been many cases like these in India, the most notable being the Bhopal gas tragedy. By the end of 2020, 32,737 km of gas pipelines is to be laid through the length and breadth of the country with a few of these pipes going through areas of dense habitation. In these cases, the prevention of leaks is mandatory and anomalies that may occur due to human negligence, and errors are not acceptable. These scenarios arise the need for a cost-effective and accurate leak detection system to avoid dangerous occurrences due to leaks in these areas. Leak detection systems use the data acquired by the sensors in the leak-prone areas as input which are fed to the algorithm which then decides the presence of a leak. Based on the kind of signals acquired for the sensors like sound, concentration, etc., an algorithm is used to decide the presence of a leak.

Leak localization techniques have also been used to determine the exact location of the leak [1]. In this paper, localization techniques that can be used have been proposed and their performances are compared and analyzed.

In [2], they have used conventional machine learning algorithms like Naïve Bayes to detect breast cancer using a predictive breast cancer dataset and given a diagnosis of how conventionally these algorithms have performed in detecting breast cancer. In this paper, we take a similar approach by using conventional algorithms that can be used to detect the leak and the performances of these algorithms have been compared and analyzed.

1.1 Related Works

The correct prediction of the leak in industries is vital due to the impact they have on both the environment and human life around it. Many industries work with explosive gases which if leaked may have hazardous consequences [3]. Acoustic based detection systems have been developed using adaptive filter technology to detect leaks in natural gas pipelines. The detection system has been simulated using LabVIEW by differentiating the leak characteristics. Another method [4] that they have used for detecting VOC (volatile organic compound) Gas leaks is by using infrared sensors. In this paper, they have created a circuit that produces an analogous time series output from the PIR sensor. After processing the data obtained and producing the required wavelet coefficients a Markov Modeling based classifier is developed to detect the gas leaks. Many solutions using machine learning and neural networks have been proposed. In [5], for predicting the leak points in the petrochemical industry, a three-level backpropagation algorithm has been proposed. This algorithm improves the response speed of the learning process and lowers the prediction time. They have also proposed a better method to decrease the learning error and improve the convergence rate of the algorithm. In [6] they have proposed a Gaussian-based model to detect small leaks in gas transportation pipes by learning the distribution of small leaks in the pipeline. They have also based the model by analyzing the acoustic signals. The model proposed takes into account the environmentally and randomly high noise.

2 Data and Pre-processing

The data used in this paper was collected from the tests performed [7] in an experiment at the Texas A&M Engineering Extension Service facility, College Station, TX, USA. The sensor network consists of 20 sensors in a 4 * 5 configuration. The sensors are placed at an elevation of 2.25 m and the leak sources are at an elevation of 0.5 m and 5.5 m respectively. 60 releases were performed during this experiment where the leak duration for each release was for 2 min. Although there were only 2 sources of release the releases varied from each other based on the source's nozzle size, the flow rate of gas. The nozzle sizes were 2, 6.35, 19 and 63.5 mm while the flow rates ranged from 1.35 to 1020 lb/h. The sensor data has a 0–2% measurement range (0–20,000 ppm). The data consist of a sensor showing a concentration reading at a rate of 1 measurement every 5 s. Initially, all the sensor readings were taken as zero and in each recorded duration a sensor value is updated. The dataset consists of 20 concentration with a sensor value updated in each instance of time and the corresponding status (leak or no leak) based on the scenario during which the readings were taken. We have only taken the rows where all the sensor values are unique from the previous one, i.e., Rows after 20 instances of time. We have also any duplicate values from the dataset.

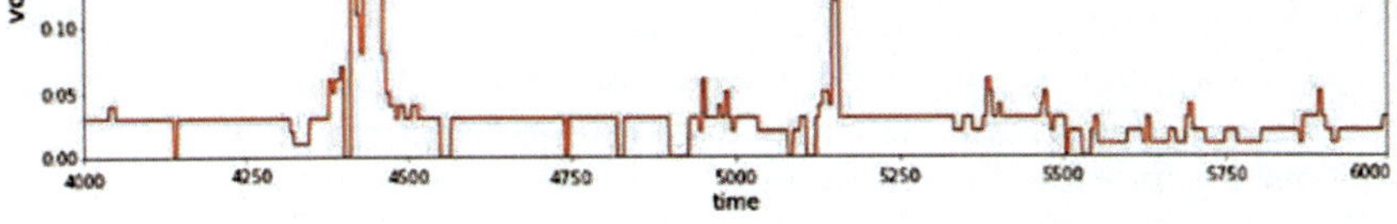

Fig. 1 This figure represents a data sample and how the values deviate during a leak. The sudden data spikes show instances of the leak

Figure 1 shows how our data show changes during a leak which is seen by the sudden rise in the volume percentage of Fig. 1 but due to the reason, the propane is heavier than air it settles in the air for some time before dispersing due to this reason we eliminate few of the data values which is shown even after the leak time has been completed. After we finalize the data from its raw to a processed state we feed the data into the algorithms discussed in Sect. 3 and also details about the data are given if they are further modified from the processed state as described in this section.

3 Description of Algorithms Used

3.1 Naïve Bayes

This Classifier is used for membership probabilities prediction of each separate class. In our model, the different classes are leak and no leak. Based on the class with the highest probability, the scenario of leak or no leak is predicted.

These classifiers are an application of the Bayes probability theorem. It is not an individual algorithm but a family of algorithms sharing a common trait; i.e., every pair of features being classified is independent of each other. Firstly, the feature matrix contains all the vectors of the dataset in which each vector consists of the value of dependent features. Secondly, the Response vector contains the value of the class variable which is the output or predicted value for each row of the feature matrix. In our paper, we have used Naïve Bayes Classifier to identify the leak and no leak conditions.

The simple Bayes theorem can be expressed as:

$$P\left(\frac{I}{J}\right) = \frac{P(\frac{I}{J})P(J)}{P(J)} \tag{1}$$

which can be theoretically represented as,

$$\text{Posterior} = \frac{\text{Likelihood} * \text{Prior}}{\text{Evidence}} \tag{2}$$

In simple words, this is a rather simple transformation, but it bridges the gap between what we want to do and what we can do.

But a downside to the algorithm is that it makes a naive assumption that all features are independent. Despite its simplicity, Naive Bayes forms a position in which the other models have been developed.

3.2 ACF Model

In this model, we initially convert all 20 sensors values in each row into a single value by determining the likelihood of leak and likelihood of no leak at each instance of time. We first prepare the probabilities of events in individual columns of the data. To determine the probability of the event of the leak, we find the individual probability of both leak ($z = 1$) and no leak ($z = 0$). After determining the probability of leak at each instance, we use the likelihood function to determine the likelihood of leak and the likelihood of no leak at a specific instance of time. The likelihood at each time step is calculated by:

$$P_t(z) = \prod_{i}^{20} p(s_i(t)/z) \tag{3}$$

P = likelihood function
z = event of a leak or no leak
i = sensor number (1–20)
t = time instance.

To determine anomalies, we use methods that are used for detecting abnormal EEG signals from normal ones. In [8], A neural network has been applied here for classifying epileptic signals from normal EEG signal detection but we use a different approach based on autocorrelation to find the extent of correlation between two values in the same dataset but during different time steps. Even neural networks can be used to find anomalies between the leak signals but we use the autocorrelation function to provide a real-time monitoring approach to the leak detection system. From [9], where autocorrelation has been used for robust artifact detection in ECG has been used in this model to detect the leak. Autocorrelation is a degree of similarity between a given time series and its former-self lagged by a factor k which depends on the user. It calculates the correlation between two-time series, except that here both are the same series one the current one and the other the lagged one. We take a time lag of 5 s. We then use cosine similarity to determine the similarity between the data points in both the time series (Figs. 2, 3 and 4).

The cosine similarity between 2 vectors is given by the equation

$$\frac{\sum_{i=1}^{n} Xi * Yi}{\sqrt{\sum_{i=1}^{n} X_i^2} * \sqrt{\sum_{i=1}^{n} Y_i^2}} \tag{4}$$

This equation is the extension of the dot product between two vectors to determine the cosine of the angle between two vectors X and Y

$$\cos \varnothing = \frac{X\,.\,Y}{||X|| * ||Y||} \tag{5}$$

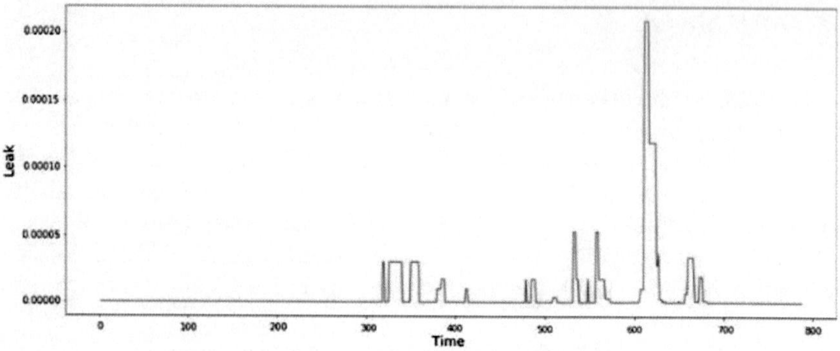

Fig. 2 Represents the likelihood of leak plot

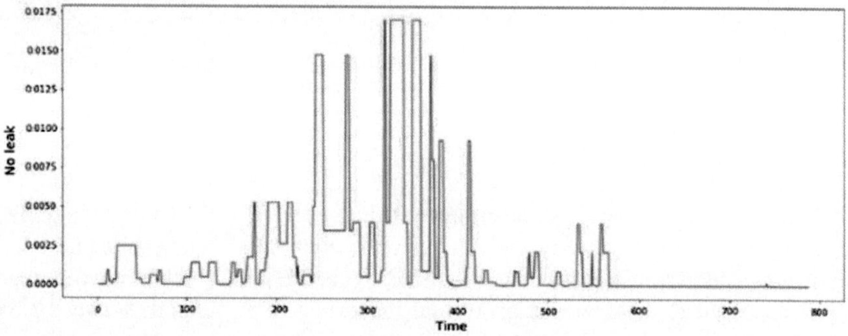

Fig. 3 Represents the likelihood of no leak plot

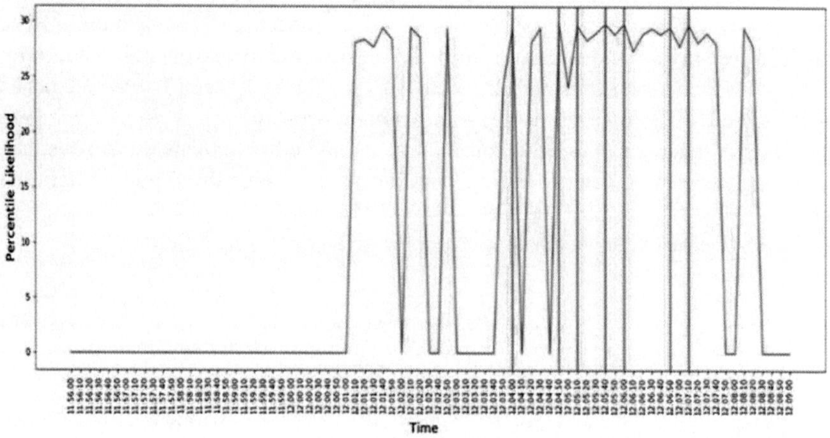

Fig. 4 The red lines represent the predicted leak points in the graph

But in our scenario, we have two lists X_i and Y_i which corresponds to the present likelihood values and the other the likelihood values before an interval of 10 s as the window size used for comparison in the autocorrelation function is 10 s. So, we determine the cosine similarity between the two lists using Eq. 4. The values range from -1 to 1 where 1 is perfectly similar and -1 is perfectly dissimilar.

Cosine similarity is used because it can determine how similar the series are irrespective of their size. It measures the cosine of angles between two data points projected in a multi-dimensional space. This captures the orientation of data in the 3d space rather than the magnitude. Sometimes although the series are far apart by the Euclidean distance (Magnitude) because of the size they may have a small angle between them making them more similar. If the data points are dissimilar more by a value greater than the provided threshold, then it is determined as a leak. The threshold used in the model was 90 percentile.

3.3 Decision Tree

We use a decision tree for decision analysis (determine the presence of leak) and to help identify the easiest strategy to reach the goal. The decision tree is supervised learning and performances well for both categorical and continuous data. It divides the data samples into two or more subsets based on the most significant differentiator in the input variable. A decision tree consists of a root node with all the samples and then the algorithm breaks down the dataset into little subsets while correspondingly an associated decision tree is developed incrementally. The final tree consists of a tree with decision and leaf nodes. The decision node has two or more subsets (branches) while the leaf node corresponds to a classification or decision, in our context leak or no leak. The samples are split from the root node into other decision nodes which in turn are divided into leaf nodes.

3.4 Random Forest

The Random forest consists of a mixture of a huge number of the individual decision tree that together work as a single classifier. Many numbers of decision trees that act together as a complete set outperform any of the individual models. From [10], in which they detected water leaks in a pipeline by using the random forest algorithm. We can see that that the algorithm works very well in conventional situations as shown in the paper where inexpensive pressure sensors have been installed to obtain an accuracy of nearly 96%. This model works very well because the data have a very low correlation to another. This algorithm uses bagging and feature randomness for building each tree and creates a forest of trees that are uncorrelated to each other. Taking into account that all sensor values are independent of each other the data fits perfectly with the model.

4 Results and Inference

18 data samples were taken and the ACF model was applied to predict the leak points in the sample. The model predicted 59 out of the 78 samples correctly with an accuracy of 75.64% (Fig. 5).

The above detection algorithms have been applied to the data and the results have been compared with the following parameters

Mean Absolute Error (MAE) and Mean Squared Error (MSE)

$$\mathrm{MAE}\left(h,\hat{h}\right)=\frac{1}{n}\sum_{i=0}^{n-1}\left|h_i-\hat{h}_i\right| \tag{6}$$

$$\mathrm{MSE}\left(h,\hat{h}\right)=\frac{1}{n}\sum_{i=0}^{n-1}\left(h_i-\hat{h}_i\right)^2 \tag{7}$$

$\hat{h}_i$ the predicted value of the ith sample
h_i the corresponding true value of the ith sample
n sample range.

$$\mathrm{Precision}=\frac{\mathrm{TP}}{\mathrm{TP}+\mathrm{FP}} \tag{8}$$

TP is True positive

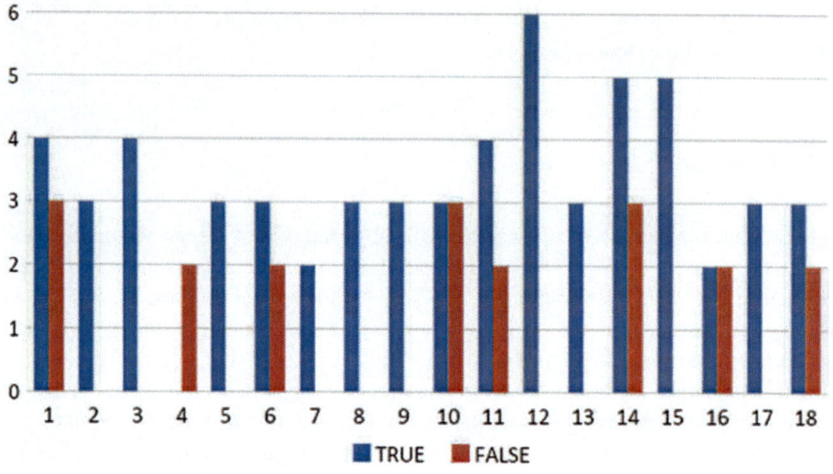

Fig. 5 Shows the number of true and false predictions that were obtained when performing each of the 18 data samples using the ACF model

FP is False positive

$$\text{Recall} = \frac{\text{TP}}{\text{TP} + \text{FN}} \tag{9}$$

where

TP is true positives
FN is false negatives

$F1$ score can be interpreted as weight harmonic mean of precision and recall

$$F1\,\text{score} = \frac{2 * (\text{precision} * \text{recall})}{\text{precision} + \text{recall}} \tag{10}$$

From Table 1, we deduce that the random forest algorithm has better accuracy when compared to the other developed models. The ACF model used shows an accuracy of 75.64%. The ACF model runs concerning real-time and depends on the quality of data it is being worked upon unlike the other three algorithms used which predict the outcome from the training we have done with the model. If more features are taken into account the accuracy of decision tree and random forest improves because they consider a lot feature and choose the best choice to make a decision The mean squared error can be used to detect large errors which cause more damage to the algorithm than an equivalent amount of small errors. We see that Naïve Bayes shows a very poor performance out of all the models and has a very low F1 score since if a unique variable present in the dataset is not seen during training occurs during prediction it just assigns a zero probability to it and a prediction is not made. After the detection has been done if the area of the leak-prone region is very large in such a way that just detection is insufficient and it also calls for localization. Localization becomes the next focus on the path to create an automated leak detection system. We can use techniques like the ones used in [11] where a gas diffusion model is proposed taking into account the possibility of gas dispersion due to wind using the weighted centroid algorithm or other techniques like [12] where the support vector machine regression algorithm has been proposed to localize single and multiple targets in an indoor environment can be used to localize the leak occurring in the pipeline which is beyond the scope of this paper and can be done as an extension of the leak detection system.

Table 1 Represents the performances of each algorithm that have been applied

Model name	Accuracy	MAE	MSE	Precision	Recall	F1 score
Naïve Bayes	77.48	0.225	0.225	0.481	0.337	0.396
Decision tree	94.25	0.059	0.056	0.849	0.898	0.873
Random Forest	97.4	0.026	0.026	0.974	0.906	0.939

References

1. Anusha KS, Ramanathan R, Jayakumar M (2019) Device free localisation techniques in indoor environments. Defence Sci J (DSJ) 69(4):378–388
2. Potdar K, Kinnerkar R (2016) A comparative study of machine learning algorithms applied to predictive breast cancer data. Int J Sci Res 5(9):1550–1553
3. Lei JC, Yuan W (2013) The research of natural gas pipeline leak detection based on adaptive filter technology. In: Proceedings of 2013 2nd IEEE international conference on measurement, information and control
4. Erden F, Birey Soyer E, Ugur Toreyin B, Enis Cetin A (2010) VOC gas leak detection using Pyro-electric infrared sensors. In: 1988 international conference on acoustics, speech, and signal processing, 1988. ICASSP-88 (2010)
5. Wang K, Zhuo L, Shao Y, Yue D, Tsang KF (2016) Toward distributed data processing on intelligent leak-points prediction in petrochemical industries. IEEE Trans Ind Inform 12(6)
6. Li J, Chen G, Liu C, Tang J (2018) Gaussian-based models for small leak identification of gas transportation pipes. In: IEEE international conference of safety produce informatization (IICSPI), Chongqing, China, pp 1–5
7. Chraim F, Erol YB, Pister K (2016) Wireless gas leak detection and localization. IEEE Trans Ind Inform 12(2)
8. Anusha KS, Mathews MT, Puthankattil SD (2010) Classification of normal and epileptic EEG signal using time & frequency domain features through artificial neural network. In: International conference on advances in computing and communications
9. Varon C, Testelmans D, Buyse B, Suykens J, Van Huffel S (2012) Robust artefact detection in long-term ECG recordings based on autocorrelation function similarity and percentile analysis. In: Proceedings of annual international conference on ieee engineering in medicine and biology society (EMBC), pp 3151–3154
10. Aymon L et al (2019) Leak detection using random forest and pressure simulation. In: 2019 6th Swiss conference on data science (SDS), Bern, Switzerland, pp 109–110. https://doi.org/10.1109/sds.2019.00008
11. Li Q, Liu Z, Wang J, Xiao X (2014) A gas source localization algorithm based on wireless sensor network. In: Proceeding of the 11th world congress on intelligent control and automation, Shenyang, pp 2514–2518
12. Anusha KS, Ramanathan R, Jayakumar M (2019) Link distance support vector regression (LD-SVR) based device free localization technique in indoor environment. Int J Eng Sci Technol (2019)

Agro Advisory System Using Big Data Analytics

Nazneen Ansari, Siddhi Martal, Namratha Bhat, and Sohan Pawar

Abstract From past decades, agriculture is remaining as a primary source of food and raw materials for human lives. Recently, the agriculture field is greatly influenced by technologies like big data and automated decision-making systems to deploy an efficient way to farm. Most of the agriculture-related data come from diverse varieties of information sources and networks. The objective of the system is to aid farmers and agriculture experts through a user-friendly website. The data is processed using the Hadoop framework, the results of which are displayed on the website by using a Tableau visualization tool. The ideology consists of data about farming and related aspects. The system has been designed by considering agriculture in India.

Keywords Big data analytics · Decision system · Agro advisory

1 Introduction

Agro advisory systems provide results approved by the agricultural experts. Big data concept is also used in smart farming where sensors are used on the farm. The amount of data generated by the sensors is huge and this data needs proper management. Big data technology gives a new perspective to farming by enhancing the speed of the process and helping in decision making. Hadoop tool is widely used in agricultural big data analytics.

N. Ansari · S. Martal · N. Bhat · S. Pawar (✉)
St. Francis Institute of Technology, Borivali, Mumbai, India
e-mail: sspawar5131@gmail.com

N. Ansari
e-mail: nazneenansari@sfit.ac.in

S. Martal
e-mail: siddhimartal98@gmail.com

N. Bhat
e-mail: bhat.namrata98@gmail.com

G. Ranganathan et al. (eds.), *Inventive Communication and Computational Technologies*, Lecture Notes in Networks and Systems 145,
https://doi.org/10.1007/978-981-15-7345-3_8

Big data can be integrated with other technologies to enhance the outcome. The proposed system can not only be used in the agriculture domain but also other mundane applications.

The paper structure is as follows. Section 2 elaborates on big data applications in agriculture. Section 3 discusses a decision-making system which thereby increases the overall production using big data techniques. Next section contains methodology that shows the implementation of the system. Finally, in Sect. 6 we discuss the results.

2 Literature Review

In the study of Zhao et al. [1], the major focus was to maximize production by extensive analysis. The agricultural automated decision-making system was introduced here. The implementation of an effective decision-making system considering various parameters was done. Here, the system uses big data analysis to interpret decisions for wheat production. The system also uses AI technology to obtain results.

Rao et al. [2] paper targets levels at which big data enhance farming fields and actionable knowledge for the practice of Smart Agriculture. It aims at research challenges, ways in which it leverages big data in the research of smart agriculture. Climate change and its impact on agricultural production are put forth by the author.

In the study of Wolfert et al. [3], the concept of smart farming was introduced. Smart Farming highlights the use of technology in the agricultural cycle. The main objective is to develop a system that can be useful for other applications. The use of smart sensors and machines is proposed through the study of literature between 2010 and 2015. The author briefs about the main factor involved in the farming business and how one can reduce the extra work with the help of innovative technologies.

In the study of Kaur et al. [4], the author expects an analytic model that identifies diseases based on symptoms similarity and recommends a solution according to the level of resemblance. Hadoop & Tableau are used for achieving the results which are represented in graphical format. Collected data is in clean and normalized form, this data is uploaded on HDFS. The result is expressed in graphical format. The author also surveys various reports of the agriculture department.

Authors 'Shriya Sahu, Meenu Chawla, Nilay Khar', [5] provides insight about agriculture as a human survival source. Here, the Hadoop framework was used for the classification of data. The data considered includes crop information. The precision of the Random Forest algorithm was calculated. The given system helps the farmers know more about their crops and increase production through accurate results. The process has some setback that needs to be improvised by adding detailed parameters.

In the study of Hirafuji et al. [6], the author briefs about a system that helps to solve the problem of generating data without any disturbance The paper gives ideas about various applications of big data in agriculture. Since the experiences of farmers are insufficient, hence decision-making applications are an essential need. Big data can modernize agriculture by contributing in various ways. The data needs to be stored securely so that it can be used in the future.

Di et al. [7] discussed the problems and solutions of farming w.r.t geography, geosciences. Characteristics of Big Data also called the 5 Vs were studied in detail. Big data applications in geo-informatics were also explored. The use of services of cloud computing for big data and how it can attend to geographical aspects is also explored in this paper.

Bendre et al. [8] presented an overview of the use of ICT services in the agricultural domain. Applications and characteristics of big data and ICT were also analyzed by the author. Algorithms like MapReduce, Kmeans, and linear regression were implemented for weather data forecasting. Different big data types are studied briefly. GPS and remote sensing was studied w.r.t big data.

Shekhar et al. [9] surveyed huge data based on agriculture. The shortcomings and other enhancements were listed. The future of big data applications was predicted and analyzed. Big data helps in decision making. For addressing the challenges and opportunities like Data sharing barriers, missing crucial data, etc. interdisciplinary collaboration is needed between research communities.

In the study of Mukesh Kumar, Prof. Mayura et al. [10], according to the authors, weather conditions play a prominent role in agriculture. It is also observed that for decision-making study of soil is equally important. Here, the use of the MapReduce algorithm is done. Precision agriculture is studied in detail. The motive of the paper is to obtain a sculptural view of the data using BI techniques and algorithms.

Varma et al. [11] did a thorough study of MapReduce. The author explains how MapReduce technology can be helpful. Also the benefits and detriments of the same were noted. Problem-solving using MapReduce is understood through the paper.

In the study of Tamilselvi et al. [12], the paper highlights the concept of performing big data analytics using Hadoop. The various characteristics of big data are discussed along with its architecture. The numerous technologies used in big data are studied thoroughly and some of the problems are listed. The applications of big data in real life as well as its importance are explained. It also elaborates on the Hadoop technology and its architecture followed by the working and advantages of MapReduce.

3 Proposed System

The proposed system is a user-friendly website. The website acts as a guide for agriculture all over India. State-wise suitable crop, soil type, as well as the ideal temperature needed for its cultivation, is given. The season, use of pesticide is also mentioned along with the intercropping species. The results are displayed on the website using a live Tableau dashboard which gives a better understanding of the results from Hadoop.

The agro advisory system aims to determine the best suitable crop and temperature for optimizing the agricultural yield, to predict the season and a suitable pesticide, to predict compatible crops for intercropping, to determine the soil type and production (Fig. 1).

Fig. 1 System design

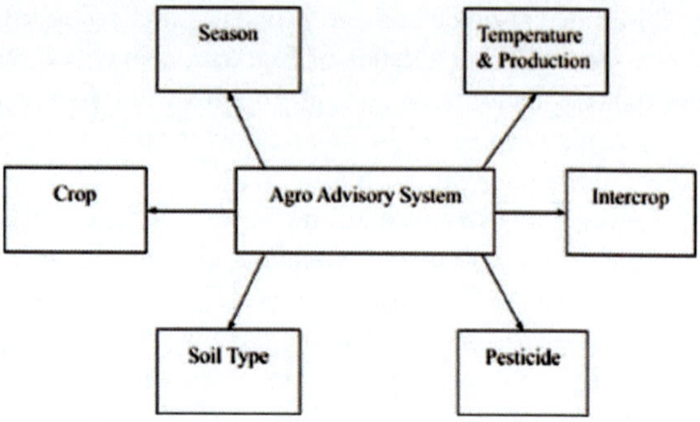

4 System Architecture

The system architecture consists of two major parts, i.e., the front end and the back end (Fig. 2).

A. Front End:

The front end is in the form of a website that is capable to interact with the Hadoop ecosystem. The front end is developed using the following:

1. User Interface:

The UI is developed using HTML, CSS, and ReactJS. ReactJS is a javascript library that allows us to build the UI components that can be reused and present data that is updated with the time.

2. Tableau tool:

Tableau is a data visualization tool in the Hadoop ecosystem. Tableau is connected with the Hadoop Hive database to fetch the Hadoop cluster. Various visualization techniques are available in the tool to plot simple graphs.

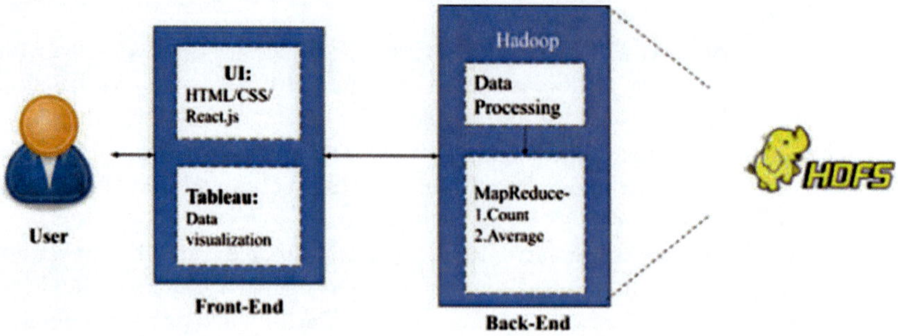

Fig. 2 The architecture of the system

B. Back End:

The back end of the website is a Hadoop ecosystem. MapReduce is a vital part of the Hadoop ecosystem which is used for computation of data. MapReduce framework is used for applications that compute the data in the HDFS. MapReduce programs being parallel are used for performing data analysis using numerous machines inside the cluster. Thus, it improves the pace and accuracy of the system.

C. HDFS:

All the data and files of the Hadoop system are stored in HDFS.

5 Methodology

Figure 3 shows the system flow of agro advisory systems. The data acts as an input to the Hadoop ecosystem. To process the data, the dataset is submitted to Hadoop.

A. Data Access

Apache Hive is used for data querying and research. Hive typically performs data summarization, querying, and analysis. HQL converts the SQL queries into MapReduce jobs, to be carried out on Hadoop.

B. Data Processing

The two applications of MapReduce, word count, and average are used for data processing in the proposed system (Fig. 4).

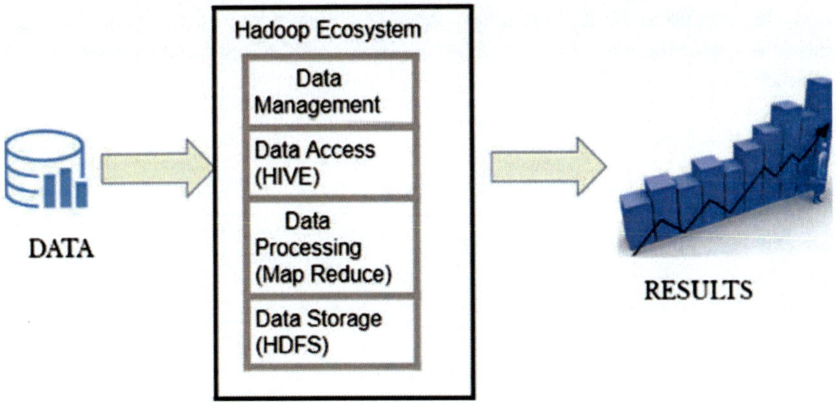

Fig. 3 System flow

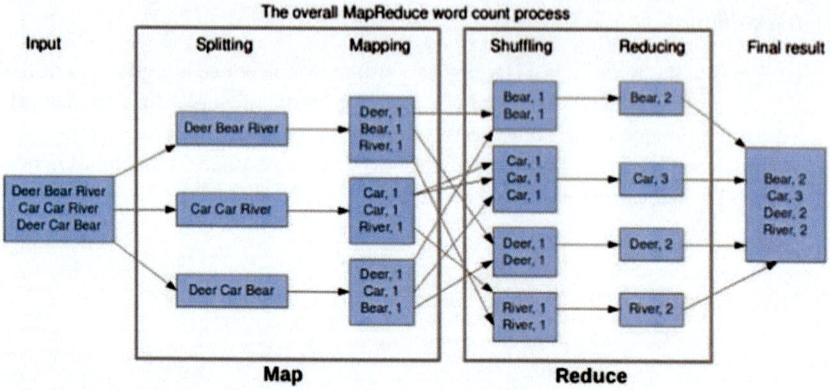

Fig. 4 MapReduce algorithm [4]

i. Wordcount algorithm:

The wordcount algorithm is executed on the dataset. The program calculates the number of times the crops in each state in India. The algorithm consists of three main sections.

1. Mapping phase

In this phase, the input dataset is first converted to key-value pairs. The key-value pair acts as an input as well as the output for the mapper function. The mapper function counts the number of distinct crops, season, intercropping species, soil type, and pesticide for each state in the dataset.

2. Shuffling phase

After the execution of the mapping phase is finished successfully, in the next phase the key-value pairs from the mapping phase act as input and then organized alphabetically.

3. Reducing phase

In this phase, all the keys are clubbed and the values for alike keys are added up to calculate the number of times a word appears. Thus the output of this phase is the count of the crop, season, intercropping species, soil type, and pesticide for each state.

Using the results of wordcount the suitable crop, season, intercropping species, soil type, and pesticide for each state are thus predicted based on maximum occurrences.

ii. Average algorithm

The execution of this algorithm is the same as that of wordcount except for the reducing phase. The reducer function in this algorithm is modified to calculate the average of the temperature and production values. The output which is in the form of key-value pairs consists of the average temperature and production of each state in India.

C. Data Storage

HDFS stores the data while MapReduce processes the data. The dataset is stored in the HDFS which is a major part of the Hadoop ecosystem.

The papers will be published in our digital library www.springerlink.com. Only subscribers to Springer's eBook packages or the electronic book series can access the full-text PDFs and references of our online publications. Meta-data, abstracts, and author e-mail addresses are freely available for all users.

6 Results

The system provides an insight into the agriculture experts about state-wise crop information which thus helps them to formulate different strategies to improve crop production. The numerous benefits of the website are-

- A single destination for all agro information.
- State-specific crop information.
- It can be accessed by users at any time.

In the case of lockdown where farmers cannot physically reach the experts, they can easily access the website for the information they need.

A. Hadoop Result

Data analysis is done using the Hadoop framework. Data processing using the MapReduce algorithm is completed. Observed results are shown in the figures below. Wordcount in MapReduce algorithm was used for finding the crop, season, soil type used, intercropping species, pesticide usage whereas average was used to find the ideal temperature.

MapReduce algorithm is applied to the dataset on Hadoop. Wordcount in the MapReduce algorithm is used for every specific condition to get the desired output (Fig. 5).

The ideal temperature was calculated using MapReduce (Fig. 6).

Estimation of Production using MapReduce, to find out the average values of the production rate of Maharashtra (Fig. 7).

```
[training@localhost Desktop]$ hadoop fs -cat /user/training/seasonresult.txt/part-r-00000
Autumn  7
Kharif  7275
Rabi    3824
Summer  1011
Whole   511
Year    511
[training@localhost Desktop]$ hadoop fs -cat /user/training/soilresult.txt/part-r-00000
AlluvialSoil    12628
LateralSoil     1
[training@localhost Desktop]$ hadoop fs -cat /user/training/pesticideresult.txt/part-r-00000
No      12588
Yes     40
[training@localhost Desktop]$ hadoop fs -cat /user/training/intercropresult.txt/part-r-00000
Arecanut        28
Maize   654
Rice    1004
TrapCrop        10941
[training@localhost Desktop]$
```

Fig. 5 Hadoop results

```
[training@localhost Desktop]$ hadoop fs -ls /user/training/teml.txt
Found 3 items
-rw-r--r--   1 training supergroup          0 2020-04-03 23:23 /user/training/teml.txt/_SUCCESS
drwxr-xr-x   - training supergroup          0 2020-04-03 23:23 /user/training/teml.txt/_logs
-rw-r--r--   1 training supergroup         59 2020-04-03 23:23 /user/training/teml.txt/part-r-00000
[training@localhost Desktop]$ hadoop fs -cat /user/training/teml.txt/part-r-00000
        Total: 313778.90020275116 :: Average: 24.912973418241457
[training@localhost Desktop]$
```

Fig. 6 Temperature results

```
bin/hadoop command [genericOptions] [commandOptions]

[training@localhost Desktop]$ hadoop fs -cat /user/training/prodresult.txt/part-r-00000
        Total: 1.2636406062E9 :: Average: 505755758.6080505762
[training@localhost Desktop]$ █
```

Fig. 7 Production average

B. User Interface

The home page of the website allows you to navigate through various tabs like crop, best practices, about, contact, team (Fig. 8).

Figure 9 shows the live tableau dashboard. These results are represented diagrammatically in the form of pie charts using the Tableau data visualization tool which is connected to HDFS of the Hadoop ecosystem.

The grid view of various crops is shown in Fig. 10.

Latest Blogs and videos related to agriculture are also made available to users (Fig. 11).

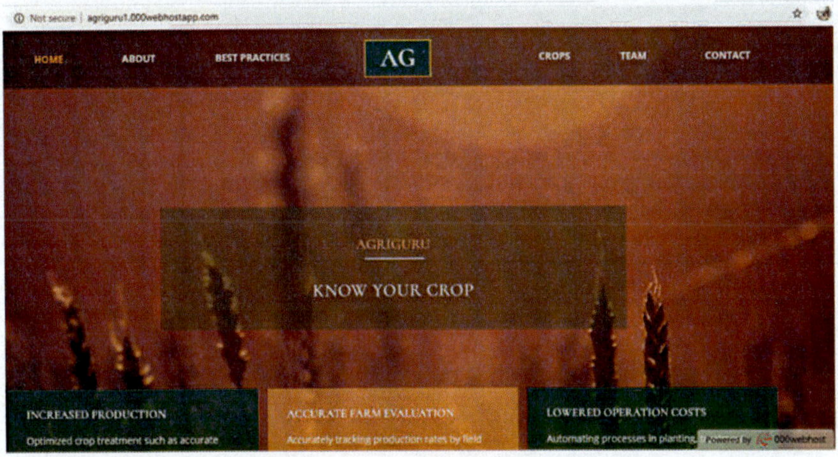

Fig. 8 Home page

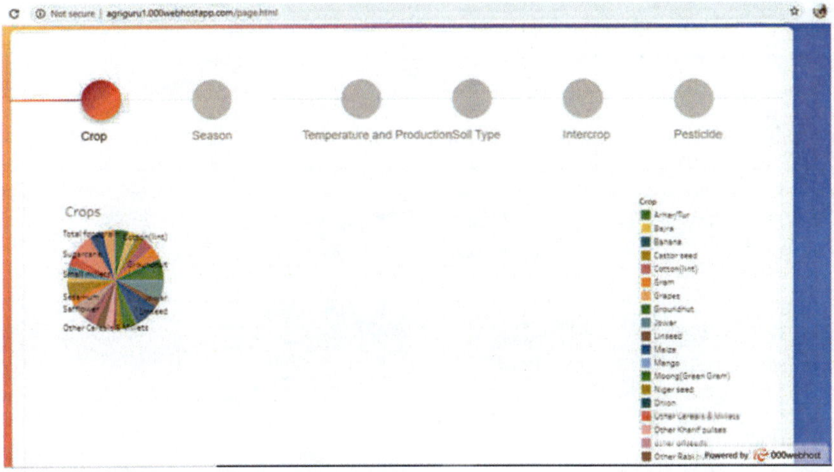

Fig. 9 Live tableau dashboard

7 Conclusion

The Agriculture and Big data together are heralded as one of the greatest advancements in the food production sector since the green revolution. With the help of the observations and detailed study conducted, a dataset is generated which has been used for the analysis and production of the crop. Another aspect of Big Data Analysis is how this data-driven enables farmers to view a complete pre and post-harvest

Fig. 10 Crops page

Fig. 11 Latest blog posts

solution. This technology is very useful for farmers that are not aware of the latest technologies and the user-friendly website also helps the farmer to understand and act.

The project is designed only for a particular state, which is Maharashtra, in the future as per the requirements and the feedback a wide variety of data can be included. Major issues such as Internet availability in village areas, the development need to be such that they can be made available at low Internet speed.

At present the website is in English, to make it comprehensible for people with diversified dialects, the future versions of the website would include a variety of other languages.

References

1. Zhao J, Guo J (2018) Big data analysis technology application in agricultural intelligence decision system. In: 2018 IEEE 3rd international conference on cloud computing and big data analysis (ICCCBDA), Chengdu, pp 209–212
2. Rao NH (2017) Big data and climate smart agriculture—review of current status and implications for agricultural research and innovation in India, Mar 25, 2017. In: Proceedings Indian National Science Academy (Forthcoming). Available at SSRN https://ssrn.com/abstract=2979349
3. Wolfert S, Ge L, Verdouw C, Bogaardt M (2017) Big data in smart farming—a review. Retrieved 5 May 2019, from https://www.sciencedirect.com/science/article/pii/S0308521X16303754
4. Kaur R, Garg R, Aggarwal H (2016) Big data analytics framework to identify crop disease and recommendation a solution. In: 2016 international conference on inventive computation technologies (ICICT), Coimbatore, pp 1–5
5. Sahu S, Chawla M, Khare N (2017) An efficient analysis of crop yield prediction using Hadoop framework based on random forest approach. In: 2017 international conference on computing, communication and automation (ICCCA), Greater Noida, pp 53–57
6. Hirafuji M (2014) A strategy to create agricultural big data. In: 2014 annual SRII global conference, San Jose, CA, pp 249–250. https://doi.org/10.1109/srii.2014.43
7. Di L (2016) Big data and its applications in agro-geoinformatics. In: 2016 IEEE international geoscience and remote sensing symposium (IGARSS), Beijing, pp 189–191. https://doi.org/10.1109/igarss.2016.7729040
8. Bendre MR, Thool RC, Thool VR (2015) Big data in precision agriculture: weather forecasting for future farming. In: 2015 1st international conference on next generation computing technologies (NGCT), Dehradun, pp 744–750. https://doi.org/10.1109/ngct.2015.7375220
9. Shekhar S, Schnable P, LeBauer D, Baylis K, VanderWaal K (2017) Agriculture big data (AgBD) challenges and opportunities from farm to table: a midwest big data hub community whitepaper, pp 340–355. https://doi.org/10.1109/ngct.2015.8284867
10. Kumar M, Nagar M (2017) Big data analytics in agriculture and distribution channel. In: 2017 international conference on computing methodologies and communication (ICCMC), Erode, pp 384–387
11. Varma C (2018) Performance analysis and challenges of the Map Reduce framework in big data analytics. In: 2018 international conference on current trends towards converging technologies (ICCTCT), Coimbatore, pp 1–5
12. Tamilselvi K, Sumithra V, Dhanapriyadharsini M (2020) Big data analytics using Hadoop technology (online) Irjet.net. Available at https://www.irjet.net/archives/V5/i1/IRJET-V5I1328.pdf. Accessed 30 Apr 2020

Big Data Technologies: A Comprehensive Survey

Varsha Mittal, Durgaprasad Gangodkar, and Bhaskar Pant

Abstract In the last decade, the digitization in every aspect of life has resulted in the explosive generation of data. Therefore, the term Big Data had drawn the attention of researchers and the corporate world. This survey paper presents the concept and definition of Big data followed by its characteristics. We provide a brief overview of the challenges of big data, its technologies, and tools that play a significant role in storing and management of big data. We also highlight the work done by various researchers in the storage and analysis of big data. A comparison of storage technologies is also presented that will help the researchers to have a fair idea to address the different challenges.

Keywords Big data · Hadoop · NoSQL databases

1 Introduction

Big data is one of the recent and prospective research frontiers. Big data is not only seeking the attention from the different social, commercial, educational, and research industry but also give a breakthrough [1]. Mckinsey, the well-known management and consulting firm assumed that every industry and modern business functions are being perforated by big data to a greater extent [2]. O'Reilly said that "The future belongs to those who convert the data into products" [3]. The way we live, work, and think are greatly influenced and changed by the values-driven from big data.

Big data is a collection of huge datasets of different data types, which is difficult to be processed using traditional processing approaches. In 2012, Gartner gave a more detailed definition of Big data [3]. He said, "Big data are high volume, high velocity, and high variety information assets that require new forms of processing to enable enhanced decision making, insight discovery, and process optimization". A dataset can be termed as Big Data when it is gigantic to be analyzed and cannot be processed

V. Mittal (✉) · D. Gangodkar · B. Pant
Graphic Era Deemed to be University, Dehradun, India
e-mail: var.aadi@gmail.com

G. Ranganathan et al. (eds.), *Inventive Communication and Computational Technologies*, Lecture Notes in Networks and Systems 145,
https://doi.org/10.1007/978-981-15-7345-3_9

using traditional software and analytic tools, require significant processing power, and cover a wide assortment of data types including text, numeric, images, videos, audios, etc.

With the increased and wide variety of data collecting options such as sensor networks, social media, and smartphones, etc. data grows exponentially. The off-shelf techniques which are designed to store and then analyze the data do not work efficiently and give satisfactory result [4]. Every stage either it is data capturing, data curation, or data visualization faces a lot of challenges. It is required to design new methods for storing, processing, analyzing, and visualizing the big data and to use it for our purposes. To address the challenges for big data representation and modeling a clear understanding of the features of big data is required.

The paper focused on discussing the evolution of big data by identifying the various challenges that exist in every stage either it is data capturing, data curation, or data visualization. The paper is organized in the following manner Sect. 2 gives a brief overview of the characteristics of Big Data. Section 3 presents the various challenges faced in representing, storing, and modeling Big data. The further section presents the various technologies used in storing, analyzing, and visualizing the Big Data.

2 V'S of Big Data

The big data is characterized by V's representing the volume, velocity, and variety. But the veracity, value, and variability have also given the new dimension to big data. The huge datasets refer to the volume, the structured, semi-structured, and unstructured nature of data refers to the variety and the continuous generation of data through various sources even without considering the relevancy shows the velocity. Among all the v's of big data, the most important is the value. It is not only difficult to extract but also quite hard to learn. Each of these V's figure out the scope of big data and are the main characteristics of it:

1. **Volume (data at rest)**: Volume refers to the magnitude of the data. Nowadays the data is being generated at high speed through various sources either the digital or through the human activity. In big data, the size can be measured in terabytes, petabytes, and higher units [5]. The main magnetism in using big data analytics is the ability to process a large amount of data. The challenge of volume can be addressed through technology like by using Hadoop and the clusters which can be created through commodity hardware.
2. **Variety (data in many forms)**: Data is being collected from various sources so structural heterogeneity of the dataset can't be ignored. As a point the data from the relational databases are structured, data collected from social media, feeds, weblogs, email, etc. is semi-structured whereas the data form video, still images or audio is unstructured [5]. This challenge is addressed at the software level.

3. **Velocity (data in motion)**: Velocity means the data generation rate and the pace at which the data should be analyzed and acted upon [6]. The generation of abundant data through several smart devices, sensors, etc. is leading to the requirement of real-time analytics. The various techniques like real-time processing and in-memory computation can be used to address this issue.
4. **Veracity (data in doubt)**: To measure the inherent untrustworthiness in some sources of data IBM coined the fourth V. Example of such sources can be data collected through social media [6]. Due to the veracity, the result derived from big data can be assigned probability. This issue is addressed through different software solutions.
5. **Value (data in highlight)**: This is a feature of big data that was introduced later by Oracle. Big data is also called "low-density data" [7]. It signifies that the value derived out of the data has low value in comparison to the volume of the data. This aspect of big data emphasizes extracting knowledge. This issue is difficult to address as value is extracted concerning the application domain or some business problems. Understanding of appropriate application domain is required by the data scientist.
6. **Variability**: Two additional dimensions of big data which increases the scope of it are variability and complexity [8]. The discrepancy in data flow rates refers to variability and complexity refers to the challenges faced in creating big data through several sources.

3 Big Data Challenges

Big data offered numerous and attractive opportunities but a lot of challenges are still there which need to be handled while dealing with big data [9]. Various hindrances are faced in capturing the data, storing the data, searching and sharing the data, and also in analyzing and visualizing it. To harness the hidden values of big data is the same as extracting the gold from its ore but we still do not have sufficient capabilities to do so.

1. **Heterogeneity and Incompleteness**: Manual analysis of data can easily tolerate and handle the heterogeneity of the data as the richness of natural language provides great help. However, the use of a machine learning algorithm for data analysis requires the data to be in a homogenous state [9]. As a consequence, the data is to be converted into a structured format before it is ready for analysis. The analysis of incomplete data may result in uncertainties in the result. Incomplete data refers to the missing values in data samples [10]. Even after performing data cleaning and error correction, it can't be assured that the data will not suffer from incompleteness and errors.
2. **Scale**: Whenever the word big data comes in the scene the very first thing that strikes is the size and managing this large and rapidly growing data is a challenging task [9]. The vast size of data cannot be managed using traditional tools. In past years, Moore's law was followed to mitigate this challenge that is to

provide more and more resources to handle the large volume of data. But now the volume of data is scaling faster than the computing resources. In the past few years, the processors with an increased number of cores are designed considering the power constraints as an alternative approach to increase the speed of processor rather than simply doubling the clock cycle [11]. Moreover, to handle the large volume of data parallelism is applied across nodes in a cluster but to achieve the desired results the intra-node parallelism is required. The techniques which can be applied for inter-node parallelism cannot be directly applied for inter-node parallelism and this is the main challenge to be addressed.

3. **Timeliness**: The size of the dataset is directly related to the time taken to process it. Large is the volume of the data more is the time required to process it. The challenging task is that the designs which are made to handle a large amount of data should also process the data at a faster rate to get the result faster. In consideration of speed not only the processing speed is important but the speed of acquiring the data is equally important [12]. Some situations require the immediate result of the analysis. Let us consider a situation, if a fraud credit card transaction is suspected, then before the completion of the transaction it is must figure it out. In such situations, the complete analysis is not possible so the partial results must be prepared in advance. And to reach the quick decision only a small amount of work with new input data is required. In a large dataset, the data meeting specified criteria are searched frequently and repetitively. To retrieve the data satisfying the search criteria from such a huge dataset will be a time-consuming process. In such cases, index structures are created to support this type of search quickly but it also suffers a limitation that these index structures are not sufficient for all types of search.

4. **Privacy**: The term Big Data is defined as a huge collection of data that cannot be ever processed, captured, and stores using any old conventional techniques. The privacy of data is another important concern that proliferates with the allusion of big data. Since data is accessible through multiple sources there is the probability of inappropriate access and use of personal data [13]. Moreover, in the real world, the data is not static and keeps growing over time there exist no prevailing techniques that assure the privacy of the personal data. The other important aspect of privacy is the security for information sharing in big data. With the growing usage of social media, the user needs to share the private information, but apart from the record level of access control, the meaning of sharing the information is not clear.
5. **Human Collaboration**: Even after much advancement in the computational analysis [11], and also the use of machine learning algorithms found great difficulty in detecting and finding the hidden pattern that can be easily identified by having the human in the loop especially in the analytics of big data. Crowd Sourcing [14] is one of the popular methods of utilizing human intelligence in analyzing and solving problems. Starbucks, Doritos, online encyclopedia are some of the best examples of crowdsourcing. The individuals may have different motives to provide some information. The information may be false or correct. But in

crowdsourcing, the error can be detected and corrected by others in the crowd. The challenge which can be identified here is the uncertainty of the data collection devices.

4 Big Data Technologies

A variety of tools can be worn in Big Data management to perform the various tasks covering all the stages of the BDA pipeline starting from data acquisition to interpretation. The majorities of these tools are the ingredient of Apache projects and are built around the Hadoop. Two core projects of Hadoop are [8]: Hadoop Distributed File system(HDFS) and MapReduce.

1. **Hadoop Distributed File System (HDFS)**: HDFS is based on the Google File System (GFS) which is designed to run on large clusters of commodity hardware [10]. It is also based on a Java based distributed file system that provides scalable and reliable storage [15, 16]. Shvachko [17] adds strength in the definition by saying that "it is designed to store very large data sets reliably, and to provide those datasets to user applications whenever required". The UNIX file system pattern is used to design the HDFS. HDFS is not suitable for interactive use rather it is used for batch processing [8, 16]. In HDFS applications, only once the files are written but they can be executed many times [12, 13, 18]; as a result, we can assure data coherency and high throughput of the system [17]. While using the HDFS file system the dedicated server, the NameNode system is used to store all the metadata and the Data Node is used to store application data.

Except for handling out large databases, HDFS has many other goals the primary objective is to store data reliably even in case of failure by using the feature of data replication. Hadoop also has the feature of scalability and load balancing which makes it suitable for processing and analyzing big data.

2. **MapReduce**: It is a programming model and is implemented to apply a distributed and parallel algorithm for a dispensation and engendering large data sets on a cluster [8] of commodity servers. It was originally developed by Google to support parallel and distributed processing. This framework supports the application-specific input data which is given by the user [15] and appropriate for semi-structured or unstructured data. The (key, value) pairs are synthesized as the output of Map Reduce. Two functions named "Map" and "Reduce" generate the "MapReduce" programming paradigm. Initially, on the input data, the Map function is applied and individual elements are broken down into tuples where each tuple consisting of (key, value) pairs; and the output from the Map is taken as input by Reduce and all intermediate keys are combined with the same intermediate values [15, 16].

In a Hadoop cluster, tasks are executed by subsequently breaking the job (i.e., a MapReduce program [7]) into small pieces. However, in many applications like

RDF/RDFS and OWL reasoning [15, 19] and also in querying structured data MapReduce has been successfully used [20–22].

5 Database Capabilities

Since from the 80s, the relational model is dominating the market but the changing trend of the data concerning its volume, variety, and the increased cost of vertical scaling forces the market and the researchers to look for the other alternatives [23]. The modern applications cannot rely on the concept of "one size fits for all". The traditional databases are not suitable to store the huge and heterogenous data called Big Data. The relational databases have a fixed and rigid schema that is to be defined in advance before the data is stored. All the traditional databases have to strictly follow the ACID properties of the transaction. An adaptive approach. NoSQL provides the solution to all the above-said problems and introduces the concept of BASE semantics which is completely suitable for a distributed environment.

Table 1 summarizes the difference between the relational databases and NoSQL databases. NoSQL databases are of four type key-value, wide column, document, and graph databases which differ in their scalability and functionality level.

1. **Key-value store**: A Key-value store does not have a fixed schema. It stores the data in the key-value pair where indexing is done on the keys. Keys use the hash table to read the associated value. These databases are a good choice when the data is not related. It provides the search for the exact match. These databases are extensively used because of their simplicity, scalability, fast retrieval, and suitable to work in a distributed environment [23]. OracleNoSQL is a good example of a key-value store. It is a scalable, distributed NoSQL database by Oracle Corporation [24]. It is premeditated to offer highly reliable, bendable data management across a set of storage nodes. For the application developer, Oracle NoSQL Database provides a very simple data model. In this data model, a unique key is used to identify each row, and also has a value, of random length, which is understood and used by the application. A single row can be manipulated (insert, delete, update, read) through a transaction performed by the application. All rows

Table 1 Comparison of SQL and NoSQL databases

Databases	Scalability and throughput	ACID complacence	Memory overhead	Data Structure Schema Design
RDBMS	Vertical & low	Strict supports	More	Pre-defined data types with the rigid schema
NoSQL	Horizontal high	BASE semantics	Less (supports memory caching)	Unstructured, unpredictable data with flexible schema

in the database can be scanned repeatedly by the application. Since it stores the data in the form of (key, value) pair so this database is suitable for storing Big Data.

2. **Column Stores**: Column store or wide column databases are inspired by Google Big table. It stores the data based on the columns rather than storing it row-wise. It provides a hybrid approach to the relational database and key-value store. It an extension of a key-value store where each column has a different structure. The structure of a column key in a column store database is r-key, column-family-store, column-value, timestamp where r-key is the row key which is used for indexing. The concept of a timestamp is used for versioning [25].

HBase is a database that is designed after Google's Big table and it is a non-relational, open-source, and distributed database. It is a database that is developed in Java and it runs on top of HDFS (Hadoop Distributed File system). It is developed as a part of the Apache Software Foundation's Apache Hadoop Project. It also provides BigTable-like competence for Hadoop [8–10]. That is, a large amount of sparse data can be stored in a reliable way using HBase (sparse data is some small information trapped in a huge anthology of empty or insignificant data, such as finding the 100 objects in a collection of 3 million records, representing less than 0.1% of a large collection).

Apache Casandra is another very popular wide column database [20]. It is an open-source, scalable distributed database management which is developed as part of the Apache Project. High availability is one of the prominent features of Casandra. It was originally developed by Facebook, which was open-sourced in 2008 afterward in 2010 it becomes the part of Apache Project. Cassandra offers strong support for those clusters which have multiple data centers, by providing masterless replication so that all clients can have low latency operations.

A high value on performance is also placed by Cassandra. In 2014, a team of researchers at the University of Toronto studying NoSQL systems concluded that "Cassandra provides the high scalability but for the maximum number of nodes, at the cost of high read and write latencies it provides the highest throughput" [17].

3. **Document Store**: Document databases are used to store, manage, and retrieve the document in the form of JSON, BSON, etc. These databases are used to store text documents like emails or XML documents. A unique key is used to store and retrieve the document. MongoDB is a popular example of a document store. It doesn't have a fixed schema and is implemented through API calls. It is designed for large datasets. It is horizontally scalable and supports replication for fault tolerance [22].
4. **Graph Databases**: Graph databases use nodes and edges to store highly interrelated data. It not only stores the data but also the relationship that exists among the data. The performance in many use cases like pathfinding problems, recommendation systems, navigating the social networks, forensic enquires is enhanced using the graph database [22].

Neo4j is an open-source graph database that stores the data in the form of nodes and edges where both nodes, as well as edges, have their properties. It provides a faster method to retrieve and store the highly connected and semi-structured data since the graph database is schema free [26]. No complex joins are required to retrieve the interrelated data.

6 Processing Capabilities

1. **Apache Hive**: On top of Hadoop, a data warehouse infrastructure is built which is named Apache Hive. It facilitates to query and manage the data stored in distributed storage. Originally it was developed by Facebook, but now many other companies such as Netflix also used and developed it. A software fork of Apache Hive is maintained by Amazon to provide web services on Amazon and it is incorporated in Amazon Elastic MapReduce [27].

The analysis of large datasets stored in Hadoop's HDFS and some compatible file systems such as the s3 file system of Amazon is supported by Apache Hive. It defines aHiveQL as a simple language like SQL that enables the user to query the database. The read and write queries on a schema can be transparently converted to Map/reduce, Apache Tez, and Spark jobs. The execution engines of Apache Tez and Spark can run in Hadoop YARN. For query optimization, the indexes are used and it also includes the use of bitmap indexes.

By default, to store metadata the client/server database like MySQL and Apache Derby database is used. Nowadays, TEXTFILE, SEQUENCE FILE, ORC, and RCFILE are the four file formats supported in Hive.

2. **Apache Zookeeper**: Apache Software Foundation has another software project named as Apache ZooKeeper. For large distributed systems it works as an open-source distributed configuration service, naming registry service, and synchronization service. Previously ZooKeeper was an associate as a part of Hadoop but now it is working as an individual a top-level project in its way.

Redundant services are used to provide high availability in ZooKeeper's architecture. If the first Zookeeper leader is unable to answer the clients can solicit another ZooKeeper leader. A Hierarchical namespace is used to store the data in ZooKeeper nodes Since the ZooKeeper has shared configuration service the read/write from/to the nodes can be done by the client easily. Updates are ordered [8, 9]. Many companies like eBay, Rackspace, Yahoo!, Odnoklassiniki, and also the open-source enterprise named as Solr uses the Zookeeper.

3. **Pig**: For analyzing large data sets, especially the data sets which use high- level language for representing the data analysis program, Apache Pig provides the platform [13, 15]. Pig Latin is the language used to work on this platform. The abstraction provided by the Pig Latin converts the Map Reduce programs in Java into a simple notation which makes it simple to use, the simplicity is the same as that of SQL provide to RDBMS systems. The UDF (User Defined Functions) written in Java, Python, Ruby, JavaScript or Groovy can be used to extend the

library of Pig Latin. These UDF can be directly called in the programs written in Pig Latin [8, 9].

To create and execute Map Reduce jobs on large data sets Yahoo Research in 2006 developed the Pig for researchers. It was included as a part of the Apache Product in 2007.

The advantage of using Pig Latin for processing is that writing script using Pig Latin only takes 5% of the time as compared to writing the script in Map Reduce which reduces the development and testing time but it increases the execution time by 50%.

4. **Chukwa**: For log collection and analysis to be used for big data an open-source data collection system is designed named Chukwa. It works at the top layer whereas in the background the Hadoop Distributed File System (HDFS) and Map/Reduce framework work. Scalability and Robustness are some of the prominent features of Chukwa. For log analysis, it provides a flexible and powerful toolkit. The tool kit can also be used for displaying, monitoring, and analyzing results [28]. Originally it was designed to aim system analysis and debugging but later on, it is found suitable to be used for applications involving lower latency in data delivery example of such application is adaptively provisioning distributed system based on the measured workout. Working with Chukwa has three important properties:

   - The system being monitored will not be imposed high overhead
   - It is capable of scaling huge volumes of data.
   - It recovers rapidly from failures

5. **Oozie**: Oozie is an open-source Java web application used for Scheduling Apache Hadoop job. It is also used to handle complicated pipelines of data processing [3, 11, 12] and is executed in the Java servlet container. All the workflows in Oozie are arranged in a control dependency graph called as DAG (Directed Acyclic Graph) and the collection of actions, i.e., Map/Reduce jobs, Pig jobs are called as the workflow [22]. The term "control dependency" signifies that without completing the first action the second action can't run. The jobs are initiated in remote systems (Hadoop or Pig) by the workflow actions. The remote systems call back Oozie after the action is completed, to inform about the action completion; so that the Oozie can start the next action in the workflow.

7 Data Integration Capabilities

1. **Apache Sqoop**: To transfer the data between the relational database and Hadoop a command-line interface application is designed named Sqoop. It is SQL to Hadoop. It is a tool written at Cloudera. It can be used to import the selected table, complete database, selected column from a table, selecting certain rows from the selected table [20]. We can imports data to populate tables in Hive or

HBase whereas to put the data from Hadoop into a relational database we can export the data. To transfer the data from Microsoft SQL Server databases to Hadoop, Microsoft uses a Sqoop-based connector.

2. **Flume**: Flume is a data ingestion mechanism to effectively handle a large amount of log data. It can efficiently collect, aggregate, and move the log data into HDFS. It is designed to transfer the log data to the HDFS [23]. It is a reliable and configurable tool. It has a simple and flexible architecture consisting of flume event and flume agent. It provides a mechanism for failover and recovery.

8 Visualization Techniques

To make valuable decisions is one of the crucial and final goals of the Big data analysis. To achieve this goal, the good and effective visualization of big data content is required. Because of this reason, the area of the visualization can't be ignored and has gained a lot of interest [3, 4], i.e., to improve, communicate and understand the result of big data analysis various techniques are used. These techniques include the method of creating diagrams, animations, or images [10].

1. **Tag Cloud**: To visually depict the text data a tool named Tag Cloud is used. It is a weighted list in visual design. To represent the reserved words, metadata on websites or to represent the free form of text the Tag Cloud is used. Tags are single words, and different font size or color is used to show the importance. To have a quick overview of the highest-flying terms used on a web page and for finding a term alphabetically the method of the Tag Cloud is used [21]. It can also be used as a navigation aid in a website by hyperlinking the terms to the items associated with the tag.
2. **History Flow**: History Flow is another visualization technique given by Mutharaju et al. [22]. To show the contributions of different authors and to show the progression of a document, this tool is used. The time is represented by the horizontal axis of a history flow and the names of the authors are represented by the vertical axis. The color code method is used to indicate the amount of text written by each author, where each author is given a color and the contribution of each author is indicated by the length of a vertical bar.

9 Conclusion and Future Scope

In today's scenario, we cannot deny the importance of the data as it affects every aspect of human life. The term Big Data had been coined to signify the importance of the data. The development in many scientific streams and the productivity of many organizations can speed up and enhanced by performing the analysis of the huge volumes of data which is collected through various sources. However, to utilize these potentials to their fullest extent certain technical challenges need to be addressed.

Various stages are there in the analysis pipeline, it includes the phase from data acquisition to result from interpretation. Each stage of big data analytics pipeline faces difficulties in handling heterogeneity of data, lack of structure, scale, error handling, privacy, timeliness, provenance, and visualization of results. Many problems like the problem of reasoning, entity linking, information extraction, consolidation, and the problem in visualizing the results can be encountered during big data management. In this paper, a deep insight into the term Big Data, its features, and the various challenges are reviewed. The paper summarizes that until now the volume is the most handled aspect through Hadoop. But now other issues like velocity, informality in data collected through various sources like web data, social data, and uncertainty needs to be handled. We also reviewed the various tools available that can be used during the various steps of the data analysis life cycle. To achieve the benefits of big data, fundamental research is forwarded towards addressing the technical challenges.

References

1. Halevi G, Moed H (2012) The evolution of big data as a research and scientific topic: overview of the literature. Res Trends 3–6
2. Krishnan K (2013) Data warehousing in the age of Big Data: the Morgan Kaufmann series on business intelligence, 1st edn. Elsevier Science
3. Gartner IT Glossary. (online). Available: http://www.gartner.com/it.glossary/bigdata/
4. Reeve A (2013) Managing data in motion: data integration best practice techniques and technologies, 1st edn. Morgan Kaufmann
5. O'Reilly Media (2014) Big data now, 2nd edn. O'Reilly Media
6. Hitzler P, Janowicz K (2013) Linked data, big data, and the 4th paradigm, Semantic web, pp 233–255 (online). Available http://blog.semantic-web.at/2012/08/09/whats-wrong-withlinked-data/
7. Bertino E, Haas L, Madden S et al (2012) Challenges and opportunities of big data, white paper, Feb 2012 (online). Available www.cra.org/ccc/files/docs/init/bigdatawhitepaper.pdf
8. Amudhavel J et al (2015) Perspectives, motivations, and implications of big data analytics. In: Proceedings of international conference on advanced research in computer science engineering technology, New York, USA, 6–7 Mar 2015, vol 9, pp 344–352
9. Suchanek F, Weikum G (2013) Knowledge harvesting in the big-data era. In: Proceedings of international conference on management of data, SIGMOD'13. ACM, New York, NY, USA, pp 933–938
10. Cuzzocrea A (2014) Privacy and security of big data: current challenges and future research perspectives. In: Proceedings of first international workshop on privacy and security of big data, PSBD
11. Big data. Nature 1–136 (2008)
12. Mayer-Schonberger V, Cukier K (2013) Big data: a revolution that will transform how we live, work, and think. Houghton Mifflin Harcourt
13. Kalil T (2012) Big data is a big deal (Online). Available http://www.whitehouse.gov/blog/2012/03/29/big-data-big-deal
14. Jagadish HV, Aggarwal D, Bernstein P (2015) Challenges and opportunities with big data. The Community Research Association
15. Loukides M (2011) What is data science? O'Reilly Media, Inc
16. White T (2009) Hadoop: the definitive guide, 1st edn. O'Reilly Media, Inc

17. Shvachko K, Kuang H, Radia S, Chansler R (2010) The Hadoop distributed file system. In: Proceedings of IEEE 26th symposium on mass storage systems and technologies (MSST), MSST'10. IEEE Computer Society, Washington, DC, USA, pp 1–10
18. Oussous A, Zahara F, Lahcen A, Belfkih S (2018) Big data technologies: a survey. J King Saud Univ Comput Inf Sci 30:431–448
19. Corbellini A, Mateos C, Zunio A, Godody D, Schiaffino S (2017) Persisting big data: the NoSQL landscape. J Inf Syst 63:1–23
20. Borthakur D (2014) The Hadoop distributed file system: architecture and design. The Apache Software Foundation, pp 1–14
21. Dean J, Ghemawat S (2008) MapReduce: simplified data processing on large clusters. Commun ACM 51(1):107–113
22. Mutharaju R, Majer F, Hitzler P (2010) A MapReduce algorithm. In: Proceedings of 15th international conference on description logics, CEUR workshop proceedings, Waterloo, Ontario, Canada, pp 464–474
23. He Y et al (2011) Rcfile: A fast and space –efficient data placement structure in map reduce based warehouse systems. In: Proceedings of IEEE 27th international conference on data engineering, ICDE'11. IEEE Computer Society, Washington, DC, USA, pp 1199–1208
24. Manyika J et al (2011) Big data: the next frontier for innovation, competition, and productivity. McKinsey Global Institute
25. Oussous A, Benjelloun F, Lahecen A, Belfkih S (2019) Big data technologies: a survey. J King Saud Univ Comput Inf Syst 30:431–448
26. Corbellini A, Mateos C, Zunio A, Godoy D, Schiaffini S (2017) Persisting big data: the NoSQL landscape. J Inf Syst 63:1–23
27. Thusoo A et al (2010) Hive—a petabyte-scale data warehouse using Hadoop. In: Proceedings of 26th international conference on data engineering. IEEE, pp 996–1005
28. Boulon J et al (2008) Chukwa, a large scale monitoring system. Cloud Comput Appl 1–5

Automatic Face Recognition and Finding Occurrence of Actors in Movies

Prashant Giridhar Shambharkar, Umesh Kumar Nimesh, Nihal Kumar, Vj Duy Du, and M. N. Doja

Abstract In this paper, artist face recognition, and all movie prediction system is proposed. It comprises of two phases. Initially, the faces in the video are recognized using an l1-minimization CNN + HOG framework, and some keyframes are selected, based on a robust measure of confidence. Then the labels are propagated from the keyframes to the remaining frames, by using transductive learning. The constraints in both feature and temporal spaces are integrated simultaneously. The output of the algorithm is tested on Indian Movie Face—Database and generated all the movies of those actors/actresses of the film.

Keywords CNN · HOG · L1 minimization · Transductive learning · JSON

1 Introduction

The launch of new unconstrained databases like PubFigs [1], LFW [2], and YouTube Celebrities [3] have amplified the interest in face-recognition. After the launch of such datasets, substantial progress has taken place in enhancing the output in recognition and verification. Such developments have sparked curiosity in recognition of

P. G. Shambharkar · U. K. Nimesh (✉) · N. Kumar · V. D. Du
Delhi Technological University, Delhi, India
e-mail: umeshkumar27101998@gmail.com

P. G. Shambharkar
e-mail: prashant.shambharkar@dtu.ac.in

N. Kumar
e-mail: nil8535@gmail.com

V. D. Du
e-mail: vuduydu1997@gmail.com

M. N. Doja
Jamia Millia Islamia, New Delhi, India
e-mail: mdoja@jmi.ac.in

G. Ranganathan et al. (eds.), *Inventive Communication and Computational Technologies*, Lecture Notes in Networks and Systems 145,
https://doi.org/10.1007/978-981-15-7345-3_10

faces in the unconstrained environment [2, 3]. The recognition of actors present in films may have several applications, such as particular actor/actress related scene retrieval, video indexing, etc. The task is difficult due to broad variation in posture, appearance/look, camera motion, face expressions, and occlusions. Many of the recent solutions tackle the issue by independently labeling the person's faces and the facial tracks, while disregarding the large number of unlabeled faces.

An efficient approach to face recognition for the movie clips has been presented by us, provided a broad collection of a labeled facial dictionary. The initial step in our approach is to classify all the faces identified in a video using our facial recognition techniques such as the Convolutional Neural Network + Histogram of Oriented Gradients. The key-frame labels which are highly positive are retained based on video clip frame features [4]. In the next step, only the faces of the video clip given are examined and view the keyframes from clip as labeled and rest of the frames in the clip as unlabeled. In the third step, all the movies of those actors/actresses are found from The Movie Database using the JSON file API, component to our algorithm. The similarities in features and temporal space are examined in the clip as propagate through frames, hence successfully utilizing the correlation among the faces in the video and between consecutive frames.

Our technique would be extremely effective in the offline annotation of a movie trailer or movies' clips. Our algorithm (HOG + CNN) can also work with criminal face dataset to generate the crime history of that particular criminal and find all details of the criminal and crime can be controlled effectively using digital technologies.

2 Related Theory

Numerous attempts are made in the last few years to recognize faces in videos. These approaches can be roughly classified into three groups: keyframe-based approach, image-set approach, and temporal model-based approach. Keyframe-based [5, 6] approaches treat video as a set of images. They depend on the identification of a single image face and perform identification on all or a selection of frames. Only a set of keyframes is selected for recognition, rather than recognizing each frame. The selection is based on several good quality heuristics which are appropriate for face-recognition. Several methods have been suggested to pick high-quality frames like relative eye and nose positions [7], reliable statistics to filter out distorted images of the face [6], etc. When the keyframes for each track are established, they use majority-voting-schemes for eventually marking the complete video. Temporal model-based [8] methods consider the relation between continuous frames in a movie clip. The various dynamics of the face like non-rigid head movements and rigid facial expressions are modeled to learn how a video sequence differs between individual faces. However, in Image set-based [9] approaches face tracks are treated as the image-sets and the face image locations are modeled in each sequence. The similarity in recognition is then measured. There may be instances where the temporal sequence of faces in the clip is not present because of detection limitations or other factors. In such cases, it may not be possible to observe the temporal coherence between the frames. A Histogram-Oriented Gradient approach [10], this approach focused on the assessment of well normalized local histograms of dense grid image gradient orientation. The basic principle is that the size, shape, or structure of the local entities can usually be well-identified through using the distribution of edge directions or local intensity gradients, often without knowing about the respective gradient or edge positions in detail. Our approach is CNN + HOG resembles keyframe-based methods and temporal methods. Unlike current keyframe based approaches which concentrate on selecting the high-quality frames, our strategy selects those frames that are identified confidently. In contrary to temporary coherence-based approaches, we have not modeled any face dynamics but consider only the temporal-proximity when doing propagation on the labels. As a major task, we predict all movies of actors/actresses present in the clip using JSON implementation in JavaScript Language.

3 Dictionary Formation of Movie Actors

A face dataset—Indian Movie Face Database (IMFDB), inspired by PubFigs [1] and LFW [2], has recently been launched to further facilitate face-recognition work in unconstrained environments. The database contains about 34,000 facial images of 100 actors/actresses from about 103 movie clips and Indian films. This involves 67 actors and 33 actresses, each actor having a minimum of 200 pictures. The IMFDB database would help recognize characters, actions, and shots to retrieve by aligning the scripts as previously done [11]. From a large and diverse range of Indian films, IMFDB has been created to record the maximum amount of variations. The faces images present in the films are diverse in detail, lighting, age, hair, pose, and voice. This is produced by carefully choosing and removing the faces present in the films based on detail, lighting condition, age, hair, pose, and voice. First, films were selected at several different phases of the career of each actor to include the differences in age. To have a diverse presence of actors, they were chosen from five distinct languages which are spoken in India: Hindi, Bengali, Malayalam, Telugu, and Kannada. The films were also from a wide period (1970–2012), thus having differences in quality of image, resolution, medium recording, and illumination. Faces were identified manually from the individual video-clips, without the aid of face detectors. The actors pose may have significant variations in films that lead to a diverse-shots in songs, action scenes, etc. It took considerable cost and human effort to detect the faces manually, however, the resulting collection of images was very diverse in terms of poses. The produced poses were new and challenging which may be hard even for the most advanced detectors to detect. After the frames were detected, only a few different sets of images (facial expression) were selected from each film for preventing any resemblance between the images. Figure 1 displays some pictures of actor Salman Khan from the IMFDB database which shows the rich diversity in expression, posture, resolution, lightening, occlusion, quality, and age. The IMFDB database offers a very detailed annotation of the pose, voice, makeup, age, occlusion, and lighting for each

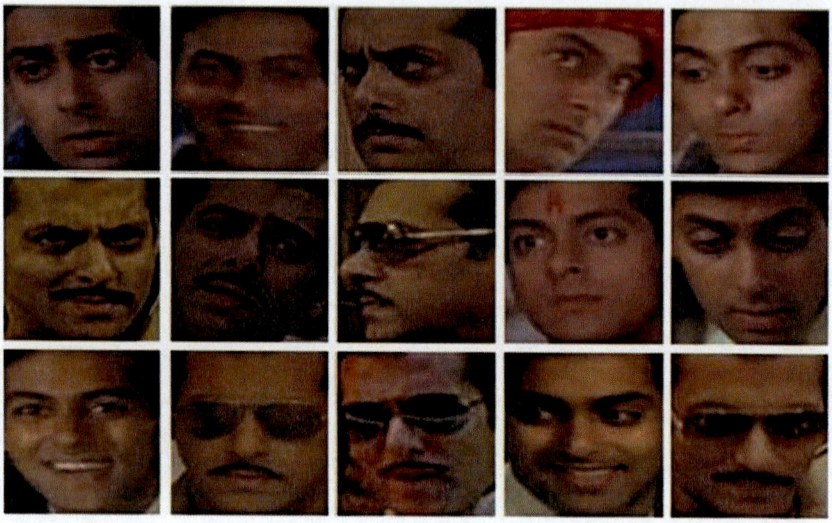

Fig. 1 Sample images obtained from different movies from the Indian actor Salman Khan's IMFDB. Faces of actors are identified and cut manually from a broad set of films chosen on various phases of their careers. Note the great differences in terms of appearance, pose, occlusion, and age

actor. In the following ways, IMFDB varies from current unconstrained databases, including LFW [2] and PubFigs [1]. Images are collected from Internet sources like Google image search in such databases.

However, IMFDB is generated from movies. Hence the images have greater differences in lighting, pose and image resolution. The Actors/Actresses images were taken from the web would have a similar public image with a very limited variety.

Also, as the images are obtained by search query, the images collected may be similar and contains minimum variations in age. Celebrities and public figures frequently maintain their personality (expressions, appearance, dress styles) when they are in public which leads to limited variations, whereas in movie actors will exhibit a very broad diversity of expression, postures, and appearances. IMFDB also encompasses broad diversity in age by careful film selection.

4 Types of Face Recognition Approach in Video

4.1 Key-Frame Based Approach

In this class, the video clips are viewed as a sequence of frames/images. The face-recognition is performed with the comparison of all or the subset of each frame/face-images against the training set of images using image-based face-recognition methods. No assumption is made about the input face image distribution. Also, they do not use the temporal coherence between the input face images. They are based on the proposition that the correlation between the videos from the test set and the training-data is represented in the similitude of the actor image from the test video clip and the training-data. The training data may be still images or videos. Satoh [12] compares two facial sequences based upon the facial correlation between the closest pair of face images between the two sequences. Such techniques can crash, as the impact of outliers is not taken into account by them [12]. A technique would be slow if it requires to compare every pair of images/samples taken from the clip from the input. This approach is used for our work.

4.2 *Temporal Model-Based Approach*

A video not only provides a large number of still frames, but it also allows us to do the characterization of the faces according to the intrinsic dynamics which cannot be done in the case of unmoving images. The characteristics of face comprise the non-rigid motion of features of the face (such as expressions on face) and the rigid motion of the entire face (such as moving head). Psychological findings show that facial features play a remarkable role in the procedure of face identification and facial information both static and dynamic is used to recognize faces in the visual system for humans. Under impaired conditions, such as low lighting, distance recognition, and, low resolution. facial dynamics are even more important. Many facts related to them have been validated in computer vision research (Fig. 2).

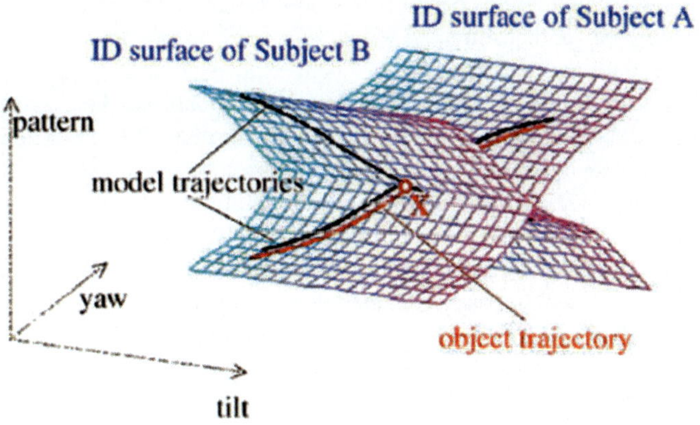

Fig. 2 Temporal model-based approach for identity surfaces in dynamic face recognition [13]

4.3 Image-Set Matching Based Approach

Although temporal dynamics may in some cases be manipulated, the collected facial images could not be temporarily successive in a more general scenario. This may be because of the functional shortcomings in the current face identification technique, for example, it is not easy to find or monitor faces from any video frame continuously, or the images are unordered and sparse examination obtained over a long period and from different viewpoints. For face recognition, instead, modeling facial dynamics is not necessary. In image-set matching techniques, face recognition is done by comparing the probe image set with the whole collection of image sets. The image set represents a single person. The temporal-coherence between successive frames are not taken into account. These methods are applicable to set of images that contain observation in the organized form obtained over sequential time stages, as well as image-sets obtained by several individuals motionless shots (e.g., collection of pictures) or discontinuous observations in long term.

5 Automatic Face Recognition in Movies

The faces in a movie wanted to be recognized, despite a labeled dictionary of actors. The method proposed is composed of two phases. During the first step, mentioned as keyframe selection, marked all the faces in a movie using the face recognition algorithm based on CNN + HOG. They maintain a few very confident keyframes labeling. In the next step, the labels are propagated from the keyframes to the rest of the frames which integrate space of time constraints and function. Our research is based on the premise that some of the faces in the films look identical to the faces present in the dictionary set in terms of facial expression and one can label them with a high level of confidence. If this labeling can be identified, then can effectively promote this label from the chosen keyframes to the rest of the frames. It can be especially useful for the shots having zooming in or zooming out of the faces of actors/actresses or in shots in which there is a moderate change of posture. Provided a labeled dictionary with Di having training examples from Class $I \in \{1, 2, 3, \ldots, c\}$, have to do the annotation a dictionary set containing N number of faces in a clip of video. In this whole paper, assumed that each D and X column has a unit of l_2—norm.

5.1 Key-Frame Selection

Here, a Convolutional Neural Network [4] is used, a very popular still image face recognition algorithm for still images, to initially label the video frames with help of the labeled dictionary. Each frame is represented as a set of linear dictionary faces

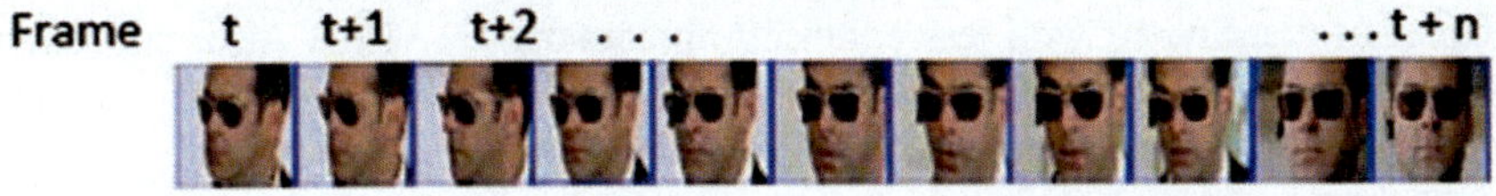

Fig. 3 A scene where an actor's pose gradually shifts. The dictionary cannot cover all actor's poses. For these cases, even though few images close to the dictionary-set have been defined with large confidence in the pose transition, their marks could be effectively transmitted to the rest of the poses [15]

by CNN. It is based upon the concept that facial-images are based on a manifold of low-dimension and provided an adequate example for the training samples, they comprise of low dimensional space [14]. Thus, represent any face-image by linearly combining the training samples of the class to which belongs to. This representation would be sparse because of the large numbers of classes. The diverse solution is got by solving the given $l1$ minimization which under certain conditions gives sparse solution.

$$\hat{\alpha} = \arg\min_{\alpha_i} ||X_i - D\alpha_i||_2 + \lambda||\alpha_i||_1 \tag{1}$$

Here λ is Lagrangian constant, which controls the trade-off between error in reconstruction and sparsity.

Finally, a class-label to a sample X_i.

$$\text{label}(X_i) = \arg\min_j ||Xi - D_j\hat{\alpha}_{ij}||_2 \tag{2}$$

where D_j represents training samples of class j and α_{ij} represents their respective weights.

After marked every frame, the key-frames are selected based on confidence. This confidence measures how well a sample of queries has been identified using the samples of a specific class. The basis of this fact is that face images with identical poses and looks like that the elements from the dictionary will have a large confidence-measure, whereas other faces with different poses, the appearance will have a low measure of confidence (Fig. 3). By marking these keyframes, labels will effectively propagate to remaining frames.

5.2 *Propagating the Key-Frames*

Once keyframes are selected and propagated the labels to the complete video [16], there are several advantages of this technique. First, as a dictionary includes a broad collection of face-images, there is a strong risk of confusion. The numbers of actors present in a film, though, is normally low. Therefore, if can confidently mark some of the frames which belong to a collection of images, then those labels could be effectively transmitted with less confusion throughout the video. Also, the actor presence in separate video frames is normally identical, which makes propagation easily than independent labeling. In the next step, considering the keyframes chosen in the previous step as a labeled-set X_l and unlabeled X_u rest of the frames. Let's say $X = [X^l \; X^u] \in \mathbb{R}^{d \times N}$ is the data matrix, $F \in R^{N \times c}$ is a non-negative matrix in which every row corresponds to data-point. F a scoring function is viewed which indicates the probability of data points to belongs to a given class. The face label $X_i, \{i = 1, 2, \ldots, N\}$ derived from F as $y_i = \arg\max F_{ij}$, where, $j = \{1, 2, 3, \ldots, c\}$. Let Y be the labeling in the matrix initially. If the faces in the matrix have been labeled, then take $Y_{ij} = 1$ for $y_i = j$ and 0 otherwise. If the faces are unlabeled take $Y_{ij} = 0$ for all $j = \{1, 2, 3, \ldots, c\}$.

Provided X, a graph h_V, E_i is created utilizing both unlabeled and labeled data-points. There is a node corresponding to every facial image, and there is a similarity between the edges E. Larger the weight of the edge, greater the resemblance of the faces. Two types of similarities have been considered here—first in function, and second in time-space. Faces that are similar to each other in temporal space need to be from the same class and therefore have broad weight. It is accomplished as per the following,

$$\sum_{i,j} V_{ij} \left\| F_i - F_j \right\|^2 \tag{3}$$

where $V_{ij} = \exp\left(-\left(t_i - t_j\right)\gamma_{iji}/2\sigma^2\right)$. t_i and t_j represent the frame no. of ith and jth frames and γ_{ij} represents absolute sum of difference between coordinate of ith and jth frames. Instinctively, the given constraint indicates the same class will belong to the neighboring frames which have an identical bounding box for face images. Similarly, the related points should have significant weights in appearance (feature space). It can be achieved as

$$\sum_{i,j} W_{ij} \left\| F_i - F_j \right\|^2 \tag{4}$$

where W_{ij} represents the appearance of weight. A popular measurement for appearance similarity is Gaussian distribution which can be represented as $W_{ij} = \exp$, here σ manages the width of the Normal distribution.

5.3 *Rejection of Unknown Faces*

The frames from the video that comprise of unidentified faces are not included in the dictionary. The algorithm must reject any such labeling of faces according to the measure of confidence. The two highest labeling scores are found to be a measure of confidence in accepting or rejecting a face label [17]. Automatically, the scoring vector F_i will have a large score corresponding to the correct class for a sample, but a much lower score for the rest of the classes indicating a single class contribution to the sample reconstruction. Labeling dominance score (LDS) can be defined on the basis of this instinct, It decides whether the labeling of the face is allowed or rejected.

$$\text{LDS}(i) = \frac{F_{ij}}{\arg\max\limits_{k,k \neq j} F_{ij}} \text{ where } j = \arg\max_{j} F_{ij} \tag{5}$$

A face is annotated if the difference between the two greatest F_{ij} scores is large.

5.4 *The Movie Database (TMDb)*

The Movie Database (TMDb) is a database of TV shows and movies that allows users to edit data. Since 2008, the users have been adding and editing the data and are using this database for fetching all movies of actors/actresses occurring in a video-clip.

The TMDb is used for Movies fetching because of several advantages given below.

1. Since 2008, the number of submissions to the database per year has been growing. TMDb has become a leading source for metadata with more than 200,000 developers and businesses using the tool.
2. TMDb also provides one of the best collections of high-resolution posters and fanart, along with comprehensive documentation for movies, TV shows, and individuals. More than 1000 photos are added each day on average.
3. TMDb is international. While it supports 39 languages officially, it also has comprehensive regional data. The TMDb is used every single day in over 180 countries.
4. TMDb is a trusted platform. Every single day, millions of people use its service as it handles more than 3 billion requests. For years, TMDb has proven that it is a service that can be trusted and relied on.

6 Experiment

The trained feature models of Convolutional Neural Network (CNN) were utilized in this paper using the keyframe based approach for face recognition in video. This approach has advantages to the following. Firstly, since a dictionary includes a large number of subjects, there is a strong risk of confusion. The number of actors in a video, though, is usually smaller. Therefore, if one could confidently mark few frames belonging to a collection of subjects, then those labels can be effectively propagated with less ambiguity through the video. Also, the actors' presence in separate video frames is essentially identical, making propagation simpler compared to individually labeling them. The model has the features of the entire labels of the face recognition systems. Against these models, the test images are validated and provide the maximum value of probability among the labels and claims to be the person/actor. Hence the architecture of the project is given in Fig. 4.

There are four main components in the architecture: Key-frame generator, Face Detector, Face Recognition component, and the component to fetch movies. The architecture also includes two databases, the first is a database of known face and embeddings, the other is the TMDb Movie Database. A video clip file is provided as an input. The input undergoes a series of steps and the actors present in the video clip and their movies are produced as output. Firstly, the Key-frame generator generates the key-frames from the input video using the key-frames approach described previously. After that, the face detection is performed using the CNN + HOG approach. Next face recognition is done using the Indian Movie Face Dataset and the database of known faces and embeddings. In the final step, the movies of the actors recognized are fetched using The Movie Database and the movies are produced as output.

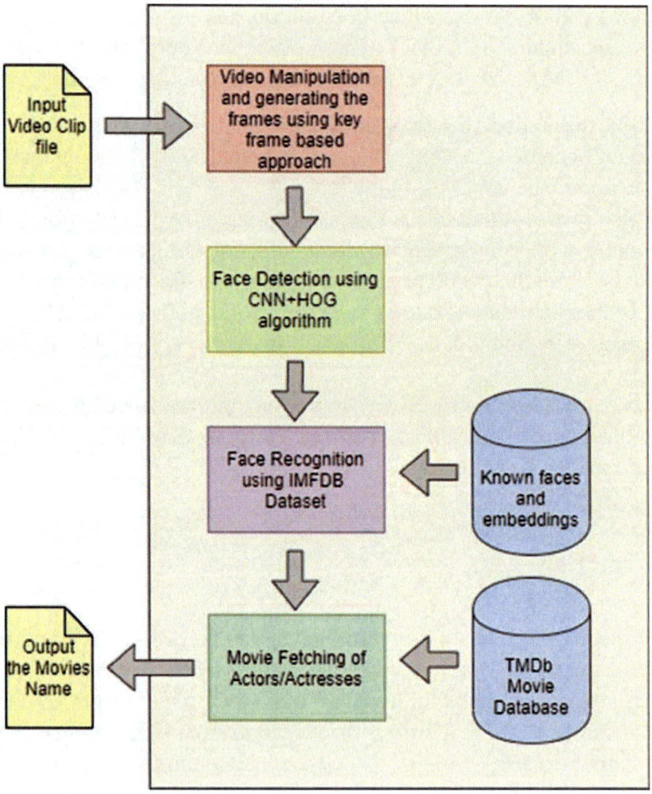

Fig. 4 The architecture of our proposed work

A video clip is loaded in this model to detect the faces in the video, and it is pre-processed from converting RGB to Grayscale to eliminate the effects of lighting in the video. For detecting faces in each video series, a cascade face detector is applied. The videos are taken for training and testing for comparison purposes so that the algorithm can identify the individual correctly. Key-points of features were extracted by the extraction method for the CNN + HOG algorithm. It distinguishes salient points in the image that are uniquely differentiated from any point. The HOG features are obtained by more number layers divided into n octaves and intra-octaves and finally by using the corner detector obtains several key-points.

6.1 HOG Feature Extraction

HOG is a standard object detection method that counts edge-directional incidences in a local image neighborhood. We have to consider all the faces in a frame in the face recognition phase. This detection function is accomplished with the application of HOG in the frame acquired. The input frame is converted from RGB into a gray-scale frame for simplicity before the application of HOG.

Its use was to extract the features. A problem arises because of variations in projection and posing, so next, we resolved that problem. The model for the classification of the frames in the video clip was trained using Convolutional Neural Networks.

6.2 Fetching of All Movies of Actors/Actresses Present in Clip

Before movies' names are fetched, the list of actor names is filtered out to discard the names that appear in a very less number of frames which are likely to be incorrect detections. To do that, the normal distribution technique is used. We calculate the mean as well as the standard deviation of the number of appearances of names. Based on that, the names that stand out three-time standard deviation are discarded. The filtered list contains only names with the largest number of votes.

All the actor names are then forwarded to the movie name fetching module. Here, each actor's name is used to find the movies in the database. The Movie Database TMDb is chosen as a reliable source which is a TV show and movie database built by the users in which data are being added by the community since 2008. Finally, the list of movies of each actor is printed to the output.

7 Result

Our Face Detection algorithm in the video clip works effectively from a good frame in the clip to the blurred frame in the clip. It correctly recognizes multiple actors/actresses in a frame. All Movies of actors/actresses collected from the clip are successfully fetched. The output has been given in Figs. 5, 6, 7, and 8.

The recognition accuracy of our algorithm is 99% with 20.35 s time in feature extraction using HOG and 17.18 ± 0.9 s in classifying using CNN approach.

Fig. 5 Correct identification of actor's face in decent frames in clip

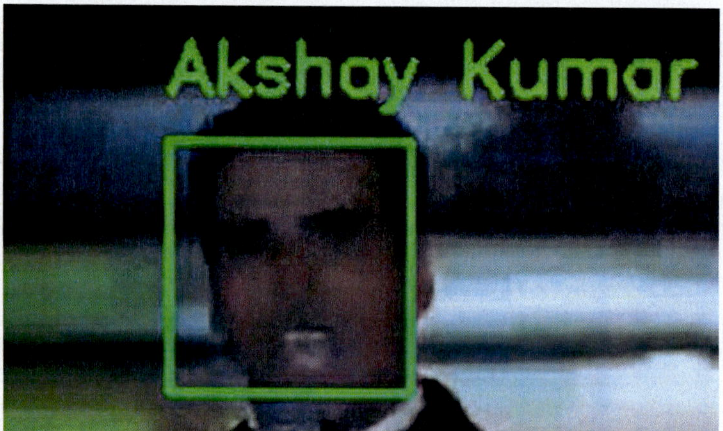

Fig. 6 Correct identification of actor's face in blurred frames in clip

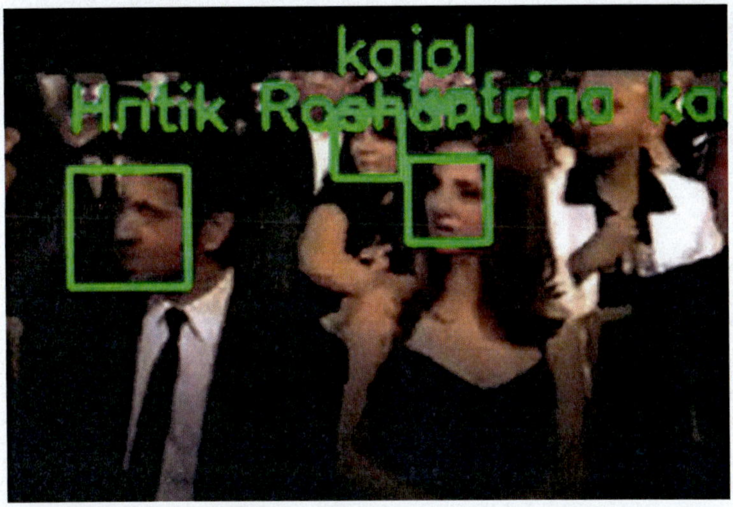

Fig. 7 Correct identification of multiple actors/actresses face in blurred frames in clip

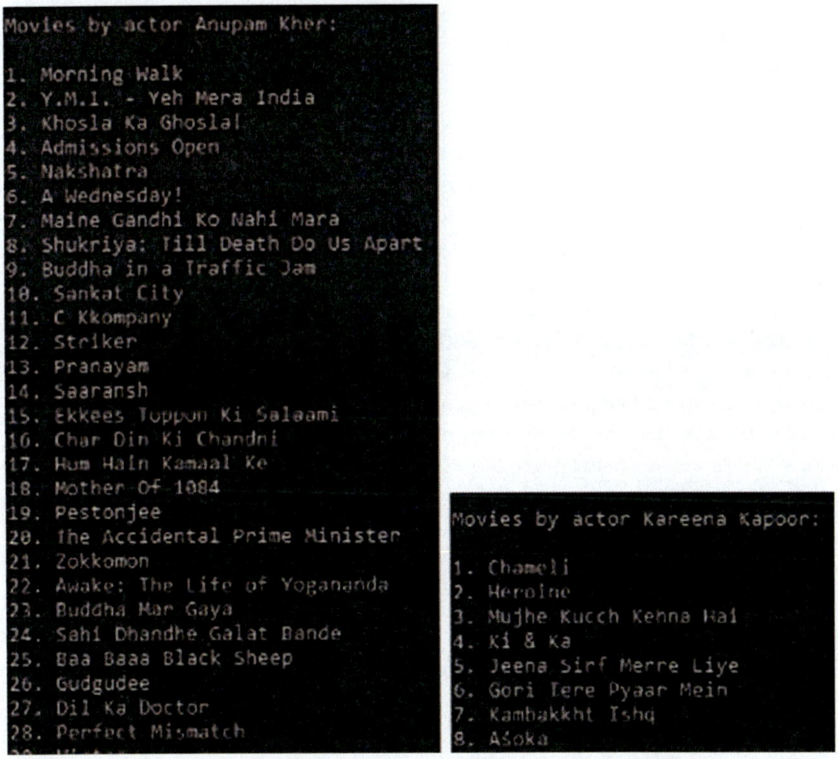

Fig. 8 Some movies of actor/actress fetched from video clip

Table 1 Performance measures of different algorithms for face recognition in a video clip

Experiments	Model implementation time (s)		Recognition accuracy (%)
	Time in feature extraction (s)	Time in classifying (s)	
IP with ANN	10.96	15.58 ± 0.7	99.50
IP with CNN		21.84 ± 0.9	99.50
IP with SVM		195.42	98.00
HOG with ANN	20.35	11.89 ± 0.7	99.00
HOG with CNN (**Our Algorithm**)		17.18 ± 0.9	99.00
HOG with SVM		214.43	96.00
BoW with ANN	30.50	10.33 ± 0.5	96.00
BoW with CNN		25.77 ± 0.8	94.00
BoW with SVM		28.82	97.00

The comparison table of different algorithms used for face Recognition is given in Table 1.

8 Conclusion

In the Actors/Actresses Face Recognition in video task, the detection of the actors/actresses faces is achieved. All the actor/actress's Movies can be found through this research. The tracking of faces in each frame of a video is achieved using HOG + CNN, thus better-identifying faces and faces within each frame. The extraction method of the HOG feature extracts interesting key points in faces that help in matching people's faces in the video. The CNN classifier achieves better results in identification, memory management and is easy to use when face detection has to be done in a video. The Video-based face identification system allows us to recognize and track the face of individuals with various poses and illuminance conditions, which contributes to better recognition results. Training of a person with more in number and detecting more people in the video can be achieved in the future along with more accuracy. Many facial recognition applications can be found in video clips.

References

1. Kumar N, Berg AC, Belhumeur PN, Nayar SK (2009) Attribute and simile classifiers for face verification. In: ICCV-2009
2. Huang GB, Ramesh M, Berg T, Miller EL (2007) Labeled faces in the wild: a database for studying face recognition in unconstrained environments. University of Massachusetts, technical report

3. Kim M, Pavlovic V, Kumar S, Rowley HA (2008) Face tracking and recognition with visual constraints in real-world videos. In: CVPR-2008
4. Wright J, Yang AY, Ganesh A, Sastry SS, Ma Y (2009) Robust face recognition via sparse representation. PAMI
5. Zhao M, Adam H, Yagnik J, Bau D (2008) Large scale learning and recognition of faces in web videos. In: FG
6. Berrani SA, Garcia C (2005) Enhancing face recognition from video sequences using robust statistics. In: Advanced video and signal based surveillance
7. Gorodnich DO (2002) On importance of nose for face tracking. In: FG-2002
8. Edwards GJ, Taylor CJ, Cootes TF (1999) Improving identification performance by integrating evidence from sequences. In: CVPR
9. Shakhnarovich G, Fisher JW, Darrell T (2002) Face recognition from long-term observations. In: Computer vision—ECCV 2002, vol 2352
10. Surasak T, Takahiro I, Cheng CH, Wang CE, Sheng PY (2018) Histograms of oriented gradients for human detection in video. In: ICBIR-2018
11. Setty S et al (2013) Indian movie face database: a benchmark for face recognition under wide variations. In: 2013 fourth national conference on computer vision, pattern recognition, image processing and graphics (NCVPRIPG), Jodhpur, pp 1–5
12. Satoh S (2000) Comparative evaluation of face sequence matching for content based video access. In: Proceedings of the fourth IEEE international conference on automatic face and gesture recognition
13. Shan C (2010) Face recognition and retrieval in video. In: Video Search and Mining 2010, 235–260
14. Georghiades AS, Belhumeur PN, Kriegman DJ (2001) From few to many: illumination cone models for face recognition under variable lighting and pose. Pattern Anal Mach Intell
15. Kumar V, Namboodiri AM, Jawahar CV (2014) Face recognition in videos by label propogation. In: ICPR-2014
16. Zhu X, Ghahramani Z (2002) Learning from labeled and unlabeled data with label propagation. In: CMU-CALD-02-107
17. Kumar V, Namboodiri AM, Jawahar CV (2013) Sparse representation based face recognition with limited labeled samples. In: ACPR-2013
18. Stiefelhagen R, Bauml M, Tapaswi M (2012) "Knock! knock! who is it?" probabilistic person identification in TV-series. In: CVPR

Signature Forgery Recognition Using CNN

Amit Chaurasia, Harsh Agarwal, Ankur Vishwakarma, Ashish Dwivedi, and Arpit Sharma

Abstract This paper presents the recognition of hardwritten signatures. This is troublesome as even the human eye does not have that much visual capacity to distinguish everything in the manually written signature. It is hard for people to recognize the original and the fashioned ones. By utilizing profound realization which utilizes the refined reproduction of human cerebrum, we can recognize the fraud done in signature with higher precision. Confirmation of signature may be achieved using either offline or online mode, depending on the program. Digital systems use contextual data from a target taken at the moment of making the target. Disconnected structures hack away at the mark's image tested. A methodology for offline signature authentication actually utilizes a number of simple geometric highlights based on form. The highlights that are considered are area, eccentricity, center of gravity, pressure, pen up/down, and inclination. Before extricating the highlights, preprocessing of a filtered picture is important to detach the marked part and to expel any fake commotion present.

Keywords CNN · Dropout · Adam · RMSprop · ReLU · Softmax

1 Introduction

Conventional bank checks, bank credits, visas, other government-issued funds, and different authoritative reports are an essential piece of the advanced economy. They are one of the major mediums from which individuals and companies pass cash and pay taxes. Indeed, even today, any one of these especially budgetary exchanges needs confirmation of our signatures. The inescapable reaction of signatures is that they can be misused to pretend an archive's validity. Thus, the requirement to inquire about productive mechanized answers for signature recognition and check has expanded as of late to abstain from being defenseless against misrepresentation.

A. Chaurasia (✉) · H. Agarwal · A. Vishwakarma · A. Dwivedi · A. Sharma
Department of Computer Engineering and Applications, GLA University, Mathura, India
e-mail: amit.chaurasia@gla.ac.in

G. Ranganathan et al. (eds.), *Inventive Communication and Computational Technologies*, Lecture Notes in Networks and Systems 145,
https://doi.org/10.1007/978-981-15-7345-3_11

Fig. 1 Types of signatures

Signature recognition is crucial to measure examination of archives. Signature is a significant apparatus for person confirmation. Each individual has a novel method for doing his/her signature, on imitation, loses its key highlights. A series of signature samples are shown in Fig. 1. Accordingly, signature confirmation turns into a significant security angle. Manually written signature acknowledgment can be worked in two different ways: online and offline relying upon how is the signature procured. In the online mode, the mark is caught while composing along these lines giving the dynamic data including area, speed, increasing speed, pen pressure, pen up, pen down, edge, and time. In the offline mode, the mark is filtered after the mark is composed consequently prompting the static picture of the mark called the checked mark. Testing a target in the disconnected mode is more difficult than in the online mode, which provides more estimates. Handwritten signatures have various sizes and shapes, and the varieties in them are so enormous now and again that it is hard to confirm the real people. In addition, the mark of an individual differs every now and then. Little varieties are innate, and furthermore, these can be endured by the validation framework. In any case, when there is a noteworthy change in the mark, the confirmation framework ought to be refreshed with the new signature database.

In the signature recognition system, the signs can be classified into three categories, and this categorization is done on the basis of the extent of similarity with the original signature. Categories are random forgery, unskilled forgery, and skilled forgery. In random forgery, the person does not have any idea about the person whose signature belongs to, not even the name or signature structure.

In plain forgery or unskilled forgery, the person knows the name of the actual person whose sign belongs to but not to whom his sign represents. However, in professional forgery, a somewhat close impersonation of the original signature is created by a falsifier who has seen and practiced on the genuine signature writing. Using a CNN takes into consideration increasingly precise example acknowledgment as it can proficiently distinguish spatial highlights to order pictures freely. At the point when such systems are applied to offline signature verification, one can all the more likely identify falsification. From a comparable perspective, our answer can likewise apply to the field of legal sciences in helping authorities distinguish the essayist

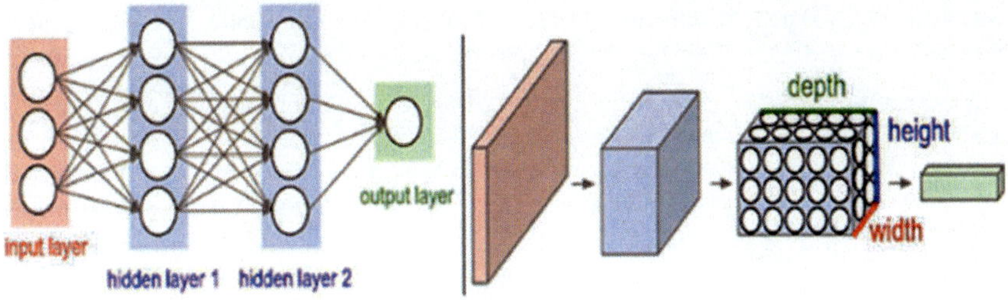

Fig. 2 Convolutional neural network

of a report, perhaps supporting legitimate cases. Tragically, ongoing endeavors to arrange frauds have been moderately fruitless with a noteworthy pace of mistake, even with bigger datasets. Also, for example, the variety in characters and strokes found in an individual's penmanship, which are generally hard to recognize, has brought about a background marked by disappointment for acknowledgement of penmanship utilizing conventional neural systems. Be that as it may, the utilization of CNNs for penmanship investigation has demonstrated increasing productivity, with an exactness rate of around 98% (Fig. 2).

2 Objective

In this paper, we are trying to make a CNN model that can distinguish between a genuine and a forged signature. Nearly all the time in the bank system you need to sign the form to withdraw the money. If a person is able to forge someone's signature which cannot be identified by human senses, the possibility of theft increases, and we want to design a model which can help systems to prevent someone breaching into the system.

3 Recent Work

Signature verification is one of most testing zones of pattern acknowledgment. Being a conduct biometric quality, which can be imitated, we face a test in planning such a framework to counter the intrapersonal and relational varieties. It is quite difficult to make a model that can differentiate among genuine and forged signature. To solve this difficulty in differentiating characteristics that can enhance performance, scholars have tried to deploy different methodologies, in which we can observe the main use of the features involved [2, 4].

The study suggested in [10] is focused on the concepts of the Langevin equation. In this analysis, it is believed that the pen-drawn signature fits the Langevin equation.

The Minkowsky handwritten signature factor is used as a parameter, and the logistic regression is used as a classifier throughout the study. Rudyi and Semyon used the notion of the Langevin equation as cited in [10]. In this, they believed that the signatures produced were provided by pen and adopted the Langevin equation as follows. For their study, they used Minkowsky element of handwritten signature as a attribute and logistic regression as a classifier. In [1], a detached brand identification utilizing worldwide highlights such as area, tallness, and distance was presented. The Euclidean separation is used to locate a connection in the database between the check signature and the image. The system brings the identification score of approximately 89%.

In [8], the Euclidean distance was used to set up a detached label validation structure. For e.g., the topological highlights are used, template incline bottom, angle proportion, uniform area, focal point of gravity of the whole mark image and line slant joining the focuses of gravity of two sections of a mark picture. As a proportion of the overlap between the two, the Euclidean distinction between the claimed symbol and the style fills in. Of chance, this distinction is not necessarily a predefined point, and the check mark is assumed to be the true mark declared as a fraud in either event. The system provides 100% precision to the organization of the true and fake signatures.

Bertolini et. al. [3] suggested, in the WI case, a writer-independent solution for hand signature test printing. This approach incorporates the principles of representation of individuality and SVMs as classifiers. Signature picture representations are studied in a user-independent format by Hafemann et al. using CNN [5]. Throughout the part learning phase, a tale description is introduced to add knowledge on talented imitations of a subset of clients. It aims to grab viewable cues distinguishing real signatures from fabrications giving no consideration to the consumer. CNN's basic aim is to segregate the change package within the customers. A executing several tasks program is suggested to push the highlights to be appropriate in understanding gifted abuse by talking of two words in the expense function for providing information. The main term pushes the model to consider specific customers, while the following term pushes the model to identify certifiable signatures and skilled produce. In the midst of training, CNN is used to delete highlight portrayals of 2048 signature dimensions from the test collection and to prepare subordinate classifiers for clients (Fig. 3).

In [7], authors set up a function dependent on Eigeneseses (enormous and tiny Eigeneseses from the upper and lower envelopes of signatures) organized using the neural system and accomplished a precision of 98.1% and a false recognition rate of 1.9%. In (Yılmaz and Yanikoglu, 2016) [12] authors demonstrated a method relying on a combination of classifiers. Our analysis on the sample produced an equal error rate of 6.97%: GPDS-160. Deep neural networks are ground-breaking ML architectures with a large range of parameters. But in these structures, overfitting is a big concern. Big networks are often hard to use, rendering overfitting impossible to handle when consolidating the predictions of a diverse variety of different neural networks at the time of research. Dropout is a technique to tackle this problem [9]. Srivastava et. al. [11] showed that dropout increases the appearance of neural systems in vision on specific learning undertakings, recognition of dialog, and report order (Fig. 4).

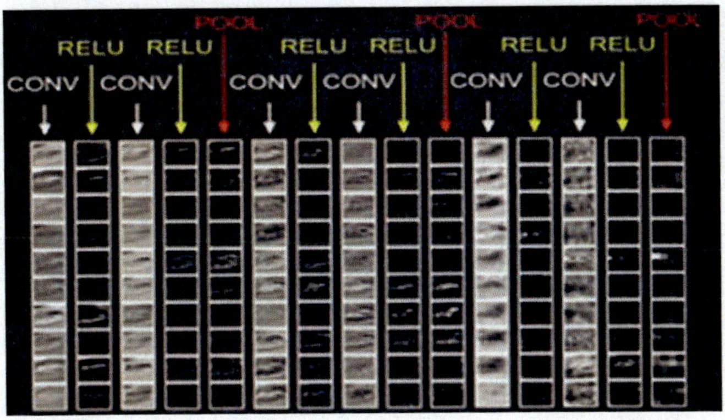

Fig. 3 ConvNet architecture

Fig. 4 Pooling layer matrix

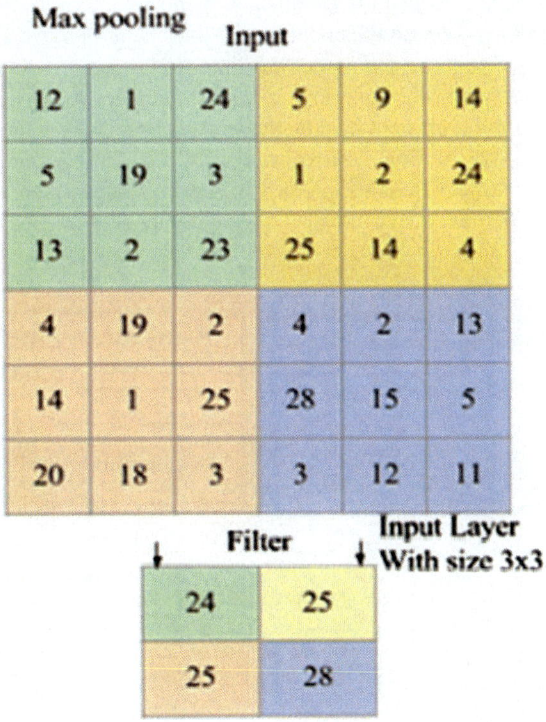

4 Proposed Model

The aim is to create a manually written signature test system that is detached to distinguish the legitimate and counterfeit signature based on features that are extracted using CNN. The current models use hand-picked features from a picture, and these are taken care of into a classifier. The models are just as solid as the picked features, and they regularly need a huge sum of effort to develop. The unambiguous characteristics comprise dimensional, graphometric, spatial, wavelet, outline, and surface properties. The figure of CNN network is shown in Fig. 2.

The biological inspiration of convolutional neural networks is in the human visual cortex which is responsible for detecting objects in the human brain. There is a variety of neurons that fires only when exposed to light at specific angles,so basically convolutional neural networks are designed to identify objects in the image. In our dataset, we have RGB image, so we have three channels in our model. So, the basic idea is to convolute the images with some kernels and applying *ReLU* on the top of them. We do this a couple of times, and then, we apply max pooling layer from Fig. 5 to make our model invariant to size, color, orientation of image. We also add padding to our input image, so that we can get the size of our output image which is similar to input image after convolution is done. This makes it one epoch, and to avoid overfitting, we use dropout [9] and continue our epoch. Our task is to optimize the kernel values and produce a more accurate output using *adam* and *rmsprop* which use exponential average decay of gradient at previous steps, and our loss function is

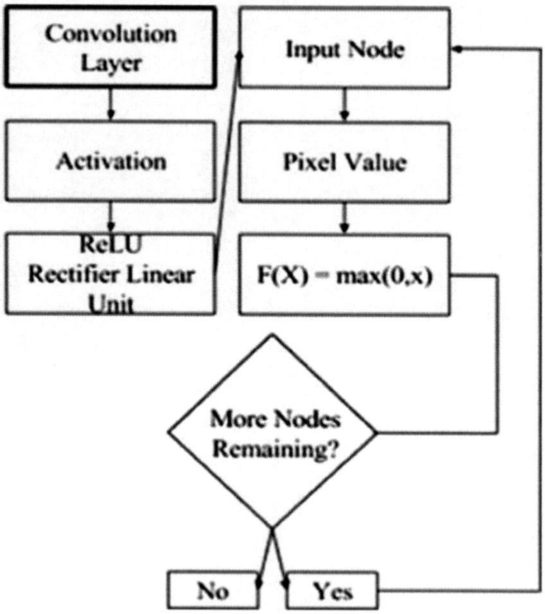

Fig. 5 ReLU function flowchart

cross entropy. After all the epochs are done, we flatten the image and fully connected output layer on which we apply *softmax* activation function which helps us to classify the images.

4.1 Architecture

Simple representation ConvNet classification model has the architecture [$INPUT - CONV - RELU - POOLFC$]. The architecture is shown in Fig. 3.

Steps of spatial arrangement

1. First, the depth of the data is called a hyper-parameter: it refers to the amount of filters we may like to use, each finding out how to scan in the details for something else.
2. Secondly, the filter must be specified which is used to slide the filter.
3. Padding the input volume using zeros around the border. This element of zero-padding is a hyper-parameter. The benefit of zero padding is that it will enable one to monitor the production volumes spatially.

We take several different masks and convoluted it with input images which are multiplication followed by addition. We use ReLU activation function from Fig. 4 because if the neuron activates. As long as it is not a dead neuron, successive updates are fairly effective. ReLU is also very quick to evaluate.

4.2 Pooling Layer

The pooling process entails moving a two-dimensional filter over each channel of the object map and summarizing the features within the area protected by the filter. The purpose is to slowly minimize the representation's spatial scale and reduce the number of parameters and calculation in the network, and thereby thus to monitor over-fitting. Around the same time, pooling layers, for example, MAX pooling will maintain valuable details while managing overfitting issue. The pooling layer operates independently on every depth slice of the input and resizes it spatially, using the (most commonly used) MAX operation. Pooling allows features to change relative to each other, resulting in effective function matching even in the face of slight distortions. There are also several other advantages of pooling, such as:

- Reduces the function map's spatial aspect.
- And the number of parameters far up the processor hierarchy is therefore popular which simplifies the difficulty of the overall layout.

Loss function for an overall number of P predictions, the network forecast gives performance y_p and its corresponding targeted values as t_p, and the mean squared error is given using Eq. 1

$$E = \frac{1}{2} \sum_{p} (t_p - y_p)^2 \tag{1}$$

We use backpropagation, and two modifications are carried out, one for the weights and the other for the deltas. We use Adam optimizer to update weights. Adam is an optimization algorithm for adaptive learning rate that has been designed mainly for training deep neural networks. It was first published in 2014.

4.3 Dropout

Dropout is a regularization method that approximates training a large number of neural networks with different architectures in parallel. During training of a model, few numbers of layer outputs are randomly disregarded or "dropped out." As a result, it has the effect of making the layer look-like and be treated-like a layer with a different number of nodes and connectivity to the prior layer. In effect, each add-on to a layer during training is performed with a different "view" of the configured layer. In simple words, dropout is a way to prevent neural networks from overfitting.

Before now, different types of standardization layers have been introduced for usage of ConvNet architectures. This standardizes data across features instead of standardizing data features through the process level of process normalization. Completely linked layers are an essential component of CNNs and have been shown to be very effective in identifying and classifying images for computer vision. Fully connected layers form the last few layers of the network. The input to the fully integrated layer is the output of the final pooling or convolutionary layer, which is flattened and then fed into the fully connected layer. A completely linked layer is targeted at taking the effects of the convolution/pooling phase and used them to organize the picture into a name.

4.4 Softmax

Softmax function, also known as soft arg max or normalized exponential function, is a wonderful activation function which transforms numbers into probabilities which add up to one. The function softmax produces a matrix representing the distributions of probability of a number of possible outcomes. It is also a key factor used in the classification of deep learning tasks. We used softmax function on last layer of our network. Softmax function works well for multi-class classification. Moreover, softmax returns the vector of probabilities of a datapoint which becong to a certain class. We can also analyze the probabilities to find out certainty about the classification.

5 Result

During our time of training, we held out a test set. The matrix of performance is evaluated on the basis of test set accuracy, and for hyper-parameter tuning, we judge accuracy on validation set. The dataset used to train the model is from the ICDAR 2009 Signature Verification Competition (SigComp2009) for more details refer [6]. The collection contains authentic signatures from 100 writers and forged signatures from 33 writers. The best model is one which gives highest accuracy on validation set, and using this model, we evaluated the accuracy on test set. Our model performs classification whether a signature is genuine or forged. We evaluated our performance using (Table 1):

$$\text{Accuracy} = \frac{(\text{true positives} + \text{true negatives})}{\text{num data}} \tag{2}$$

The snapshot for the simulation is depicted in Fig. 6. Higher the accuracy, the better the model. However, the difference between the accuracies of validation set and test set should not have much difference. A model is said to overfit if our model shows higher accuracy on training set and lower accuracy on test set. We used a customized dataset in which we have combination of forged and genuine signature of different authors. Our test accuracies were little bit lower than validation set, we tuned many hyper-parameters using the validation dataset like learning rate, dropout rate, number of layers in model, and regularization constant. We tried to choose the best parameters according to performance given by validation set but to some extent hyper-parameters fitting noise in validation set. Our best model used dropout for regularization with a dropout rate of 0.5. And for optimization, we used Adam with a learning rate of 0.0001. If we do not use regularizer, we will certainly overfit the model because there are a lot of kernels to be updated. We measured accuracy at the end of tenth epoch on validation set. We choose Adam for faster convergence. Our model gives a loss of 0.04, accuracy of 98% on training set, and accuracy of 84% on the test set.

Table 1 Performance of model after tenth epoch

Train accuracy	98.04%
Loss for train set	4.1%
Validation accuracy	83.33%
Loss on validation set	5.4%

```
Epoch 1/10
51/51 [==============================] - 16s - loss: 2.6048 - acc: 0.0392 - val_loss: 4.3524 - val_acc: 0.1667
Epoch 2/10
51/51 [==============================] - 17s - loss: 8.0346 - acc: 0.3333 - val_loss: 6.4420 - val_acc: 0.5000
Epoch 3/10
51/51 [==============================] - 17s - loss: 6.1385 - acc: 0.4510 - val_loss: 4.7921 - val_acc: 0.3333
Epoch 4/10
51/51 [==============================] - 17s - loss: 4.1047 - acc: 0.5294 - val_loss: 3.3684 - val_acc: 0.6667
Epoch 5/10
51/51 [==============================] - 17s - loss: 3.1750 - acc: 0.6667 - val_loss: 3.0212 - val_acc: 0.6667
Epoch 6/10
51/51 [==============================] - 17s - loss: 0.6488 - acc: 0.8039 - val_loss: 0.3667 - val_acc: 0.6667
Epoch 7/10
51/51 [==============================] - 16s - loss: 0.3661 - acc: 0.8824 - val_loss: 0.8382 - val_acc: 0.8333
Epoch 8/10
51/51 [==============================] - 17s - loss: 0.0540 - acc: 0.9804 - val_loss: 0.8631 - val_acc: 0.8333
Epoch 9/10
51/51 [==============================] - 17s - loss: 0.0864 - acc: 0.9804 - val_loss: 0.5829 - val_acc: 0.8333
Epoch 10/10
51/51 [==============================] - 17s - loss: 0.0410 - acc: 0.9804 - val_loss: 0.5441 - val_acc: 0.8333
```

Fig. 6 Snapshot of simulation

6 Conclusion

Neural networks have shown their achievements in a lot of applications because of their capacity to take care of certain issues without hardly lifting a finger of utilization. One of the principle highlights, which can be ascribed to CNN, is its capacity to learn nonlinear issues disconnected with specific preparation, which can prompt adequately exact reaction. Additionally, the utilization of enormous databases is not required to show the capacity of learning for this kind of issue, and we have picked just five certified signatures and three fashioned ones for preparing, and we get generally excellent outcomes. Anyway for genuine practice use, bigger preparing information can build the strength of the framework. Our model gives accuracy of 98% on validation set; however, it can be increased.

References

1. Anand H, Bhombe D (2014) Relative study on signature verification and recognition system. Int J Innov Res Adv Eng (IJIRAE) 1(5). ISSN: 2163–2349
2. Bashar A (2019) Survey on evolving deep learning neural network architectures. J Artif Intell 1(02):73–82
3. Bertolini D, Oliveira LS, Justino E, Sabourin R (2010) Reducing forgeries in writer-independent off-line signature verification through ensemble of classifiers. Pattern Recogn 43(1):387–396
4. Gideon SJ, Kandulna A, Kujur AA, Diana A, Raimond K (2018) Handwritten signature forgery detection using convolutional neural networks. Procedia Comput Sci 143:978–987
5. Hafemann LG, Sabourin R, Oliveira LS (2017) Learning features for offline handwritten signature verification using deep convolutional neural networks. Pattern Recogn 70:163–176
6. ICDAR-2009: Signature (2009) http://www.iapr-tc11.org/mediawiki/index.php?title=ICDAR_2009_Signature_Verification_Competition_(SigComp2009). Data retrieved from SigComm2009

7. Jagtap AB, Hegadi RS (2016) Eigen value based features for offline handwritten signature verification using neural network approach. In: International conference on recent trends in image processing and pattern recognition, pp 39–48. Springer
8. Jana R, Saha R, Datta D (2014) Offline signature verification using euclidian distance. Int J Comput Sci Inf Technol 5(1):707–710
9. Kingma DP, Ba J (2014) Adam: a method for stochastic optimization. arXiv preprint arXiv:1412.6980
10. Rudyi SS, Vovk TA, Rozhdestvensky YV (2019) Signature identification by minkowski dimension. Chaos Interdisc J Nonlinear Sci 29(5):053110
11. Srivastava N, Hinton G, Krizhevsky A, Sutskever I, Salakhutdinov R (2014) Dropout: a simple way to prevent neural networks from overfitting. J Mach Learn Res 15(1):1929–1958
12. Yılmaz MB, Yanıkoğlu B (2016) Score level fusion of classifiers in off-line signature verification. Inf Fus 32:109–119

Image Sentiment Analysis Using Deep Learning

Vipul Salunke and Suja Sreejith Panicker

Abstract Determining the image sentiment is a tedious task for classification algorithms, owing to complexities in the raw images as well as the intangible nature of human sentiments. Classifying image sentiments is an evergreen research area, especially in social data analytics. In current times, it is a common practice for majority people to precise their feelings on the web by substituting text with the upload of images via a multiplicity of social media sites like Facebook, Instagram, Twitter as well as any other platform. To identify the emotions from visual cues, some visual features as well as image processing techniques are used. Several existing systems have already introduced emotion detection using machine learning techniques, but the traditional feature extraction strategies do not achieve the required accuracy on random objects. In the entire process, normalization of image, feature extraction, and feature selection are important tasks in the train module. This work articulates the newest developments in the field of image sentiment employing deep learning techniques. Also, the use of conventional machine learning techniques is compared along with deep learning algorithms. It is indicative that a combination of fast recurrent neural networks and CNN may produce high accuracy with minimum time complexity. It is noted from the survey that existing researchers reflect CNN provides around 96.50% average accuracy for sentiment classification on the flicker image dataset.

Keywords Deep learning · Machine learning · DCNN · Image processing · Image sentiment analysis · Social data analytics first section

V. Salunke (✉) · S. S. Panicker
School of Computer Engineering & Technology, MIT World Peace University, Pune, Maharashtra, India
e-mail: salunke.vipul26@gmail.com

S. S. Panicker
e-mail: suja.panickar@mitwpu.edu.in

G. Ranganathan et al. (eds.), *Inventive Communication and Computational Technologies*, Lecture Notes in Networks and Systems 145,
https://doi.org/10.1007/978-981-15-7345-3_12

1 Introduction

Nowadays, people share a lot of data on social media in the form of images and videos—be it personal, daily scenes, or their opinions in the form of memes. Internet is a huge platform for communication and for sharing information globally and instantaneously, thus providing users with a good collection of people's perspectives and their sentiments regarding a huge spectrum of topics [1]. Several posts on social media rarely contain any textual caption but are rather flooded with images. Thus, primarily contributing to a variety of opinions and emotions being conveyed quite implicitly merely through visual content.

One can express sentiments through text, image, or videos. Although several works in the past have employed techniques to decipher sentiments from user posts, especially on social media [2, 3], image sentiment analysis is yet to be explored. In the current scenario, there is increasing use of social media for expressing sentiments, hence it is an important research area. Current developments aim at improving accuracy. There are many algorithms and techniques proposed for image sentiment analysis.

Planning to find out, if applying CNN to the study of visual emotion, provides advantages. It could be possible to conduct learning during backpropagation only in top layers since they have less basic features for the dataset. The experimental findings indicate this domain-specific fine-tuning is successful in enhancing neural network efficiency for heterogeneous image classification.

Overview of Deep Learning
Deep learning plays a massive role in image sentiment analysis for providing various techniques like Convolutional Neural Network, Deep Neural Network, Region Neural Network, and Deep Belief Network to get optimum results [4]. The major problem occurs in situations where found incompatible emotions which is express through image and text [5].

Rest of the paper is organized as follows. Section 2 gives a brief overview of the latest research, Sect. 3 explains proposed work, Sect. 4 covers observations, Sect. 5 presents research contribution, Sect. 6 elaborates on applications of image object detection, Sect. 7 presents future scope, Sect. 8 covers the conclusion.

These are mainly classified into lexicon-based algorithms and machine learning-based algorithms. Lexicon based algorithms include semantic-based as well as statistical techniques, while machine learning based algorithms include neural network, Bayesian network, support vector machine, naïve Bayes, and maximum entropy.

1.1 Background

Recently, a whopping number of people worldwide are increasingly using images and (video/audio) recordings to express their feelings freely on social media. Recognition of visual matter emotions on such a large scale can help improve a client's

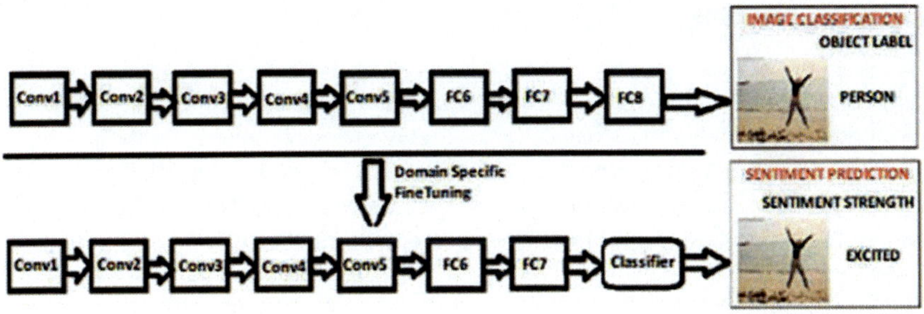

Fig. 1 System basic overview [1]

assessment of opportunities or topics, for example, in image tweets. The assumptions models that help this are bounteous: incredible dazzling images frequently contain rich information that helps to view the effectively interface with those images. With the proliferation of internet-based life an expanding number of individuals utilize images to express their feelings, opinion, and fatigue via web-based networking media stages like Flickr and Instagram. Programmed induction of the sentiment and estimation data from such consistently developing, huge measures of client produced images is of expanding significance to numerous applications in medicinal services, human sciences, correspondence studies, advertising, and many sub-territories inside software engineering, for example, computer vision [1].

Figure 1 shows the framework for visual sentiment prediction. The Convolution Neural Network was used on a large dataset—ImageNet for classification of images. The parameters of previously trained layers are transferred to the sense prediction layer in order to produce the image representation using fine-tuning specific to the field. Individuals share a great deal of substance via web-based networking media as images—be it individual, or ordinary scenes, or their sentiments portrayed as kid's shows or images. An important aspect of image sentiment algorithms is to break down the images (obtained from social media such as Flickr, Twitter, Facebook, and so forth.) into individual objects from which sentiments are deciphered. This can be used for the further general estimation of individuals about a state of mind too. Likewise, it is valuable to comprehend the sentiment an image delineates to and consequently anticipate the class label. As a chunk of this task, it intends to give a sentiment-based class to an image. Images conventionally fall under five classifications—love, happiness, sadness, violence, and fear [1].

2 Literature Survey

Several recent trends in this field are surveyed and tabulated the techniques, datasets used, and research gap in Table 1.

3 Proposed Works

The proposed research work on image sentiment classification employs a deep learning approach. This work basically illustrates various feature extraction as well as selection techniques from image objects and builds the training knowledge accordingly. Various feature selection methods have been used to extract the different parameters, like shape, texture, alpha, density, and color features. Sometimes text meta-data

Table 1 Brief overview of survey

Reference No.	Year	Technique	Dataset	Advantages	Research gap
[20]	2019	DCNN and R-CNN	Not mentioned	Small overhead is easy to train and normalize	Reduced accuracy for very deep neural network
[16]	2019	CNN using Deep Learning	Twitter, photo tweet, multiview and for each dataset	Good accuracy for RGB as well as CMYK image model	Generate high time complexity when multi CNN has used
[13]	2017	RCNN	Not mentioned	Low cost required for entire execution due to regional features extraction	when extract region base features it can effect on accuracy
[18]	2016	ImageNet library has used for deep image learning	Twitter	System achieves better performance on flicker dataset. Provides good accuracy to real time as well as synthetic dataset	Single dataset has used only and Default ADAM optimized has used which can eliminate relevant feature during execution
[7]	2016	DNN, PNN, RNN base deep learning algorithms	Sina weibo dataset	System carried out execution like supervised learning like already trained module while unsupervised earning like pre-trained module.	The system is not able to detect multiple objects in a grid and loss accuracy rate. Sometimes it detects objects many times. unable to localize small objects
[8]	2015	Deep CNN	Twitter and Flickr	Easy training less complexity than RCNN with acceptable accuracy	More and more computing resources are required simultaneously

(continued)

Table 1 (continued)

Reference No.	Year	Technique	Dataset	Advantages	Research gap
[6]	2015	Deep learning with CNN	Flickr and Instagram	System generate background training knowledge based on image as well as text features which increased the classification accuracy	Maximum focus for generate background knowledge on text data, sufficient amount of high-level features were not generated for small items
[1]	2015	Deep Learning algorithm with CNN	Flickr	Good accuracy for object classification results. And various features extraction done Fast RNN	Large time complexity when unknown emotions have extracted during training
[11]	2015	Aspect mining and sentiment classification	Flickr and Instagram	All levels of sentiments are valuable	To reduce accuracy remove the top-down connections
[19]	2014	NLP and ML algorithms with supervised learning	Twitter	Best accuracy for structure as well as semi-structured dataset. System also predicts the positive-negative sentiment, respectively	System works only large text data not image audio and video sets
[9]	2014	NLP with supervised learning algorithms	Flickr	It works like text sentiment analysis and classification based on features on large text data	No provision for image sentiment classification Machine learning algorithms should be taking large processing time
[10]	2013	Deep learning base visual features extraction using multi CNN layers	ImageNet and ILSVRC	No need for predefined anchors much faster than basic CNN	More time is spent generating corners and evaluating the base networks

(continued)

Table 1 (continued)

Reference No.	Year	Technique	Dataset	Advantages	Research gap
[12]	2012	Deep CNN	ImageNet	Very high accuracy than other deep learning modules even system deals with pre-trained environment	API dependency for train as well as test system. Some realistic feature has eliminated by ImageNet library

also used to identify the respective image sentiment. Normalize data set should be most impactful to achieve better classification accuracy (Fig. 2).

In the training phase, various features have been extracted from the training data set and a training model is built accordingly. A similar feature extraction strategy has been applied to the testing data set which extracts each image feature accordingly. The weight calculation process is identifying the similarity between testing and training features. It is the subprocess of similarity evaluation between two features sets. The weight factor is evaluated with desire threshold values and defines sentiment labels accordingly. 0 is initial weight while threshold can be user-defined. Details are as under:

1. Image resizing: Evaluate each image height and width accordingly and change as per the required size.

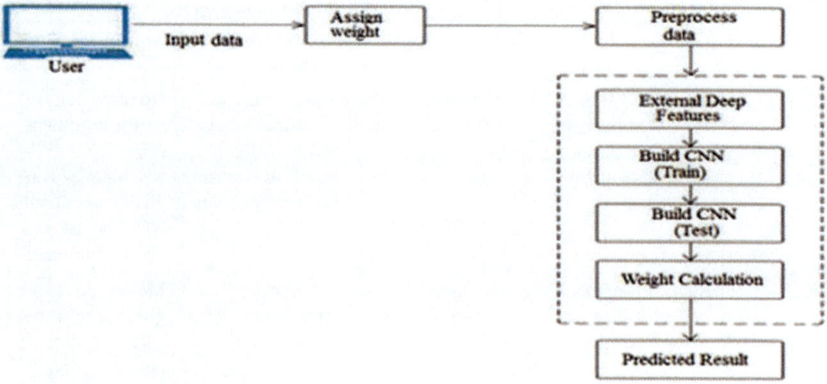

Fig. 2 Proposed system overview

2. Sometimes image contains some noise or specific input image already contains a different kind of noise. Using the Gaussian noise filter eliminated noise from images.

The proposed deep learning module having the ability to identify such features using image Net library. The DCNN classifier has used to detect the sentiment class of the entire test dataset.

- System initially works with a flicker image dataset for sentiment classification with a supervised learning approach.
- It first generates 5-fold, 10-fold and 15-fold cross-validation, respectively.
- The data is already preprocessed in trained module so the system directly builds a train module for respective scaled images.
- For test data, performing on real-time as well as synthetic dataset, respectively, shall be performed.

3.1 System Overview

Module 1: Data collection: To collect real-time data from flicker image dataset is proposed [14].
Module 2: The classification is performed using CNN The steps in train and test phases are as follows:

a. Data preprocessing and visualization
b. Pass the train data and test data as input to VGGImage Net model.
c. Networks stores the activation and second to last fully connected layer as feature vectors.
d. CNN classifier train data for each sentiment category.
e. For each image in test data, mark the highest count from CNN algorithm as predicted classification of image in labels.

Module 3: Analysis The accuracy of proposed system is demonstrated and evaluated with other existing systems.

3.2 Datasets Used

To evaluate the proposed system planned to use real-world datasets from a Kaggle website Flicker [14] and Twitter [15]. These are publicly available.

Dataset name	Description and size	Train	Test
Flicker	2048 * 2048, 1600 * 1600	8850	1000
ImageNet	1024 * 1024	7000	3000
Twitter	1500 * 1500	1402	622

Flickr [14]
For current work proposed to use the Flicker dataset and dynamically select 70–75% images from the half-million Flickr dataset as training data. This shall be used to train the module according to selected features. The ensuing 30–25% images shall be considered as test dataset. We propose to train the Convolutional Neural Network using various iterations and each iteration shall hold '*n*' number of images, respectively.

Twitter [15]
A real-time image dataset is built from the various social media web application as well as image tweets. Image tweets are the tweets that contain some input image and are posted by individual users. A large set as test images is proposed for entire research to validate system accuracy with real-time dataset.

4 Observations

- According to the above literature review, many systems deals with deep convolutional neural network, different kind of boosting methods have used to achieve a better time as well as space complexity.
- The region-based convolutional neural network and other neural network algorithms like PNN, RNN, and DCNN have used to identify the sentiment of the heterogeneous image dataset.
- ImageNet library has used to extract the features and build train model accordingly. Sometimes it is hard to achieve better accuracy than CNN using PNN, RNN, etc.
- Most frequently used datasets are Flickr dataset, Twitter testing dataset, and imageNet dataset.
- The combination of fast recurrent neural networks and CNN is suggested to produce highest accuracy with minimum time complexity.
- It takes a very long time for CNN when multi convolutional NN has generated, even it also takes large data when the system deals with heterogeneous datasets. Feature extraction methods generate high dimensionality issues with irrelevant features extracted.

- Typically several of the existing research on CNN provides around 96.50% average accuracy for sentiment classification on flicker image dataset.

5 Research Contribution

- The recent developments is articulated in the field of image sentiment analysis employing traditional machine learning and deep learning techniques. This shall be very useful for fellow researchers.
- Also performed gap finding based on a recent survey.

These gaps shall serve as important pointers for future research.

6 Applications

There are various fields with numerous application that uses image sentiment analysis and image object detection technique:

- There is a wide range of applications of Image sentiment analysis that can be used in both Industry as well as academics like Political forecasting, social network analysis, predicting prices of stock. Also useful in face analysis, brain signals, posture analysis, and behavior analysis [16].
- Prediction and recognition of user behavior. The knowledge obtained from such systems can be used in a variety of applications like product or services recommendation or prediction modeling, marketing, etc. [18].
- Tag predictions for images uploaded on social sites as well as in the classification of emotion during and after an election [17].
- Most recent and widely use case is self-driving cars that use the technique of object detection in order to identify objects like trees, vehicles, humans and other things present on the road. Another area that is used to collect images of earth called Remote sensing uses detection of objects to capture images of an oasis in deserts, forest fires, etc. In medical field, it is used to detect tumors. Some of the use cases are given below:

    a. Face Detection: One of the primary use lies in detecting human faces in the image. Convolution Neural network one of the techniques of Deep learning is widely used for this purpose. CNN can automatically identify the important features, train network, and then identify them in other images. Social media is a platform that gives users the ability to share and express their feelings often use this technique to identify the faces present in the image uploaded.
    b. Security and Surveillance: Anti-social activities have increased with large amount. Detection of intruders, explosives, weapons, etc. along with remote

sensing comes under this category. Research work is still going to increase the performance of such detection and automate the process [19].

7 Future Work

Our observational study suggests that Deep Learning algorithm gives promising results on image data along with the classification of emotion. Future work includes running experiments on a large-scale processing environment. Video scenes can be accommodated along with images to further classify sentiments into different genres like happy, thriller, funny, romance, etc.

8 Conclusion

Our study explores the techniques of deep learning in systems of classifying image sentiment. Images have several applications in programmed labeling with sentimental sections in sequence, consequently classifying video groups with emotions category and automatically classifying video scenes into thrillers, comedy, romance, and more. As per the survey performed, it is noted that CNN can be generated with different confusion matrix parameters when the system deals with different datasets. Highest accuracy obtained is with DCNN using imageNet library with more than 96% average classification accuracy.

References

1. Jindal S, Singh S (2015) Image sentiment analysis using deep convolutional neural networks with domain specific fine tuning. In: International conference on information processing (ICIP). IEEE, pp 447–451
2. Kunte A, Panicker S (2019) Using textual data for personality prediction: a machine learning approach. In: International conference on information systems and computer networks (ISCON), Mathura, India
3. Kunte A, Panicker S (2019) Personality prediction of social network users using ensemble and XGBoost classifiers. In: 2nd international conference on computing analytics and networking (ICCAN), Bhubhaneshwar
4. Mittal N, Sharma D, Joshi ML (2018) Image sentiment analysis using deep learning. In: International conference on web intelligence (WI). IEEE, pp 684–687
5. Kumar A, Jaiswal A (2017) Image sentiment analysis using convolutional neural network. In: International conference on intelligent systems design and applications (ICISDA). Springer, pp 464–473
6. Wang Y, Hu Y, Kambhampati S, Li B (2015) Inferring sentiment from web images with joint inference on visual and social cues: a regulated matrix factorization approach. In: International conference on web and social media (ICWSM)
7. Yuhai Y, Hongfei L, Meng J, Zhao Z (2016) Visual and textual sentiment analysis of a microblog using deep convolutional neural networks. Algorithms 9:2

8. You Q, Luo J, Jin H, Yang J (2015) Robust image sentiment analysis using progressively trained and domain transferred deep networks. In: Twenty-ninth AAAI conference on artificial intelligence
9. Yang Y, Jia J, Zhang S, Wu B, Chen Q, Li J, Tang J (2014) How do your friends on social media disclose your emotions? In: Proceedings of AAAI conference on artificial intelligence AAAI
10. Frome A, Corrado GS, Shlens J, Bengio S, Dean J, Ranzato MA, Mikolov T (2013) DeViSE: a deep visual-semantic embedding model. In: Proceedings of advances in neural information processing systems (NIPS), pp 2121–2129
11. Wang Y, Li B (2015) Sentiment analysis for social media images. In: International conference on data mining workshop (ICDMW). IEEE, pp 1584–1591
12. Krizhevsky A, Sutskever I, Hinton GE (2012) Imagenet classification with deep convolutional neural networks. In: Advances in neural information processing systems, pp 1097–1105
13. Mandhyani J, Khatri L, Ludhrani V, Nagdev R, Sahu S (2017) Image sentiment analysis. Int J Eng Sci
14. https://www.kaggle.com/hsankesara/flickr-image-dataset. Last accessed 29 Nov 2019
15. https://www.kaggle.com/paoloripamonti/twitter-sentiment-analysis. Last accessed 29 Nov 2019
16. Ragusa E, Cambria E, Zunino R, Gastaldo P (2019) A survey on deep learning in image polarity detection: balancing generalization performances and computational costs. Electronics 8:783
17. Gajarla V, Gupta A (2015) Emotion detection and sentiment analysis of images. Institute of Technology, Georgia
18. Islam J, Zhang Y (2016) Visual sentiment analysis for social images using transfer learning approach. In: International conferences on big data and cloud computing, social computing and networking, sustainable computing and communications (SustainCom) (BDCloud-SocialCom-SustainCom). IEEE, pp 124–130
19. Anjaria M, Guddeti RMR (2014) Influence factor based opinion mining of Twitter data using supervised learning. In: Sixth international conference on communication systems and networks (COMSNETS). IEEE, pp 1–8
20. Bhagya C, Shyna A (2019) An overview of deep learning based object detection techniques. In: 1st international conference on innovations in information and communication technology (ICIICT). IEEE, pp 1–6

Fake News Detection Using Passive-Aggressive Classifier

Saloni Gupta and Priyanka Meel

Abstract People can get infected with fake news very quickly with misleading words and images and post them without any fact-checking. The social media life has been used to distribute counterfeit data, which has a significant negative influence on individual consumers and on a wider community. The fake news problem is tackled using a machine learning algorithm. Different classifiers are used for the purpose of identifying fake news. In this paper, Passive-Aggressive Classifier is implemented for this purpose. The approach is implemented on two datasets of fake and real news. After performing the experiment, it is observed that Passive-Aggressive Classifier provides an accuracy of 97.5%. The performance of the proposed model is compared with the existing methods. The Passive-Aggressive Classifier provides the best result compared to others.

Keywords Fake news detection · Machine learning model · Classifier · Passive-Aggressive Classifier · TfIdfVectorizer

1 Introduction

Misinformation is defined as inaccurate or misleading information. Unintentionally, it could spread because of honest reporting errors or incorrect interpretations. Disinformation, on the other hand, is misleading knowledge which is intentionally circulated for confusing or promoting a prejudicial purpose. Like misinformation, the falsification is intentionally meant to confuse people; in consistency, it is characterized as amusing and nasty. Satirical news is published mainly with the intent of amusing and offending the audience, but it can be harmful if spread out of context, similar to hoaxes. False news for different financial and political reasons has emerged

S. Gupta (✉) · P. Meel
Delhi Technological University, Delhi 110042, India
e-mail: salonigupta140398@gmail.com

P. Meel
e-mail: priyankameel@dtu.ac.in

G. Ranganathan et al. (eds.), *Inventive Communication and Computational Technologies*, Lecture Notes in Networks and Systems 145,
https://doi.org/10.1007/978-981-15-7345-3_13

with the development of social networks and has become popular in the online world. The universe is now seeing many elaborate scandals; outlets for social networks that encourage the circulation of misinformation even further. The proliferation of fake news in social media, make a fall in general confidence and faith in conventional mass media. Two key reasons for the growth in misleading sites are one in a commercial context that generates substantial advertisement profits from viral news stories and another in a more ideological manner since fake news outlets tend to manipulate public sentiment on such subjects. Indeed, journalism is often filled with fake reporting.

In this paper, Passive-Aggressive Classifier is used for fake news detection which is one of the classification techniques. A comparative study is also done in which the other classifiers are evaluated and used for knowing the classifier with the best accuracy. The other classifiers are Naïve Bayes Classifier, Support Vector Machines (SVM), K-Nearest Neighbors, Decision tree, and Random forest. Apart from classifiers, several other machine learning and deep learning algorithms are applied to the same problem. Convolutional Neural Networks (CNN), Recurrent Neural Network (RNN), N-Gram Analysis, Deep Neural Networks have also been implemented in this field. Artificial intelligence is also believed to address this problem as these algorithms work better to classify many things like face recognition, voice identification [1]. Thus, different articles provide different techniques that can be used for this problem.

2 Related Work

The task of rumor detection, gossip detection, satire detection, is closest to several interesting problems like detection of news which are not true. Because any person should be able to understand these similar concepts intuitively, each research takes on its own meaning. The purpose of this analysis is to detect the output of news material that could be checked by using Passive-Aggressive Classifier. Research in the area of machine learning and deep learning helps in figuring out how to distinguish counterfeit news have demonstrated execution upgrades over customary strategies because of their improved capacity to separate significant features. Initial methods have sought to spot false news by using linguistic features derived from the data of news reports. One of the simple experiments was conducted with Naïve Bayes Classifier [2] in order to identify fake content. This paper used a Facebook news dataset and was implemented as a software system. A structure is recommended that recognizes and distinguishes Twitter messages utilizing both Convolutional Neural Networks (CNN) and long-short term Recurrent Neural Network (RNN) models [3]. Other than the explicit features, there additionally exist some concealed patterns in the words and pictures utilized in counterfeit news, which can be caught with a lot of latent features removed by means of the different convolutional layers in the model named as TI-CNN (Text and Image data based Convolutional Neural Network)

[4]. Another research paper proposed different classifiers on the problem of classifying fake news with real news. It was implemented using different classifiers of machine learning: Support Vector Machines (SVM), K-Nearest Neighbors, Decision tree, Random forest [5]. Deep neural networks have shown tremendous success in image processing and text representation. These are implemented effectively to different activities such as image summarization, answering graphic questions, and identifying fake news. A framework is proposed, a false news identification platform using n-gram method and machine learning techniques [6]. This experiment uses Term Frequency-Inverted Document Frequency (TF-IDF) for extracting features and Linear Support Vector Machine (LSVM) for classification. A model is proposed as a start to finish arrange, Multimodal Variational Autoencoder (MVAE) [7], which utilizes bimodal variational autoencoder combined with a binary classifier for the errand of misleading information recognition. This approach contains three principle parts, an encoder, a decoder, and a fake news detector. The variable autoencoder is programmed to learn stochastic idle vector models by streamlining the limits of the periphery or marginal likelihood of the knowledge observed. Fake news detector uses multimodal images collected from the bimodal self-encoder to check the news as fake or real. Event Adversarial Neural Network (EANN) [8] has been suggested to assess symmetric highlights from occasion to occasion and consequently benefit from the recognition of disinformation on recent occasions. This contains three primary components: the multi-modular extractor, false news detector, and the discriminator of events. The multi-modular extractor will isolate the linguistic and graphic highlights from the articles. It allows the false news detector to familiarize itself with the selective image of fake news. Event discriminator is used to evacuate the occasion of explicit highlights. The feedback of the news mentioned in microblogs can also be used for this problem [9]. A naïve approach is used to solve the problem of fake news detection. The Passive-Aggressive algorithms [10] uses a basic but efficient randomized procedure in order to determine if the label of a category should be queried. The state-of-the-art algorithm is implemented to exploit the labeled instance by updating it in case of wrong answer. Passive-Aggressive Classifier is a new classifier that hasn't been used till now. It is used for training a model, making parameter changes only when necessary, while removing all updates that do not upset the balance. Thus, it is used in the domain of fake news detection.

3 Proposed Work

3.1 Dataset

The experiment of classifying news as real or fake is performed in Python using two datasets. The first dataset is taken from Kaggle. It contains two files, one .csv file and one .h5 file. The size of this dataset is 149 MB having 3983 unique tuples. It has 4 attributes: URLs (source), Headline (title), Body (content), and Label (fake or real). 3186 tuples are used for the training set and 797 are used for the test set while performing passive-aggressive classifier. Another dataset used is a dataset of size 100 MB containing approx. 25,000 tuples with 54 attributes and it is available online [11]. This dataset contains several attributes like title/headline, body/text, image_url, site_url, title-length, text_length, sentiment words count and other metadata (54 columns) for almost 25,000 news with 50% real and 50% fake news. Some important attributes are author (author of the news), title (headline of the news), site_url (url of the site of the news), type (fake or real). Out of 25,000 news, 10,000 tuples were used for classification out of which 2000 were test tuples and the rest 8000 tuples were training set.

3.2 Methodology

Nowadays, machine learning has become the solution to each and every problem. It is used in almost every field, domain, and research. It is having the power of giving solutions that humans can't think of easily. The machine-learning algorithm uses mathematical operations and is commonly used. Dataset having numerical values can be handled easily by applying machine learning algorithm. But, dataset with categorical data, requires some transformations before the application of these algorithms. To detect the news credibility is one of the applications of Machine Learning Algorithm. The algorithm of machine learning used in this paper is Supervised Learning. The learning is called supervised when the model is trained using a dataset which has both parameters of input and output. One of the Machine Learning algorithms is classification. Classification is the mechanism by which the class of the data objects class is predicted. Classes are referred to as labels or categories. A classifier is an algorithm that sort data into designated classes or categories. Classifiers are a practical understanding of pattern recognition in certain types of computer processing. Predictive classification is the function of approximating (f) mapping from input (X) variables to distinct output variables (y). An algorithm that sorts information into named classes or knowledge groups is a classifier. For certain types of computer processing, classifiers are a practical understanding of pattern recognition. Passive-Aggressive Classifier is one such classifier that is used for solving the problem of detecting fake news. For this application, the following steps are implemented:

- Cleaning the Dataset

Before applying any model on the dataset, it is advised to clean the dataset. Cleaning the dataset is also known as pre-processing the dataset. Bad or incorrect knowledge leads to false results. The accuracy of our results is strongly determined by how much clean, pre-process, and understand the data. Data cleaning requires various problem-specific strategies. For each one with its own trade-offs, various approaches can be implemented. Incorrect data can be erased, rectified, or imputed. The required fields of the dataset are tested for null-value cells.

- Train Set and Test Set

The dataset is divided into two parts. The first part is known as the training set. It is used for training the algorithm and the model uses this part for learning. The second part is known as the test set and it is used for providing the result of the model applied on the training set. The model is evaluated using the test set. Test set is used only after the model is trained completely. The training and the test should not be biased because it will cost accuracy. Accuracy of the model depends on how the model has been trained. Thus, training or the learning of the model should be done properly. The dataset used in this paper is divided into train set and test set. To provide good amount of training for the model, the training is done on the 80% data of the dataset. Rest 20% of the data becomes the test set and is used for calculation of the accuracy.

- TfIdfVectorizer

Term frequency (TF): It measures the frequency of a term in a document, the term frequency is divided by the length of the document.

Inverse document frequency (IDF): It detects the relevance of a term. It is the logarithm of the count of all documents in a corpus divided by the count of documents having that term in it.

The TfIdfVectorizer is used to convert a group of raw documents into a matrix of TF-IDF features. It can calculate word count, IDF, TF-IDF values together. When there is a need to calculate TF-IDF value on documents within the training set, TfIdfVectorizer is used. Thus, it is applied on the training set (80% of the dataset). The output obtained after applying TfIdfVectorizer is a matrix of TF-IDF features.

- Passive-Aggressive Classifier

The Passive-Aggressive algorithms are online algorithms used for large-scale learning. This algorithm remains passive if correct result is obtained after classification, and turns aggressive if there is any miscalculation or incorrect result. It doesn't converge, unlike most other algorithms. The goal is to correct the error, resulting in very little improvement in the weight vector norm. The PA algorithm uses the margin to adjust the present classification [12]. A classification upgrade is carried out by solving an optimization problem that is limited: and wants the new classification to be as similar to the existing one as possible and to reach at least a unit margin on the latest cases. Forcing a unit margin in the presence of noise may prove to be too aggressive. The input of the Passive-Aggressive Classifier is a matrix of TF-IDF Features. Thus, a model is formed which is trained on the data of training set and will be applied on the test set to evaluate the performance of this classifier (Fig. 1).

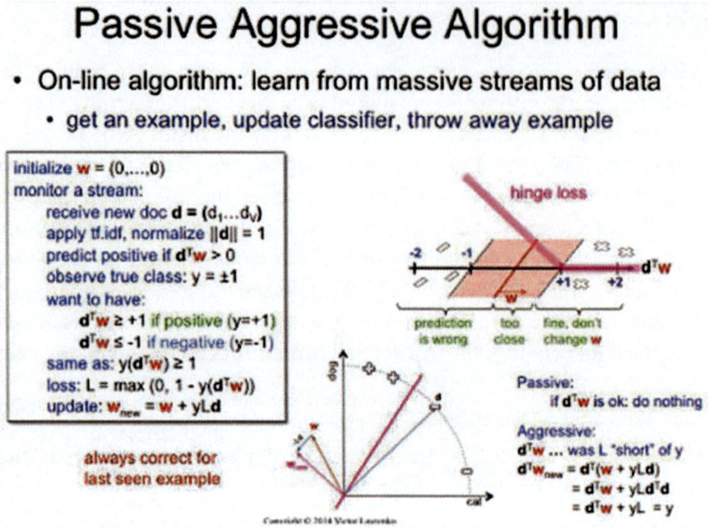

Fig. 1 Passive-Aggressive algorithm. *Image source* https://www.youtube.com/watch?v=uxGDwyPWNkU

4 Result Analysis

The performance of the Passive-Aggressive Classifier can be evaluated after applying on the test set. The result is calculated using some metrics and then compared with the result of other classifiers. To analyze the result, performance measures like Accuracy is considered and confusion matrix is drawn. Confusion Matrix (Fig. 2) is a 2×2 matrix comparing the actual classification to the predicted classification. This matrix itself defines some important metrics and also forms the basis for the other types of metrics. The metrics of confusion matrix are as follows:

- True Positives: Predicted value—YES and Actual value—YES.
- True Negatives: Predicted value—NO and Actual value—NO.

Fig. 2 Confusion matrix

Confusion Matrix

	Actually Positive (1)	Actually Negative (0)
Predicted Positive (1)	True Positives (TPs)	False Positives (FPs)
Predicted Negative (0)	False Negatives (FNs)	True Negatives (TNs)

- False Positives: Predicted value—YES and Actual value—NO.
- False Negatives: Predicted value—NO and Actual value—YES.

Accuracy for the classifier is calculated using the formula:

- Accuracy = (True Positives + True Negatives)/Total Number of Samples

The accuracy of the Passive-Aggressive Classifier can be calculated using the above formula after training the model. The experiment is done on the test set. For the calculation of accuracy, the confusion matrix is formed with the four parameters—True Positives, True Negatives, False Positives, and False Negatives. Then, the accuracy can be calculated by dividing the correct number of predictions by the total number of instances of the test set.

The first dataset used for classification achieved 98.87% accuracy with the confusion matrix (Fig. 3) having True Positives = 417, False Positives = 6, False Negatives = 1 and True Negatives = 373. The second dataset used for classification achieved 96.25% accuracy with the confusion matrix (Fig. 4) having True Positives = 1002, False Positives = 30, False Negatives = 45, and True Negatives = 924.

The results of this paper can be compared with the other machine learning models used for classifying fake news. The comparison of the other classifiers with the Passive-Aggressive Classifier used in this paper can be seen in Table 1. The table represents the classifiers used for fake news detection with the papers in which they are implemented. It compares all the classifiers using the measure as to accuracy. From this, the better performance of the Passive-Aggressive Classifier can be seen amongst all the classifiers used so far in this domain.

The graph (Fig. 5) plotted to check the performance of the classifiers with classifiers on the x-axis and accuracy on the y-axis can also be used in the purpose of comparing the classifiers with each other. The graphical comparison can be considered to check that the best classification is provided by the Passive-Aggressive

Fig. 3 Confusion matrix of dataset—1

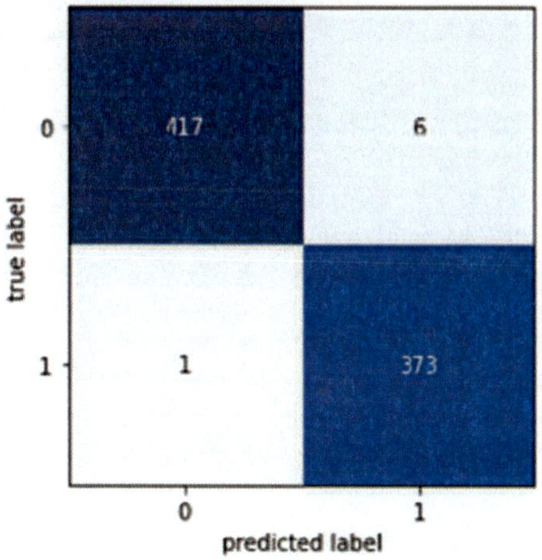

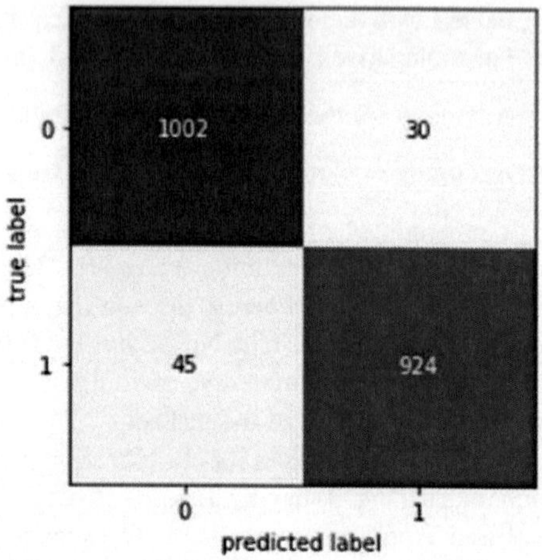

Fig. 4 Confusion matrix of dataset—2

Table 1 Different classifiers with their accuracy for detecting fake news

Paper reference	Classifiers used for fake news detection	Accuracy (%)
Granik and Mesyura [2]	Naïve Bayes classifier	74
Lakshmanarao et al. [5]	SVM classifier	75.5
Lakshmanarao et al. [5]	K-Nearest Neighbor	79.2
Lakshmanarao et al. [5]	Decision tree classification	82.7
Lakshmanarao et al. [5]	Random forest classification	90.7
	Passive-Aggressive classifier	97.56

Classifier. All the other classifiers have lower bars than the Passive-Aggressive Classifier.

The best accuracy for this problem is obtained by the Passive-Aggressive Classifier. The result analysis reveals that the Passive-Aggressive Classifier has the ability to improve standard classification methods to identify fake news for certain data substantially within the tested machine learning algorithms.

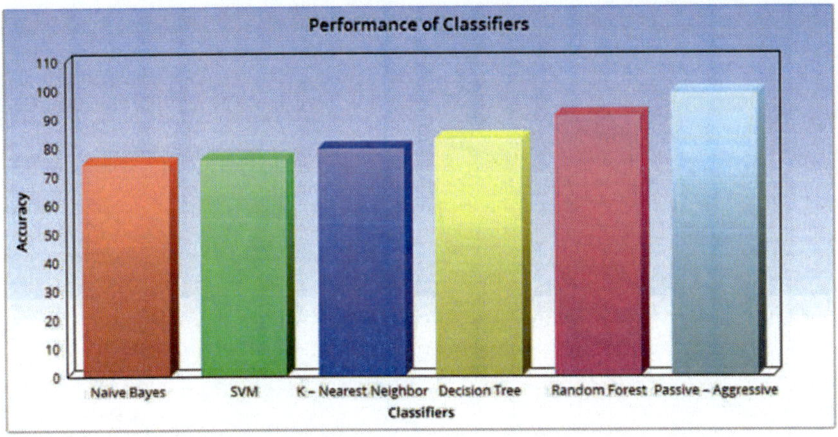

Fig. 5 Graph of classifiers with their accuracy for detecting fake news

5 Conclusion

With the expanding notoriety of social media life, an increasing number of individuals consume news from web-based life rather than conventional news media. In any case, internet-based life has likewise been utilized to distribute misinformation, which has solid negative effects on singular clients and more extensive community. The problem of detecting fake news is handled using Machine Learning algorithm in this research. Thus, Passive-Aggressive Classifier is used as it provides the best accuracy amongst all the classifiers with accuracy of 97.5%. Thus, it is suggested to battle against the subject of fake news detection.

References

1. Metz C (2016) The bittersweet sweepstakes to build an AI that destroys fake news, Dec 2016 (Online). Available https://www.wired.com/2016/12/bittersweet-sweepstakes-build-ai-destroys-fake-news/
2. Granik M, Mesyura V (2017) Fake news detection using naive Bayes classifier. In: 2017 IEEE first Ukraine conference on electrical and computer engineering (UKRCON), Kiev, Ukraine
3. Bhowmik D, Zargari S, Ajao O (2018) Fake news identification on twitter with hybrid CNN and RNN models. In: Proceedings of the 9th international conference on social media and society
4. Zheng L, Zhang J, Cui Q, Li Z, Yang PS, Yang Y (2018) TI-CNN: convolutional neural networks for fake news detection. arXiv preprint arXiv:1806.00749
5. Lakshmanarao A, Swathi Y, Kiran TSR (2019) An efficient fake news detection system using machine learning. Int J Innov Technol Exploring Eng (IJITEE) 8(10)
6. Ahmed H, Traore I, Saad S (2017) Detection of online fake news using n-gram analysis and machine learning techniques. In: International conference on intelligent, secure, and dependable systems in distributed and cloud environments ISDDC 2017

7. Khattar D, Goud JS, Gupta M, Varma V (2019) MVAE: multimodal variational autoencoder for fake news detection. In: The web conference-2019, San Francisco
8. Wang Y, Ma F, Jin Z, Yuan Y, Xun G, Jha K, Su L, Gao J (2019) EANN: event adversarial neural networks for multi-modal fake news detection. In: 24th ACM SIGKDD international conference on knowledge discovery & data mining, London
9. Markines B, Cattuto C, Menczer F (2009) Social spam detection. In: 5th international workshop on adversarial information retrieval on the web, 2009
10. Lu J, Zhao P, Hoi SCH (2016) Online passive-aggressive active learning. https://doi.org/10.1007/s10994-016-5555-y
11. Available https://drive.google.com/file/d/0B3e3qZpPtccsMFo5bk9Ib3VCc2c/view (Online)
12. Crammer K, Dekel O, Keshet J, Shalev-Shwartz S, Singer Y (2006) Online passive-aggressive algorithms. J Mach Learn Res 7:551–585

Recurrent Neural Network-Based Character Recognition System for Tamil Palm Leaf Manuscript Using Stroke Zoning

Suganya Athisayamani, A. Robert Singh, and A. Sivanesh Kumar

Abstract Tamil is one of the ancient Indian languages which has a vast collection of literature in the form of palm leaf, stones, metal plates, and other materials. Palm leaf manuscript was a broad tool to narrate medicines, literature, drama, and many more. Recognition of the characters written in palm leaf manuscripts is still an open task because of the need for digitization and transcription. In this paper, the recurrent neural network (RNN) is used to train the features of the characters extracted from the palm leaf. This method contains a preprocessing method to eliminate noise; then the character is segmented from the image and trained using the bidirectional long short-term memory (BLSTM) network. A feature vector with nine zones of character strokes is used to train and test the characters. A rich set of characters are used to train the features of the characters. This method provides better recognition accuracy than the other neural network-based character recognition.

1 Introduction

Tamil is the ancient Indian classical language that has 247 letters with 12 vowels (Uyir Ezhuthu), 18 consonants (Mei Ezhuthu), 216 compound letters (Uyirmei Ezhuthu), and one unique character (Aytha Ezhuthu). One of the historical documents [3] in the Tamil language is palm leaf manuscript. These manuscripts narrate about music, literature, grammar, novels, astrology, medicines, and astronomy. The writing style of these letters changes through different centuries. Such manuscripts are found across the globe and need automatic transcription [14]. Bidirectional long

S. Athisayamani
School of Computing, Sastra Deemed to be University, Thanjavur, India

A. Robert Singh (✉)
School of Computing, Kalasalingam Academy of Research and Education, Krishnankoil, India
e-mail: robertsinghbe@gmail.com

A. Sivanesh Kumar
Department of Computer Science and Engineering, Saveetha School of Engineering,
Saveetha Institute of Medical and Technical Sciences, Chennai, India

G. Ranganathan et al. (eds.), *Inventive Communication and Computational Technologies*, Lecture Notes in Networks and Systems 145,
https://doi.org/10.1007/978-981-15-7345-3_14

short-term memories recurrent neural network (BLSTM-RNN) is widely used for various recognition applications like face recognition [1], speech recognition [2], protein pattern prediction, and character recognition. The handwritten letters can be recognized by pattern recognition [15] using different image processing methods. Hidden Markov model (HMM) is the general approach for handwriting recognition [16]. HMM has the disadvantage of failing to model the contextual parts of the characters.

The proposed method has a RNN with BLSTM learning method. The feature maps include nine stroke zone details of each character for training and testing.

2 Related Works

Kiruba [8] summarized the available Tamil character recognition methods for palm leaf manuscripts. Panyam et al. [9] proposed a Telugu character recognition in palm leaf manuscript using a 3D feature recognition. The combination of 2D DWT, 2D FFT, and 2D DCT is used for feature extraction. It is a hardware-based model that estimates the X, Y, and Z coordinates to measure the depth of the impression in 3D. Valy et al. [10] proposed two methods for Khmer palm script manuscript character recognition. The first method uses character recognition using the neural networks like CNN, LSTM-RNN, and their combination. The second method proposes simultaneous character recognition using 1D and 2D RNN. Vinoth et al. [11] proposed an optical character recognition for Vettezhuththukkal, an ancient Tamil writing style. This method has a series of operations including noise reduction, image enhancement, segmentation, character extraction, and character recognition using a neural network with a single hidden layer.

3 Proposed Palm Leaf Character Recognition

In this paper, the identification of palm leaf characters is done by dividing the character into nine zones called center, left top, left bottom, left, top, right top, right bottom, right, and bottom. The RNN learns the pattern of strokes with BLSTM recognition. The HMM contains many univariate internal states that can give a few amounts of information. The BLSTM can access the contextual information in both directions from the input and output layer. The proposed network has two hidden layers that are learning in forwarding and backward directions that are connected to the same output layer. The detailed process of character recognition is shown in Fig. 1.

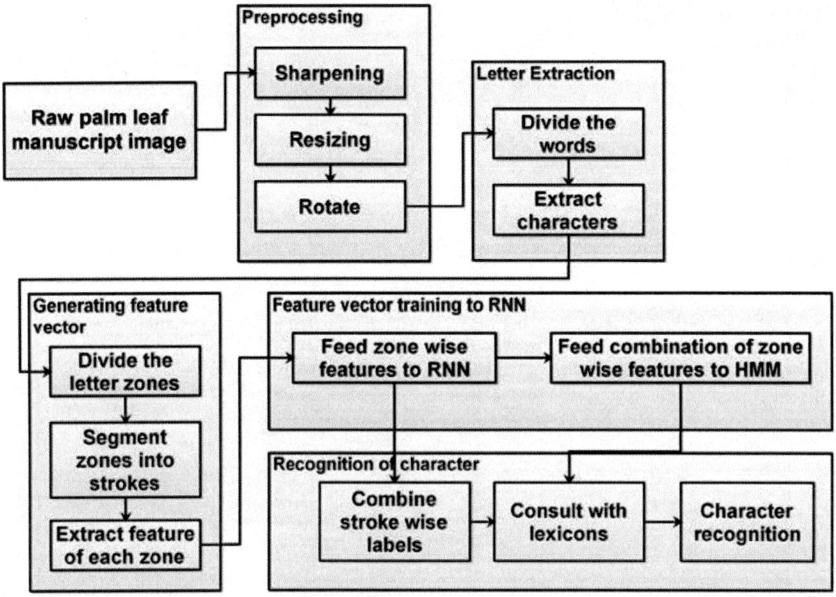

Fig. 1 Architecture of palm leaf recognition

3.1 Preprocessing

Preprocessing is an important process that makes the raw palm leaf image into a recognition ready version [13]. Palm leaf is inscription of characters using a sharp-edged pen-like device. The surface of the leaf may be soft or rough. The image capturing may cause some rotated contents. These issues of the raw palm leaf images can be addressed by a sequence of preprocessing methods. In this paper, sharpening—to enhance the quality of the image, resizing—to create a uniform character size for processing, and rotate—to keep a uniform orientation of characters are applied. Sharpening is the process of enhancing the quality of input image in spatial domain. The linear transformation of pixel intensity is applied to get the enhanced image. The linear transformation is given in Eq. (1).

$$s = 255 \times \sin(r \times \frac{\pi}{255}) \tag{1}$$

where r is the intensity value of each pixel in the raw image. The output of such enhancement is shown in Fig. 2. Resizing will ensure the constant size of each character that can easily process the recognition. In this paper, the size is fixed as 256 × 512 pixels for each leaf that can resize the image without large amount of loss in resolution. Rotation is performed by mapping the subsamples of the image with a constant

Fig. 2 **a** Original image, **b** enhanced image

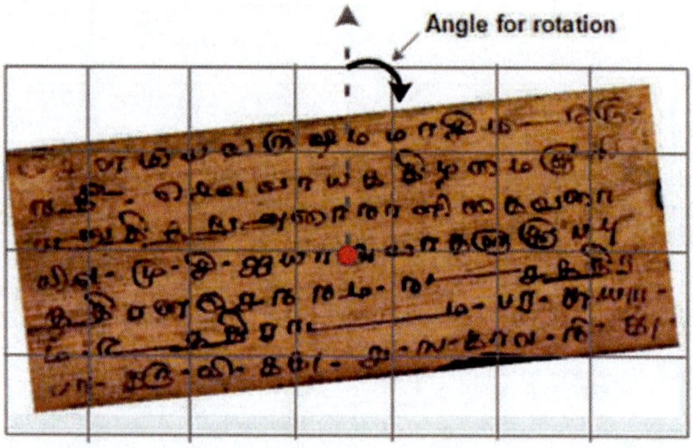

Fig. 3 Rotation angle calculation

size. For example, the image given in Fig. 3 is divided into grids and the angle between the perpendicular line from the center point of the image and the outline of the image is considered as the angle for rotation.

3.2 Letter Extraction

The enhanced image contains a sequence of characters. Tamil palm leaf manuscripts have specific set of symbols and space to separate characters and words. In general, space is used to separate characters where dot (.), vertical line (|), and horizontal line (-) are used to separate words. Such separators are shown in Fig. 4. These separators should be removed before separating the characters. The local binary filter (LBF) is used to recognize these separators. The given space is mapped into a 3 × 3 window and if a match is found, then they are eliminated. The LBF patterns for the separators are shown in Fig. 5.

3.3 Feature Vector Generation for Each Character

The neural network trains the feature map of the characters. The feature map is identified by the strokes of the characters in different zones of the character. Figure 6 shows the different zones of a character. The character to be tested is mapped into

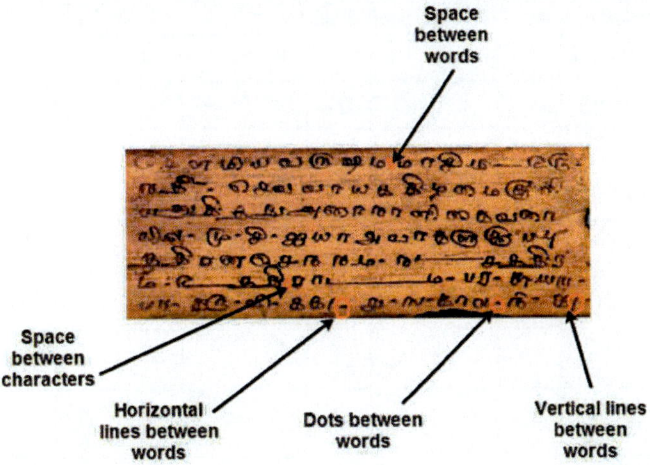

Fig. 4 Word and character separators

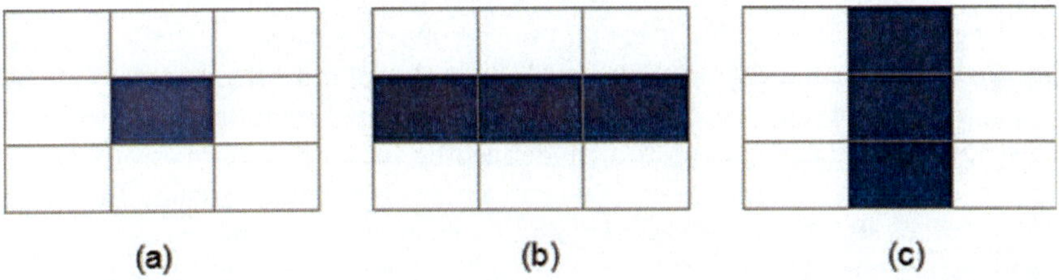

Fig. 5 LBF for separators **a** dot, **b** horizontal line, **c** vertical line

Fig. 6 Zoning of character

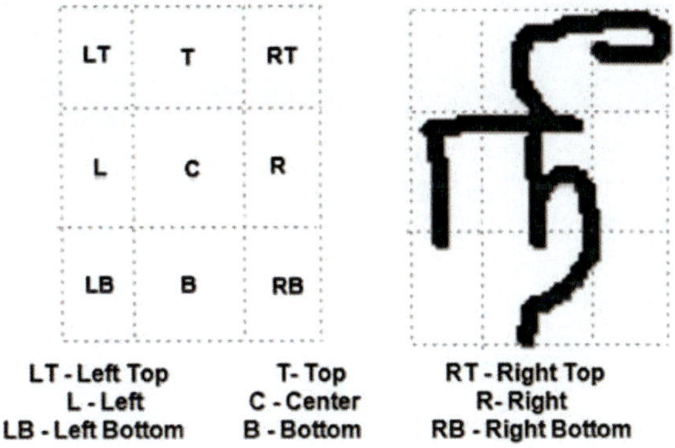

Table 1 Strokes of zones of different Tamil characters

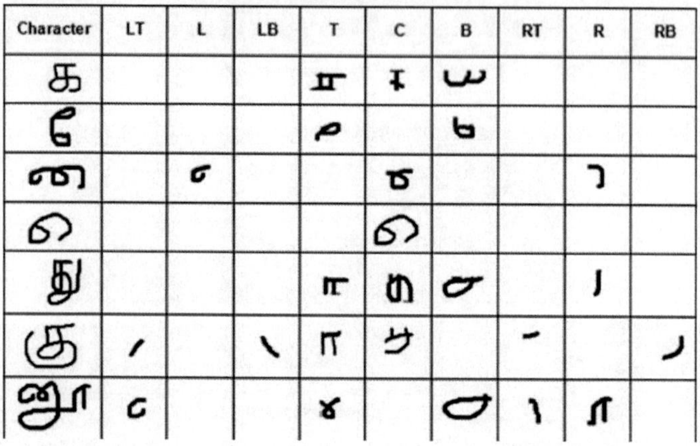

a rectangle. The rectangle is further partitioned into nine parts by equally dividing through horizontal and vertical directions. The strokes of different zones for various Tamil characters are listed in Table 1. The feature vector contains the nine zones and the shape in the zones. The shapes are defined by four different types of strokes. Slope [18], curliness, linearity [19], and curvature [17] are used to define the shape. The part of the stroke is known as the vicinity V under process with the starting point S and ending point E. S is considered as the left topmost valid pixel in the given stroke area. Slope of the vicinity V is computed by finding the cosine of the angle θ between the straight line from the point E and the last vicinity of the character L. Curliness U is calculated by the deviation of the vicinity V from the straight line as given in Eq. (2).

$$U = \frac{l}{\text{Max}(dx, dy)^2} \tag{2}$$

Linearity (LI) is calculated by the average square distance between each point in the vicinity V and the straight line joining S and E of V as given in Eq. (3).

$$LI = \frac{1}{N}\sum_i D_i^2 \tag{3}$$

where N is the total number of points in V. The curvature of the vicinity V is evaluated by the sine and cosine of the angle β with the two neighbors that are not immediate next of V as given in Eqs. (4) and (5).

$$\cos\beta = \cos\theta(p-1) \times \cos\theta(p+1) + \sin\theta(p-1) \times \sin\theta(p+1) \tag{4}$$

$$\sin\beta = \cos\theta(p-1) \times \sin\theta(p+1) + \sin\theta(p-1) \times \cos\theta(p+1) \tag{5}$$

3.4 Training of Feature Vector Using RNN

The basic strokes are labeled for RNN-based character recognition with connectionist temporal classification (CTC). These sample strokes under each zone are assigned with a unique class value. Training is done by extracting the different strokes for a large scale of samples. In this paper, 5736 Tamil palm script pictures from 24 bundles of Noolaham online archive [6] and the palm leaf manuscripts from the Agama academy digital library [4] are used to generate the training set, and the testing images are obtained from [4] and the Tamil Heritage Web archive [5].

3.4.1 Architecture of Proposed RNN

LSTM is the commonly used architecture that can train a rich set of data and to test on raw data. In this method, a bidirectional LSTM is used. This architecture uses forward and backward layers to merge outputs to the next level of training. The general architecture of the proposed method is given in Fig. 7. It contains nine neurons in the input layer to process nine different zoning data of the character. Each neuron contains the slope, curliness, linearity, and curvature that are the features for learning. Finally, an activation layer is used to apply sigmoid between the previous output and the current output. This output gives a probability of the label generated by the training process. For each character, such a probability model is created and the output is generated as given in Eq.(6). If the probability is maximum for the given test character, then the label of the trained character will be returned as output.

$$P(x)_\pi = \Pi_{i=1}^{9} p_\pi^t, \forall \pi \tag{6}$$

where x is the input, π is the possible sequence of output, and p_π^t is the probability of a label from the sample π at t.

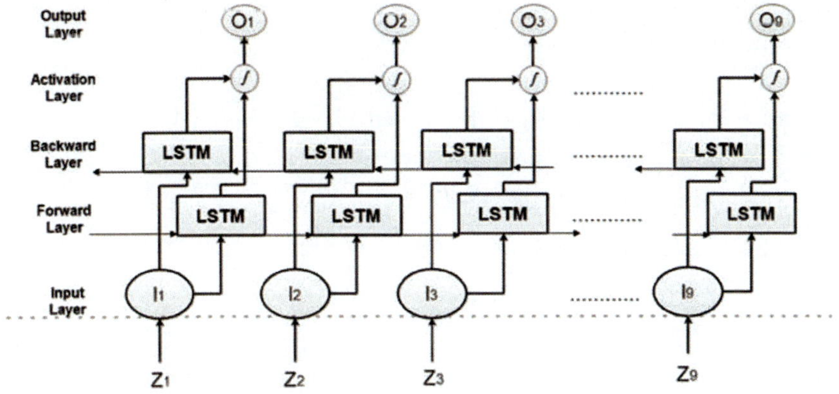

Fig. 7 Architecture of proposed BLSTM-RNN

4 Result Analysis and Discussion

The performance of the proposed method is tested for various quality metrics. The proposed method is implemented in Python 3.0 along with OpenCV for image processing operations. The system configuration is Intel Pentium 4 CPU 2.80GHz processor, 2GB RAM, and Windows 7 Enterprise Service Pack 1 operating system. The results are evaluated using the target performance metrics against the existing research method with varying input values.

4.1 Accuracy

The correctness of the proposed model is defined by accuracy and calculated by Eq. (7). Consider the parameters Tp—true positive, Tn—true negative, Fp—false positive, and Fn—false negative sentiments.

$$\text{Accuracy} = \frac{\text{Tp} + \text{Tn}}{\text{Tp} + \text{Tn} + \text{Fp} + \text{Fn}} \tag{7}$$

4.2 Sensitivity

Sensitivity is the ability of a test to correctly identify the characters with the training characters. Sensitivity, also called the true-positive rate, the recall, or probability of detection in some fields, measures the proportion of positives that is correctly identified.

$$\text{Sensitivity} = \frac{\text{Tp}}{\text{Tp} + \text{Tn}} \tag{8}$$

4.3 Specificity

Specificity is a metric to test the proposed method without any training set.

$$\text{Specificity} = \frac{\text{Fn}}{\text{Fn} + \text{Fp}} \tag{9}$$

4.4 Precision

The precision is the metric to estimate the relationship between real and fake samples. It is the ratio between the true positive and sum of true positive and false positive.

$$\text{Precision} = \frac{\text{Tp}}{\text{Tp} + \text{Fp}} \tag{10}$$

4.5 F-Measure

In image classification, F-measure is used to summarize precision and specificity. F-measure combines precision P and recall R by

$$F = \frac{2PR}{P + R} \tag{11}$$

Table 2 summarizes the performance of the proposed method on different Tamil compound letters that are shown in terms of accuracy, sensitivity, specificity, precision, and F-measure. The characters with more number of zone strokes like ‘ka,’ ‘tha,’ ‘thu,’ and ‘gna’ are achieving more accuracy percentage than other letters. Sensitivity, specificity, precision, and F-measure are depending on the data set trained. These parameters can be improved by using more number of training samples.

The proposed method is compared with deep belief network (DBN) [7]-based Tamil vowel recognition and sequence to sequence RNN (RNN-StoS) method [12] for offline character recognition. Epoch is one of the most important training parameters that determines the number of passes of the algorithm through the entire data set to adjust the internal weight. The number of epochs in a range of 5–65 is examined for average test error rate for DBN, RNN-StoS, and BLSTM-RNN as shown in Fig. 8. The result shows that the proposed method registers minimum test error due to zoning of different parts of the letter. With the epoch count 30, the proposed method converges and the average test error reaches zero. Thus, the proposed architecture performs well with fewer number of runs through the given data set than the state-of-the-art methods. RNN-StoS and BLSTM-RNN have automatic weight adjustment in each run of the learning on the data set. DBN fails to adjust the weights automatically that make it converge at 50th run on the entire data set.

Table 3 compares the accuracy for various characters. The results show that the proposed method outperforms in more number of characters for recognition than the state-of-the-art methods. RNN-StoS is a recurrent learning method that could recognize the characters like ‘sa,’ ‘nu,’ and ‘moo’ with better accuracy than BLSTM-RNN. DBN is a general neural network that fails to adjust the learning weight. So, the recognition accuracy of DBN is lesser than both the RNN methods.

Table 2 Performance comparison for proposed method

Letter	Accuracy	Sensitivity	Specificity	Precision	F
	97	0.2458	0.28909	92	0.90685
	97	0.19348	0.64024	81	0.99489
	97	0.63542	0.8941	92	0.70106
	91	0.95406	0.18741	84	0.49034
	97	0.96914	0.53774	86	0.57367
	91	0.1595	0.67009	81	0.77025
	91	0.51991	0.64302	83	0.79842
	94	0.48346	0.15319	86	0.68371
	94	0.95053	0.26354	82	0.27129

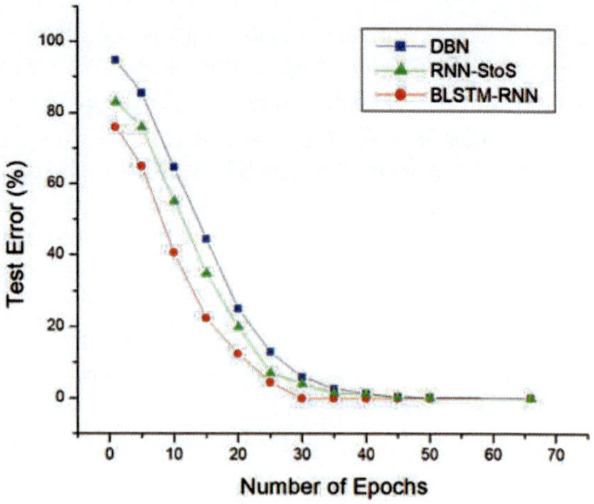

Fig. 8 Comparison of epoch versus test error

Table 3 Comparison of accuracy BLSTM-RNN versus RNN-StoS (bold values are the best)

Letter	BLSTM-RNN	RNN-StoS	DBN
	**97**	94	95
	**97**	94	93
	**97**	95	94
	91	**93**	92
	**97**	94	95
	91	**94**	92
	91	**93**	90
	**94**	92	91
	**94**	93	90

5 Conclusion

Ancient Tamil character recognition is the most concentrated research field in the real world which focuses on recognizing the ancient Tamil characters so that much information can be learnt. Specifically, ancient medicine system like Siddha can be seen in the ancient Tamil scripts only through the medium of palm leaves. In the proposed research, recognizing and prediction of the ancient Tamil characters are done by implementing BLSTM-RNN. This method performs in a better way and produces better result by correcting the prediction error value in every layer at run time. The experimental evaluation of the proposed research method is proved to provide better result than the existing research method in terms of accurate recognition of the ancient Tamil characters.

Acknowledgements Hereby the authors acknowledge Agama Academy for providing a rich set of digital archive of Tamil palm leaf manuscript for this research work.

References

1. Ko BC (2018) A brief review of facial emotion recognition based on visual information. Sensors 18(2):401
2. Graves A, Schmidhuber J (2005) Framewise phoneme classification with bidirectional lstm and other neural network architectures. Neural Netw 18(5):602–610
3. Kumar DU, Sreekumar G, Athvankar U (2009) Traditional writing system in southern India–palm leaf manuscripts. Des Thoughts 7:2–7
4. Shivachariar RV (2019) Agama academy. Online. Accessed 4 July 2019
5. Subashini (2017) Tamil heritage. http://thfcms.tamilheritage.org/category/palm-leaf/. Online. Accessed 4 July 2019
6. Noolaham Foundation (2015) Noolaham, http://www.noolaham.org/wiki/index.php/. Online. Accessed 4 July 2019
7. Ram Prashanth N, Siddarth B, Ganesh A, Kumar VN (2017) Handwritten recognition of Tamil vowels using deep learning. In: IOP conference series: materials science and engineering
8. Kiruba KB (2016) A survey of methods and strategies in Tamil palm script recognition. Int J Comput Sci Inf Technol 7(2):972–977
9. Panyam NS, TR VL, Krishnan R (2016) Modeling of palm leaf character recognition system using transform based techniques. Pattern Recogn Lett 84:29–34
10. Valy D, Verleysen M, Chhun S, Burie J (2018) Character and text recognition of khmer historical palm leaf manuscripts. In: 16th International conference on frontiers in handwriting recognition (ICFHR), pp 13–18
11. Vinoth YPR, Rajesh R (2017) Intelligence system for tamil vattezhuttu optical character recognition. Int J Comput Sci Eng Technol 8(04S):22–26
12. Shkarupa Y, Mencis R, Sabatelli M (2016) Offline handwriting recognition using LSTM recurrent neural networks. In: The 28th Benelux conference on artificial intelligence
13. Saxena LP (2014) An effective binarization method for readability improvement of stain-affected (degraded) palm leaf and other types of manuscripts. Curr Sci 107(3):489–496
14. Antara Kesiman MW, Burie JC, Ogier JM, Grangé P (2018) Knowledge representation and phonological rules for the automatic transliteration of balinese script on palm leaf manuscript. Computación y Sistemas 21
15. Chamchong R, Fung CC (2012) Text line extraction using adaptive partial projection for palm leaf manuscripts from Thailand. In: International conference on frontiers in handwriting recognition, pp 586–591
16. Gilloux M (2004) Hidden Markov models in handwriting recognition., In: Impedovo S (ed) Fundamentals in handwriting recognition. NATO ASI series (Series F: computer and systems sciences), vol 124
17. Chiang Y, Knoblock CA (2011) Recognition of multi-oriented, multi-sized, and curved text. In: International conference on document analysis and recognition, pp 1399–1403
18. Hafiz AM, Bhat GM (2014) Handwritten digit recognition using slope detail features. Int J Comput Appl 93:14–19
19. Hamrouni S, Cloppet F, Vincent N (2014) Handwritten and printed text separation: linearity and regularity assessment. In: Noise modelling in parallel magnetic resonance imaging: a variational approach, pp 387–394

Microcontroller Based Smart Grinder for Automatic Batter Collection and Grinder Cleaning

B. N. Neethu, D. Vijai Srinivas, S. Jayanthy, and J. Judeson Antony Kovilpillai

Abstract An automated grinding device is developed for domestic and commercial purpose that grinds, collects batter and cleans the grinder without human involvement. This system converts the rice into batter and after the completion of grinding, the collecting and cleaning is done automatically. The smart grinder is interfaced with an embedded system based on ARM LPC2148 and the prototype is validated. For grinding rice and other food items, AC Induction Motor coupled with mechanical arrangement is used. The DC Motor controls both the wiper movement and valve mechanism. The knob is used to decide which process (wiper or valve) is to be performed. Valve and Wiper helps in the automated batter collection process. Finally the cleaning is done. The status of the system is displayed in the LCD.

Keywords ARM LPC2148 based microcontroller · Smart grinder · Automation · Sensors · DC motors · Stepper motor · Motor drivers · LCD

1 Introduction

Rice is the most important food crop in the world. Rice flour is produced from broken rice by different processes which is used to produce various kinds of foods and snack items, baby foods, candies etc. Usually, there are three method which is

B. N. Neethu (✉) · S. Jayanthy · J. Judeson Antony Kovilpillai
Department of Electronics and Communication Engineering, Sri Ramakrishna Engineering College, Coimbatore, India
e-mail: neethunair159@gmail.com

S. Jayanthy
e-mail: jayanthy.s@srec.ac.in

J. Judeson Antony Kovilpillai
e-mail: judeson@srec.ac.in

D. Vijai Srinivas
VTEC Engineering, Coimbatore, India
e-mail: info@vtec.co.in

G. Ranganathan et al. (eds.), *Inventive Communication and Computational Technologies*, Lecture Notes in Networks and Systems 145,
https://doi.org/10.1007/978-981-15-7345-3_15

used for preparing the rice flour: wet, semidry, dry grinding. Of these wet grinding is the traditional method for preparing the rice batter which involves process such as washing, soaking, grinding, and the preparation include use of machines and manpower [1]. The wet grinders are still the practical method for the preparation of batter. For improving the existing wet grinders into new types which is suitable for domestic kitchen, various attempts are made [2]. Some of the existing wet grinders are as follows.

1.1 Conventional Wet Grinders

Conventional wet grinders need a lot of space due to its size, which is not a portable kitchen appliance. These type of wet grinders are used in small restaurants and commercial kitchens. It is made of stainless steel, aluminium and stone based mechanical elements driven by a motor. This model wet grinder is easy to use and is of high capacity but lack modern features and mainly used in commercial application.

1.2 Tabletop Wet Grinders

A tabletop wet grinder is compact and requires less space which can be easily placed on a table, platform or on a counter and is small-sized, portable kitchen appliance. In this model its drum cannot be tilted backwards or forwards but can only be removed/disengaged from the motor mechanism after grinding.

1.3 Tilting Wet Grinders

It is a burden to remove the drum and collect the batter from the tabletop wet grinders. Hence another type of wet grinder was developed and named as tilting wet grinders, which requires less space compared to conventional models. In this model the drum is attached by using adaptors to the main assembly which can tilt the drum forwards and backwards.

These models are good based on its basic functionality but these models don't have advanced features that help in easy operation and maintenance. To overcome these drawback in existing wet grinders that are available in today's market, the new "Smart Grinder" [3] is developed which transfer the batter automatically to a vessel after grinding and then automatically clean all the parts involved in grinding without human intervention.

In this paper "smart grinder" which is a automated wet grinder, is developed for automatic grinding the rice and dhal into batter both for domestic and commercial use. This invention automates the wet grinder which involves less human involvement in its operation. Smart grinder has electronic, electrical and mechanical components for its several operations.

2 Related Work

Numerous research works have been undertaken on automation of these wet grinders to reduce the human involvement and as a result few models have been developed.

In selecting components and mechanism for automation different types of methods are opted.

In [4] the author have designed an improved version of wet grinder which is a tilting type, which will overcome the difficulty in lifting and then transferring of batter by tilting the drum. But the disadvantage of this model is that, the cleaning of the grinder is difficult because the drum cannot be detached from the main assembly. Cleaning the drum takes more time and effort. Another disadvantage is that it difficult to maintain because this model comes with suspension equipment or springs and also due to this the system may cause issues with screws or wearing out of spring and hence increase the repair cost.

In [5] the author developed a table top wet grinder which is portable and is of small size when compared to the conventional wet grinders, which have the advantage of removing the drum for collecting batter and cleaning. The disadvantage is that this system does not provide automation for collecting batter and cleaning the grinder.

In [6] authors have designed a unipolar unidirectional stepper motor circuit. This circuit has low costs, stable performance and provides high accuracy. But the disadvantage of the system is that when this is compared to the bipolar stepper motor, unipolar stepper motor provides less torque.

In [7] authors have used an old DC compound motors which are upgraded with DC converter without the modification of the mechanical assembly by achieving better speed response and less losses.

In [8] authors have developed a Brushless DC Motor Controller which can control the speed of the motor by using the PWM technique which gives a higher torque.

In [9] authors have designed a system to achieve DC motor speed control using PWM for the operation such as forward, reverse, speed up, slow down, stop.

In [10] the authors have implemented the digital alarm clock, which also displays an external text. When the input is given by the user, it is displayed as text along with alarm which can help the users to remind that the work is completed. The disadvantage of the system is that this type of LCD only displays the text, and to display customized character or images, graphical LCD is needed.

In all the above works human intervention is required for collecting the batter and cleaning the grinder. Manual monitoring of the grinding process is required. In this paper an embedded system based automation of wet grinder is developed using ARM microprocessor which is interfaced with motors and display unit. The automation involves collecting the batter and cleaning the parts automatically without human involvement. The status of the system is shown on the LCD which can be viewed by the user. Sensors can be fixed for monitoring the process. Prototype is developed and validated.

3 Proposed System

The proposed system is implemented using ARM Microprocessor-LPC2148 [11]. The grinder converts the rice/dhal to batter using electrical and mechanical components. For the grinding process AC induction motor which is coupled with the mechanical components is used and then after grinding, the device automatically collects the batter and cleans the grinder using the wiper valve mechanism and the batter is collected in the collecting compartment. Figure 1 represents the block diagram of the System. The DC motor is connected to the microcontroller via the motor driver [12] which is used to drive the mechanism of wiper and valve. The DC Motor used for these mechanisms is Johnson motor and to drive these Johnson motor VNH2SP30 motor driver is used. The Johnson motor is connected via the VHN2SP30 motor driver to the wiper and valve mechanism performs forward, reverse, start and stop operations. The single motor is used for both the mechanisms. The two mechanisms are selected by a KNOB which is controlled by the stepper motor for its operation. This stepper motor is connected to the ARM LPC2148 microcontroller via the stepper motor driver [13]. DRV8825 motor driver is used to drive the stepper motor. The stepper motor performs forward and reverse rotation to control the KNOB. The Limit Switch is used to detect the stop and start of the motor. The collecting of batter is automatically done using these mechanisms after the grinding of the rice/dhal items into batter. Finally cleaning of grinder is also done automatically. The status of the entire process is displayed through the ILI9341 TFT LCD Display.

Figure 2 shows schematic diagram of the proposed system. The DC motor, DC motor drivers, Stepper motor, Stepper motor driver and LCD are interfaced with ARM-LPC2148 microcontroller.

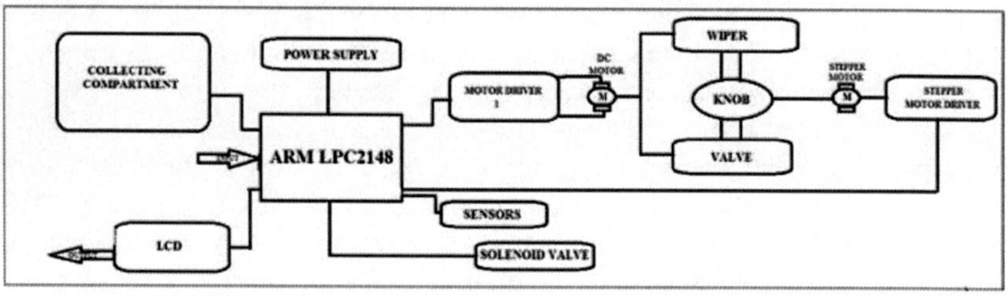

Fig. 1 Block diagram

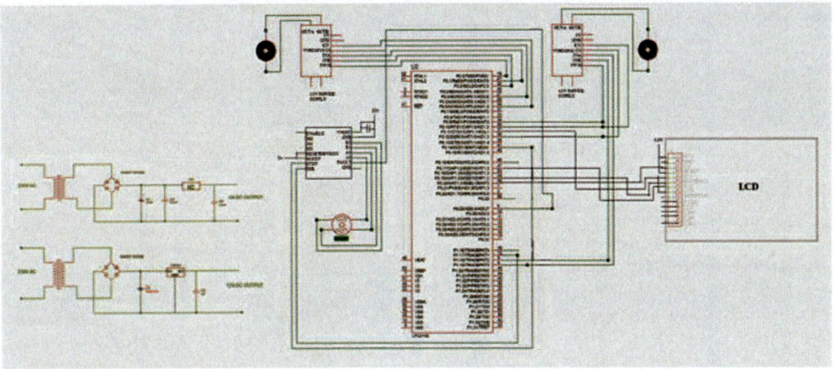

Fig. 2 Schematic diagram

3.1 *Hardware Description*

3.1.1 ARM-LPC2148 Microcontroller

ARM7 based LPC2148 is an Advanced Reduced Instruction set computer (RISC) machine, which has a 32-bit processor architecture. The automation of the grinding device is carried out by interfacing various components of the device to this controller.

3.1.2 Johnson Motor (DC Motor)

This simple DC Motor is of compact size and is coupled with metal gearbox to deliver more torque/power. The Johnson motor driven by VNH2SP30 motor driver controls the wiper and valve mechanism.

3.1.3 VNH2SP30 Driver (DC Motor Driver)

VNH2SP30 is a full bridge motor driver, widely used for automotive and industrial applications. These motor drivers are used for driving the Johnson DC motor by connecting the motor via driver to the microcontroller.

3.1.4 Stepper Motor

Stepper motor is a brushless DC electric motor. This motor divides the full rotation into number of equal steps. The stepper motor is of two types, unipolar stepper motor and bipolar stepper motor. In this system high torque unipolar stepper motor is used, which has six lead that controls the knob by motor rotation. The KNOB position

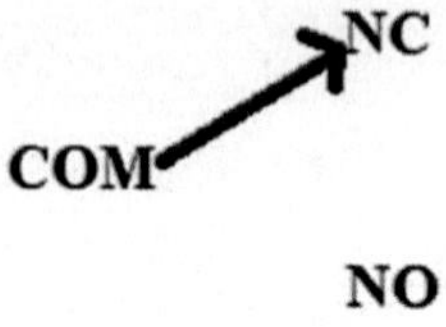

Fig. 3 Normally closed

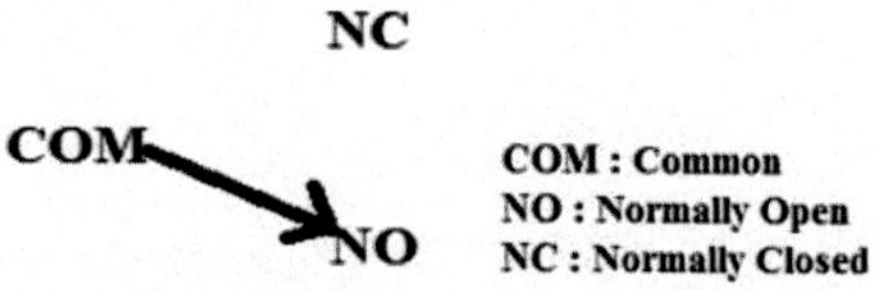

Fig. 4 Normally open

decides for which mechanism (wiper or valve) the Johnson motor is to be used. This stepper motor is connected to the microcontroller via the DRV8825 driver board.

3.1.5 DRV8825 Driver(Stepper Motor Driver)

DRV8825 is a micro-stepping stepper motor driver, which has over temperature and over current protection with adjustable current limiting capability. This driver has six micro-step resolutions (i.e.) down to 1/32 step. This motor driver is used for driving the stepper motor by connecting the motor via driver to the microcontroller.

3.1.6 Limit Switch

Limit switch is used for controlling the electrical devices. Limit switch is a mechanical switch, where the actuator senses the pressure of any physical object which is moving towards it and provides the electrical output. The limit switch that is used has three terminals NO (Normally Open), NC (Normally Close) and COM (Common). Normally when no pressure is applied the switch will be in NC (Normally Closed) Position, and when the pressure is applied this will be switched to NO (Normally Open) position shown in Figs. 3 and 4. In this system when the limit switch is pressed the motor stops and when the pressure is released the motor starts its rotation.

3.1.7 Solenoid Valve

Solenoid valve is an electromechanical device. These valves allows to autonomously control the flow of fluid without the need of human involvement, by connecting to the controller for opening and closing the valves. These valves are of two types NO (Normally Open) and NC (Normally Close). When the electric current is passed through the coil, these generate the electromagnetic field and the valve opens which result in flow of fluid, and as soon as flow of electric current stops magnetic field disappears and the valve goes back to its Normally Closed Position. In this system, the cleaning of grinder is carried out by opening and closing of valves along with the wiper mechanism.

3.1.8 ILI9341 TFT LCD Display

The 2.8 in. SPI TFT display has a resistive touch screen with 240 * 320 pixels and an individual RGB Pixel control. This is connected to the microcontroller to display the status of the entire process.

3.2 Hardware Specification

Table 1 shows the Hardware Specification for the proposed model. The components shown are used for the automation of the proposed model.

4 Methodology

First the rice is added to the grinding drum, and the device is turned ON. The grinding of rice is carried using the AC single phase induction motor coupled with mechanical components. After grinding the batter will be ready to be removed from the grinder. During the grinding process, the wiper will be at an angle allowing the batter to pass forward the wiper aiding in grinding. The valve will be in closed position. In the next step after the batter is prepared, the collection of batter is done automatically by opening the valve as the wiper moves towards the center. The wiper and valve mechanism is operated using the Johnson motor, (i.e.) to open the valve, the motor will rotate in forward direction and once the valve comes to the open position the limit switch stops the motor. Then the KNOB switches the motor for wiper mechanism using the stepper motor. Once the KNOB engages the motor with the wiper mechanism, again the motor starts rotating in the forward direction (due to treading this will be considered as reverse direction) and once the wiper comes towards center, again the limit switch senses and stops the motor. When the wiper is at the center position the valve will be in open position. The prepared batter is collected by rotating the grinding drum slowly in the collecting compartment. After the collection process gets completed, the valve and wiper will go back to its initial position by the same above mechanism, but the motor rotation will be in the reverse direction. Then the cleaning is done by spraying water through nozzles, by opening and closing of the solenoid valve. After cleaning the valve opens, wiper moves towards the center and the grinding drum rotates slowly and then the system is turned off. The status of these entire operation will be displayed on the LCD [14] (Fig. 5).

Table 1 Hardware specification

Components	Specification
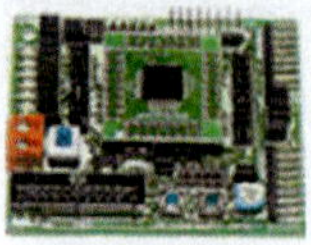 ARM-LPC2148	Operating voltage: 3.3 V Clock speed: 12 MHz Flash memory: 512 KB On-chip static Ram: 40 KB 2 UARTS, 2ADC and 1DAC 60 MHz maximum CPU clock Up to 5 V tolerant I/O pads
 Johnson motor (DC motor)	Motor type: High torque geared motor Operating voltage: 12 V DC RPM: 60 RPM POWER: 20 W No-load current: 0.8 A Load current: 7.5 A(MAX)
 VNH2SP30 (DC motor driver)	Voltage range: 5.5–16 V Maximum current rating: 30 A Practical continuous current: 14 A Current sense output proportional to motor current Maximum PWM frequency: 20 kHz Thermal shutdown
 Stepper motor	Motor type: Unipolar Operating voltage: 12–37 V Rated voltage: 4.2 V Current: 1.2 A Holding torque: 0.317 Nm Step angle: 1.8° No of lead: 6 N°
 DRV8825 (stepper motor driver)	Minimum operating voltage: 8.2 V Maximum operating voltage: 42 V Continuous current per phase: 1.5 A Maximum current per phase: 2.2 A Minimum logic voltage: 2.5 V Maximum logic voltage: 5.25 V Microstep resolutions: full, 1/2, 1/4, 1/8, 1/16, and 1/32.
 AC induction motor	Phase: Single phase induction motor Voltage: 180–240 V, AC Power: 0.2 HP Frequency: 50 Hz Wattage: 180 W
 ILI9341 LCD display	Type: TFT Touch type: Resistive Input voltage: 3.3–5 V Current: 80 mA Pixel resolution: 240 * 320

(continued)

Table 1 (continued)

Components	Specification
 SMPS power supply	Input voltage: 110–240 AC Frequency: 50 Hz Output voltage: 12 V Output current: 10 A Power: 150 W

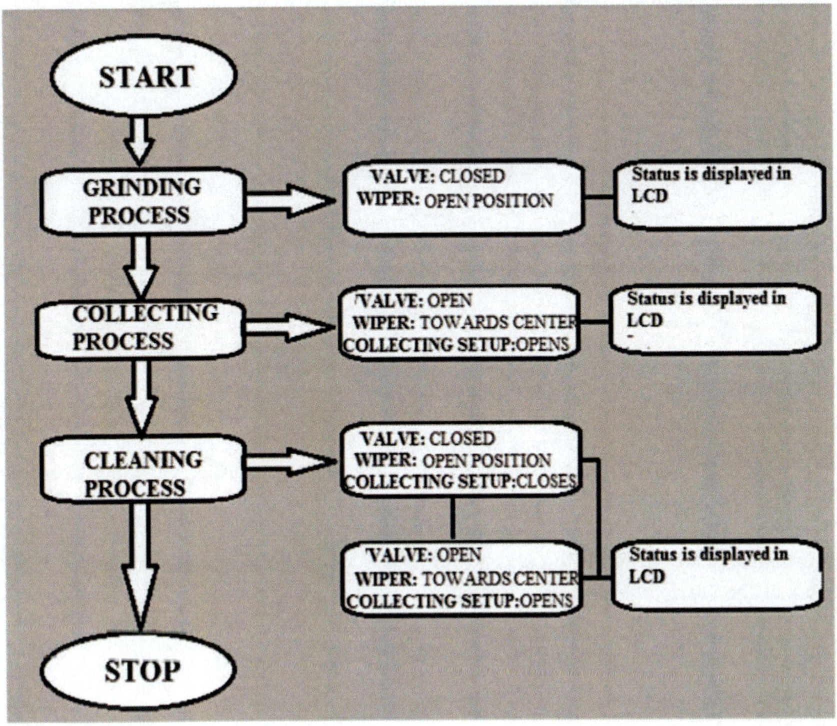

Fig. 5 Flow chart

5 Experimental Results

The device shown in Fig. 6 carries out three process namely grinding, collecting batter and cleaning the system. The grinding shown in Fig. 8 is carried out by the mechanical elements present in the grinder. For collecting and cleaning process the wiper and the valve is automated. The wiper, valve automation is done using the motor controlled through a micro processor which is shown in Fig. 7. Wiper is used

Fig. 6 Smart grinder

for directing the batter towards Valve and also to assist in cleaning. Valve controls the flow of batter to collecting vessel after the completion of grinding shown in Fig. 9. After the collection of batter, the wiper and valve assists in the cleaning of the device automatically by spraying the water and also by using the above mechanism. The status of the entire operation is displayed in the LCD which is shown in Fig. 10.

6 Conclusion

A Smart Grinder is proposed using ARM LPC2148 microprocessor which automates the collecting and cleaning process, which is absent in the existing wet grinder models. After grinding the rice/dhal into batter, the grinder automatically collects the batter and cleans the grinder. In commercial kitchen or large kitchen application the grinder can be incorporated with features to monitor its batter consistency using sensors and can also be monitored from a centralized location.

Fig. 7 Hardware connection of motor

Fig. 8 Grinding

Fig. 9 Collecting

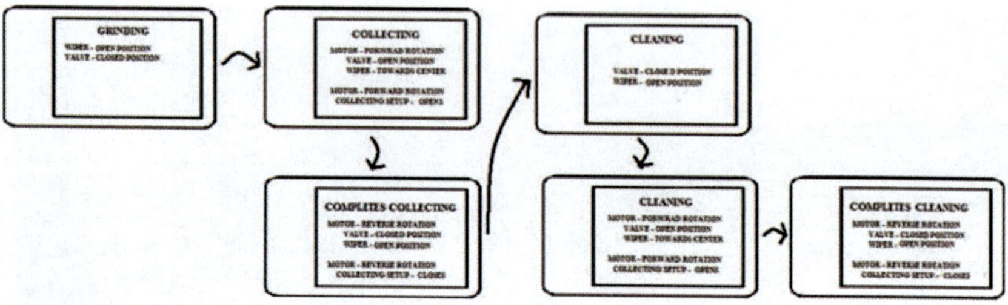

Fig. 10 LCD status display

References

1. Ngamnikom P (2011) The effects of freeze, dry and wet grinding processes on rice flour properties and their energy consumption. J Food Eng
2. Sharma P (2008) Grinding characteristics and batter quality of rice in different wet grinding systems. J Food Eng
3. Vijai Srinivas D (2017) Smart grinder: Indian patent application for smart grinder
4. Ramachandran G (2009) Wet grinder: Indian patent application for wet grinders
5. Doraiswamy RN (2008) A table top wet grinders: Indian patent application for table top wet grinders
6. Benhua Z, Chenghua L, Shiming S, Lu G (2011) Design on a unipolar and unidirectional stepper motor circuit. In: Proceedings of 2011 international conference on electronic & mechanical engineering and information technology, pp 1795–1797. https://doi.org/10.1109/emeit.2011.6023452
7. Soressi E (2012) New life for old compound DC motors in industrial applications. In: 2012 IEEE international conference on power electronics, drives and energy systems (PEDES), pp 1–6. https://doi.org/10.1109/pedes.2012.6484440
8. Lee C-C, Chiou G-J, Chen J-Y, Tung Y-C, Juang F-S (2017) Implementation of a novel brushless DC motor controller. In: 2017 12th IEEE conference on industrial electronics and applications (ICIEA), pp 1641–1645. https://doi.org/10.1109/iciea.2017.8283102
9. Sun J, Sun Q (2015) Design and simulation of PWM DC motor speed regulator based on proteus. In: 2015 international conference on fluid power and mechatronics (FPM), pp 1210–1213. https://doi.org/10.1109/fpm.2015.7337304
10. Sandilya KS, Eswari M, Remadevi G, Bini B, Madhavan A (2018) Design for a programmable alarm clock using alphanumeric display (PACAD). In: 2018 IEEE international conference on computational intelligence and computing research (ICCIC), pp 1–4. https://doi.org/10.1109/iccic.2018.8782289
11. Chaure R, Pande NA (2016) Design and implementation of ethernet based embedded network controller using ARM 7(LPC248) processor. In: 2016 world conference on futuristic trends in research and innovation for social welfare (startup conclave), pp 1–4. https://doi.org/10.1109/startup.2016.7583924
12. Mulay G, Yembarwar A, Raje S (2014) A DC motor driver consisting of a single MOSFET with capability of speed and direction control. In: 2014 IEEE 6th india international conference on power electronics (IICPE), pp 1–3. https://doi.org/10.1109/iicpe.2014.7115811
13. Aranjo B, Soori PK, Talukder P (2012) Stepper motor drives for robotic applications. In: 2012 IEEE international power engineering and optimization conference Melaka, Malaysia, pp 361–366. https://doi.org/10.1109/peoco.2012.6230890
14. Wang Y, Ma L, Jiang H (2019) Detecting conductive particles in TFT-LCD with U-MultiNet. In: 2019 8th international symposium on next generation electronics (ISNE), pp 1–3. https://doi.org/10.1109/isne.2019.8896458

Cardiovascular Disease Classification Using Different Algorithms

Rahul, Monika, Pranav Ray, Roshan Bapurao Kharke, and Saurav Singh Chauhan

Abstract The coronary heart stroke rates are increasing rapidly in people of all ages and gender. Cardiovascular diseases are posing a crucial and critical challenge and also the inaccurate prediction may lead to fatality. Contemporary prediction techniques like machine learning have been a useful approach in predicting these attacks with the help of the healthcare industry. In this paper, different methods are suggested to find a good-sized feature set by applying various prediction techniques which leads to enhancement of accuracy. The predictive model is delivered with various machine learning strategies.

Keywords Prediction · Heart disease · Symptoms · Machine learning · Neural network · Adaboost · Random forest · LightGBM · Gradient boosting · Gaussian Naïve bayes

Rahul · P. Ray · R. B. Kharke (✉) · S. S. Chauhan
Department of CSE, Delhi Technological University, Delhi 110042, India
e-mail: roshan.kharke1111@gmail.com

Rahul
e-mail: rahul@dtu.ac.in

P. Ray
e-mail: om.rai1997@gmail.com

S. S. Chauhan
e-mail: sauravchauhan00@gmail.com

Monika
Department of CS, Shaheed Rajguru College of Applied Sciences for Women University of Delhi, Delhi 110096, India
e-mail: monika.siwaliya@gmail.com

G. Ranganathan et al. (eds.), *Inventive Communication and Computational Technologies*, Lecture Notes in Networks and Systems 145,
https://doi.org/10.1007/978-981-15-7345-3_16

1 Introduction

The complexity of heart diseases and their associated problems affects the personal and social life of an individual and must be dealt very carefully. The heart strokes generally occur in people of age 50 or above is a misconception. In India, 50% of heart attacks occur to those who are below 50 years and 50% of these people are in the age group of 40–50 [1]. Lack of exercise, eating unhealthy food or junk food regularly increases the chance of getting heart strokes and hence these are the main reasons why people in urban areas are thrice more likely to become victims of heart attack than the people living in the rural areas. Eating junk foods will lead to obesity, leads to an increase in bad cholesterol and increases the chance of heart strokes.

The patients' previous data is calculated through clinical exams such as ECG and CT scan. There are many sensors to detect the needed factors and they are easily available nowadays in smartwatches, smartphones etc. and the data from these sensors will act as our record. After preprocessing the data, the dataset is divided into train and test. Next, a model is trained and checked the efficiency of our trained model and will run this model on the test data created earlier. This will give accuracy and the higher the accuracy means, the model will predict a patient's chance of getting a stroke more correctly. Now, multiple models are trained to get which model or algorithm works best.

2 Related Work

Plenty of research has been done to find out cardiovascular disorders in different patients. Statistical procedures and applications of information mining processes are some techniques being used by the researchers. The causes of risk elements for coronary heart disease as acknowledged by conducting statistical methods are total cholesterol, smoking, age, blood pressure, diabetes, hypertension, obesity and absence of exercises. Monitoring heart sickness is important for the prevention and healthcare of people with heart problems. Over time there has been a lot of research done in this field. Priyanka N and Dr. Pushpa Ravi Kumar in 2017 implemented Naïve Bayes and Decision Trees on UCI dataset [2]. In the place of missing values, they stored mean. In 2018 Aditi Gavhane et al. wrote an MLP classifier for predicting heart disease [3]. In 2019 Ahmed M. Alaa and et al. worked on a dataset of 423,604 patient's records where each record had 473 features. They developed an ML-based model using Auto Prognosis to improve the accuracy of already existing models [4]. Rahul and et al. in 2019 implemented feed-forward neural network and achieved an accuracy of 84% [5]. In 2019, Senthil Kumar Mohan et al. created a hybrid machine learning model called HRFLM (Hybrid Random Forest Linear Model) for cardiovascular disease classification and achieved 88.4% accuracy [6]. K. Subhadra and B. Vikas in 2019 took some models and compared their accuracy, specificity, sensitivity and precision [7]. Also in 2019, A. Golande and T. Pavan Kumar used different methods to predict heart disease (KNN, Decision Tree, AdaBoost and K means clustering) [8]. The data that determines the heart disease of a person is collected. During the early stages of the research main focus was put on the factors having a greater influence on the health of the heart. It was found that some of the factors affecting heart problems were unalterable like family background, sex and age etc. and then there were some factors which can be kept in check or control by following some rules or taking some preventive measures. These factors were diastolic pressure, heart rate and BMI etc. People who did not have a normal range of alterable factors were advised by doctors. The advice mainly consisted of taking healthy food, doing exercises regularly etc. Now of all the attributes playing part in the heart strokes, not all possessed a major threat or had a greater stake [9]. Those attributes are:

- **Hyper Cholesterol**
- **Blood Pressure**
- **Diabetes**
- **Heart Disease**
- **Body Mass Index (BMI)**
- **Age**
- **Sex**

Most of the researchers in this field work on the Cleveland dataset which can be obtained from the UCI Library [10] and the same has been used by us. This dataset uses a total of 76 parameters which collectively describes the condition of one's heart. All the parameters or the attributes in the dataset mentioned above obtained by conducting clinical exams which are CT scan, ECG etc. Of the plethora of the given attributes, not all are needed and only 13 of them are needed for the prediction system trying to design. Total of 303 data points or information of 303 patients were given. Out of these 303 records, there were incomplete records of six patients. In the case of missing values, the missing value can be replaced with the mean or median or mode of that particular attribute. Another thing that can be done in the case of missing values is to remove the data points or the records of those patients completely. Then there were some variables which were categorical which means that those variables or attributes or parameters were classified in some categories. Now, these categories were numbered like 1, 2, 3 and 4. The problem faced with such is that these may or may not be in precedence.

3 Proposed Approach

The Block Diagram shows the workflow of our model and the organization of our data (Fig. 1).

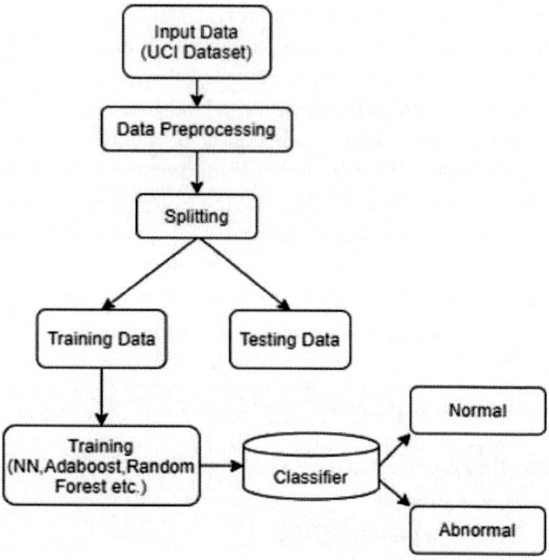

Fig. 1 Proposed data classification system

3.1 Data Preprocessing

The 6 records which had missing values were removed. In the dataset, some attributes have categorical values like Chest Pain (cp), Slope, Thal, and RestECG. In our model, used one-hot encoding to deal with categorical Values (Table 1).

3.2 Splitting

After preprocessing our data and before training our dataset is divided into two parts: train and test set. As the name suggests train, this set is employed for training the model. After training the model the model is tested by running it on a test set to check how accurate the model is and have 70% of the dataset for the train and 30% for tests and before splitting the dataset is shuffled.

3.3 Classification

The training dataset from above is trained using different techniques of machine learning, i.e., **Neural Network, Random Forest, Adaboost, Gradient Boosting, LGBM, Gaussian Naïve Bayes**. These were implemented on python using Jupyter Notebook with sci-kit learn, python library [11].

Table 1 Dataset description [10]

Features	Description	Type
Age	Age of person in years	Numeric
Sex	Gender of patient (male = 1, female = 0)	Nominal
Cp	Types of chest pain. There are 4 values of chest pain:	Nominal
	(1) Typical angina	
	(2) Atypical angina	
	(3) Non-anginal pain	
	(4) Asymptomatic	
trestbps	Blood pressure level at rest mode (in mm Hg when admitted to hospital)	Numeric
chol	Serum cholesterol in mg/dl	Numeric
fbs	If Fasting blood sugar >120 mg/dl value = l(true) else value = 0(false)	Nominal
restecg	Values from electrocardiogram while the patient is at rest represented by 3 values. Value 0 = Normal, Value 1 = Abnormality in ST-T wave (which is because of T wave inversions or/and ST depression or elevation of >0.05 mV	Nominal
thalach	Measurement of maximum heart rate that was achieved.	Numeric
exang	Angina caused by exercise 1 = yes; 0 = No	Nominal
oldpeak	ST depression caused by exercise correlative to rest	Numeric
slope	Gradient of the ST segment during exercise at peak. It has 3 values. 1 = Upsloping, 2 = Flat, 3 = Downsloping	Nominal
ca	Total count of major vessels coloured by doing fluoroscopy. Range 0–3	Numeric
thai	Condition of heart depicted by 3 values. Value3 = normal; Value 6 = fixed defect; Value 7 = reversible defect	Nominal
target	Diagnosis of the cardiovascular disease. Value 1 = Positive (has disease); Value 0 = Negative	Nominal

3.3.1 Neural Network

Neural Networks are mostly effective in medical predictions [12], Artificial Neural Networks contain the algorithms inspired by the structure of the brain and form the base for deep learning [13]. The core processing units of the neural network are the layers of neurons. The network, has an input layer, output layer and in the middle of the several hidden layers. Most of the computations required by the network are completed in that area. The Nodes (Neurons) of one layer are connected to the nodes of the other layers and each connection is assigned a weight which needs to be tuned as our training of network proceeds. After providing input values to the first layer a weighted sum is calculated and added along with the bias, this value is then passed through an activation function. The node with maximum value is involved in the output. Finally, the output layer represents the probability of the predicted class, i.e., one with maximum value. This Process is called Forward Propagation.

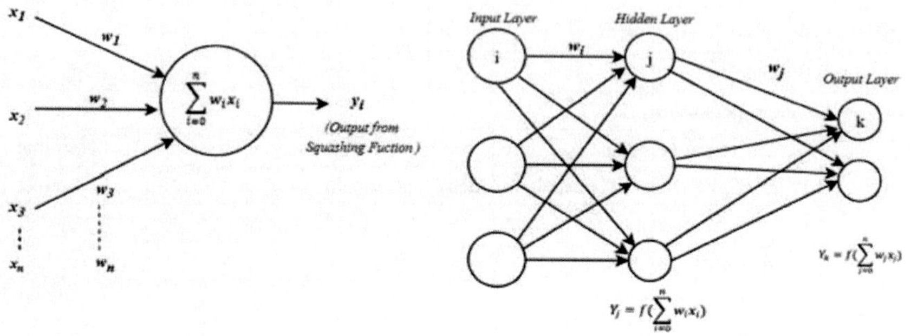

Fig. 2 Structure of neuron, neural network architecture

After this error is calculated and weights adjusted and sent back, this is called Back Propagation. The following formula is used for calculating the node value:

$$Y_k = f\left(\sum_{j=0}^{n} w_j x_j\right)$$

where f is the activation function (Fig. 2).

3.3.2 Random Forest

Random Forest merges the simple nature of decision trees and flexibility which increases the accuracy to a great extent. In the random forest method, many trees are grown instead of one single tree. Every tree gives a prediction result and then the trees vote for that class. The output from each tree in the random forest is evaluated and the class with maximum votes is known as our model's prediction (Fig. 3).

Syntax:

```
RandomForestClassifier(bootstrap=True, ccp_alpha=0.0, class_weight=None,
                       criterion='gini', max_depth=None, max_features='auto',
                       max_leaf_nodes=None, max_samples=None,
                       min_impurity_decrease=0.0, min_impurity_split=None,
                       min_samples_leaf=1, min_samples_split=2,
                       min_weight_fraction_leaf=0.0, n_estimators=50,
                       n_jobs=None, oob_score=False, random_state=0, verbose=0,
                       warm_start=False)
```

3.3.3 AdaBoost

In Random Forest, a full-size tree is made whereas in AdaBoost trees has a single node and two leaf nodes which are also called Stump. The stump can only use one

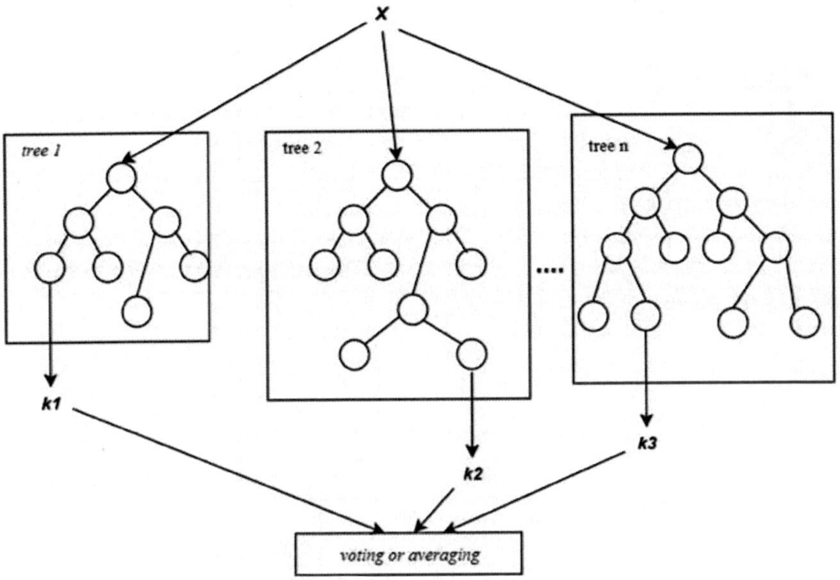

Fig. 3 Visualization of a random forest model making a prediction

variable for decision making thus, Stumps are weak learners. AdaBoost combines these individual stumps to increase the performance of the algorithm [14]. Each tree in Random Forest has an equal vote in final classification whereas in AdaBoost different stumps have different contributions or vote to the final classification. In AdaBoost order of the stumps are important, error made by the first stump influences how the second stump is made. Each sample has a sample weight, initially which is given by:

$$\text{weight}(x_i) = \frac{1}{n}$$

n is the total count of samples and x_i = ith instance of sample. Sample weight is manipulated according to the error.

Syntax:

```
AdaBoostClassifier(algorithm='SAMME', base_estimator=None, learning_rate=0.09,
                   n_estimators=20, random_state=None)
```

3.3.4 Gradient Boosting

Gradient Boosting enhances the performance of weak learners. In Gradient Boost first, a single leaf is made instead of tree or stump. The leaf represents an initial guess of weights of all the samples, then gradient boost builds the tree.

In this algorithm, trees are built which are of fixed size and the current tree considers the error of the previous tree. In this sense, it is similar to Adaboost. The part where gradient boost is different from Adaboost is that the size of the tree which is more than one root and two leaf nodes only. The shortcomings can be identified by using the gradient in the loss function ($y = mx + c + E$, E is the error term). The loss function is used to define how good our model's coefficients are.

Syntax:

```
GradientBoostingClassifier(ccp_alpha=0.0, criterion='friedman_mse', init=None,
                           learning_rate=0.5, loss='deviance', max_depth=3,
                           max_features=None, max_leaf_nodes=None,
                           min_impurity_decrease=0.0, min_impurity_split=None,
                           min_samples_leaf=1, min_samples_split=2,
                           min_weight_fraction_leaf=0.0, n_estimators=100,
                           n_iter_no_change=None, presort='deprecated',
                           random_state=None, subsample=1.0, tol=0.0001,
                           validation_fraction=0.1, verbose=0,
                           warm start=False)
```

3.3.5 LGBM

LightGBM is a fast and high-performance tree-based algorithm because the main time-consuming step in Decision tree-based algorithms is to find the best split points, LightGBM uses a histogram-based algorithm to find these split points since it is more efficient in both the speed of training and memory usage. LightGBM is faster than XGBoost and other GBDT Algorithms [15].

Syntax:

```
LGBMClassifier(boosting_type='gbdt', class_weight=None, colsample_bytree=1.0,
               importance_type='split', learning_rate=0.1, max_depth=1,
               min_child_samples=20, min_child_weight=0.001, min_split_gain=0.0,
               n_estimators=32, n_jobs=-1, num_leaves=31, objective=None,
               random_state=None, reg_alpha=0.0, reg_lambda=0.0, silent=True,
               subsample=1.0, subsample_for_bin=200000, subsample_freq=0)
```

3.3.6 Gaussian Naive Bayes

Gaussian Naive Bayes is a classification algorithm that employs Bayes Theorem. In Naive Bayes classifiers, every feature is independent of each other. In Gaussian Naive Bayes the distribution of our dataset is based on Gaussian distribution and it is easier than others because only needed to calculate the mean and standard deviation from our dataset during training [16].

$$P(M|x) = \frac{P(x|M)P(M)}{P(x)}$$

Syntax:

```
GaussianNB(priors=None, var_smoothing=1e-09)
```

Confusion Matrix Confusion matrix is a layout in tabular form which shows the statistics of output of our prediction method and denotes and helps us to graphically visualize the performance of our model.

Implementation
Random Forest Classifier, Gradient Boosting Classifier, AdaBoost Classifier methods were imported from sklearn.ensemble API while Gaussian NB was imported from sklearn.naive_bayes API. For using LGBM Classifier had to install lgbm first which was done using.

```
conda install -c conda-forge lightgbm
```

Model.fit() and model.score() was used for every model to train and get accuracy, respectively.

Neural Networks was implemented using tensorflow, and added 1 input, 2 hidden, and 1 output layers. Then, compiled the model, trained, and got accuracy using model.compile(), model.fit(), model.evaluate(), respectively.

4 Result and Discussion

After implementing the algorithms on our dataset, the models are tested on the test dataset and obtained the following result and can see from the results that Gaussian Naive Bayes, AdaBoost and Neural Network are the best classifiers for this task. The reason why Naïve Bayes is doing so well is that the parameters are independent and the size of the dataset is less and Naïve Bayes is a classifier which works great with fewer data. The Neural Network model here has 2 hidden layers and does have backward propagation because of which weights are being updated. Adaboost works better because it combines different stumps (trees with one root node and two leaf nodes) with different weights and current stump takes previous one's mistake into account.

Confusion Matrices showing calculations of the proposed algorithms (Tables 2, 3, 4, 5, 6, 7 and 8; Fig. 4).

Table 2 Neural network

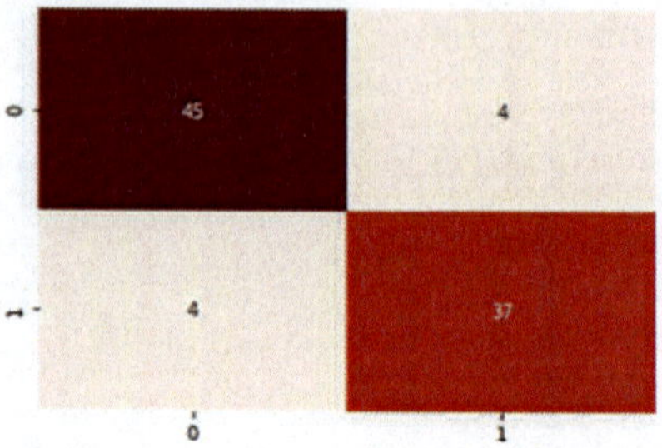

Table 3 Random forest

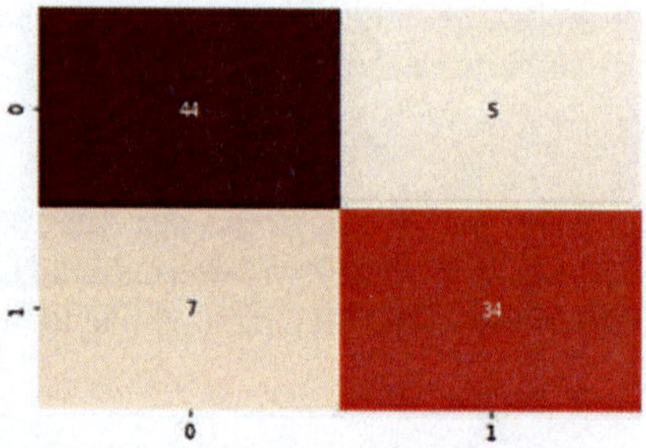

Table 4 AdaBoost classifier

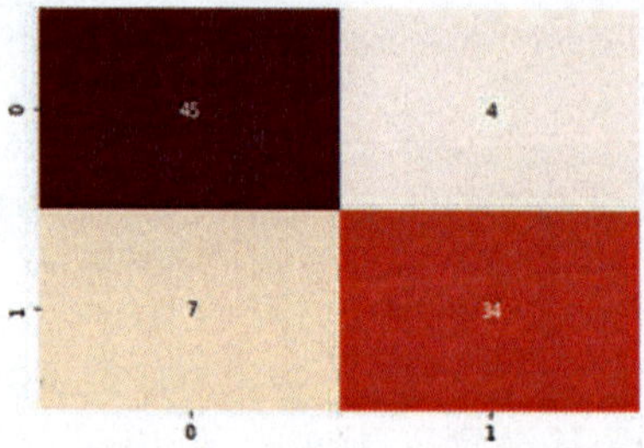

Table 5 Gradient boosting

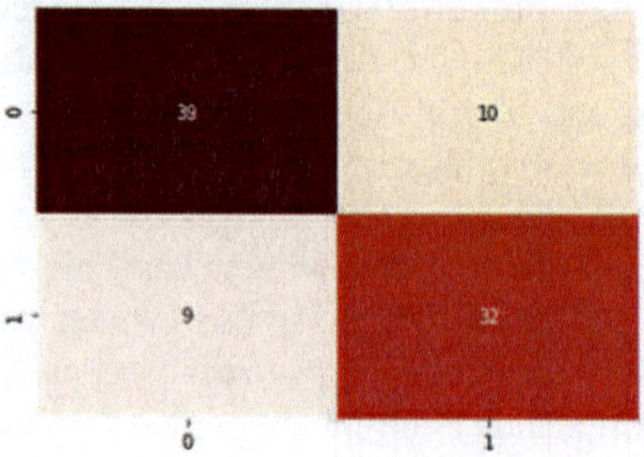

Table 6 LGBM

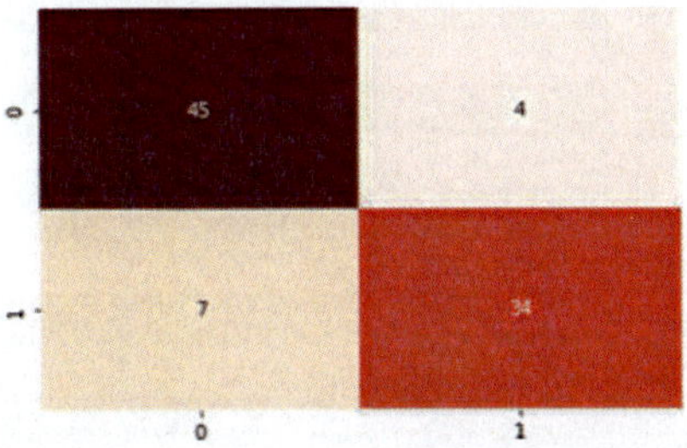

Table 7 Gaussian Naïve Bayes

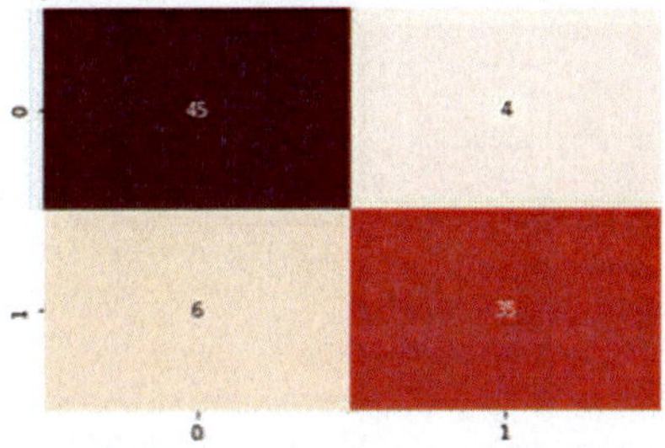

Table 8 Table for sensitivity, specificity and precision for different models

	Neural Network	*Random Forest Classifier*	*AdaBoost Classifier*	*Gradient Boosting Classifier*	*LightGBM Classifier*	*Gaussian Naïve Bayes*
Specificity	0.92	0.92	0.92	0.8	0.92	0.92
Sensitivity	0.9	0.83	0.78	0.83	0.83	0.85
Precision	0.9	0.89	0.76	0.89	0.85	0.9

5 Conclusion

In this paper, different prediction techniques that can predict heart disease is studied using a dataset. The processing of data will help save the lives of people by detecting heart conditions early. Heart stroke prediction is a challenging yet important task in the field of medicine. Evolved techniques like Adaboost, Random Forest etc. have been used. The accuracy of this model can be increased by using different composition of machine learning approaches by tuning the parameters. The death rate can be gradually decreased if more and more people try to use this system.

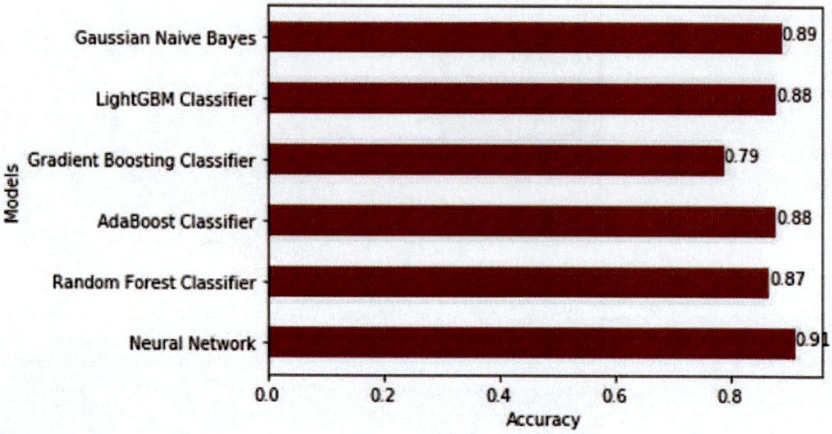

Fig. 4 Graphical representation of accuracy

6 Future Work

Similar prediction and classification systems for other diseases such as Cancer, Diabetes etc. can be built using recent Machine learning technologies. New algorithms can also be proposed to improve prediction accuracy. Android applications and devices can be made which employ these models to quickly give results based on input data from the user. This can act as a portable and fast Heart disease detector.

References

1. Vaideeswar P, Tyagi S, Singaravel S (2019) Pathology of atherosclerotic coronary artery disease in the young Indian population. Forensic Sci Res 4(3):241–246. https://doi.org/10.1080/20961790.2019.1592315
2. Priyanka N, Ravikumar P (2017) Usage of data mining techniques in predicting the heart diseases—Naïve Bayes & decision tree. In: Proceedings of IEEE international conference on circuit, power and computing technologies ICCPCT 2017. https://doi.org/10.1109/iccpct.2017.8074215
3. Aditi Gavhane KD, Kokkula G, Pandya I (2017) Prediction of heart disease using supervised learning algorithms. Int J Comput Appl 165(5):41–44. https://doi.org/10.5120/ijca2017913868
4. Alaa AM, Bolton T, Di Angelantonio E, Rudd JHF, van der Schaar M (2019) Cardiovascular disease risk prediction using automated machine learning: a prospective study of 423,604 UK Biobank participants. PLoS ONE 14(5):1–17. https://doi.org/10.1371/journal.pone.0213653
5. Rahul, Bansal H, Monika (2020) Classification techniques used in sentiment analysis & prediction of heart disease using data mining techniques: review, pp 1–6. https://doi.org/10.1109/icict46931.2019.8977707
6. Mohan S, Thirumalai C, Srivastava G (2019) Effective heart disease prediction using hybrid machine learning techniques. IEEE Access 7:81542–81554. https://doi.org/10.1109/ACCESS.2019.2923707

7. Subhadra K, Vikas B (2019) Neural network based intelligent system for predicting heart disease. Int J Innov Technol Explor Eng 8(5):484–487
8. Golande A, Pavan Kumar T (2019) Heart disease prediction using effective machine learning techniques. Int J Recent Technol Eng 8(1) Special Issue 4:944–950
9. Heart-Health Screenings|American Heart Association. https://www.heart.org/en/health-topics/consumer-healthcare/what-is-cardiovascular-disease/heart-health-screenings
10. UCI machine learning repository: heart disease data set. http://archive.ics.uci.edu/ml/datasets/Heart+Disease
11. scikit-learn: machine learning in Python—scikit-learn 0.22.2 documentation. https://scikit-learn.org/stable/
12. Karaduzovic-Hadziabdic K, Köker R (2015) Diagnosis of heart disease using a committee machine neural network, vol 1, pp 351–360. https://doi.org/10.14794/icai.9.2014.1.351
13. Vieira S, Pinaya WHL, Mechelli A (2017) Using deep learning to investigate the neuroimaging correlates of psychiatric and neurological disorders: methods and applications. Neurosci Biobehav Rev 74:58–75. https://doi.org/10.1016/j.neubiorev.2017.01.002
14. Wu P, Zhao H (2011) Some analysis and research of the AdaBoost algorithm. Commun Comput Inf Sci 134(1):3–5. https://doi.org/10.1007/978-3-642-18129-0_1
15. Ke G et al (2017) LightGBM: a highly efficient gradient boosting decision tree. In: Advances in neural information processing systems, vol 2017-Decem, no. Nips, pp 3147–3155
16. Gayathri BM, Sumathi CP (2016) An automated technique using Gaussian naive Bayes classifier to classify breast cancer. Int J Comput Appl 148(6):16–21

Empirical Test Design Strategies Using Natural Language Processing

T. S. Abishek, Adithya Viswanathan, Akash Kumar Pujari, and V. S. Felix Enigo

Abstract With the rise in the role played by computers and their ubiquity, both developers and consumers share responsibility for improvements in the digital platform. Thus, there is a need to reduce the load on both ends with the help of automation. This proposed system, provides a simplistic interface, where users can submit issues with applications and developers are provided with test cases before writing off on the solutions to these issues. It aims to intelligently identify preexisting similar issues, if any, to reduce redundancy, identify system requirements unsatisfied by these issues as well as identifying the criticality of the issues. It also seeks to generate test cases for functional requirements. The system uses a 'Deep Averaging Network (DAN)' model developed by Google for sentence similarity to detect similar issues, providing an accuracy of upwards of 89% for an in-house prepared test set of more than 100 entries. It also uses sentence dependencies and a 'Satisfiability Modulo Theories (SMT)' solver, for identifying parts of speech and generates appropriate test cases.

Keywords Test case generation · Natural Language Processing · Natural language understanding · Dependency parsing · Sentence similarity · Requirement analysis · Deep averaging network

T. S. Abishek · A. Viswanathan · A. K. Pujari (✉) · V. S. Felix Enigo
Department of CSE, SSN College of Engineering, Chennai, India
e-mail: akash16007@cse.ssn.edu.in

T. S. Abishek
e-mail: abishekshyamsunder@gmail.com

A. Viswanathan
e-mail: adithya16002@cse.ssn.edu.in

V. S. Felix Enigo
e-mail: felixvs@ssn.edu.in

G. Ranganathan et al. (eds.), *Inventive Communication and Computational Technologies*, Lecture Notes in Networks and Systems 145,
https://doi.org/10.1007/978-981-15-7345-3_17

1 Introduction

It is common knowledge that software development is an iterative process, with the developers constantly adding functionalities, fixing bugs, analysing usage and holistically improving the product with each iteration. Almost every digital service provider such as Apple, Mozilla Foundation, Google, Amazon, etc. all diagnose issues submitted by their users and products aimed at refinement. Thus, one can understand that a combined effort from both users as well as developers is essential. The problem, however, doesn't end there. Every time a new version of a software/application is released, it is imperative that it is subject to rigorous application testing, which is aimed at finding bugs in the software as well as in its dependencies.

The testing phase also includes the creation of a test plan, the process of Test Case Estimation as well as Test Case Generation explained in detail in the work done by Myers [1] On an average, the development of a test case takes about 10 min, and in general, development of a test plan without test cases and its review can take about 2–3 days. A cumulation of these factors creates a necessity for a system, that if not removes, at least reduces the time and energy spent on this.

Thus, the system is developed with the following objectives in mind:

- The system must notify the user submitting an issue if another similar issue already exists in the database.
- The system must automatically gauge the criticality of the issue being submitted by the user based on keywords.
- The system must map the issue being submitted, to the requirement that it violates.
- The system must generate test cases for functional requirements, to ease the workload of the developers.

The last seven decades have seen a rapid increase in the importance of Natural Language Processing and its various applications. Sentence Similarity is an important subset of NLP which itself has several applications today. It evolved from the computation of extended cosine distance between the vectorised forms of sentences to the use of various neural networks to group similar sentences [2]. The current trend is the use of LSTM, to measure the similarity between sentences [3]. A comprehensive study of various deep learning neural network architectures utilised in various applications is given in the work done by Bashar [4].

However, there are a few disadvantages to the use of such Recurrent neural networks. They need a large amount of training time. As the dimensionality increases, the computations at each node of the parse tree become expensive. They also require consistent syntax between training and test data due to their reliance on parse trees and thus cannot effectively incorporate out of domain data.

There is also a steady increase in the use of NLU, which is a subset of NLP to create entities such as chatbots, to provide tailored responses to inputs. A theoretical interpretation of Natural Language Understanding can be obtained from the work done by Shank [5] as early as 1972. Rasa [6], which provides an open-source conversational AI framework for building contextual assistants is one of the industry's leading service providers. Their core functionalities can be ported for obtaining responses based on keywords, by training with developer provided data corpus.

Every word in a sentence is grammatically linked to other words in the sentence. This phenomenon is known as sentence dependency and is very useful in extracting valuable information both semantically and structurally. Dependency parsing is strongly linked to POS tagging, which is also a complicated task because the contexts of the same words vary with sentences. An overview of increasing the accuracy of tagging words is given in the work done by Manning [7].

Satisfiability Modulo Theories plays a pivotal role in this work. The language of SMT solvers is First-Order Logic(FOL) where the symbols are divided into logical symbols and non-logical symbols or parameters. In the scope of this research, SMT solvers are used to providing a model that satisfies a set of asserted constraints. The work done by Dwarakanath and Sengupta [8] provides a comprehensive method for the generation of test cases using link grammar. However, the collusion of sentence dependencies and SMT solver will serve as a powerful tool to generate test cases for scenarios with multiple dependant constraints.

The work done by Dwarakanath and Sengupta [8] provides a comprehensive method for the generation of test cases using link grammar. However, the collusion of sentence dependencies and SMT solver will serve as a powerful tool to generate test cases for scenarios with multiple dependant constraints.

1.1 Structure of Recommendation by the System

The system provides its recommendation to the user in the form of three simple sentences.

- The first sentence notifies the user of any similar issues identified.
- The second sentence informs the user of the requirement unsatisfied by the issue being submitted.
- The third sentence automatically recommends the criticality that is to be associated with the issue being submitted.

The user submits their issues via a form in an HTML based GUI. Thus, these recommendations are automatically projected in the interface before the user submits the form.

1.2 Figures and Tables

The figures in this chapter, serve as a visual aid to complement the information presented in the work. Figure 1 gives us an insight into the logical structure of the system. Figures 2 and 3 help us visualise how text is converted to a vectorised form. Figure 4 highlights a similar issue matched by the system. Figure 5 gives a flowchart

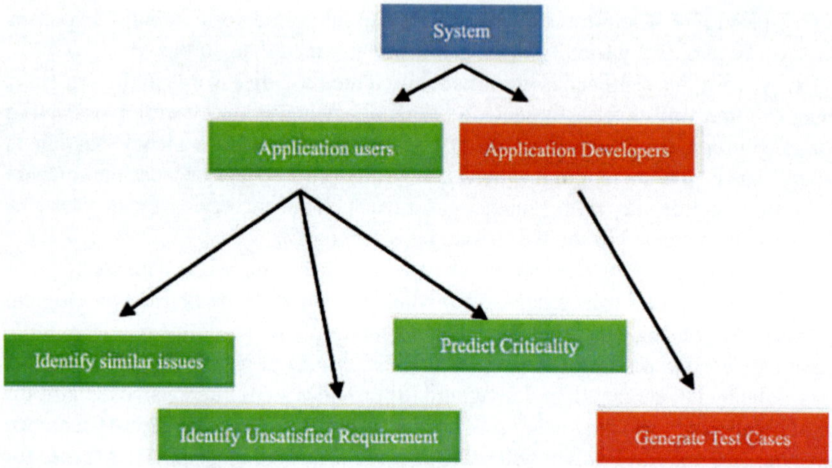

Fig. 1 Abstract system architecture

Fig. 2 One hot encoding

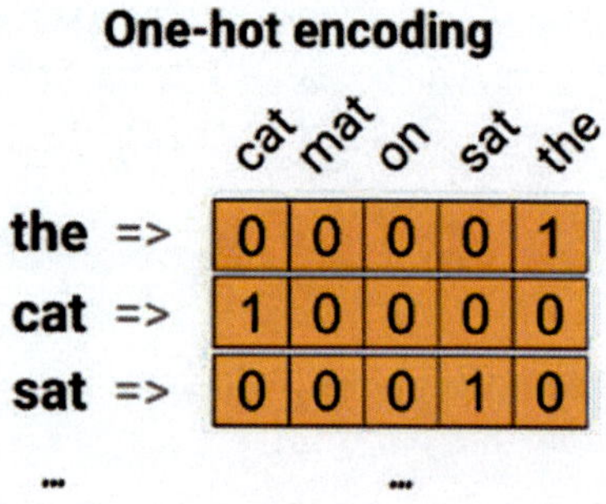

Fig. 3 Word embedding

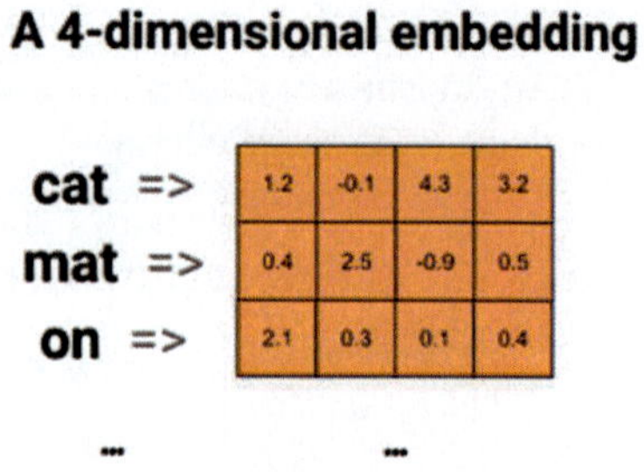

of how the response to a message is obtained in RASA. Figure 6 gives us a sample of a criticality recommendation. Figure 7 gives us a parsed and tagged sentence which is dissected in Sect. 4.

Table 1 gives us a list of POS tags while Table 2 provides information about the accuracy of universal sentence encoder model for relevant test cases.

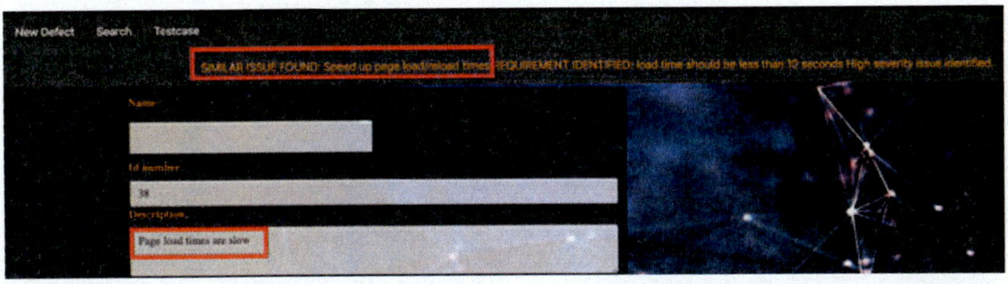

Fig. 4 Sample issue recommendation

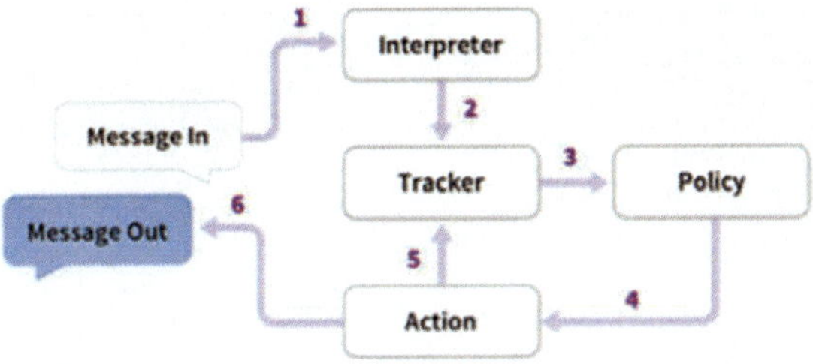

Fig. 5 RASA architecture

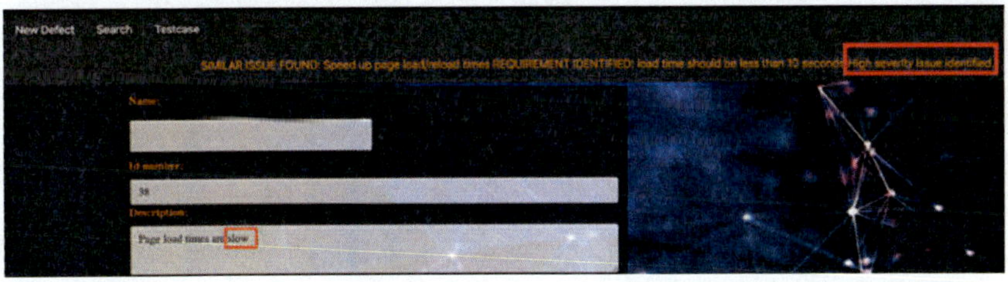

Fig. 6 Sample criticality recommendation

2 Sentence Similarity

Sentence Similarity or Semantic Textual Similarity is a measure of how similar two pieces of text are, or to what degree they express the same meaning. This is broadly useful in obtaining good coverage over the numerous ways that a thought can be expressed using language without needing to manually enumerate them. It is widely used in the domain of Natural Language Processing. The process of identifying similar sentences usually entails that the text is first converted into high dimensional vectors. This process is known as embedding.

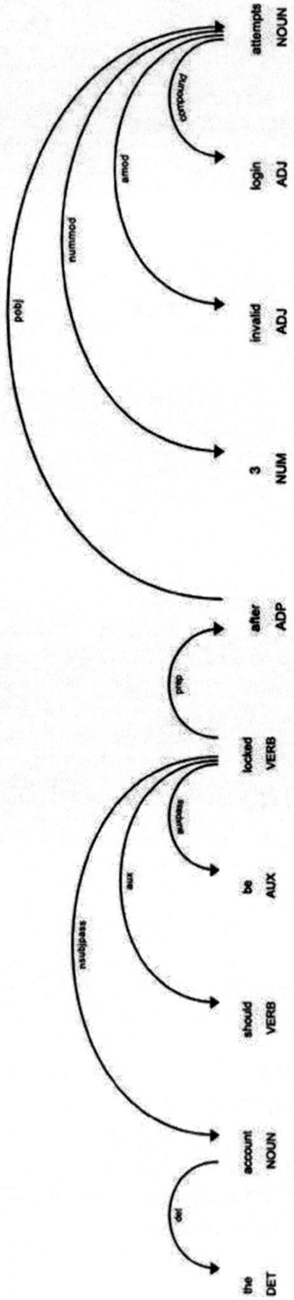

Fig. 7 Example dependency

Table 1 Universal POS tags

Open class words	Closed class words	Other
ADJ	ADP	PUNCT
ADV	AUX	SYM
INTJ	CCONJ	X
NOUN	DET	
PROPN	NUM	
VERB	PART	
	PRON	
	SCONJ	

Table 2 Accuracy table

Person ID synonymous sentences matched accuracy in percent			
	Written	Correctly	Age (%)
1	28	27	96.43
2	28	26	92.86
3	28	27	96.43
4	28	23	82.14
5	28	23	82.14
Total	140	126	90.00

2.1 Embedding of Text

There are three strategies that one can use to embed text

- One-hot encoding—Here, to represent each word, we will create a zero vector with length equal to the vocabulary, then place a one in the index that corresponds to the word. This process is very inefficient.
- Encode each word with a unique number—In this approach, each word is encoded using a unique number. Thus, instead of a sparse vector, now have a dense one. However, this is not recommended as well because it does not properly represent the relationships between the words and is also very difficult to interpret.
- Word Embeddings—This gives us a way to use an efficient, dense representation in which similar words have a similar encoding. The most important factor to be considered here is that do not have to manually specify the encodings by hand. These can be trainable parameters (weights learned by the model during training).

The model used here is the universal sentence encoder model developed by Google [9].

2.2 Understanding Deep Averaging Networks

A basic idea of DAN can be obtained from the work done by Iyyer et al. [10]. The intuition behind feedforward neural networks is that each layer learns a more abstract representation of the input than the previous one. Thus, the model is expected to identify sentences that are 'closer' to one another in meaning, despite their differences in the word embedding average being small. Consider a scenario where given three sentences, 'I love eating pizza', 'I like eating pizza' and 'I hate eating pizza' named s1, s2 and s3, respectively. The vector averages of these three sentences will most definitely be identical. However, the averages associated with sentences s1 and s2 will be slightly more similar to one another when compared to s3 because s1 and s2 convey a similar meaning.

The word embedding model used was trained with unsupervised data sourced from a variety of web sources which was then augmented with supervised data from the Stanford Natural Language inference corpus.

DAN is a model that classifies a group of words in a documented sentence into various classes. It represents the words as a continuous bag of words, i.e. Each word is represented as an embedding which is an *n*-dimensional vector.

$$\text{CBOW}(w_1, \ldots w_N) = \sum_i E[W_i]$$

$$z_0 = \text{CBOW}(w_1, \ldots, w_N)$$

DAN takes this representation of the document and applies a linear transformation to it, followed by passing it through multiple layers of non-linearities, element-wise to get the final output.

$$z_1 = g(W_1 \cdot z_0 + b_1)$$
$$z_1 = g(W_1 \cdot z_0 + b_1)$$
$$z_1 = g(W_1 \cdot z_0 + b_1)$$

Practical use of this module requires a download of a 1 GB file to be downloaded the first time it is instantiated. Thus, the time taken to load may vary with the user's network speed. However, because the modules are cached by default, reloads do not consume as much time. Furthermore, once the module is loaded onto memory, inference time should be relatively fast.

In general text preprocessing involves steps such as tokenisation, removal of HTML tags, removal of extra white spaces, conversion of accented characters to ASCII characters, expansion of contractions, removal of special characters, normalisation (converting text to lower case or upper case), convert number words to numeric form, removal of numbers, removal of stop words and Lemmatisation. However, this module performs best effort text input preprocessing. Thus, it is not required to pre-process the data before applying it to the module.

2.3 Recommendation by the System

When a user uses our system, they try to submit issues/bugs which they have identified while using a particular software. There is a chance that another user may have already submitted an issue regarding this problem. Thus, the system must use sentence similarity models to identify these similar issues and prompt the user asking them if they still want to submit. The system also uses sentence similarity to identify requirements that have not been satisfied by these issues. The system based on DJANGO uses AJAX calls to dynamically prompt helpful information to the user. From the example given in 4, when an issue 'Page Load times are slow', the system matches it with an already existing issue, 'Speed up page load/reload times' even though there is a considerable difference in the way both the sentences are structured.

There also exists an option for the users to register if the predictions by the system were right, which can then be stored and used later of analytical purposes.

3 RASA and Natural Language Understanding

To understand why Natural Language Processing was used to create this part of the system, one must understand the necessity that the system was built upon. When a user inputs an issue, the system had to intelligently categorise the issue into levels of criticality. This required the extraction of specific keywords and taking a course of action according to these.

3.1 Rasa

Incidentally, chatbots performed this very same task, just for a different use case and thus their functionality was ported into our project to serve a different purpose. RASA provides an open-source machine learning framework to help build contextual AI assistants and chatbots and also enable scaling with their enterprise-grade platform.

The steps involved in how a response to a message is obtained are:

1. The message is received and forwarded to the Interpreter, which translates it into a python dictionary, including the original text, the purpose, and any entities that have been defined. This part is being handled by the NLU.
2. Tracker is the tool that keeps track of the state of conversation. It receives information that a new message is coming in.
3. The strategy is focused on the current state of the tracker.
4. The strategy must determine which action to take next.
5. The selected activity is signed into the server.
6. The answer will be sent to the user.

The basic work performed by rasa is known as intent classification, which takes a sentence such as, 'I am looking for a Mexican restaurant in the centre of town' and returns structured data such as,


```
{
  "intent": "search_restaurant",
  "entities": {
    "cuisine" : "Mexican",
    "location" : "center"
  }
}
```


3.2 *Mechanism Behind RASA*

Working with intents can be described by two principal components which are featurizers and intent, classification models. Featurizers such as CountVectorsFeaturizer or SpacyFeaturizer take words separately as tokens and encode them as vectors to represent them as numeric encodings based on multiple attributes. EmbeddingIntentClassifier and SklearnIntentClassifier are different intent classification models that are present On the other hand the intent classification models use the output from the featurizer and uses them to make a prediction about which intent matches the corresponding message. The output is represented as a ranked list of intent predictions. Rasa's domain is an essential component of a Rasa dialogue management model. It controls the following requirements including

- What meaning does the user convey?
- What responses the model can provide?

The domain contains Intents, Stories, Actions, Templates, Entities and Slots Using an NLU model, a section named Intents is created to define a list of intents that the Rasa can understand.

Actions are the things that the bot runs in response to the input.

The templates sections are where the text responses are defined for Rasa to respond using the utterance predicted by the dialogue management model. More than one template can be used for an utterance.

Entities are useful and influence how the Rasa responds to the users' input, they take into account the location and facility type, etc.

Slots work as the memory of Rasa and are used to remember the important details of the inputs specified. They act as a key-value pair to store information critical to the inputs. They can be text, bool, categorical, float, list, unfeatured

The training for Rasa is by default set to 100 epochs.

As an example, consider a project which has three levels of criticality for its issues, moderate, high and critical, respectively. By associating keywords such as 'improve', 'slow' and 'crash' to the three categories, respectively in the NLU markdown file, associate specific words with intents. Consider the same example given in Sect. 2.3, represented by Fig. 6. Given that the issue being input has the word slow, the predicted criticality is 'high'.

4 Dependency Parsing

Dependency parsing is the task of extracting a dependency parse of the sentence that represents its grammatical structure and defines the relationship between 'head' words and words which modify those heads. A Dependency parser transforms a sentence into a dependency tree. A dependency tree is a structure that can be defined as a directed graph with 'V' nodes corresponding to the words and 'A' arcs corresponding to the syntactic dependencies between them. Labels are also attributed to dependencies called relations. These relations give details about the dependency type.

A very important part of NLP that must be addressed before moving on to the working of sentence dependencies is POS tagging. This refers to Parts Of Speech Tagging and it is the process of marking up a word in a corpus to a corresponding port of a speech tag, based on its context and definition. This task is not straightforward, as a particular word may have a different part of speech based on the context in which the word is used. For example, in the sentence 'Give me your answer', the answer is a Noun, but in the sentence 'Answer the question', the answer is a verb. In the case of the spaCy python module, the fastest syntactic parser in the world [11], which is used in the system, a statistical model comes in, which enables spaCy to predict which tag or label most likely applies in this context.

A model consists of binary data and is produced by showing a system enough samples for it to make predictions that generalise across the language. Table 1 gives us a sample of commonly used universal POS tags.

The scope of the system is limited to the generation of test cases for functional requirements which have numerical parameters to be satisfied. With this in mind, it is easy to understand that the most important dependency is the nummod dependency which stands for the numeric modifier. Let us first define a few terms which will help us better understand how information is extracted from a sentence using dependencies.

- Variable—An entity that is constrained by the parameter
- Parameter—Numerical value which serves as the start of the search
- Subject—The subject of the sentence
- Object—The object of the sentence
- Verb/Auxiliary term—The action word that is linked to both the subject and the object.

The number in the sentence is identified and set it as the parameter while the head term is the nummod of along with those linked to it, delimited by the verb are together considered to be a single entity known as the variable.

In all sentences, the action word that is linked to the subject is considered as the verb and the sentence may or may not have an object. This can be clearly understood with an example:

Consider the sentence 'The account should be locked after three invalid login attempts' represented by Fig. 7. Here the verb is locked, and it is linked to the word account with the dependency 'nsubpass' and thus the word account is considered to be the subject. 3 is the numerical modifier of the word attempts which in turn is linked to the words invalid and login, thus 3 is taken as the parameter while the words 'invalid login attempts' are taken as the variable.

In this example, the sentence has no object. Thus, any word linked with a dependency containing sub-strings 'sub' and 'obj' to the verb is the subject and object, respectively.

This information from the sentence is explicitly extracted because once this information is passed to an SMT solver and the output is received, a legible test case must be constructed to be submitted to the developers.

Given that the probability of representing numbers as words while writing the requirements is very high, the system has the capability of identifying these and converting them to numeric form, before feeding it to the SMT solver, discussed in Chap. 6. For example, when parsing the sentence:

'The battery percentage must be greater than thirty-three' the system will recognise thirty-three to be number 33. This, in turn, works with the python module 'word2number' used along with in the SMT solver.

It is also important to note that there may be requirements for which the system is not capable of generating test cases. Thus, the system identifies these cases and notifies the user of the same.

5 SMT Solver

5.1 Z3

In computer science and mathematical logic, the Satisfiability Modulo Theories problem is a decision problem for logical formulas concerning combinations and background theories expressed in classical first-order logic with equality. There are many available SMT solvers and the system uses the Z3 [12], which has built-in theories such as empty theory, linear arithmetic, non-linear arithmetic, bit-vectors, arrays, data types, quantifiers, strings, etc. It also has APIs in C/C++, .NET, Python, Java, etc. It is developed by Microsoft research and is used in many applications such as software/hardware verification and testing, constraint solving, analysis of hybrid systems, security and geometrical problems.

In terms of the use cases that the system will be subjected to, an SMT solver is required to provide a solution for the set of asserted constraints. The solution is a model for the set of asserted constraints. A model is an interpretation that makes each asserted constraint true.

5.2 Test Case Generation

Given that test cases must be generated for all possible scenarios such as correct inputs, corner cases, etc. even though the constraint is extracted from the input sentence using sentence dependency, the test cases are generated for all permutations and combinations. The output of the SMT solver is an assignment. This must then be used to construct a test case that can easily be understood by the developer.

One might then wonder why to go to the trouble of extracting the constraints and various other information using dependency parsing. When the test cases are presented in the interface, they are divided into two sets. One containing the positive and corner test cases while the other containing the negative ones.

However, if one understands the workings of Z3, they will be able to see that for a given set of constraints, the model created will always be the same. This drawback was overcome by using a random selection of multiple generated test cases. Rather than adding constraints to the z3 solver with multiple iterations, where each constraint imposed a new rule where the model could not generate the same solution as the previous iteration, the solver was set to create a model for a fixed number of iterations, where each iteration had different constraints altogether. From the final set of models created by this process, the required number of positive, negative and corner cases are chosen.

6 Results and Analysis

Accuracy of the model is defined here as the number of sentences matched correctly to their corresponding similar sentence, divided by the total number of input sentences. The total accuracy of the proposed system is 90% With 126 out of 140 sentences matched correctly.

It is important to note that the test sentences were extracted from the bug list of an anonymous project. To maintain the diversity of sentence structure and writing style in this set, the similar sentences were sourced from five people from different backgrounds, and the split up is given in Table 2.

The system provides the users with the functionality to define their levels of the criticality of issues, based on certain keywords. For this, the system uses 'Rasa', a chatbot that generates responses based on the pre-defined keywords matched in the parsed sentence. The responses generated are modified in the code to return the severity of the issue. Rasa also has an added advantage of a very small training time required.

7 Conclusion and Future Work

This paper provides an extensive report on four deliverables that have been satisfied by the system as mentioned in Sect. 1.

The system uses a DAN model over an LSTM model mainly because DAN's perform better in scenarios where the training sentences and testing sentences vary in structure. This is highly relevant to this project as there is a high probability that it will be employed in highly technical scenarios with a diverse input corpus.

There are also many ongoing research works on the use of Capsule neural networks for classification due to their enhanced performance as mentioned in the work done by Vijayakumar T [13]. Albeit, the work covers the prowess of CapsNet in image processing, future research may pave the way for challenging NLP applications.

The combination of sentence dependency parsing and SMT provides a powerful tool that can take in multiple related constraints and still provide relevant test cases. However, the system is currently limited to generating test cases for functional requirements that have numeric dependencies only.

Development of a module that generates test cases for non-functional requirements as well is quintessential and will ensure maximum usability and reach of the product.

References

1. Myers GJ, Sandler C, Badgett T (2011) The art of software testing. Wiley
2. Mikawa K, Ishida T, Goto M (2011) A proposal of extended cosine measure for distance metric learning in text classification. In: 2011 IEEE international conference on systems, man, and cybernetics. IEEE
3. Mueller J, Thyagarajan A (2016) Siamese recurrent architectures for learning sentence similarity. In: Thirtieth AAAI conference on artificial intelligence
4. Bashar A (2019) Survey on evolving deep learning neural network architectures. J Artif Intell 1(02):73–82
5. Schank RC (1972) Conceptual dependency: a theory of natural language understanding. Cogn Psychol 3(4):552–631
6. Bocklisch T et al (2017) Rasa: open source language understanding and dialogue management. arXiv preprint arXiv:1712.05181
7. Manning CD (2011) Part-of-speech tagging from 97% to 100%: is it time for some linguistics? In: International conference on intelligent text processing and computational linguistics. Springer, Berlin, Heidelberg
8. Dwarakanath A, Sengupta S (2012) Litmus: generation of test cases from functional requirements in natural language. In: International conference on application of natural language to information systems. Springer, Berlin, Heidelberg

9. Cer D, Yang Y, Kong SY, Hua N, Limtiaco N, John RS, Constant N, Guajardo-Cespedes M, Yuan S, Tar C, Sung YH (2018) Universal sentence encoder. arXiv preprint arXiv:1803.11175
10. Iyyer M, Manjunatha V, Boyd-Graber J, Daume III H (2015) Deep unordered composition rivals syntactic methods for text classification. In: Proceedings of the 53rd annual meeting of the association for computational linguistics and the 7th international joint conference on natural language processing (volume 1: long papers)
11. Choi JD, Tetreault J, Stent A (2015) It depends: dependency parser comparison using a web-based evaluation tool. In: Proceedings of the 53rd annual meeting of the association for computational linguistics and the 7th international joint conference on natural language processing (volume 1: long papers)
12. De Moura L, Bjørner N (2008) Z3: an efficient SMT solver. In: International conference on tools and algorithms for the construction and analysis of systems. Springer, Berlin, Heidelberg
13. Vijayakumar T (2019) Comparative study of capsule neural network in various applications. J Artif Intell 1(01):19–27

Sign Language to Text Conversion Using Deep Learning

P. V. S. M. S. Kartik, Konjeti B. V. N. S. Sumanth, V. N. V. Sri Ram, and P. Prakash

Abstract Sign Languages also popularly called Signed Languages. This is the only language by which deaf and dumb can communicate with each other and one of the most unexplored areas. So the proposed paper represents a model that can be used to translate given sign language to simple English text so that it could be used for dumb people. By late proceedings in the field of profound learning, there exist many applications that neural network systems can solve one such problem is this sign language. To meet the requirements, the proposed model uses deep learning techniques to solve the problem. One such technique is Convolutional Neural Networks for image depiction and classification. The model designed in the proposed paper has a dataset that consists of 87,002 images. The dataset has been split into training data and test data in the ratio of 9:1. The proposed model was trained on 78,300 images and the testing was performed on 8700 images which were classified in 29 classes. The model has produced a training accuracy of 98.67%.

Keywords Image processing · Deep learning · Convolutional neural network (CNN) · Tensorflow layers

P. V. S. M. S. Kartik · K. B. V. N. S. Sumanth · V. N. V. S. Ram · P. Prakash (✉)
Department of Computer Science and Engineering, Amrita School of Engineering, Coimbatore, Amrita Vishwa Vidyapeetham, Coimbatore, India
e-mail: p_prakash1@cb.amrita.edu

P. V. S. M. S. Kartik
e-mail: cb.en.u4cse17033@cb.students.amrita.edu

K. B. V. N. S. Sumanth
e-mail: cb.en.u4cse17035@cb.students.amrita.edu

V. N. V. S. Ram
e-mail: cb.en.u4cse17068@cb.students.amrita.edu

G. Ranganathan et al. (eds.), *Inventive Communication and Computational Technologies*, Lecture Notes in Networks and Systems 145,
https://doi.org/10.1007/978-981-15-7345-3_18

1 Introduction

The natural language used for communication without sound is called sign language. Sign language has its own set of rules called grammar and a set of lexicons. Generally, people who are not dumb never try to learn sign language. They simply show some gestures to communicate with dumb people. American Sign Language (ASL) is one such sign language with complete grammar. This language is expressed by the movements of the hands. Like other languages, it has a set of rules for pronunciation, semantics for word formation, and a list of sequences for word order. This language can be communicated using single-handed gestures. So, it is not complex to learn and understand. Other sign Languages are a little complex since they use both hands and other body parts for communicating.

Hand gestures techniques are split into two categories such as static and dynamic hand gestures. Static gestures require only single input for processing. The image sequences are needed to get the dynamic gestures. American Sign Language has 24 static hand sign postures and two dynamic gestures to compose the alphabets (A–Z). This proposed system is also able to recognize other hand gestures like delete, space and nothing (no hand gesture) [1].

The objective of the proposed system is to convert sign language to text obtaining a high accuracy producing model for the American Sign Language Dataset.

The organization of the remaining sections is as follows as Sect. 2 starts with surveys on existing research whose inferences are aligned with our work. Section 3 explains the dataset features and preprocessing techniques employed. Section 4 describes the setup used for the comparative study and evaluation for different methods. The results obtained from the deep learning model have been furnished in Sect. 5. The concluding remarks are presented in the last section.

2 Related Works

Various works related to this sign language recognition have been done over the past and some eminent research right now referenced.

Bheda and Dianna Radpour [2] proposed a method using Deep Learning Convolutional Networks to recognize the for American Sign Language. There are around 22 hand figures that compare 26 letters as set named as the alphabet, and ten digits can be signed with the help of a hand. The design included three clusters of two then a max-pool followed by a dropout layer, and two clusters of fully connected layers than a dropout and output layer, respectively. The experiment resulted in an accuracy of 82.5% for alphabets and 97% for validation set accuracy on digits.

Lopes et al. [3] proposed a sign language interpretation using deep learning. The dataset used, contains a sampled image set of all-American Sign Language alphabets Aa-Bb signs and 1-10-digit signs. The most common type of convolution that is used is the 2D convolutional layer. Conv2D filters are used only in the initial layer of a CNN. The experiment resulted in an accuracy of 95%.

Agnisarman Namboodiri et al. [4] proposed Automatic Sign Language Finger Spelling Using Convolution Neural Network [5]. This paper focuses on the recognition of static gestures of American Sign Language which are collected from Kinect sensors. They used the PDNN implementation of CNN. The architecture consists of a single convolutional layer with 20 feature maps, the local filter having a size of 5 × 5, and a pooling size of 2 × 2. The proposed work attained an efficiency of 94.6774%.

Xie et al. [6] proposed an American Sign Language Recognition model using Computer Vision and Machine Learning. The main objective is to make a dream based application that converts sign language into text using American Sign Language Dataset. The video sequences are fed into the model and extract essential features from it. The architecture used CNN for recognizing spatial features and RNN to train on the temporal features. This architecture used an Inception module with 1 × 1, 3 × 3, 5 × 5 convolutions and 3 × 3 max-pooling. It also used a wider RNN network with 512 LSTM units and a deeper RNN network with three layers of 64 LSTM units. The proposed work attained an efficiency of 91%.

Sahoo et al. [7] proposed a technique for perceiving Indian Sign Language in the space of numerals precisely so the less blessed individuals will have the option to speak with the outside world without the need of a mediator out in the open spots like railroad stations and so on. Here they made a numeric sign database containing 5000 signs, 500 pictures. Here they had direct pixel value and progressive hierarchical centroid procedures were utilized to remove wanted highlights from sign pictures and utilized systems like neural networks and kNN classification methods to classify the signs. The result of this experiment achieved an accuracy of up to 97.10%.

Andersson et al. [8] proposed a model that utilizes a self-fabricated dataset and a second dataset named 'Hand Gesture Recognition Database'. It has a dataset comprising 8000 pictures as a training set and 1600 pictures as a test set. They depicted the significance of Convolutional Neural Networks (CNN) for image representation and classification. After applying several preprocessing techniques on the dataset and fed into the model. Their proposed model has two convolution layers, two max-pooling layers, two fully connected layers and an output layer. The experiment had an accuracy of up to 97.12%. There exist some cloud computing services like Amazon Web Services (AWS) which help us to do image classification but there is no kind of deep learning service which helps us to convert image to text. Suresh et al. [9] explained the image classification using AWS services.

However, the above-proposed works used datasets containing less than 10,000 images with a maximum accuracy of 97%. Also, the feature extraction by manual procedure has a few downsides. Strategies for programmed communication via gestures acknowledgement couldn't utilize the profundity detecting innovation that is as generally accessible today. Past works utilized extremely fundamental camera innovation to produce datasets of basically pictures, with no profundity or form data accessible, simply the pixels present.

Our proposed architecture involves one of basic supervised learning using an American Sign Language Dataset having more than 80,000 images. Our objective is to classify the images using deep learning techniques and predict every letter in ASL. The inputs were not of fixed size and were resized to 100 by 100. Our architecture used two convolutional layers with two max-pooling layers and an optimization function used is ADAM optimizer. The experiment resulted in an accuracy of 98.68% for alphabets and 90% for validation accuracy.

3 Exploratory Data Analysis

Here, this section provides the description of the dataset, hidden layers used, layers of Convolutional Neural Network architecture that have been used. The process flow of methodology is shown in Fig. 1. The methodology and the proposed technique involved include the data gathering, minor preprocessing, hidden layers used, constructing the configuration of Convolutional Neural Network and constructing the model.

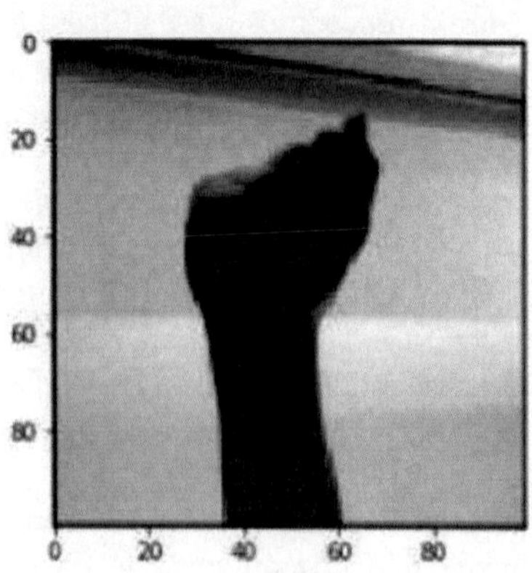

Fig. 1 Sample image after preprocessing

3.1 *Input and Training Data*

Images that are to be trained are selected randomly from the American Sign Language Dataset as an input for the training data. It is accepted that the information pictures contain precisely one hand gestures made with the left hand. Here, the location of the hand is not exactly at the centre, the hand covers all the possible angles that the alphabet can represent like the hand representing the alphabet can be at the top-right portion of the image or it may be at the bottom portion of the image. The training images contain all possible angles of the alphabet so that the prediction of the alphabet will be more accurate. The prediction procedure will be not so mind-boggling but rather more proficient if the foundation is less intricate and the differentiation is high on the hand. In this way, it is accepted that the foundation of the pictures was less unpredictable and accurate.

3.2 *Minor Data Remodelling*

A base preprocessing is applied to the dataset to decrease calculation entanglements and to accomplish better proficiency. Right off the bat, the pictures were changed over to grayscale pictures. Since grayscale pictures contain just one shading which will be simple for CNN to learn. Then the images are normalized by dividing with 255 so that the images lie between 0 and 1 to have a better computation for the prediction. Now the images are resized to the required size since the required size is 100*100 pixels the images are resized accordingly and fed to CNN for training.

3.3 *Dataset*

The proposed system classifies the whole dataset [10] into 29 classes (A, B, C, ..., Z, space, del, empty) to recognize. The data set contains 3000 images with an 80:20 ratio for training and testing, respectively. So, this makes it a total of 78,000 for training and 8700 for testing. Here all these images are pre-processed and then fed to the proposed CNN model. The sample image is shown in Fig. 1.

4 Proposed Architecture

The proposed model comprises two convolutional layers, with two max-pooling layers and an optimization function here used is ADAM optimizer which is shown in Fig. 2.

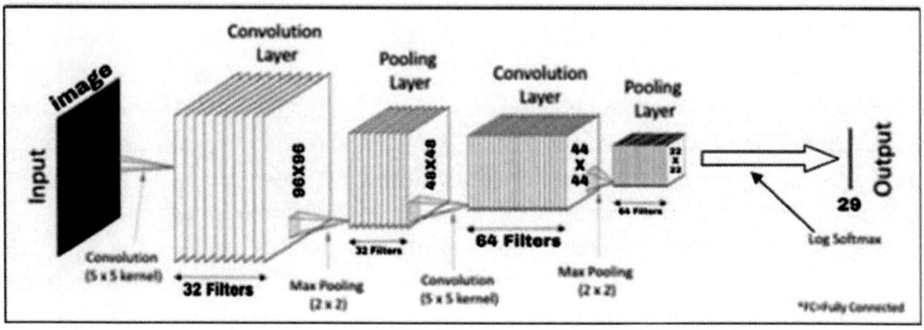

Fig. 2 The proposed model

The first convolution layer has 32 different filters with a kernel size of 5 × 5. The activation function used is the Rectified Linear Unit (ReLU). From numerous instances, from the earlier studies, it is conclusive that the Rectified Linear activation function gives better results than other activation functions for similar architectural models. As in r the first layer the size of the input image is specified. The input image size is 100 × 100 × 1 which demonstrates that a grayscale image size of 100 × 100 is provided to the network. This can be noted by seeing the third dimension which turns out to be one. The output of the current layer is extracted and passed on to the trailing layer which turns out to be the pooling layer.

The first CNN of the proposed model has a max-pooling layer of 5 × 5 size which extracts the maximum value from a window of 5 × 5 size. The spatial size of the portrayal is decreased bit by bit as the pooling layer takes just the max values which are produced and forget about the rest. This layer enables the system to comprehend the pictures all the better as it chooses the significant features out of it.

The successor layer is again a convolution layer and now the number of filters is 64 but the same kernel size of 5 × 5 remains the same along with the default stride. Again, ReLU activation function is used in this layer. This is followed by another max-pooling layer which is of size 2 × 2.

The output layer has 29 nodes each relating to one class of American sign language. This layer uses SoftMax as the activation function which gives the output which is the expected value for each class.

The compilation of the proposed model is done with ADAM optimizer function with a default learning rate of 0.001. Finally, the metric of accuracy was stated to make a note of the assessment procedure. The skeleton of the proposed model is provided in Table 1.

Table 1 Model summary

Model configuration	Attributes
First convolution layer	32 filters with kernel size 5 × 5, ReLU, input size 100 × 100
First pooling layer	Max-pooling—Size 2 × 2
Activation function	ReLU
Second convolution layer	64 filters with kernel size 5 × 5, ReLU, input size 100 × 100
Second pooling layer	Max-pooling—Size 2 × 2
Activation function	ReLU
Output Layer	29 nodes corresponding to 29 classes with a SoftMax activation function used.
Optimizer	ADAM Optimizer
LR	0.001
Evaluation metrics	Accuracy

5 Results

The proposed approach has detailed an innovative deep learning approach for classifying the American sign language images into text. We likewise perceive that the portrayal itself has an enormous effect on the presentation of calculations like our own, so we want to locate the best portrayal of our information with the best accuracy. Some of the results of previously published papers related to this field with accuracy metrics are provided in Table 2 and the statistics and metrics of the proposed model are provided in Table 3. An insight comparison of the accuracy of the proposed model with other models in the form of Bar graph is visualized in Fig. 3. The accuracy versus loss lines of the model as the number of epochs increases, i.e. as the training progresses is visualized in Fig. 4.

Table 2 Comparison of the proposed model with other models

Name	Accuracy (%)
Model 1 [2]	90
Model 3 [4]	94.6774
Model 2 [3]	95.20
Model 4 [6]	91

Table 3 Statistics of the proposed model

Metrics	Value
Training accuracy	98.68%
Training loss	0.0485
Validation accuracy	90%
Validation loss	0.0654

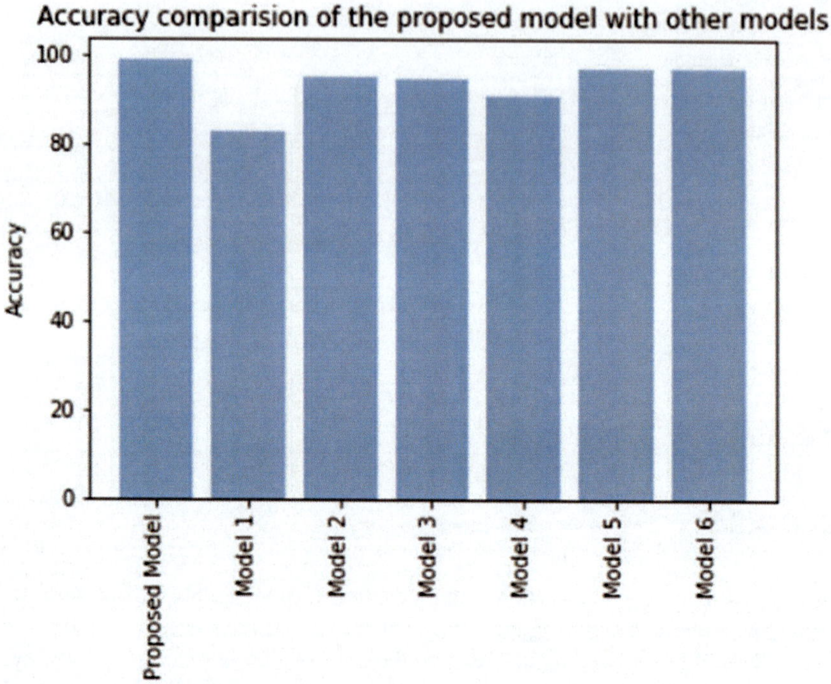

Fig. 3 Accuracy comparison of the proposed model with other models

6 Conclusion

Based on the results obtained from the proposed model, it has been demonstrated that it is possible to effectively convert sign language to text using proper neural network architectures. The scope of this paper is that it does not include direct conversion from video to text continuously. To do this, video preprocessing and frame-wise splitting of data into images are required. This improvement could help in real-time applications like video conferences which eventually could include people who cannot speak. The proposed model could add up as a boost to numerous other developments in the architecture and technique which can, in turn, be a useful product on a real-time basis.

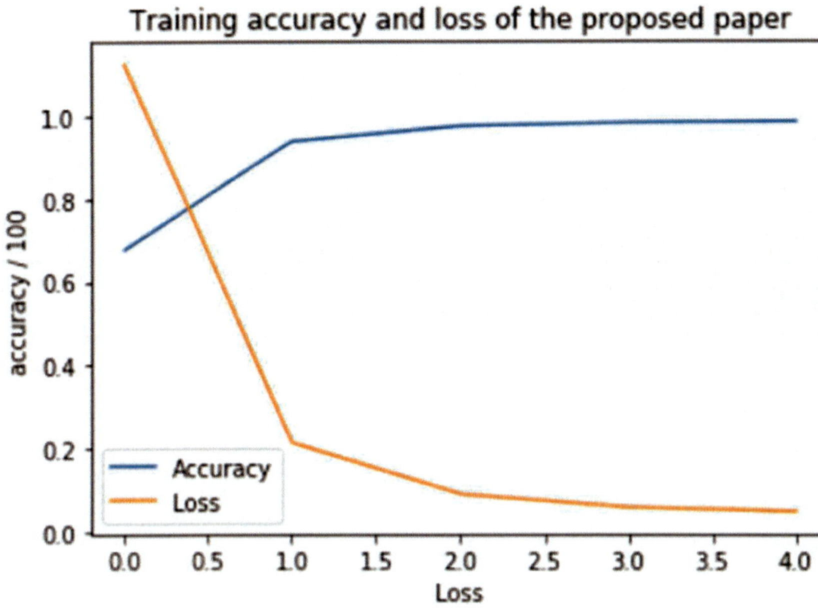

Fig. 4 Accuracy versus Loss with an increasing number of epochs

References

1. Agarwal A, Thakur M (2013) Sign language recognition using microsoft kinect. In: IEEE international conference on contemporary computing
2. Bheda V and Dianna Radpour N (2017) Using deep convolutional networks for gesture recognition in american sign language. CoRR, abs/1710.06836
3. Dias J, Patil D, Raut P, Lopes M (2019) Sign language interpretation using deep learning (IOSRJEN)
4. Beena MV, Agnisarman Namboodiri MN (2017) Automatic sign language finger spelling using convolution neural network: analysis. Int J Pure Appl Math 117(20):9–15
5. . http://deeplearning.net/software/theano/
6. Bantupalli K. Xie Y (2018) American sign language recognition using machine learning and computer vision. In: IEEE international conference on big data
7. Sharma M, Pal R, Sahoo AK (2014) Indian sign language using neural networks and kNN classification techniques. J Eng Appl Sci 9(8):1255–1259
8. Islam MZ, Hossain MS, Ul Islam R, Andersson K (2019) Static hand gesture recognition using convolutional neural network with data augmentation. In: 8th International conference on informatics, electronics & vision (ICIEV), At Washington, USA
9. Suresh R, Prakash P (2018) Deep learning based image classification on amazon web service. J Adv Res Dyn Control Syst 10:1000–1003
10. Dataset. https://www.kaggle.com/grassknoted/asl-alphabet/data#G1006.jpg

Classification of Banana Leaf Diseases Using Enhanced Gabor Feature Descriptor

N. Ani Brown Mary, A. Robert Singh, and Suganya Athisayamani

Abstract Banana leaf disease classification requires the application of machine learning and image processing on the images obtained from the field. A lot of diseases like banana black sigakota (BBS), banana bunchy top virus (BBTV), and banana bacterial wilt (BBW) have been reported on banana plantation. In this paper, the merits of both Gabor filter and 2D log Gabor filter have been used to construct an enhanced Gabor filter, which extracts features from the images of diseased plant. Further, KNN classifier is used in the classification of diseases. Classification results with diseased datasets show that the proposed method achieves better when compared to SIFT and SURF feature descriptors.

Keywords KNN · Gabor filter · Log Gabor filter · Classification · Canny edge detector

1 Introduction

Banana plant, Musa paradisiaca, is an outstanding supply of potassium. A single banana supplies 23% of potassium for the users. Banana leaves have wonderful health benefits and medicinal uses. Banana leaf body wrap treatments are used to keep the body warm. Still burns due to fire are treated using banana leaf. Another name for banana leaf is natural eraser because it gives relief from skin irritation, rashes, poisonous insect bites, spider bites, and bee stings. Banana leaf diseases spoil the banana plant and its growth.

N. Ani Brown Mary
Sarah Tucker College, Tirunelveli, Tamil Nadu, India

A. Robert Singh (✉)
School of Computing, Kalasalingam Academy of Research and Education, Krishnankoil, India
e-mail: robertsinghbe@gmail.com

S. Athisayamani
School of Computing, Sastra Deemed to be University, Thanjavur, India

G. Ranganathan et al. (eds.), *Inventive Communication and Computational Technologies*, Lecture Notes in Networks and Systems 145,
https://doi.org/10.1007/978-981-15-7345-3_19

Fig. 1 Leaf affected by **a** BBTV, **b** BBW, **c** BBS, and **d** healthy leaf

Many banana leaf diseases have been reported, but black sigakota and yellow sigakota are the most widespread diseases. Sigakota disease is caused by a fungus, which is an airborne disease. Black sigakota is widespread in lowlands while yellow sigakota is widespread in highlands. Black sigakota and yellow sigakota are characterized by the appearance of small streaks which then develop into brown streaks. The streaks then enlarge and fuse to form black patches which then turn to bea disease symptom as spots in banana leaves. Banana bunchy top virus (BBTV) disease exhibits bunchy leaves on the top of banana plant. Diseased plant leaves will be narrow and erected upright and will stop producing fruits. Removal of diseased leaves from plants will help to minimize infection in plantation. Diseases in plantation spoil the banana plants and their growth. Figure 1 shows some differences between diseased leaves and healthy leaves.

There is not much work reported in the literature for the classification of diseases in banana leaves. Godliver et al. [1] have classified banana leaf disease using six classifiers, namely nearest neighbor, random forest, naive Bayes, support vector classifier, decision tree, and extremely randomized trees and have shown that extremely randomized trees provide excellent results over the other classifiers. Banana leaf disease features are extracted using thresholding technique for the analysis of connected components in terms of morphological features. This has yielded 96% accuracy. Vipinadas et al. [2] proposed classification of banana leaf disease using support vector machine (SVM) along with radial basis function, and features are analyzed using discrete wavelet transform (DWT). They have reported an accuracy of 95%.

The difficulty of classifying diseases is a tedious one. Farmers stay for time for the disease symptoms to detect the disease in the crops. However, since the plantation is large and clumsy, it becomes tedious to discover it at an early stage. Therefore, this study intends at early disease classification. The indications are able to be seen in the stem, male bud, leaves, and fruit. The disease commences with any leaf and makes them to brown, yellow, and then it wilts. Young diseased plants become underdeveloped and may not make any fruits. Therefore, there is a need for an automated system to classify banana leaf diseases.

Gabor filter is generally used in for texture feature extraction from the images [3, 4] and has been very proficient. Gabor filter can tune the scale and direction for texture analysis [5]. 2D Gabor functions enhance valleys as well as edge contours and ridge contours of the diseased banana leaf image. This corresponds to enhancing borders and diseased edges, which are supposed to be the main clue for disease recognition. Having feature maps specialized for each banana leaf disease in database makes it feasible to keep overall leaf data while enhancing local characteristics. Gabor filters are band pass filters that are mostly used for feature extraction technique in image processing. But, the real part of the complex 2D Gabor filters is not strictly band pass filter. So, the performance of the method is hindered by the background brightness. This difficulty can be overcome by 2D log Gabor filter, so that the background brightness does not influence the extraction of banana leaf features. Thus, the merits of both Gabor filter and 2D log Gabor filter have been combined to develop an enhanced Gabor filter to compromise the drawback of one with the other.

Contribution in this work can be summarized as follows:

- Development of a novel feature descriptor termed as enhanced Gabor filter which helps to improve classification accuracy.
- An efficient automated system using the proposed feature descriptor for the classification of diseases in banana plantation.

2 System Description

The proposed work is based on disease identification, from banana leaf image using enhanced Gabor as texture descriptor. KNN classifier is trained to perform the classification task. The architecture of the feature extraction technique is shown in Fig. 2. The subsequent subsections describe the main steps of the approach. Banana leaves considered are in RGB color space, but problem in RGB color space could not characterize color in terms of human interception [2]. So, banana leaves are transformed from RGB to YCbCr where Y, Cb, and Cr represent luma component, blue-difference, and red-difference chroma components, respectively. Here, Y component alone is considered for feature extraction as given in Fig. 3.

2.1 Segmentation

The diseased banana leaf region has to be segmented from an image, and a group of points from the contour is extracted to represent the shape of the leaf. This makes the segmentation problem tough because banana leaf disease dataset comprises leaf images with uniform pattern or diseased pattern. Usually, K-means algorithm is used for segmenting leaf images. Each cluster is populated with the data points that are closer to the cluster center. Then, for each cluster, it updates the cluster center

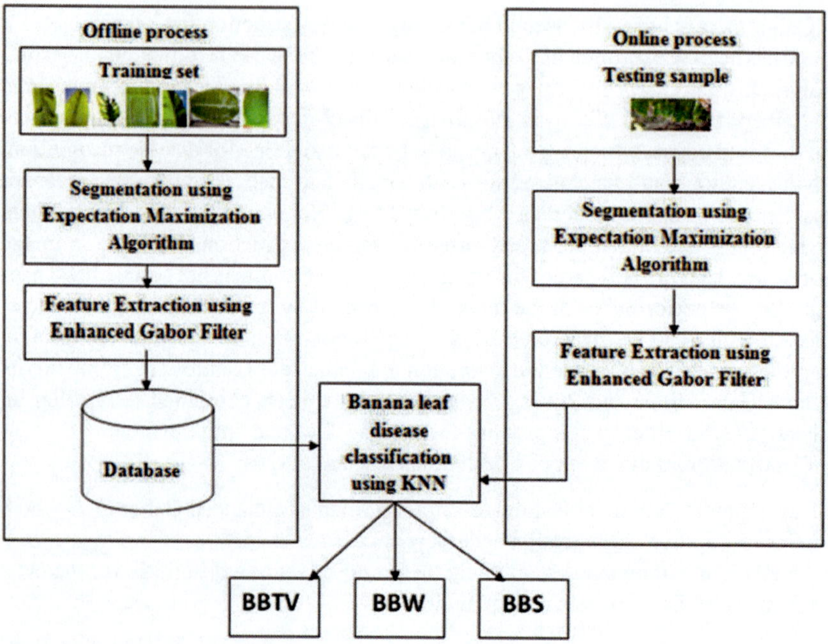

Fig. 2 Architecture of the proposed feature extraction technique

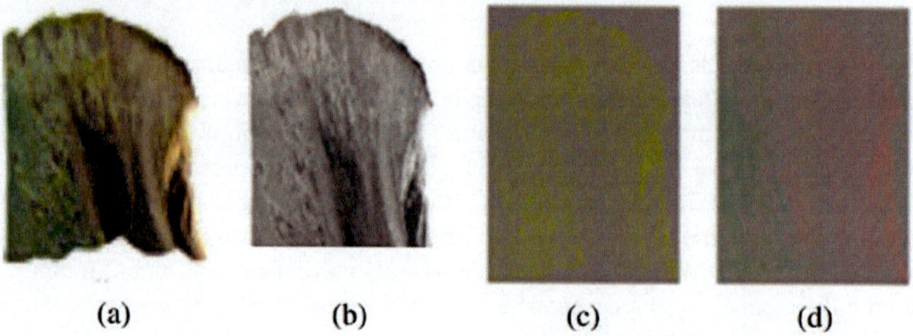

Fig. 3 **a** Diseased leaf, **b** Y component, **c** Cb component, and **d** Cr component

according to the data points assigned to it. These two steps are repeated until the cluster centers are changed. The demerit of K-means is that it gets stuck when local optimal problem occurs. To overcome these difficulties, expectation-maximization (EM) algorithm is used. In EM algorithm, each data point associates to a cluster with particular probability, and finally, it will be assigned to a cluster with maximum probability [6]. Figure 4 shows the result of segmentation.

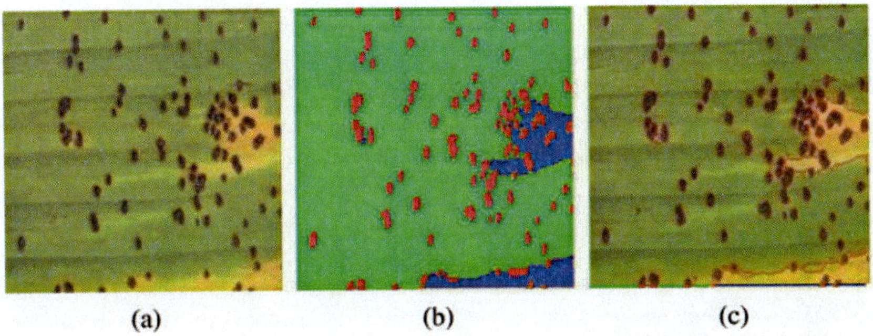

Fig. 4 **a** Input image, **b** segmentation with EM algorithm, and **c** segmented boundary image

2.2 *Feature Extraction*

2.2.1 Scale-Invariant Feature Transform (SIFT) Detector

SIFT has four important levels: (a) detection of scale space, (b) localization of key point, (c) assignment of orientation, and (d) descriptor for key point. Difference-of-Gaussian (DoG) function is used in the first level to identify potential interest points [7, 8], with constant scale and orientation. DoG has the advantage of high computational speed over Gaussian [8–10].

$$Z(u, v, \delta) = (GV(u, v, n\delta) - GV(u, v, \delta)) \times F(u, v) \quad (1)$$

$$= C(u, v, n\delta) - C(u, v, \delta)) \quad (2)$$

where $\times$ is the convolution operator, $GV(u, v, \delta)$ is a variable scale Gaussian, $F(u, v)$ is the input image, $Z(u, v, \delta)$ is DoG with n scale. In the key point localization step, orientation assignment is done by forming a histogram of gradient orientation of sample points within a region around the keypoint [8]. So, this results in SIFT descriptor using 4×4 array of histograms with eight orientation bins each with $4 \times 4 \times 8 = 128$ dimensions.

2.2.2 Speed-Up Robust Feature Detector (SURF) Detector

SURF is an algorithm that extracts key points and narrative method that gives a comparable alternative to SIFT and obliges much fewer processing time for perceiving similar key points. SURF is visualized to ensure fast execution of the feature detection steps say recognition, narrative, and matching. There are 64 dimensions in SURF descriptor, and a 4×4 Haar wavelet responses on oriented square sub-regions are used to generate it.

SIFT and SURF operators are almost the same feature extraction techniques. SIFT and SURF algorithms make use of some what diverse ways of detecting features. The difference between them are: (i) SURF uses square-shaped filters and SIFT uses cascaded filters. (ii) Image pyramids are built by filtering each layer with Gaussians of increasing sigma values and taking the difference. At the same time, a stack is created by SURF without 2:1 downsampling for pyramid's higher levels in that causes images with common resolution [11]. (iii) Salient image regions are used to extract the SURF key points. SIFT key points are rotation-invariant, which means, even if the image is rotated, it seems to have the same corners. (iv) There are 64 bins in SURF descriptor, which is half of that of SIFT descriptor [12]. (v) SURF is the best, and SIFT is the least in their execution speed, but SIFT targets more number of descriptors [7]. This is because SURF uses "Fast-Hessian" detector [7].

SURF and SIFT are texture-based matching algorithms, but they will get confused with the images with textures, because they are liable to rotation and differences in illumination. This degrades the image recognition performance. Problems of these descriptors are overcome by enhanced Gabor filter by using scales and orientation.

2.3 Enhanced Gabor

Feature extraction methods depend on calculating local areas on leaves and signifying corresponding information in an well-organized way. But, selecting exact feature locations and the corresponding values is extremely critical for the performance of a banana leaf disease classification. Gabor filters are tunable band pass filters that satisfy the lower-most leap of the time-spectrum resolution. It is attained by modulating sinusoid with Gaussian distribution [13]. It is a multi-scale, multi-resolution filter with adaptive orientation, spectral bandwidth, and spatial extent. A 2D Gabor filter is defined as

$$GF_{w,\theta,\mu,\sigma,\beta}(u,v) = exp(-\frac{u'^2+\beta^2v'^2}{2\sigma^2})exp(i(2\pi wu'+\mu)) \quad (3)$$

where, $u' = ucos\theta + vsin\theta$, $v' = -usin\theta + vcos\theta$ w denotes the spatial frequency of the sinusoidal factor, θ denotes the orientation of the normal to the parallel stripes in Gabor function, μ denotes the phase offset, σ denotes the Gaussian kernel's standard deviation, and β denotes the spatial aspect ratio representing the ellipticity of the support of the Gabor function. The output of Gabor filter has real and imaginary components. They are convolved with the input image to produce the Gabor-filtered image.

2D log Gabor filters are created in the polar coordinate system of frequency domain, and inverse Fourier transform can be used to construct in spatial domain. Log Gabor filters extend information equally across the channels. The log axis is the optimum method to represent spatial frequency response. It is represented as in Eq. (4).

$$\mathrm{LGF}(\tau,\alpha) = \exp\left\{\frac{(-\log(\frac{\tau}{\tau_0})^2)}{(2\log(\frac{q}{\tau_0}))^2}\right\}\exp\left\{\frac{(-(\alpha-\alpha_0)^2)}{(2RR(\Delta\alpha)^2)}\right\} \tag{4}$$

where τ_0 symbolizes the midpoint, q symbolizes the bandwidth in the radial path, α_0 symbolizes the direction position, RR is the sizing feature, and $\Delta\alpha$ is the direction spacing.

2D Gabor filter in Eq. (3) and 2D log Gabor filters in Eq. (4) are combined to form enhanced Gabor filter as demonstrated by Peng Yao and Jun Li for iris recognition [14]. In this work, an attempt is made to utilize the merits of Gabor filter and 2D log Gabor filter to form an enhanced Gabor for banana leaf disease feature extraction. The process to apply the enhanced method is as follows:
Step 1: The frequency term of 2D log Gabor filter is transformed, and enhanced Gabor filter in the Cartesian coordinate system is constructed as follows,

$$\mathrm{EGF}_{mn} = \exp\left\{\frac{-\log(\frac{m_1}{m_0})^2}{2\log(\frac{q}{m_0})^2}\right\}\exp(\frac{-(n_1)^2}{2\delta_n^2}) \tag{5}$$

where $m_1 = m\cos\theta + v\sin\theta$, $n_1 = -m\sin\theta + n\cos\theta$, θ denotes the direction, m_0 denotes the center frequency, q denotes the bandwidth in the m_1 direction, and the filter's bandwidth in m_1 direction is δ_n. Step 2: Consider an image $f(x, y)$ with size $P \times Q$, the convolution is specified by

$$\mathrm{GC}_{mn} = \sum\sum f(X-c, Y-v)\mathrm{EGF}_{mn}(c, v) \tag{6}$$

where $c \times v$ is the size of filter mask size that is taken as 60×60. m and n are scale and orientations of feature vector, respectively.
Step 3: Enhanced Gabor filter of different orientations is applied with varying scales on the image to the array of magnitudes as given by Eq. (6):

$$\mathrm{EG}(m, n) = \sum_x\sum_y |\mathrm{GC}_{mn}(X, Y)| \tag{7}$$

Step 4: Mean μ_{mn} and standard deviation σ_{mn} of magnitudes of the transformed coefficient are calculated as in Eqs. (7) and (8):

$$\mu_{mn} = \frac{EG(m, n)}{P \times Q} \tag{8}$$

$$\sigma_{mn} = \sqrt{\frac{\sum_x\sum_y(|\mathrm{EGF}_{mn}(X, Y)| - \mu_{mn})^2}{p \times Q}} \tag{9}$$

where $P \times Q$ represents the size of the image.

(a) (b)

Fig. 5 Example of banana leaf image response to enhanced Gabor filters **a** input image and **b** enhanced Gabor filters which corresponds to five spatial frequencies and eight orientations

Step 5: Five different scales and eight orientations are used to form a feature vector, FV, as given in Eq. (9):

$$FV = (\mu_{00}, \sigma_{00}, \cdots \mu_{00}, \sigma_{07}, \mu_{01}, \sigma_{00} \cdots \mu_{01}, \sigma_{07}, \cdots \mu_{04}, \sigma_{00} \cdots \mu_{04}, \sigma_{07}) \quad (10)$$

Thus, a diseased image can be characterized by enhanced Gabor wavelet transform allocating the explanation of both spatial frequency structure and spatial relations. Enhanced Gabor consists of 80 bins which is lower than the size of SIFT descriptor which has 128 bins. In Fig. 5, an input banana leaf image and the amplitude of enhanced Gabor filter responses are publicized.

2.4 *Classification*

The KNN classifier is used for classification [15] with Euclidean distance [16] as the distance metric is given in Eqs. (11) and (12).

$$d(x, u) = \sqrt{(x_1 - u_1)^2 + (x_2 - u_2)^2 + \cdots (x_n - u_n)^2} \quad (11)$$

$$d(x, u) = \sqrt{\sum_{i=1}^{n} (x_i - u_i)^2} \quad (12)$$

where $x = (x_1, x_2, \ldots, x_n)$ and $u = (u_1, u_2, \ldots, u_n)$ are the feature vectors.

3 Experiments Carried Out

In this section, the classification performance of the proposed enhanced Gabor for feature extraction are accessible for two banana leaf disease datasets, namely Scotnelson [17] and Godliver [18] that are discussed in Table 1. Implementation is carried out using MATLAB 2013a.

3.1 Results with Discussion

The input image that is considered for segmentation is the Y component of YCbCr. It is the biased summation of gamma-compressed RGB components of an image that is the brightness in an image. Y color component for banana leaf diseases provides good results compared to RGB image.

3.2 Accuracy

In this paper, two datasets of diseased banana leaf are used to compare the performance of feature extraction techniques for automated classification approaches. Table 2 compares different qualitative classification parameters with different implementation metrics. Table 2 discusses accuracy rate for different datasets. 98 % accuracy is obtained by the proposed method for Scotnelson dataset, that is 10% superior than the state-of-the art techniques. 97% accuracy is obtained for Godliver dataset, that is 8% higher than state-of-the-art methods. Finally, KNN classifier with Euclidean distance provides good results during classification (Figs. 6 and 7).

3.2.1 Computational Complexity

In the existing approaches and the proposed method, the complexity of an algorithm is estimated using various constraints. In SIFT and SURF algorithms, complexity depends on descriptor matching [19]. Let *N*, *M*, and *D* be the number of features in the test image, number of features in the training dataset, and the dimension of the feature vector, respectively. The complexity of proposed method is O(NMD). The complexity of enhanced Gabor is directly propositional to the size of the convolution mask [20]. The complexity to calculate the filter response for a single point is $O(M^2)$, where M is the width and height of the mask [21]. For applying filtering on the entire image with size $N \times N$, complexity is $O(M^2N^2)$.

Table 1 Evaluation of feature extraction techniques on various banana leaf datasets where (Prec-Precision, Reca-Recall and Fm-Fmeasure)

Dataset	Feature descriptors	Division of samples								
		Testing-10% Training-90%			Testing-25%, Training-75%			Testing-50%, Training-50%		
	Metrics	Prec	Reca	Fm	Prec	Reca	Fm	Prec	Reca	Fm
Scotnelson	Enhanced Gabor	97.5	96.4	96.94	85.4	84.54	84.96	78.59	80.1	79.33
	Gabor	93.9	94.07	93.98	83.67	80.43	82.01	76.08	77.88	76.96
	Log Gabor	94.9	96.32	95.6	86.9	83.29	85	79.1	77.48	78.28
	SHIFT	85.6	86.91	86.25	79.67	78.4	79.02	71.9	69.03	70.43
	SURF	83.9	84.99	84.44	78.54	79.43	78.98	71.16	69.99	70.57
Godliver	Enhanced Gabor	98.2	99.1	98.64	85.5	87.43	86.45	79.9	77.7	78.78
	Gabor	95.9	93.76	94.81	84.8	82.3	83.53	77.3	74.45	75.84
	Log Gabor	96.4	93.12	94.73	83.97	82.41	83.18	77.58	74.25	75.87
	SHIFT	87.98	85.7	86.82	78.54	76.9	77.71	70.9	68.2	69.52
	SURF	86.5	85.23	85.86	76.45	75.2	75.81	69.4	67.91	68.64

Table 2 Feature extraction techniques with overall accuracy

Feature extraction techniques	Overall accuracy	
	Scotnelson [17]	Godliver [18]
Enhanced Gabor	98.12	97.43
Gabor	95.76	96.56
Log Gabor	96.6	96.73
SHIFT	85.99	89.43
SURF	84.43	88.9

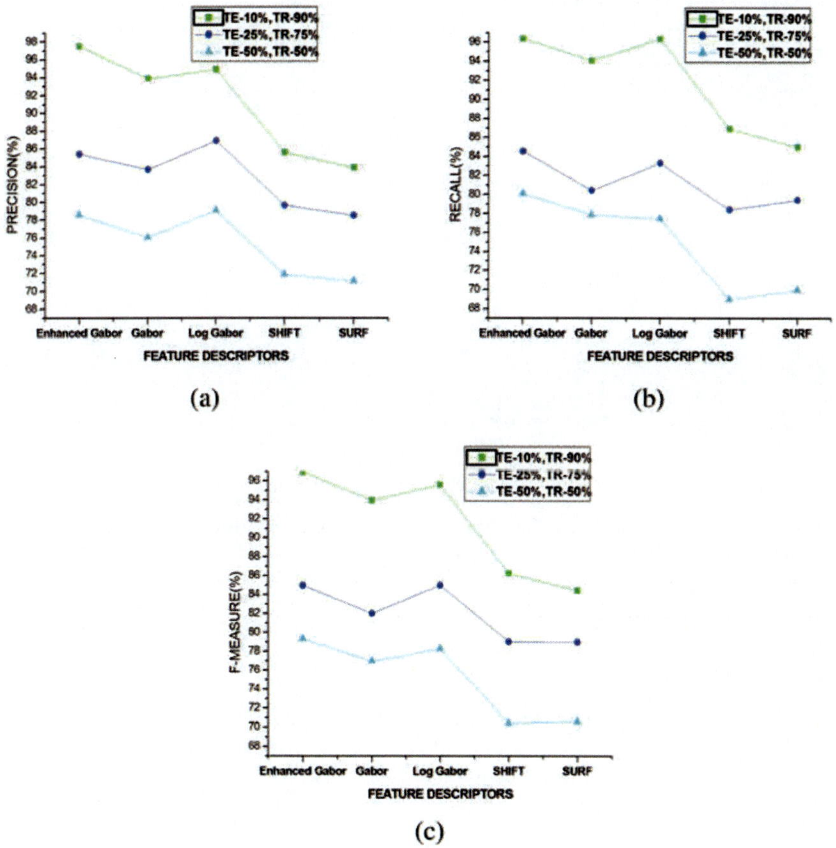

Fig. 6 **a** Precision, **b** recall, and **c** F-measure analysis using ScotNelson dataset

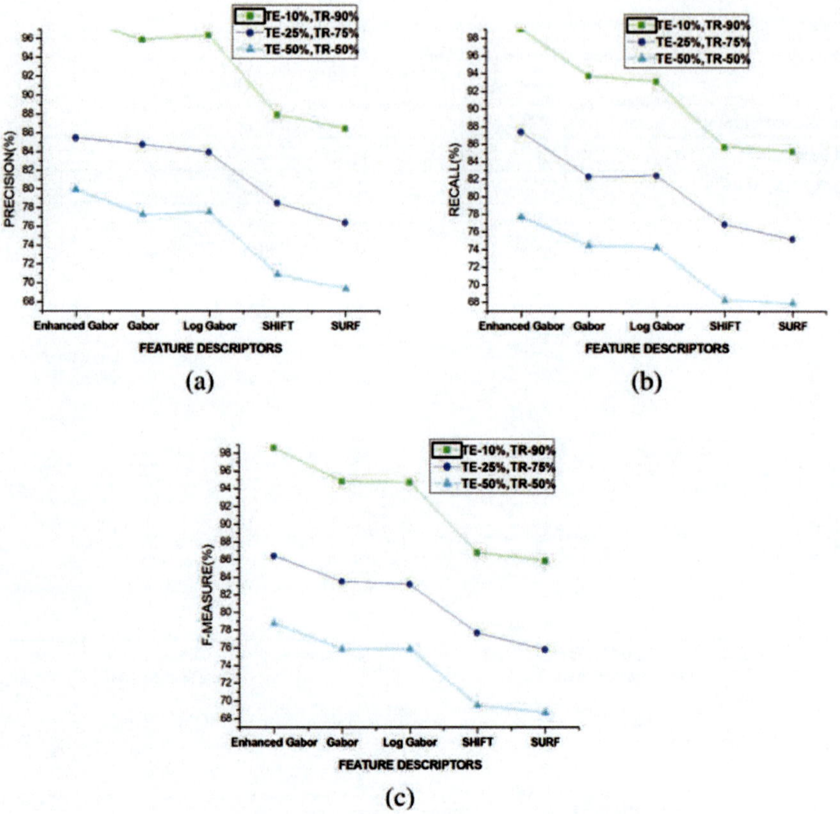

Fig. 7 **a** Precision, **b** recall, and **c** F-measure analysis using GODLIVER dataset

3.2.2 Time Complexity

SIFT and SURF have four important time-consuming levels. But compared to SIFT, SURF is several times faster. But enhanced Gabor feature extraction technique consumes a small amount of memory [21], and so it greatly reduces the time complexity of the recognition system as given in Table 3.

4 Conclusion

An enhanced Gabor filter is proposed in this paper for extracting features to classify banana leaf disease with good classification accuracy when compared to other existing feature extraction approaches. When compared with the existing descriptors, enhanced Gabor filter improves accuracy while reducing time complexity. This

Table 3 Feature extraction techniques with time complexity

Feature descriptors	Total time (ms)	
	Scotnelson [17]	Godliver [18]
Enhanced Gabor	32.6	41.3
Gabor	33.2	42
Log Gabor	34	41.6
SHIFT	49.1	59.8
SURF	42	51

proposed method has good feasibility, high efficiency, and provides results with high robustness. It is noted that the observed accuracy of 98% may be considered as a satisfactory estimate compared to other descriptors. Compared to Gabor, enhanced Gabor produces 3% more accuracy.

References

1. Owomugisha G, Quinn JA, Mwebaze E, Lwasa J (2014) Automated vision-based diagnosis of banana bacterial wilt disease and black sigatoka disease. In: 1st International conference on the use of mobile ICT in Africa
2. Vipinadas MJ, Thamizharasi A (2016) Banana leaf disease identification technique. Int J Adv Eng Res Sci 3(6):120–124
3. NishantShrivastava and VipinTyagi (2014) An efficient technique for retrieval of color images in large databases. Comput Electr Eng 46:314–327 Elsevier
4. Manjunath BS, Jens-Rainer O, VasudevanVinod V, Akio Y (2001) Color and texture descriptors. IEEE Trans Circ Syst Video Technol 1(6):703–15
5. Sebe N, Lew MS (2000) Wavelet based texture classification. In: Proceedings on international conference on pattern recognition, vol 3, pp 959–962
6. Chen D (2008) Expectation-maximization algorithm and image segmentation
7. El-gayar MM, Soliman H, Meky N (2013) A comparative study of image low level feature extraction algorithms. Egypt Inf J 14:175–181
8. David GLOWE (2004) Distinctive Image features from scale-invariant keypoints. Int J Comput Vis 60(2):91–110
9. Wang X-Y, Wu J-F, Yang Hong-Ying (2010) Robust image retrieval based on color histogram of local feature regions. Multimed Tools Appl 49:323–345 Springer
10. Takacs G, Chandrasekhar V, Tsai S, Chen D, Grzeszczuk R, Girod B (2010) Unified real-time tracking and recognition with rotation-invariant fast features. In: IEEE computer society conference on computer vision and pattern recognition, San Francisco, CA, pp 934–941
11. Yang Z, Guo B (2008) Image mosaic based on SIFT. In: International conference on intelligent information hiding and multimedia signal processing, Harbin, pp 1422–1425
12. Sergieh HM, Egyed-Zsigmond E, Döller M, Coquil D, Pinon J, Kosch H (2012) Improving SURF image matching using supervised learning. In: Eighth international conference on signal image technology and internet based systems, pp 230–237
13. Sinha A, Banerji S, Liu C (2012) Novel gabor-phog features for object and scene image classification. In: SSPR and SPR 2012. LNCS 7626, pp 584–592
14. Yao P, Li J, Ye X, Zhuang Z, Li B (2006) Iris recognition algorithm using modified log-gabor filters. In: The 18th IEEE international conference on pattern recognition (ICPR'06), pp 1–4

15. Omranpour H, Ghidary SS (2016) A heuristic supervised Euclidean data difference dimension reduction for KNN classifier and its application to visual place classification. In: Neural computing and applications. Springer
16. Partridge M, Jabri Marwan (2002) Hierarchical feature extraction for image recognition. J VLSI Sign Process 32:157–167
17. http://www.flickr.com/photos/Scotnelson/albums
18. https://github.com/godliver/source-code-BBW-BBS.git
19. Ramisa A, Vasudevan S, Aldavert D, Toledo R, Mantaras R (2009) Evaluation of the SIFT object recognition method in mobile robots, pp 9–18. https://doi.org/10.3233/978-1-60750-061-2-9
20. Amayeh G, Tavakkoli A, Bebis G (2009) Accurate and efficient computation of gabor features in real-time applications. In: Bebis G et al (eds) Advances in visual computing. ISVC 2009 Lecture Notes in Computer Science, vol 5875. Springer, Berlin, Heidelberg, pp 243–252
21. Ilonen J, Kämäräinen J-K, Kälviäinen H (2005) Efficient computation of gabor features. Lappeenranta University of Technology

A Tool to Detect Plagiarism in Java Source Code

Swati Srivastava, Akshit Rai, and Mahima Varshney

Abstract The act of plagiarism occurs when an author uses other author's intellectual ideas without his/her permission. In academics, scholars used to submit assignments now and then in the form of codes or text documents. In this work, the primary focus will be on program codes. In the current scenario, a major perspective is that the scholars may facsimile the codes from a source record without appropriately referencing the original writer or programmer. As there is a wide range of programming languages like C, C++, Java, Python, and many more, in this paper, we have dealt with Java code files. The objective of this work is to estimate the percentage of plagiarism in the given input programming code.

Keywords Plagiarism · Java code · Normalization · Levenshtein distance · Similarity index

1 Introduction

The act of copying one person's effort without taking permission is referred to as plagiarism. It is like the act of stealing a car, watch, cell phone, and a variety of gazettes of others which is liable to be punished by law. Certainly, the act of stealing an intellectual's idea is considered unlawful. Though, this does not signify that scholars should not observe diverse works or sources for references. To enhance knowledge by taking opinions and ideas from experts is a good thing. Nevertheless, most prominent is to ensure that the basis of the sources and references are accordingly credited.

S. Srivastava (✉) · A. Rai · M. Varshney
Department of Computer Engineering and Applications, GLA University, Mathura, Uttar Pradesh, India
e-mail: swati.srivastava@gla.ac.in

A. Rai
e-mail: akshit.rai_cs16@gla.ac.in

M. Varshney
e-mail: mahima.varshney_cs16@gla.ac.in

G. Ranganathan et al. (eds.), *Inventive Communication and Computational Technologies*, Lecture Notes in Networks and Systems 145,
https://doi.org/10.1007/978-981-15-7345-3_20

Plagiarism in coding is not a completely novel experience. This concern has been studied earlier by researchers to recognize the rigorousness of the problem [1, 2]. Plagiarism in programming assignment, not only engrossed the replication of source code but comments and input data are also considered as plagiarism. There are many reasons for students of getting involved in plagiarism like sometimes they feel lazy to write their code. Usually, plagiarism in coding is firm to sense since similar coding is used for the same application. Plagiarism in coding is straightforward to do but tricky to detect. Scholars facsimile all or part of a program from a source or different sources and put forward the fake as their work. This includes students who act as a team and present analogous work. Such plagiarism is felt to be ordinary, even though the true similarity level is hard to assess. When a teacher in a programming course gives a common problem to all scholars then all have to work on the same problem. Consequently, some scholars may inscribe the source code of a problem on their own. While other scholars just obtain the code and change the variable names, the order of statements, functions, and variables of a class. Such modifications in source code are complicated to seize. There are two categories of source code variation: lexical change and structural change. Lexical change can be done without any prior programming knowledge. Structural changes need prior knowledge of programming language. Change in the number of iterations, conditional statements, the order of statements, a procedure to function, and vice versa, adding comments are structural changes.

For the code in Fig. 1, one can use the same logic devoid of considering this code. For sure, this is not considered plagiarism. Such a scenario can be handled by putting some constraints over the size of the code. The constraints may be like that if *n* consecutive lines are similar in two codes then it will be considered as stealing. We need a system to calculate the similarity percentage of code between two Java files. We proposed a plagiarism detection system based on a novel normalization process, to identify the uniqueness of the scholar's code by comparing the input code with the original code. It may be used by the teachers to detect whether the student committed plagiarism or not. This is possible when the plagiarism is estimated for two Java files. If the percentage of plagiarism is less than the specified threshold, then the input code is acceptable otherwise not.

Fig. 1 Sample code

```
package Implementation;
public class test
{
public static int add(int a, int b)
{
return a+b ;
}
public static void main(String[] args)
{
int c;
c= add(2,3);
System.out.println("Sum = " +c);
}
}
```

The rest of the paper is organized as Sect. 2 represents the previous work on plagiarism detection. Section 3 presents the proposed work. The results are discussed in Sect. 4. Section 5 concludes the proposal.

2 Related Work

Many researchers have given methods for plagiarism detection in text and programming code [3, 4, 5, 6]. While some researchers gave a comparison among different plagiarism detection tools [7, 1, 2]. Nurhayati and Busman [8] intended the Levenshtein Distance (LD) algorithm for plagiarism detection in the document. They developed software for Android smartphones. One way to measure the distance is a string metric which is the result of the LD algorithm. In [9], the authors created an application using the LD algorithm to identify similarity in Java codes. A technique for uncovering the plagiarism between C++ and Java codes based on semantics has been projected in [10]. It is a multimedia-based e-Learning and smart estimation method. Input code transformed into tokens to determine semantic comparison token by token. Then it estimated the semantic similarity for the whole input code. In literature, there exist many similarity detection algorithms. Based on these algorithms, the researchers developed a similarity detection system referred to as SCSDS [11]. SCSDS was slower than existing methods. By the fusion of various similarity detection algorithms, the speed and performance of SCSDS became even worse. SCSDS required speed and performance improvement. In [12], the plagiarism detection system considered only text documents for plagiarism tasks. No consideration was given to the syntactical structure of formal programming language. They used normalization of commonly used identifiers to detect a pair of programs that have the same objective. They proved that removal of these normalized operations improves the system.

3 Proposed Method

The proposed system aims to estimate the plagiarism percentage in the given input code. Initially, the user needs to give an input code that has to be checked for plagiarism. The already available codes are called here as original codes that are used for comparison. These two codes are stored in separate variables. After that, the code stored in these two variables is converted to a form that can be easily used for detecting plagiarism. This is done in the normalization step. Following steps are performed to normalize the code:

- Removing white spaces
- Removing comments
- Removing all the keywords

- Removing all the operators
- Replacing all the identifiers with **identifier**
- Sorting.

Removing white spaces

Generally, there are white spaces before and after any operator to enhance the readability. If the code is copied from any online platform then users generally take care of these extra spaces because it looks like it has been copied. So, there is no need for extra spaces as it will increase the length of our string. As the length of the string increases, it will reflect on the LD algorithm as its complexity is $O(n^2)$.

Removing comments

As comments do not affect the actual functioning of code, it is merely there for understanding code in case of complex and long code. We are removing comments because someone can add an extra comment or edit the copied comment. Since the LD algorithm checks similarity character by character, it will affect the result of our plagiarism detection tool. The following regular expression is used to detect the comments.

replaceAll("(?:/*(?:[^*]|(?:*+[^*/]))**+/)|(?://.*)","")

Removing all the keywords

This is the most significant step. It involves removing all the keywords that belong to a language. In our proposal, we check plagiarism only in Java code, so we removed all the keywords that belong to Java language. We are removing keywords because the code of the same program will generally have some type of data types and inbuilt functions. Therefore, they are generally increasing the length of our string which will again reflect the complexity as $O(n^2)$. So, to save time and space we remove keywords. Sometimes users come around with some hack and use different data types and functions to complete the code. Although the code is copied, as he/she understood the copied code, he/she edited it to avoid plagiarism. Removing all the keywords will help in detecting the genuine similarity index.

Removing all the operators

Generally, codes of the same program used the same type and the same number of operators even if they are not copied. They are only increasing the time and space complexity of our code. To get away from this, we remove all the operators.

Replacing all the identifiers with **identifier**

Users generally change the name of identifiers involved in a code to dodge plagiarism. So, we are renaming all the identifiers in both the codes that mean original code and the code to be checked by "**identifier**".

Sorting

Sort both the strings containing original code and the code to be checked alphabetically. A user can change the position of copied code (function, class, etc). Sometimes user also changes the position of statements. Therefore, we need to sort both the strings. The result of sorting is stored separately for original code as well as code to be checked to detect plagiarism even if the user has changed the position of copied code. This completes the normalization step.

After performing all these steps, we get normalized code that again can be stored in a variable. Now, we simply apply the LD algorithm [8]. After that, we store the result of the LD algorithm in a variable. Now, we calculate the plagiarized value using the result of the LD algorithm.

Levenshtein Algorithm

The LD algorithm [8] is used to find the distance which is used for measuring the dissimilarity between two progressions. This distance is referred to as Levenshtein distance or edit distance. It may also denote a larger family of distance metrics. It gives a minimum number of single-character alterations, essential to change one word into the other, between two terms.

Calculating Plagiarism

After performing normalization, we get normalized codes in the form of string both for original code and code to be checked. The original code is referred to as source string (δ). The code to be checked string is referred to as the target string (ε). After this, we fed these two strings to the LD algorithm. It gives us a numeric value which corresponds to the difference between these two strings. This is called LD distance (đ) and is defined as:

$$đ = |\delta - \varepsilon| \tag{1}$$

Now, using plagiarized value formula, we can calculate plagiarism between these two stings. The plagiarized value (ƥ) can be calculated as:

$$ƥ = \frac{1 - đ}{\max(\delta, \varepsilon)} \times 100 \tag{2}$$

where đ is the LD distance, δ represents the original code, ε is code to be checked for plagiarism, max(δ, ε) is maximum length between δ and ε. Figure 2 shows the working of the proposed plagiarism detection system.

Fig. 2 Framework of the proposed plagiarism detection system

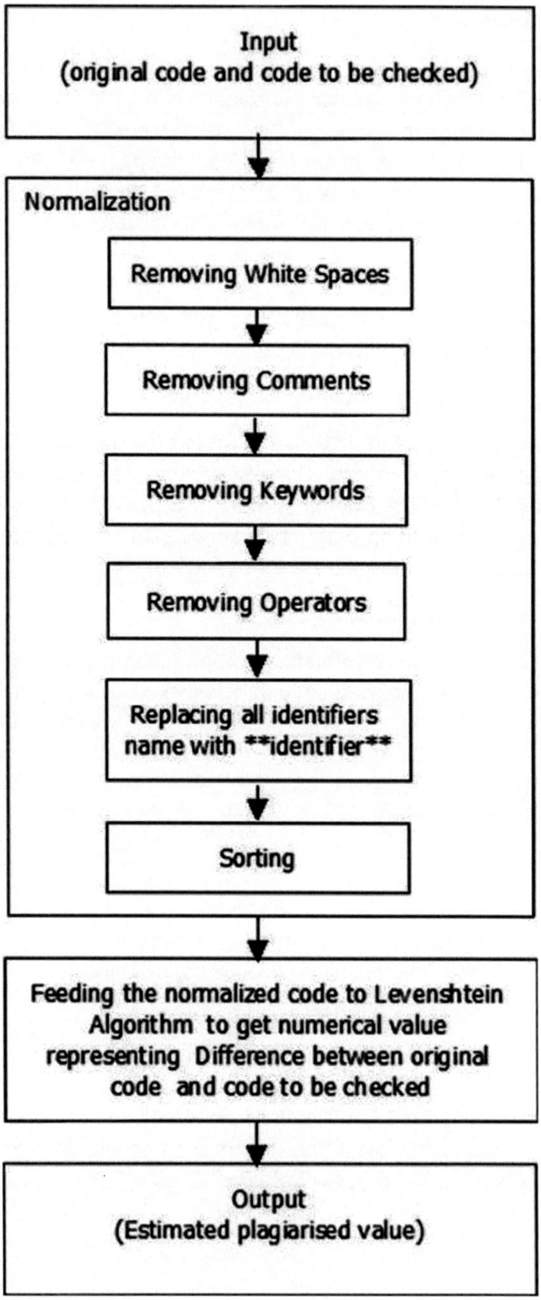

4 Results and Findings

To estimate the plagiarism percentage of the given input code, first, the user needs to give input code that has to be checked for plagiarism along with the original code. Figures 3 and 4 show the samples of the original code and code to be checked, respectively. This code is injected into the normalization step which results in normalized code. Now, the LD algorithm [8] is applied to the normalized code. Then, using the result of the LD algorithm, the plagiarized value can be estimated. Figure 5 shows the user interface of the proposed system. Figure 6 shows the interface after filling the code in the specified area. Figure 7 shows the estimated plagiarism by clicking on the check fraud button. From Fig. 8, it can be observed that the standard plagiarism detection software is not suitable to detect the originality of a Java programming code. Since there are common keywords in a programming language used by the programmers. Therefore, merely the detection of the same words is not the correct criteria to investigate the originality of source code. As can be seen from Figs. 7 and 8, standard software (Turnitin) gives the similarity index of 78% whereas the proposed system gives the similarity index of 51% for the same code. The similarity index calculated by the proposed method and standard software can be compared from Table 1. The above comparison can also be seen in Fig. 9. Thus, it can be stated that the proposed system is more suitable for Java codes than other software for originality detection of source code.

Fig. 3 Sample original code

```
public class PrimeExample{
public static void main(String args[]){
int i,m=0,flag=0;
int n=3;
m=n/2;
if(n==0||n==1){
System.out.println(n+" is not prime number");
}else{
for(i=2;i<=m;i++){
if(n%i==0){
System.out.println(n+"is not prime number");
flag=1;
break;
}
}
if(flag==0){System.out.println(n+"is prime number");}
}
}
}
```

Fig. 4 Sample code to be checked

```
public class Prime {
public static void main(String[] args) {
int num = 29;
boolean flag = false;
for(int i = 2; i <= num/2; ++i)
{
if(num % i == 0)
{
flag = true;
break;
}
}
if (!flag)
System.out.println(num + " is a prime number.");
else
System.out.println(num + " is not a prime number.");
}
}
```

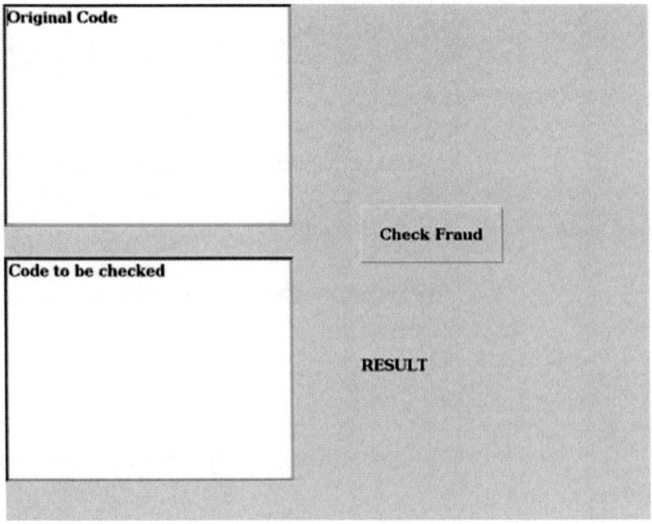

Fig. 5 User interface

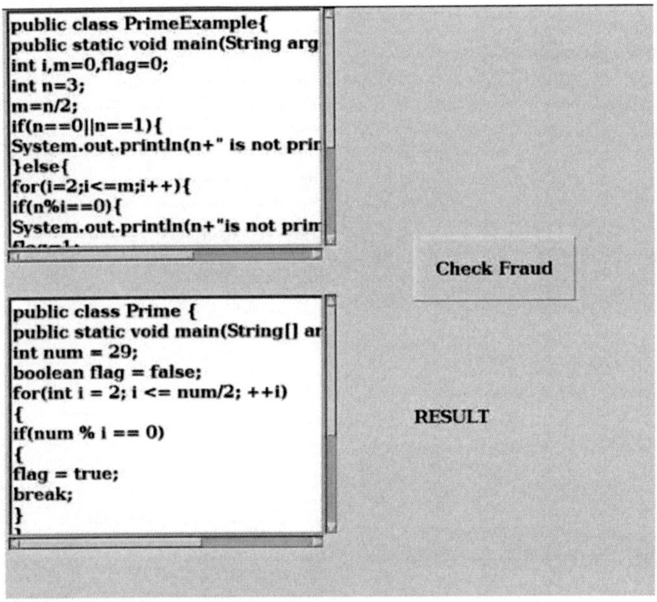

Fig. 6 After filling both the text areas accordingly

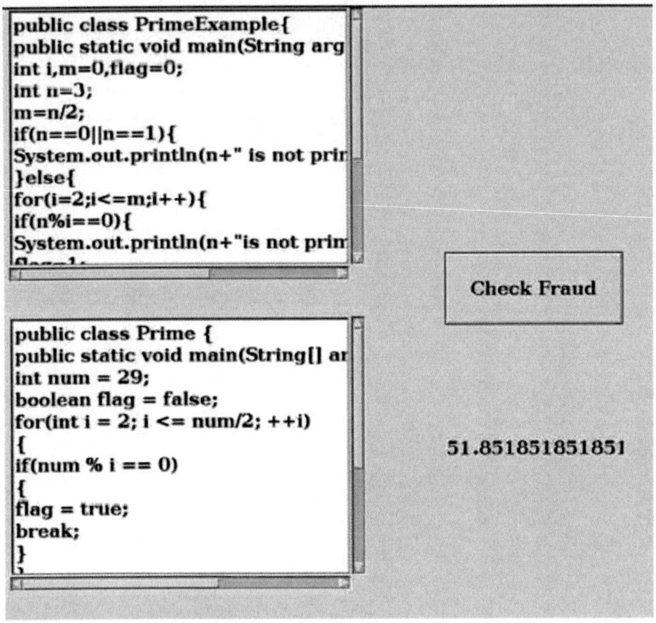

Fig. 7 After clicking on check fraud

```
public class Prime {
public static void main(String[] args) {
int num = 29;
boolean flag = false;
for(int i = 2; i <= num/2; ++i)
{
if(num % i == 0)
{
flag = true;
break;
}
}
if (!flag)
System.out.println(num + " is a prime number.");
else
System.out.println(num + " is not a prime number.");
}
}
```

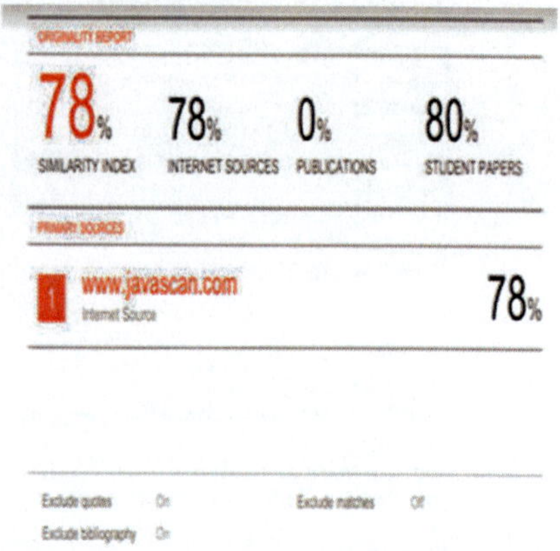

Fig. 8 Plagiarism report of a standard plagiarism detection software

Table 1 Comparison of similarity indexes of proposed system and existing software

Input	Similarity index (proposed system) (%)	Similarity index (existing software) (%)
Code 1	51.85	7
Code 2	54.76	80
Code 3	57.29	83
Code 4	53.26	81

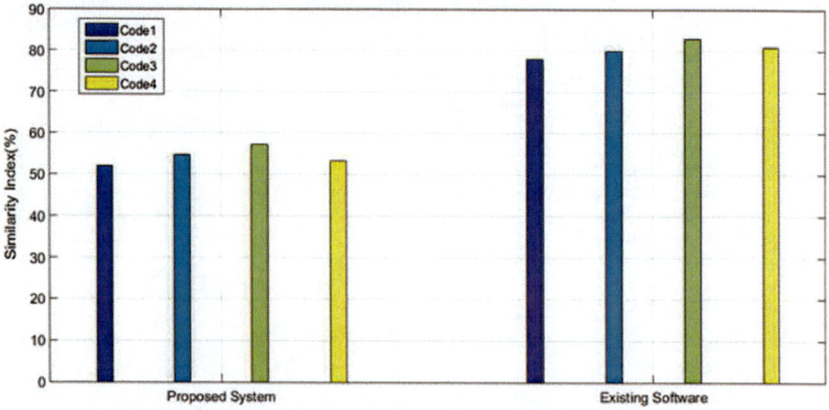

Fig. 9 Comparison of similarity indexes of proposed system and existing software

5 Conclusion

We have proposed a tool that can efficiently be used to check whether the input Java code is plagiarized or not. To carry out plagiarism detection, first, the code is preprocessed through normalization. Normalization of code consists of various steps: removing white spaces, removing comments, removing all the keywords, removing all the operators, replacing all the identifiers with **identifier**, sorting. Then the normalized code is fed into the LD algorithm to obtain LD distance. The value returned by the LD algorithm is used to calculate the plagiarized value. The proposed tool only works on Java source code. Further, it could be extended to work on all programming languages. Plagiarized value has been calculated for 4 codes through the proposed system as well as the existing system. From the results, it can be concluded that the proposed system is more suitable for Java codes than the existing system for originality detection of source code.

References

1. Foltýnek Tomáš, Meuschke Norman, Gipp Bela (2019) Academic plagiarism detection: a systematic literature review. ACM Comput Surv (CSUR) 52(6):1–42
2. Naik RR, Landge MB, Mahender CN (2015) A review on plagiarism detection tools. Int J Comput Appl 125(11)
3. Ghanem B, Arafeh L, Rosso P, Sánchez-Vega F (2018) HYPLAG: hybrid Arabic text plagiarism detection system. In: International conference on applications of natural language to information systems. Springer, Cham, pp 315–323
4. Jadalla Ameera, Elnagar Ashraf (2008) PDE4Java: plagiarism detection engine for java, source code: a clustering approach. IJBIDM 3(2):121–135
5. Alzahrani SM, Salim N, Abraham A (2011) Understanding plagiarism linguistic patterns, textual features, and detection methods. IEEE Trans Syst Man Cybern Part C (Appl Rev) 42(2):133–149
6. Sulistiani Lisan, Karnalim Oscar (2019) ES-Plag: efficient and sensitive source code plagiarism detection tool for academic environment. Comput Appl Eng Educ 27(1):166–182
7. Ali AM, Abdulla HM, Snasel V (2011) Overview and comparison of plagiarism detection tools. In: DATESO, pp 161–172
8. Nurhayati B, Busman B (2017) Development of document plagiarism detection software using levensthein distance algorithm on Android smartphone. In: 2017 5th International conference on cyber and IT service management (CITSM), pp 1–6
9. Liaqat AG, Ahmad A (2011) Plagiarism detection in java code
10. Ullah F, Wang J, Farhan M, Jabbar S, Wu Z, Khalid S (2018) Plagiarism detection in students' programming assignments based on semantics: multimedia e-learning based smart assessment methodology. In: Multimedia tools and applications, pp 1–18
11. Đurić Zoran, Gašević Dragan (2013) A source code similarity system for plagiarism detection. Comput J 56(1):70–86
12. Heblikar S, Sharma P, Munnangi M, Bankapur C (2015) Normalization based stop-word approach to source code plagiarism detection. In: FIRE workshops, pp 6–9

Improved Skip Algorithm for Single Pattern Searching

K. Padmaveni and D. John Aravindhar

Abstract In this paper, we present a novel idea for a single pattern matching in strings. The idea is named skip search (SS). The pattern is checked for presence in the given text by accessing only half the memory locations even for the worst case. The search is a modified version of the Naïve search algorithm by accessing only half the characters in the given string and by checking if those characters are the starting or ending character of the pattern. By doing this, we could obtain the skip length of the pattern size and hence reduce the search time. Since there is no preprocessing phase or no extra memory space required, the algorithm is good in terms of space complexity. The time taken for execution is compared with Naïve algorithm and Knuth–Morris–Pratt (KMP) algorithm, and the skip algorithm gave a better performance for most of the test cases. The algorithm has a big difference in execution time when the pattern is present toward the end of the text or if the pattern is not present in the text. The algorithm can be recommended for any pattern matching projects.

Keywords Algorithm · Pattern matching · Searching

1 Introduction

The pattern searching algorithm aims to find one or all occurrences of a pattern in the given text. In this paper, we proposed an exact pattern matching algorithm called skip algorithm which is based on a Naïve search algorithm with modification in the skip positions. The algorithm returns the position of the first occurrence of the pattern in the text.

K. Padmaveni (✉) · D. John Aravindhar
Hindustan Institute of Technology and Science, Chennai, India
e-mail: kpadmaveni@hindustanuniv.ac.in

D. John Aravindhar
e-mail: jaravindhar@hindustanuniv.ac.in

G. Ranganathan et al. (eds.), *Inventive Communication and Computational Technologies*, Lecture Notes in Networks and Systems 145,
https://doi.org/10.1007/978-981-15-7345-3_21

A literature review of exact pattern matching algorithms starts from the Knuth–Morris–Pratt algorithm. Most of the literature survey focuses on reducing the number of comparisons and processing time. In this paper, we compare the results of the skip algorithm with the Naïve algorithm and Knuth–Morris–Pratt algorithm. The complexity is compared with many other algorithms referred to in the literature survey.

A brief description of the existing string matching algorithms is presented in Sect. 2. The basic concept and working are explained in Sect. 3, and the result is shown in Sect. 4.

2 Literature Review

The Naïve search or brute-force algorithm compares the pattern with every possibility of the text by moving in one direction starting from one end. There is no preprocessing phase in the Naive search algorithm. The time complexity of the Naïve search algorithm is $O(mn)$ where m is the text size and n is the pattern size [1].

The Knuth–Morris–Pratt (KMP) algorithm works by improving the length of shift. This speeds up the procedure by using the previous knowledge obtained from comparisons. This requires preprocessing of the pattern. This works well when there is similarity within the pattern [2].

One of the efficient algorithms published by Boyer–Moore (BM) in 1997 performs the comparisons from right to left of the pattern. The algorithm is considered as one of the most efficient strings matching algorithms. Different shifting rules are used. The algorithm had a preprocessing phase and required additional time in the worst case. The preprocessing here has the time complexity of $O(m + |\sum|)$ and the worst case time is $O(nm + |\sum|)$ [3]. The Boyer–Moore–Horspool (BMH) algorithm uses the occurrence heuristic to increase the shift length in the text. The preprocessing here has the time complexity of $O(m + |\sum|)$ and the worst case time is $O(nm)$ [4]. Another popular search algorithm is the quick search (QS) algorithm. The searching here is done from left to right and the shifting criteria are based on bad character shift rule. The algorithm has the worst. Case complexity is derived similarly to the Horspool algorithm [5]. The algorithm by Boyer–Moore–Smith (BMS) [6] is derived from BMH as the reference and the property of bad character shifting. The preprocessing complexity here is $O(m + |\sum|)$ and the searching time is $O(mn)$. Karp et al. presented an algorithm to find a match with the help of mathematical calculations and finding approximate matches to a pattern in a string [7]. Ziad et al. [8] proposed a skip algorithm for multiple pattern matching and have reduced the number of comparisons for multiple pattern search. Commentz Walter [9] presented an algorithm for multiple pattern matching. This is a modified algorithm of the Boyer–Moore algorithm and showed a reasonable increase in searching speed. Rami et al. [10] presented three algorithms having occurrence search as a common concept. They preprocessed the given pattern and created the occurrence list and utilized it or pattern matching. Hussain et al. [11] proposed an algorithm for bidirectional searching. The authors have used the concept of improving the shift decision. Matching is done by comparing the rightmost and mismatched character of partial text window to the left of the pattern at same shift length. The complexity of the algorithm is $O(mn)/2$. But it needs a preprocessing phase with additional time.

The faster approach by preprocessing and modification of the Naïve search algorithm was proposed [12–15]. The occurrence algorithm was proposed by Mohammed et al. [16]. The algorithm is an improvement of brute force, and this includes preprocessing of the pattern as well as the text.

A comparison is done based on the preprocessed input in the algorithm by Reverse Colussi (RC) [17]. The preprocessing phase has a complexity of $O(\mathrm{m})$ whereas the searching phase is $O(n)$.

3 Algorithm

3.1 Concept

The aim is to provide a faster search. Similar to most of the commonly used searching algorithms like KMP, BMH, BMS, etc., the proposed skip search also searches the pattern from left to right. The uniqueness of this algorithm is that the index is incremented by skipping unwanted elements.

It is accomplished by dividing the text into three groups.

1. The part of the string which contains elements that can be only the starting point of the pattern
2. The part of the string which contains elements that can be the starting as well as the ending point of the pattern
3. The part of the string that can be only the ending point

For example,

If the pattern is of length 5 (stored in index 0 to 4) and the text is of length 40 (stored in index 0 to 39), the string index from 0 to 3 can only be the starting point of the pattern. This falls under Group 1. The string index from 4 to 35 can be either the last character of the pattern or the starting character of the pattern. This falls under group 2. The string starting from index 36 to 39 can only be an ending character of the pattern. This falls under Group 3.

The memory accessing time plays an important role in the search time. The lesser is the number of memory locations accessed, the lesser will be the execution time. The search time could be drastically reduced if the number of memory locations accessed is less. To optimize the search time, the starting character is alone checked for a match in Group 1. The Group 2 search checks for both the starting and ending character of the pattern. The search enables us to reduce 50% of memory locations accessed in the worst case. The last character match is alone checked for the Group 3 elements. The working of the algorithm is shown as example. Let the text S be

0	1	2	3	4	5	6	7	8	9
a	a	b	a	c	b	a	b	c	b

10	11	12	13	14	15	16	17	18	19
c	c	a	c	c	a	c	c	b	b

20	21	22	23	24	25	26	27	28	29
c	a	c	b	c	b	c	a	c	a

30	31	32	33	34	35	36	37	38	39
c	c	a	b	c	b	c	a	c	a

40	41	42	43	44	45	46	47	48	49
a	a	b	b	c	c	c	a	c	a

50	51	52	53	54	55	56	57	58	59
b	c	a	b	a	b	c	a	c	b

And pattern P be

0	1	2	3	4
a	b	c	a	b

The string S is split into three

Group 1

0	1	2	3
a	a	b	a

Group 2

4	5	6	7	8	9
c	b	a	b	c	b

10	11	12	13	14	15	16	17	18	19
c	c	a	c	c	a	c	c	b	b

20	21	22	23	24	25	26	27	28	29
c	a	c	b	c	b	c	a	c	a

30	31	32	33	34	35	36	37	38	39
c	c	a	b	c	b	c	a	c	a

40	41	42	43	44	45	46	47	48	49
a	a	b	b	c	c	c	a	c	a

The elements that are referred from memory for comparison are highlighted in the following figure.

50	51	52	53	54	55
b	c	a	b	a	b

Group 3

56	57	58	59
C	a	c	b

0	1	2	3	4	5	6	7	8	9
a	a	b	a	c	b	a	b	c	b

10	11	12	13	14	15	16	17	18	19
c	c	a	c	c	a	c	c	b	b

20	21	22	23	24	25	26	27	28	29
c	a	c	b	c	b	c	a	c	a

30	31	32	33	34	35	36	37	38	39
c	c	a	b	c	b	c	a	c	a

40	41	42	43	44	45	46	47	48	49
a	a	b	b	c	c	c	a	c	a

50	51	52	53	54	55	56	57	58	59
b	c	a	b	a	b	c	a	c	b

For the elements in Group 1, the elements are checked if it could be the starting character. The colored entries are accessed during search under worst case. In best case, all the white entries gets skipped. All those substrings which match the initial character are subjected to further checking using search algorithm shown as Algorithm 1.

```
Algorithm SEARCH
Input:
String S of length l. Stored in an array of index 0 to l-1
Pattern P of length m. Stored in an array of index 0 to m-1
Output:
Print the starting index in the string if the pattern is present (can be between 0 to l-
m).
Return l if the index does not exist
  1. F=P[0];L=P[m-1];icount=m-1//Store the first and last character of the
     pattern in F and L
  2. For i=0 to i=m-2// Elements of Group 1
      If(s[i] =F) SEARCH_FOR _MATCH(i,i+m-1)
  3. I=i+(m-1)//Initial position of Group 2
  4. While i<=l-m //Elements of Group 2
     a. If(s[i] =F) SEARCH_FOR _MATCH (i,i+m-1)
     b. If(s[i] =L) SEARCH_FOR _MATCH (i-m+1,i)
     c. i=i+1;
     d. icount=icount+1;
     e. if (icount == m-1)
           a. i=i+m-1;//This enables to skip alternate m-1 elements in
              group 2
           b. icount=0;
  5. For i=l-m+1 to l-1// Elements of Group 3
     If(s[i] =p[m-1]) SEARCH_FOR _MATCH (i-m+1,i)
```

Algorithm 1: Search Algorithm For the elements in Group 2, the elements are checked if it could be the starting character or ending character. All those substrings which match either first or last character are subjected further checking using the SEARCH_FOR_MATCH algorithm which is direct and one to the comparison of elements. The checking is done only for an alternate substring of $m - 1$ elements, i.e., 4. For the elements in Group 3, the elements are checked if it could be the ending character.

4 Evaluation

The time taken for the execution of the algorithm depends upon the number of memory locations accessed.

If the text that has the same subpatterns appear more than once, the KMP algorithm performs faster than this proposed algorithm. Skip algorithm works better where the input characters are randomly placed.

The result is compared with search algorithms namely

1. Naive search algorithm
2. Knuth–Morris–Pratt (KMP) algorithm.

Naïve search algorithm is chosen for comparison because it has no preprocessing phase as well as it does not require any extra space.

Knuth–Morris–Pratt algorithm is chosen for comparison because it has a minimal preprocessing phase complexity compared to other algorithms, i.e., $O(m)$ and the searching phase is $O(n)$.

The comparison was done for the time taken for execution.

The execution was done in an intel i7 machine with 16 GB RAM. The execution was repeated 10 times and the average execution time was taken.

The test cases were characterized based on

- Text size (The number of characters in the text)
- Pattern size (the number of characters in the pattern)
- Alphabet size (the number of different characters used in the text).

Data set 1

The first test data was generated with all possible combinations of

- Text size 'L' (characters)
- Pattern size 'M' (characters)
- Alphabets 'A' (characters).

The details regarding the input parameters are shown in Table 1.

The average time taken for execution for data set 1 is shown in Table 2.

The result of data set 1 was compared with the Naïve algorithm (Fig. 1).

Table 1 Input parameters for data set 1

Input	Combinations
Text size 'L' (characters)	1 to 15, 16 to 63, 64 to 255, 256 to 2048, >2048
Pattern size 'M' (characters)	1 to 255 depending upon text size
Alphabets 'A' (characters)	Digits alone (0, 1, 2,…, 9) Only three alphabets 3 (a, b, and c)

Table 2 a Time taken for two alphabet input. b Time taken for numerals s

Three alphabets	
Algorithm	Average time(ms)
(a)	
Naive search	68.36
Skip algorithm	66.69
(b)	
Naive search	63.32
Skip algorithm	60.8

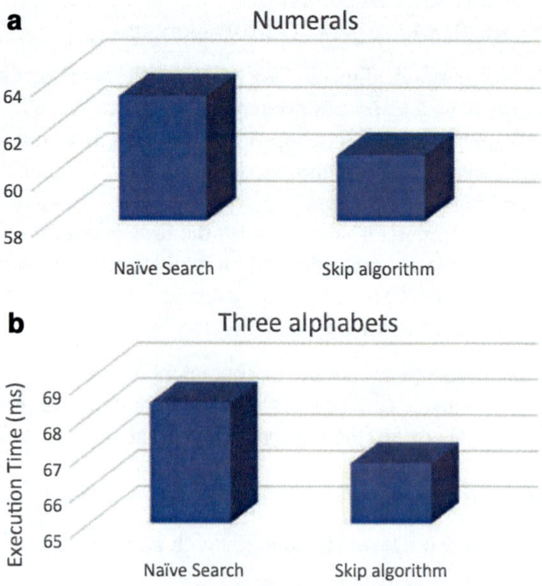

Fig. 1 **a** Skip algorithm versus Naïve for numerals. **b** Skip algorithm versus Naïve algorithm for three alphabets

Table 3 Input parameters for data set 2

Input	
Pattern size 'M' (characters)	16, 32,64, 128, 256,512,1024
Position of the pattern in text	512, 1024, 2048, 4096, not available

Data set 2

The second test data was generated with all possible combinations of pattern sizes and positions as in the following table for a constant input size of 8062 characters. The parameters used are shown in Table 3.

The result obtained are plotted in Fig. 2a–e (Fig. 3; Table 4).

The time complexity is directly proportional to the number of elements accessed from the given string. The average time complexity is $O(mn/2)$. The best case comes when the pattern to be searched is in the beginning which is $O(1)$. The worst case when the pattern is not found. Here too the average time complexity is $O(mn/2)$.

The space required for the algorithm is only the space for pattern and string. There is no extra space required. Hence, it is $O(m + n)$. This is less when compared to other algorithms and is shown in Table 5.

In all the test cases, the accuracy was 100%. The test case 1 was to verify the algorithm performance for three alphabet inputs as well as numerals. Skip algorithm gave a better performance than the Naïve search algorithm. Since the practical usage of the algorithm involves random text, test case 2 was used. The skip algorithm performed better than the Naïve search as well as the KMP algorithm. The time taken for execution decreased gradually as the search position was longer. Hence, the best performance of the skip search algorithm is when the search pattern is not found in the given text. The skip search converges quicker than the other algorithms. The $m - 1$ elements before the last $m - 1$ could be skipped in some cases. This can be identified, and further time can be saved. The algorithm can be further developed for multiple patterns multiple search problems. Conceptual graphs are used here because they enable humans to quickly understand what they should do being a victim during the disaster. The instances in disaster are linked to each other with some relations. The link connects whenever the relation exists.

5 Conclusion and Future Improvement

The data involved in pattern matching is drastically improving day by day along with the increase in storage capacity. An algorithm that performs search faster is always a need. The algorithm can be applied in search and replace operations. We have proposed a pattern matching algorithm which is a modified form of the Naïve search algorithm. Our algorithm does not require any preprocessing on the pattern of the text. The algorithm works with the logic of minimizing the number of locations accessed from the given text by checking if the element accessed is the first or last

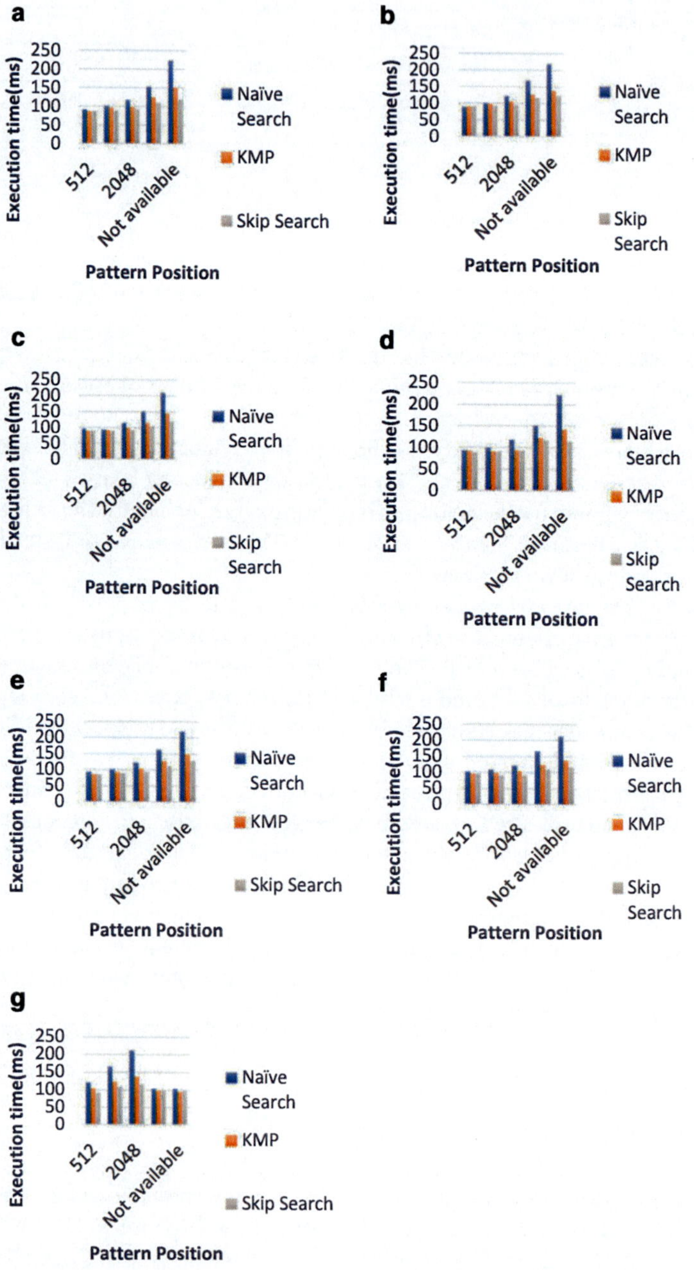

Fig. 2 **a** Pattern size 16. **b** Pattern size 32. **c** Pattern size 64. **d** Pattern size 128. **e** Pattern size 256. **f** Pattern size 512. **g** Pattern size 1024

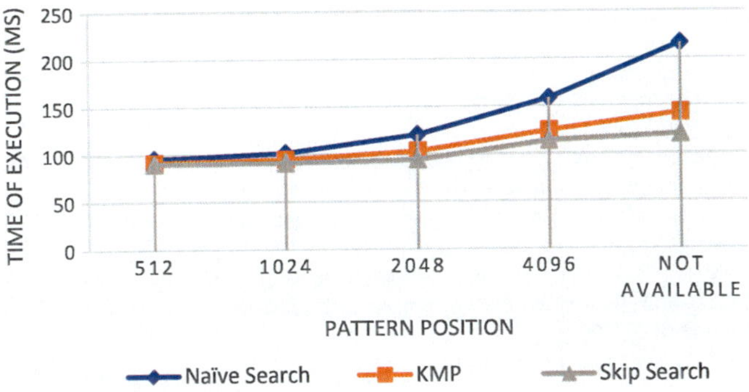

Fig. 3 Overall performance

Table 4 Average performance of all patterns

Pattern position	Naive search	KMP	Skip search
512	95.84	91.44	90.34
1024	101.74	94.99	91.22
2048	120.27	102.91	94.37
4096	157.98	124.17	113.62
Not available	215.21	142.65	120.18

Table 5 Complexity analysis

Algorithms	Preprocessing phase	Searching space	Extra space
Naïve search	–	$O(mn)$	–
Knuth–Morris–Pratt (KMP)	$O(m)$	$O(n)$	$O(m)$
Boyer–Moore (BM)	$O(m+\|\sum\|)$	$O(mn)$	$O(m+\|\sum\|)$
Boyer–Moore–Horspool (BMH)	$O(m+\|\sum\|)$	$O(mn)$	$O(\|\sum\|)$
Quick search (QS)	$O(m+\|\sum\|)$	$O(mn)$	$O(\|\sum\|)$
Boyer–Moore–Smith (BMS)	$O(m+\|\sum\|)$	$O(mn)$	$O(\|\sum\|)$
Reverse Colussi (RC)	$O(m^2)$	$O(n)$	$O(m+\|\sum\|)$
Bidirectional	$O(m+\|\sum\|)$	$O(mn)/2$	$O(m)$
Skip search	–	$O(mn)/2$	–

character in the given pattern. This can be further improved by accessing one element of text and checking that is a part of a pattern in any of the positions. Hence, the time can be further reduced.

The proposed algorithm can be used in applications like intrusion detection, plagiarism detection, bioinformatics, digital forensics, text mining research, string-based name retrieval, etc.

References

1. Cormen TH, Leiserson CE, Rivest RL (1990) Introduction to algorithms. MIT Press, Cambridge, pp 853–885
2. Knuth D, Morris JH, Pratt V (1977) Fast pattern matching in strings. SIAM J Comput 6(2):323–350. https://doi.org/10.1137/0206024
3. Boyer RS, Moore JS (1977) A fast string searching algorithm. Commun ACM 20(10):762–772
4. Horspool RN (1980) Practical fast searching in strings. Softw Pract Experience 10(3):501–506
5. Sunday DM (1990) A very fast substring search algorithm. Commun ACM 33(8):132142
6. Smith PD (1991) Experiments with a very fast substring search algorithm. Softw Pract Experience 21(10):1065–1074
7. Karp R, Rabin M (1987) Efficient randomized pattern-matching algorithms. IBM J Res Dev 31:249–260
8. Alqadi ZAA, Aqel M, El Emary IMM (2007) Multiple skip multiple pattern matching algorithm (MSMPMA). IAENG Int J Comput Sci
9. Commentz-Walter B (1979) A string matching algorithm fast on the average. In: Proceedings of 6th international colloquium on automata, languages, and programming, pp 118–132
10. Mansi RH, Odeh JQ (2009) On improving the Naïve string matching algorithm. Asian J Inf Technol (Medwell Joumals) 8(1):14–23
11. Hussain I, Kausar S, Hussain L, Khan MA (2013) Improved approach for exact pattern matching (bidirectional exact pattern matching). Int J Comput Sci Issues (IJCSI) 10(3):59–67
12. Lecroq T (1995) Experimental results on exact string matching. Softw Pract Experience 25:727–765
13. Sleit A, AlMobaideen W, Baarah AH, Abusitta AH (2007) An efficient pattern matching algorithm. J Appl Sci 7(18):269–2695
14. Ahmed M, Kaykobad M, Chowdhury RA (2003) A new string matching algorithm. Int J Comput Maths 80(7):825–834
15. Hudaib A, Al-Khalid R, Suleiman D, Itriq M, Al-Anani A (2008) A fast pattern matching algorithm with two sliding windows (TSW). J Comput Sci 4(5):393–401
16. Mohammad Ababneh, Saleh Oqeili, Abdeen Rawan A (2006) Occurrences algorithm for string searching based on brute-force algorithm. J Comput Sci 2(1):82–85
17. Colussi L (1994) Fastest pattern matching in strings. J Algorithms 16(2):163–189

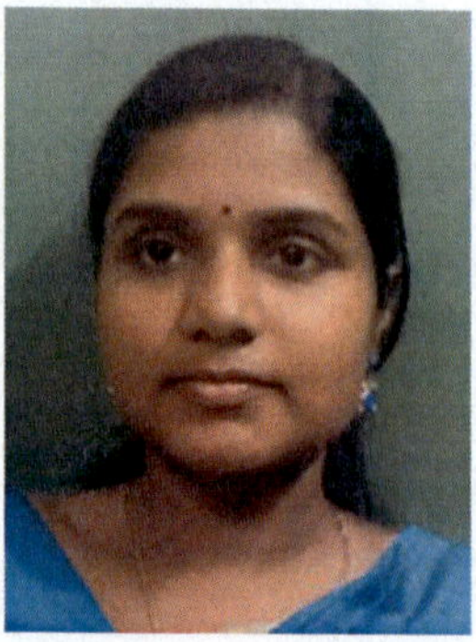

Dr. K. Padmaveni received Ph.D. degree in cloud computing from Hindustan Institute of Technology and Science. She is currently an Associate Professor in Hindustan Institute Of Technology and Science and her main interest are algorithms, virtualization and scheduling in cloud.

Dr. D. John Aravindhar received Ph.D. degree in Data mining from Hindustan Institute of Technology and Science. He is currently a Professor in Computer Science Department of Hindustan Institute of Technology and Science. His area of interest are Algorithms, Data mining and cloud.

Classification of Plant Leaf Using Shape and Texture Features

A. Sujith and R. Neethu

Abstract Plants are mainly classified based on their characteristics of plant components such as leaves, flower, stem, root, seed, etc. Feature or characteristics is an essential fact for plant classification. A good feature extraction technique can help to extract quality features that give clear information to discriminate against each class. Computer engineers can help botanists to identify plants and their species through advanced computational techniques with the stipulated time. The proposed method gives efficient hybrid feature extraction using the PHOG, LBP, and GLCM feature extraction techniques. The fused feature vector is normalized and reduced size by Neighborhood Components Analysis (NCA). The efficient feature extraction and feature selection techniques have helped to improve the classification performance and reduced the model complexity. Two benchmark plant dataset Flavia and Swedish Leaves used to evaluate the proposed work. The primary contributions of this paper arc introducing a multi-feature fusion shape and texture method for plant leaf image classification. The experimental result shows the average accuracy of the proposed method is 98.23%, and the average computational complexity is 147.98 s.

Keywords Plant classification · Pyramid histogram of oriented gradients · Local binary pattern · Gray level co-occurrence matrix · Neighborhood components analysis

A. Sujith (✉)
Department of Computer Science, Research Centre, University of Kerala, Thiruvananthapuram, Kerala, India
e-mail: sujdcbin@gmail.com

R. Neethu
Department of Computer Science, University Institute of Technology, Mukhathala, Kollam District, Kerala, India

G. Ranganathan et al. (eds.), *Inventive Communication and Computational Technologies*, Lecture Notes in Networks and Systems 145,
https://doi.org/10.1007/978-981-15-7345-3_22

1 Introduction

The plant plays a vital role in the survival of every living organism on earth. They help to maintain and protect the ecosystem. Wäldchen et al. [1], traditional way, the plant species classification is a difficult task which is time-consuming and frustrating for novices. Wet lab process is done by experts having good knowledge about the plants. In the modern era, plant classification and identification from images are made by computer engineers through advanced computational techniques. Develop a prediction model for the classification of the plant species. The main challenge of the prediction model is to enhance the accuracy of the model. The main factor is extracted useful features from the image, choose relevant features using feature selection algorithm, and choose the correct classifier. The automatic plant classification and recognition system consist of two main modules: (1) Feature extraction module, (2) Recognition modules. Here we concentrate on the feature extraction module to extract efficient features for improving the efficiency of the prediction model. The plant features can take from leaf, flower, seed, bark, root, etc. Existing Feature extraction techniques have merits and demerits. A combination of these feature extraction techniques may enhance the feature quality and improve the performance of the model. Texture, shape, color, and geometrical features are an important feature for using any recognition model. The proposed model used combined feature extraction techniques based on texture and shape feature extraction methods to extract efficient features for the classification model. The potential application of plant classification and identification model may use in the field of agriculture, pharmaceutical, production of cloth, and spices.

In recent years numerous research works have shown that shape and texture feature has an important role in plant classification and recognition. Kaya et al. [2] used geometric, LBP, HOG features, and comparison of Extreme Learning Machine (ELM), SVM classifier for Oak classification. Geometric features with ELM classifiers achieve 75%, and with SVM achieves 75.60% classification accuracy. HOG features with ELM give 88.60% with SVM gives 86.10%. LBP features with an ELM classifier gives 90.40% accuracy and with SVM classifier gives 89.20% accuracy. Weighted score fusion of LBP and HOG based ELM classifier achieves 92.17% classification accuracy with Oak leaves dataset. Janahiraman et al. [3] used LBP and HOG feature combinations with SVM classifier. The prediction model achieves 90.22% in the Swedish dataset and 77.33% in the Flavia dataset. Rajapaksa et al. [4] classify crop lodging using GLCM and LBP texture feature, Gabor filters with SVM classifier. The dataset contains drone-based images of canola and wheat. Sharma et al. [5] used the HOG feature extraction method and to classify and identify leaf by using KNN and backpropagation ANN. ANN classification has better performance with 97% than KNN with 92.3% classification accuracy.

The subsequent section of the paper has been categorized; Sect. 2 illustrates the proposed methodology; Sects. 3 exemplifies the results and discuss the details of the results. The final Sect. 4 concludes the outcome and future enhancement of the proposed work.

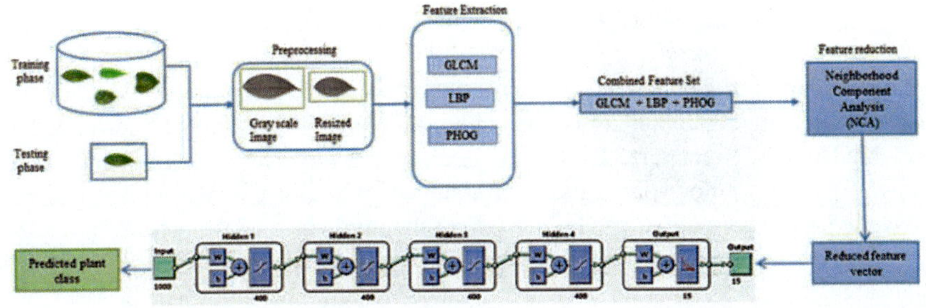

Fig. 1 The architecture of the proposed system

2 Methodology and Experiment Analysis

2.1 Proposed Model

The proposed work consists of five stages—preprocessing, feature extraction, feature fusion, feature selection, and reduction, and classification. The combination of three feature extraction techniques, PHOG, LBP, and GLCM used to build the proposed model. This proposed model provides better results than state of the art methods. Figure 1 shows a diagrammatic representation of the proposed work. In the preprocessing step, the color image is converted to grayscale and resized in a fixed size. Next extract features using PHOG, LBP, and GLCM techniques. The extracted features fused and get a new feature vector. The combined feature extraction technique has a vital role in getting better classification performance. Relevant feature selection is made with the help of neighborhood components analysis (NCA). Selected feature vector inputted to classifier for classification. The proposed system uses two benchmark datasets Swedish Leaves and Flavia. The system has three phases training, validation, and testing, and its data division ratio is 80:10:10, respectively.

2.2 Preprocessing

The captured images are usually messy or noisy. To feed the image to machine learning algorithms, they need to be standardized, clean up, and mold to our application. Preprocessing used to do these steps that reduce complexity and increase accuracy. Leaf image accusations and preprocessing are reflected in the accuracy of classifier Rzanny et al. [6]. Preprocessing enhances image quality, and it plays a significant role in various applications Hamuda et al. [7], by filtering the images. In the proposed work, we used two benchmark datasets, and images do not have the same size and tiff file format. This color image is converted into a grayscale image for reducing the complexity and unwanted information. This grayscale image has resized into a fixed size of 150×200.

$$\text{Gray} = 0.2989 * R + 0.5870 * G + 0.1140 * B \tag{1}$$

According to this equation, red has contributed 30%; Green has 59%, which is higher in all three colors, and blue has to contribute 11%. Equation 1 uses called a weighted method to convert RGB to a grayscale image and gives the correct grayscale image. Different color spaces have different weights. A true-color image can be converted to a grayscale image by preserving the luminance of the picture.

2.3 Feature Extraction

After preprocessed, we obtained noise-free images. Feature extraction techniques apply to this image. Standard features such as shape, texture, color, and geometrical features can be extracted with the help of computation algorithms. Feature extraction techniques can extract features and make it a vector form. Jiang [8] shows the analysis of various feature extraction techniques organized in four categories human expert knowledge-based methods, image local structure-based approach, image global structure-based techniques, and machine learning-based statistical approaches. This paper analyses four feature extraction methods HOG, PHOG, GLCM, and LBP. Finally, we combine a better combination among these used in our model. In our model, we fused multiple feature extraction techniques (PHOG, LBP, and GLCM combination).

Histogram of Oriented Gradients (HOG)

Xiao et al. [9] treat HOG as a new representation of shape used for plant leaf classification. HOG feature calculated using gradient magnitude and gradient angle. If a pixel in leaf images is described as $P(x, y)$ where (x, y) is, respectively, the horizontal and vertical coordinates of the pixel, the gradient value can be given by:

$$g(x, y) = \sqrt{\Delta x^2 + \Delta y^2} \tag{2}$$

The gradient direction is given by

$$\theta(x, y) = \arctan \frac{\Delta_x}{\Delta_y} \tag{3}$$

Compute all the gradient values and magnitude value from the given patch of the image and build a histogram Gradient direction nine bins for each block calculated the feature vector size should be 9. Concatenate the feature vector value of four adjacent blocks and make it as a single vector. The normalization process applies after each single vector creation. Similarly, one block shift to the right of the row then does this process till the grid reaches the bottom right corner of the image.

Pyramid Histogram of Oriented Gradients (PHOG)

The Pyramid Histogram of Gradients is an extended version of the HOG feature descriptor. This spatial shape descriptor was introduced by Dalal and Triggs [10]. In this paper, PHOG preserves the spatial layout of the image by dividing the image into different sub-regions at multiple resolutions and apply the HOG descriptor in each sub-region. To compute the PHOG features, first convert all images in grayscale. Extract edge counter using a canny edge detector. This image divide into different sub-regions. The spatial pyramid is created with three levels, and set orientation bins are 8 in the range angle is 0° to 360°. HOG is calculated for all eight bins in each level. The final PHOG vector for an image is the concatenation of all the HOG vectors at each level.

Gray Scale Co-occurrence Matrix (GLCM)

A gray-level co-occurrence matrix (GLCM) is a popular statistical texture analysis model. The texture is a spatial arrangement of the pixel value. Using a single pixel, we can't define its texture. GLCM does not consider the pixel neighborhood relation and considers the probability of a particular gray level in that location. In the proposed model, compute GLCM following way. Choose a 2D square matrix value of the given image and compute the 2D co-occurrence matrix. Next, calculate the symmetric GLCM with the summation of the 2D co-occurrence matrix and its transpose. Symmetric GLCM has to be normalized. This normalized GLCM matrix used to calculate various statistical features such as contrast, entropy, variance, correlation, etc. Chaki et al. [11] show 11 statistical features calculated by the GLCM matrix.

Local binary patterns (LBP)

The local binary pattern is a texture feature extraction technique proposed in Ojala et al. [12]. Local binary pattern feature extraction technique computes a local representation of texture. This local representation is constructed by comparing each pixel with its surrounding neighborhood of pixels. Here we select neighborhood size is eight, and we have a total of 2^8 or 256 possible combinations of LBP code. The following step perform in this proposed model for extract LBP feature. First, convert color image to grayscale image then calculate the LBP value of each pixel from this image. Third, we store this LBP value into2D array. Compute a histogram over the 2D array of LBP codes.

2.4 Feature Fusion

LBP texture feature can reflect the intensity relation between a pixel and their neighbor. But spatial location and gray level information is lost LBP capture local pattern and uses eight directions of each pixel. GLCM may consider the spatial relationship of pixels and efficiently extract surface information of an image. Shang and Li [13] GLCM-based approach has good performance in terms of computational complexity and CPU processing time. HOG uses only one direction for each pixel is compared to LBP. PHOG is an extended version of HOG and less processing time. In the proposed model, concatenate PHOG, GLCM, and LBP features for plant classification. The major problem of this type of fusion is the size of the feature vector. Reducing the size of the feature vector, we used the NCA feature selection and dimensionality reduction technique.

2.5 Feature Selection and Reduction

This section selects relevant features used to build a good machine learning model. For the automatic selection of most important features, choose feature selection techniques such as PCA, NCA, etc. In this proposed model, use the Neighborhood Component Analysis algorithm (NCA).

Neighborhood Component Analysis (NCA)

Manit and Youngkong [14] NCA is used in the k-nearest neighborhood classification algorithm and capable of learning low dimensional labeled data with Mahalanobis distance measure. NCA is a supervised metric learning algorithm, and it transfers data points in new space. The distance between two data points in the same class is small, and it is large in a different class. This metric learner use to learn distance with the help of Mahalanobis distance measure. The weights of the unwanted features should be close to zero. NCA can do linear dimensionality reduction with low-rank distance. In this case, the learned metric will be of low rank. The selected feature set ranked based on maximum feature weights and chose maximum weighted 1500 elements to form a new feature vector.

2.6 Classification

Che et al. [15] backpropagation neural network to train a feed-forward neural network have better outcomes than the genetic algorithm to train a feed-forward neural network. The proposed model classifies plant leaves by using Feed Forward back propagation Neural Network classifier. The combined feature vector will be the input of the Feed Forward Neural Network, which reduces the output layer through four hidden layers. We selected 1500 features out of 2354 by using NCA so that the input layer formed with 1500 neurons. After the evaluation with a different number of neurons, we found that each hidden layer with 400 neurons gives better classification accuracy in less computational time.

Performance evaluation metrics: This metric helps to evaluate the efficiency and performance of the proposed plant classification model. Precision is the measure of accuracy for the specified class as predicted. The recall is the measure of the prediction model to select instances from a certain class. F1 score is a useful metric when data is imbalanced. It is the harmonic mean of recall and precision. Cross entropy loss helps to measure the performance of the predicted model. The predicted probability value reaches one, then cross-entropy value towards zero.

$$\text{Acccuracy} = \left(\frac{\text{No. of correctly recognized samples}}{\text{total no. of samples}}\right) \times 100\% \tag{4}$$

$$\text{Precision} = \frac{\sum \text{True Positive}}{\sum \text{Predicted Condition Positive}} \times 100 \tag{5}$$

$$\text{Recall} = \frac{\sum \text{True Positive}}{\sum \text{Condition Positive}} \times 100 \tag{6}$$

$$F1\text{Score} = 2 \times \frac{\text{Precision} \times \text{Recall}}{\text{Precision} + \text{Recall}} \tag{7}$$

$$\text{Cross entropy loss} = -\sum_{c=1}^{M} y_{o,c} \log(P_{o,c}) \tag{8}$$

where M—Number of classes, log—the natural log, y—Binary indicator (0 or 1) if class label c is the correct classification for observation o, P—Predicted probability observation o is of class c.

3 Result and Discussion

3.1 Dataset

The introduced technique is trained and tested with two publically available benchmark data set Swedish Leaves and Flavia. The Swedish data set in Söderkvist [16], consists of 15 tree species with 1125 plant 24-bit RGB images with white background (Fig. 2). Each class contains 75 images and file various dimensions with the tiff file format. The proposed model resizes the image to 150 × 250. Both data set divided into 80% for training, 10% for validation, and 10% for testing. Flavia dataset in Wu et al. [17] consists of 32 plant classes with 1907 RGB plant images with white background. Each class contains a different number of images—all images jpg file format with 1600 × 1200 file dimensions. The proposed model used 1600 images; each class consists of 50 images. Images have to resize with 250 × 150 file dimensions.

Fig. 2 Swedish Leaves dataset

3.2 Experimental Results Analysis

This section analyses various texture and shape feature extraction methods such as GLCM, LBP HOG, and PHOG, respectively, by using the feed-forward with back-propagation classifier with the Swedish and Flavia dataset. Our model outperforms with the Swedish dataset.

Analysis 1: Analyses shape feature and texture feature using HOG, PHOG GLCM, and LBP feature extraction methods.

Conclusion 1: Table 1 shows the classification performance metrics of each feature extraction method. From this analysis, we conclude the LBP and PHOG feature extraction techniques are higher classification accuracy and speed of the model. The study of Table 2 shows LBP achieve 97.34% classification accuracy with 125.08 s. The average cross-entropy loss is 0.0010. Table 4 shows PHOG achieve 93.09% classification accuracy with 69.78 s.

Analysis 2: HOG, PHOG, GLCM, and LBP feature extraction techniques with possible combination analysis.

Conclusion 2: Combined feature extraction technique helps to improve the model classification performance and minimize computational time. Table 2 shows a combination of PHOG, GLCM, and LBP achieve 98.23% classification accuracy with 147.98 s.

Figure 3 shows the result of a confusion matrix after testing the data. This matrix helps to evaluate the correctness of the proposed classification model. The *X*-axis represents the predicted class, and *Y*-axis represents an actual class. Diagonal elements represent true positive (TP) values, which are the total numbers of correctly classified of each class. Other main terminologies false negative (FN), false positive (FP) and true negative (TN). From this matrix, we can calculate model performance metrics such as accuracy, precision, recall, and F1-Score. The bottom left corner value 98.23% is the overall accuracy of the model, and 1.8 indicates the overall misclassification rate. FN indicates the sum of the corresponding row value of each class, excluding the TP value. FP indicates the sum of the corresponding column

Table 1 GLCM, LBP, HOG, and PHOG feature extraction analysis with Swedish Leaves

	Precision	Recall	F1 score	Accuracy (%)	Time (s)	Loss
GLCM	0.8143	0.8035	0.8089	80.53	73.74	0.0286
	0.7552	0.7495	0.7523	74.33	38.68	0.0369
	0.8143	0.8035	0.8089	80.53	74.09	0.0285
	0.8303	0.8507	0.8404	83.18	44.97	0.0305
	0.7969	0.7751	0.7859	77.87	61.68	0.0301
	0.7448	0.7477	0.7462	75.22	39.15	0.0400
	0.7873	0.7807	0.7840	78.76	66.94	0.0332
	0.8272	0.8332	0.8302	83.18	67.20	0.0211
	0.8258	0.8062	0.8159	83.18	45.64	0.0321
	0.8290	0.8351	0.8320	84.07	50.38	0.0271
Mean values	0.8025	0.7985	0.8004	**80.08**	**56.24**	0.0308
LBP	0.9683	0.9683	0.9683	97.34	125.32	0.0012
	0.9814	0.9759	0.9786	97.34	121.57	0.0007
	0.9870	0.9837	0.9853	98.23	157.12	0.0002
	0.9483	0.9592	0.9537	95.57	115.76	0.0013
	0.9666	0.9750	0.9708	97.34	125.08	0.0009
	0.9783	0.9735	0.9759	97.34	104.02	0.0014
	0.9628	0.9655	0.9641	96.46	122.49	0.0018
	0.9865	0.9833	0.9849	98.23	128.61	0.0003
	0.9828	0.9745	0.9786	97.34	155.03	0.0004
	0.9865	0.9822	0.9843	98.23	107.64	0.0015
Mean values	0.9666	0.975	0.9708	**97.34**	**125.08**	0.0010
HOG	0.8773	0.8672	0.8722	86.73	697.87	0.0061
	0.8609	0.0061	0.8617	85.84	758.51	0.0061
	0.7963	0.8362	0.8158	80.53	749.19	0.0080
	0.8681	0.8435	0.8557	84.96	751.35	0.0071
	0.9458	0.9252	0.9354	92.92	717.42	0.0043
	0.9476	0.9663	0.9569	96.46	1059.71	0.0024
	0.9018	0.9086	0.9052	90.27	1050.98	0.0973
	0.9193	0.9283	0.9238	92.04	1156.87	0.0057
	0.9000	0.8823	0.8911	88.50	1052.18	0.0068
	0.8717	0.8674	0.8695	87.61	943.46	0.0057
Mean values	0.8889	0.8031	0.8887	**88.58**	**893.75**	0.0150
PHOG	0.9376	0.9409	0.9393	93.80	69.21	0.0044
	0.9242	0.9287	0.9264	92.03	78.31	0.0035
	0.9469	0.9463	0.9466	93.80	68.70	0.0044

(continued)

Table 1 (continued)

	Precision	Recall	F1 score	Accuracy (%)	Time (s)	Loss
	0.9394	0.9318	0.9356	93.80	70.76	0.0045
	0.8945	0.8951	0.8948	91.15	66.20	0.0060
	0.9497	0.9497	0.9497	94.69	69.89	0.0045
	0.9011	0.9181	0.9096	92.03	61.13	0.0085
	0.9405	0.9492	0.9448	95.57	73.58	0.0019
	0.9362	0.9374	0.9368	93.80	67.87	0.0051
	0.8976	0.9225	0.9099	90.26	72.18	0.0049
Mean values	0.9268	0.9320	0.9293	**93.09**	**69.78**	0.0048

Table 2 Performance metrics for combination of feature extraction with Swedish Leaves

Dataset	Algorithm	Precision (Mean)	Recall (Mean)	F1 score (Mean)	Accuracy (%) (Mean)	Time (s) (Mean)	Cross entropy
Swedish Leaves	PHOG + GLCM	0.9466	0.9466	0.9466	94.10	86.15	0.0035
	PHOG + LBP	0.9738	0.9707	0.9723	96.99	158.39	0.0009
	PHOG + GLCM + LBP	0.9821	0.9817	0.9820	**98.23**	**147.98**	0.0008
Flavia	PHOG + GLCM	0.9410	0.9312	0.9360	93.13	154.84	0.0017
	PHOG + LBP	0.9646	0.9637	0.9641	96.25	379.18	0.0008
	PHOG + GLCM + LBP	0.9686	0.9686	0.9686	**96.66**	**212.90**	0.0012

value of each class excludes TP value.TN indicates the sum of all column and row values, excluding corresponding class column and row values.

Figure 4 shows the result of the Receiver Operating Characteristics (ROC) curve after training, validation, and testing the data. ROC curve visualizes the performance of a proposed classification model. The *X*-axis represents the False Positive Rate (FPR), and *Y*-axis indicates the True Positive Rate (TPR). The ROC curve fall in the top left corner area indicates a good performance level. A diagonal line or baseline of the ROC curve represents a purely random classifier. The curve always above the baseline (towards the top left corner) indicates the model classifier is good. The Swedish Leaves dataset has 15 tree species; each curve represented single species with a different color.

Figure 5 shows the network performance of the proposed model. This graph deter-

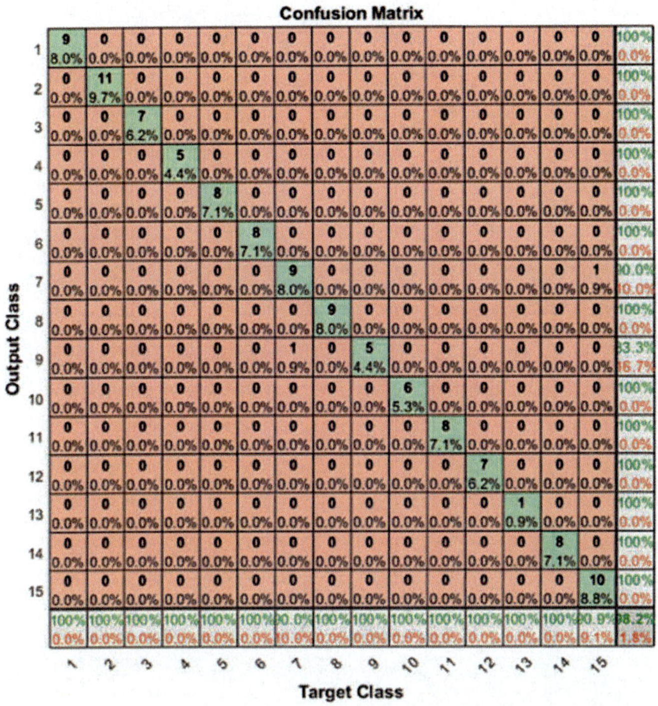

Fig. 3 Confusion matrix obtained for testing with Swedish Leaves

mines to check whether any changes have to be made in the training process, the network architecture, or the dataset. These three lines indicate training, validation, and testing have to be done significantly better performance. The training, validation, and test curves are diminishing and hence do not indicate overfitting. The best validation performance is 0.0014296 was obtained at epoch 66. When it reaches the minimum validation value, it can stop validation. Here uses a cross-entropy loss function for evaluating the proposed model. The average cross-entropy loss of 0.0 is a perfect model.

4 Conclusion and Future Enhancement

In this paper, four feature extraction methods PHOG, HOG, LBP, and GLCM used to analyses the features from the plant image. The computational complexity of the HOG feature extraction method is higher among these four. After the analysis, we found the combined feature set has better results than the individual feature set. We combined PHOG, LBP, and GLCM to build our proposed model. To reduce the size of the feature vector and for feature selection, we used the NCA technique.

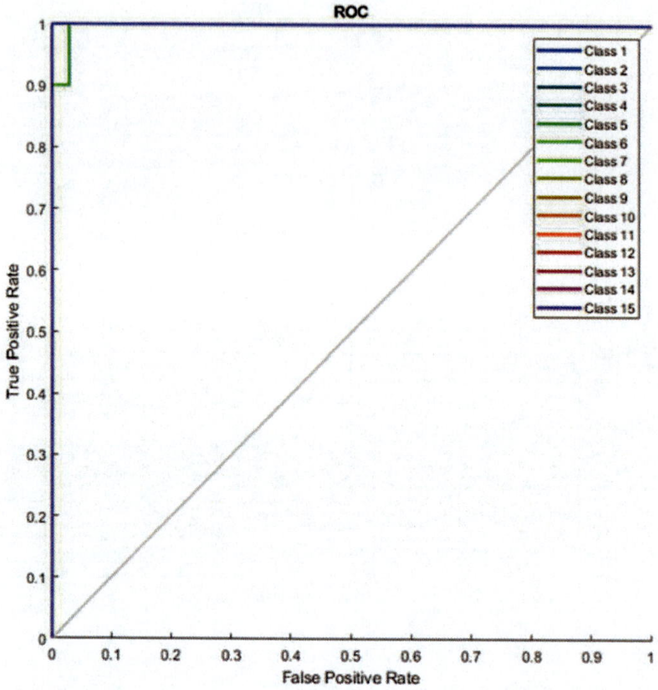

Fig. 4 Receiver operating curve for the testing with Swedish Leaf dataset

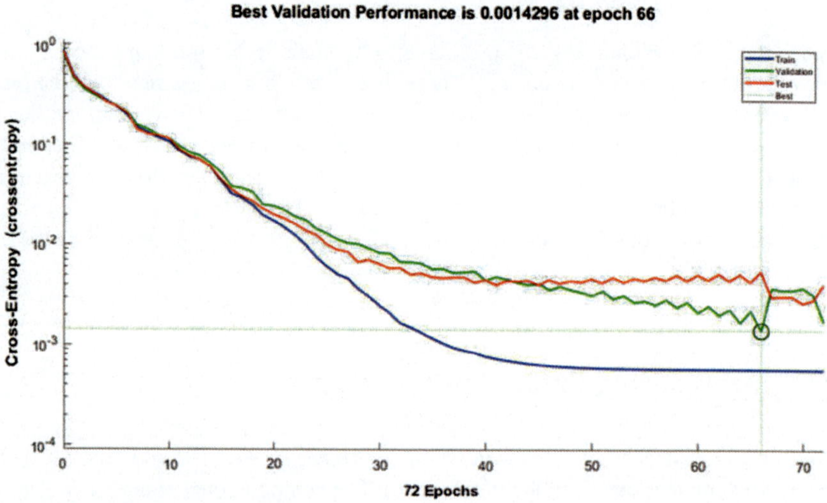

Fig. 5 Network performance analysis

NCA selects the relevant feature from the feature set. Dataset Swedish Leaves and Flavia are used for training, validating, and testing. The proposed model gives better results in Swedish leaves than the Flavia dataset. The individual feature set such as GLCM, HOG, and PHOG and LBP produce classification accuracy 80.08% in 56.24 s, 88.58% in 893.75 s, 93.09% in 69.78 s, 97.34% in 125.08 s, respectively. Combined feature extraction techniques such as PHOG and GLCM, PHOG and LBP, PHOG, LBP, and GLCM produce classification accuracy 94.10% in 86.15 s, 96.99% in 158.38 s, and 98.23% in 147.98 s, respectively. The performance of the combined feature set with the latest classification technique such as deep learning, transfer learning techniques will be used to be exploring in the future.

References

1. Wäldchen J, Rzanny M, Seeland M, Mäder P (2018) Automated plant species identification—trends and future directions. PLoS Comput Biol 14(4):e1005993
2. Kaya H, Keklık İ, Ensarı T, Alkan F, Bırıcık Y (2019) Oak leaf classification: an analysis of features and classifiers. In: 2019 Scientific meeting on electrical-electronics and biomedical engineering and computer science (EBBT). IEEE, pp 1–4
3. Janahiraman TV, Yee LK, Der CS, Aris H (2019) Leaf classification using local binary pattern and histogram of oriented gradients. In: 2019 7th International conference on smart computing& communications (ICSCC). IEEE, pp 1–5
4. Rajapaksa S, Eramian M, Duddu H, Wang M, Shirtliffe S, Ryu S, Josuttes A, Zhang T, Vail S, Pozniak C, Parkin I (2018) Classification of crop lodging with gray level co-occurrence matrix. In: 2018 IEEE winter conference on applications of computer vision (WACV). IEEE, pp 251–258
5. Sharma P, Aggarwal A, Gupta A, Garg A (2019) Leaf identification using HOG, KNN, and neural networks. In: International conference on innovative computing and communications. Springer, Singapore, pp 83–91
6. Rzanny M, Seeland M, Wäldchen J, Mäder P (2017) Acquiring and preprocessing leaf images for automated plant identification: understanding the tradeoff between effort and information gain. Plant methods. 13(1):1–1
7. Hamuda E, Glavin M, Jones E (2016) A survey of image processing techniques for plant extraction and segmentation in the field. Comput Electron Agric 1(125):184–199
8. Jiang X (2009) Feature extraction for image recognition and computer vision. In: 2009 2nd IEEE international conference on computer science and information technology. IEEE, pp 1–15
9. Xiao X-Y, Hu R, Zhang S-W, Wang X-F (2010) HOG-based approach for leaf classification. In: International conference on intelligent computing. Springer, Berlin, Heidelberg, pp 149–155
10. Dalal N, Triggs B (2005) Histograms of oriented gradients for human detection. In: 2005 IEEE computer society conference on computer vision and pattern recognition (CVPR'05), vol 1. IEEE, pp 886–893
11. Chaki J, Parekh R, Bhattacharya S (2015) Plant leaf recognition using texture and shape features with neural classifiers. Pattern Recogn Lett 58:61–68
12. Ojala T, Pietikäinen M, Harwood D (1996) A comparative study of texture measures with classification based on featured distributions. Pattern Recogn 29(1):51–59
13. Shang Z, Li M (2016) Combined feature extraction and selection in texture analysis. In: 2016 9th international symposium on computational intelligence and design (ISCID), vol 1. IEEE, pp 398–401
14. Manit J, Youngkong P (2011) Neighborhood components analysis in sEMG signal dimensionality reduction for gait phase pattern recognition. In: 7th International conference on broadband communications and biomedical applications. IEEE, pp 86–90

15. Che ZG, Chiang TA, Che ZH (2011) Feed-forward neural networks training: a comparison between genetic algorithm and back-propagation learning algorithm. Int J Innov Comput Inf Control 7(10):5839–5850
16. Söderkvist O (2001) Computer vision classification of leaves from Swedish trees
17. Wu SG, Bao FS, Xu EY, Wang YX, Chang YF, Xiang QL (2000) A leaf recognition algorithm for plant classification using probabilistic neural network. In: 2007 IEEE international symposium on signal processing and information technology. IEEE, pp 11–16

Value-Based Behavioral Analysis of Users Using Twitter

Surbhi Kakar, Deepali Dhaka, and Monica Mehrotra

Abstract In daily lives, the way people use words depicts their beliefs, opinions, and values. Each individual possesses some set of value system or other. Different geographical regions may experience varying value systems. The decisions a person takes in everyday life are influenced by these value systems. This work uses the Twitter social network to collect tweets from different regions and proposes a value-based lexicon to identify the behavior of users in terms of their value systems. The objective of this work is to inspect the prominent value systems possessed by the majority of people in these regions with the help of a value-based lexicon method. The analysis demonstrated that the majority of users in the Delhi region fall in the Self-Transcendence value system and that of Washington's fall in the Hedonism value system. This indicated that users in the Delhi region are more inclined towards social welfare whereas users in the Washington region are involved more in self-indulgence.

Keywords Social networks · Twitter · Value systems

1 Introduction

Social networks are the platform where users can disseminate information irrespective of the physical barrier between them. Recently, social networks have become a source of the humongous amount of data. One of the examples of such a social network is Twitter. Twitter is a micro-blogging platform where users share short

S. Kakar (✉) · D. Dhaka · M. Mehrotra
Department of Computer Science, Jamia Millia Islamia, New Delhi 110025, India
e-mail: Kakar.surbhi3@gmail.com

D. Dhaka
e-mail: deepali.dhaka@gmail.com

M. Mehrotra
e-mail: drmehrotra2000@gmail.com

G. Ranganathan et al. (eds.), *Inventive Communication and Computational Technologies*, Lecture Notes in Networks and Systems 145,
https://doi.org/10.1007/978-981-15-7345-3_23

messages by using the Tweet button [1]. The platform provides the following actions that a user can perform:

- The reply is an action invoked to reply to a specific post/message.
- Retweet: action is performed to spread a message further. A person retweeting a specific post signifies the interest of that user in the post.
- Like: action denotes that the post is liked by the user who hit the Like button.

The content shared by the user is an indicator of user behavior [2].

One of the major contributions done in the area of studying behaviors in the form of value systems has been done by [3]. The work proposed a basic human value system across nations. According to Schwartz, values are beliefs that people hold about themselves or other people and circumstances. Values are one of the promising and important features to study user behavior and the choices he/she may take in the future [4]. Value systems can be broadly categorized into the following types [5, 6].

1.1 Self-transcendence

People belonging to this value system mainly encompass two characteristics:

- Universalism: These people believe in inner harmony, peace, wisdom, protecting the environment, and social justice. They are interested in the well-being of other people.
- Benevolence: Such people believe in being spiritual, responsible, honest, and care about their close relationships and family.

1.2 Self Enhancement

People in this category are more involved with their interests. Following are the two significant facets of this category:

- Power: Such people talk about power and authority. They are more concerned about their wealth and social recognition and are supposed to be influential.
- Achievement: These people are concerned about their success and achievements.

1.3 Hedonism

This category consists of values of self-indulgence, enjoyment, and pleasure. People in this category talk more about sex, pleasure, and other enjoyment activities.

1.4 Openness to Change

This value system comprises the below human values:

- Stimulation: People in this category are daring, and have a desire to live an exciting and varied life.
- Self-Direction: Such people are self-directed. They believe in independence and have their independent thought. They are also curious and creative.

1.5 Conservation

This value system is composed of the below given human values:

- Tradition: Such people believe in customs, traditions, and religions.
- Conformity: People in this category believe in conforming themselves to the rules and regulations they believe in. They are polite, honor elders, and self-disciplined. They refrain from upsetting other people as they believe in looking good to the outer world.
- Security: Such people are more concerned about their security, their family's security, and national security. They also believe in social order and cleanliness.

This work uses these value systems and intends to perform a content-based analysis of the messages people share on Twitter, thereby, labeling the users with their values. The objectives of this paper can be stated as follows:

1. Building a lexicon for each type of value system defined above.
2. Labeling the users on Twitter with their corresponding value systems based on their geographical region. The geographical regions considered for this work are Delhi and Washington. The two regions are specifically taken up to study value systems of people residing in a developing versus a developed region. As per [5], it has been observed that different regions may experience varying patterns of value systems. This served as an inspiration for this work to inspect the value systems across different geographical regions.
3. Extracting the prominent value systems in each region.
4. Validating the results generated from the lexicon.

The benefits of this work are multi-fold.

(i) This work can enable social activists who are interested in targeting relevant people in a social network for initiating and promoting their campaign. People belonging to the category of Self-Transcendence are often inclined towards the well-being of the public. This lexicon can help identify such users and can be useful in targeting them.
(ii) Knowing the value system of a user, one can also use it to predict what decisions will a user take in the future, like predicting, if a user will retweet a certain

content or not. Retweet Prediction is a problem where one can predict if a user will retweet a given Tweet or not [7].

(iii) Identifying users who post sensitive content on social media is another application of this work [8]. People, who fall in the hedonistic category of value systems, talk more about sex and are more involved in pleasure-seeking activities. Such users are more suspected to post sensitive content on social media. The lexicon proposed can help in identifying such users with high values under this category.

2 Literature Review

An extensive amount of work has been published around this area, studying the various value systems and its applications. Some of them can be referred to in [9–12].

Reference [9] applied Schwartz's value system theory to study user attitudes and behavior. Reference [10] studied value systems to further infer socially responsible consumer behavior in the context of environmental attitudes.

Another work proposed by [11] examined the relationship between religiosity and value systems. It was concluded by the results that more religious people believe in valuing certainty, conformity, and refrainment from offending others. Few researchers also inspected the relationship between the levels of trust of people in institutions with their respective value systems [12].

Some of the researchers in [6, 13–16] show the significance of content-based analysis of value systems. Reference [13] proposed in ongoing research that human values can be classified automatically performing a content-based analysis, however, the results are not conclusive yet.

Another work by [14] demonstrated a technique for automatically classifying Schwartz's values to each text spoken in a public debate. However, this analysis was based on a restricted topic of public debate. Also, the data was collected using Amazon Mechanical Turk. There is a probability of inauthentic responses from such services as the volunteers are paid money for recording their responses on this platform. This can lead to a non-serious/inauthentic response recording from the volunteers. Another such work was suggested by [15] where they collaborated human annotators and machines to identify values in a text. A limitation of their approach was that the data used for manual analysis is very small.

Reference [16] investigated some of the feature words associated with the category of people who are conservative and liberal. They also used a classifier to classify incoming posts in the former categories. Reference [6] analyzed word use and the user's value system. They used LIWC to correlate various LIWC categories with each value system [17]. They also confirmed that the content used by people on social media can be a promising predictor of people's value systems.

Recent works exploiting personal values include those in [18–22].

Reference [18] study the various value systems of managers and affirm that these value systems shape their leadership styles in a company. They used a survey method for collecting data from respondents to identify value systems.

Another work suggested by [19] examined bipolar value dimensions and how they impact on travel decisions made by young adults. They also used a survey method to predict young adult's behavior.

It has been noted that personal values also shape foreign policy orientations, voting preferences, and healthcare [20–22]. In addition to it, [23] explored value systems with a dimension of concern for animals.

Apart from working on value systems to study behaviors, major works have also been proposed on studying personality traits [24–28]. However, several studies prompt that studying value systems is a more promising predictor of behavior than studying personality traits [29–31]. Hence, this manuscript is focused on studying value systems.

The research methodology is inspired by works demonstrated in [6, 16]. The research works defined in the literature above either uses a survey method to collect data or works on a restricted topic to identify value systems. Survey methods as discussed by [13] have their own limitations. To the best of our knowledge, this work is the first to propose a value-based lexicon called, ValueDict and present a comparative analysis of content-based data from two different regions based on value systems. Also, this work is not limited to just studying tweets on a single topic. This work crawls tweets for different users which incorporates their tweets on various topics, thereby not confining to a single topic.

The rest of the paper is organized as follows. Section 3 shows the approach for data collection. Section 4 discusses the methodology adopted for this research. In Sect. 5, results are shown. Section 6 the approach for validation is presented. Section 7 summarizes the conclusion and future scope of this work.

3 Data Collection

Data for the experimentation purpose was crawled using the Twint API. It is an advanced Twitter scraping tool that has been written in the Python language [32, 33].

Initially, 60 users from each region, with a total of 120 users, were manually analyzed to build the lexicon. For each user, 600 tweets were crawled. Therefore, a total of 72,000 tweets were manually analyzed to build the lexicon. The next step involved fetching a list of 50 followees for each of the base users from each region. For each followee, a total of 1000 tweets were crawled. Hence, a total of 6,000,000 tweets and 6120 users were crawled. However, the followees were then filtered based on the location of Delhi and Washington, resulting in a total of 441 users from Delhi and 432 users from Washington.

4 Methodology

To fulfill the first objective of our work, a lexicon was generated. This lexicon was created using a team of three people. However, the words used to create the lexicon were also discussed with experts from fields of psychology and behavioral studies as mentioned later in the paper. The sample of words in the ValueDict lexicon is shown in Table 1. The github link to the lexicon is given in [34]. The strength of the dictionary is being updated using similar words generation from word2vec model [35].

Initially, a total of 120 users (60 users from each region) were manually analyzed. In this manual analysis, the description of each of the users that they mentioned in their profile was recorded and then the tweets were analyzed to see which kind of value system the user could be categorized into.

The words each user used in their tweets were inspected to see which kind of value system could be best reflected by these words. These words were then recorded in our lexicon under their respective value systems. The final value system of a person was recorded based on what they said about themselves in the description and on what the majority of the tweets reflected about them.

Once the lexicon was constructed, the next step was to run this lexicon over new users fetched from these regions. The list of users manually inspected was taken as seed users for each region. As our next objective was to inspect the prominent value systems in each of the regions, a list of 50 followees (people, these users follow), for each of the seed user was extracted. The list of followers was then filtered based on the geographical location of the seed users. Tweets for each of such users were crawled and labeled using the lexicon. Algorithm 1 was used to label the users with their respective value systems. The algorithm takes as an input, the list of users to be labeled, and the lexicon containing words belonging to each value system category. For each user, the list of tweets was inspected. Each Tweet was then checked and the count of words belonging to each value system was maintained as a key-value pair where the key represented the value system and the value denoted the counts associated with it. These counts were updated as more tweets were inspected for

Table 1 Sample of words in ValueDict lexicon

Hedonism	Self-transcendence	Self enhancement	Conservation	Openness to change
Beach life	Deforestation	Power	Cleanliness	Artwork
Cinema	Donate	Control	Customs	Condemn
Club	Divine	Achievement	Discipline	Creative
Hangout	Employment	My wealth	Elders	Curiosity
Holidays	Education	i lead	Honor	Daring
Vacation	Empowerment	i win	Humble	Excited
Pleasure	Equality	My money	Temple	Freedom
Safari	Harmony	Subscribe	Self-disciplined	Independent
Travel	Humanity	Watch me	Safety	Self-respect

that user. Finally, the value system with the maximum count of words across all the tweets was assigned to this specific user.

Data given:

```
USERS: List of users to be labeled
LEX: Lexicon containing words and respective value systems

INIT (USER, LEX):
USER_LABEL = {}
 for tweet in USER [TWEETS]:
    L = LABEL_TWEET(tweet, LEX)
    USER_LABEL = MERGE_LABEL(USER_LABEL, L)
USER_VALUE_SYSTEM=MAX(USER_LABEL.key)
 return USER_VALUE_SYSTEM

MERGE_LABEL(L1, L2):
    L3 = {}
  FOR each key in L1, L2:
    L3[key] = 0
  FOR each key in L1, L2:
    L3[key] = L1[key] + L2[key]
  return L3

LABEL_TWEET (tweet, LEX):
  LABEL_MAP = {}
  FOR each key in LEX:
    LABEL_MAP[key] = 0
  FOR each key, values in LEX:
    FOR each value in values:
      IF tweet contains value:
        count = LABEL_MAP[key]
        LABEL_MAP[key] = count + 1
  return LABEL_MAP

 for USER in USERS:
      USER_VALUE_SYSTEM=INIT(USER,LEX)
```

5 Results

The labeling of the users with their respective value systems in each of the regions yielded the following results:

As per Figs. 1 and 2, it can be seen that 58% of users in the Washington region reflect the Hedonism value system. Whereas, 46% of users in the Delhi region fall under the category of Self- Transcendence value system. Therefore, it can be concluded that users from the Delhi region are more inclined towards the welfare of the society whereas users from the Washington region reflect self-indulgence.

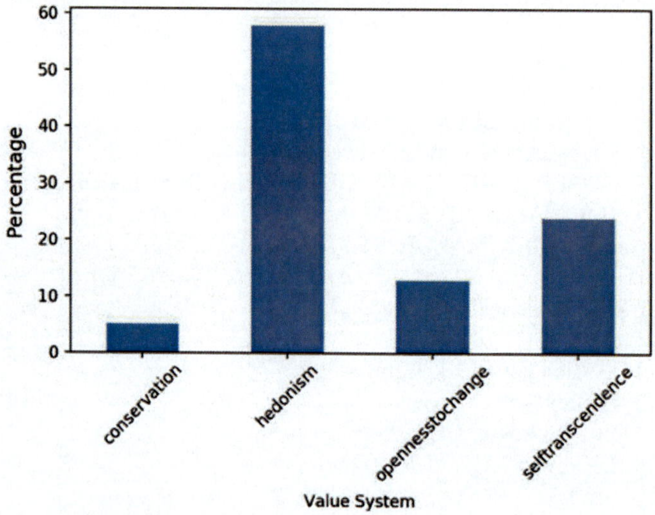

Fig. 1 Distribution of users across value system in the Washington users

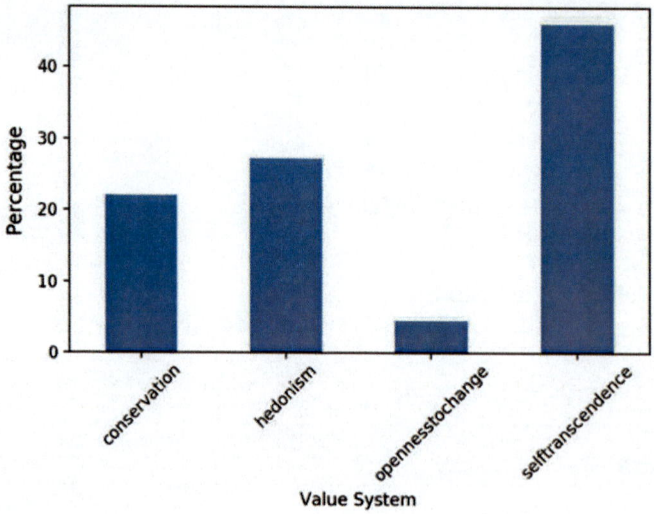

Fig. 2 Distribution of users across value system in Delhi region

Another interesting insight to observe that users in the Delhi region seem to be more conservative as compared to users in the Washington region. It can be observed that none of the users fall in the Self Enhancement category. The possible reason behind this may be that not many users were identified in this category during the manual analysis of the users.

6 Validation

To validate the lexicon, a total of 250 users across all value systems were taken. To create the ground truth, the description of the profile of the users was inspected to see what kind of value system they can be categorized into. For example, a person who states in their description that they are a social worker, can be considered to fall in the value system of Self-Transcendence. This strategy was used to create ground truth for users across other value systems as well. These users were then validated with our lexicon, to see if the results conformed to what people said in their description. Figure 3 shows the accuracy of our validated results across the value systems. The accuracy is higher when the results from the users' description in their profile match with the results generated by our lexicon, thereby, validating the lexicon.

According to the results, the value systems, "Hedonism" and "Self-Transcendence" seem to perform the best as compared to the others.

Regarding the Self Enhancement category, since a substantial number of users were not identified during the manual analysis in this category, the validation of this value system is being pursued as future scope of this work.

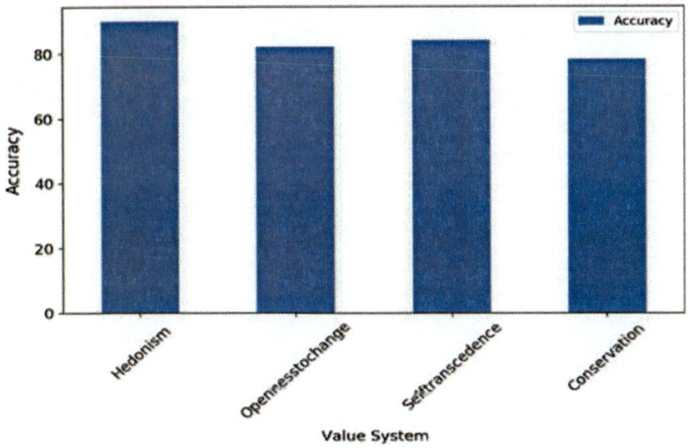

Fig. 3 Accuracy across value systems

7 Conclusions and Future Scope

A tremendous quantity of data is prevailing over the internet, as social networks like Twitter supports millions of users. Therefore, content-based data in the form of user's tweets are easily available. In this manuscript, we have analyzed these tweets to know about the behavior of Twitter users. We analyzed their tweets to know what value system they possess and constructed a value-based lexicon called ValueDict. Also, we aimed to determine the behavior of users in terms of the value system, across two different regions. Hence, the data collected for this research was crawled from the regions of Delhi (developing) and Washington (developed) region.

Through our findings, it can be seen that the prominent value system across the Delhi region is Self-Transcendence and that of Washington is Hedonism. Hence, the users in Delhi are more inclined towards social welfare whereas the users in the Washington region are more involved in self-indulgence.

The future scope of this work includes improvising the lexicon for the Self Enhancement category and using value systems as a feature to predict decisions of a user. This work can also be extended to finding pornographic users in social networks.

Acknowledgements We deeply acknowledge the following people who supported us in moving this work forward. These people helped us with our questions regarding the building of the lexicon.

- Prof. Shalom H. Schwartz, Department of Psychology, University of Jerusalem, Jerusalem.
- Prof. Sibnath Sarma, Department of Philosophy, Gauhati University, Gauhati.
- Dr. Sarmishtha Sarma, Ph.D. in Consumer behavior, Indraprastha University, Delhi.

References

1. Naveed N, Gottron T, Kunegis J, Alhadi AC (2011) Bad news travel fast: a content-based analysis of interestingness on twitter. In: Proceedings of the 3rd international web science conference. ACM, Koblenz Germany, pp 1–7
2. Boyd D, Golder S, Lotan G (2010) Tweet, tweet, retweet: Conversational aspects of retweeting on twitter. In: 2010 43rd Hawaii international conference on system sciences. IEEE, Honolulu, HI, USA, pp 1–10
3. Schwartz SH (2003) A proposal for measuring value orientations across nations. In: Questionnaire package of the European social survey, vol 259, no 290, p 261
4. Sagiv L, Roccas S, Cieciuch J, Schwartz SH (2017) Personal values in human life. Nat Hum Behav 1(9):630–639
5. Schwartz SH (1994) Are there universal aspects in the structure and contents of human values? J Soc Issues 50(4):19–45
6. Chen J, Hsieh G, Mahmud JU, Nichols J (2014) Understanding individuals' personal values from social media word use. In: Proceedings of the 17th ACM conference on Computer supported cooperative work and social computing. ACM, Baltimore, Maryland, USA, pp 405–414
7. Huang D, Zhou J, Mu D, Yang F (2014) Retweet behavior prediction in twitter. In: 2014 Seventh international symposium on computational intelligence and design, vol 2. IEEE, Hangzhou, China, pp 30–33

8. Singh M, Bansal D, Sofat S (2016) Behavioral analysis and classification of spammers distributing pornographic content in social media. Soc Netw Anal Min 6(1):41
9. Puohiniemi MA (1997) Values, consumer attitudes and behaviour: an application of Schwartz's value theory to the analysis of consumer behaviour and attitudes in two national samples 0694–0694
10. Grunert SC, Juhl HJ (1995) Values, environmental attitudes, and buying of organic foods. J Econ Psychol 16(1):39–62
11. Schwartz SH, Huismans S (1995) Value priorities and religiosity in four Western religions. Soc Psychol Q, pp 88–107
12. Devos T, Spini D, Schwartz SH (2002) Conflicts among human values and trust in institutions. Br J Soc Psychol 41(4):481–494
13. Fleischmann KR, Oard DW, Cheng AS, Wang P, Ishita E (2009) Automatic classification of human values: applying computational thinking to information ethics. In: Proceedings of the American society for information science and technology, vol 46, no 1, pp 1–4. Wiley Online Library
14. Templeton TC, Fleischmann KR, Boyd-Graber J (2011) Simulating audiences: automating analysis of values, attitudes, and sentiment. In: 2011 IEEE third international conference on privacy, security, risk and trust and 2011 IEEE third international conference on social computing. IEEE, pp 734–737
15. Ishita E, Oard DW, Fleischmann KR, Cheng AS, Templeton TC (2010) Investigating multi-label classification for human values. In: Proceedings of the American society for information science and technology, vol 47, no 1, pp 1–4. Wiley Online Library
16. Dehghani M, Gratch J, Sachdeva S, Sagae K (2011) Analyzing conservative and liberal blogs related to the construction of the 'Ground Zero Mosque'. In: Proceedings of the annual meeting of the cognitive science society, vol 33, no 33
17. Pennebaker JW, Boyd RL, Jordan K, Blackburn K, Austin, TX (www. liwc. net) (2007) The development and psychometric properties of LIWC2007. [Software manual]. Austin, TX
18. Ali S, Katoma V, Tyobeka E (2015) Identification of key values and behaviours influencing leadership orientation in Southern Africa. J Emerg Trends Educ Res Policy Stud 6(1):6–12
19. Ye S, Soutar GN, Sneddon JN, Lee JA (2017) Personal values and the theory of planned behaviour: a study of values and holiday trade-offs in young adults. Tour Manage 62:107–109
20. Rathbun BC, Kertzer JD, Reifler J, Goren P, Scotto TJ (2016) Taking foreign policy personally: personal values and foreign policy attitudes. Int Stud Q 60(1):124–137
21. Kaufmann E (2016) It's NOT the economy, stupid: Brexit as a story of personal values. British politics and policy at LSE
22. Mazzi MA, Rimondini M, van der Zee E, Boerma W, Zimmermann C, Bensing J (2018) Which patient and doctor behaviours make a medical consultation more effective from a patient point of view. results from a European multicentre study in 31 countries. Patient Educ Couns 101(10):1795–1803
23. Dietz T, Allen S, McCright AM (2017) Integrating concern for animals into personal values. Anthrozoös 30(1):109–122
24. McCrae RR, John OP (1992) An introduction to the five-factor model and its applications. J Pers 60(2):175–215
25. Komarraju M, Karau SJ, Schmeck RR (2009) Role of the Big Five personality traits in predicting college students' academic motivation and achievement. Learn Individ Differ 19(1):47–52
26. Gerber AS, Huber GA, Doherty D, Dowling CM, Ha SE (2010) Personality and political attitudes: relationships across issue domains and political contexts. Am Polit Sci Rev 104(1):111–133
27. Wang SS (2013) "I share, therefore I am": Personality traits, life satisfaction, and Facebook check-ins. Cyberpsychol Behav Soc Netw 16(12):870–877
28. Blackwell D, Leaman C, Tramposch R, Osborne C, Liss M (2017) Extraversion, neuroticism, attachment style and fear of missing out as predictors of social media use and addiction. Personality Individ Differ 116:69–72

29. Caprara G, Vecchione M, Schwartz SH (2009) Mediational role of values in linking personality traits to political orientation. Asian J Soc Psychol 12(2):82–94
30. Caprara GV, Schwartz S, Capanna C, Vecchione M, Barbaranelli C (2006) Personality and politics: values, traits, and political choice. Polit Psychol 27(1):1–28
31. Dirilen-Gümüş Ö, Cross SE, Dönmez A (2012) Who voted for whom? Comparing supporters of O bama and M c C ain on value types and personality traits. J Appl Soc Psychol 42(12):2879–2900
32. Bonsón E, Perea D, Bednárová M (2019) Twitter as a tool for citizen engagement: an empirical study of the Andalusian municipalities. Gov Inf Q 36(3):480–489
33. Zacharias C (2017) Twint-twitter intelligence tool
34. Kakar S, Dhaka D (2020) https://github.com/deepalidhaka/ValueDict
35. Mikolov T, Chen K, Corrado G, Dean J (2013) Efficient estimation of word representations in vector space. CoRR abs/1301.3781

Patient Health Monitoring and Diagnosis Using IoT and Machine Learning

Vishal Gupta, Akshay Ingle, Dhanashree Gaikwad, and Mahesh Vibhute

Abstract Including rapid and accurate up-gradation in medical technology, disclosure (detection), and monitoring of patients is not an obvious issue. Nowadays sustained and trend healthcare monitoring devices are available in the market. This paper elaborates on the concept of monitoring and diagnosis of patient health conditions. The main objective of this proposed system is to target heart patients and provide them an integrated portable health monitoring device. The proposed idea helps patients as well as cardiologists. In this system, Raspberry-pi has been used along with communication protocols i.e. MQTT and Machine learning algorithms which make it efficient, affordable, and a reliable prototype.

Keywords Internet of things (IoT) · Raspberry-pi pulse rate sensor · Hemoglobin sensor · ECG sensor · Thingspeak · Machine learning · K-means clustering · Logistic regression

1 Introduction

In recent years Internet of Things based health care devices plays a vital role in embedded systems and communication technologies. It has a significant and valuable contribution to the elevation of Health care systems. The real-time monitoring,

V. Gupta (✉) · A. Ingle · D. Gaikwad · M. Vibhute
School of Electrical Engineering, MIT Academy of Engineering Alandi, Pune, India
e-mail: vgupta@mitaoe.ac.in

A. Ingle
e-mail: asingle@mitaoe.ac.in

D. Gaikwad
e-mail: dpgaikwad@mitaoe.ac.in

M. Vibhute
e-mail: mcvibhute@entc.mitaoe.ac.in

G. Ranganathan et al. (eds.), *Inventive Communication and Computational Technologies*, Lecture Notes in Networks and Systems 145,
https://doi.org/10.1007/978-981-15-7345-3_24

tracking, and diagnosis of the patient's health care activities are remaining as a challenging task. With the development of the world, Health monitoring device is being used in almost every fields such as hospital, biomedical, and so on. This healthcare monitoring system can also be used for heart patients over 60 years of age who need daily check-up [1]. In day to day life, it is a very tedious job to keep track of abnormalities in heartbeat count, ECG signal variation, and hemoglobin level for the patient itself manually. There are various devices available in the market to keep track of internal body parameter changes.

In this proposed concept Raspberry-pi is used to design the system. Different medical sensors like Pulse Rate sensor, ECG sensor, and hemoglobin sensor are used for monitoring the heart condition which is integrated on a single platform to build wearable and portable devices [2]. If an emergency condition occurs, then notification via the cloud can be sent to doctors it helps them to take immediate action towards it. This alert notification may help to save patients life from the further critical conditions which would arise in the future. The second issue is that people may not be aware of the heart condition during sleeping hours. To overcome this situation, this paper proposes a system that monitors and diagnoses the patient's health condition at a regular interval of time. The heart rate sensor and ECG sensor would help us analyze the patient's current health condition. These main sensors are used to monitor the heart condition of the patients, especially at the hospital at equal intervals. These monitored details are stored in the cloud platform "ThingSpeak". The doctor can monitor the patient's health data and can generate reports by applying machine learning algorithms like K Means clustering and Logistic regression to detect the abnormalities in heart condition. This concept will help patients and doctors to minimize risks of tracking patient's condition randomly.

The rest of the paper is categorized as follows. Literature survey in Sect. 2. Section 3 describes the architecture of the proposed system in detail. Section 4 provides the result of the proposed system. Section 5 implies future scope and conclusion of the proposed idea.

2 Literature Survey

Several innovations are focusing on the development of wireless systems which continuously monitors the health activity of patients and provides a detailed graphical representation of health conditions. Many inspiring ideas can be implemented to upgrade the new trend technologies like the Internet of Things. Using the concept of IoT many health care systems contribute a vital and significant role in human life [3].

Nyani [4] focused on integrating several sensors into one device. Biomedical sensors like ECG and EEG records patient heart-related information. Then the obtained data can be further processed by Bluetooth connection for doctor purpose. Using Bluetooth connection, it gives connectivity up to 9 meters only which can be removed by using the Wi-Fi module in our proposed idea.

Sharma [5] gives an idea about the important parameter of the human body is hemoglobin which focused on developing an optical sensor for measurement of hemoglobin by non-invasively technique. This work deal with a technique like photoplethysmography. It is low cost and easy to operate the device to determine real-time hemoglobin levels of the heart patient. Using PIC controller, IR- LED and photodetector output displayed on LCD.

Vasava [2] proposes an embedded approach to Portable Health Monitoring System. This paper explores the idea of integration of sensors to raspberry-pi and data acquisition can be done using a communication protocol. It gives valuable and significant importance in designing an efficient embedded system.

Chakravorty [1] paper present an idea for people over 60 years of age and need special health treatment in day-to-day life. This system conveys detailed information on health parameters like Body temperature, Heartbeat rate, and blood pressure, actuating the output and displayed on an LCD. It can further process for diagnosis purposes.

Existing methodologies in patient monitoring devices focus on providing better healthcare facilities to the patient but it is very difficult to make a quick decision and action in critical condition. This proposed idea eliminates the hurdles such as distance and improves access to medical services.

3 Architecture

Our model does not contain ADC as an inbuilt component so we used an external ADC which is acting as a bridge between a chain of sensors and Raspberry-pi. Cloud transfers the data to the doctor's end as well because of the diagnosis part that needs to be handled by the doctor. This proposed system collects real-time data from patients and delivers to doctors or nurse. This autonomous system replaced all the traditional methods to collect parameters regularly by the nurse. It also helps to reduce human error while collecting the reading of parameters. We have used protocol for transferring data via the Wi-Fi module to the cloud platform "Thingspeak" to establish communication between patients and doctors [6]. Detailed data may help to plot a graph across a standard value using machine learning algorithms.

Block diagram (Fig. 1) The system has been composed of three stages (1) Sensing unit (2) Actuation of data and (3) Transmission of data. Block diagram explains each stage of the system, where sensors are used to collect data from the patients and the obtained data can be further acquired by the processing unit. ADC has been used to convert analog values into digital. Data can be sent through the Wi-Fi module which is inbuilt in the Raspberry-pi processor. Thingspeak is a cloud platform from which data can be obtained for diagnostic purposes for doctors. Machine learning Algorithms applied on data to form a cluster and report will be generated as per algorithms it helps the doctor to identify the critical condition of the heart. The detailed report can be sent by the nurse/doctor to the member associated with the patient via smart device and accordingly actions can be taken.

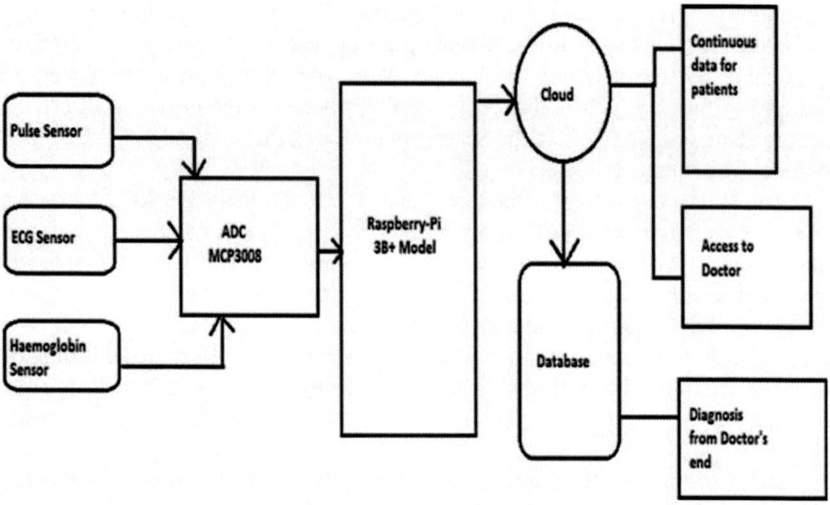

Fig. 1 Block diagram of proposed system

In this proposed system three main sensors have been used, they are (1) Pulse sensor (2) ECG sensor [AD8232] and (3) Hemoglobin sensor to check oxygen level. Raspberry-pi has processed sensing data and process further for diagnosis purpose. To establish communication between patient and doctor, the MQTT communication protocol has been used. Machine learning algorithms process data to get results.

Following are the details of the system:

3.1 Pulse Sensor [SEN-11574]

A pulse rate sensor is an electronic device that is used to measure the heartbeat rate. Usually, the source of light in a pulse rate sensor would be an IR-LED and the detector is any photodetector like a photodiode. The LED on the front side of the sensor is placed over a vein in the human body. Light emits through when the heart is pumping. This minor change is received by light is analyzed over time to determine the heartbeat of the patient [7].

3.2 ECG [Ad8232]

AD8232 is an ECG sensor. This is used to measure the electrical activity of the heart. This electrical activity can be charted as Electrocardiogram and output as an analog reading. AD8232 single lead heart rate monitor acts as an op-amp to help obtain a clear signal from the PR and QT intervals easily [8].

3.3 *Hemoglobin Sensor*

Hemoglobin is an important parameter in the human body, and it is responsible for the transportation of oxygen from the lungs to the rest of the body. This optical measurement method is based on radiation of red and near-infrared light, emitted by LED in the ranges of 600–1400 nm [5]. In this project, we have used circuitry consist of LDR and resistor of 270 K, 10 K in a series, and parallel form. Simple LDR is used as a photodetector. It helps us to calculate intensity from which wavelength of reflected light can be measured and thus final Hb level can be determined using the formula.

$$Hb = (\varepsilon_{Hb} * \lambda_{(red)} * C_{Hb})/(\mu H_2O * \lambda_{(IR)} * 64{,}500 \text{ g/dl}) \quad (1)$$

3.4 *Raspberry-Pi*

Raspberry pi is a miniature processor that can perform all the functions. We are giving sensors input to the Raspberry-pi it processes data and passes to the cloud using inbuilt Wi-Fi module.

4 Hardware Implementation

4.1 *Communication Protocol*

MQTT stands for Message Queuing Telemetry Transport. It is a lightweight publish and subscribe system protocol. Where we can publish and receive messages. So, it's the perfect solution for this proposed system. The inbuilt Wi-Fi module in Raspberry helps to send data over the MQTT protocol.

4.2 *System Working*

Pulse rate sensor works on the principle of photodetector diode. This sensor either be placed on fingertip or on-ear tips. Now the LED emits light which will fall on vein directly, the veins will have blood flow inside them. If the flow of blood is detected then the ambient light sensor will pick up lighter since they will be reflected by the blood. It has three pins consisting of GND, VCC, and a signal pin. Pins are connected to ADC to get digital values.

Collected output processed by Raspberry-pi and transmission can be done via the Wi-Fi module. Cloud platform "thingspeak" acquire real-time data on the same channel after every 15 s and provided a graphical representation of obtained data. Database formed on "thingspeak" which further utilized by machine learning algorithms to form a cluster. This result is getting by comparing standard data with sensing data. A detailed graph can be concluded in the situation of a heart patient (Fig. 2).

The proposed system works on direct data transfer which became possible by the use of "thingspeak" as a cloud platform, the data is then transmitted to user's database easily it can also be exported as .csv format so that on that datasheet various ML algorithms can be applied and put on and thus using certain calculations we can find out the result in the form of prediction whether the person can suffer any heart disease. The doctor's role becomes very important in such a case that's why the system can also perform various alert systems that can be implemented later on [9].

Fig. 2 Flow chart of the proposed system

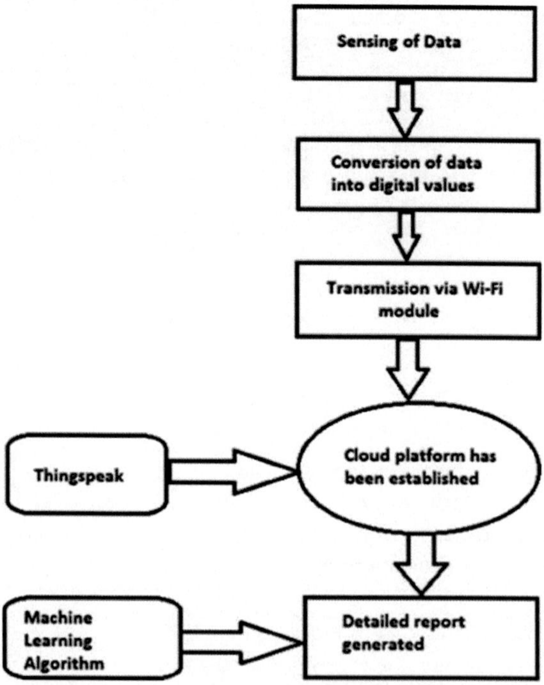

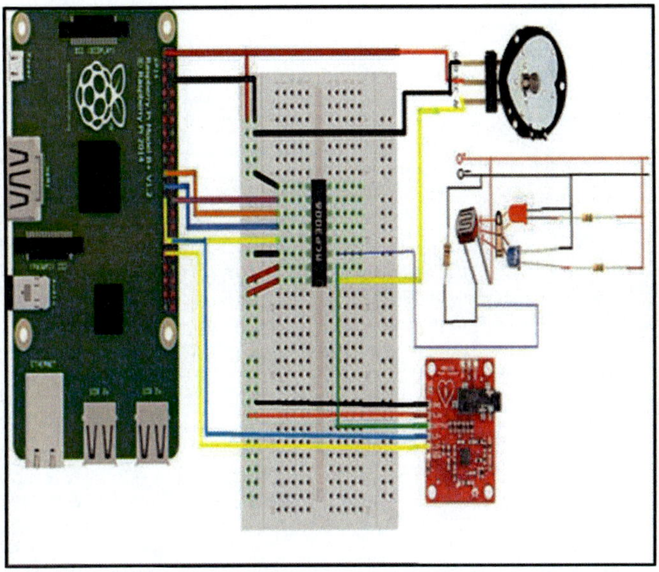

Fig. 3 Circuit diagram of the proposed system

4.3 Circuit Diagram

An ECG [AD8232] is a 20 pin IC. It is an integrated circuitry having signal conditioning, amplification blocks for heart rate monitoring. It consists of a specialized instrumentation amplifier (IA), an operational amplifier (A1) (Fig. 3; Table 1).

Hemoglobin sensor is readily available in the market but the integration of such sensors with others is difficult. So, in this proposed system we have designed a hemoglobin sensor on the principle of non-invasive technique. A designed sensor can be placed on a finger, ear, or toe, where pulses can be detected. Obtained output can be given to the ADC channel.

Pin no. 2 is positive Input typically connected to the left arm (LA) Pin no. 2 Negative Input is typically connected to the right arm (RA). Output pin no. 10 given to ADC to convert analog input to digital. Pin no. 16 and 18 provide GND and VCC to the circuitry.

4.4 ML Algorithms

Machine learning consists of various algorithms, this idea uses K Means clustering that analyzes data and used for classification, which attempts to find the natural clustering of the data and divide the obtained data into a group. It helps to find

Table 1 Pinout table of each connection

Raspberry Pi	MCP3008	Sensor connections
Pin 1 (VCC)	Pin 15, Pin 16(Vref)	AD8232, Hb, Pulse sensor +ve terminals
Pin 6 (GND)	Pin 13, Pin 9	AD8232, Hb, Pulse sensor −ve terminals
Pin 24 (CE0) Pin 21 (SPI-MISO) Pin 23 (SCLK) Pin 19 (SPI-MISO)	Pin 10 (CE) Pin 12 (MISO) Pin 13 (SCLK) Pin 11 (MOSI)	-NIL-
Pin 30 (GND) Pin 31 (GPIO 6)	-NIL-	ECG- LO- ECG- LO+
-NIL-	Pin 1 (CH 0) Pin 2 (CH 1) Pin 3 (CH 2)	Pulse sensor (Signal) AD8232 (OUTPUT) Hb sensor (OUTPUT)

the abnormalities of the heart patient. Using K means clustering cluster formation, prediction, and using logistic regression on the data gathered accuracy of the system can be generalized.

5 Results

5.1 Monitoring Section

The sensors and the Raspberry-pi controller are connected according to the block diagram. The output can be display on the monitor display by the sensor as bits per minute (Figs. 4 and 5).

Thingspeak update data in real-time, data can be sent by the Wi-Fi module and it gives the graphical representation of sensor data. The graph shows the variation in BPM after every second. Results are displayed on the RPi monitor and are transferred on the Thingspeak website using standard API. Following is the graphical representations.

Of the hemoglobin sensor values. This graph is also formed by the ThingSpeak cloud platform. The graph shows a different set of values after a certain interval of time (Fig. 6).

Figure 7 demonstrates the ECG plot formed on the data gathered by the AD8232 sensor and this plot can be viewed by doctors and diagnose the patient.

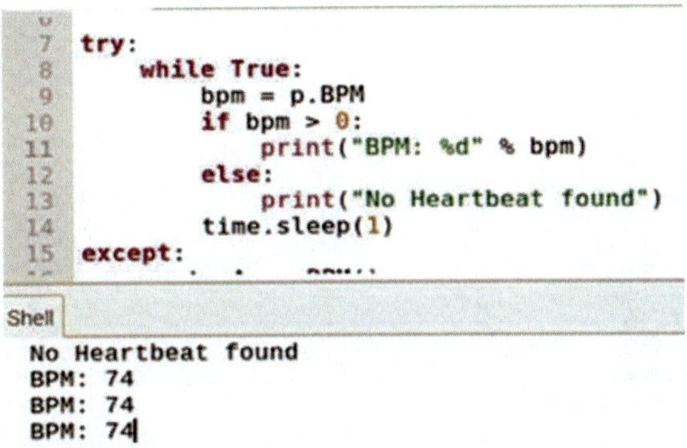

Fig. 4 Result of the pulse sensor

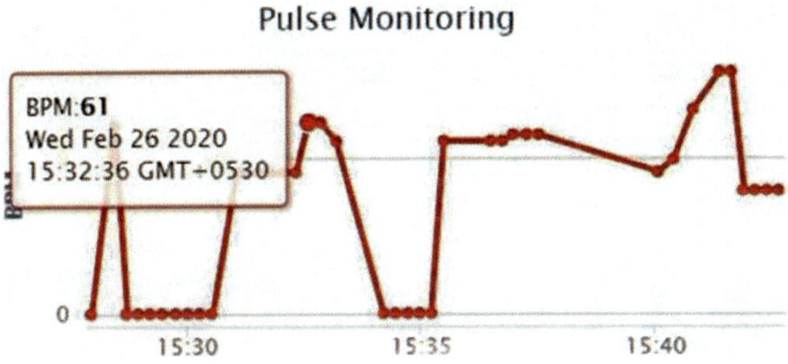

Fig. 5 Graph of pulse sensor on thingspeak

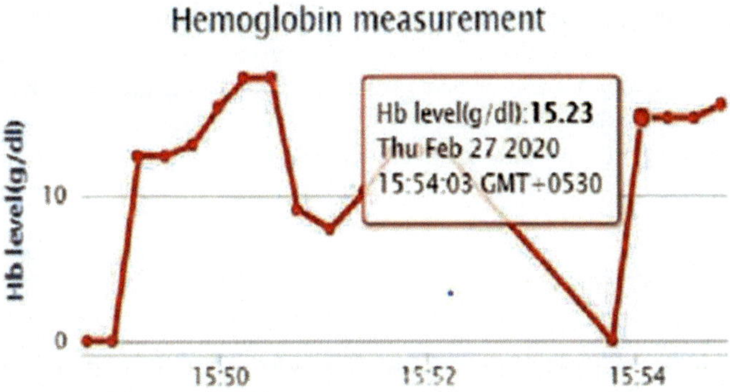

Fig. 6 Graph of hemoglobin sensor on thingspeak

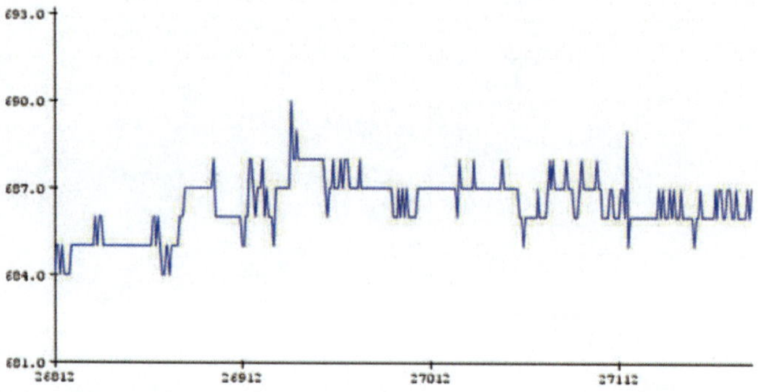

Fig. 7 A plot of ECG sensor

Table 2 Result comparison table

S. No.	Clinically data		System data		Error %	
	Heart rate	Hb gm/dl	Heart rate	Hb. gm/dl	Heart rate	Hb. gm/dl
1.	70.5	15.4	73.2	15.75	2.17	2.12
2.	81.2	14.2	84.2	13.22	3.65	6.9
3.	69.8	13.6	67.7	11.54	2.9	15.23
4.	74.2	12.8	73.6	12.32	0.82	4.62
5.	69.5	10.9	64.7	9.33	6.8	14.40

Table 2 shows the comparison between clinically obtained data and system data. The proposed system gives approximately accurate data in case of pulse sensor however in case of hemoglobin sensor it may give deviated data which can be misleading but in most cases, the data is close to clinical data and it can be improved by using good qualities of the photodiode, this data can be further used by doctors in critical condition of the patient. It helps to take quick actions against the patient.

5.2 *Diagnosis Section*

After getting the data such as heart rate, hemoglobin, ECG on the cloud from all of the sensors we have to move with the diagnosis part of the heart patient. Here we are using supervised (logistic regression) and unsupervised (clustering) machine learning algorithms to predict the output [10].

In the diagnosis part firstly, we are applying a k-means clustering algorithm to label our data. According to feature engineering here we are taking some of the important features such as age, hemoglobin, heart rate and sex to predict the output. After analyzing the feature, we are plotting the data using these parameters.

After applying K-Means at initial level data are classified and the heart patient is denoted by a green dot and if the patient is not a heart patient that is denoted by a red dot. The centroids are assigned randomly in the cluster. According to Euclidean distance, the centroid is shifted [11].

Figures 8 and 9 describe the plotting of data right from raw plotting to cluster formation with shifted centroids, this segregation helps the system to predict which particular patient belongs to critical state and which is in normal state.

Figure 10 is the Elbow plot which means the point of inflection on the curve. This technique is used to run K-Means clustering on the dataset and compute the exact value of k.

Figure 11 describes some of the datasets and the count of the heart patients.

Figure 12 describe the accuracy of the system right from gathering target values to the labelling of each of the parameters and finding accuracy using python commands.

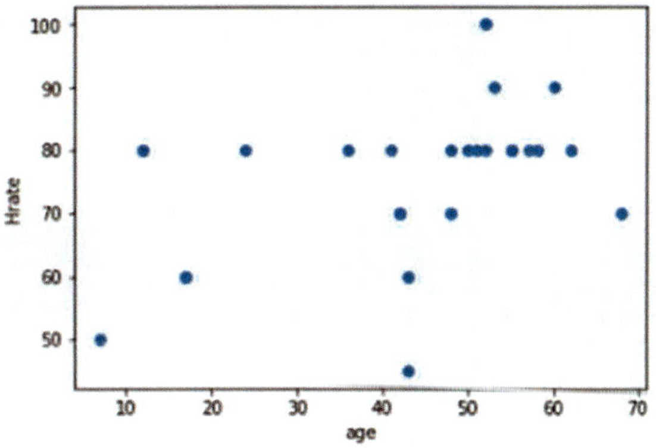

Fig. 8 Plotting of data gathered from sensors

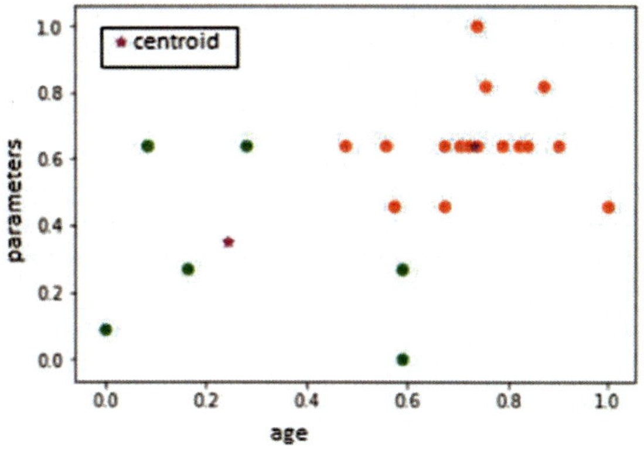

Fig. 9 Cluster formation of shifted centroids

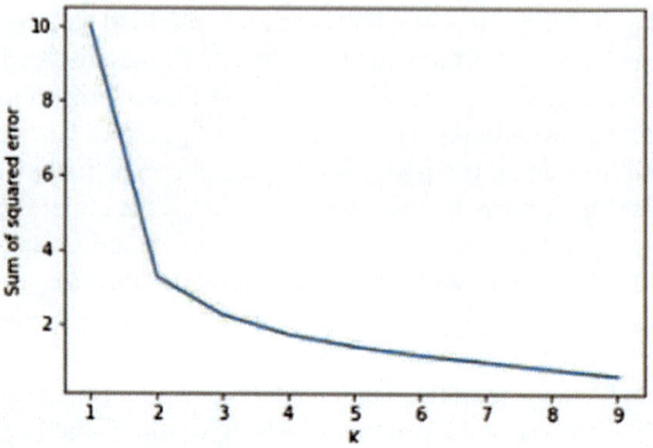

Fig. 10 Elbow plot for finding the value of K

	age	Hrate	hemo	sex	target
0	48	80	15.4	1	1
1	7	50	11.3	1	0
2	62	80	9.6	0	1
3	48	70	11.0	1	1
4	51	80	11.6	1	1

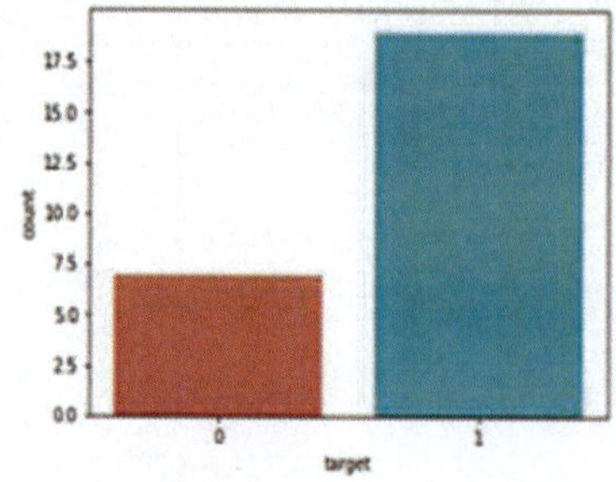

Fig. 11 Labelling and target values using logistic regression

Fig. 12 Accuracy of data on the proposed testing of gathered data

```
print('Accuracy:%d', (logmodel.score(x_test,y_test)))

('Accuracy:%d', 0.8333333333333334)
```

6 Conclusion and Future Scope

This proposed idea focused on real-time healthcare monitoring systems using IoT and cloud computing services which is more beneficial for elders and heart patients. The current methods which are available in the market are not fully survived so the challenging part of realization is also highlighted. The system gives near about 80% accurate report and will be able to send notifications to the patient as well as the doctor. The system optimizes an available set of medical resources and minimizes the overall cost and time while monitoring the patient. It also helps to increase inpatient life by taking action in critical condition.

In the future, the system can be improved by making it wearable and integrating it as a small device. Also focusing on appropriate medicines that can be suggested based on the diagnosis provided by doctor on a web-based application and precautions alerted by system.

References

1. Chakravorty D, Islam S, Rana TK (2018) IoT based patient guidance system using raspberrypi. IEEE paper 978-1-5386- 5550-4/18/$31.00 ©2018
2. Vasava DS, Vasavada NM, Sametriya DP (2017) Design and working of portable health monitoring system: an embedded approach. In: 2017 International conference on wireless communications, signal processing and networking (WiSPNET). IEEE 978-5090-4442-9/17/$31.00©
3. Chouffani (2016) Can we expect the Internet of Things in healthcare? http://internetofthingsagenda.techtarget.com/feature/Can-we-expect-the-Internet-of-Things-in-healthcare
4. Nyni KA, Vincent LK, Varghese L, Liya VL, Johny AN, Yesudas CV (2017) Wireless health monitoring system for ECG, EMG and EEG detecting. In: 2017 International conference on innovations in information, embedded and communication systems (ICIIECS)
5. Sharma P, Kumar A, Kumar G, Balwant R, Sanjana K, Estimation of haemoglobin using optical sensor based system. https://doi.org/10.15662/ieee.2018.0704052
6. Huang PC, Lin CC, Wang YH, Hsieh HJ (2019) Development of health care system based on wearable devices. 978-1-7281-0329-7/19/$31.00 ©2019 IEEE
7. Hodge A, Humnabadkar H, Bidwai A (2018) Wireless heart rate monitoring and vigilant system. In: 2018 3rd International conference for convergence in technology. 978-1-5386-4273-3/18/$31.00 ©2018 IEEE
8. Hesar HD, Mohebbi M, An adaptive Kalman filter bank for ECG denoising. IEEE J Biomed Health Inf. https://doi.org/10.1109/jbhi.2020.2982935
9. da Costa CA, Pasluosta CF, Eskofier B, da Silva DB, da Rosa Righi R (2018) Internet of health things. Artif Intell Med 89:61. PMID 29871778
10. Zhao X, Zeng X, Koehl L, Tartare G, de Jonckheere J, Song K (2019) An IoT-based wearable system using accelerometers and machine learning for fetal movement monitoring. 978-1-5386-8500-6/19/$31.00@2019 IEEE
11. Oiwa R, Ito T, Kawahara T (2017) Timber health monitoring using piezoelectric sensor and machine learning. 978-1-5090-4253-1/17/$31.00 ©2017 IEEE
12. Mahmood SN, Ercecbi E (2018) Development of blood pressure monitor by using capacitive pressure sensor and microcontroller. In: Presented at 2018 international conference on engineering technologies and their applications (ICETA), Islamic University. 978-1-5386-7858-

Handwritten Devanagari Character Classification Using CNN

Addepalli Kavya, Nunna Vivek, Maddukuri Harika, and Venkatram Nidumolu

Abstract Deep Convolutional Neural Network (CNN) has been effectively outlined acknowledgement of the written by hand character in Devanagari content. Convolutional Neural Networks build up a framework that can be set up to perceive the morphology of pictures and perform class estimate and concentrate the property of pictures. Written by hand character acknowledgement is a huge undertaking as it tries to perceive the right class for client autonomous manually written digits. Preparing of the model is finished utilizing MNIST dataset This challenge is much more difficult with a heavily impressionistic, morphologically complicated, and inherently juxtapositional character consisting of dialect such as Devanagari. A dataset of 73,600 samples were trained and tested in this proposed CNN that are separated to perform both training and testing which provided quite impressive results. 97% exactness was gotten for the test information when the system was tried with a Devanagari character dataset.

Keywords Deep neural networks · Devanagari characters · Handwritten character recognition · Machine recognition · Convolutional neural networks

A. Kavya (✉) · N. Vivek · M. Harika · V. Nidumolu
Department of Electronics and Computer Engineering, Koneru Lakshmaiah Education Foundation, Vaddesswaram, AP, India
e-mail: addepallikavya58@gmail.com

N. Vivek
e-mail: nunnavivek3@gmail.com

M. Harika
e-mail: harikamaddukuri@gmail.com

V. Nidumolu
e-mail: dean.academics@kluniveristy.in

G. Ranganathan et al. (eds.), *Inventive Communication and Computational Technologies*, Lecture Notes in Networks and Systems 145,
https://doi.org/10.1007/978-981-15-7345-3_25

1 Introduction

The pattern recognition tasks are all almost done with the help of Deep Learning Techniques. The traditional methods for pattern recognition tasks are Fuzzy based Classification [1], Support Vector Machine (SVM), Hidden Markov Model [2], etc. Almost all the Deep Learning techniques have got the highest performance metrics. The pattern recognition jobs incorporate estimation of the human posture [3], acknowledgement of the human face [4], picture acknowledgement [5, 6], and distinguishing proof of characters [7, 8]. Profound Learning systems had repealed the old and existing techniques. The other advantage of Deep Learning is that it does not require much hand-engineering. The most noteworthy target of each visual affirmation development for a machine is to isolate the features from the image [9, 10]. The element extraction is done consequently in Deep Learning.

One of the huge premiums in Deep learning is the "written by hand character acknowledgement" due to its wide extent of employments. The applications are writer affirmation, examining the postal locale, taking care of the bank checks, seeing the substance on the number plate, etc. [11]. Handwritten character identification is constantly a test to accomplish because the characters fluctuate as indicated by the various authors, their style of composing, and the commotion in their composition [12]. There are various variations for perceiving the written by hand characters: (1) Optical Character Recognition (OCR). This is known as an offline mode where the written by hand input is given as a picture. (2) Online mode where the input is a recorded list that portrays the handwritten characters [13]. The past works related to Devanagari character acknowledgement are not a lot of because of the morphological multifaceted nature of the digits.

In the field of Digital Image Processing, Convolutional Neural Networks are known as the "state of the art". CNN came into existence through the IMAGENET challenge of 2011 [12]. CNN's are best known for the image type of inputs because of their insensitivity to both scale and translation variances of the images [11]. Picture structure can be spoken to plainly with the assistance of CNN. CNN is also very useful in analyzing the documents [14]. The strategies of weight sharing and local connectivity help in good image structuring.

In our system, CNN is utilized to achieve a better than average precision rate for the affirmation of Devanagari characters. The remainder of the paper is composed as given underneath: Sect. 2 deals with the past reviews made in the field of character affirmation. Section 3 deals with the methodology followed in this paper. Section 4 covers the observations and experimentation results. Segment 5 gives a finish of our work. Finally, last section contains the references that helped us in completing our paper (Fig. 1).

Fig. 1 Devanagari character dataset. *Source* Kaggle

2 Literature Survey

Nishal Ancelette Pereira, Prajwal Rao. Akshay K Kallianpur and KG Srinivasa [15] developed a system that uses the discrete artificial bee colony for the feature extraction and identify both the offline and online characters of Kannada, English, and Hindi and they got the desired accuracy.

Rajput and Suryakant Baburao [16] also worked on the same three languages Kannada, Hindi, and English for text recognition from a handwritten document and they also used KNN classifier and the feature of scale invariance in their system to achieve better results.

Pasha and Padma [17] structured a framework that is equipped for distinguishing the two characters just as numerals of the Kannada language. The classifier utilized in their framework is ANN alongside wavelet change which gives the best outcomes for worldwide element extraction. They achieved an accuracy of 97%.

Chackoa and Dhanya [18] worked on Malayalam Character recognition and they proposed a multiple classifier system. The features involved in their system are density and gradient-based, And with the help of the product rule combination scheme, they got a recognition capacity of 81.82%.

Vijayaraghavan and Sra [19] have worked on Tamil character recognition systems using CNN. They used the ConvNetJS Library for learning the highlights. They accomplished a precision of 94.4% when tried over the IWFHR-10 dataset.

Acharya et al. [20] concocted a Large scope Handwritten Devanagari Character Recognition framework which depends on Deep Learning. The CNN count is used to describe the characters. They tried two models, model A will be a shallow system though Model B comprises of a solitary completely associated layer and three convolutional layers. The precision of Model An and Model B wound up being 0.9826 and 0.9847 individually.

Sonawane and Shelke [14] used a transfer learning mechanism to identify the Devanagari characters. They also utilized a CNN named AlexNet to achieve their target. They got 95.46% test accuracy and also 94.49% validation accuracy.

Arora et al. [21] likewise concocted an arrangement of various classifiers with the assistance of second invariants and chain code histogram for the acknowledgement of Devanagari characters.

3 Methodology

The fundamental explanation for CNN is feature building not required. Before CNN, ample amounts of energy is invested in feature choice (calculation). At the point when high quality highlights and CNN is contrasted, CNN execution well and it gives better precision. It additionally takes in various features from pictures for better accuracy.

Convolutional Neural Networks or ConvNets are mostly used to recognize and classify the images and share their parameters. For applying CNN, a large dataset of training images is required. A set of learnable filters composes a convolutional layer which is the core building block of CNN. Kernel, K represents a set of learnable filters whose main purpose is to detect features and create various feature maps (See Eq. 1). These component maps are further useful in the distinguishing proof and grouping of pictures.

$$s[t] = (x * w)[t] = \sum_{a=\infty}^{a=\infty} x[a]w[a+t] \tag{1}$$

In Eq. (1), $s[t]$ indicates feature map, x indicates input and w indicates kernel.

According to the depicted generalized architecture, the CNN model consists of multiple layers. Apart from input and output layers, there are three important layers which are namely convolutional, pooling, and fully connected.

The information layer is the spot the image is given as data which is used to expel the features from it using described channels. These diverts are regularly present in *convolutional layers* which will, as a rule, spare the pixels of the data picture and pass it to extra layers. This is done by taking a small number of squares of an input image. This helps to extract features in a more detailed manner (Fig. 2).

The limit of the *pooling layer* is to decrease the spatial estimations which are important for compelling computations with less parameters to set up a model on. This results in the lower possibilities of the over-fitting of the model. A pooling layer

Fig. 2 Generalised architecture of CNN

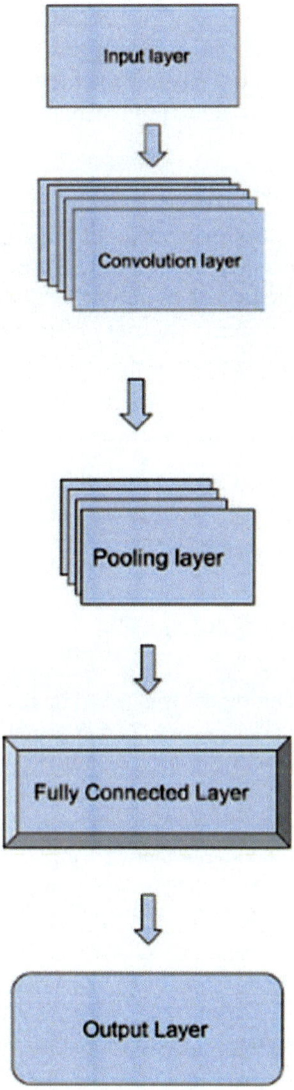

is generally of three kinds where each one is used for different purposes. For our situation, MAX_POOLING is utilized with the necessary steps of the information grid.

A fully connected layer is just a thick multi-layered neural framework that performs direct assignments which is, by and large, used to smooth the neural framework. This layer does the conjecture and gatherings of the neural framework according to the important number of classes.

Rectified Linear Units (ReLU) activation function (see Eq. 2) is mostly used across the hidden layers. Here hidden layers refer to the convolution layers which are placed before the fully connected layer. The output is equivalent to include if the info is more prominent than zero and the other way around.

$$\mathrm{ReLU}(x) = \max(x, 0) \tag{2}$$

Softmax activation function (see Eq. 3) isolates each output into fluctuated classes like the categorial likelihood dispersion and demonstrates the probabilities of the predefined number of classes and the classes which have the most elevated likelihood are taken forward to the output layer.

$$\sigma(x_j) = \frac{e^{x_j}}{\sum_i e^{x_i}} \tag{3}$$

In various terms, the ReLU work is used by its snappier convolution rate however Softmax work is used to get the essential finish of the neural framework.

4 Results

Our model was tested on the public dataset in kaggle comprising of 92,000 images in our dataset, 73,600 samples were trained and 18,400 samples were validated. A model can get complex in times when the data is limited and with hidden layers. The model will in general dive itself into overfitting where a machine gains from the commotion of the preparation set. To avoid this dropout technique is used where a random number of samples are dropped to zero. Thus dropout can be referred to as a preventive measure for any model to be efficient.

A model needs to be trained and tested as well which is why the dataset is divided into training, validation, and test sets. The readiness data is given as a commitment to the layers model and the endorsement set is used to test the model how far it is fit. Graphs are plotted for training and validation accuracy over the total number of epochs to measure the overall accuracy. Similarly, loss function is likewise determined by plotting training and validation loss. The portrayed charts are delineated underneath. Loss can be viewed as a separation between the genuine estimations of the issue and the qualities anticipated by the model. More noteworthy the loss is, progressively colossal is the blunders you made on the information. Accuracy can be viewed as the quantity of errors you made on the information. Here training loss is plotted during training of data and validation loss is plotted when the model is trained for validation and likewise with accuracy (Figs. 3 and 4).

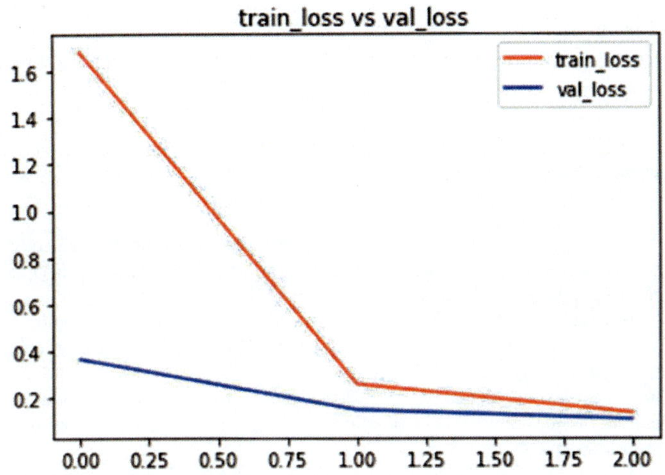

Fig. 3 Simulation graph of training loss versus validation loss

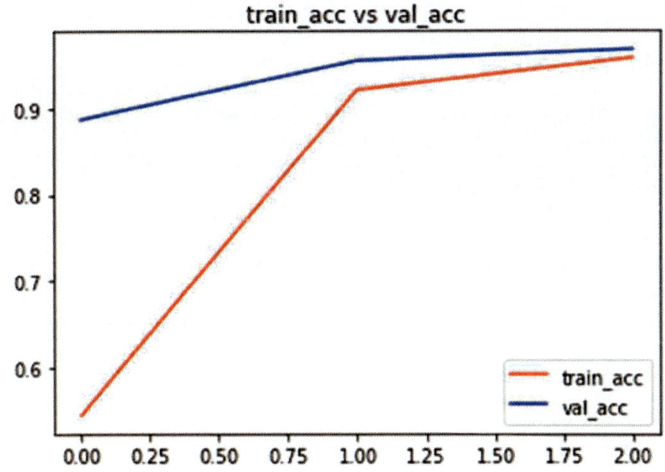

Fig. 4 Simulation graph of training accuracy versus validation accuracy

5 Conclusion

One of the most effective mechanisms to recognize the handwritten characters is "deep convolutional neural network." The biggest challenge for the Devanagari Character set is the proximity of their shapes. The proposed strategy utilizes profound CNN for both extraction and grouping of Devanagari characters. With the extension in the restriction of the frameworks, can get dynamically indisputable features. The best possible regularization procedure helps in staying away from the overfitting issue. CNN is currently known to be the best count for all the vernaculars and thusly gave the most imperative exactness pace of 97% for Devanagari characters too.

References

1. Shelke S, Apte S (2015) A fuzzy based classification scheme for unconstrained handwritten devanagari character recognition. In: International conference on communication, information & computing technology (ICCICT)
2. Shaw B, Parui S, Shridhar M (2008) Offline handwritten devanagari word recognition: a holistic approach based on directional chain code feature and HMM. In: International conference on information technology, pp 203–208
3. Tompson JJ, Jain A, LeCun Y, Bregler C (2014) Joint training of a convolutional network and a graphical model for human pose estimation. In: Advances in neural information processing systems, pp 1799–1807
4. Taigman Y, Yang M, Ranzato M, Wolf L (2014) Deep face: closing the gap to human-level performance in face verification. In: IEEE conference on computer vision and pattern recognition. IEEE, pp 1701–1708
5. Krizhevsky A, Sutskever I, Hinton GE (2012) Imagenet classification with deep convolutional neural networks. In: Advances in neural information processing systems, pp 1097–1105
6. Szegedy C, Liu W, Jia Y, Sermanet P, Reed S, Anguelov D, Erhan D, Vanhoucke V, Rabinovich A (2014) Going deeper with convolutions. arXiv preprint arXiv:1409.4842
7. Simard PY, Steinkraus D, Platt JC (2003) Best practices for convolutional neural networks applied to visual document analysis. In: 2013 12th International conference on document analysis and recognition, vol 2. IEEE Computer Society, pp 958–958
8. Ciresan DC, Meier U, Gambardella LM, Schmidhuber J (2011) Convolutional neural network committees for handwritten character classification. IEEE, pp 1135–1139
9. Trier ØD, Jain AK, Taxt T (1996) Feature extraction methods for character recognition-a survey. Pattern Recogn 29(4):641–662
10. Lauer F, Suen CY, Bloc G (2007) A trainable feature extractor for handwritten digit recognition. Pattern Recogn 40(6):1816–1824
11. Purkaystha B, Datta T, Islam MS (2017) Bengali handwritten character recognition using deep convolutional neural network. In: 2017 20th International conference of computer and information technology (ICCIT), 22–24 Dec 2017
12. Nair PP, James A, Saravanan C (2017) Malayalam handwritten character recognition using convolutional neural network. In: International conference on inventive communication and computational technologies (ICICCT 2017)
13. Asha K, Krishnappa HK (2018) Kannada handwritten document recognition using convolutional neural network. In: 3rd IEEE international conference on computational systems and information technology for sustainable solutions
14. Sonawane PK, Shelke S (2018) Handwritten devanagari character classification using deep learning.In: 2018 International conference on information, communication, engineering and technology (ICICET), Zeal College of Engineering and Research, Narhe, Pune, India, 29–31 Aug 2018
15. Pereira NA, Rao P, Kallianpur AK, Srinivasa KG (2017) Discrete artificial bee colony algorithm based optical character recognition. IEEE
16. Taigman Y, Yang M, Ranzato M, Wolf L (2014) Deepface: Closing the gap to human-level performance in face verification. In: IEEE conference on computer vision and pattern recognition. IEEE, pp 1701–1708
17. Pasha S, Padma MC (2015) Handwritten Kannada character recognition using wavelet transform and structural features. In: 2015 International conference on emerging research in electronics, computer science and technology (ICERECT)

18. Chackoa AMMO, Dhanya PM (2014) Multiple classifier system for offline malayalam character recognition. In: International conference on information and communication technologies (ICICT)
19. Vijayaraghavan P, Sra M, Handwritten tamil recognition using a convolutional neural network. MIT Media Lab
20. Acharya S, Pant AK, Gyawali PK (2015) Deep learning based large scale handwritten Devanagari character recognition. In: 2015 9th International conference on software, knowledge, information management and applications (SKIMA)
21. Arora S, Bhattacharjee D, Nasipuri M, Basu DK, Kundu M (2008) Combining multiple feature extraction techniques for handwritten Devnagari character recognition. In: 2008 IEEE Region 10 and the third international conference on industrial and information systems, ICIIS, pp 1–6

Performance Analysis of Machine Learning Algorithms in Credit Card Fraud Detection

Anupam Yadav, Vinod Jain, and Anuj Kumar

Abstract Credit cards are very useful to make payments nowadays. Security is a major issue in making transactions using credit cards. Lots of frauds happen in different ways using credit cards. Machine learning algorithms are very useful in making predictions. In this paper, machine learning algorithms are used to analyze and predict the fraud in a credit card transaction. Different machine learning algorithms such as decision tree, random forest, linear regression, support vector machine, XG-Boost are applied on a standard data set. The prediction accuracy of these algorithms is measured and compared to each other. It is concluded that machine learning algorithms are very useful in predicting frauds using credit cards. The prediction accuracy of XG-Boost and the random forest is best among all.

Keywords Artificial intelligence · Machine learning · Credit card frauds

1 Introduction

Present time is Technology time. Due to this, all people are surrounded by online activities like shopping, transactions, chatting, and many more. Some issues will also occur with these facilities like frauds in money transaction on online platforms. Different techniques were used in past time to prevent these online frauds, where credit card fraud is one of them. In large groups, money matters of business and monetary cheats are very pivotal matters. Moreover, monetary cheats create unreliability in wealth of any business. It also affects the basic amount of people which they

A. Yadav · V. Jain (✉) · A. Kumar
Department of Computer Engineering and Applications, GLA University Mathura, Mathura, India
e-mail: vinod.jain@gla.ac.iv

A. Yadav
e-mail: anupam.yadav@gla.ac.in

A. Kumar
e-mail: anujkumar.gla@gla.ac.in

G. Ranganathan et al. (eds.), *Inventive Communication and Computational Technologies*, Lecture Notes in Networks and Systems 145,
https://doi.org/10.1007/978-981-15-7345-3_26

spend on daily needs. They are different cheats are hold in different area like automobile securities, monetary statements, credit cards. Debit cards, internet banking, etc. security issues arise here in all sectors. In this paper, machine learning algorithms are applied on the detection of fraud in credit cards. In this section, Credit card fraud detection techniques are focused. Classification, clustering, prediction, outlier detection, regression, and visualization are techniques that are used in machine learning concepts. For credit card fraud detection, these data mining techniques of machine learning algorithm is required. These approaches of data Mining techniques are shown in Fig. 1. In this generation, such machine learning methodologies is needed which can work on large dataset Because it not easily possible for human beings. Machine learning algorithms are very useful in prediction, clustering, etc. Machine learning is applicable on large number of research problems. Supervised learning and unsupervised learning are the two classifications, which contains all the concepts of machine leaning Supervised learning requires prior classification to anomalies. Several supervised learning algorithms used during the last few years in detecting credit card fraud. Supervised Learning is further divided into two parts A. Classification and B. Regression. Unsupervised learning used clustering Approach. Overall process flow in two steps with the help of two types of data like categorical and numerical data, in first step data sets which is in original form with categorical data. Data cleaning and pretreating mechanism can be used for primary data. In the first step, the data comes from a group that can be modified into numerical data and then a suitable method is putted to do the assessment. Further, categorical data is used in machine learning techniques to find the best solution finding algorithm. Supervised learning is a machine learning in which a well-labeled data is provided to machine and train the machine based on labeled data after which it tries to predict the outcome of the new data by analyzing the training data. Classification falls in the category of supervised learning where our source code program learns from the inputted data and this type of learning machine will have used for grouping of new things (objects).

Classification algorithms are very useful in making predictions, many researchers applied classification for prediction and forecasting [1, 2].

In unsupervised machine learning, don't have the teacher or critic which can direct us. It's up to the algorithm to find the commonalities in data and make features. Unsupervised learning is useful in finding unknown patterns in the data.

Fig. 1 Types of machine learning algorithms

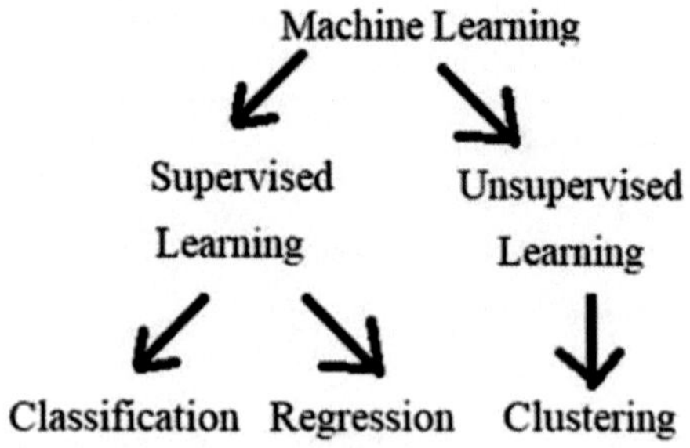

Clustering is type of unsupervised learning method. In clustering, basically, clusters are groups which contain the objects in which they are divided into conceptually meaning [3]. Two types of cluster groups in which those objects containing the same feature are in same cluster. While dissimilar objects do not become the part of that cluster; it transfers to another cluster (group of objects) [2, 3].

2 Related Work

Many researchers applied machine learning algorithms to solve different problems of science and engineering [4–6]. Here are some important contributions to machine learning algorithms in this area. In the paper [7], the random forest machine learning algorithm makes a forest of decision trees. These trees are used to votes for different classifications of the attributes. The classification having the maximum number of votes is selected for classification.

The logic regression is useful to predict values that are binary, i.e., either true or false. It uses linear model [8]. The support vector machine SVM algorithm uses all the data points as features in n dimensional space. This algorithm is also very useful in predicting values. Decision tree is a well-known supervised machine learning algorithm. It is very useful for both types of data that are categorical data and continuous data. The algorithm divides the data into two sets. The most significant features in the data set are used to make different groups. Tree divided into the subset until it stored into mutually exclusive subgroups eventually. It is also referred to as classification and regression tree (CART) [3]. XGBoost [9] is a well-known machine learning algorithm that performs better in many situations. It is based upon ANN Artificial Neural Network. It works better for a large amount of data which is presented in that article [9]. Detection of Fraud in credit card by using RF, SVM and LR is examined in [5]. Genetic algorithm was also used for prediction of credit card frauds [10]. Genetic Algorithm (GA) and scatter search were two algorithms applied by the author for deduction in the rate of incommensurable. Further at last author found that our model increased two times our performance as compared with previous results [10]. For identity of Monetary statement cheats in organization or groups two algorithm were used by author Decision Trees (DT) and Bayesian Belief Networks (BNN) [11]. Raj et al. [12] apply recurrent neural networks and nonlinear prediction in SVM. The paper highlight the use of recurrent NN using SVM for prediction. Many other researchers applied machine learning algorithms in different areas in recent years [13–16]. After going through this literature it is concluded that machine learning algorithms are very useful in making predictions. The power of machine learning algorithms can also be used to predict and stop the frauds in credit cards. The next section discusses some machine learning algorithms that are applied in detection of credit card frauds.

3 Proposed Work

In this work, machine learning algorithms are used to predict fraud in credit card transactions. The five machine learning algorithms are used and applied on a data set of credit card frauds.

Figure 2 is showing the proposed model for prediction of credit card frauds using machine algorithm.

The proposed methodology of credit card fraud detection using machine learning algorithm applies different ML algorithm on a data set of credit card frauds. The technique first applies the ML algorithm by training the ML algorithms on 70% of the data set. It will create a model for the ML algorithm. Then the remaining 30% of the data is used to test the accuracy of ML algorithm.

The details of the applied machine learning algorithms are as follows:

Logistic Regression—The logic regression is a mathematical model that fit the points on logic functions. Because of the logic function, the algorithm is called logical regression. The logic regression is useful to predict values that are binary, i.e., either true or false. So this algorithm is useful to predict fraud occurs or not in a credit card transaction.

Support Vector Machine—The support vector machine SVM algorithm uses all the data points as features in n-dimensional space. Every feature is corresponding to a dimension in the n-dimensional space. The coordinates of the points work as support vectors. This algorithm is also very useful in predicting values.

Decision Tree—Decision tree is a well-known supervised machine learning algorithm. It is very useful for both types of data that is categorical data and continuous data. The algorithm divides the data into two sets. The most significant features in the data set are used to make different groups. This algorithm is also very useful in prediction.

Fig. 2 Proposed model for credit card fraud detection using machine learning

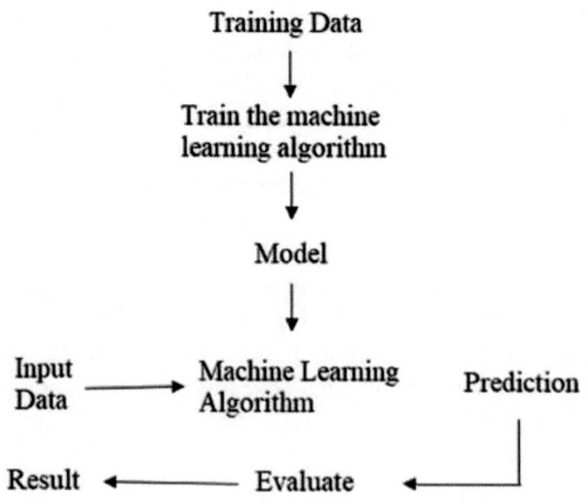

Table 1 Prediction accuracy of machine learning algorithms

S. No.	Machine learning algorithm	Prediction accuracy
1	Logistic regression	99.838
2	Support vector machine	99.860
3	Decision tree	99.923
4	Random forest	99.957
5	Xgboost	99.962

Random Forest—The random forest machine learning algorithm makes a forest of decision trees. These trees are used to votes for different classifications of the attributes. The classification having the maximum number of votes is selected for classification. This machine learning algorithm is also very useful in prediction.

Xgboost—It is a very new and very powerful machine learning algorithm. The outcoming results of Xgboost algorithm outperforms the tree or forest-based machine learning algorithms. This algorithm is also very useful in prediction.

After going into depth of all these machine learning algorithms it is proposed that machine learning algorithms will play a very important role in the detection of credit card frauds.

The next section discusses the results of applying these algorithms on a data set of credit card frauds.

4 Result Analysis

The proposed work is implemented on a data set of credit card frauds and the prediction accuracy of these algorithms is calculated. Table 1 is showing the calculated prediction accuracy of the proposed five machine learning algorithms. From this table, it is clear that the machine learning algorithms can be easily used to predict the fraud in credit card transactions. The prediction accuracy of Xgboost algorithm is 99.962% and it is best among all the applied machine learning algorithms.

Figure 3 is showing a pictorial representation of the prediction of different machine learning algorithms. The accuracy of logistic regression is least which is 99.838% and Xgboost algorithm is most which is 99.962% for the given data set of credit card fraud detection. The next section discusses the conclusions, limitations, and future scope of this work.

5 Conclusion and Future Scope

This paper applied machine learning technique in predicting fraud in credit card transactions. Different machine learning algorithms are applied to a data set. It is concluded that the prediction accuracy of different machine learning algorithms is different. The

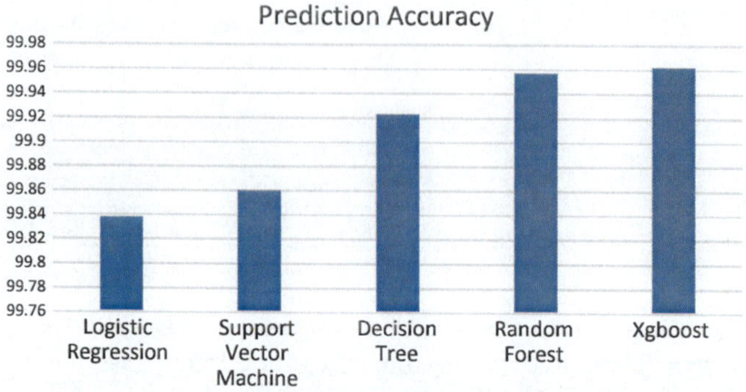

Fig. 3 Comparison of prediction accuracy of machine learning algorithms

prediction accuracy of Random Forest and XG-Boost machine learning algorithms are best among all the five algorithms applied. The summary of the achievement of this work includes that the logic regression machine learning performs least and XGBoost machine learning algorithm performs best as compared to Support Vector Machine, Decision Tree, Random Forest in credit card fraud detection.

Some of the limitations and future scope of this paper are as follows:

1. Other machine learning algorithms can be applied on the same data set and their prediction accuracy can be calculated and compared.
2. The machine learning algorithms and their prediction accuracy can be tested on different data set having more data.
3. Other application areas of machine learning in cyber security can be explored where this technique may play some role.

References

1. Han J, Kamber M (2006) Data mining: concepts and techniques. 2nd ed, Morgan Kaufmann Publishers, pp 285–464
2. Ravisankar P, Ravi V, Rao G, Bose I (2011) Detection of financial statement fraud and feature selection using datamining techniques. Decision Support Syst 491–500
3. Kirkos E, Spathis C, Manolopoulos Y (2007) Data mining techniques for the detection of fraudulent financial statements. Expert Syst Appl 995–1003
4. Popat RR, Chaudhary J (2018) A survey on credit card fraud detection using machine learning. In: 2018 2nd international conference on trends in electronics and informatics (ICOEI), Tirunelveli, pp 1120–1125. https://doi.org/10.1109/icoei.2018.8553963
5. Bhattacharyya S, Jha S, Tharakunnel K, Westland JC (2011) Data mining for credit card fraud: a comparative study. Decision Support Syst 50(3):602–613
6. Yue X, Wu Y, Wang Y, Li Y, Chu C (2007) A review of data mining-based financial fraud detection research. In: International conference on wireless communications sep, networking and mobile computing, pp 5519–5522

7. Kumar MS, Soundarya V, Kavitha S, Keerthika ES, Aswini E (2019) Credit Card fraud detection using random forest algorithm. In: 2019 3rd International conference on computing and communications technologies (ICCCT), Chennai, India, pp 149–153. https://doi.org/10.1109/iccct2.2019.8824930
8. Ngai E, Hu Y, Wong Y, Chen Y, Sun X (2011) The application of data mining techniques in financial fraud detection: a classification framework and an academic review of literature. Decision Support Syst 559–569
9. Jidong L, Ran Z (2018) Dynamic weighting multi factor stock selection strategy based on XGboost machine learning algorithm. In: 2018 IEEE international conference of safety produce informatization (IICSPI), Chongqing, China, pp 868–872. https://doi.org/10.1109/iicspi.2018.8690416
10. Duman E, Ozcelik MH (2011) Detecting credit card fraud by genetic algorithm and scatter search. Expert Syst Appl 38(10):13057–13063
11. Kirkos E, Spathis C, Manolopoulos Y (2007) Data mining techniques for the detection of fraudulent financial statements. Expert Syst Appl 32(4):995–1003
12. Raj JS, Ananthi JV (2019) Recurrent neural networks and nonlinear prediction in support vector machines. J Soft Comput Paradigm (JSCP) 1(01):33–40
13. Feng C, Wu S, Liu N (2017) A user-centric machine learning framework for cyber security operations center. In: 2017 IEEE international conference on intelligence and security informatics (ISI), Beijing, pp 173–175. https://doi.org/10.1109/isi.2017.8004902
14. Farooq HM, Otaibi NM (2018) Optimal machine learning algorithms for cyber threat detection. In: 2018 UKSim-AMSS 20th international conference on computer modelling and simulation (UKSim), Cambridge, pp 32–37. https://doi.org/10.1109/uksim.2018.00018
15. Abdullah MS, Zainal A, Maarof MA, Nizam Kassim M (2018) Cyber-Attack features for detecting cyber threat incidents from online news. In: 2018 Cyber resilience conference (CRC), Putrajaya, Malaysia, pp 1–4. https://doi.org/10.1109/cr.2018.8626866
16. Kadoguchi M, Hayashi S, Hashimoto M, Otsuka A (2019) Exploring the dark web for cyber threat intelligence using machine leaning. In: 2019 IEEE international conference on intelligence and security informatics (ISI), Shenzhen, China, pp 200–202. https://doi.org/10.1109/isi.2019.8823360

An Empirical Evaluation of Bitcoin Price Prediction Using Time Series Analysis and Roll Over

N. M. Dhanya

Abstract Bitcoin has attracted considerable attention in today's world because of the combination of encryption technology along with the monetary units. For traders, Bitcoin leads to a promising investment because of its highly fluctuating price. Block chain technology assists in the transactions of documentation. The characteristics of the bitcoin which is derived from the blockchain technology has led to diverse interests in the field of economics. The bitcoin data is selected from 2013 to 2018, over a period of 5 years for this analysis. Here a new roll over technology is applied where new data is obtained over time which will close out the old information during machine training. This mechanism will help in incorporating new information in the short-term learning. The results show that the rollover mechanism improves the time series prediction accuracy.

Keywords Bitcoin · Time series analysis · Price prediction · Roll over mechanism

1 Introduction

Time series information in the real world involves the properly ordered and observed sequence of values of any real-world objects, or process or domain information or the values coming straight from the sensor fixed to capture the underlying information. These time series information data come at periodic intervals with the same or different frequencies so that it enables the user to analyse the information from the captured data.

These data therefore offer double advantage to the user, as it helps in past information analysis and also to capture the essence of the future data. Hence modelling of the time series information helps to analyse, predict and solve many prediction problems that might help in various fields of economic growth.

N. M. Dhanya (✉)
Department of Computer Science and Engineering, Amrita School of Engineering, Coimbatore, Amrita Vishwa Vidyapeetham, India
e-mail: nm_dhanya@cb.amrita.edu

G. Ranganathan et al. (eds.), *Inventive Communication and Computational Technologies*, Lecture Notes in Networks and Systems 145,
https://doi.org/10.1007/978-981-15-7345-3_27

Hence proper regression algorithms can be employed to analyze the time series data, and to capture the essential information. So, it is necessary to build a proper regression based modeling technique to evaluate the upcoming sensor data. Subsequently, to assess the best model for temperature forecast utilizing the time arrangement information, the model ought to be sufficiently healthy to stay away from commotion; it ought to likewise be exceedingly solid when working with information influenced with expectation inclination or with scaling mistake. The chosen model ought to be adaptable, so that the model can be utilized to contrast the forecast that comes about and distinctive datasets.

1.1 Need for Bitcoin Prediction

Bitcoin has of late pulled in significant consideration in the fields of financial markets, cryptography, and PC science because of its inalienable nature of consolidating encryption, innovation and money related units. Bitcoin uses the Blockchain technology which provides transparency to the transactions made and hence has become a popular means of transaction across the globe. However, the extreme volatility in the Bitcoin values is a reason for concern for investors as well as regulatory authorities. Hence a reliable prediction model for bitcoin price movements is the need of the hour. The observed results is compared with other straight and non-direct benchmark models on predicting bitcoin price. Figure 1 shows the bitcoin price fluctuations from 2012 to 2020 [1].

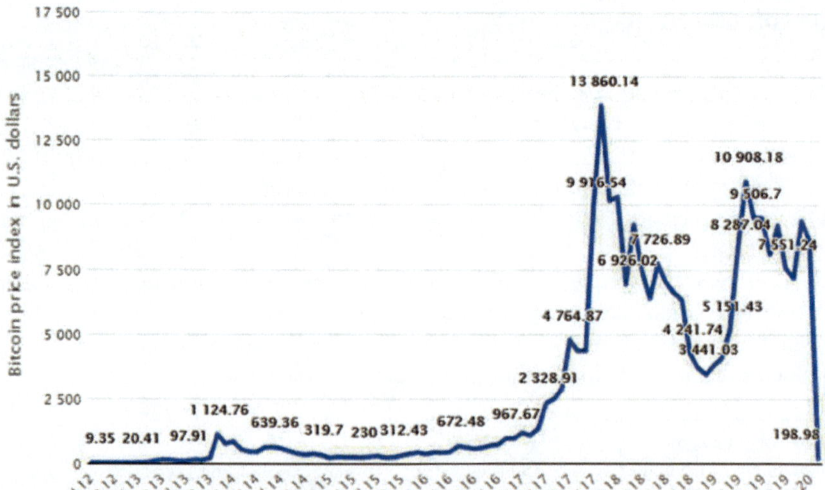

Fig. 1 Bitcoin price changes

Bitcoin is an effective figure cash brought into the money related market in view of its one of a kind convention and Nakamoto's orderly basic detail [2]. Dissimilar to existing fiat monetary standards with national banks, Bitcoin points to accomplish finish decentralization. Inborn attributes of Bitcoin inferred from Blockchain advancements have prompted different research interests in the field of financial aspects as well as in cryptography and machine learning. A machine trained just with Bitcoin value list and changed costs displays poor prescient execution [3]. Our model looks at the precision of anticipating Bitcoin cost through linear Regression, Support Vector Regression, Random forest regression algorithms, ARIMA model and Deep learning approaches.

2 Literature Survey

Existing work on forecasting bitcoin prediction techniques are: Gaussian Process; Linear Regression; Sequential minimal optimization (SMO) regression and Multilayer Perceptron. Gaussian Process [2] implements a classifier function without hyper-parameter tuning for regression. This method however is computationally expensive. Linear regression [4, 5] models the relationship between one or more explanatory variables and a scalar response. It has two main drawbacks i.e., it is limited to linear relationships and it looks for only the dependant variable's mean. Multilayer perceptron [3] is to forecast the webpage views and it makes use of back-propagation method. After the network is built, during the training time it can be monitored and modified. It is quite painful to select the suitable architecture of the network. SMO regression [6] where it has potential for speeding up the forecasting process and also it scales linearly with the size of the training set. But it fails to handle the large-scale training problems because of memory issues. Li and Moore [7] presented an algorithm called Elastic Smooth Season Fitting (ESSF) algorithm which derives the seasonality employing residual sum of squares minimization by smoothness regularization. ESSF accuracy improves significantly over other methods that ignore the yearly seasonality. Jung and Lee [8] created a bitcoin prediction system using block chain information. The authors applied a Bayesian neural network for predicting the price. Li et al. [9] used an LSTM architecture for predicting the Bitcoin price. Block chain statistics is also used for prediction. Various machine learning and deep learning algorithms can be used for prediction [10] and Table 1 shows the functions and limitations of the existing time series models. Sriwiji and Primandari [11] used a Bayesian regularization network for predicting the bitcoin price. They employed a subset selection technology to reduce the number of features and were able to get an accuracy of 91%. Sin and Wang [12] proposed an ANN based ensemble approach based on Genetic algorithm based selective neural network. The next day price of the bitcoin is predicted from the past 50 days observation over 200 features. This strategy generated 85% of the returns.

The following Table 2 shows the already existing time series models for prediction.

Table 1 Functions and limitations of time series models

Models	Function	Cons
ARIMA	Combination of auto regressive, integrated moving average	Noise not considered and parametric
ANN	Linear neuron model, learning by delta rule	Not applicable for non-linear relationships
SVM	Linear learning using decision boundary	Cannot deal with non-linearly separable and noisy data
Recurrent (MLP)	Non linear models, supervised learning by feedback based back propagation	Memory is needed to store temporal information

Table 2 Existing bitcoin prediction techniques

Models	Function
Cryptocurrency price prediction using news and social sentiments [4]	Analyses the ability of news and social media data to predict price fluctuations
Bayesian regression and Bitcoin [6]	Analyses the effect of Bayesian technique for predicting bitcoin fluctuations
Predicting fluctuation in cryptocurrency transactions based on user comments and replies [7]	Trains prediction model based on crawling data from online cryptocurrency communities
Predicting bitcoin price fluctuation with twitter sentiment analysis [14]	Uses sentiment analysis and option mining for prediction
Automated bitcoin trading via machine learning algorithms [15]	Studies the Bayesian neural network algorithm for predicting prices

3 DataSet

For data modelling and bitcoin prediction, time series bitcoin dataset is chosen each with 4000 rows (Time series bitcoin price data from 2011 to 2018) and 24 columns. Preliminary pre-processing of data is done and missing data is filled. This dataset has the following features [13].

- Date: Date of the bitcoin price observation.
- btc_market_price: Average market price of bitcoin.
- btc_total_bitcoins: Total number of bit coins mined.
- btc_market_cap: The total value of the bitcoin in circulation.
- btc_trade_volume: The total value of trading volume of bitcoin.
- btc_blocks_size: Total size of all headers and transation in the block chain.
- btc_avg_block_size: The average block size in MB.
- btc_n_orphaned_blocks: Total number of mined blocks which are not attached to the blockchain.
- btc_n_transactions_per_block: The average number of transactions per block.

- btc_median_confirmation_time: The time for a bitcoin transaction to accept into a mined block.
- btc_hash_rate: The estimated number of tera hashes per second the Bitcoin network is performing.
- btc_difficulty: A measure on the difficulty in finding a new block.
- btc_miners_revenue: The total rewards and fees paid to miners.
- btc_transaction_fees: The total value of all transaction fees paid to miners.
- btc_cost_per_transaction_percent: miners revenue as percentage of the transaction volume.
- btc_cost_per_transaction: miners revenue divided by the number of transactions.
- btc_n_unique_addresses: The total number of unique addresses used on the Bitcoin blockchain.
- btc_n_transactions: The number of daily confirmed Bitcoin transactions.
- btc_n_transactions_total: Total number of transactions.
- btc_n_transactions_excluding_popular: The total number of Bitcoin transactions, excluding the 100 most popular addresses.
- btc_n_transactions_excluding_chains_longer_than_100: The total number of Bitcoin transactions per day excluding long transaction chains.
- btc_output_volume: The total value of all transaction outputs per day.
- btc_estimated_transaction_volume: The total estimated value of transactions on the Bitcoin blockchain.
- btc_estimated_transaction_volume_usd: The estimated transaction value of bitcoin.

A feature engineering mechanism of XGBoost with Bayesian optimization is applied on the features to reduce it and we selected top 15 contributing features. The feature importance is given in Fig. 2.

4 Architecture and Roll Over Mechanism

The below flow chart in Fig. 3 describes the process flow of the proposed work. And Fig. 4 describes the overall workflow of the models applied.

5 Prediction of Error Rate by Rollover

The prediction analysis is done with the traditional models such as linear regression, random forest and SVM and time series models such as ARIMA and LSTM. A roll over mechanism is used to improve the accuracy of the methods. Since the variations in bitcoin price is very high, this rollover mechanism will allow us to iterate our model with latest information in time thus making the model more dynamic and current context aware. The rollover mechanism will work as follows. A time frame

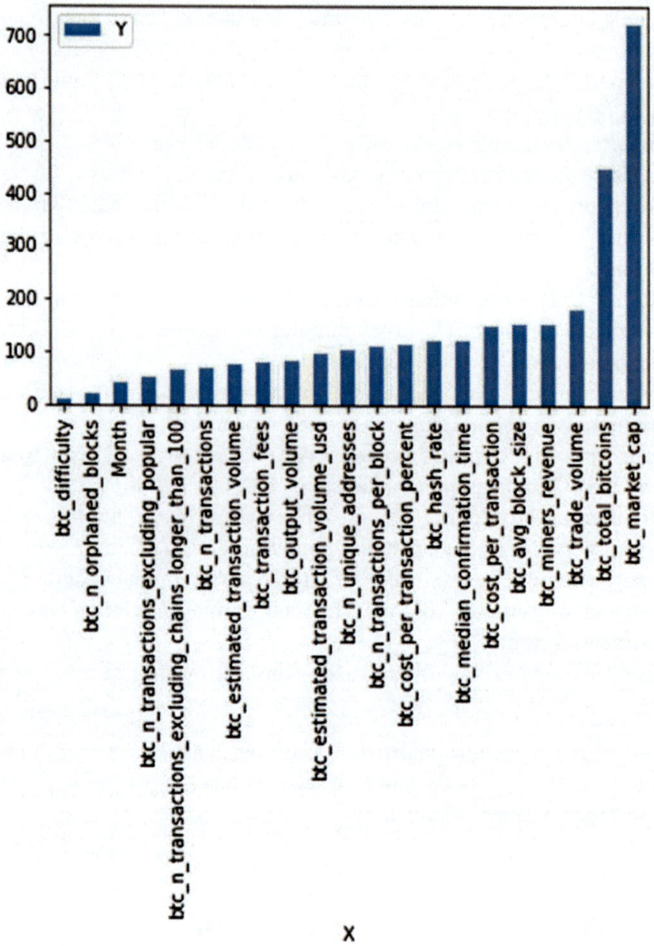

Fig. 2 Feature selection

is set for rolling over in such a way that the old data is closed over time and new data is acquired for rollover. The schematic definition of this method is given in Fig. 5.

The training of the framework will start with N training samples N_{train}, and the prediction performance is tested with the testing data N_{test}. After a time frame of t'-t from time frame t, the model is trained again with a training data N_{train} from time t' and updation to the old model is done. Again, the testing is preformed using test data N_{test}. Similarly training and updation is done through the entire dataset.

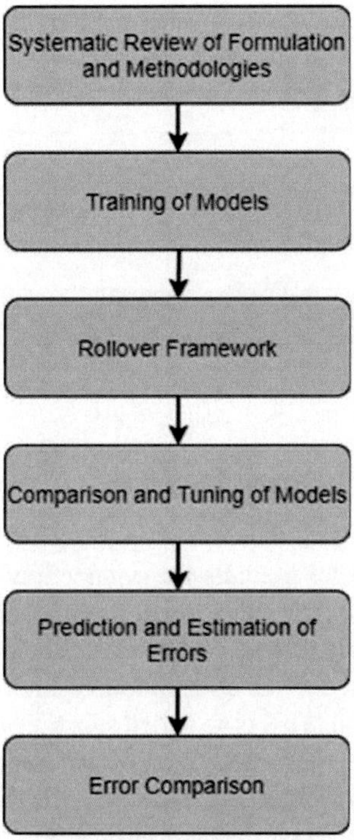

Fig. 3 Overall design

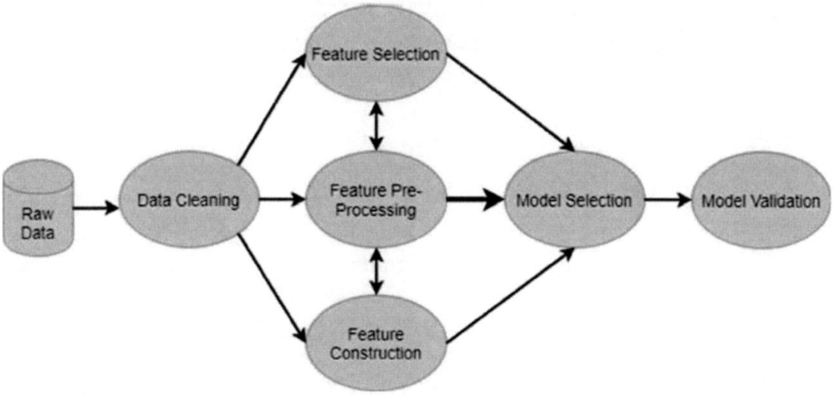

Fig. 4 Prediction model

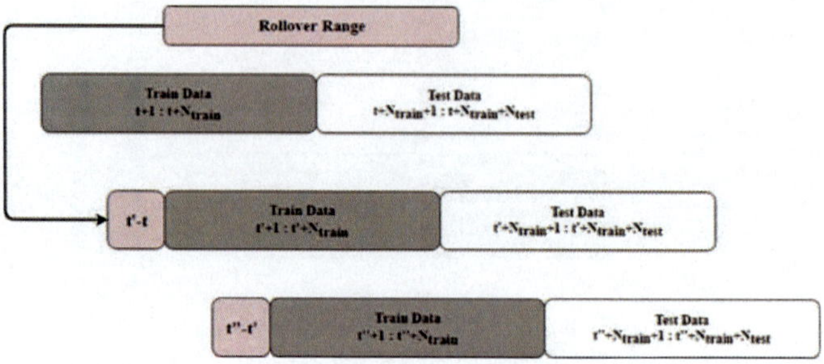

Fig. 5 Rollover Framework

6 Prediction Using Classical Regression Algorithms

Linear Regression: This model predicts the connection between the indicator and reaction factors utilizing straight indicator capacities which are extricated from the information. The parameters for straight relapse are recognized utilizing relationship, and the components which are much corresponded shapes the indicator factors.

Random Forest Regression: This is an added substance model, where information is anticipated by consolidating the aftereffects of different choices obtained from the easier base models. Henceforth the last yield show is the aggregate of all the less difficult models. This empowers to accomplish preferred outcomes over alternate methods.

Support Vector Regression: It is the enhancement-based relapse system where in each stage, the inclination work relating to the information is improved to yield preferable outcomes over the past stage. Improvement is being performed on the negative angle side where the comparing relapse tree is fitted.

7 Prediction Using Time Series Models

ARIMA is a classic time series analysis algorithm. For applying ARIMA the data should be stationary. From Fig. 6, it can be observed that bitcoin price has an exponential trend, the confirmation of this is one with Augmented Dicky Fuller Test, where the p value is >0.05. Hence, log of the data is taken to make it stationary, but the data appears to be still seasonal and as the last step differencing is applied to remove the trend and seasonality. Figure 4 represents all these transformations. The differencing is done automatically by the ARIMA model.

The ARIMA results are shown in the Fig. 7, which depicts the actual and predicted value of the bitcoin price.

LSTM-based prediction model is created and evaluated in various historical window sizes and network parameters. The window size of 30 was giving the minimum RMSE. A simple LSTM model is created with the following layers and

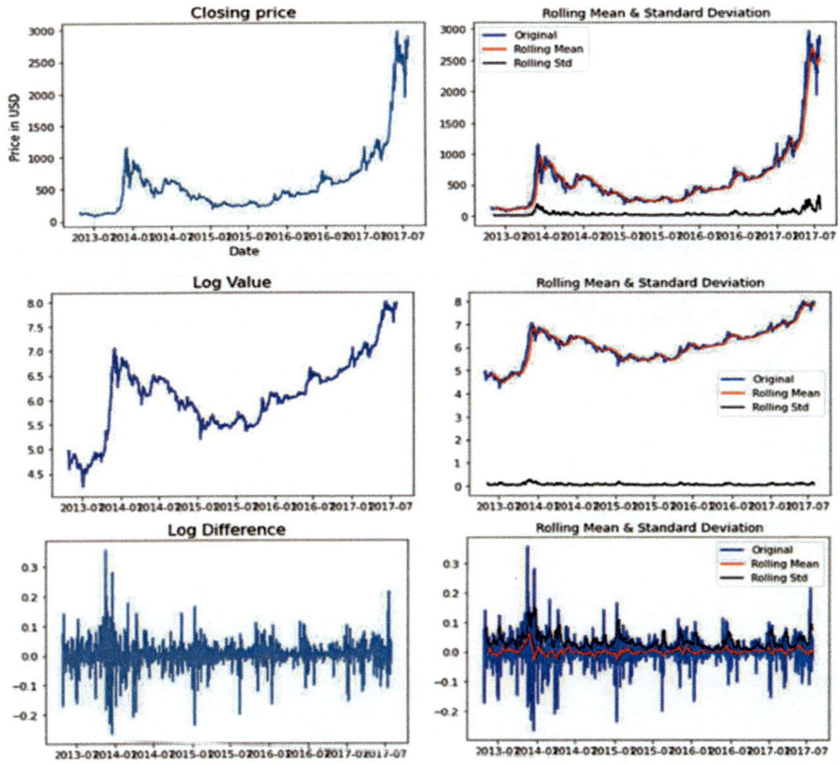

Fig. 6 Original data, log and log-differencing

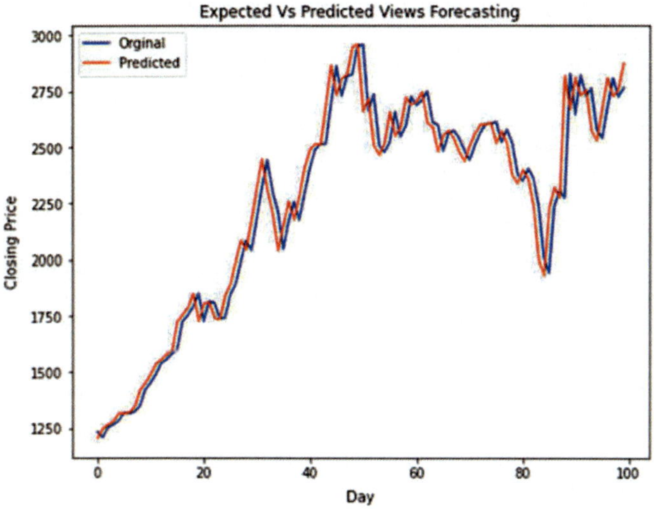

Fig. 7 ARIMA prediction result

```
Model: "sequential_4"
_________________________________________________________________
Layer (type)                 Output Shape              Param #
=================================================================
lstm_5 (LSTM)                (None, 50)                10400
_________________________________________________________________
dense_3 (Dense)              (None, 1)                 51
=================================================================
Total params: 10,451
Trainable params: 10,451
Non-trainable params: 0
_________________________________________________________________
```

Fig. 8 Neural network model

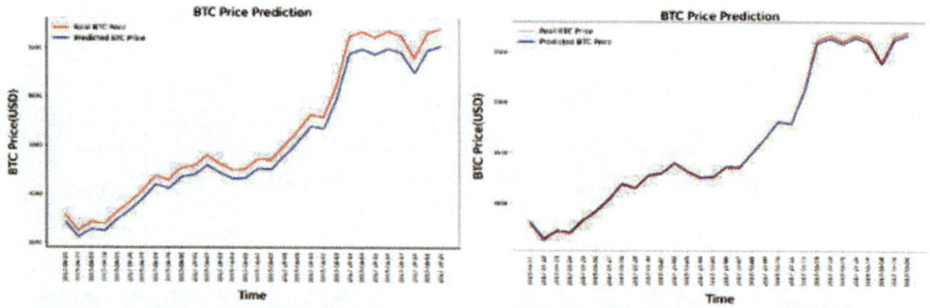

Fig. 9 LSTM and GRU prediction

the result is shown in Fig. 8. A similar architecture is tried for a GRU also and, it is giving much better prediction as LSTM. Figure 9 describes the results of the deep learning model. As shown in the result, it is clear that GRU gives better prediction accuracy than LSTM model.

8 Performance Metrics for Prediction

Each of the algorithms is tested for the following metrics as a result of rollover framework run.

Mean Squared Error (MSE): It is the metric that is computed by taking the normal distinction between the square of the anticipated qualities and the real qualities. It says how the anticipated qualities are near the relapse line.

$$\mathrm{MSE} = \mathrm{Average}\left(y - y'\right)$$

where, y is the actual values and y' is predicted values.

Root Mean Squared Error (RMS Error): It is characterized as the square foundation of Mean Squared Error. With the end goal of correlation of models, Normalized RMS estimates are utilized here. RMS = (MSE).

$$\text{Normalised RMS} = \frac{\sqrt{\text{MSE}}}{(y\text{max} - y_{\text{min}})}$$

Variance: This metric measures how far the qualities got strayed from the mean estimate.

9 Methodology and Execution

- Choosing the Data record: The dataset with 15 features and 4000 rows fills in as the contribution for the accuracy calculation from CSV document.
- Dependent and the independent factors are appropriately inputted to the model.
- Generation of Training and Testing information: The given dataset is 10 fold cross-validated to get the preparation and testing information with the proportion of 80:20.
- Model forecast: The Selected information is tested with the various models.
- Each calculation is executed in python where the parameters for the models are kept steady for all the datasets.
- For SVM regression, number of estimators is settled at 25, minimum leaves is set to 5, and number of arbitrary states is set at 3.
- For Random forest regression, number of estimators is settled at 25, learning rate is set to 0.2, maximum profundity is set at 5 and number of irregular states is set to 3.
- ARIMA is applied with $p, d, q = (2, 1, 0)$.
- LSTM and GRU is tried with 50 nodes and for 100 epochs.
- Then the algorithm is adjusted for rollover framework with all models accessible and the parameters of the ideal model is settled.
- The above model setting is rehashed for all the run and the measurements estimated.

10 Results and Discussions

The following Table 3 shows the execution results of classic algorithms run on the dataset.

From Table 3 results, application of Rollover Framework using deep learning techniques executes better than the other classical techniques ultimately resulting in minimization of error rate.

Table 3 Execution results

Models	Metric	Without rollover	With rollover
Linear	RMSE	0.4823	0.3863
	Variance	0.2663	0.2655
Random forest	RMSE	0.0039	0.5039
	Variance	0.0138	0.0124
SVM	RMSE	0.3201	0.2801
	Variance	0.0428	0.0098
ARIMA	RMSE	0.0875	0.0359
	Variance	0.9314	0.9445
LSTM	RMSE	0.0825	0.0325
	Variance	0.9147	0.9436
GRU	RMSE	0.0725	0.0315
	Variance	0.9185	0.9523

11 Conclusion

Bitcoin is a cryptocurrency mechanism which is extensively studied. In this paper, the Bitcoin prize is analysed using time series analysis. A linear model, random forest and SVM is applied and the results are analysed. The GRU model with Rollover is giving the highest accuracy in predicting the closing price. The LSTM and GRU model can be improved by hyperparameter tuning such as dropout and other regularization techniques.

References

1. https://www.statista.com/statistics/326707/bitcoin-price-index/
2. Lane ND, Bhattacharya S, Georgiev P, Forlivesi C, Kawsar F (2015) An early resource characterization of deep learning on wearables, smartphones and internet-of-things devices. In: Proceedings of the 2015 international workshop on internet of things towards applications—IoT-App 15
3. Spuler M, Sarasola-Sanz A, Birbaumer N, Rosenstiel W, Ramos-Murguialday A (2015) Comparing metrics to evaluate performance of regression methods for decoding of neural signals. In: 37th Annual international conference of the IEEE engineering in medicine and biology society (EMBC)
4. Ahmed NK, Atiya AF, Gayar NE, El-Shishiny H (2015) An empirical comparison of machine learning models for time series forecasting. Technical Report
5. Veerakumar S, Dhanya NM (2018) Performance analysis of various regression algorithms for time series temperature prediction. J Adv Res Dyn Control Syst 10(3):996–1000
6. Anufriev M, Hommes C, Makarewicz T (2012) Learning to forecast with genetic algorithms. Working Paper
7. Khadka M, Popp B, George KM, Park N (2010) A new approach for time series forecasting based on genetic algorithm. In: CAINE, pp 226–231

8. Jang H, Lee J (2018) An empirical study on modeling and prediction of bitcoin prices with bayesian neural networks based on blockchain information. IEEE Access 6:5427–5437
9. Li L, Arab A, Liu J, Liu J, Han Z (2019) Bitcoin options pricing using LSTM-based prediction model and blockchain statistics. In: 2019 IEEE international conference on Blockchain (Blockchain)
10. Dhanya NM, Harish UC (2018) Sentiment analysis of twitter data on demonetization using machine learning techniques. In: Lecture notes in computational vision and biomechanics, vol 28, pp 227–237
11. Sriwiji R, Primandari AH (2020) An empirical study in forecasting bitcoin price using bayesian regularization neural network. In: Proceedings of the 1st international conference on statistics and analytics, ICSA 2019, 2–3 Aug 2019, Bogor, Indonesia
12. Sin E, Wang L (2017) Bitcoin price prediction using ensembles of neural networks. In: 2017 13th International conference on natural computation, fuzzy systems and knowledge discovery (ICNC-FSKD)
13. Kaggle. https://www.kaggle.com/c/bitcoin/data
14. Dobslaw F (2010) A parameter tuning framework for metaheuristics based on design of experiments and artifcial neural networks. In: Proceedings of the international conference on computer mathematics and natural computing (WASET '10)
15. https://docs.microsoft.com/en-us/azure/machine-learning/machine-learning-algorithm-choice, http://scikit-learn.org/stable/modules/cross_validation

Ensure the Validity of Forensic Evidence by Using a Hash Function

K. Aishwarya Lakshmi, Prasad B. Honnavali, and S. Rajashree

Abstract A hash function is a mathematical function used to identify and authenticate the file data. Data corruption in forensic evidence will tamper the digital evidence and could mislead the investigation. To ensure information security, various identity and access management tools are necessary. Hashing plays a key role in such scenarios and can be used to check the integrity of the data. Ensuring the validity of such evidence becomes crucial. Validity can be performed at various stages, viz. collection, preservation, and at the time of analysis.

Keywords Forensic evidence · Hashing · Integrity · FTK imager · SHA1

1 Introduction

The rapid, widespread adoption of new technology often outpaces society's development of a shared ethic governing, its use and the ability of legal systems to deal with it. The handling of digital evidence is a perfect example [1]. The integrity of a data source and the data is especially important. Moreover, integrity is one principle component of the security triad, i.e. the CIA triad. Ensuring the original data value during storage, transmission, and at receiving phase is very crucial.

Hashing is a technique through which the integrity of data can be checked, which may be in the form of a file, filesystem format, image file (.iso), or even a picture. Hashing can be defined as the method or process of converting the data into a unique string (maybe a hexadecimal string). As stated earlier, it is used to check for the

K. Aishwarya Lakshmi (✉) · P. B. Honnavali · S. Rajashree
Department of Computer Science, PES University, Bangalore, India
e-mail: aishwaryalakshmi359@gmail.com

P. B. Honnavali
e-mail: prasadhb@pes.edu

S. Rajashree
e-mail: rajashrees@pes.edu

G. Ranganathan et al. (eds.), *Inventive Communication and Computational Technologies*, Lecture Notes in Networks and Systems 145,
https://doi.org/10.1007/978-981-15-7345-3_28

integrity of the data, if any unauthorized changes/manipulations have been made to the data under consideration as evidence in the investigation.

A crucial property of a hash function is the collision resistance property. Collision resistance is the property of cryptographic hash function such that it becomes impracticable to find to input that has the same output [2]. The most common standard design of the hash function is based on the Merkel--Damograd Construction. The key of this method was that, if the completion function is collision-resistant so will the hash function derived from it [2].

In this paper, MD5 and SHA1 verify the values of an image file. Even though the use of MD5 and SHA1 has been questioned based on the degree of collision resistance, the chance of hash collisions occurring randomly is improbable due to the significantly large numbers involved [3].

MD5 (Message Digest 5) [4, 5] is a cryptographic hashing algorithm produces a 128-bit hash value for any arbitrary length input. SHA1, under the family of SHA160 (which includes SHA0), produces a 160-bit hash value for any arbitrary length input [6, 7].

2 Problem Definition

To analyze the evidence for its integrity using the hash value, (SHA 1) is performed on FTK Imager.

3 Design and Implementation

3.1 Access Data FTK Imager 3.1.1.8

FTK® Imager is a data preview and imaging tool that lets you quickly assess electronic evidence to determine if further analysis with a forensic tool such as AccessData® Forensic Toolkit® (FTK) is warranted [8]. It is one of the most popular forensic tool kits used for image analysis and verification purpose. Other products can be used to recover deleted file and perform an operation on them. EnCase is another tool used instead of FTK toolkit.

Step 1: Installing FTK Imager 3.1.1.8 on Windows (Fig. 1).

Step 2: Creating a disk image.

The disk image is being adopted for this assignment (Fig. 2).

Step 3: Describing the evidence details (Fig. 3).

Step 4: Checking the MD5 and SHA1 hash values of the files created (Fig. 4).

Step 5: Checking the contents of the created text typed file (Fig. 5).

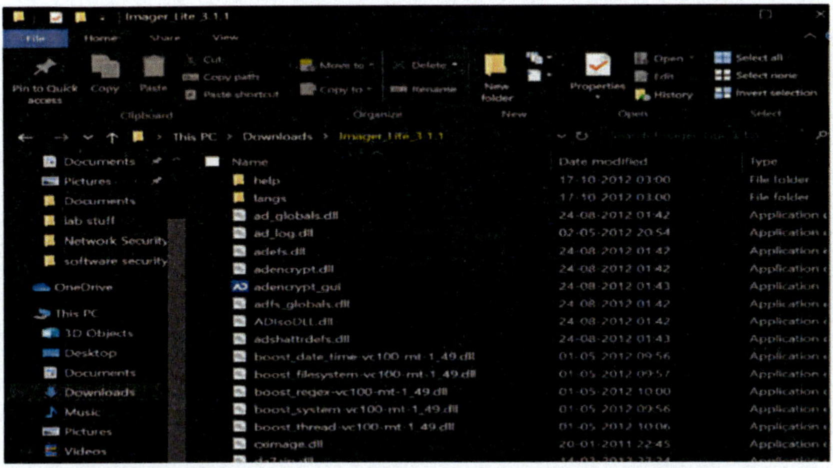

Fig. 1 Installation of FTK Imager

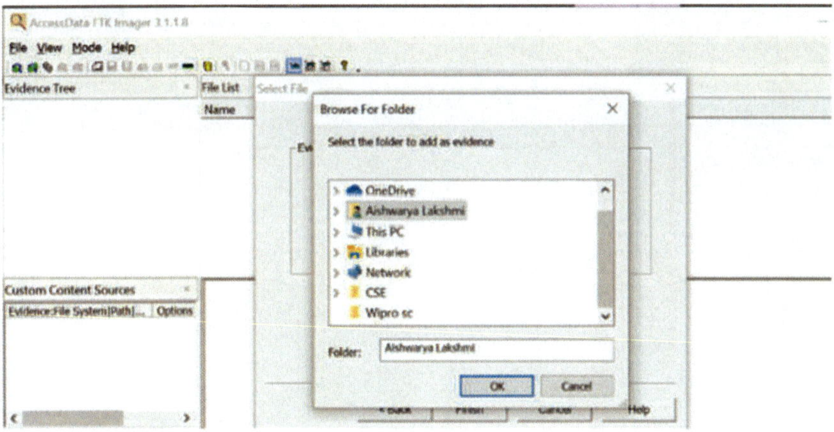

Fig. 2 Creating disk image on FTK Imager

Step 6: Exporting the hash value to CSV file for later comparison (Fig. 6).

Step 7: Modifying the file and then exporting the hash value to a CSV file (Fig. 7).

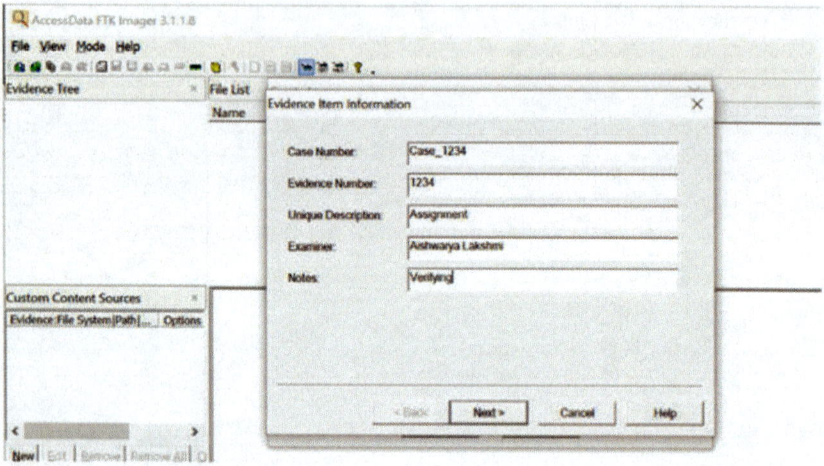

Fig. 3 Recording the entry of the evidence

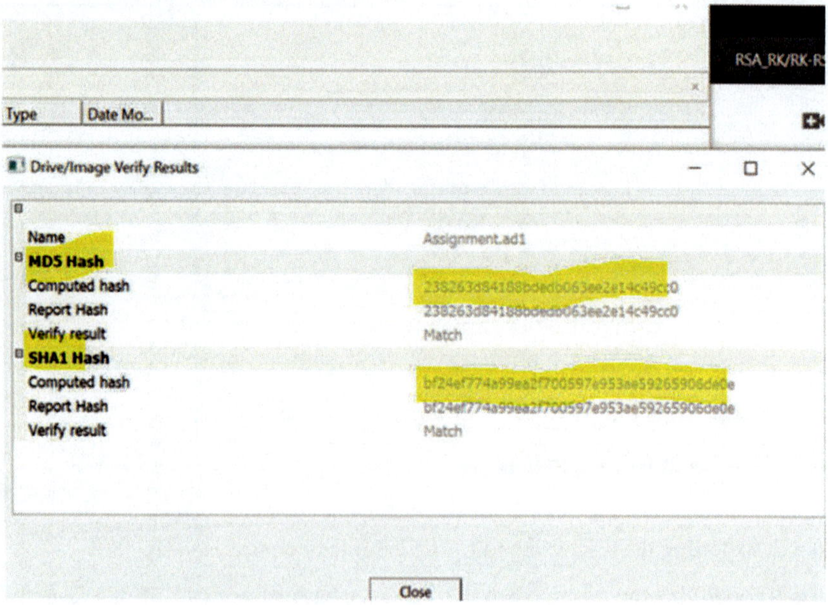

Fig. 4 MD5 and SHA1 hash value of evidence

```
Created By AccessData® FTK® Imager 3.1.1.8

Case Information:
Acquired using: ADI3.1.1.8
Case Number: Case_1234
Evidence Number: 1234
Unique Description: Assignment
Examiner: Aishwarya Lakshmi
Notes: Verifying

--------------------------------------------------------------

Information for C:\Users\kechi\OneDrive\Desktop\Assignment.ad1:
[Computed Hashes]
 MD5 checksum:    238263d84188bdedb063ee2e14c49cc0
 SHA1 checksum:   bf24ef774a99ea2f700597e953ae59265906de0e

Image information:
 Acquisition started:   Wed Apr 01 20:29:01 2020
 Acquisition finished:  Wed Apr 01 20:29:01 2020
 Segment list:
  C:\Users\kechi\OneDrive\Desktop\Assignment.ad1

Image Verification Results:
 Verification started:  Wed Apr 01 20:29:01 2020
 Verification finished: Wed Apr 01 20:29:01 2020
 MD5 checksum:    238263d84188bdedb063ee2e14c49cc0 : verified
 SHA1 checksum:   bf24ef774a99ea2f700597e953ae59265906de0e : verified
```

Fig. 5 Contents of the disk image file created

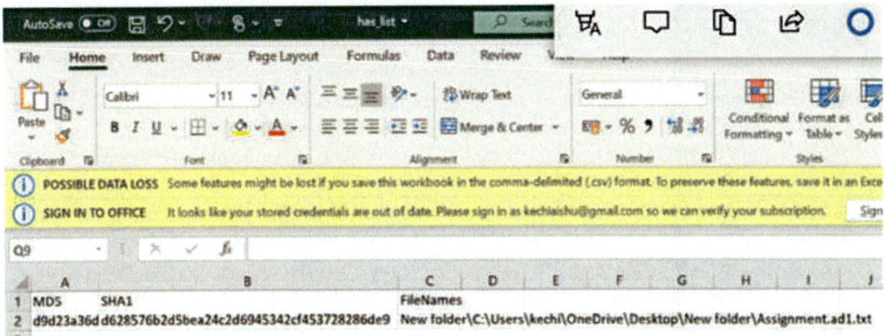

Fig. 6 Hash values of the evidence before the modification exported to CSV file. *Note* SHA1 hash value before modifying the file d628576b2d5bea24c2d6945342cf453728286dc9

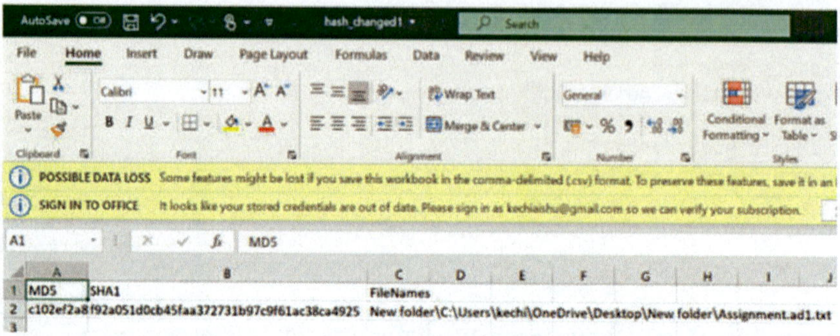

Fig. 7 Hash values of the evidence after the modification exported to CSV file. *Note* The SHA 1 hash value of the file after modification f92a051d0cb45faa372731b97c9f61ac38ca4925

4 Conclusions

The SHA1 hash value in step 6 (before modifying the file) was: d628576b2d5bea24c2d6945342cf453728286de9 and, the SHA1 hash value in step 7(after modifying the file) was: f92a051d0cb45faa372731b97c9f61ac38ca4925.

The two values are not the same and justify the fact that any changes to the file will change the hash value of the function and the same can be verified. This makes it evident that the significance of the hash function in the field of digital forensics is valuable.

References

1. Digital evidence in the courtroom: a guide for law enforcement and prosecutors, U.S. Department of justice office of justice programs national institute of justice, Jan 2007
2. Cryptography: a crisis revealed a resolution solved. A case study from Isaac Newton institute for mathematical science
3. Schmitt V, Jordaan J (2013) Establishing the validity of Md5 and Sha-1 hashing in digital forensic practice in light of recent research demonstrating cryptographic weaknesses in these algorithms. Int J Comput Appl 68(23):0975–8887
4. Kumar K (2012) Significance of hash value generation in digital forensic: a case study. PEC University of Technology, Chandigarh, India, July 2012
5. Crypto Case 3: ensure the validity of forensic evidence by using a hash function
6. Rodenny McCamish "When is digital evidence forensically sound?"
7. Gopalakrishnan A (2018) The art of piecewise hashing: a step toward better evidence provability. Amrita School of Engineering, Amrita Vishwa Vidyapeetham, Coimbatore, India, Jan 2018
8. Access Data, Imager 3.1.1, release notes. https://accessdata.com/products-services/forensic-toolkit-ftk/ftkimager

Enhanced Opinion Classification Using Nature-Inspired Meta-Heuristics for Policy Evaluation

Abhilasha Sharma, Nikhil Arora, and Paridhi Sachdeva

Abstract In today's competitive and result-driven environment, every decision needs to be carefully weighed before implementation. The government faces a similar problem while evaluating its schemes and forming policies. The goal of this study is to recommend a suitable and optimal automated tool for a highly accurate process of sentiment analysis on government schemes. This paper studies the effects of adding a feature selection phase to the conventional opinion mining model by analyzing the impact on the accuracy of the different models. For this purpose, swarm evolutionary algorithms, namely binary cuckoo search algorithm and firefly algorithm, have been used and analyzed, coupled with TF-IDF-based feature extraction for an optimized opinion classification process. Digital India, the flagship campaign of the Indian government, has been selected as the topic of study for this research due to its significant impact on Indian society in recent years. This paper aims to examine the success of the Digital India program, while at the same time, determine the most appropriate model for future assessment of government schemes.

Keywords Government schemes · Sentiment analysis · Swarm evolutionary algorithms · Optimized opinion classification · Digital India

A. Sharma · N. Arora (✉) · P. Sachdeva
Delhi Technological University, Delhi, India
e-mail: nikhilarora986862@gmail.com

A. Sharma
e-mail: abhi16.sharma@gmail.com

P. Sachdeva
e-mail: paridhisachdeva98@gmail.com

G. Ranganathan et al. (eds.), *Inventive Communication and Computational Technologies*, Lecture Notes in Networks and Systems 145,
https://doi.org/10.1007/978-981-15-7345-3_29

1 Introduction

In a democratic nation like India, the governing body of a nation is that entity which is elected to power by the citizens of the country to run the nation. Some of the many roles of the government include ensuring that justice prevails in the country, domestic order is maintained, rights of individuals are protected, and public goods and services are accessible to all [1]. To fulfill the above-mentioned roles, the government launches many schemes, policies, campaigns and programs. They cover a wide array of domains, such as criminal and justice, educational, economic, environmental, health and welfare. [2]. These range from income tax deduction laws and application of Goods and Services Tax (GST), to the provision of food at subsidized rates for the underprivileged and regulations for the protection of nature. According to the Oxford dictionary, a policy is defined as "A course or principle of action adopted or proposed by an organization or individual" [3]. In other words, any decision taken by the government is reflected in the form of a policy or a scheme. Hence, accessing the performance of the government also requires one to assess the outcome of government schemes. The success of a scheme is a direct indication of the success of the leaders who launched it. The effectiveness of policy can, in turn, be measured through various techniques such as gathering statistics about its impact or comparing the state or situation before and after the implementation of the scheme. Another method, that appears to be more efficient, is to collect the opinion of the target group of the policy. A policy that is meant for the welfare of a particular section of the society would not be considered a success if that section of the society does not benefit from it or finds faults in it. Therefore, gathering and analyzing the sentiments of the target group is often considered an important and effective step in measuring the success of a policy.

One standard but the inconvenient approach of gathering public opinion involves going door to door to ask for feedback of a policy. Since this is a very time-consuming and cumbersome process, people often turn toward automated tools for data collection. Feedback forms or online surveys can also be used for this purpose. However, one needs to explicitly create, send out and get the forms filled, which is another method that requires time and efforts of both the person gathering the feedback and the one providing it. A better course of action is to make use of the huge amounts of data already available online. Also known as big data, this raw and unstructured data lies unused in large quantities and can be exploited to mine patterns and trends to give interesting results [4]. Twitter is an example of such a data repository, where people from all over the world post their views on various topics. This micro-blogging platform can be used free of cost for sharing content, information or feelings with other users worldwide. It is fast, accurate and has a wide reach, which makes it an appropriate choice of tool for collecting public sentiments on any topic. Due to the aforementioned reasons, Twitter has been selected as the source of data collection for this study.

1.1 Digital India

Digital India is a flagship program, set in motion on July 1, 2015, by the Prime Minister of the country, Narendra Modi, to transform the country into a digitally connected and empowered nation [5]. The idea of the campaign was to ensure that all government services and facilities are easily accessible to every individual equally by enhancing the online infrastructure, improving technological connectivity in remote areas and increasing digital literacy throughout the country. It is an umbrella program that covers various domains categorized into nine pillars, i.e., e-Governance, Broadband Highways, Information for All, Universal Access to Mobile Connectivity, Electronics Manufacturing, Early Harvest Programs, Public Internet Access Program, e-Kranti-Electronic Delivery of Services and IT for Jobs [6]. Each of these departments further has many schemes under it. Some of the most successful services under the Digital India campaign include:

- **MyGov.in.** This is an online platform for partnership between the government and the citizens, to improve the quality of governance by crowdsourcing innovative ideas from the citizens themselves [7].
- **Wi-Fi hotspots.** The government has provided high-performance Wi-Fi hotspots throughout the country for public use. The number of public Wi-Fi hotspots in the country is expected to reach 21 Lakh by 2021.
- **eHospital.** This online facility has enabled citizens to book doctor appointments, pay hospital fee, get a clinical diagnosis and check availability of blood groups in blood banks online.
- **DigiLocker.** This cloud storage space is a facility for individuals to store and access their essential documents, such as pan card, voter ID card, driving license and access them on the go [8].
- **BHIM (Bharat Interface for Money).** BHIM is a mobile application for easy transfer of cashless money directly from one bank account to another through Unified Payment Interface (UPI) [9].
- **Himmat app.** This mobile application is an emergency service for female victims, launched by the Delhi police. It allows the person to send distress calls or SOS alerts to the Delhi police and specific close contacts [10].

Due to the widespread impact and relevance of this government program, Digital India has been selected as the topic of study for carrying out this research and validating the suggested model.

The remaining paper is structured as specified: Sect. 2 explains the detailed step-wise methodology undertaken in this research work. Section 3 analyzes and summarizes the results obtained from the study and Sect. 4 concludes the paper.

2 Methodology

The complete workflow of the opinion mining and sentiment classification process undertaken in this research to assess the performance of Digital India campaign has been depicted in Fig. 1. The following subsections describe every stage of the process in detail.

2.1 *Data Collection and Pre-processing*

"Tweets," or content posted on Twitter, are short messages with or without media that are openly visible to anybody worldwide. Users often end their tweets with a hashtag (#), followed by a certain keyword which categorizes the tweet. By clicking on a hashtag, a user is easily able to see all the tweets posted by other users containing the same hashtag in the message. This helps in increasing the visibility of the tweet. After legally procuring, an authentication key from Twitter, all the tweets containing the hashtag "#DigitalIndia" and posted till December 31, 2019 were extracted using the Tweepy API [11]. The first appearance of such a tweet was on July 1, 2015, also the launch date of the flagship program. A total of 3323 tweets were acquired in this manner over a time frame of approximately five years.

The data collected was raw and unstructured and needed to be cleaned, to convert it into a usable form. For this purpose, raw data was pre-processed using a six-step cleaning process to eliminate noise from the data and improve its overall quality. The six stages of data pre-processing were as follows [12]:

- **Redundancy removal**. Removal of duplicate data or re-tweets to ensure unbiased results.
- **Filtering**. Removal of URLs, special symbols (e.g.!,?,;, @, etc.) and emoticons, as they are insignificant to the results.
- **Tokenization**. Segmentation of tweets into a bag or array of words by omitting punctuation marks and spaces.
- **Stop word removal**. Removal of insignificant, but frequent terms such as "the", "and", "is" from the data set.

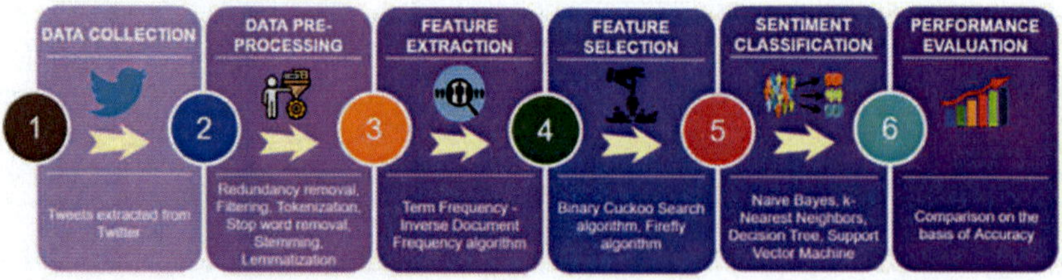

Fig. 1 Detailed functional flow of the proposed model

- **Stemming**. Reduction of complex words to their stems by eliminating common suffixes, e.g., “brilliantly” → “brilliant”.
- **Lemmatization**. Reduction of complex words to their dictionary-meaning root forms, e.g., “happiness” → “happy”.

2.2 Feature Extraction and Feature Selection

Term frequency–inverse document frequency (TF-IDF) algorithm has been used to extract relevant features in this study [13]. This algorithm assigns weights to every term in all the data sets or documents, which are directly proportional to the importance a term has to a particular data set or document. Term frequency (TF) equals the number of occurrences of a term (t) in a document (d), whereas inverse document frequency (IDF) measures how concentrated a term is in a given set of documents. The higher the value of TF for a term in a document, the more suitable it is for inclusion in selected features as it directly implies the importance of the word in that document. At the same time, the terms that rarely occur in many documents are considered more important than those that occur frequently in different documents. Hence, terms with low values of IDF are given higher preference for inclusion in selected features as per the TF-IDF mechanism of feature selection. So, for better results, the two are taken together and multiplied to rate the terms in order of their importance for inclusion in selected features’ list. It is a measure to determine how concentrated a term (t) is in a particular document (d) and is calculated as follows:

$$\mathrm{TF}(t, d) = \frac{\text{frequency of } t \text{ in } d}{\text{total number of terms in } d} \tag{1}$$

$$\mathrm{IDF}(t) = \log \frac{\text{total number of documents}}{\text{number of documents that contain } t} \tag{2}$$

$$\mathrm{TF} - \mathrm{IDF}(t, d) = \mathrm{TF}(t, d) \times \mathrm{IDF}(t) \tag{3}$$

In the conventional model, once the significant features are extracted, the feature matrix can be input into the classification models along with the training data to obtain the results. However, an alternate approach has been proposed in this paper which has shown to give extremely promising results. Instead of using all the features to train the model, feature selection can be applied to eliminate those features whose removal results in an optimized performance by improving the accuracy of the model. Feature subset selection is an NP-hard problem since it is not possible to obtain an optimal subset of N features in polynomial time, as there would be a total of 2^N such candidate subsets. Therefore, presently there does not exist any specific feature selection algorithm that has proven to give the best results on all kinds of data sets. In our study, nature-inspired meta-heuristics, also known as swarm evolutionary (SE) algorithms are applied, for this purpose, due to their wide acceptability and varied application area. Furthermore, these algorithms make use of particular stopping criteria to limit the number of iterations, and at the same time, ensure that the final state is the most optimized one. Nature-inspired swarm evolutionary algorithms, namely binary cuckoo search (BCS) and firefly algorithms have been used for this purpose in this study, due to the impressive results shown by these algorithms in [14, 15], respectively. These algorithms have been applied with the accuracy of the sentiment classification model as the fitness function so that it improves with every iteration of the algorithm. The description and details of the application of the algorithms are as follows:

- **Binary Cuckoo Search (BCS) Algorithm.** The binary cuckoo search algorithm is a modified version of the conventional cuckoo search algorithm which comes from the behavior of cuckoo birds. This bird looks for host nests of other birds to lay its eggs into. If the host bird spots the faulty egg, it either discards the egg or abandons its nest and builds a new one. Since feature selection is essentially a dual-class problem where the features included are represented by 1 and those not included are represented by 0, the output of cuckoo search has to be discretized. In this variation of conventional cuckoo search (CS) algorithm that employed to solve the above-mentioned problem, the search space is represented as an n-dimensional Boolean lattice, where the solutions are modeled and updated at the corners of the hypercube [16]. In this technique, Sigmoid function is used to limit the values assumed by the eggs, as shown by Eq. 4. Since the eggs (terms) are supposed to assume only two values, 0 and 1, another function is applied to convert it to a specific value of 0 or 1, as shown by Eq. 5.

$$s\left(x_i^j(t)\right) = \frac{1}{1 + e^{-x_i^j(t)}} \tag{4}$$

$$x_i^j(t+1) = \begin{cases} 1 \text{ if } S\left(x_i^j(t)\right) > \sigma \\ 0 \text{ otherwise} \end{cases}, \tag{5}$$

  $\sigma \in (0, 1)$ and $x_i^j(t)$ represent the value of jth egg present at ith nest at time t.
- **Firefly Algorithm.** The firefly algorithm is inspired by fireflies and their flashing lights that help in attracting mates. The attractiveness of a firefly is directly proportional to the brightness of its light, which further depends on the distance between this firefly and the viewer firefly. A less-brighter firefly (firefly i), having position x_i moves toward a more-brighter one (firefly j), having position x_j, as per the following equation [17]:

$$x_i = x_i + \beta_o e^{-\gamma r_{ij}^2}\left(x_i - x_j\right) + \alpha(\text{rand} - 0.5) \tag{6}$$

  In Eq. 6, the second additive term of the RHS is added because of the attraction between the two fireflies where $\gamma \in [0, \text{infinity}]$ is the light absorption coefficient and r_{ij} is the Euclidean distance between the two fireflies i and j. This term is therefore responsible for the exploitation of the search space. The third additive term of the RHS gives a random number which is added to the search space to explore the region. Here, $\alpha \in [0, 1]$ denotes the noise and rand is a random number between the range (0, 1).

2.3 *Sentiment Classification and Performance Evaluation*

The primary goal of this research is to determine that model which correctly labels tweets into three categories—"Positive" or +1, "Neutral" or 0 and "Negative" or −1. For this, the data set is split into a 70:30 training data: testing data ratio. The training data, along with the selected features are used to train each of the classifiers that are then tested on the testing data to compare and analyze the performance of each of these classifiers.

The reason for choosing the below-mentioned algorithms for opinion classification is because they show promising results, as validated by various other authors in the past. Furthermore, unlike complex ensemble learning algorithms, the implementation of these algorithms is relatively easier. Multiple ML algorithms in combination with swarm techniques were used for this process, to test various models and carefully compare and analyze them to find the most suited one (Table 1).

Once the sentiment classification process is complete, the performance of all the models is compared based on accuracy, which is one of the most important standard evaluation measures. Accuracy is the closeness of the predicted labels with the actual labels. Since the quality of labeling and classification depends on the accuracy, our results are compared and established based on this parameter. It is the ratio of correctly labeled instances and the total number of instances.

3 Results and Findings

3.1 *Assessment of Digital India Program*

The models proposed in this study act as an automated tool for sentiment classification of acquired tweets, based on the sentiments of these tweets. 3323 tweets were collected across the time frame, each of which has been labeled as positive, negative or neutral according to its polarity. Table 2 summarizes the results obtained, while Fig. 2 gives a graphical representation of the same.

As can be seen from Fig. 2, 46.61% of the people had a positive response and praised the program, while 24.56% of the citizens found faults with the campaign and therefore had a negative opinion about it. The remaining 28.83% of the tweets portrayed neither of the two extreme sentiments and were merely factual or stating statistics related to the initiative.

Table 1 Description of Machine Learning algorithms used for sentiment classification

Algorithm	Description
Naive Bayes (NB)	This is a probabilistic method which calculates the probability of every possible label being the correct label. The one with the highest probability is chosen as the correct label for the given class instance or data [18].
k- Nearest Neighbors (kNN)	In this method, all the data points are hypothetically plotted on a Cartesian plane. Only the closest k points from the given class instance are taken into consideration. This distance is calculated using Euclidean formula, i.e. the shortest straight-line distance between two points. Finally, the label that the majority of these k particles have is given to the class instance in question [19].
Decision Tree (DT)	This algorithm employs the use of a tree structure for deciding the labels. The distinguishing attributes or features form the internal nodes of the decision tree, the presence or absence of these features are the branches and the labels are the leaf nodes. The traversal starts from the root of the tree, and the leaf node at which the traversal ends become the determined class label for the class instance [20].
Support Vector Machine (SVM)	This method of classification uses hyperplanes or decision boundaries to separate various classes. The label for the class instance is chosen on the basis of the region in which the instance lies, as defined by the maximal margin hyperplanes surrounding the region [21].

Table 2 Polarity of tweets across the entire time frame

Class	Number of tweets
Positive	1549
Neutral	958
Negative	816
Total	3323

Fig. 2 Graphical representation of polarity of tweets across the entire time frame

3.2 Analysis and Comparison of Classification Models

This research was inspired by the work of the authors in [18] who applied particle swarm optimization (PSO) and artificial bee colony (ABC) algorithms for optimized sentiment classification. The results obtained in this paper supersede the results obtained by the authors in the aforementioned research. Twelve different combinations of classification models (with and without optimization) have been analyzed and compared based on:

- The extent of feature reduction on the application of feature selection.
- Improvement inaccuracy of the models on the application of feature selection.

Feature Reduction 886 features were obtained on applying the TF-IDF algorithm on the pre-processed data. These features were further reduced using BCS and firefly algorithms, the results of which have been summarized in Table 2.

According to Table 4, the maximum feature reduction of 49.66% by BCS was shown with kNN, while the least was 13.32% with SVM. On the other hand, applying a firefly algorithm for feature selection with NB led to a feature reduction of 44.58%, whereas only 17.95% of the features were reduced with SVM.

Figure 3a, b represents the average percentage of features reduced by BCS and firefly algorithms, respectively. As can be seen from the graphs, the average feature reduction with BCS was 31.54%, whereas using firefly algorithm resulted in a 31.15%

Table 3 Result of feature selection on the reduction of features across different models

Algorithm	TF-IDF	TF-IDF + BCS		TF-IDF + Firefly	
		Features selected (#)	Features reduced (%)	Features selected (#)	Features reduced (%)
NB	886	484	45.37	491	44.58
kNN	886	446	49.66	514	41.99
DT	886	728	17.83	708	20.09
SVM	886	768	13.32	727	17.95

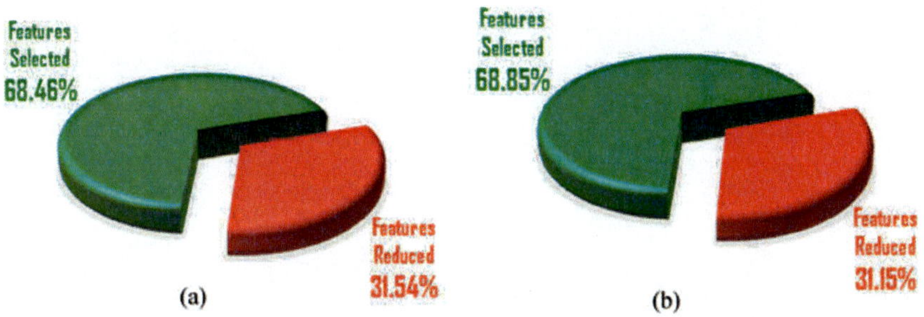

Fig. 3 Average feature reduction using. **a** BCS and **b** firefly algorithm

average reduction in the number of features. Since the time taken to train a model is directly dependent on the size of the data set and many selected features, a lesser number of features would mean lesser computational time. Hence, BCS outperformed firefly algorithm in this regard.

Improvement in Accuracy For this research, accuracy has been used as the parameter of comparison of the performance of various non-optimized and optimized supervised learning models. Table 4 depicts the change in accuracy of the different models on applying swarm optimization, while Fig. 4 graphically encapsulates the same results.

Without applying feature selection, the maximum accuracy of 87.60% was shown by SVM classifier. However, the optimized opinion classification models show better results. Although the maximum accuracy gain (24.14%) was shown by a combination of BCS and NB, applying BCS for feature selection with SVM for classification has proven to be the most accurate (92.14%). Overall, applying BCS for feature subset selection resulted in an average improvement of 11.51%, whereas firefly algorithm

Table 4 Result of feature selection on the accuracy of different models

Algorithm	TF-IDF	TF-IDF + BCS			TF-IDF + Firefly		
	Accuracy (%)	Accuracy (%)	Accuracy improvement	Accuracy gain (%)	Accuracy (%)	Accuracy improvement	Accuracy gain (%)
NB	72.56	90.08	17.52	24.14	89.44	16.88	23.26
kNN	78.82	89.56	10.74	13.63	90.28	11.46	14.54
DT	74.45	87.68	13.23	17.77	89.21	14.76	19.83
SVM	87.60	92.14	4.54	5.18	90.03	2.43	2.77

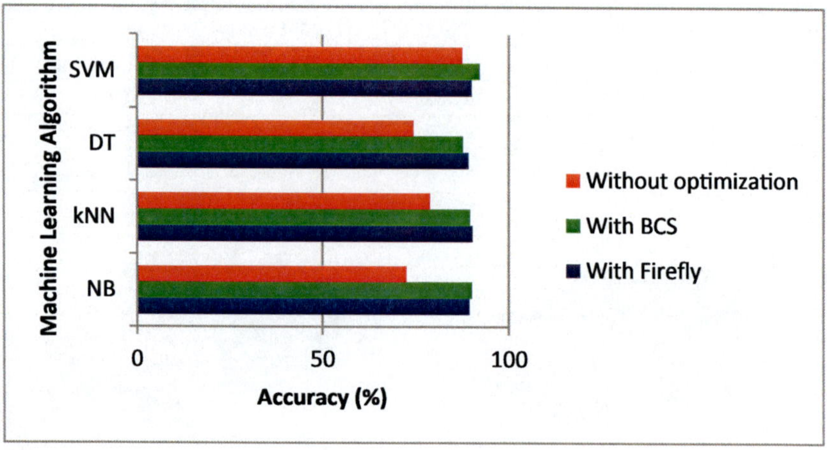

Fig. 4 Graphical representation of the result of feature selection on the accuracy of different models

led to an average improvement of 11.38%. Hence, the BCS algorithm outdid firefly algorithm in both respects.

4 Conclusion

Digital India was put into action to remove the technological barriers between the government and the citizens to ensure that all government services are equally available to every individual. Like every other scheme, initially, the public was apprehensive about this one as well. However, as the effects of the venture started surfacing, the citizens too started appreciating the program. One important goal of this paper was to evaluate the success of Digital India campaign by analyzing the perception of the public. A positive response from the public points toward the success of a scheme, while a large number of negative views does not indicate a good impact. As shown by the results, the majority of the citizens are satisfied with the program and in complete support of it.

However, the key objective of this paper was to find and recommend a model which can be used for future appraisal of all government schemes, while at the same time, is highly optimized and accurate. The model hence proposed differs from the conventional sentiment analysis model by the addition of a feature selection phase. In conclusion, for maximum accuracy, this study suggests the use of binary cuckoo search algorithm for feature subset selection, in combination with support vector machine classifier for sentiment classification for best results. Additionally, BCS performed better in terms of average feature reduction and an average improvement in accuracy across all the different swarm evolutionary algorithms for feature selection.

References

1. Friedman M (1955) The role of government in education. Rutgers University Press New Brunswick
2. Public policy of the United States. https://en.wikipedia.org/wiki/Public_policy_of_the_United_States
3. Policy|Definition of policy by Lexico. https://www.lexico.com/en/definition/policy
4. Russom P (2011) Big data analytics. TDWI best practices report, fourth quarter. 19:1–34
5. Kedar MS (2015) Digital India new way of innovating india digitally. Int Res J Multi Stud 1:34–49
6. Thomas PN (2012) Digital India: understanding information, communication and social change. SAGE Publications India
7. Lamba A, Yadav D, Lele A (2016) Citizenpulse: a text analytics framework for proactive e-Governance—a case study of Mygov. In: Proceedings of the 3rd IKDD conference on data science, association for computing machinery, pp 1–2
8. Petare P, Mohite P, Joshi M (2015) Digilocker (digital locker-ambitious aspect of digital india programme). Ge-Int J Manage Res 3:299–308

9. Sharma A, Agarwal H (2018) Study of recent developments related to cashless commerce in India. J Commer Trade 13:66
10. Joshi D, Kulkarni CM (2016) Protection circuit for girls. Int J Eng Trends Technol 33:246–247
11. Handayanto RT, Setiyadi D, Retnoningsih E (2019) Corpus usage for sentiment analysis of a hashtag twitter. In: 2019 fourth international conference on informatics and computing (ICIC), IEEE, pp 1–5
12. Sharma A, Arora N, Sachdeva P (2018) Machine learning based social big data mining for communal welfare. Int J Inf Syst Manage Sci 1
13. Kadhim AI (2019) Term weighting for feature extraction on twitter: a comparison between BM25 and TF-IDF. In: 2019 international conference on advanced science and engineering (ICOASE), IEEE, pp 124–128
14. Kumar A, Jaiswal A, Garg S, Verma S, Kumar S (2019) Sentiment analysis using cuckoo search for optimized feature selection on kaggle tweets. Int J Inf Retrieval Res 9:1–15
15. Kumar A, Khorwal R (2017) Firefly algorithm for feature selection in sentiment analysis. Computational İntelligence in Data Mining, Springer, pp 693–703
16. Rodrigues D, Pereira LAN, Almeida TNS, Papa JP, Souza AN, Ramos CCO, Yang XS (2013) BCS: a binary cuckoo search algorithm for feature selection. In: 2013 IEEE international symposium on circuits and systems (ISCAS), IEEE, pp 465–468
17. Emary E, Zawbaa HM, Ghany KKA, Hassanien AE, Parv B (2015) Firefly optimization algorithm for feature selection. In: Proceedings of the 7th balkan conference on informatics conference, Association for Computing Machinery, pp 1–7
18. Sharma A, Sachdeva P, Arora N (2020) Swarm optimized opinion classification model for policy assessment. Int J Eng Adv Technol 9
19. Rish I (2001) An empirical study of the naive Bayes classifier. IJCAI 2001 workshop on empirical methods in artificial intelligence, pp 41–46
20. Tan S (2006) An effective refinement strategy for kNN text classifier. Expert Syst Appl 30:290–298
21. Safavian S, Landgrebe D (1991) A survey of decision tree classifier methodology. IEEE Trans Syst Man Cybern 21:660–674
22. Keerthi S, Shevade S, Bhattacharyya C, Murthy K (2000) A fast iterative nearest point algorithm for support vector machine classifier design. IEEE Trans Neural Netw 11:124–136

Automatic Distributed Gardening System Using Object Recognition and Visual Servoing

D. Ruth Anita Shirley, K. Ranjani, Gokulalakshmi Arunachalam, and D. A. Janeera

Abstract The proposed work outlines the architecture and implementation of an autonomous style of gardening, which operates by itself with the support of autonomous robots. The robot operates inside the garden with the help of sensors to monitor and maintain a database of the plants such as soil content, nutrients, environmental conditions and fruits location. In this paper, the architecture of the system along with experimental results for object recognition, navigation, and manipulation is presented and discussed. The work is carried out using cherry tomatoes which are fitted with sensors to keep track of the plant's well-being. The proposed work reduces manual labour and increases the efficiency of the system.

Keywords Autonomous greenhouse · Robotized garden · Garden monitoring · Distributed autonomous gardening

D. Ruth Anita Shirley (✉) · D. A. Janeera
Department of Electronics and Communication Engineering, Sri Krishna College of Engineering and Technology, Kuniyamuthur, India
e-mail: ruthshirley06@gmail.com

D. A. Janeera
e-mail: janeerada@gmail.com

K. Ranjani
Department of Electronics and Instrumentation Engineering, SNS College of Technology, Coimbatore, India
e-mail: ranjani.ece.snsece@gmail.com

G. Arunachalam
Department of Bio Medical Engineering, SNS College of Technology, Coimbatore, India
e-mail: gokulalakshmiravi@gmail.com

G. Ranganathan et al. (eds.), *Inventive Communication and Computational Technologies*, Lecture Notes in Networks and Systems 145,
https://doi.org/10.1007/978-981-15-7345-3_30

1 Introduction

The main objective of this work is to build an autonomous greenhouse with intelligent robots. To aid the function of the robots, the plants and pots in the greenhouse are equipped with sensing, computation and communication [1]. This collaboration of plants, pots and robots will transform nutrients, water and energy and produce into fruits. To enhance this arrangement, precision agriculture system provides nutrients and water to the plant as and when needed, while the fruits are harvested at the optimum time. The robot operates inside the garden with the help of sensors that are placed on the plants to monitor and maintain a database about the details of the plants such as soil content, nutrients, environmental conditions and fruits location. In general, the cultivation of specific vegetables and fruits will need a considerable about of manual labour in comparison with the Broadland crops [2].

1.1 Project Structure

The proposed work has been designed using a spiral design to analyse the experiment through the sub-systems output. Many design revisions implemented over the period to fine-tuning the hardware and software components leading to the final system design are outlined in this work.

1.2 Related Work

Automation in commercial agriculture is a booming field that has grown to enhanced automation over the past few years. However, it is noted that only Broadland crops are using automated gardening. This includes the use of autonomous combiners. Likewise, application of fertilizers and pesticides is also available commercially with the help of satellite images that make identification of the plants easier. In recent years, many autonomous solutions have been built for crops such as cucumbers [3], mushrooms [4], cherry [5] and apples [6]. Moreover, in [7] and [8], a system based on distributed WSN has been outlined. This work is focused on extending this distributed system to the precision agriculture in which the robots are controlled by a network system inside a controlled environment. In some other cases, there is a need for actuators and sensors to guide the robot using an embedded system in the pre-defined surroundings such as a home companion [9, 10].

The proposed work is briefed in the following manner.

Section 2 will describe the distributed robot garden architecture followed by a detailed explanation of the individual robotic system components in Sect. 3.

- Section 4 deals with objection recognition
- This is followed by visual servoing, grasping and task allocation in Sect. 5

- Finally, the results are documented in Sect. 6, and the conclusion is drawn at the end.

2 System Archıtecure

The architecture of the robot garden is given in Fig. 1 which comprises of the plants and robots. The architecture holds the sub-systems namely plants and robots.

2.1 Network Architecture

With the help of the optimized link state routing algorithm, it is possible to create an ad hoc mesh network. This algorithm is used to connect a large area of wireless sensor networks, working with many nodes at the same time. The nodes are linked together, and information of every node is passed on to its neighbouring node till all the nodes and links are discovered. The shortest local path is computed, so that the node travels to the end node. The location is broadcasted to the IPs mentioned in the kernel routing table. The hop count is maintained using the routing table. Hence, only one-hop communication limit is used in this approach.

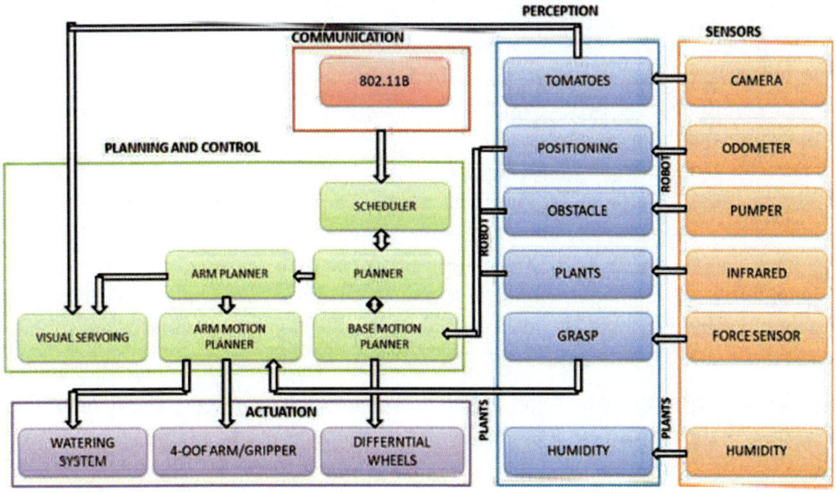

Fig. 1 System architecture

2.2 Plant Architecture

The basic architecture of the plant works on the following components.

- To serve and update the details on maturity and location of fruits-PHP interpreter and lighttpd
- To issue HTTP request-wget tool
- To check the status of humidity periodically-Linux cron daemon.

The information gathered is saved in JSON format and further transmitted to the destination. The plant architecture is fit with sensors that are used to track and monitor the progress of the plant using various technologies and interaction between the sensors, and the system is established to instruct the robot accordingly.

2.3 Robot Architecture

Figure 1 gives an overall representation of the software architecture. The planner and scheduler are the core processes.

- The scheduler is used to keep track of all the tasks assigned to the robot and queuing it based on the priority. The tasks can be assigned either by the supervisor or by the plants. To receive commands from the supervisor, the common gateway interface is used.
- The planner gets the task from the scheduler and decomposes it into a stream of events that can be completed by the individual models and further sees through the execution of the jobs. To do this, there is a need for interfacing with the visual servoing and navigation process.
- The robot operating system [11] is the IPC framework used and all modules communicate via it locally. This is an open-source software and gives interfacing with many programming languages like C, Octave, LISP and Python. ROS has many advantages as it enables many open-source algorithms and further offers integration with visualization and management tools. Thus, an HTTP protocol can be called via the Web services, and information is accessed from the remote machine, establishing plant-robot and robot-robot communication.

With the help of the plant's and robot's server, through common gateway interface, the output of the performance and the execution of requests are made using the HTTP protocol.

3 Automation and Operation

3.1 Plants

- Cherry tomato shrubs which carry fruits (red and green tomatoes) and flowers as they mature and continue to grow is observed. Dedicated pots are used to grow these shrubs
- A docking station to docking the robot is present with which provides about 130 Wh current using a battery pack is provided with an infrared field which gives the robot the ability to dock autonomously
- A wireless sensor node works on a radio-on-chip (AR2315 Atheros)
- Humidity sensor
- A mesh network represents the unique identities of every plant using the IPs.

Sensors are embedded on the plants giving the ability for computation, sensing and communication resulting in many positive aspects. Using the sensors, the humidity of the plant is sensed to update when the plant runs out of water. The fruit inventory results that are tracked by the robot are stored in a central server, thereby giving a distributed system which results in a robust and scalable solution which will have an economical cost price. The plants by themselves are used to coordinate plant-specific which implies that there is no need for a centralized system for task allocation. Hence, it will result in improving the robustness and scalability of the system. The plants that are monitored will hold the functionality given below:

- A plant will remove the details of fruit from its database, on request.
- The maturity and location of the fruits of the plant are reported to the local coordinate frame, on request.
- The humidity sensor of the plant is read periodically, and when the water level is less than the threshold value, it requests watering.
- Watering the plant.
- Docking at the pot to which it navigates based on a request from a specific plant.
- Pick the fruits that mature on the plant.
- Using the on-board camera to give an inventory of the system.

For these functions to operate, TCP? IP connection is necessary as they rely on leveraging standard tools and Web services.

3.2 Robots

- A computer system that works on Ubuntu Linux.
- An iRobot Create gives an infrared sensor, wall sensor, four cliff sensors and bumper sensors at the base of the robot. To improve odometry, rear-end caster wheel is retrofitted into the robot. An external battery is used to power the

robot while the peripherals are attached with laser-cut structure. A USB-to-serial connection is used to connect and control the robot.

- For global localization, a sensor is used to locate the plant and the robot based on the four markers mounted.
- Video monitoring is done by connecting a camera to the system.
- The system is powered using a 19.2 V, 130 Wh battery pack.
- Pump and water tank.
- The resistive force sensor and arm gripper.
- A servo board that controls the robot's arm with the help of A-D converter.
- The radio card Atheros PCMCIA.

The use of the computer system to visualize, debug and program made it easier and helps to speed up the process. Moreover, the provision of a USB helps to connect with all the components of the robot.

The use of Crustcrawler arm serves to be an economical benefit as it costs about 1/20th of the other advanced arms in the package. Also, for autonomous charging, a single battery pack is used to power all the components of the system. To fit the voltage requirements of the various components, the power is regulated using the computer system. Achievable frame rate is possible with the Logitech Quickcam along with a Firewire camera.

4 Navigation and Path-Planning

To locate the position of the plant and travel to that specific plant, Create platform gives the basic navigation structure. However, the location of the plant is done using the odometry which is not enough to guide the robot along the navigating passage which is said to be only slightly wider than the width of the robot. Moreover, this system does not give means of error recovery. Hence, Hagisonic Stargazer is used for global localization, giving a sufficiently accurate orientation and location using the infrared markers that are fixed to the ceiling. Using transformation and rotation, the position of the plant is computed in the global coordinate frame. The odometry updates information on the position at 50 Hz which Hagisonic Stargazer conveys the information at 2 Hz. To get a highly accurate position at a short distance, odometry will be the right fit. Hence, both the sensors are used, and the information gathered is fused asynchronously. When data from the sensor is received, the robot will either update the global positioning information or integrate odometry. The ROS MapServer will provide a configuration-space map for the robot. A uniform arrangement of the pots is recommended which can be further expanded to the use of the large scale of pots. Using the Create's built-in docking algorithm, the robot will plan the path to be taken to reach the docking station of the plant. At every service request, the actual location of the docking is instructed to the robot. If a collision occurs, the robot will load a low-level avoidance behaviour and will trace a safe position to re-plan. If a collision occurs, the robot will load a low-level avoidance behaviour and will trace a safe position to re-plan. To determine if the robot is near the dock, a combination of force field detection, green buoy and the red buoy is required.

5 Object Recognition

The main aim of this section is to identify the object, in our case, to identify the green and red tomatoes. On object identification, the coordinates of the fruit are found and communicated with the robot. This will serve as a visual servo for the robots to help in grasping the fruit when it is ripen. The coordinates serve as a location to guide the robot to harvest the tomatoes. The major challenge faced is in recognizing the tomatoes and the place since the foliage has complex geometry, and the lighting present can also affect the recognition in many ways. Though the tomatoes are considered to be round in general, there may be variation in their shape and size. Hence, the fruit might be partially obscured by leaves or even by peer tomatoes. In the proposed work, feature-based resource-intensive classifiers are used to detect an object through MATLAB.

6 Visual Servoing, Grasping and Task Allocation

To harvest the fruit, there are three steps involved:

1. Identification of the fruit
2. Reaching for the ripen tomato and grasping it
3. Removing the fruit from the plant.

From the previous section, the object recognition output is fed as the input to the visual servoing algorithm. This works on information on the location and object recognition based on colour to instruct the robot's function. As it is possible to locate the fruit only approximately, design a closed-loop control algorithm which will coordinate the fruit and the gripper, to grasp the fruit. The algorithm for visual servoing and fruit grasping presented in Fig. 2.

Here, the input is a series of images of the fruit, and the output is the grasping of the fruit.

Task allocation is another important activity which requires a high-level of coordination to ensure that all tasks are completed. The tasks and requests (e.g., harvesting, inventory, watering, etc.) are kept track of using the task allocation algorithm. Each task is assigned to one robot, with a unique ID, such that there is load balancing between the robots during the execution of the tasks. The robots are allocated the tasks over a mesh network which are replied to by the robots with the task's cost. The robot's ability, the task queue length and distance to the plant determine the cost function.

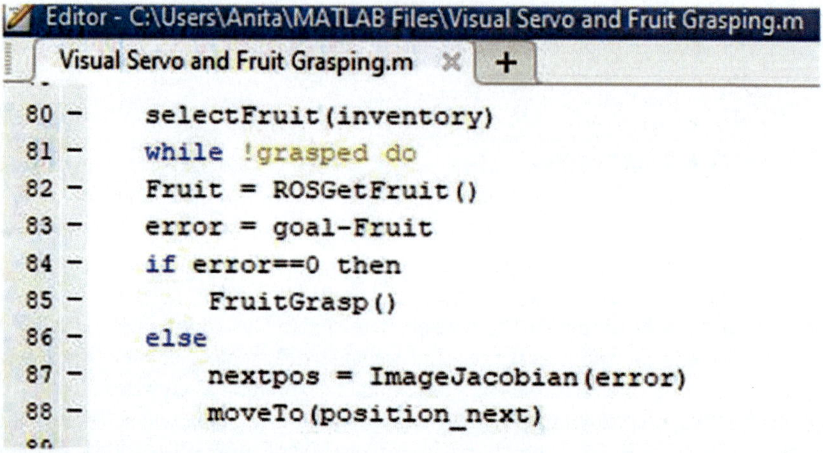

Fig. 2 Algorithm for visual servoing and fruit grasping

7 Results

The proposed algorithm is implemented using Python, PHP, C++ and C and is connected with Web interfaces and ROS messages. The results of the reliability of both the plant sensory operation and that of the robot's operation is observed and recorded, the effectiveness of task allocation algorithm and the ability to coordinate multiple robots. Based on the network load and distance, the data rate at which the message reaches the robot has been tested. Accordingly, a mission is said to be 'successful', if the message sent has been properly received by the robot and vice versa. Table 1 shows the distances at which the messages were sent and the time received by the receiver. It was found that almost all the messages (over 95%) reached within 0.05 s.

The navigation performance is evaluated, and the robot's position is examined within a square of 90 cm length. However, it was found that odometry did not perform as well as expected with about 10% off the actual positive in over ten iterations. It was found by including an external localization system, and there were no errors in more than 20 loops in the assigned square. The use of Stargazer sensor plays a

Table 1 Data rate transmission between network and robot

Experiment trial	Distance between router and robot (m)	Distance between laptop and router (m)	Status
1	1	<1	Successful
2	13	>1	Not successful
3	27	<1	Successful
4	32	<1	Successful

significant role in supporting the poles if it gets stuck when the global positioning information is absent. The rate of watering the plants was successful in over 100 trials, and the time taken to navigate to the dock closest ranges between 79:25 ± 10.5 s and 92.30 ± 10.3 s.

With the help of a boosting algorithm, a pre-defined training set with 20 images have been fed into the system and are further classified from the robot's live images. Many image are captured, and each image consists of one or more tomatoes. It was observed that using feature extraction, the shape of the tomatoes which were closer to being round was recognized with high accuracy when compared to the other tomatoes with varied shapes. As the detector had to take into account about 25 most important features, about 15 s were required to run every image on the computer. The level of accuracy measured was about 77% to identify red and green tomatoes. Using a filter-based approach, the right colour of the tomato is identified, and instructions are given to the robot for visual servoing and grasping. Object recognition plays a key role in bringing out the success rate of the algorithm. Fig. 3, represents the detection of the tomato grown based on object recognition.

8 Conclusion

The proposed work establishes a distributed robot garden which uses plant sensors networks and mobile manipulators. This paper demonstrates a system which can coordinate robot activities and plant requests to accomplish tasks like fruit harvesting, fruit inventory and plant watering. However, this faced difficulty in increasing the robustness of the system due to the limitation of ARM that has chosen and workspace. In particular, the changes in lighting conditions affected robustness of the system, and navigation errors recovery was difficult due to the global localization mechanism. The current challenge that focused on is the autonomous operation of the distributed gardening system in terms of many weeks.

Fig. 3 Object recognition to detect tomatoes (red and green)

References

1. Reed JN, Miles SJ, Butler J, Baldwin M, Noble R (2001) AE—automation and emerging technologies: automatic mushroom harvester development. J Agric Eng Res 78(1):15–23

2. Nicola S, Tylecek R, Blaich M, Petkov N, Biber P, Hemming J, van Henten E et al (2018) Trimbot2020: an outdoor robot for automatic gardening. In: ISR 2018; 50th international symposium on robotics, VDE, pp 1–6
3. Zhang W, Kantor G, Singh S (2004) Integrated wireless sensor/actuator networks in an agricultural application. In: Proceedings of the 2nd international conference on Embedded networked sensor systems, pp 317–317
4. Van Henten EJ, Hemming J, Van Tuijl BAJ, Kornet JG, Meuleman J, Bontsema J, Van Os EA (2002) An autonomous robot for harvesting cucumbers in greenhouses. Autono Robots 13(3):241–258
5. Grift T, Zhang Q, Kondo N, Ting KC (2008) A review of automation and robotics for the bio-industry. J Biomechatronics Eng 1(1):37–54
6. Tabb AL, Peterson DL, Park J (2006) Segmentation of apple fruit from video via background modeling. In: 2006 ASAE annual meeting, American society of agricultural and biological engineers, p 1
7. Kim Y, Evans RG, Iversen W, Pierce FJ (2006) Instrumentation and control for wireless sensor network for automated irrigation. In: 2006 ASAE annual meeting, American society of agricultural and biological engineers, p 1
8. Agostini A, Alenya G, Fischbach A, Scharr H, Woergoetter F, Torras C (2017) A cognitive architecture for automatic gardening. Comput Electron Agric 138:69–79
9. Ismail INB, Iskandar AHB, Eqwan MR, Zuhdi AWM, Mohamad D, Isa MR, Zahari NM et al (2018) Design and development an automatic plant pot prototype. In: AIP conference proceedings, vol 2030, no 1. AIP Publishing LLC, PP 020201
10. Abraham C (2019) RGBD analysis for finding the different stages of maturity of fruits in farming. J Innov Image Process (JIIP) 1(02):111–121
11. Chanya P, Srinitiworawong P, Samerjai W, Sunetnanta T (2016) DIY sensor-based automatic control mobile application for hydroponics. In: 2016 Fifth ICT international student project conference (ICT-ISPC), IEEE, pp 57–60

Time and Energy-Efficient Load Balancing Algorithm Toward Green Cloud Computing

P. Geetha and C. R. Rene Robin

Abstract The objectives of green cloud computing are to deal with the force and vitality effectiveness, decision of eco-neighborly equipment and programming, and reusing the material to build the item's life. Load balancing needs to consider the heterogeneous sort of assets in cloud server farm alongside its present state while choosing the allotment of client assignments to the asset. Cloud computing spins around Web-based securing and arrival of assets from a data center. Being Web-based dynamic processing, distributed computing likewise may experience the ill effects of over-burdening of solicitations. Be that as it may, it despite everything faces a couple of difficulties, for example, asset usage in a cloud server farm and nature of administration to the end-clients because of inappropriate outstanding task at hand balances among accessible assets. In this paper, a new approach of time and energy-efficient load balancing (TELB) is introduced. This algorithm is based on time parameters and to use effective load balancing and scheduling of resources to consume more energy as well as within a minimum duration while executing the millions of tasks from various regions. This proposed algorithm has been actualized and found to give good outcomes of accurate, predictable, and reliable one.

Keywords Load balancing · Cloud computing · Load manager · Resource scheduler · Energy consumption

P. Geetha (✉)
Department of Computer Science and Engineering, Dhanalakshmi College of Engineering, Chennai, Tamil Nadu, India
e-mail: geetha_mails@yahoo.com

C. R. Rene Robin
Department of Computer Science and Engineering, Jerusalem College of Engineering, Chennai, Tamil Nadu, India
e-mail: crrenerobin@gmail.com

G. Ranganathan et al. (eds.), *Inventive Communication and Computational Technologies*, Lecture Notes in Networks and Systems 145,
https://doi.org/10.1007/978-981-15-7345-3_31

1 Introduction

Green cloud can be utilized to create novel arrangements in checking, asset designation, remaining task at hand booking just as advancement of correspondence conventions and system foundations. Cloud load balancing is the way toward appropriating outstanding tasks at hand and figuring assets in a distributed computing condition. Burden adjusting permits undertakings to oversee application or outstanding task at hand requests by designating assets among numerous PCs, systems, or servers. Green cloud load balancing generally keeps up the dynamic servers as per current interest, which brings about low vitality utilization than the moderate methodology of over-provisioning. Besides, high use of server brings about more force utilization, server running at higher use can process more remaining task at hand with comparative force use. Cloud computing is among the most recent developing standards in information and communication technological area where administrations are given over the Web to the client on request. Burden adjusting; guarantee high accessibility is one the most testing zones in Cloud Computing.

Green cloud computing is basic for guaranteeing that the future development of cloud figuring is economical. Something else, cloud figuring with progressively unavoidable front-end customer gadgets associating with back-end server farms will cause a tremendous heightening of vitality utilization. Load balancing quantitative measurements of ward parameters are throughput, migration time, make span, energy consumption, power consumption, turnaround time, and resource use factor. Also, autonomous parameters are reaction time, execution time, VM creation; find the quantity of over-burden hubs and under-loaded hubs. Load balancing process requires different trade of data, for example, holding up time of occupations in line, handling intensity of CPU, appearance pace of changed employments, and so forth. Disappointment of any of these data of the heap balancers may prompt genuine repercussion, and losing the information is one of them.

Another region of worry inside a data center is its cooling framework that adds to right around 35% of aggregate vitality utilization. Air-based and water-based cooling frameworks are vital that they straightforwardly cool the hot gear as opposed to whole room region. In this way, more up to date vitality proficient cooling frameworks are proposed dependent on fluid cooling, nanoliquid cooling frameworks, and in server, in-rack, and in-push cooling by organizations. With load adjusting setup, the servers in the cloud can perform better in this manner improving accessibility of cloud and adaptability. Adaptability alludes to the capacity to serve expanded number of customers without debasing execution. Accessibility alludes to the measure that demonstrates the accessibility of server in a given year.

In this manner, it gets basic to build up a calculation which can improve the framework execution by adjusting the remaining task at hand among virtual machines. There are different burden adjusting calculations accessible, for example, round robin, weighted cooperative effort, dynamic burden adjusting, equally spread current execution (ESCE) algorithm, first come first serve, subterranean insect colony calculation, and throttled calculation. The most every now and again utilized booking strategies for a non-preemptive system are first in first out. CloudSim-3.0.3 is the recreation condition for the distributed computing research works. It bolsters framework and conduct demonstrating of cloud framework segments, for example, server farms, has, virtual machines (VMs), what is more, asset provisioning arrangements.

The goal is to upgrade the presentation of virtual machines utilizing the blend of static and dynamic burden adjusting by distinguishing the length of the occupations, resource capacities, interdependency of different errands, viably foreseeing the underutilized VMs, and maintaining a strategic distance from the over-burden on any of the VMs. This extra parameter of "work length" thought can help plan the occupations into the privilege VMs at any minute and can convey the reaction in a very least execution time. The viable planning on this calculation will likewise limit the over-burden on a VM and in this manner it will likewise limit the assignment movements.

The remaining paper is structured as follows. Section 2 describes the review study of load balancing in green cloud computing. Section 3 describes proposed work of the given problem statement, TELB- algorithmic steps with its architecture. Section 4 follows simulation environments and its results. Section 5 shows the comparative results of load balancing algorithms into green cloud computing. Section 6 give the conclusion and future work of given proposed algorithm.

2 Review Study

Aarti Singha, Dimple Junejab, Manisha Malhotraa et al. [1, 2], Autonomous Agent-Based Load Balancing Algorithm (A2LB), in this paper, have proposed an independent specialist-based burden adjusting component which gives dynamic burden adjusting for cloud condition. Significant commitment of this component is proactive burden count of VM in a DC and at whatever point heap of a VM comes to approach limit esteem, and load operator starts to look for an applicant VM from other DCs. Keeping data of competitor VM already lessens administration time. Whenever a VM gets over-burden, the specialist organization needs to disseminate the assets in such a way, that the accessible assets will be used in an appropriate way and burden at all the virtual machines will stay adjusted.

Rekha P M, Dakshayini M et al. [3, 4], In this paper, heuristic based methodology called dynamic cost-load (DCLASB) mindful help handling arrangement is proposed to decrease in general preparing time, Cost of Virtual Machine what's more, reaction time by relegating client demands in proficient method. Rule-based heuristic calculation is executed to locate every single accessible asset in which assets are assigned by the client necessities. The proposed calculation steps depicted underneath by orchestrating VMs concerning MIPS, and server farm with least handling time is chosen.

Nguyen Xuan Phi, Cao Trung Tin, Luu Nguyen Ky Thu, and Tran Cong Hung et al. [5, 6] proposed throttled modified algorithm (TMA) to improve the reaction time for the client (UserBase) and handling time of server farm by viable reallocation of the undertakings. It has demonstrated efficiencies when the quantity of VMs increments: lessening the reaction time and handling time of cloud server farms.

D. Chitra Devi and V. Rhymend Uthariaraj et al. [7–11], IWRR algorithm, say that the cloud processing needs to dole out the computational errands to the most appropriate virtual machines from the dynamic pool of the VMs by thinking about the prerequisites of each errand and the heap of the VMs. The solicitations from the customers are coordinated to any of the server farms in the cloud. On the other hand, similar solicitations are coordinated by the server farm to the most appropriate VMs dependent on the cloud and the board approaches rely upon the heap of the person VMs.

P Geetha, Rene Robin, et al. [12, 13], in this paper, proposed the diverse subjective measurements that are viewed as significant for load adjusting in distributed computing, and the performance parameters are throughput, migration time, fault tolerant, resource utilization, response time, and scalability. Green cloud computing (GCC) arrangements diminish these organization and operational expenses, and in this way, spare vitality, henceforth, decreases antagonistic natural effects.

3 Problem Statement

There is need of a calculation which can offer most extreme asset use, greatest throughput, least reaction time, dynamic asset planning with adaptability and unwavering quality. The load balancing calculation needs to consider the heterogeneous kind of resources in cloud data center alongside its present state while choosing the portion of client assignments to the resource. Task execution is a reliable procedure of doing arranged undertakings and creating expected outcomes by one individual or a gathering of people joint into a group, in consistence with preset prerequisites and desires such an arrangement.

Task scheduling consists of the VMs and hosts and both can be appointed by moment freedom premise.

Just the VMs are appointed by the moment premise yet the hosts are doled out on the freedom premise.

Just the VMs are appointed by the freedom premise yet the hosts are doled out on the moment premise.

Both the VMs and hosts are appointed by the freedom premise.

Resource scheduling doles out the exact and precise errand to CPU, system, and storage. Efficiency mindful asset booking communicates the measure of assets expended for handling, contingent on the focused on assets to improve the productivity. It is an assortment of administration execution that shows the level of fulfillment of cloud client for the IaaS assets or administrations, which is a want of cloud clients to get a help that ought to be increasingly capable monetarily and proficiently. The accompanying targets are regularly considered for the ideal asset planning for cloud figuring those are cost, time, make span, QoS, vitality, load adjusting, accessibility, dependability, disappointment rate, and so forth.

4 Proposed Work

From the writing survey, obviously constrained work has been accomplished for load balancing in cloud computing condition and those current components do have confinements that should be tended to. In this manner, there is need of a calculation which can offer most extreme asset usage, greatest throughput, least reaction time, dynamic asset booking with adaptability and dependability. This work proposes a time and energy-efficient load balancing algorithm (TELB) for green cloud computing to address the above issues.

Cloud burden adjusting or load balancing is the development of distributing workloads across numerous computing properties. And it distributes the loads among internal servers of data centers. Overall, it reduces the response time and execution time in multiple tasks with effective increasing of energy consumption.

4.1 Telb Algorithm—Steps

1. Collect the customers' requests.
2. Customer's requests have been stored in a table which contains size, id, and time of registered.
3. The cloud usage customers requests have given to the primary load manager.
4. Primary load manager of task scheduler has been maintained and monitored the table continuously.
    a. Performs the collecting the requests, arranging it in order and identifying it.
    b. Primary load manager also maintained the list of customers from various regions.
    c. Task scheduler schedules the task based on first in first out (FIFO) manner.
    d. FIFO principle based on the time of requests has been sent by the customers.
    e. Arrange, schedule it, and send it to the secondary load manager. Goto Step 4.
    f. If it is necessary, task migration is taken place.
    g. Update, reschedule if it is necessary.
5. Secondary load manager of resource scheduler which will allocate the resources based on the customer's requests.
    a. Identifying the VM based on the customers size of requests.
    b. Secondary load manager will collect a list of VMs.
    c. Resource scheduler will create a list of VMs and segregated depends on the size.
    d. Create a list of PMs and store it in a table and arrange it in order based on its memory size.
    e. If the PM reaches its maximum capacity, then it is set as OFF state otherwise ON state.

f. Arrange the VMs based on its capacity.
g. In the list of VMs, half of the VMs are in ON state and rest in OFF state with its capacity.
h. Performs the initialization, mapping, balancing, and running the VM.
i. If ON state VMs are reached the condition—filled to capacity (overfull), then the VMs executed the tasks of allocated resources successfully based on the time constraints. If not reallocate the VM again instructed by the resource scheduler. Remaining VMs are ready to set again. Then, the list of VMs is ready for further process—ON state.
j. Then, the remaining VMs are set it as OFF state to ON state. The rest of VMs are doing the process of executing the tasks. If not reallocate the VM again.
k. Both the states of VMs are updated and monitored continuously by secondary load manager of resource scheduler. Goto Step 5.
l. If it is necessary, VM migration is taken place.
m. Update and reallocate if it is necessary.

6. Assign the data centers based on the size of VMs. Split it by half of VMs.
7. Check the size and time of user request and compare with VM capacity then the data center will allocate.
8. Data centers will monitor the ON/OFF state of VMs and PMs already stored in the secondary load manager table.

   a. Check the size of user request and compare with VM capacity of the data center through monitor table.
   b. On the off chance that a VM builds the utilization of swap, to diminish it, then will increment memory random access memory (RAM) (if it necessary)
   c. If the physical machine presents high burden, to diminish the heap, then will move the VM with all the more preparing to another physical machine.
   d. If the data center presents high burden, to diminish the general burden, then will turn on increasingly physical machines.
   e. Allocate the PMs
   f. Run the requested jobs and maintain the table for VMs. Goto step 6
   g. Update and reallocate if it is necessary.

9. Energy manager will calculate the energy how it will be consumed more energy through data center.

   a. If both the PM and VM are in OFF state, then the energy consumption is 0Ws
   b. If PM is in ON state and VM is in OFF state, then the energy consumption is 200Ws
   c. If both the PM and VM are in ON state, then the energy consumption is 215Ws
   d. Update the status of VMs and PMs.

10. After completing the requests given by customers, the VM informs to the secondary load manager.
11. Then, the secondary load manager de-allocates the VM and updates the list.
12. Then the primary load manager reallocates the requests based on new customers.
13. Repeat step 1 to 12 as whenever a new request come.

This proposed algorithm has been actualized and found to give good outcomes of accurate, predictable, and reliable one. While executing the tasks, the time parameters will produce accurate results to predict future results when increasing more number of tasks different regions. And the results should be reliable because scheduling of tasks as well as balancing the loads is through PMs, VMs, migration among tasks, and VM migration in Data Centers.

Real-time applications of load balancing algorithms of green cloud computing are business organizations. There are some constant uses of cloud computing. Cloud computing is empowering organizations to exploit the most recent innovations without spending fortunes on exorbitant programming, equipment, and IT services. Emails are probably the most famous specialized techniques that organizations use today. This administration is advancing at an exceptionally quick rate getting progressively dependable and faster. Through distributed computing, Webmail customers can utilize distributed storage while giving investigation encompassing email information from any area all around. Cloud computing has made it simpler for workers, customers, and organizations to team up easily. Sharing records and reports has been made simpler by distributed computing. This has improved associations that are simple and less tedious. However, it provides complete, specific tasks that take just hours instead of months to accomplish. Organizations are likewise utilizing cloud-based SaaS applications to empower access to big business data in a flash from any area. In a perfect world, cloud computing has made it simpler for organizations to official inward procedures easily. It has improved correspondence and data spread over all organization offices all through an association.

4.2 Design of Telb Algorithm

Cloud usage customer: Consider the customers of Flipkart, a cloud-based application of social Web sites. As per the requests, we distributed the products among 5 or 6 states in India (Fig. 1).

Let us assume, 100 millions of customers from Tamilnadu, 85 millions of customers from Karnataka, 95 millions of customers from Kerala, 125 millions of customers from Andhra, 110 millions of customers from Maharashtra, and 50 millions of customers from Odissa. Requests are prepared by the server dependent on the equalization of burden on the servers. It means that demands that originate from client are sent to various segments dependent on the heap level of the servers associated with the segments. There is no arrangement for cloud segment rules.

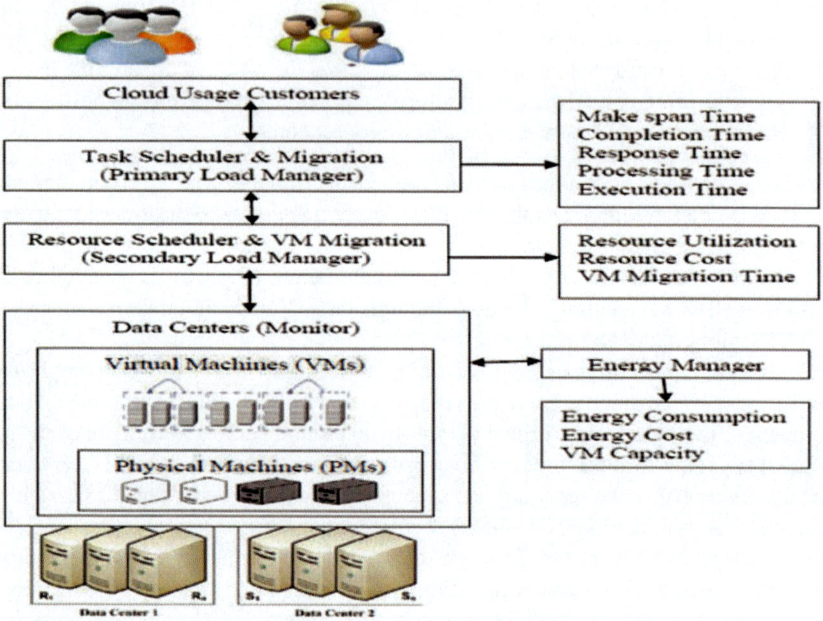

Fig. 1 Design of telb

Primary load manager—task scheduler: It collects the customer's requests and stored it in table which contains size, id, and time of registered, once the collected requests and arranged it in order. Task scheduler will schedule the task based on FIFO of time basis. Once the allocation over, then rearrange it, schedule it, and then send it to the secondary load manager. It performs task scheduling and task migration.

(1) *Task scheduling*—The undertaking scheduler partitions the errand into subtasks and doles out it to various entertainer servers. The entertainer servers play out the subtasks relegated to them. As the errand is separated into subtasks and relegated to various entertainer servers, the all out time required for task culmination is decreased contingent on the quantity of entertainer servers utilized. Therefore, the time required for task finishing is decreased furthermore; the heap on the server is additionally decreased. If there should arise an occurrence of two entertainer servers, the time required for task fulfillment procedure will be decreased to half and the heap on the servers will likewise be diminished to half.
(2) *Task migration* will take place, if the data center performs the task if it is free. If the data center is busy in performing the OFF (previous task), then the current task is put into the queue. When it completes then the task is released from the queue.

Secondary load manager—resource scheduler: It ensures the format of requests and also allocates the VMs through data center monitor. Find the most appropriate VM for the customers and assign the job to the VM depending on the size. Resource scheduler will create a list of VMs and segregate depending on the size and its memory capacity. And it will create/delete a list of PMs and store it in a table and arrange it in order based on its memory size. In the list of VMs, half of the VMs are in ON state and rest in OFF state with its capacity. It will allocate the PMs to the VMs and maintain the list based on ON/OFF state. Scheduling policies are also applied whether it may be moment-based or freedom-based allocation of VM. It can perform two processes, likely resource scheduling and VM migration

(1) *Resource scheduling:* Resource scheduling doles out the exact and precise errand to CPU, system, and storage. Efficiency mindful asset booking communicates the measure of assets expended for handling, contingent on the focused on assets to improve the productivity. It is an assortment of administration execution that shows the level of fulfillment of cloud client for the IaaS assets or administrations, which is a want of cloud clients to get a help that ought to be increasingly capable monetarily and proficiently. The accompanying targets are regularly considered for the ideal asset planning for cloud figuring those are cost, time, make span, QoS, vitality, load adjusting, accessibility, dependability, disappointment rate, and so forth.
(2) *VM migration*: Migration is used for load balancing and optimization of VM consumption in data centers. VM migration is used to copy the memory from the source to the destination, without stopping the execution of the VM. It does the launch of VM on the goal yet with a similar state as suspended source.

Data centers—monitor: Data center monitor controls the input and output of physical machines (PMs) and virtual machines (VMs) and also controls the size of requests. It is also used to identify the VM and allocate the customer requests. Data center monitor monitors the load of VM and PM continuously and forward it to load managers based on moment and result based. The moment-based means when a task is detached from a VM or allocated to the VM then the monitor will be updated. In result-based means continuously monitors the resources after a particular time interval.

Virtual machines (VMs): The list of VMs is monitored as well as maintained by data centers. Data center will allocate the job to the VMs. Then, the VM utilizes two diverse undertaking execution instruments like moment and freedom. In freedom method, the assignments will be executed consistently. It infers that just one assignment for every CPU/center is executed in its CPU. The rest of the errands appointed to that VM ought to be in the holding up line. In moment method, the tasks are executed concurrently in a time-sliced manner which resembles the execution of tasks in parallel mode. So, almost 95% of tasks are completed successfully. By default, 50% of the VMs are in ON state, the rest of 50% of VMs are in OFF state. Once the 50% of VMs are allocated fully, then VMs are set in OFF state. The rest of VMs are activated by ON state. The VM performs the initialization, mapping, balancing, and running the instances.

Physical machines (PMs): The list of PMs is maintained as well as monitored by data centers. If the physical machine presents high burden, to diminish the heap, then will move the VM with all the more preparing to another physical machine. If the data center presents high burden, to diminish the general burden, then will turn on increasingly physical machines.

Energy Manager (EM): Energy manager will calculate the energy how it will be consumed more energy through data center. If both PMs and VMs are in OFF state, then no energy will be consumed. If anyone PM or VM are in ON state then the energy will be consumed above 200 Ws. The data center will decide PM or VM based on the energy consumed.

5 Simulation Environments and Results

Because of the trouble of repeating tests in genuine conditions and with the objective of performing controlled and repeatable analyses, we picked to approve the proposed situations utilizing reenactment. For this undertaking we utilized the CloudSim framework. To empower the examination of the proposed approach, the tests model two groups (data centers) with 50 physical machines each. Each group hub has: one CPU center with execution equal to 1000, 2000 or 3000 MIPS; 8 GB of RAM; and 1 TB of capacity limit. The dynamic remaining burden was displayed as per data on top burden periods separated from a Web server. The pinnacle load periods are arbitrary and do not present any cycle. It has API for making server farms, cloud representatives, virtual machines, and asset provisioning-related things, as CloudSim empowers us to have diverse asset provisioning and booking approaches.

This is the simulation environment for time and energy efficient load balancing in green cloud computing data centers. Let us assume, the customers' requests from various regions likely

$CR = \{CR_1, CR_2, CR_3 \ldots CR_n\}$ of n requested job per customer,

$VM = \{VM_1, VM_2, VM_3 \ldots VM_v\}$ of v number of Virtual Machines.

Then the processing time of jobs to the VM is computed by

(1) Processing time (PT) = Finishing time of simulation − starting time of simulation. (jobs of VMs)(ms)
(2) Data center processing time (DPT) = Average processing time of all VMs.(ms)

Energy consumption is computed by convention of energy while moving data in both the physical machines and the data center in the ON/OFF mode, send and receive the packets from the two closures.

(3) Energy consumption (EC) = Average of energy consumed by each node/total energy (Ws)

On average, data center energy consumption is around 430kWh. Energy cost is the process of measuring, monitoring, and modeling energy use in data centers. In this algorithm, estimate the energy cost on a yearly basis approximately by 440$.

Total cost is the cost of computing the data transmission cost with the allocated virtual machine cost. This algorithm estimates the total cost of 50 simulations over 6 VMs of 2 data centers in a year exactly 880$, comparatively less than the existing approaches of 884$.

(4) Total cost (TC) = Transmitted size of data cost + virtual machine cost ($)

Response time is the time taken by the framework to react to the client request. It is alluring to react all the more rapidly to give worthy nature of administrations to the cloud customers.

(5) Response time (RT) = Exit time – entry time + transmission wait time (ms)

Where entry time is the start of transmission, exit time is the finishing time of requests given by the customers, and transmission wait time is to compute the requesting time from the customers to the allocation of VM.

Make span is to compute the maximum completion time of given allocated resource submitted to the system.

(6) Make span time (MST) = Max of completion time (CT) (jobs)(ms)

Completion time (CT) of the machine has the expected time of a task completion. Execution time depends on the executed period of task completion. The execution time or CPU time of a given assignment is characterized as the time spent by the framework executing that task, including the time invested executing run energy or framework benefits for its sake.

(7) Execution time (ET) = Finishing time of task execution – beginning time of task execution (ms)
(8) Average utilization (AV) = Completion time of allocated resources by VMs/make span * no.of VMs.(ms)

Discover the limit of virtual machine (asset) to realize that what number of virtual machine are over-burden conditions and under-stacked condition. In the event that any virtual machine is over-burden, at that point move the undertaking to under-stacked virtual machine, on the off chance that huge number of virtual machine are in under-stacked condition, at that point check the condition and decrease the virtual machine.

(9) VM capacity = No. of dealing out elements in the VM * processing time of VM (MIPS)

Migration time alludes to the aggregate sum of time required to move a virtual machine at source to data center without influencing its accessibility. VMs can be moved to another VM without shutting down.

(10) VM migration time = Aggregated sum of moving of resources from one VM to a data center.

6 Comparative Results Discussion

The difficulties of the load balancing calculations are investigated right now request to recommend increasingly effective load adjusting strategies in future. Comparatively, TELB will differ time within 0.5 ms of existing load balancing algorithms. On average, data center energy consumption is around 430kWh. Energy cost is the process of measuring, monitoring, and modeling energy use in data centers. In this algorithm, estimate the energy cost on a yearly basis approximately by 440$.

Total cost is the cost of computing the data transmission cost with the allocated virtual machine cost. In this algorithm, estimate the total cost of 50 simulations over 6 VMs of 2 data centers in a year exactly 880$, comparatively less than the existing approaches of 884$.

6.1 Parameter Table

This table shows the parameters used in the TELB algorithm (Table 1).

Case 1: The results show the comparative study on load balancing approaches in reviewed articles (Fig. 2,3).

Discussion of the existing load balancing algorithms with proposed TELB algorithm based on time and energy parameters, i.e., make span is reduced from 10% to 8%, response time is reduced from 13% to 12%, processing time is reduced from 15% to 13% and utilization of resources from 12% to 11%, and energy will be consumed from 8% is reduced to 6%. The overall TELB algorithm performance is better than the existing load balancing algorithms. And it may take the load in terms of millions of different regions then it is scalable one.

Table 1 Parameters used in this TELB proposed algorithm

S. no.	Parameter name	Parametric value
1	No. of requests from customers (Region)	5–10 millions
2	No. of data center	02
3	No. of physical machines (PMs) per VM	25
4	No. of virtual machines (VMs) per data center	06
5	Memory	In GB
6	Cost	$
7	Time	Milliseconds (ms)
8	Energy	kWs

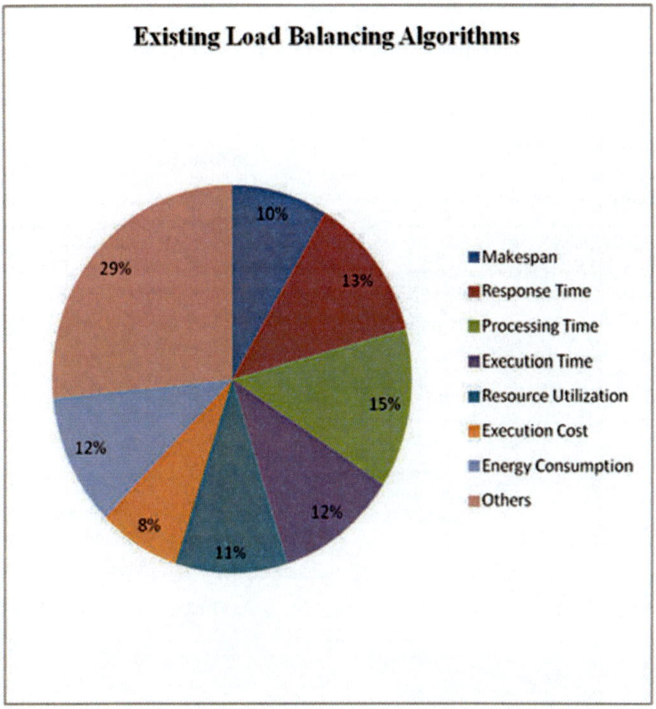

Fig. 2 Shows the results of load balancing parameters of existing algorithms

Case 2: The results of experiments in terms of execution time for given jobs in the presence of different number of virtual machines are presented.

TELB algorithm will produce the time-based parameters in the best way among existing algorithms. Likely

Make span time is reduced comparatively with the existing LB algorithms within 0.5 ms, execution time among the tasks 1.8 ms, average response time will differ by 0.32 ms, and data center processing time will differ by 0.42 ms (Fig. 4).

Case 3: It shows the result of cost-aware load balancing with TELB algorithm.

This algorithm estimates the total cost of 50 simulations over 6 VMs of 2 data centers in a year exactly 880$, comparatively less than the existing approaches of 884$, 887.41$, 885.27$ (Fig. 5).

7 Conclusion

This work centers on load balancing in distributed computing condition, i.e., it is to keep up the load to each processing element with the end goal that all the handling components become neither over-burden or inert or under-stacked that implies each

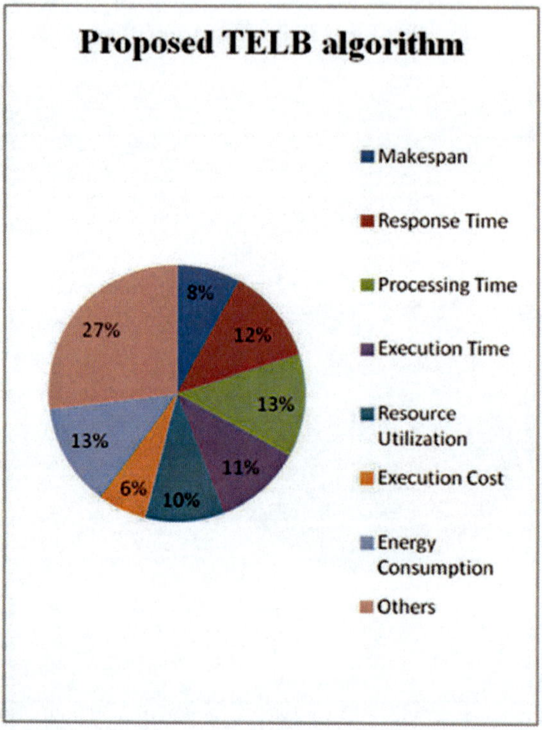

Fig. 3 Shows the results of load balancing parameters of TELB algorithm

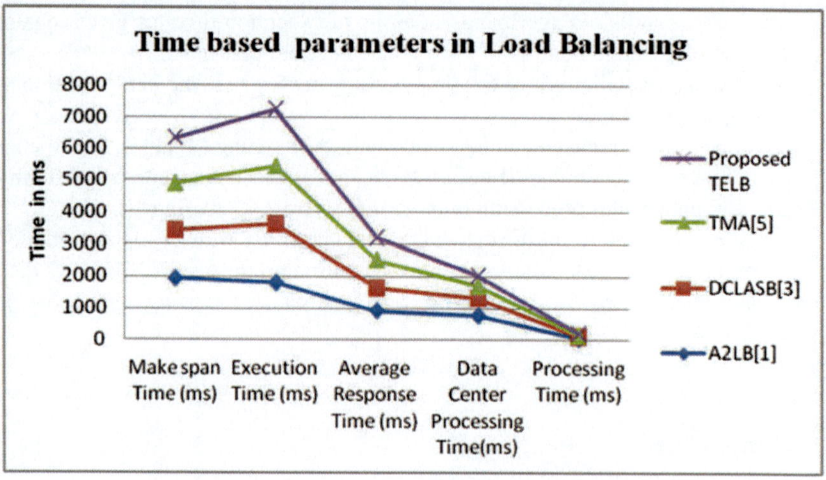

Fig. 4 Load balancing algorithms differ with time parameters versus TELB

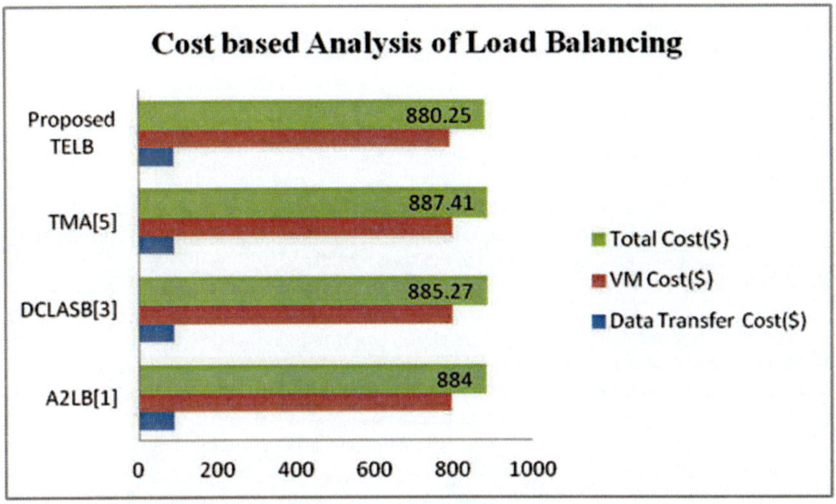

Fig. 5 Load balancing algorithms differ in cost versus TELB

processing element in a perfect world has equivalent burden at any snapshot of time to get the most extreme presentation of the system (minimum execution time, maximum energy consumption). Load balancing in distributed computing has been disregarded, yet fast development in number of cloud clients has raised interest for load adjusting systems. Also, in case of any crash during the task execution, the task is transferred to any other VM to improve the reliability of the system. Surely, it will give accurate task execution results. The better forecasts of undertaking portion progresses the adjusting of the heap just as it upgrades the framework make span and also cost of VM and response time by passing on user requests in proficient method. Result acquired through usage demonstrated that this calculation works agreeably. In future, the work will be on firm cost assessment by moving number of parameters for the various figurings and applying green deciding for various server living courses of action to spare cost and reduce the expense of Load.

References

1. Aarti Singha, Dimple Junejab, Manisha Malhotraa, Autonomous Agent Based Load Balancing Algorithm in Cloud Computing, International Conference on Advanced Computing Technologies and Applications (ICACTA-15), Procedia Computer Science 45 (2015) 832 – 841
2. Gaurang Patel, Rutvik Mehta, Uphendra Bhoi, Enhanced Load Balanced Min-Min algorithm for static meta task scheduling in Cloud Computing, 3rd International Conference on Recent Trends in Computing 2015
3. Rekha P M, Dakshayini M, Dynamic Cost-Load Aware Service Broker Load Balancing in Virtualization Environment, International Conference on Computational Intelligence and Data Science (ICCIDS 2018)

4. Monika Chaturvedi, Prof. Deepak Agrawal, Optimal Load Balancing in Cloud Computing by Efficient Utilization of Virtual Machines, International Journal of Innovative Research in Computer and Communication Engineering, Vol. 5, Issue 12, December 2017
5. Nguyen Xuan Phi, Cao Trung Tin, Luu Nguyen Ky Thu and Tran Cong Hung, Proposed Load Balancing Algorithm to Reduce Response Time and Processing Time, On Cloud Computing, International Journal of Computer Networks & Communications (IJCNC) Vol.10, No.3, May 2018
6. Badshaha P Mulla, C. Rama Krishna, and Raj Kumar Tickoo, Load Balancing Algorithm for Efficient VM Allocation in Heterogeneous Cloud, International Journal of Computer Networks & Communications (IJCNC) Vol.12, No.1, January 2020
7. D. Chitra Devi and V. Rhymend Uthariaraj, Load Balancing in Cloud Computing Environment Using Improved Weighted Round Robin Algorithm for Non-preemptive Dependent Tasks, Hindawi Publishing Corporation, The Scientific World Journal
8. Shahbaz Afzal and G. Kavitha, Load balancing in cloud computing – A hierarchical taxonomical classification, Journal of Cloud Computing: Advances, Systems and Applications
9. Shalini Agarwal, Juhi Singh, Performance Analysis of Load Balancing Algorithms for Cloud Computing, International Journal of Recent Technology and Engineering (IJRTE) ISSN: 2277–3878, Volume-7, Issue-6S4, April 2019
10. Japman Kaur Dhaliwal, Mohd Naseem, Aadil Ahamad Lawaye, Ehtesham Husain Abbasi Fibonacci Series based Virtual Machine Selection for Load Balancing in Cloud Computing, International Journal of Engineering and Technology
11. Sambit Kumar Mishra, Bibhudatta Sahoo, Priti Paramita Parida, Load Balancing in Cloud Computing: A big Picture, Journal of King Saud University - Computer and Information Sciences
12. Mohit Kumar, S.C. Sharma, Deadline constrained based dynamic load balancing algorithm with elasticity in cloud environment, Computers and Electrical Engineering 0 0 0 (2017) 1–17
13. P Geetha, Dr C R Rene Robin, A Comparative-Study of Load-Cloud Balancing Algorithms in Cloud Environments, International Conference on Energy, Communication, Data Analytics and Soft Computing (ICECDS-2017),https://ieeexplore.ieee.org/document/8389549

Machine Learning Techniques and Cloud Computing to Estimate River Water Quality—Survey

M. Ranjithkumar and L. Robert

Abstract The quality of the water for drinking purpose is an important aspect to be considered to the well-being of nature and humans. River water is the major source for drinking, agribusiness, and hydroelectric power plants, and as the river flows through different areas, assessment of change in quality has to be done by the water authorities regularly. This paper displays an overview of arrangements of different machine learning techniques to recognize contamination in waterway water. In this survey, a helpful examination of different river water dataset and categorizer utilized for classifying of river water is done successfully.

Keywords Water pollution · Machine learning · Cloud computing · Internet of things

1 Introduction

Rivers, surface water, and lakes in India are getting contaminated to a great extent in many regions. The motivation of the survey is acquired from the Ganga cleaning process put forth by the Indian government. The perineal water source of India has contaminated and has become a danger to consume [27]. Specialists have arranged water into two classes: groundwater and surface water. The survey highlights the method of classifying the contaminated surface waters. Data about surface water is accessible in the type of contamination control board. A study has been done on the most proficient method to use the dataset for investigation of the surface water. Researchers detailed the accompanying three issues during river water classification:

M. Ranjithkumar (✉) · L. Robert
PG & Research Department of Computer Science, Government Arts College (Autonomous), Coimbatore, Tamil Nadu, India
e-mail: ranjithphdcs@gmail.com

L. Robert
e-mail: robertatgac@gmail.com

G. Ranganathan et al. (eds.), *Inventive Communication and Computational Technologies*, Lecture Notes in Networks and Systems 145,
https://doi.org/10.1007/978-981-15-7345-3_32

(1) it is hard to get the dataset from the diverse source. Consequently, the determination of the sensor requires a cautious examination [5]. (2) The waterway extraction from the dataset is likewise testing assignment. (3) The extricated stream water information from the individual office should be labeled for the order of the waterway water. There is no prenamed dataset accessible for stream water dataset comprising water contamination levels or contamination classes.

In this paper, Sect. 1 contains the introduction of the river water quality status and machine learning techniques. Section 2 gives literature reviews about river water and machine learning techniques. Section 3 gives the comparison of machine learning classifications algorithms for river water quality classification. Section 4 contains the conclution of this Survey.

2 Literature Review

Genç et al. [7] utilized the KNN and the SVM to estimate the contamination level in small water streams. Javan et al. [12] utilized the ANN to examine the river quality. Kasiviswanathan et al. [14], in this paper, utilized three extraordinary strategies for evaluating the expectation interim in ANN models. The demonstrated method exhibited in this paper is delineated through stream anticipating utilizing the information gathered from Kolar bowl, India. The expectation interim was evaluated utilizing the measures, for example, level of inclusion and normal width. The examination between these strategies shown that PI strategy has come about in generally less forecast and parameter vulnerability, other than the improved model execution. Moreover, the PI technique created an exact forecast of hydrograph top, which is a general worry in ANN models.

Khaki et. al. [15] proposed new strategies, ANNs and ANFIS, with an adaptable numerical structure which can decide complex non-direct connections among information and yield information in examination with other traditional demonstrating approaches. The appropriateness of new machine learning drawn near, including ANN and ANFIS, for groundwater quality evaluation was created and applied in this examination, in five diverse observing stations. The presentation assessment criteria, in particular, the MSE and the relationship coefficient were broke down on a six years database of water quality parameters. Additionally, results effectively speak to the system's recreated incentive to groundwater quality in all perception information with an MSE of 0.031–0.096 m^2. The results for the preparation, testing, and checking of information sets clearly show the ability of the ANN and ANFIS model to reenact the estimations of TDS and EC well indeed, using the estimations of the other existing water quality parameters that are exhibited as inputs. The ANFIS model with summed up ringer work with an MSE of 0.031–0.054 m^2 indicated the best execution and it has been decided as the best fitting model contrasted with the exhibition of different models.

Lan [16], the author, utilizes the SVM in estimating the river water quality. An estimation technique for the water spending plan had been dismissed because it requires a lot of data on the impact of components, for example, inflow, precipitation, vanishing, also, counterfeit water withdrawals, as model data sources. The method shows higher efficiency compared to the other models such as the RBF as well as the polynomials.

Lin et al. [17] proposed the exhibition of ANN (330) models which was only a little superior to SVM (330) models. In any case, this propose that ANN is consistently the best technique to utilize.

Mao et al. [20], in this study, proposed a managed PNN structure assurance calculation. A critical component of this managed learning calculation is that the necessities on the system size and order mistake rate are legitimately fused during the time spent deciding to organize the structure. As a result, the proposed calculation frequently prompts a genuinely little system structure with good order exactness.

Aggarwal et al. [1] In this paper, determining of stage and release was done in a period arrangement structure utilizing three models: FNNN, SVM and the PM and found that the SVM was more convincing.

Ashwini et al. [2], in this paper, plan to structure and build up an ease framework for the constant checking of water quality utilizing the Internet of things (IoT) and machine learning (ML). The physical and synthetic parameters of water, for example, temperature, level, dampness, stickiness, and perceivability, are estimated utilizing separate sensors. ESP8266, the center controller, is utilized to process the deliberate qualities from the sensors. The information gained from sensors is sent to the Django server. Irregular forest (RF) and K-nearest neighbors (KNN) calculation are utilized in the investigation and forecast of water quality. K-nearest neighbors (KNN) model is more precise than random forest (RF).

Mim et al. [21] examined the water streams using the methods RF—random forest, SVM—support vector machine, and DT—decision tree and summarized that RF was convincing compared to the other two.

Lodhi et al. [19] In this paper, we have taken information about 12 Australian waterways for making their water quality forecast. The undertaking comprises of two stages. The main stage furnishes bits of knowledge of the dataset with the assistance of API and diagram libraries. This stage does the perception of the bits of knowledge we get from the information. The subsequent stage is an expectation model. It predicts the class to which the water quality has a place. The various classes that are utilized are as per the following: excellent, fair, good, very good, poor, marginal, and worst. At first, our information does not contain any classifications. Along these lines, for classification K-implies is utilized and for expectation process decision tree J48 algorithm is utilized. The general precision of the forecast model is 99% and have additionally built up an Android application for showing the consequences of the undertaking.

Barzegar et al. [3], in this paper, concluded that the present investigation affirms the reasonableness of the AI models applied for improving the first DRASTIC powerlessness technique for GCR appraisal. The recently proposed multi-model group (i.e., ANN board-based AI model) may be effectively utilized as a successful instrument by researchers, partners, and leaders, toward spring assurance; in this manner, natural manageability will be profoundly encouraged by creating precise and hearty

GCR maps, and consequently mostly advances natural maintainability through an increasingly sound administration of groundwater assets.

Cao et al. [4], in this paper, have finished the foundation of multi-class LS-SVS water quality evaluation model, which joins the versatile transformation PSO calculation to finish and quicken the improvement of parameters. Through the exploratory reproduction results, we can see that the improved PSO calculation can improve the precision of the outcomes by 1.5% and the combined speed is essentially quicker when contrasted and fundamental PSO calculation.

Du et al. [6] This paper focus in Taihu Lake. With everything taken into account, this investigation improves the exactness of TP reversal by consolidating the new characterization calculation with DRF. This is huge for the advancement of class-based water quality parameter reversal calculations for water shading remote detecting, and this methodology can be applied in the powerful the executives and control of the eutrophication of lake conditions. Three sorts of models, DRF, BP, and RF, were built to analyze the exhibition of TP reversal. The outcomes show that the exactness of the three techniques is improved after water grouping contrasted and the first information. Likewise, in the three techniques, DRF performs best, with the most elevated exactness in various kinds of waters. The wonder of underestimation exists in the approval results due to the disintegrated phosphorus in the water.

Kamyab-Talesh et al. [13] the inconstancy in the water quality was identified using the SVM. Haghiabi et al. [8] applied ANN and the SVM was determining the condition of the surface waters.

Hoa et al. [9] "In this paper contemplated the utilization of choice tree model for water quality record expectation in a tropical domain. The month-to-month water quality information from the Klang River for ten years (2001–2010) was used in this exploration. A choice tree calculation was created to anticipate the WQI of the Klang River by considering a few situations, each of which utilized a fluctuating number of water quality parameters as demonstrating inputs. Three distinct situations were analyzed utilizing the choice tree model, viz. those with five, four, and three water quality parameters as the model contribution, with the WQI class as the objective yield for every situation. In this investigation, the best forecast precision for the main situation is 84.09% when NH3-N was overlooked from the information factors. In the subsequent situation, the best forecast precision of 81.82% was accomplished when NH3-N and pH were precluded from the information factors, and an expectation exactness of 77.27% was accomplished when NH3-N, pH, and SS were precluded as information factors in the third situation. The three outcomes accomplished an expectation precision that is higher than the benchmark of 75% expectation exactness. This investigation has demonstrated that the quantity of water quality parameters in a checking procedure can be decreased. Every one of the three situations has indicated that NH3-N, pH, and SS have a less significant impact on the anticipated WQI since the expectation precision of the model stayed over the 75% benchmark at the point when these parameters were precluded from the information factors. These discoveries could change the way WQI class is anticipated and checked in the future, in this manner taking into consideration better water assets the board by lessening the expense and the time engaged with the checking procedure."

Victoriano et al. [25], in this examination, intend to foresee the contamination level that undermines the Marilao River, situated in the area of Bulacan, Philippines. This study used the data mining technique-based predicted model using the random forest that has scored 91.75% in terms of correctly classified instances and was able to generate 0.8115 Kappa values which indicate that the model used to produce a strong level of agreement.

Muharemi et al. [23] This work additionally examines and proposes an answer for certain difficulties when managing time arrangement information. The accompanying models are applied to water quality information: strategic relapse, straight discriminate investigation, "using SVM, ANN and LSTM" the outcomes show that all calculations are helpless even though SVM, ANN, and calculated relapses will, in general, be somewhat less helpless, while DNN, RNN, and LSTM are entirely powerless.

Prakash et al. [24]. The water tests have been characterized (great, normal, and terrible quality) given the mineral substance present in the tests. A similar investigation of arrangement strategies was done dependent on perplexity framework, the precision of arrangement and receiver operating trademark (ROC) and utilized the "SVM, KNN and DT" to sort out the quality, and found SVM to be better.

Liu et al. [18] utilized the surrogate structures prepared by the SVM and the ANN to have an exact categorization of water.

3 Machine Learning Classification Algorithms

Modaresi et al. [22] and Dezfooli. et al. [5] uses "PNN, SVM and KNN" Sarkar et al. [25] This paper exhibits the utilization of the artificial neural network (Table 1).

4 Conclusion

The paper is about the survey on the river quality extraction applying the machine learning and the cloud computing, numerous specialists have dealt with river water characterization and it is progressively troublesome contrasted with straightforward grouping gave various classes of the river water. For preprocessing, thresholding procedure is generally utilized and for the most part, utilized highlights are shading and surface highlights. For river water arrangement, most analysts have utilized distinctive arrangement strategies, for example, SVM, RF, what is more, KNN. It is seen from the review that KNN provides great precision in extracting rivers features, numerous analysts have utilized distinctive assessment measures. General execution assessment measures are accuracy precision, and recall. Further recommends the utilization of deep learning as the potential answer in analyzing the contamination of rivers.

Table 1 Comparison of machine learning classification algorithms in river water

Author	Algorithm used	Advantages	Didvantages	Reported accuracy
Aggarwal et al.	– Perseverance model – Feedforward neural system (FFNN) model – Support vector machine (SVM) model	– SVMs method works equally well – SVM utilized as effective model	– The models are applied to daily data collected from a river site in India	The SVM is more accuracy compared to other two
Ashwini et al.	– Random forest (RF) K-nearest neighbors (KNN) algorithm	– Low cost – Real time getting data – No data loss – Monitor the quality of water without any human intervention	– This method is inefficient, as water samples cannot be concurrently collected from all areas – Human intervention is required to monitor the quality of water	K-Nearest Neighbors (KNN) algorithm gives 97% accuracy
Mim et al.	– Support vector machine(SVM) – Decision Tree - Random Forest	– Easily to monitoring the river water – Low cost – No man power	– Waterbodies is not constant each year	Random forest shows 92% accuracy
Lodhi et al.	– K-means algorithm for prediction process – decision tree J48 algorithm for the overall accuracy of the prediction model	– The increasing consumption of water has led to water scarcity, – various efforts are being made to conserve water for future generation	– Dataset collection process is very tedious job – It gets more time for data collection	– The overall accuracy of the prediction model is 99%
Cao et al.	– Least square support vector machine (LS-SVM)	– Algorithm are faster in training speed and higher in accuracy	– Water quality interval variation in the next three days can be Predicted	– PSO algorithm can improve the result accuracy by 1.5% compared with basic PSO Algorithm

(continued)

Table 1 (continued)

Author	Algorithm used	Advantages	Didvantages	Reported accuracy
Du et al.	– Data regression analysis and fitting (DRF), – Backpropagation neural network (BP) – Random forest(RF)	– Spectral information would be lost when the hyperspectral reflectance is simulated	– Only one parameter used (dissolved phosphorus)	– Data regression analysis and fitting (DRF), algorithm gives the best performs
Kamyab-Talesh et al.	– Support vector machine (SVM)	– water quality index could be improved by other statistical and intelligent models	Water samples were collected by plastic bucket and were transported to the laboratory—Taking more time manpower needed	It got 87% accuracy
Amir Hamzeh Haghiabi et al.	– Artificial neural network (ANN), support vector machine (SVM), group method of data handling (GMDH)	– Structure of SVM showed that the best accuracy – data dispersion was less in SVM, and SVM model is more reliable	– time-consuming	SVM showed that the best accuracy
Jun Yung Ho et al.	Decision tree algorithms	– "More efficient process-cost-effective approach for the computation and prediction of WQI"	– "The traditional method for computing WQI is always associated with errors – Experimental testing is very high"	"Proposed model has worthy shown appropriate prediction accuracy"
Jayson M Victoriano et al.	Random forest classification (RF)	– Produced a strong level of agreement	– Underestimation of the water parameter results – Incorporating additional data to the training process	It got 91.75% accuracy

(continued)

Table 1 (continued)

Author	Algorithm used	Advantages	Didvantages	Reported accuracy
Fitore Muharemi et al.	– logistic regression, linear discriminant analysis, support vector machines(SVM), artificial neural network (ANN), deep neural network (DNN) recurrent neural network (RNN), and long short-term memory (LSTM)	– "Improved the prediction while using SVM classifier"	– Weakness of machine learning algorithms when applied to a highly imbalanced dataset	SVM classifier best compared to other algorithms
Ramya Prakash et al.	– Decision tree (DT), K-nearest neighbor (KNN), support vector machine (SVM)	– "SVM is a better classification model than -KNN and DT models"	– "The data acquisition interval is a time span of 1 year"	97.1%
Chuankun Liu et al.	– Artificial neural network (ANN) – Support vector machine (SVM)	– Greatly decreased computational costs	– Higher expected costs	SVM, ANN gives same accuracy
Onur Genç et al.	– Artificial neural networks (ANNs) , support vector machine (SVMs), and K-nearest neighbor algorithms (k-NN)	– Reliable -Low cost	– Speed appropriation in little streams.	k-NN algorithm getting more accuracy compare to other algorithm
Kazem Javan et al.	– Hydrological simulation program , Fortran (HSPF) – artificial neural networks (ANNs)	– Good training process – Suitable algorithms – The prediction is more accurate	– HSPF model, less data is required	ANN model have better performances

(continued)

Table 1 (continued)

Author	Algorithm used	Advantages	Didvantages	Reported accuracy
Fereshteh Modaresi et al.	– Support vector machine (SVM), probabilistic neural network (PNN), k-nearest neighbor (KNN)	– Most accurate results—no errors in calibration and validation phases	– Time-consuming process Reduce the computational time	– “SVM algorithm presents the best performance”
Donya Dezfoolil et al.	– Probabilistic neural network (PNN) -k-nearest neighbor (KNN) - Support vector machine (SVM)	– Reduce the sampling costs and computation time	“Great error might occur in water quality classification”	90.70% accuracy
Archana Sarkar et al.	– Artificial neural networks (ANN)	– Very limited efforts—Efficient approach for water quality modeling—“It does not require any assumption about the range of flow discharge, temperature, BOD and DO”	– Water nature of a stream at any area is a monotonous work	– Noticeable exactness by creating high connections (up to 0.9

References

1. Aggarwal SK, Arun Goel, and Vijay P. Singh. “Stage and discharge forecasting by SVM and ANN techniques.” Water resources management 26.13 (2012): 3705–3724
2. Ashwini K et al (2019) Intelligent model for predicting water quality. Int J Adv Res Ideas and Innovations Technology 5(2):70–75
3. Barzegar, Rahim, et al. “Mapping groundwater contamination risk of multiple aquifers using multi-model ensemble of machine learning algorithms.” Science of the total environment 621 (2018): 697–712
4. Cao, Sheng, Shucheng Wang, and Yan Zhang. “Design of River Water Quality Assessment and Prediction Algorithm.” 2018 17th IEEE International Conference on Machine Learning and Applications (ICMLA). IEEE, 2018
5. Dezfooli, Donya, et al. “Classification of water quality status based on minimum quality parameters: application of machine learning techniques.” Modeling Earth Systems and Environment 4.1 (2018): 311–324
6. Du, Chenggong, et al. “Estimation of total phosphorus concentration using a water classification method in inland water.” International journal of applied earth observation and geoinformation 71 (2018): 29–42
7. Genç Onur, Dağ Ali (2016) A machine learning-based approach to predict the velocity profiles in small streams. Water Resour Manage 30(1):43–61

8. Haghiabi, Amir Hamzeh, Ali Heidar Nasrolahi, and Abbas Parsaie. "Water quality prediction using machine learning methods." Water Quality Research Journal 53.1 (2018): 3–13
9. Ho, Jun Yung, et al. "Towards a time and cost effective approach to water quality index class prediction." Journal of Hydrology 575 (2019): 148–165
10. Raj Jennifer S (2020) Machine Learning Based Resourceful Clustering With Load Optimization for Wireless Sensor Networks. Journal of Ubiquitous Computing and Communication Technologies (UCCT) 2(01):29–38
11. https://en.wikipedia.org/wiki/Pollution_of_the_Ganges. (Pollution Of The Ganges [Online]. Available)
12. Javan, Kazem, Mohammad Reza Fallah Haghgoo Lialestani, and Majid Nejadhossein. "A comparison of ANN and HSPF models for runoff simulation in Gharehsoo River watershed, Iran." Modeling Earth Systems and Environment 1.4 (2015): 41
13. Kamyab-Talesh, Forough, et al. "Prediction of Water Quality Index by Support Vector Machine: a Case Study in the Sefidrud Basin, Northern Iran." Water Resources 46.1 (2019): 112–116
14. Kasiviswanathan KS, Sudheer KP (2016) Comparison of methods used for quantifying prediction interval in artificial neural network hydrologic models. Modeling Earth Systems and Environment 2(1):22
15. Khaki, Mahmoud, Ismail Yusoff, and Nur Islami. "Application of the Artificial Neural Network and Neuro-fuzzy System for Assessment of Groundwater Quality." CLEAN–Soil, Air, Water 43.4 (2015): 551–560
16. Lan Yingying (2014) Forecasting performance of support vector machine for the Poyang Lake's water level. Water Sci Technol 70(9):1488–1495
17. Lin Jian-Yi, Cheng Chun-Tian, Chau Kwok-Wing (2006) Using support vector machines for long-term discharge prediction. Hydrol Sci J 51(4):599–612
18. Liu, Chuankun, et al. "Optimizing the Water Treatment Design and Management of the Artificial Lake with Water Quality Modeling and Surrogate-Based Approach." Water 11.2 (2019): 391
19. Lodhi, Pooja, Omji Mishra, and Gagandeep Kaur. "WQVP: An API enabled Open Data Machine Learning based Solution for Water Quality Visualization and Prediction." Journal of Telecommunication, Electronic and Computer Engineering (JTEC) 10.2 (2018): 61-72
20. Mao, Ke Zhi, K-C. Tan, and Wee Ser. "Probabilistic neural-network structure determination for pattern classification." IEEE Transactions on neural networks 11.4 (2000): 1009–1016
21. Mim, Mahbina Akter, and KM Shawkat Zamil. "GIS-Based Analysis of Changing Surface Water in Rajshahi City Corporation Area Using Support Vector Machine (SVM), Decision Tree & Random Forest Technique." Machine Learning Research 3.2 (2018): 11
22. Modaresi Fereshteh, Araghinejad Shahab (2014) A comparative assessment of support vector machines, probabilistic neural networks, and K-nearest neighbor algorithms for water quality classification. Water Resour Manage 28(12):4095–4111
23. Muharemi, Fitore, Doina Logofătu, and Florin Leon. "Machine learning approaches for anomaly detection of water quality on a real-world data set." Journal of Information and Telecommunication (2019): 1–14
24. Prakash, Ramya, V. P. Tharun, and S. Renuga Devi. "A Comparative Study of Various Classification Techniques to Determine Water Quality." 2018 Second International Conference on Inventive Communication and Computational Technologies (ICICCT). IEEE, 2018
25. Sarkar Archana, Pandey Prashant (2015) River water quality modelling using artificial neural network technique. Aquatic procedia 4:1070–1077
26. Victoriano, Jayson M., et al. "Predicting Pollution Level Using Random Forest: A Case Study of Marilao River in Bulacan Province, Philippines." International Journal of Computing Sciences Research 3.1 (2019): 151–162
27. Waste Water Generation And Treatment In India,[Online].Available: http://www.mediaforrights.org/infopack/englishinfopack/443-waste-water-generation-and-treatment-in-india
28. Water Sanitation Health," [Online]. Available: http://www.who.int/water_sanitation_health/takingcharge.html

Low Transition Dual LFSR for Low Power Testing

Navya Mohan, M. Aravinda Kumar, D. Dhanush, J. Gokul Prasath, and C. S. Jagan Sai Kumar

Abstract Low transition dual Linear Feedback Shift Register (LFSR) offers maximum fault coverage with comparatively less test data storage and less power dissipation than the traditional single LFSR. One of the serious problems faced during IC testing is higher power dissipation, which affects reliability and efficiency of manufacturing systems. The proposed research provides a possible solution for reducing power dissipation with reduced transitions and a new reseeding algorithm. Reduction in transitions is obtained by implementing the properties of AND and OR operations to the TPG (Test Pattern Generation). Vast experiments were conducted on all the benchmark circuits in ISCAS 85 circuits using simulation tools (ModelSim, Vivado, HOPE) to check the possibility and effectiveness of the proposed research idea. Results obtained from the proposed solution shows power reductions up to 20% than the traditional single LFSR during IC testing.

Keywords LFSR · Reseeding · BIST · Switching activity

N. Mohan · M. Aravinda Kumar · D. Dhanush (✉) · J. Gokul Prasath · C. S. Jagan Sai Kumar
Department of Electronics and Communication Engineering, Amrita School of Engineering, Amrita Vishwa Vidyapeetham, Coimbatore, India
e-mail: cb.en.u4eie16010@cb.students.amrita.edu

N. Mohan
e-mail: m_navya@cb.amrita.edu

M. Aravinda Kumar
e-mail: cb.en.u4eie16003@cb.students.amrita.edu

J. Gokul Prasath
e-mail: cb.en.u4eie16014@cb.students.amrita.edu

C. S. Jagan Sai Kumar
e-mail: cb.en.u4eie16022@cb.students.amrita.edu

G. Ranganathan et al. (eds.), *Inventive Communication and Computational Technologies*, Lecture Notes in Networks and Systems 145,
https://doi.org/10.1007/978-981-15-7345-3_33

1 Introduction

Progress in the field of VLSI is getting enormous nowadays. Development also leads to complexity and expensive external test equipment. Primary goal during IC testing is to cover utmost faults with lesser test data available [1]. Testing of these circuits became more difficult because of the instantaneous power surge due to large test data volume and transition between patterns. Conventional testing methods are not feasible for increased device counts and data [2]. Built in self-tester (BIST) represents a reliable solution for this problem [3, 4].

In a BIST framework on chip LFSRs create test patterns and determine the response to the provided details by reducing channel limitations and the expensive test equipment specifications [3–10].

A reliable BIST structure must guarantee maximum fault-coverage while reducing the test application time, hardware overhead and test data volume storage. Several strategies for test pattern generation were suggested previously to cover failure with different compensations between the parameters listed above [5]. These techniques range from deterministic pattern generation [6] technique to pseudo-random test pattern technique [5]. Both of these techniques have their own advantages and disadvantages. Deterministic pattern generation technique produces full fault coverage with significant amount of test storage whereas pseudo-random pattern generation technique consumes only negligible amount of test storage producing lesser fault coverage.

Making a trade off in fault coverage, pseudo-random pattern generation techniques is considered to concentrate more on lesser test data storage. Pseudo-random patterns cover all the easily detectable faults but do not cover certain random pattern resistant faults. To achieve a higher fault coverage including the random pattern resistant faults, a reseeding algorithm is proposed which helps in selection of initial seeds to the test pattern generator (LFSR). Reseeding algorithm is incorporated in this system to determine the initial test seed which covers most of the test faults [11, 12]. Any modifications in the test set will require a whole BIST hardware resynthesis, as they depend more on the test set despite the quality patterns provided. To reduce transitional power, Dual-LFSR is introduced with multi polynomials through which the transitions between successive patterns is reduced [13, 14]. BIST deals primarily with structures based on offline testing which comprises LFSRs. Reseeding scheme is further used to reduce the transitions between the successive patterns [14].

This paper presents a Low Transition Dual LFSR along with a reseeding algorithm which aims to overcome higher power loss caused by external test equipment during IC testing. The Fig. 1 shows the proposed architecture of LTD-LFSR. The proposed approach reduces power dissipation and reaches optimum fault coverage in a short time. The proposed reseeding method proved to be an effective method in finding the initial seeds for generation of test patterns which achieves maximum fault coverage with minimum patterns. Selected the initial seeds, the output of two multi-polynomial LFSR are operated parallelly and combined using AND or OR depending upon the number of 1 and 0 s in the test patterns [10]. A series of 25%

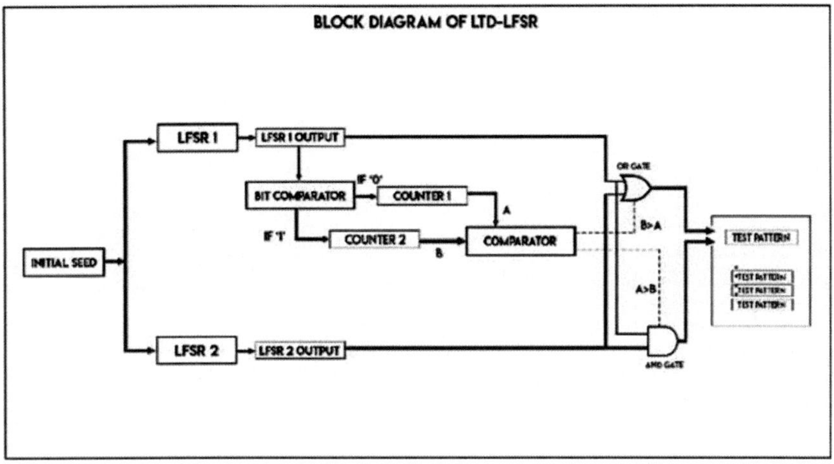

Fig. 1 Proposed architecture

fewer transformations than single LFSRs is created. A large number of tests were conducted using simulation tools in benchmark circuits to analyze the efficiency of the technique proposed for test pattern generation.

The next section deals with the related work used for this method. Section 3 deals with the methodology and the following section is about the design of Low Transition Dual-LFSR and the experimental results are given in the last section.

2 Background Work

2.1 Reseeding Algorithm

LFSR reseeding is one of the different methods for generating testing patterns. Pseudo random patterns are generated in this reseeding system. Figure 2 explains the proposed reseeding algorithm. Initially a seed is randomly selected and given to the LFSRs, then the LFSRs are made to run parallelly to generate test patterns. These test patterns are appended to the Pattern List and the fault coverage calculation is done. Once the desired fault coverage is obtained the algorithm stops. If the obtained fault coverage is not sufficient, the next seed is selected uniquely which is not present in the Pattern List. After selecting the successive seed, the LFSRs will generate the test patterns.

In this paper, trade-off the fault coverage to a certain extent to optimize both the power dissipation during testing and test data storage. Lesser number of test patterns can be used for the fault coverage if reseeding method is used efficiently.

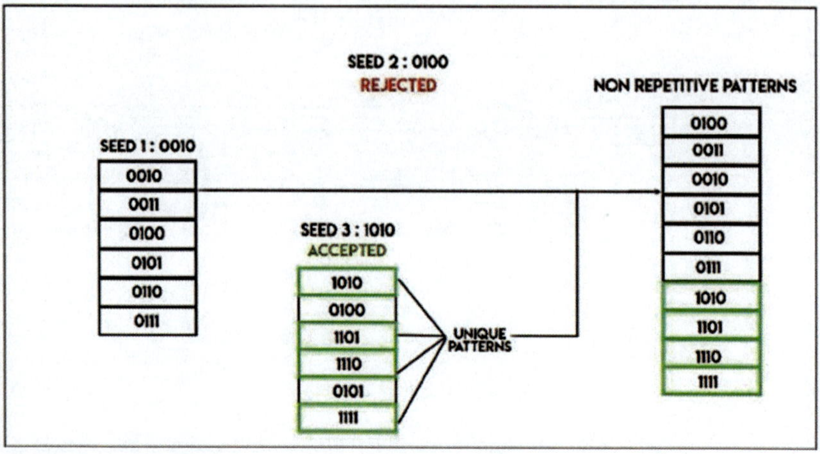

Fig. 2 Reseeding schema

2.2 *LFSR Clocking*

Linear feedback shift register is a type of register which shifts bit data and generates test patterns by a feedback loop. LFSR is mainly used to generate pseudo-random patterns using parallel input and parallel out method and to store bit values. Found the initial seeds using the reseeding algorithm, it is loaded onto its flip-flops. A LFSR clock creates a random output series according to the polynomial function of the LFSR. By modifying input taps on an LFSR or the initial seeds the generated performance can be modified. Figure 3 represents Implementation of 5-bit Low Transition Dual LFSR (LTD-LFSR).

The bit values from chosen D flip-flops are feedback to a common flip-flop of the register after the values are implemented through XOR gate [13].

3 Methodology

Step-1: Design of a n-bit single LFSR.

Step-2: Design of a n-bit dual LFSR.

Step-3: Reseeding algorithm for Dual LFSR.

Step-4: Fault coverage for ISCAS-85 benchmark using the patterns obtained from the Low Transition Dual LFSR.

Step-5: Power analysis comparison between single and low transition dual LFSR.

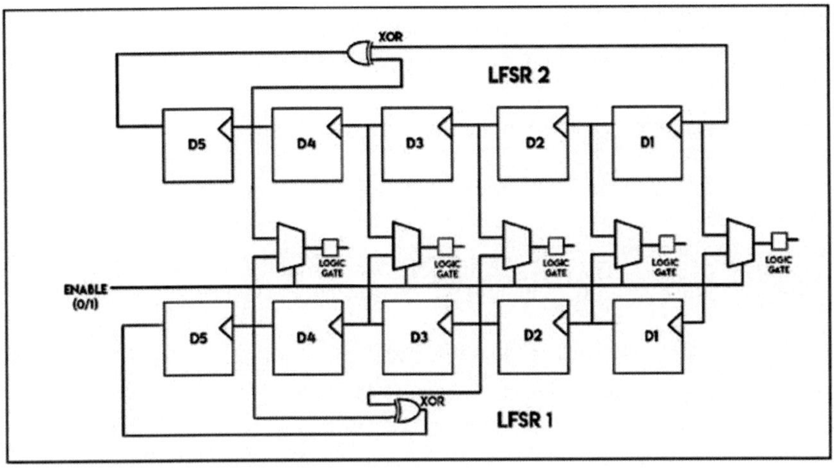

Fig. 3 Implementation of 5-bit LTD-LFSR

4 Low Transition Dual-LFSR

LFSR generated bits have a large amount of transitions between them because of their pseudorandom nature. So, when LFSR bits are given as input to a combinational circuit these transitions directly reflect on the input of the circuit and causes high switching activity in the combinational circuit. So, known properties of AND and OR gates is used to reduce the switching activity in the scan chain during shift [10].

Let us consider a binary signal '*S*' which means the signal can either be '1' or '0'.The sum of probabilities of getting a '0' or a '1' is always equal to 1, and the probability of getting a '0' or a'1' is always 0.5.

$$P_0(S) + P_1(S) = 1$$
$$P_0(S) = P_1(S) = 0.5$$

The transition probability of a signal is found by using the common formula

$$P_{tr}(S) = P_1(S) * P_0(S) + P_0(S) * P_1(S)$$

So, for a random binary signal

$$P_{tr}(S) = 0.5 * 0.5 + 0.5 * 0.5 = 0.5$$

Considering two random mutually independent signals S_A and S_B.

$S_{AND} = S_A$ AND S_B. So, the probability of getting 1 in an AND gate is calculated by $P_1(S_A)*P_1(S_B)$.

$$P_1(S_{AND}) = 0.5 * 0.5 = 0.25$$
$$P_0(S_{AND}) = 1 - 0.25 = 0.75$$

So, the transition probability of a AND gate is given by

$$P_{tr}(S_{AND}) = 2 * P_1(S_{AND}) * P_0(S_{AND}).$$
$$P_{tr}(S_{AND}) = 2 * 0.75 * 0.25 = 0.375$$

Similarly, the transition probability of OR gate can be calculated in the same way.

$$P_0(S_{OR}) = 0.5 * 0.5 = 0.25$$
$$P_1(S_{OR}) = 1 - 0.25 = 0.75$$
$$P_{tr}(S_{OR}) = 2 * P_1(S_{OR}) * P_0(S_{OR}) = 0.375$$

From the above findings it is understood that the activity of AND and OR of two separate random signals is 25% lower than the initial signals. Thus, the switching activity can be significantly reduced using AND or OR combinations in BIST test pattern generators, which, in turn, reduces power dissipation in the under-test circuit.

The key issue is that check trends arising from AND or OR activity should be sufficiently unpredictable to mask faults easily detected over time. This problem can be solved by ensuring that the two composite behaviours are equally exclusive. Another concern is that AND or OR operations will produce test patterns for random patterns tolerant faults to ensure maximum fault coverage with pre-computed test cubes.

LFSR reseeding can therefore be used for the creation of an appropriate test pair for AND or OR operations in order for the switching activity in the scan chain to be minimized thus maintaining the specific bits of the test patterns and thus optimizing fault coverage.

In Fig. 4 two separate LFSR isusedin which one is the main LFSR and other is the secondary-LFSR. These two LFSR have different feedback combinations so that the patterns do not match during every clock cycle. The initial seed is common for both the LFSR and it is made to work parallelly. For every clock cycle the output form the main LFSR is taken into consideration and the number of 0 bits and 1 bit are counted. Depending on the most dominant bit from the output the AND or OR operation is done in the following way. If the most dominant bit is '1' the output from both the LFSR is combined using OR logic. If the most dominant bit is '0' the output from both the LFSR is combined using AND logic.

Example: Let us consider a 5-bit test pattern 11010. The primitive polynomials for the main LFSR and secondary LFSR are $x^5 + x^3 + 1$ and $x^5 + x^4 + 1$ respectively. Solving the set of linear equations for calculating the initial seeds we get 10101 and 10100 respectively for the primary and secondary LFSR. Later these seeds are given to its respective LFSR to get two outputs. Considering the pattern from the main LFSR, it is evident that the number of 1-bits to be greater, so the output from two

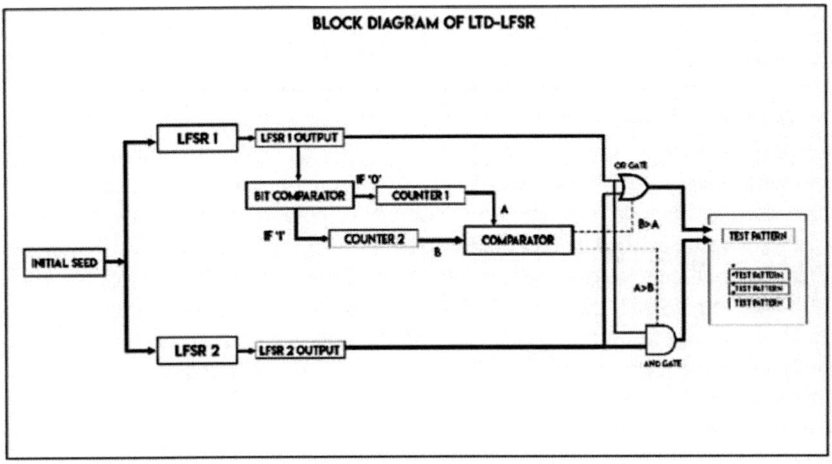

Fig. 4 Working of LTD-LFSR

LFSR is combined using an OR logic. This gives the output as 10101. Running the setup for the desired time to obtain patterns which have low transition between them.

Here the system should be smart enough to decide the operation that it is going to process in the next clock cycle. This can be done only if the number of 1 and 0 s present in the main LFSR is known. For this purpose, a certain number of 1-bit comparators and two counters is used to find the number of 0 and 1 s. The number of bit comparators required is equal to the length of the LFSR because at maximum be can have all1s or all 0 s. A bit comparator is used to find whether the output is either 1 or 0. If the output is 1 counter-1 is incremented by 1, else counter-2 is incremented by 1. After each and every bit of the LFSR is compared, counter-1 and counter-2 will have the number of 1 and 0 s. Now these values are given to another comparator which compares the number of zeroes and ones. If the number of 0 s is greater AND logic is applied to the output of 2 LFSRs, else OR logic is applied to the output of 2 LFSRs. After this all the comparators, counters are reset to 0 for their operation during the next clock cycle. Thus we get the low transition test patterns is got which applied to various benchmark circuits to get better efficiency.

The proposed logic was applied to several ISCAS-85 benchmark circuits and the switching activity was reduced up to 20% depending on the level and the complexity of the circuits.

5 Experimental Results

Various experiments were performed on ISCAS85 benchmark circuits to analyse the fault coverage achieved and to find the reduction in switching power when compared to traditional LFSR. The reseeding algorithm is used to produce low transition test patterns and optimize fault coverage while preserving the lowest possible number of test patterns. LFSRs were designed using VHDL coding platform and synthesized using Vivado software. Fault coverage was done using HOPE fault simulator.

For a 5-bit C17 benchmark circuit various LFSR with different feedback combinations and same initial seeds were utilized to find which feedback combination gives high fault coverage while keeping the switching activity as low as possible. The particular combination was then taken into consideration and various initial seeds were given to the particular combination to find the number of test patterns required to get the desired fault coverage. For a 5-bit circuit the desired fault coverage is always 100%. So, to get 100% fault coverage, the number of test patterns required for each initial seed is taken into consideration and the seed which requires the minimum number of test patterns is finalized. The fault coverage was found using HOPE fault simulator tool. For a 5-bit circuit when the initial seed was 10,000 the fault coverage of 100% was achieved when 17 patterns was used, but when the initial seed was 00110, the desired fault coverage of 100% was achieved using 10 test patterns only. Table 1 demonstrates the effectiveness of the proposed approach by calculating the fault coverage using the software HOPE simulation tool [15].

Likewise, LFSR has been checked with various validation schemes, with a range of input variations and feedbacks are selected with the highest average fault coverage and the

original seed offers optimum fault coverage with minimal number of test patterns. The design of the proposed Low Transition Dual-LFSR was done using VHDL coding platform and the design was synthesized using Vivado software. The value change dump file was generated from Vivado software to get the switching activity of the

Table 1 Fault coverage for ISCAS' 85 benchmark circuits

Circuit	Inputs	Test patterns	Fault coverage (%)	Collapsed faults	Detected faults	Undetected faults	Simulation time (s)
C17	5 BITS	10	100	22	22	0	0.126
C432	16 BITS	792	99.237	524	520	4	0.267
C499	41 BITS	369	98.945	758	750	8	0.341
C880	60 BITS	2029	98.832	942	931	11	0.517
C1355	41 BITS	1700	99.111	1574	1560	14	0.267
C1908	13 BITS	3043	98.776	1879	1856	23	1.567
C3540	50 BITS	3099	95.74	3428	3282	146	2.6
C6288	32 BITS	80	99.535	7744	7708	36	0.267

Table 2 Power comparison table

Circuit	S-LFSR	LT-DLFSR	% Reduction
C17	0.014	0.013	7.1428
C432	0.049	0.041	16.3265
C8S0	0.076	0.066	13.1578
C1355	0.054	0.043	20.3703
C1908	0.045	0.036	20.00
C2670	0.270	0.216	20.00
C3540	0.071	0.060	15.4929
C6288	0.050	0.041	18.00

proposed model which was then changed to saif file using synopsis tool to find the switching power dissipation during the time of testing. The number of test patterns needed for optimum fault coverage is directly proportional to the complexity and number of gates on the circuit, as the complexity of the test circuit decreases the fault coverage declines so that the number of patterns necessary and the switching operation are reduced.

The switching power comparison in Table 2 between the conventional LFSR and the proposed low transition dual-LFSR was given in the following table. For lower order circuits the switching power does not make much difference, but when the complexity and inputs increase the reduction in the switching activity gets reduced to up to 20%. This shows that the proposed method reduces the switching activity during the testing of a VLSI circuit.

6 Conclusion and Future Scope

The work presented in this paper proposes a reliable solution for IC testing with lower test data storage and dissipation power. It is evident that by employing AND/OR operation, the patterns produced by the LFSR is of lesser transitions. Applying AND/OR operation to the dual LFSR, the transition probability reduces from 0.5 to 0.375. Lesser transitions contribute to lesser power dissipation and reseeding scheme helps in reducing test data storage and higher fault coverage. The proposed TPG (Test Pattern Generator) structure achieves significant power reduction, lesser test data storage than the traditional single LFSR scheme. Experimental results have illustrated power reduction up to 20% and fault coverage for ISCAS'85 benchmark circuits.

This technique can be improved by implementing mixed mode BIST which has two types pattern generation technique; pseudo random pattern generation technique and deterministic pattern generation technique. This technique produces 100% fault accuracy. Although they produce higher fault coverage, they occupy significant amounts of test data storage. A new efficient reseeding algorithm or scheme must be implemented to overcome higher test data storage in case of mixed mode BIST.

References

1. Mohan N, Krishnan M, Rai SK, MathuMeitha M, Sivakalyan S (2017) Efficient test scheduling for reusable BIST in 3D stacked ICs. In: 2017 International conference on advances in computing, communications and informatics (ICACCI), Udupi, pp 1349–1355
2. Rinitha R, Ponni R (2016) Testing in VLSI: a survey. In: 2016 International conference on emerging trends in engineering, technology and science (ICETETS), Pudukkottai, pp 1–6
3. Wang S, Gupta SK (2002) DS-LFSR: a BIST TPG for low switching activity. IEEE Trans Comput Aided Des Integr Circuits Syst 21(7):842–851
4. Singh B, Khosla A, Bindra S (2009) Power optimization of linear feedback shift register (LFSR) for low power BIST. In: 2009 IEEE international advance computing conference, Patiala, pp 311–314
5. Ying J, Tseng W, Tsai W (2018) Bipolar Dual-LFSR Reseeding for Low-Power Testing. In: 2018 IEEE Conference on dependable and secure computing (DSC), Kaohsiung, Taiwan, pp 1–7
6. Pomeranz I (2015) Computation of Seeds for LFSR-Based Diagnostic Test Generation. IEEE Trans Comput Aided Des Integr Circuits Syst 34(12):2004–2012
7. Bushnell ML, Agrawal VD (2000) Essentials of electronic testing for digital, memory and mixed-signal VLSI circuits. Kluwer, Norwell, MA
8. Agrawal VD, Kim CR, Saluja KK (1993) A tutorial on built-in self-test, part 1: principles. IEEE Des Test Comput 10(I):73–82
9. Nourani M, Tehranipoor M, Ahmed N (2008) Low-Transition test pattern generation for bist-based applications. IEEE Trans Comput 57(3):303–315
10. Yarraya K, Rao KC (2014) The power optimization of linear feedback shift register using fault coverage circuits. Int J Sci Eng Res 5(9):917–922
11. Rosinger P, Al-Hashimi BM, Nicolici N (2003) Dual multiple-polynomial LFSR for low-power mixed-mode BIST. IEE Proc comput Digit Tech 150(4):209–217
12. Roy, Annu, and J. P. Anita. (2017) Pattern generation and test compression using PRESTO generator. In: International symposium on security in computing and communication. Springer, Singapore
13. Yang MH, Kim Y, Park Y, Lee D, Kang S (2007) Deterministic built-in self-test using split linear feedback shift register reseeding for low-power testing. IET Comput Digit Tech 1(4):369–376
14. Kalaiselvi M, Neelukumari KS (2013) LFSR-reseeding scheme for achieving test coverage. In: 2013 International conference on information communication and embedded systems (ICICES), Chennai, pp 1209–1213
15. Lee HK, Ha DS (1996) HOPE: an efficient parallel fault simulator for synchronous sequential circuits. IEEE Trans Comput Aided Des Integr Circ Syst 15(9):1048–1058

High Speed and Low Power Buffer Based Parallel Multiplier for Computer Arithmetic

N. S. Kalyan Chakravarthy, O. Vignesh, and J. N. Swaminathan

Abstract In Digital Signal Processor (DSP) the arithmetic elements are playing a major vital role in processing the data in processors. The multiplier is the most complex part of the arithmetic circuits which is used in DSP applications. The partial product generation is reducing the speed of the overall multiplier operation in the processor. The computation of partial products for the multiplier is a critical part of hardware implementation. In this paper, we reduce the computational complexity in the proposed buffer based parallel multiplier which occurs in partial product generation. The proposed 16×16 multipliers is designed using a tri-state buffer, adder, shifter and multiplexer. The power and delay analysis are observed and compared with existing multipliers. The main and sub-modules are simulated using the Altera EDA tool and implemented using the Altera Cyclone II FPGA family.

Keywords Parallel multiplier · Tri state buffer · Shifter · High speed · Low power · Altera FPGA

1 Introduction

Plenty of research works are going towards high-speed operation for efficient hardware architecture. The drastic development in silicon technology, millions of transistors can be fabricated in a single chip. The arithmetic units are fabricated with overall system architecture as in System on Chip (SoC) concepts. In general, two

N. S. Kalyan Chakravarthy (✉) · O. Vignesh · J. N. Swaminathan
QIS College of Engineering and Technology, Ongole, Andhra Pradesh 523272, India
e-mail: chairman@qiscet.edu.in

O. Vignesh
e-mail: vicky6058@gmail.com

J. N. Swaminathan
e-mail: sammbuddy@gmail.com

G. Ranganathan et al. (eds.), *Inventive Communication and Computational Technologies*, Lecture Notes in Networks and Systems 145,
https://doi.org/10.1007/978-981-15-7345-3_34

major types of multiplications are used in DSP processor one is sequential multiplication and another one is parallel multiplication. The sequential multiplication method is consuming low static power dissipation and large latency due to the computational complexity. Parallel multiplication is used to produce less latency but the static power consumption is the critical parameter to reduce.

Most of the high speed and high-performance applications are implemented using the parallel multiplication method. In this paper, we present low power and high-speed parallel multiplier. Data-path reduction and rearrangement methods and also adding a small data-path element for modifying of data-path circuits are used to reduce static power consumption and overall delay in the processor. In DSP processor applications, the data-path reduction and modification methods are used to reduce a huge amount of computational complexity. It is used in high-speed data communication applications. The hardware architecture of both existing and proposed parallel multipliers are implemented using Altera Field Programmable Gate Array (FPGA).

1.1 Related Previous Works

The sequential multiplication method is consuming less amount of static power dissipation, at the same time the output of the multiplier is given to the iteration loop. The timing complexity is increased due to the sequential iteration process.

In [1], the radix based sequential multiplier is proposed. The partial products of a sequential multiplier are implemented using radix-16. In this work, components of the radix based sequential multiplier split into high and low components. The latency of the radix based sequential multiplier lower than the other sequential multiplier and the energy-delay product also improved. For optimization of power, delay, and area in the processor, some of the data-path circuits can be designed using approximate concepts. The approximate multipliers can be implemented for error-tolerant applications only but not in all DSP applications [2]. An unsigned truncated approximate sequential multiplier is proposed for the efficient hardware architecture of the processor [3].

The hardware efficient shifting based approximate multiplier is proposed for error-tolerant DSP applications [4]. In accuracy-based applications are required the exact multiplication operation to build the high-performance DSP processor. The floating-point multipliers are used to build a more accurate DSP processor. Mantissa multiplier is the most complex part in the floating-point multiplier and also the high precision like double and quadruple floating-point multipliers are required a higher bit size of the mantissa multiplier. The complexity of the floating-point multiplier can be reduced using some Vedic multiplication methods. In recent days most of the multiplier designs are computed using Vedic mathematics [5]. The multiple-precision support floating-point multiplier is implemented using Karatsuba Vedic mathematics [6]. In Parallel multiplier, all the inputs of the multiplier can be feed into multiplication processes at a time. The area of the parallel multiplier depends on the number of input bits. The size of the parallel multiplier is higher than the normal and sequential multiplier [7]. So, the fast multiplication process is used to increase the overall speed of the DSP processor. The parallel processing concept is implemented in all the data-path circuits to attain high-speed performance. In Booth and Modified Booth multiplier radix based parallel concepts are applied to reduce the latency in DSP processor [8]. The parallel multiplier has been implemented using optimized Boolean networks. In this work, three different fixed-point multipliers are implemented using FPGA. The decimal parallel multiplier [9] has been implemented using BCD (8421), (5221), and Excess 6 code conversion. The parallel multiplier was implemented based on the counter control circuit for the factorial circuit [10]. A pre-stored register-based parallel multiplier has been implemented for factorial circuit design [11]. All the possible inputs are pre-stored in the register and then the inputs are controlled by the tri-state buffer to perform the multiplication process.

2 Proposed Shift/Buffer Based Parallel Multiplier

The new architecture for the parallel multiplier is proposed. It contains four levels and each level has one sub-module (Block). The Blocks are designed with tri-state buffer, shifter, adder, and multiplexer. The main complex part of the parallel multiplier is the partial product generation (PPG). The block architectures are used to compute the partial product generation of parallel multiplier with less amount of latency and power consumption. The proposed parallel multiplier is skipped the partial product operation based on the carry generation of the inputs. So the switching power is reduced and also latency can be minimized. The 16-bit parallel multiplier is designed and the inputs are A[15:0] and B[15:0]. Figure 1 shows the architecture of the proposed parallel multiplier.

2.1 Block Architectures

At the first level, block 1 architecture contains 1-bit shifter, 17-bit Tristate buffer 'D1', 16-bit Tristate buffer 'D2', 17-bit adder, and 4:1 multiplexer, and it is shown in Fig. 2. There are eight block1 architectures are utilized for 16 × 16 parallel multiplier. Block 1 comprises input signal 'A' 16-bit, select lines for 4:1 multiplexer, which enable signal for both tri-state buffers. The select lines are taken from the input signal 'B'. The enable signal 'G1' is computed from AND operation of select lines of 'Bn' and 'Bn + 1'. The 4:1 multiplexer is used to choose the input signal to output concerning the select lines. The four inputs of the multiplexer are 18 bits of zeros, the input signal (A), one-bit Shifted input (A'), and the addition of input (A) and shifted input (A'). The output of the block1 architecture is 18 bits [Pn]. There are eight outputs (Pn, where n is 0–7) that are taken from eight-block 1 architecture.

At the second level, block 2 architecture contains 2-bit shifter, 20-bit Tristate buffer 'D3', 18-bit Tristate buffer 'D4', 20-bit adder, and 4:1 multiplexer, and it is shown in Fig. 3. There are four block2 architectures are utilized for 16 × 16 parallel multiplier.

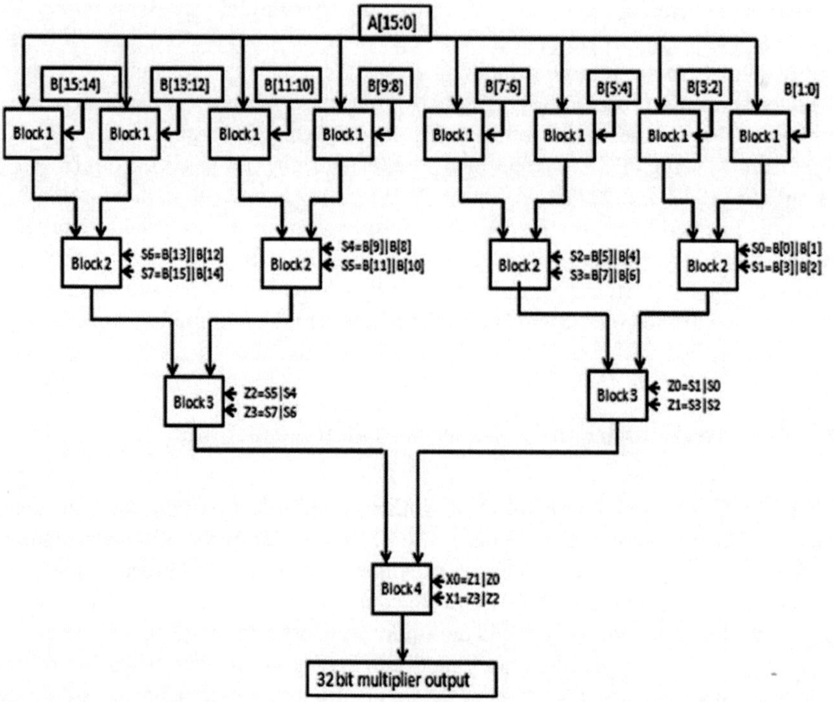

Fig. 1 Proposed shift/buffer based parallel multiplier

Block 2 comprises the input signal as the output of block 1 ‘Pn’ 18-bit, select lines for 4:1 multiplexer, an enable signal for both tri-state buffers ‘G2’. The select lines are computed from the OR operation input signal ‘B’. The Boolean expression of the select lines is Sn=Bn OR Bn + 1. The enable signal ‘G2’ is computed from AND operation of select lines (Sn and Sn + 1). The 4:1 multiplexer is used to choose the input signal to the output concerning the select lines. The operation 4:1 multiplexer is working as same as block 1 architecture’s multiplexer but the number of bits is varied according to the output of block 1 architecture. The output of the block 2 architecture is 21 bits [Qn, where n is 0–3].

At the third level, block 3 architecture contains 4-bit shifter, 25-bit Tristate buffer ‘D5’, 20-bit Tristate buffer ‘D6’, 25-bit adder, and 4:1 multiplexer, and it is shown in Fig. 4. There are two block-3 architectures are utilized for 16 × 16 parallel multiplier. Block 3 comprises an input signal as the output of block 2 ‘Qn’ 21-bit, select lines for 4:1 multiplexer, which enable signal for both tri-state buffers. The select lines are computed from the OR operation input signal ‘B’. The Boolean expression of the select lines is Zn=Sn OR Sn + 1. The enable signal is computed from the AND operation of select lines (Zn and Zn + 1). The 4:1 multiplexer is used to choose the input signal to output with respect to the select lines. The output of the block2 architecture is 26 bits [Rn, where n is 0–1].

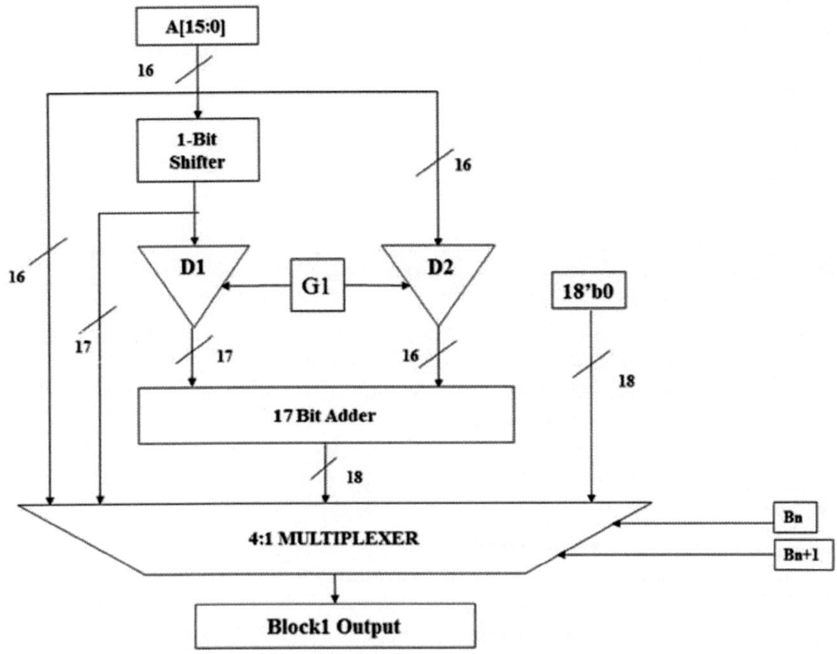

Fig. 2 Block 1 architecture of parallel multiplier

At last level, block 4 architecture contains 8-bit shifter, 34-bit Tristate buffer 'D7', 26-bit Tristate buffer 'D8', 25-bit adder, and 4:1 multiplexer, and it is shown in Fig. 5. One block 4 architecture is utilized for a 16 × 16 parallel multiplier. The operation of block 4 as the same as like block 1, 2, 3 but the number of bits are varying from other blocks and the figure shows block 4 architecture. The final output of the parallel multiplier is taken from truncated last 32 bits of block 4 architecture. The output from the multiplexers of each block is shown in Table 1.

2.2 Submodules of Block Architectures

Each block architecture has one shifter module. The shifter is used to append zeros at the Least Significant Bit (LSB) of the inputs. The number of zeros appends at LSB is fixed at each level. The first level appends 1 bit, the second level append 2 bits, and similarly third and fourth level append 4 and 8 bits. In multiplier, an adder is a vital block to compute and also consuming higher power. The computational complexity can be reduced by skipping the addition operation using the multiplexer and shifter. Tri-state buffer is used to either enable the data to addition or disable the data flow to addition, which is used to avoid unnecessary switching power dissipation on that

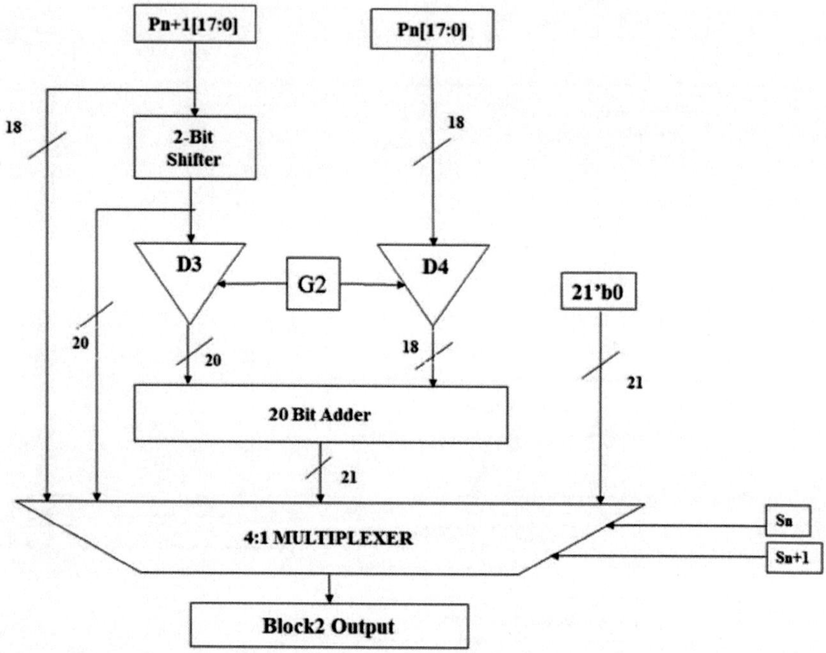

Fig. 3 Block 2 architecture of parallel multiplier

data-path. Each block architecture has two tri-state buffers. Based on the select lines of the multiplexer, both tri-state buffers enabled signals are getting active. Four adder circuits are used to design a 16 × 16 parallel multiplier. Each level has one adder block is used. Four different bit size adders like 17, 20, 25 and 34-bit adders are utilized. Four 4:1 multiplexer circuits are used to choose the outputs from the four different options based on select lines. In the proposed parallel multiplier, the critical path is skipped and latency can be improved.

3 Result and Implementation

Altera Quartus II EDA tool is used to implement all existing and proposed main and submodules of the parallel multiplier. The power analysis is simulated form the Altera Powerplay power analyzer. In Altera Quartus, the Time Quest timing analyzer is used to measure the worst-case critical path in all the modules of the parallel multiplier.

The Altera FPGA Cyclone II family is used to implement the parallel multiplier. It is one of the low-cost digital signal processing based application board. The parameters analysis of parallel multiplies shown in Table 2.

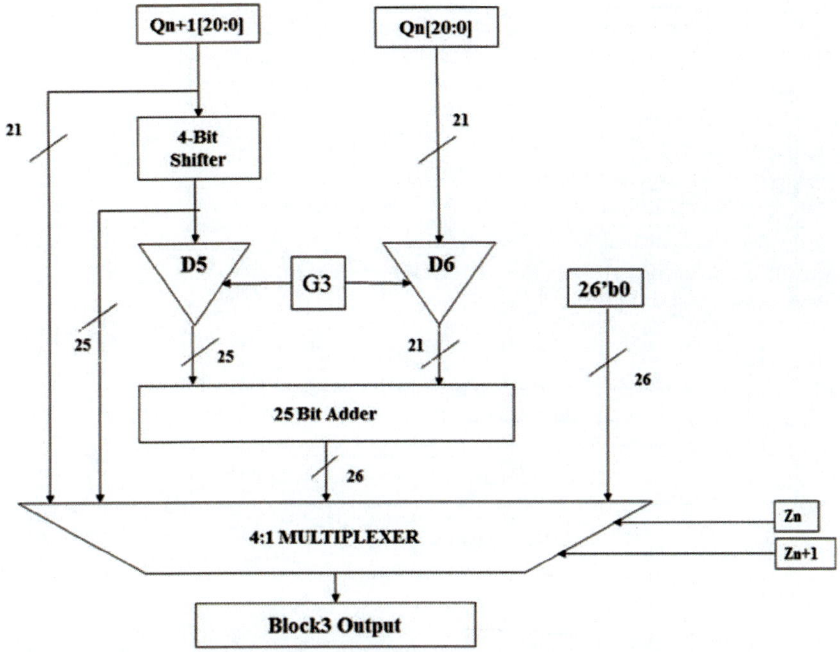

Fig. 4 Block 3 architecture of parallel multiplier

The power of the proposed parallel multiplier is 24.96% and 22.19% minimized when compared to Existing Parallel Multiplier [12] and existing Buffer Parallel Multiplier [13], respectively. The speed of the proposed parallel multiplier is improved 33.22% and 28.72% when compared existing Parallel Multiplier [14] and Existing Buffer Parallel Multiplier [15–19], respectively. But the number of logic elements is slightly increased due to the addition of some small data path circuit which is used to low power and high speed.

4 Conclusion

In this paper, low power and high-speed parallel multiplier have been designed using shifter and tri state buffer. The data path rearrangement techniques without changing the performance of the overall circuit are one of the best techniques to attain high-performance computing. Here, the same technique is used in the proposed parallel multiplier to attain less computational complexity for computer arithmetic. Further, the proposed parallel multiplier can be improved by using reconfigurable computing based on the compact scheduling algorithm. The repeated data path circuits in the

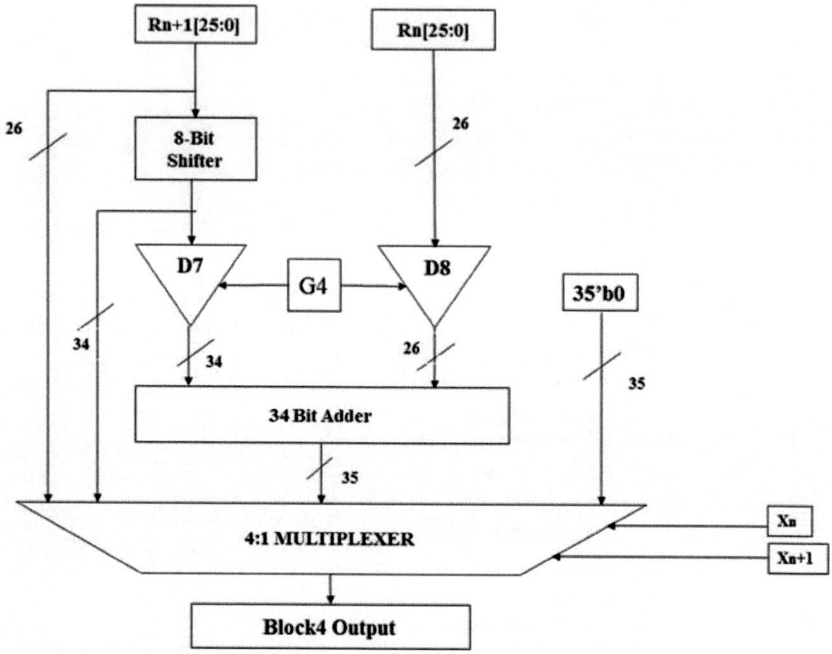

Fig. 5 Block 4 architecture of parallel multiplier

Table 1 Outputs of block architectures

Select1	Select0	Block 1 output	Block 2 output	Block 3 output	Block 4 output
0	0	18′b 0	21′b 0	26′b 0	35′b 0
0	1	A	Pn	Qn	Rn
1	0	A′	Pn + 1′	Qn + 1′	Rn + 1′
1	1	A + A′	Pn + Pn + 1′	Qn + Qn + 1′	Rn + Rn + 1′

Table 2 Different parameters comparative details (existing method 1 and 2 versus proposed method)

Parallel multipliers	Area (No. of LE's)	Power (mW)	Delay (ns)	Frequency (MHz)	PDP (J $\times 10^{-9}$)
Existing parallel multiplier [12]	465	123.18	16.316	61.29	2.009
Existing buffer parallel multiplier [13]	546	118.95	15.287	65.41	1.818
Proposed buffer and shifter based parallel multiplier	572	92.56	10.896	91.77	1.008

overall architecture of the parallel multiplier can be eliminated using schedule based reconfigurable computing.

Acknowledgements We the authors are pleased to acknowledge Dr. N. S. Kalyan Chakravarthy, Chairman and Correspondent, QIS Group of Institutions, Ongole, Andhra Pradesh on the successful completion of this research work.

Funding Details The research work is carried out in Product-Research and Development Lab, QIS College of Engineering and Technology, Ongole, Andhra Pradesh, funded by Sri Nidamanuri Educational Society (File No: QISCET/S&C/02/04/1255).

References

1. Amanollahi S, Jaberipur G (2017) Fast energy efficient radix-16 sequential multiplier. IEEE Embed Syst Lett 9(3):73–76
2. Osorio RR, Rodríguez G (2019) Truncated SIMD multiplier architecture for approximate computing in low-power programmable processors. IEEE Access 7:56353–56366
3. Tabei SM, Nikmehr H (2017) An unsigned truncated sequential multiplier with variable error compensation. Microprocessors Microsyst 49:9–17
4. Malek S, Abdallah S, Chehab A, Elhajj IH, Kayssi A (2017) Low-Power and high-speed shift-based multiplier for error tolerant applications. Microprocess Microsyst 52:566–574
5. Paliwal P, Sharma JB, Nath V (2019) Comparative study of FFA architectures using different multiplier and adder topologies. MicrosystTechnol 1–8
6. Cui X, Dong W, Liu W, Swartzlander EE, Lombardi F (2017) High performance parallel decimal multipliers using hybrid BCD codes. IEEE Trans Comput. 66(12):1994–2004
7. Imaña JL (2018) Fast bit-parallel binary multipliers based on type-I pentanomials. IEEE Trans Comput 67(6):898–904
8. Del Barrio AA, Hermida R, Ogrenci-Memik S (2019) A combined arithmetic-high-level synthesis solution to deploy partial carry-save radix-8 booth multipliers in datapaths. IEEE Trans Circuits Syst I Regul Pap 66(2):742–755
9. Moss DJM, Boland D Leong PHW (2019) A two-speed, radix-4, serial–parallel multiplier. IEEE Trans Very Large Scale Integr (VLSI) Syst 27(4):769–777
10. Akbari O, Kamal M, Afzali-Kusha A, Pedram M (2017) Dual-Quality 4:2 compressors for utilizing in dynamic accuracy configurable multipliers. IEEE Trans Very Large Scale Integr (VLSI) Syst 25(4):1352–1361
11. Bahar AN, Wahid KA (2019) Design of QCA-Serial Parallel Multiplier (QSPM) with Energy Dissipation Analysis. IEEE Trans Circ Syst II Express Briefs. https://ieeexplore.ieee.org/abstract/document/8902172
12. Saha P, Banerjee A, Dandapat A, Bhattacharyya P (2011) ASIC design of a high speed low power circuit for factorial calculation using ancient vedic mathematics. Microelectron J 42(12):1343–1352
13. Vignesh O, Mangalam H (2019) Low power binomial coefficient architecture for unused spectrum detector. Analog Integr Circ Sig Process 99(3):599–606
14. Khurshid B (2017) Technology-optimized fixed-point bit-parallel multipliers for FPGAs. J Signal Proc Syst 89(2):293–317
15. Boppana NVVK, Kommareddy J, Ren S (2019) Low-Cost and high-performance 8 × 8 booth multiplier. Circ Syst Signal Proc 38(9):4357–4368
16. Jaiswal MK, So HKH (2019) Design of quadruple precision multiplier architectures with SIMD single and double precision support. Integra VLSI J 65:163–174

17. Vestias M, Neto H (2019) Improving the area of fast parallel decimal multipliers. Microprocess Microsyst 61:96–107
18. Reddy KM, Vasantha MH, Nithin Kumar YB, Dwivedi D (2019) Design and analysis of multiplier using approximate 4-2 compressor. AEU Int J Electron Commun 107:89–97
19. Ahmeda SE, Varmaa S, Srinivasb MB (2018) Improved designs of digit-by-digit decimal multiplier. Integra VLSI J 61:150–159

Classification of Power Transmission Line Faults Using an Ensemble Feature Extraction and Classifier Method

Ani Harish and M. V. Jayan

Abstract This paper proposes an ensemble of feature extraction techniques for extracting features and an ensemble machine learning algorithm for classification of transmission line faults from voltage, voltage angle, and frequency signals. Transmission line protection is an important facet of a reliable power system. Many measures have been adopted by utilities worldwide for transmission line protection. Phasor Measurement Units (PMU) have been deployed in power grids throughout the world. PMU data can be used for detection, classification, and localization of faults in transmission lines. The characteristics of voltage, current, and frequency signals changes on different kinds of faults. The PMU data can be analyzed for signal characteristics, and these can be extracted as features. Maximum overlap discrete wavelet packet transform (MODWPT), Autoregressive coefficients, and Wavelet variance methods are used for feature extraction from power system signal data. A Machine learning algorithm for classification, ensemble bagging with a tree as a weak learner is used for classification of transmission lines faults. The performance of the ensemble classification algorithm is compared with other widely used machine learning classification algorithms.

Keywords Phasor measurement units (PMU) · Data · Fault · Classification · Transmission line · Feature extraction · Machine learning

Ani Harish (✉)
Rajiv Gandhi Institute of Technology, Kottayam, Kerala, India
e-mail: ani.ramachandran@gmail.com

M. V. Jayan
Government Engineering College, Thrissur, Kerala, India
e-mail: jayan@gectcr.ac.in

G. Ranganathan et al. (eds.), *Inventive Communication and Computational Technologies*, Lecture Notes in Networks and Systems 145,
https://doi.org/10.1007/978-981-15-7345-3_35

1 Introduction

Transmission line protection is a very important facet of a reliable and self-healing smart grid. Intelligent methods for detection, classification, and localization of faults on transmission lines is vital towards achieving the aspiration of a reliable smart grid. With the advancement of technology and computing power, many ıntelligent electronic devices (IEDs) are deployed on the power system to provide timely information about the state of the grid. PMUs are IEDs that capture voltage, voltage angle, current, current angle, frequency, and rate of change of frequency values on different buses. There is an increasing number of PMUs that are deployed in power systems throughout the world. The PMU readings are time-synchronized using the Global Positioning System (GPS). So these values are reliable and can provide an effective real-time understanding of the entire power system or the grid. The GPS synchronized measurements provided by the PMUs can be used to derive knowledge about the state of the system. The PMU data can be utilized for detecting, classifying, and localizing the short circuit faults on transmission lines. Effective algorithms can derive insights from PMU data, and thus PMUs can be key components of a power grid back up protection system.

Machine learning algorithms based on pattern matching can be used for the classification of transmission line faults [1–3]. The knowledge of line and system parameters is not required while using machine learning algorithms for fault classification on PMU data. Feature extraction techniques like phaselet transform, discrete wavelet transform (DWT), S-Transform, maximum overlap discrete wavelet transform (MODWT) have been applied to extract features for transmission line fault classification [4–8].

This paper proposes an ensemble of feature extraction techniques, MODWPT, Wavelet variance, and Autoregressive coefficients for feature extraction from power system voltage, voltage angle, and frequency signals. The features extracted are then classified using an Ensemble classifier algorithm. A comparative study of the proposed classification algorithm with other machine learning classification algorithms like support vector machines (SVM), K nearest neighbor (KNN), and decision tree (DT) is done, and the performance of the algorithm is analyzed.

The structure of this paper is as follows. Section 2 describes the methodology adopted. Section 3 discusses the results and performance analysis, and Sect. 4 is the Conclusion.

2 Methodology

The flowchart for the data-driven approach for classification of faults in transmission lines is as shown in Fig. 1.

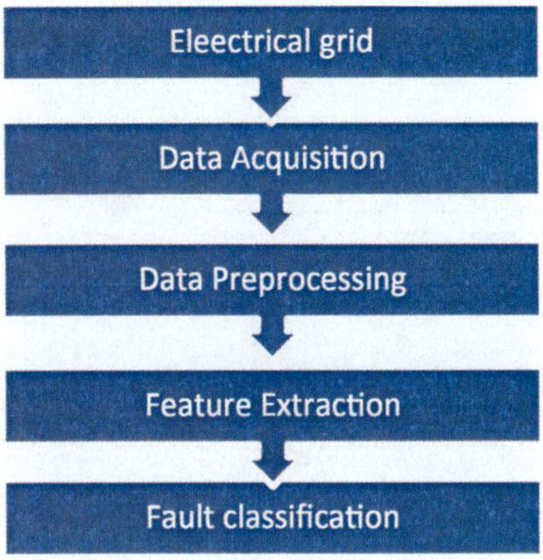

Fig. 1 Flow chart for classification of transmission line faults using machine learning algorithms

2.1 Data Acquisition

The PowerWorld simulator is a simulation package capable of simulating "high voltage power system operation within a time frame ranging from several minutes to several days" [9]. This power flow simulation engine package is capable of solving cases for up to 250,000 buses. The PowerWorld simulator has an add-on Transient stability tool. This enables transient stability studies on multi-bus systems. Transient stability studies help in analyzing the system's response to disturbances, such as faults. System response a few seconds post fault can be analyzed by transient stability studies. The PowerWorld Simulator for educational purposes capable of power flow studies for 40 buses is used here for the data simulation [9].

A PowerWorld compatible model of an IEEE 39 bus system (Fig. 2) is downloaded from [10] and is simulated on the PowerWorld Simulator. Contingency records are inserted in the transient stability window. Single line to ground fault (SLG), Line to Line fault (LL), Double line to ground fault (DLG), and Three Phases Balanced (3 PB) fault are applied to different buses. Simulation is run for 5 min with a sampling frequency of 60 Hz. The simulation results are exported and stored in a CSV file. The Dataset is a time series data with voltage, voltage angle, and frequency values of the 39 buses for pre-fault, during fault and post fault conditions. The Dataset generated has 18,403 samples of 117 signals. Oversampling of the minority classes is done to get a comparatively balanced Dataset. Dataset created with oversampling has 66,412 samples. Predictor variables are the bus voltages, bus voltage angles, and frequency measurement at each bus.

The PowerWorld Simulator generates and exports transient stability data in a C37.118 (the IEEE standard for PMU data transmission) compliant form. The data is exported and saved to a Comma Separated Values (CSV) file.

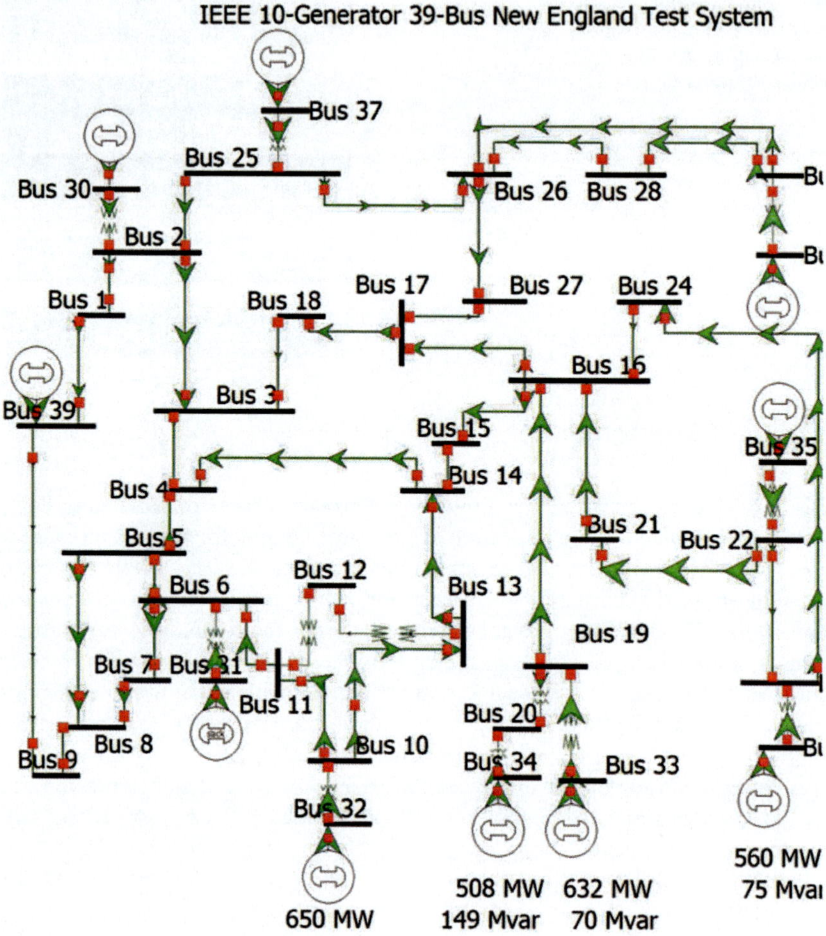

Fig. 2 One-Line diagram of IEEE 39 bus New England test system (*Source* PowerWorld Simulator)

2.2 Data Preprocessing

The data in the CSV file is checked for any inconsistencies. The Blank fields are replaced with 'NaN.' The Events data from the PowerWorld Simulator is exported as a CSV file. The Events data has the time of occurrence of faults along with the bus number. A column is inserted in the transient stability data CSV file as labels. The time instances are labeled according to the Events data. The labels are '0' for No-Fault, '1' for SLG Fault, '2' for LL Fault, '3' for DLG Fault, and '4' for 3 PB Fault. The Dataset is imported into MATLAB workspace as a 66,412 by 118 numeric matrix. Then the Dataset is randomly split into training and test sets. Take 70% of the labeled Dataset as a training Dataset and 30% as a test Dataset. Training data is a 46,488 × 117 numeric matrix, and training labels is a 46,488 × 1 matrix with categorical data. Test Data is a 19,924 × 117 numeric matrix, and Test labels are a 19,924 × 1 matrix with categorical data.

2.3 *Feature Extraction*

Feature extraction from the Dataset is done using an ensemble of three different techniques. The methods used for features extraction from the data are

Autoregressive coefficients

An Autoregressive process is a stochastic process where the output variable Y is dependent on the input variable X and the past values of Y. This model can be used in forecasting outputs of Time series. In [11] 'Burg method 'is used for calculating the Autoregressive coefficients. The same method as in [11] is used for calculating the Autoregressive coefficients of the time series data. Autoregressive coefficients of order 4 are found for the Dataset and taken as features.

Shannon entropy values for Maximal overlap discrete wavelet packet transform (MODWPT)

The 'modwpt' function of MATLAB (2019b) returns the wavelet packet tree, with each row of the matrix representing the sequence, ordered wavelet coefficients at each level. Shannon Entropy of the wavelet coefficients at level 4 is calculated and considered as features [12].

Wavelet variance

The 'modwtvar' function of MATLAB is used for estimating the wavelet variance of the signals. The variability of the signals over the octave frequency bands is determined by wavelet variance.

The features obtained by all the above methods are concatenated into a feature matrix. The feature matrix for train data is a 46,488 × 22 numeric matrix, and the feature matrix for the test data is a 19,924 × 22 numeric matrix.

2.4 *Fault Classification*

Machine learning methods, SVM, Decision Trees, KNN, and Ensemble Bagging classification methods are used for training the classification models. The features are given as inputs (predictors) and the class labels as responses to the classifier algorithms. The Hyperparameters for the different classifiers are optimized by Bayesian optimization, The optimizable Hyperparameters for the KNN classification algorithm are

- Number of neighbors
- Distance metric
- Distance weight.

The optimizable Hyperparameters for the SVM classification algorithm are as follows:

- Multi-class method
- Box Constraint Level
- Kernel Scale
- Kernel Function.

The optimizable Hyperparameters for Decision Tree classification algorithm are

- Maximum number of splits
- Split Criterion.

The optimizable Hyperparameters parameters for Ensemble classification algorithm are

- Ensemble Method
- Maximum number of splits
- Number of Learners
- Learning Rate.

Table 1 gives the range of values of the Hyperparameters for KNN, SVM, Decision tree, and Ensemble classifiers. The optimal values of the Hyperparameters for the different classifiers are found out using Bayesian optimization, and the values are

Table 1 Optimizable Hyperparameters search range

Classifier	KNN	SVM	Decision tree	Ensemble
Hyper parameters search range	Number of neighbors: 1–881 Distance metric: cityblock, chebyshev, correlation, cosine, euclidean, hamming, jaccard, mahalanobis, minkowski, spearman Distance weight: equal, inverse, squared inverse	Multi-class method: One-Vs-One, One-Vs-All, BoxConstraint Level: .001-1000, Kernel scale: 0.001–1000, kernel function: Gaussian, linear, quadratic, cubic	Maximum number of splits: 1–1361, Split criterion: Gini's diversity index, twoing rule, maximum deviance reduction	Ensemble method: bag, adaboost, rusboost, maximum no. of splits: 1–1361, No. of learners: 10–500, learning rate: 0.001–1

Table 2 Hyperparameter values for transmission line fault classifiers

Classifier	KNN	SVM	Decision tree	Ensemble
Optimized parameters	The number of neighbors: 274 Distance metric: Cosine Distance weight: Squared Inverse	Kernel Function: Gaussian, Kernel Scale: 02819, Box Constraint level: 12.553 Multi-class Method: One-vs-One	The maximum number of splits: 46006 Split Criteria: Maximum Deviance Reduction	Ensemble method: Bag, The maximum number of splits: 45,297, No. of Learners: 319

as in Table 2. The training feature matrix with the train labels is given as inputs to the KNN classifier, decision tree, SVM, ensemble bagging classifiers. The training is done with 5-fold cross-validation.

The transmission line faults classification is a multi-class classification problem with the five classes as follows:

1. No-Fault (NF) labeled as '0'.
2. SLG labeled as '1'.
3. LL labeled as '2'.
4. DLG labeled as '3'.
5. 3 PB labeled as '4'.

3 Results and Discussion

The performance measures, training accuracy, misclassification cost and training time of the different classifiers are as in Table 3. Ensemble Bagging, SVM and KNN give the best fivefold cross-validation classification accuracy, 99.3%, for training data. The misclassification cost of the KNN Classifier is the least, and the Decision Tree classifier has the least training time.

The confusion matrix shows the number of observations classified into different classes. The confusion matrix for the test data by the trained classifier models is as shown in Tables 4, 5, 6 and 7. Table 4 shows the confusion matrix of Ensemble Bagging Classifier. 18,260 instances of NF class data is correctly classified, and eight

Table 3 Performance measures of the different classifiers

Measures	KNN	SVM	DT	Ensemble Bag
Accuracy (%)	99.3	99.3	96.6	99.3
Total misclassification cost	331	342	1598	342
Training time (s)	2899.8	72113	79.62	3791.5

Table 4 Confusion matrix of Ensemble Bagging classifier

Confusion matrix ensemble bagging

Output class	Target class					
		NF (0)	SLG (1)	LL (2)	DLG (3)	3 PB (4)
	NF (0)	18260	3	1	3	1
	SLG (1)	29	282	0	0	0
	LL (2)	18	0	542	0	0
	DLG (3)	12	0	0	439	0
	3 PB (4)	12	0	0	0	322

Table 5 Confusion matrix of decision tree classifier

Confusion matrix decision tree

Output class	Target class					
		NF (0)	SLG (1)	LL (2)	DLG (3)	3 PB (4)
	NF (0)	18020	64	62	75	47
	SLG (1)	53	234	10	11	3
	LL (2)	68	4	443	22	23
	DLG (3)	52	6	8	379	6
	3 PB (4)	22	11	4	8	289

Table 6 Confusion matrix of KNN classifier

Confusion matrix KNN

Output class	Target class					
		NF (0)	SLG (1)	LL (2)	DLG (3)	3 PB (4)
	NF (0)	18,268	0	0	0	0
	SLG (1)	311	0	0	0	0
	LL (2)	560	0	0	0	0
	DLG (3)	451	0	0	0	0
	3 PB (4)	334	0	0	0	0

Table 7 Confusion matrix of the SVM classifier

Confusion matrix SVM

Output class	Target class					
		NF (0)	SLG (1)	LL (2)	DLG (3)	3 PB (4)
	NF (0)	18,268	0	0	0	0
	SLG (1)	311	0	0	0	0
	LL (2)	560	0	0	0	0
	DLG (3)	451	0	0	0	0
	3 PB (4)	334	0	0	0	0

Table 8 Precision, recall, and F1 score

	Precision	Recall	F1 score
Ensemble	99.277	96.2327	97.80
Decision tree	82.288	84.71	83.408

instances are incorrectly classified. Of the eight incorrectly classified instances, three instances are classified as SLG, three classified as DLG, one instance is classified as LL, and one as 3 PB. Of SLG class data, 282 are correctly classified as SLG, and 29 are incorrectly classified as NF. The 542 instances of LL are correctly classified, and 18 cases of LL are incorrectly classified as NF. Four hundred thirty-nine samples of DLG are correctly classified, and 12 cases of DLG are incorrectly classified as NF. Three hundred twenty-two instances of 3 PB are correctly classified, and 12 cases of 3 PB are incorrectly classified as NF, 99.6% of NF class, 90.7% of SLG class, 96.8% of LL class, 97.3% of DLG class, and 96.4% of 3 PB class cases are correctly classified. The overall accuracy of the Ensemble Bagging classifier is 99.6%.

The testing accuracy of The KNN and SVM classifier models is 91.69%, and both these classifier models predict test data only for the majority class or the NF class. The testing accuracy of the Decision Tree classifier model is 97.19%, and of the Ensemble classifier is 99.60%. The Dataset is an unbalanced Dataset with the No-Fault class as the majority class. The classifier accuracy will be high if a classifier classifies the majority class data correctly. Apart from overall accuracy, performance measures like Precision, Recall, and F1 score can be considered for evaluating the performance of a classifier.

$$\text{Precision} = \frac{\text{Sum of True Positives}}{\text{Sum of True Positives} + \text{Sum of True Negatives}} \tag{1}$$

$$\text{Recall} = \frac{\text{Sum of True Positives}}{\text{Sum of True Positives} + \text{Sum of False Negatives}} \tag{2}$$

$$\text{F1 Score} = \frac{2 * \text{Precision} * \text{Recall}}{\text{Precision} + \text{Recall}} \tag{3}$$

The Precision, Recall and F1 score for the decision tree and ensemble classifier is analyzed, and the values are as in Table 8. Ensemble Classifier model shows high accuracy and high F1 score, and this makes it best suited for transmission line faults classification.

4 Conclusion

With the deployment of IEDs in the WAMS, a huge amount of data is getting captured and stored. Insights and knowledge can be derived from these data. One such knowledge discovery is the transmission line fault detection and classification, which plays an important role in a backup power system protection system. In this paper, an IEEE 39 bus test system is considered, and the Dataset has voltage, voltage angle, and frequency measurements from all the 39 buses. The data captured is processed and transformed using an ensemble of feature extraction techniques. The features thus extracted are used for classification. SVM, KNN, DT, and Ensemble classification algorithms are considered for this multi-class classification problem. It is observed that even though the training accuracy of cross-validated SVM and KNN classifier models are high, these models fail to classify the different faults during testing. The Decision Tree classifier and the Ensemble model with base learners as tree perform well in classifying the different faults. Accuracy, Precision, Recall, and F1 score are the performance measures considered for the classifiers. Considering all the performance measures, the Ensemble classifier model with the Bagging Ensemble method is the best classifier algorithm for transmission line faults.

Acknowledgements AH, thanks Government of Kerala, for the financial support provided through Higher Education Fellowship.

References

1. Arouche Freire JC, Garcez Castro AR, Homci MS, Meiguins BS, De Morais JM (2019) Transmission line fault classification using hidden markov models. IEEE Access. 7:113499–113510. https://doi.org/10.1109/access.2019.2934938
2. Prasad A, Belwin Edward J, Ravi K (2018) A review on fault classification methodologies in power transmission systems: part—I. J Electr Syst Inf Technol 5:48–60. https://doi.org/10.1016/j.jesit.2017.01.004
3. Chen K, Huang C, He J (2016) Fault detection, classification and location for transmission lines and distribution systems: a review on the methods. High Volt 1:25–33. https://doi.org/10.1049/hve.2016.0005
4. Swain KB, Mahato SS, Cherukuri M (2019) Expeditious situational awareness-based transmission line fault classification and prediction using synchronized phasor measurements. IEEE Access 7:168187–168200. https://doi.org/10.1109/ACCESS.2019.2954337
5. Ashok V, Yadav A, Abdelaziz AY (2019) MODWT-based fault detection and classification scheme for cross-country and evolving faults. Electr Power Syst Res 175:105897. https://doi.org/10.1016/j.epsr.2019.105897
6. Jain, A., Archana, TC, Sahoo, MBK: A Methodology for Fault Detection and Classification Using PMU Measurements. 2018 20th Natl. Power Syst. Conf. NPSC 2018. (2018). https://doi.org/10.1109/NPSC.2018.8771757
7. Che J, Park J, Park G, Park T (2019) A new fault location identification method for transmission line using machine learning algorithm. In: Proceedings—2019 3rd International Conference on Smart Grid and Smart Cities, ICSGSC, pp. 81–84 (2019). https://doi.org/10.1109/ICSGSC.2019.00-14
8. Yu JJQ, Hou Y, Lam AYS, VOK L (2019) Intelligent fault detection scheme for microgrids with wavelet-based deep neural networks. IEEE Trans Smart Grid 10:1694–1703. https://doi.org/10.1109/TSG.2017.2776310
9. PowerWorld online help. https://www.powerworld.com/WebHelp/
10. Electric grid test case repository. https://electricgrids.engr.tamu.edu/electric-grid-test-cases/ieee-39-bus-system/

11. MATLAB help. https://in.mathworks.com/help/releases/R2018a/wavelet/examples/ecg-classification-using-wavelet-features.html
12. MATLAB help. https://in.mathworks.com/help/wavelet/ref/modwpt.html

A Retrospection on Selective Forwarding Attacks in WSN

A. Anitha and S. Mythili

Abstract In recent years, Wireless Sensor Networks (WSNs) are being a torrid research topic due to its fastest communication path via the networks. At the same time, it faces several security attacks. In particular, the selective forwarding attack is an attack where the attacker or the malicious node can selectively drop packets or selectively forward the packets, which will leave the entire network at risk. Typically, this selective forwarding attack occurs at the network layer and also becomes difficult to discover and avoid this occurrence. This will inevitably influence the working routine of the network. Hence, the following retrospection helps to acquire an interpretation of the various strategies for detecting and preventing the selective forwarding attacks.

Keywords Selective forwarding attacks · Sensor node · Sensor network · Attackers

1 Introduction

Wireless Sensor Networks (WSNs) comprise an enormously distributed, self-directed, small-powered tiny device called sensor nodes and those tiny nodes are dispersed in an ad hoc manner. A collection of sensor nodes collects the information from the environment to complete particular application objectives. WSN comprises many wireless sensors where it acts as nodes and also as a sink or base station (BS) and also it has a gateway node. The sensor node can communicate with each other or route the data to other sensors or back to the base station. Each sensor node contains

A. Anitha (✉)
Ph.D Research Scholar, Department of Computer Science, Kongunadu Arts and Science College, Coimbatore, Tamil Nadu, India
e-mail: anithaaruchamy91@gmail.com

S. Mythili
Associate Professor and Head, Department of Information Technology, Kongunadu Arts and Science College, Coimbatore, Tamil Nadu, India
e-mail: smythili78@gmail.com

G. Ranganathan et al. (eds.), *Inventive Communication and Computational Technologies*, Lecture Notes in Networks and Systems 145,
https://doi.org/10.1007/978-981-15-7345-3_36

three types of subsystems. The sensing unit is used for sensing the environment, the processing unit is used for calculating the confined permutations of the sensed data and the communication unit is used to perform the exchange of processed information among neighboring sensor nodes. WSN contains self-organizing capabilities of protocols and algorithms. It is a subclass of Wireless Ad hoc Network (WANET) which is typically utilized for observing either somatic or surrounding circumstances alike heat, resonate, heaviness, or comparative moistness. The main objective of WSN is cooperatively passing its information via the system to a primary region. This paper gives an overview of the different techniques in handling the selective forwarding attack.

2 SFAs (Selective Forwarding Attacks)

SFAs (Selective Forwarding Attacks) are a practice of network layer attack which ensues in the WSNs. Generally, in this type of network, the source sensor nodes will onward the data packets either to its immediate sensor node or to its nearby sensor nodes which will maintain the belief aspect that is the data packets which onwards by it to its desired destination sensor nodes which are located at the other extent in the network [18]. In the other case, this selective forwarding attacks (SFAs) are implemented in such a way that the attackers will configure the attacker sensor nodes which pretends like the normal sensor nodes in the network. Then, these kind of attacker nodes will impulsively drop the received data packets and onward only the selected data packets to the immediate sensor node. Meanwhile, when the attacker nodes drop out the entire received data packets instead of forwarding it to the neighbor nodes. Then this form of attack is called the black hole attack. Therefore, these attacks have a greater impact on the network performance when the attackers act externally during the data communication between the sensor nodes. While considering black hole attack and selective forwarding attack, the black hole attack can be identified with the ease when all the data packets transmitted are dropped which will provide an alert about the presence of attacker nodes in-network but at the same time, selective forwarding attacks are difficult to identify because only it results in selective data packet loss which creates an illusion as it happens in the network data transmission process.

3 Related Work

In this paper, the existing schemes are classified based on their working nature which is been introduced by various researches for discovering and avoiding the selective forwarding attacks (SFAs) which occurs in the wireless sensor networks(WSNs).

3.1 Distributed

The authors [1] introduced the "Light-Weight Defense Scheme against Selective Forwarding Attack" which will make use of the nearest sensor nodes and those sensor nodes are considered to be the detector sensor nodes. The nearest sensor nodes will observe the data packets at the time of data communication. During this process when some of the data packets are dropped then by considering the position of the attacker the packet is retransmitted. The nearest sensor nodes will observe the event data packets' communication. This aids in discovering the selective forwarding attacks (SFAs) and it can be notified and in case of any packet drop that packet will be resent to the respective destination. The routing algorithm is used to calculate the routing path for the event packet transmission from the source sensor node to the destination sensor node. The event packets are transmitted via the midway nodes or relay nodes. The selective forwarding attack is detected with the help of a monitor node and when such an attack occurs then the event packet is retransmitted. When such retransmission occurs due to packet drop then it is informed to the remaining sensor nodes which are located in the network by producing a piece of alert information about the attacker. And it is used by the other nodes to protect themselves from the attacker. This scheme made few assumptions such as the network topology are static and packet monitoring is done by one hop and the attack is triggered only at the time of transmitting the event packet.

Limitations.

- Probability-based routing is not always optimal.
- It does not support dynamic topology.
- There is no detection mechanism imposed for the monitor node when it suffers from the attack.
- It is cost-effective due to the presence of GPS.

The authors [2] presented the "CheckPoint-Based Multi-Hop Acknowledgement Scheme (CHEMAS)" [17]. This scheme utilizes the "Checkpoint Nodes" for dispatching and obtaining the acknowledgments allying the source sensor node and the destination sensor node. This method strived in enhancing the detection of the anomalous data packet within the wireless sensor networks. The presumption is made that whichever resolute nodes were not capable of comprising the vigilant packets along to malignantly implead the remaining sensor nodes. Thereupon accumulating the confirmation to exemplify regardless of whether that node is the attacker node or not. The source sensor nodes will regulate the location of the distrustful sensor node under the emplacement. Nevertheless, no allowance of guarantee is provided for reliable data communication though the attacker node is been positioned via the acknowledgment.

Limitations.

- It requires a high storage space.
- High energy consumption.
- The reliability of the data is very low.

The authors [3] developed the "Light-Weight Security" procedure for discovering the SFAs in the surroundings of the wireless sensor networks. This procedure uses the "Multi-Hop Acknowledgment" to initiate the consternation by acquiring the reply from the sensor nodes which are detected in the midway of the routing paths. The authors also considered this procedure as it may recognize the attacker nodes. Therefore, every sensor node that is presented in the data routing path is in charge of discovering the attacker sensor nodes. When the attacker sensor node is discovered an alarm is been send for indicating the existence of the attack. The identification of suspect nodes is stated through a midway node. When the mischievous data packet exists been spotted then the divulge data packet is generated with the transmission of those packets to sink node through the node by node. Consequently, the sink node will receive the alert data packet and it will onward that to the multiple sensor nodes from where the data communication is initiated.

Limitations.

- Great communication.
- Delay overheads.
- Low scalability.

The authors [4] are the first persons to converse the selective forwarding attacks (SFAs) along with suggested the multi-path forwarding to defend against these SFAs. The data packets are directed over with the multiple routing paths which are outrightly the disassociate and safeguarded up against the SFAs relating to most of the "N Compromised Nodes" it provides few possible security measures in the act of "K Nodes" that are endangered. In this data forwarding technique, the sensor nodes are permitted to selects its succeeding node for the travel of data packets as of the potential sensor nodes with decreasing the likelihoods of the attacker to get an advantage of the thorough regulator over the stream of data. While several routing paths ultimately intersect in the nearby province of sink node accordingly when the sensor nodes which are located all over the sink nodes are conceded then the SFAs are appropriate because of the absence of nearby sensor node observation.

Limitations.

- No notification is initiated about the malicious node.
- High energy consumption.
- This method results in communication overhead.
- There is no mechanism of a particular strategy for the identification of the attacker.
- The security is low.

3.2 Centralized

The authors [5] suggested a lightweight detection mechanism. This mechanism works on the centralized structure of clusters in which it utilizes two-hop neighborhood node information and listening technique. In the application layer, the sensor nodes are provided with the detection module. The routing rules are framed by the sensor nodes and the alert packets are generated based on the knowledge of two-hop neighborhood information. The value of the sensor nodes is estimated using "Malicious Counter" and when that value goes beyond the specified "Threshold Value." Then that sensor node is considered to be the misbehaving node or the attacker node. Hence, that attacker node will be removed straight away from its neighboring sensor nodes list. The authors presumed that none of the sensor nodes are trustworthy. At the same time, the neighboring sensor nodes should have an acquaintance in defense and reliability during the deployment period along with the constant topology construction. The "Pair-Wise Keys" are disseminated in advance between the sensor nodes which is necessary for avoiding the occurrence of external attacks. The authors also recommended "Two Routing Rules" to create an effective observation system. Hence, "the first rule is to determine if the destination node forwards the packet along the path to the sink. It generates an alert packet with the malicious factor to the sender node or the source node." And "the second rule governs that the monitor node waits and detects the packet that was already forwarded along the path to the sink. It verifies the two-hop neighbor knowledge to assess whether the destination node is on the right path to the sink. If not, it generates an alert packet with the malicious factor to the sender node or source node." The "Detection Module" is responsible for indirectly discovering the SFAs on the nearest sensor nodes.

Limitations.

- It does not detect the attack when it occurs in the centralized node or the monitor node.
- It works only for a pre-fixed topology.
- The reliability of data transmission is not assured.

The authors [6] proposed the methodology called cumulative acknowledgement based detection (CADE). It works based on cumulative acknowledgment. It is accomplished with pre-distributed keys between them. This scheme comprises three different phases. The first phase begins with the topology construction and the process of route selection. The second phase is data transmission, and the third phase is the detection process. It works only for static topology.

Limitations.

- The topology is static and not suits for a dynamic environment.
- It consumes more energy in each of its phases.

The authors [7] proposed the Support Vector Machines (SVMs) and it a form of an algorithm. It merely identifies the attacks. The local direction-finding data in association with sink node and it is used to generate alarms based on two-dimensional features such as bandwidth and hop count. The data patterns are classified based on the "one-class SVM classifier." The "MTE (Minimum Transmission Energy)" is used by the subsequent step which is taken when the communication dynamism depleted by the distribution sensor node is minimized and also in an attempt to extend the lifetime of all individual nodes. It is used for detecting two forms that are SFAs and black hole attacks. They used anomaly detection and point out the discovery of these attacks. It works on the apriori method. That is the attack training set has experimented which occurs in the network during data transmission.

Limitations.

- Attacker nodes are not detected.
- No substitute routes are discovered in the course of SFAs.
- This algorithm experiences the sole sensor node devastation issue. In other words, it can state like when the integrated sensor node is conceded formerly the entire network will be in trouble and the network operations are affected completely.

3.3 Secure Algorithm

The authors [3] projected the "Secure Routing Algorithm against Selective Forwarding Attacks" in WSNs. The focal notion of this process is described by launching the structure of the cell contained by the network. Hence "W > 1" sensor node shows the observing protagonist in every single cell which observes events of the other sensor nodes happening in that cell. Using an eavesdropping stream of traffic surrounded by the cell and the spectator sensor nodes will discover whether those sensor nodes exist malevolently dropped the packets. Indecision the spectator sensor nodes diagnose that a sensor node is mischievous and make sure of not dispatching any received data packet to the destination sensor node instead it will change the routing path of the data packets. The reminiscence overhead is been enacted by the presented procedure to the nodes trepidations solitary to the neighbor routing table. Reminiscence directly above which is equivalent to "2d." The message directly above because it executes on the observing sensor nodes aimed at eavesdropping action in this method. Conversely, it is evident that eavesdrop action devises diminutive overhead and minor liveliness likened to transferring the data packet. The overhead dispensation of this method proceeds with its second level that is every sensor node has to organize its neighboring sensor nodes in a sliding direction besides scrutinizes whether those sensor nodes are in the "W List of the observing sensor node." Presuming that "d" is considered to be the "average number of neighbor sensor nodes."

Limitations.

- Communication overhead is high.
- Processing overhead is high.

3.4 Watermark Technology

The authors [9] suggested "Secure Data Transmission (SDT)" for identifying the SFAs. The "watermark technology" is used for discovering attacked nodes. The watermark procedure will estimate the loss of data packets and also helps in discovering the attacker node. They used the reliance assessment in the direction-finding nominated procedure. It practices the topographical dispatching procedure by merging the reliance assessment through remoteness to elected finest data progressing route. After the confirmation of the optimal routing path, the transmission of data will be initiated. The original watermark message contains the K bits binary sequence and it is created by the sink and it is one of a part in the packets. The sink makes a comparison between the extracted watermark and the original watermark message for discovering the SFAs. The assumption is made in such a way that the base station does not suffers from the attack and it is always in a secured state. In the network model, all individual nodes contain a trust value. All nodes trust value is maintained in the base station to find the reliability of the node. During the process of network deployment, all nodes are deployed with a similar trust value, the trust value changes accordingly with time. It is assumed that the attacker drops only a few packets during transmission.

Limitations.

- Data transmission is not possible after the attack.
- Additional overhead because multiple attackers cannot be notified at the same time.

3.5 Traffic Monitor Based

The authors [10] developed a "Lightweight Approach." The "Extra Monitor (EM)" is used for eavesdropping and observing the entire stream of traffic during message communication. It is further employed Received Signal Strength Indicator "(RSSI)" to identify a sink-hole attack. The assessment of this received signal strength indicator was done based on "4 EM sensor nodes" that are organized to create locations for each node wherein the sink node is located on "(0, 0)." It focuses on effect with the location-dependent SFAs because the misbehaving sensor node will selectively drop the data packets and this function works on location information of the source sensor node. The network model works on certain per-fined rules such as the topology of the network is fixed, after the deployment of the nodes, it becomes static. The attacker sensor node contains the ability for modifying the sensor node's location from its actual position to make a confusing circumstance.

Limitations.

- Not suitable for a dynamic environment.
- The correctness of this scheme is inadequate.

- It does not handle the attack which occurs in the centralized node and it leads to the network failure.

3.6 Channel Aware Detection

The authors [11] developed the "Channel Aware Detection (CAD)" method. The working of this method depends upon two processes. They are the "Channel Estimation" and "Traffic Monitoring." The first process is channel estimation as the name itself implies that it is used for calculating the usual data loss ratio which is caused either by medium virtue or impingement. The second process is observing the stream of traffic for calculating the confirmed data loss ratio. The attacker is node is identified when the traffic observing ratio in a particular rush goes beyond the calculated data loss ratio. The basic standard is that each relay sensor node concurrently designates the route that enacts "Upstream" and "Downstream" observations. The "downstream" observation is the process of observing that the node is altering or dropping the messages. "upstream" observation done by perceiving the upstream node's performance by computing the loss rate. The detection of malicious nodes is carried out by the upstream and downstream observations concerning its thresholds. In this scheme, each node contains the history of the packet counts. And it includes the amount of data packets collected out of "upstream" sensor node and amount of data packets which is eavesdropped out of the "downstream" sensor node that is dispatched. By considering such investigations all the sensor nodes keep up the chances of having mistrustfulness of its "downstream" sensor node. This scheme contains more than one type of data packets. They are the "PROBE packet" and "PROBE ACK packet" which is used for detecting misbehavior activities. The source node transmits the "PROBE packet" onto its desired destination sensor node to find whether that path is secure or not. The PROBE packet is transmitted via the nodes which are presented in the available path. Each of those nodes will append its traffic monitoring parameters to the PROBE packet. The most important process of the destination node is to verify whether there is any packet loss or if any other distrust is happening in the path and it is found with the available information of the received PROBE packet from each node. Based on the received PROBE packet the decision is made by the destination sensor node whether to disseminate "negative PROBE ACK" or "positive PROBE ACK" to its source sensor node. "Negative PROBE ACK" is sent when malicious activity is detected. Otherwise, "Positive PROBE ACK" is sent by insisting that there is no packet loss in the destination. While getting "positive PROBE ACK" the source sensor node carries on its normal data transmission.

Limitations.

- Additional data packets are used in communication for discovering the attack and this leads to extra overhead.
- Produces a high traffic rate.

- Every activity should be observed during both upstream and downstream monitoring.

3.7 Multipath Data Flow Topologies

The authors [12] developed the strategy called "Multi-Flow Topologies" which is applicable in detecting the SFAs. Herein this strategy is deployed by partitioning whole sensor network into several discrete data topologies. Thus, the sensor node belonging to one discrete topology can do interact with the remaining sensor nodes which are located within that same discrete topology. All these processes are carried out at the course of the deployment period. The sink node makes use of the locality data for discovering the attacker nodes. This scheme does not require additional hardware or software for detecting the attack. It also results in a high packet delivery ratio.

Limitations.

- More expenditure is needed for implementing such a network strategy.
- The lifetime of the entire network is compromised.

3.8 Heterogeneous Sensor Network Model

The authors [13] introduced a network model that is used for detecting SFA with the "Heterogeneous Sensor Network (HSN) model." This model comprises of two different types of sensor nodes. The first type of sensor node is "High-End Sensors" and it has great potential. The second type of sensor node is enormously used and called "Low-End Sensors." After deploying all the sensor nodes, the next step was the cluster formation. One high-end sensor node will be elected to perform the role of the cluster-head. Each time when there is the action of packet dropping takes place at a sensor node than that low-end sensor nodes will provide the information regarding packet dropping to its respective cluster-head in this model it is the elected high-end sensor node. Depending upon the information acknowledged high-end sensor node goes for the examination then decides whether that sensor node is an attacker node or a non-attacker node. Furthermore, there is dual-threshold value are used in this network model for stating the activities of the sensor node through added assurance. Network security is delivered by way of using encryption and decryption techniques. The low-end sensor nodes will perform the process of encryption wherein the high-end sensor nodes will perform the process of several decryptions because it acts as a cluster-head from a group of low-end sensor nodes.

Limitations.

- When the centralized node or the cluster head suffers from the attack then it leads to the degradation of the network.

- No data transmission is initiated after the packet drop.

3.9 Multi-layers

The authors [14] developed the "Selective Forwarding Detection (SFD) approach" using multi-layers for detecting the SFAs. The SFAs are one of the most rigorous attacks which disturb the data transmission performed by the sensor nodes of the network and this happens by disconnecting the transmission links. This detection approach works in multiple layers. The structure of this approach contains three types of layers where all those layers are maintained by the various procedures. The main layer is called "Pool of MAC IDs" and its corresponding procedure is used for validating the inward stream of traffic to govern whether that sensor node is an authentic sensor node or an attacker sensor node. The next layer is called as "Rule-Based Processing Algorithm" and usage of this layer is to verify the traffic flow comparatively with the presumed rules which are specified in the rule list. The last layer is called as "Anomaly Detection" procedure which is used for tracing the unidentified outbreaks that seem to be like direct alert, deceitful rejections, and discard the stream of traffic. In this model, three presumptions are taken into concern about its working. The first presumption is each sensor node should ensure identical qualifications. The second presumption is every sensor node in the network should contain identical energy levels in the beginning stage. The third presumption is that those sensor nodes are homogeneously dispersed indiscriminately. Only after launching the selective forwarding, the packet should be dropped otherwise the packet should not be dropped. During the deployment process, the nodes cannot be attacked by the adversary. The communication exists between the sensor nodes and the sink node. The mechanism which is used for estimating the remoteness of the sink node with the sensor node is called a received signal strength ındicator (RSSI).

Limitations.

- All nodes have the same specifications that are assumed which leads to a problem when the node specifications change.
- A malicious node may drop the packet at any time.

3.10 Per-Hop Acknowledgement

The authors [15] proposed the "Per-Hop Acknowledgement (PHACK) Based" system is used during every data packet communication for detecting SFAs and recovering failed route. The acknowledgement (ACK) data packets are produced by sink node and sensor nodes along with forwarding path for each received packet to make sure that usual data packet communication is in progress. To increase the resilience against the attacks all the ACK data packets are been reverted to the source sensor node along with various data forwarding routes. It is used for preventing the attackers from the attacker nodes in the arrival of the direction-finding route. If not is been interrupted by the ACK packets of those sensor nodes. The "PHACK" is used for detecting the anomalous data packet damage besides detects suspicious sensor nodes equally with improved flexibility in contrast to attacks.

Limitations.

- Hard to detect powerful attacker nodes which will have a probability to combine with former attackers for promoting new attacks.
- The time taken to send and receive the acknowledgment along with the different routing paths leads to overhead.

3.11 Zero-Sum Game Approach

The authors [16] have framed the attack-defense game. Two players evolved in the attack model, they are the intruder and the detection system. In the intrusion detection system, each node maintains the table containing the information such as the packet drop rate, the alternate paths selection, and the security levels. The payoff function of the node is used to identify whether the node is compromised or node. While transmitting the messages from the source sensor node to the destination sensor node at that time payoff function gets executed and calculated. All such compromised nodes are been identified and segregated from the network and it is done through the cluster-head or sink node which is used for monitoring all the nodes at that state. The malicious node is removed and isolated from its cluster and further, there will no transmission amongst isolated sensor nodes with the remaining sensor nodes located in the network.

Limitations.

- Undergoes packet dropping.
- There is no proper congestion control mechanism.

4 Conclusion

This paper portrays the different schemes about the discovery and prevention of SFAs. Security is considered to be the most important basis for any sensor systems as a result of its constrained capacities and giving protection is a troublesome assignment. All current schemes have impediments in this manner an extremely attentive; viability is required for handling selective forwarding attacks.

References

1. Xin-sheng W, Yong-zhao Z, Shu-ming X, Liangmin W (2009) Lightweight defense scheme against selective forwarding attacks in wireless sensor networks, pp 226–232
2. Xiao B,Yu B, Gao C (2007) CHEMAS: Identify suspect nodes in selective forwarding attacks. J Parallel Distrib Comput 67(11):1218–1230
3. Yu B, Xiao B (2007) Detecting selective forwarding attacks in wireless sensor networks. In: Parallel and distributed processing symposium
4. Karlof C, Wagner D (2003) Secure routing in wireless sensor networks: attacks and countermeasures. Elsevier Ad hoc Netw J 1(2):293–315
5. Hai T, Huh E-N (2008) Detecting selective forwarding attacks in wireless sensor networks using two-hops neighbor knowledge. In: Seventh IEEE international symposium on network computing and applications, pp 325–331
6. Kim YK, Lee H, Cho K, Lee DH (2008) CADE: cumulative acknowledgement based detection of selective forwarding attacks in wireless sensor networks. In: Third international conference on convergence and hybrid information technology, pp 416–422
7. Kaplantzis S, Shilton A, Mani N, Sekercioglu YA (2007) Detecting selective forwarding attacks in wireless sensor networks using support vector machines. In: 3rd International conference on intelligent sensors, sensor networks and information, pp 335–340. IEEE
8. Kamari S, Jamshidi M (2016) A secure algorithm against selective forwarding attack in wireless sensor networks. Int J Comput Technol 3(8). ISSN: 2348–6090
9. Deng H, Sun X, Wang B, Cao Y (2009) Selective forwarding attack detection using watermark in wireless sensor networks. In: International colloquium on computing, communications control, and management, pp 109–113
10. Shila DM, Cheng Y, Anjali T (2010) Mitigating selective forwarding attacks with a channel-aware approach in WMNS. IEEE Trans Wireless Commun 9(5):1661–1675
11. Sun H-M, Chen C-M, Hsiao Y-C (2007) An efficient countermeasure to the selective forwarding attack in wireless sensor networks, pp 1–4
12. Brown J, Du X (2008) Detection of selective forwarding attacks in heterogeneous sensor networks. In: ICC, pages 1583–1587
13. Marti S, Giuli TJ, Lai K, Baker M (2000) Mitigating routing misbehavior in mobile ad hoc networks. In: Proceedings of international conference on mobile computing and networking, Boston, MA
14. Alajmi N, Elleithy K (2015) Multi-layer approach for the detection of selective forwarding attacks
15. Liu A, Dong M, Ota K, Long J (2015) PHACK: an efficient scheme for selective forwarding attack detection in WSNs, 9 Dec (2015)
16. Reddy YB, Srivathsan S (2009) Game theory model for selective forward attacks in wireless sensor networks. In: 17th Mediterranean conference on control and automation makedonia palace, Thessaloniki, Greece, pp 458–463, June 24–26 (2009)
17. Das AK, Sharma P, Chatterjee S, Sing JK (2012) A dynamic password-based user authentication scheme for hierarchical wireless sensor networks. J Netw Comput Appl
18. Bysani LK, Ashok KT (2001) A survey on selective forwarding attack in wirelesssensor networks. In: International conference on devices and communications (ICDeCom)

Automatic Railway Gate Control System Using GPS

B. Subramanian, A. S. Selvakumar, M. Sachithanantham, T. Saikumar, and Anisha Radhakrishnan

Abstract Nowadays, the errors caused due to the manual operation at the railway gate level crossing have been increased due to various reasons like the receiving of data about the exact location of the train through various means. In these situations, it would be a boon for the people crossing the railway gate if there are an automatic opening and closing system of the railway gate for an effective means of automation of the railway gate. This paper proposes to introduce an active gate opening system that uses live GPS data collected from GPS sensors on the train. This GPS location data is sent from the GSM transmitter to the receiver at the railway gate. The data is compared with the location of the current railway gate and if both matches with each other, the gate is closed automatically at the receiver end, and by the same means it is opened again once the train crosses the gate.

Keywords GPS sensor · GPS receiver · GSM transmitter

1 Introduction

Our focus on automation of Railway Gate Control is mainly on reducing errors of mankind, avoiding accidents by adding additional alarm techniques to alert the public and we can also have a reduction in time for opening and closing the gate. Normally, the manual gate also called as Boom Barriers, the closing operation is done with the mechanism of hydraulic motors or electromechanical system where the gate is opened and closed manually with the information received by nearest railway station or level crossing to the gatekeeper, where he manually closes and opens the gate and he also locks the gate to confirm the safety of gate from being opened. The automation

B. Subramanian (✉) · A. S. Selvakumar · M. Sachithanantham · T. Saikumar · A. Radhakrishnan
Department of Computer Science and Engineering, Amrita School of Engineering, Amrita Vishwa Vidyapeetham, Coimbatore, India
e-mail: santhoshsubramani473@gmail.com

A. Radhakrishnan
e-mail: r_anisha@cb.amrita.edu

G. Ranganathan et al. (eds.), *Inventive Communication and Computational Technologies*, Lecture Notes in Networks and Systems 145,
https://doi.org/10.1007/978-981-15-7345-3_37

is done with the help of GPS sensor [1, 2], GSM, stepper motor integrated with IC L293d driver and Arduino UNO [3, 4]. Railway gate crossing at the receiver end is automated and controlled from GPS signals that are received from train to receiver side Arduino [5, 6]. Gates are automatically operated once the train enters and leaves the range of 2 km (radius before and after the corresponding railway gate). The location is received by the server (via GSM to Arduino) [7] at frequent intervals and once the location is within the range, the server (Arduino) gives a signal to the driver—IC L293d which operates the stepper motor and operates the gate [8].

2 Existing Method

Wankhede SA [9], uses a module, which includes a webcam, alert mechanism, and a gate driving, webcams are placed on the side of the track (at a height of 5 m from the ground) about 2 km on either side of the level crossing. As the train passes through the webcam (in), it takes a sequence of images and compares it with those stored in the database to operate the railway.

Advantages It has superior image-based detection and control; it can also give a quick response, database support. Human attention and intervention-free are the merits of the system.

Disadvantages Webcam can be damaged as it's exposed out and external factors may affect the image being taken by the webcam like dust, rain, birds.

Aswani Y [10], Raju B [11] use a controlled system set components that use microcontroller GSM modem, GPS module, lighting system, alarm. Location is received using the PIC8067 that is near the railway crossing. Once the train is located at a distance of 2 km away from the level crossing, it gives a signal to the pedestrians by giving Alarms and signals, though the system provides more security than other system and can also be accessed from the remote places, practically it's inconvenient because PIC uses RISC architecture that is difficult in its implementation.

Al-Zuhairi [12] describes the automation of railway gate using IR sensors. A microcontroller can be compared to a small stand-alone computer which is a very powerful device and is capable of executing a series of pre-programmed tasks and interacting with other hardware devices. The IR sensors are placed about 2 km on either side of the level crossing. Once the train has obstructed the path of the light of sensors (in), the sensor sends a signal to the PIC which in turn operates the motor that is placed in the nearest level crossing. Then the railway gate closes with the help of DC motor that is operated by PIC.

Advantages This avoids railway accidents in remote areas, reduces man-power, elimination of human error, the time delay due to manual work.

Disadvantages It can't identify whether the object that obstructed is a train or not. DC motors get worn out easily. Programming in PIC is very long because of RISC architecture.

3 Proposed Method

In this system we use GPS sensor to find the location of the train, GPS coordinates are transmitted to the railway gate side (receiver side) with the help of GSM and Arduino UNO acts as an interface between Stepper motor and the data received from GSM, once the train enters or leaves the 2 km (the radius for closing and opening the gate) Stepper motor opens or close based on the instructions received from Arduino. Components used in our proposed method are Arduino UNO boards, GSM receiver, GSM transmitter, GPS sensor, Stepper motor, Power supply, IC Driver (L293d), A Laptop, connecting wires and Arduino IDE is the software used to run the code.

4 Architectural Design

(a) Mounted with Simcard and antenna connected

Once the GPS sensors start generating the co-ordinates, GSM can start transmitting the GPS-DATA to the receiver side with the help of Arduino UNO. GSM is mounted with a sim card, and Arduino is coded in such a way that GSM is changed to text mode (Figs. 1 and 2).

(a) Pair wires
(b) Center taps
(c) Mounted with Simcard and antenna connected.

The Arduino UNO at the receiver end retrieves the data from the GSM (sender). It compares the data present in the Arduino code (default GPS coordinates) with the coordinates of the railway gate stopping in which the Arduino is mounted. Once the co-ordinates of the gate match with the coordinates in the Arduino code then the gate is automatically closed (or open for leaving) by the operation of Stepper motor.

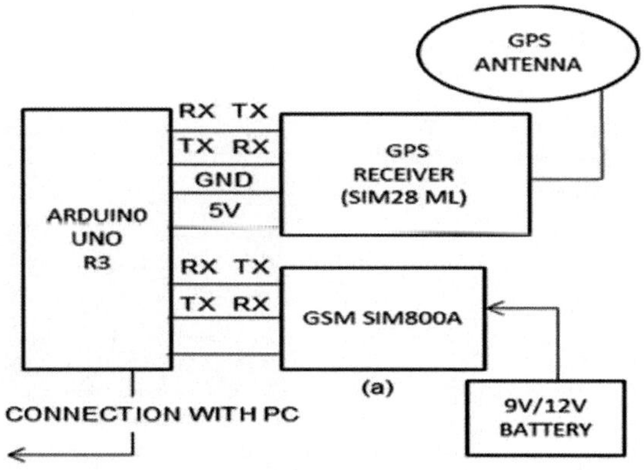

Fig. 1 Architectural diagram at sender side

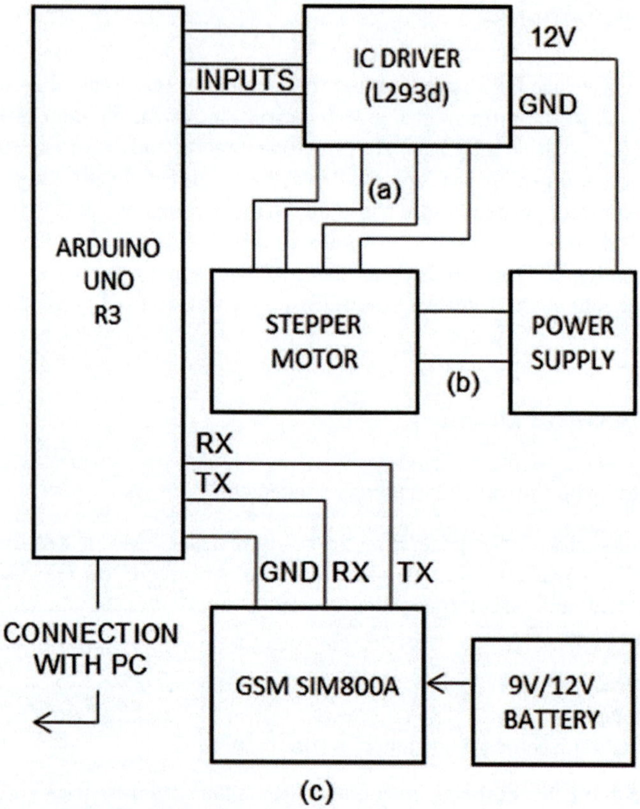

Fig. 2 Architectural diagram at receiver side

Figure 3 is the connection setup of Arduino UNO with GPS. This module needs four pins: ground, power (3.3 V or 5 V), TX, and RX. So it communicates using a serial port. TX (Transmitter) and RX (Receiver) is the pin configuration used in both modules to send and receive data's where GPS module is connected with an antenna

Fig. 3 Sender side (Arduino UNO connection with GPS sensor)

Fig. 4 Receiver side (connection of stepper motor with Arduino UNO and IC driver)

to receive co-ordinates at a particular area the coordinates are captured by Arduino UNO from TX (of GPS) to RX (of Arduino) (Fig. 4).

This module needs 6 pins for IC L293d driver: ground, 4 Input pins connected with Arduino, power (5 V or 3.3 V). 6 pins for Stepper motor: Center taps (2 wires) are connected with power supply, Pair wires (common wires of center taps) are connected to IC L293d driver's every motor. For further details about the set-ups of the sender and receiver refer Architectural design.

(Note: Use GSM in both sender and receiver side for sending and receiving messages, GSM needs 3 pins for both sender and receiver: ground, TX and RX).

5 Implementation

5.1 Algorithm for Sender Side

- Set the GSM module in text mode.
- Read the GPS data and separate the data that starts with "$GPRMC" in string GPS.
- Segregate the latitude and longitude from the GPS data (co-ordinates) from all other strings obtained (12 strings).
- When the input is passed, the current segregated GPS data is transferred to the receiver side (server in the railway gate).

5.2 Algorithm for Receiver Side

- Default GPS co-ordinates (for entering and leaving the station) is given to the Arduino code.
- For each data received, server cross-checks with the default co-ordinates stored.

- Stepper motor is integrated with IC L293d for motor operation.
- If the co-ordinates match with each other, then Arduino instructs for the Stepper motor to open and close the gate.

GPS data is generated spontaneously but it's not sent to the receiver side continuously, only if the user gives the corresponding input to the server (sender side), then the GPS data at the current time is segregated and sent to the receiver to operate the railway gate.

6 Result

GPS data that are directly displayed (without any segregation done) in the Arduino UNO monitor, where one can take any separate set of data from it (Fig. 5; Table 1).

The GPS data is modified to make it just display Latitude and Longitude in serial monitor with the help of the Arduino code (Fig. 6).

GPS data that are captured by Arduino UNO is sent from the sender to another mobile as a segregated GPS (only latitude and longitude) for testing, with the help of the GSM module (Fig. 7).

IC L293d driver is an integrated module for the working of the stepper motor. It is a 16-pin Dual-in-line Package, where 4 pins get input from Arduino UNO and other 4 pins (as pairs) from the stepper motor for motor rotation. The result that is displayed in the serial monitor of the receiver side where it shows the working of stepper motor based on the particular GPS data received from the sender side (Fig. 8).


```
$GPRMC,062758.000, A, 1059.6298,N,07657.1907,E,0.54,266.53,240317,,,A*62
$GPVTG,266.53,T,,M,0.54,N,1.01,K,A*38
$GPGGA,062759.000,1059.6297,N,07657.1905,E,1,6,1.62,431.3,M,-93.1,,*7A
$GPGLL,1059.6297,N,07657.1905,E,062759.000,A,A*5F
$GPGSA,A,3,16,08,11,23,22,26,,,,,,,1.85,1.62,0.89*00
$GPGSV,4,1,13,03,67,260,,22,62,200,12,16,51,034,24,23,38,346,15*70
$GPGSV,4,2,13,27,33,119,,08,28,161,21,26,22,034,16,07,19,281,*76
```


Fig. 5 GPS data received from the GPS sensor

Table 1 Segregated GPS—consecutive latitude and longitude

Latitude	1059.632
Longitude	7657.193
Latitude	1059.632
Longitude	7657.193
Latitude	1059.632
Longitude	7657.193

Fig. 6 Segregated GPS—testing segregated GPS latitude and longitude values from sender to mobile phone

```
Latitude =
1054.4147Longitude =
07653.8654

Latitude =
1054.4100Longitude =
07653.8704
```

Fig. 7 IC L293d driver

Fig. 8 Stepper motor working in the output screen

```
GSM RECEIVING MESSAGE

Latitude=1053 , Longitude=07653
Latitude=1053
Longitude=7653

*****STEPPER REVOLUTION****
Counter Clockwise
```

Serial monitor displaying the result of working of the stepper motor based on the instruction given by Arduino UNO when it confirms the matching of its default GPS co-ordinates with the received GPS coordinates (Fig. 9).

Fig. 9 The output of message sending from the server (sender side)

```
AT+CMGF=1
OK
AT+CMGS= "+919994308341"
>&1054.3507#07653.8
+CMGS: 34

OK
```

Latitude and longitude are sent to receiver from sender GSM with '$' for latitude and '#' for longitude, so that receiver server will find it easy to identify the latitude and longitude separately.

7 Conclusion

In this paper, we segregated the GPS data with the help of Arduino UNO code and using GSM module we transferred the latitude and longitude from sender side (from the train) to receiver side (railway gate server) and thereby operating the Stepper motor to open and close the gate when the train enters and leaves the radius of 2 km by declaring default GPS co-ordinates of the radius to receiver side Arduino UNO. This paper provides railway gate operation by using GSM to GSM message transfer.

8 Limitations and Future Work

Our paper is restricted only to the city and to a limited number of trains. Implementing this on a larger scale would require a very strong database that can store a lot of values from many trains. This application can be further developed by replacing the GSM to the GSM message transfer system with a database system, where the GPS data is pushed into the database using SQL from the sender side and at the receiver side the data is retrieved from the database for matching with the current GPS coordinates.

References

1. Costanzo A (2013) An Arduino based system provided with GPS/GPRS shield for real time monitoring of traffic flows. In: 2013 7th international conference on application of information and communication technologies (AICT). IEEE
2. Pham HD, Drieberg M, Nguyen CC (2013) Development of vehicle tracking system using GPS and GSM modem. In: 2013 IEEE Conference on Open Systems (ICOS). IEEE
3. Sharad S, Sivakumar PB, Anantha Narayanan V (2015) A novel IoT-based energy management system for large scale data centers. In: Proceedings of the 2015 ACM sixth international conference on future energy systems. ACM
4. Megalingam RK et al (2014) Smart, public buses information system. In: 2014 international conference on communications and signal processing (ICCSP). IEEE
5. Barrett SF (2013) Arduino microcontroller processing for everyone! Synth Lect Dig Circuits Syst 8(4):1–513
6. www.arduino.cc
7. Badamasi YA (2014) The working principle of an Arduino. In: 2014 11th international conference on electronics, computer and computation (ICECCO). IEEE
8. Minns PD (2013) C Programming for the PC the MAC and the Arduino microcontroller system. Author House

9. Wankhede SA, Matsagar MB (2016) Advanced railway gate control system using webcam and MATLAB. Int J Comput Appl (0975–8887). National Conference on Digital Image And Signal Processing
10. Aswani Y, Sudhakar Rao P, Rajendra Prasad VVGS (2014) Automated alarm based railway gate crossing based on GPS and GSM. Int J Prof Eng Stud 2(1)
11. Raju B, Sreenivas B (2013) Alarm system of railway gate crossing based on GPS and GSM. Int J Sci Eng Res (IJSER) 1(1). ISSN (Online): 2347-3878. Young M (1989) The technical writer's handbook. University Science, Mill Valley, CA (1989)
12. Al-Zuhairi ASM (2013) Automatic railway gate and crossing control based sensors & microcontroller. Int J Comput Trends Technol (IJCTT) 4(7)

Secure Decentralized Public Key Infrastructure with Multi-signature in Blockchains

M. J. Jeyasheela Rakkini and K. Geetha

Abstract Decentralized public key infrastructure (DKPI) with permissioned/permissionless blockchains has gained tremendous popularity due to the inherent transparent, decentralized, and tamper-proof nature of blockchains. Digital certificates with transport layer security/secure socket layer (TLS/SSL) which are issued by a certificate authority (CA) and which associates the public key to an individual/organization is highly centralized and is held by a small group of multinational organizations. The centralized nature of TLS/SSL certificate generation can be decentralized with blockchain technology. We propose a decentralized public key infrastructure (DPKI), with a set of nodes in the blockchain which act as certificate authorities and are used to sign and validate the public key of the user requesting the certificate, thus facilitating the issue of transparent and tamper-proof TLS/SSL certificates. In this paper, we have implemented a group signature with m nodes out of n nodes to generate a digital certificate that ensures decentralization in the blockchain.

Keywords Decentralized public key infrastructure (DPKI) · Proof of authority (PoA) · Proof of elapsed time (PoET)

1 Introduction

Blockchain which is a decentralized, distributed, and tamper-proof ledger with an append-only log of records can be used for decentralized public key infrastructure (DPKI) with multiple participants and with no trusted third party. The blocks in the blockchain have a hash of all the transactions and are added to the blockchain by

M. J. Jeyasheela Rakkini (✉) · K. Geetha
SASTRA Deemed to be University, Thanjavur, India
e-mail: jeyasheelarakkini@cse.sastra.edu

K. Geetha
e-mail: geetha@cse.sastra.edu

G. Ranganathan et al. (eds.), *Inventive Communication and Computational Technologies*, Lecture Notes in Networks and Systems 145,
https://doi.org/10.1007/978-981-15-7345-3_38

the miners on a consensus method. The two types of blockchains which are permissioned blockchains and permissionless blockchains can be used for implementing decentralized public key infrastructure (DKPI). Smart contracts in Ethereum, which is a permissionless blockchain that has many complex data structures can be used for the creation of digital certificates. Smart contracts validate transactions between the sender and the receiver and can monitor their proper streamlined behavior with event triggers and functions. Transparency and tamper-proof nature are the quintessential security features that are the driving force behind the implementation of a decentralized PKI. The public key identity management is also done with a smart contract in blockchains.

The nodes in the blockchain verify the public key of a particular user/organization with multi-signatures. The centralized public key infrastructure for authentication and distribution of public key for secure data transfer is susceptible and vulnerable to many attacks such as denial of service/distributed denial of services (DOS/DDOS), spoofing attacks, and compromise of certificate issuing servers. In a traditional public key infrastructure, the Web application which runs on a server is requested by the client. The root certificate authority, which governs all certificate authorities (CA), issues certificates, which validates a public key against an authenticated user, maintains revocation of certificates, and logs all the activities with the transparency of the certificate. The architecture of the proposed system is shown in Fig. 1. There is a lack of transparency between the certifying authority and the verifying authority in centralized public key infrastructure. There is also a lack of transparency between the certifying authority and governing authority. We have implemented the blockchain in Python, which accelerates blockchain development for digital certificate generation

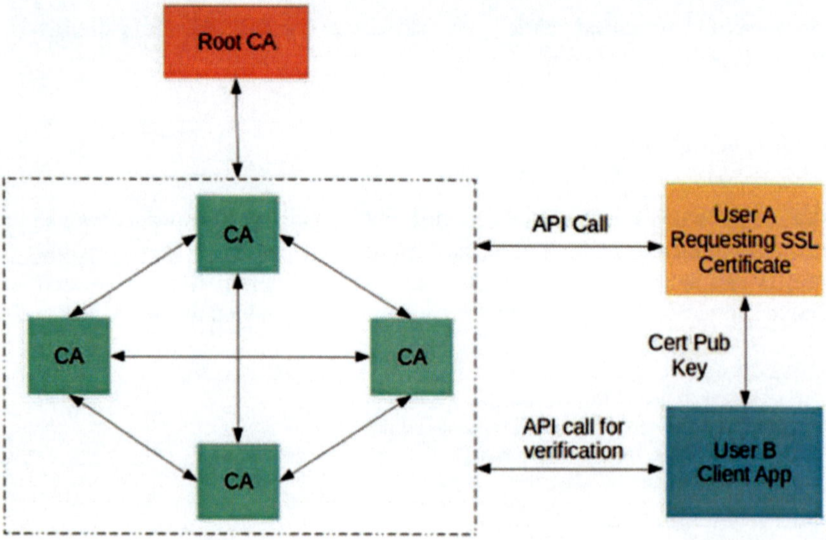

Fig. 1 Decentralized PKI

in a decentralized way and the verification is done by elliptic curve cryptography with an elliptic curve digital signature algorithm (ECDSA) for the validation and message digest of the certificate is done by the hash functions. ECDSA offers the same security with lesser key size than RSA. So, we have preferred ECDSA. There are n peer nodes in our proposed blockchain and m out of n nodes have to sign to trigger a transaction. Here, a single signature to work on behalf of m peer nodes to validate a message and reach a consensus in the blockchain. The digital signature promises integrity, authentication, and non-repudiation of any message. The total of 'n' key pairs are generated, where m out of n public keys are required to sign to generate and authorize a transaction. The smart contracts are written in Python and have the wrapper API for blockchain.

2 Related Work

Sukhodolskiy et al. [1] proposed a set of secure, private keys generated by cryptographic protocols, which facilitates privacy preserving traits and role based access control of datasets in the untrusted cloud environments. Wenbo Jiang et al. [2] exploits k anonymous clients with their privacy preserving characteristics facilitate a blockhain based decentralized public key infrastructure. Yin W et al. [3] exploits the random space of lattice and generates the master public/private key pair from which a set of public/private keys are generated for ensuring the authenticity of transactions in blockchain. A secure escrow address is generated jointly by peers [4] in the bitcoin network who want to transact, who generate an ephemeral key, and mixing of the output address is done by mixing operations which assure unlinkability, anonymity, and protection from DOS and Sybil attacks. The security challenges of domain name systems (DNS) and certificate authority namely identity management, name resolution, and decentralized storage [5] concerning blockchain-based solutions such as Ethereum name services, Blockstack, Namecoin, EmsDNS, and DNSchain. Singla et al. [6] compared Emercoin, Ethereum with remote node, Ethereum with light sync mode for PKI for IoT concerning features such as time taken, the storage required, the trust required, and cost. Friebe et al. [7] discussed decentralized ID, with four smart contracts, namely registry contract, root identity contract, shared identity contract and attribute contract with identity locator file, attribute locator file and attribute data stored in the distributed hash table. Yeow et al. [8] discussed the state-of-the-art decentralized consensus systems for IoT which is more edge centric, the data structures in blockchain, scalable ledgers both permissioned and non- permissioned, tokenized and non-tokenized transaction models. Isirova et al. [9] asserted that security in a decentralized PKI exceeds the PKI in centralized systems. The Crypviser app acts as a block in a blockchain and this generates ID and private key for each user. The public key is generated from this private key. The digital signature that is generated with the user's private key is verified based on the user's public key. After this initial phase, the value of the hash of the public key and ID of the user is matched with the hash of the public key and status; if equal then verification is found to be successful. Here, the status refers to the certificate status register in the transaction block. The sender's certificate can be traced back to its publication. Existing users (nodes) can check the new user's certificate register.

Won et al. [10] asserted the fact that IoT device owners can generate the name, value pair by themselves by sending coins to an address, and a private key is generated according to the address. The public key, the private key generated by the device manager, and the NVS nodes form a blockchain. The NVS blockchain resembles Emercoin and Namecoins. Anada et al. [11] discussed a decentralized, multi-authority, and anonymous authentication has five algorithms, the first algorithm generates public parameter values based on an input security parameter, the second algorithm generates public parameter values and authority index a, the public key is generated. The third algorithm generates the private secret key of a prover. The fourth and fifth algorithms are the prover and verification algorithms based on which user is authenticated and the verification of them. Coin join transactions are tracked and the money laundering transactions with coin join are detected [12].

Yao et al. [13] discussed homomorphic hash and blockchain-based authenticated key exchange protocol, where blockchain transactions of multiple unfamiliar parties authenticate common secret holders and publish the hash value of the common secret along with attributes related to the secret. Key chain [14] binds ID to public key and stores as a table in each node where the byzantine fault is resolved by delegated proof of stake. Li et al. [15] concealed the metadata that has information about the location of data in decentralized blockchain-based storage. Aitzhan et al. [16] require multiple independent parties to sign a transaction so that it becomes valid and the tokens held by them are spent for energy transactions. Suma et al. [17] discussed the security concerns and issues in blockchain as it throws a plethora of issues owing to its decentralized and immutable nature.

3 Proposed Work

3.1 *Phase 1*

3.1.1 Domain Requesting New Certificate or Renewal

The certificate authority network has a root node/governing body which sets the compliances and creates the genesis block. The root node is generated and is seen in Fig. 1. The first node is added by the root CA. New nodes are added to the network after being approved from root CA and a minimum of n nodes out of m nodes must have signed in the blockchain. If a domain owner requests a new digital certificate or if it is for renewal of an old digital certificate, the domain control identification record and domain information are sent to the CA network along with the set of tokens. After validation of the domain control identification record with the validate address API of the proposed system, the digital certificate is issued (again in the case of renewal) along with multi-signatures from CA nodes that are n nodes out of m nodes should sign for the approval. The CA network creates a hash of the public key and the hash is available for verification using API available in the blockchain.

$$\text{Multi-signature} = \text{Aggregate}\ (s_1,\ s_2,\ s_3, \ldots,\ s_m).$$

where $s_1, s_2, s_{3,} \ldots, s_m$ are the signature of m nodes, respectively.

Hash = SHA256 (public key of the domain owner)

The function GetDNSRecord in the smart contract is used to generate a unique key pair with ECDSA which is related to a particular domain address and this function writes this unique key, zone signing key (ZSK) found in the DNS security version of DNS, namely DNSSEC for the domain name in the blockchain. The X.509 format is used when we write a DNS record in the blockchain. The unique key pair (public key, private key) is given to the domain owner.

Algorithm 1: GetDNSRecord (domain Name)

Function GetDNSRecord (domain Name)

1. Generate a unique key for the specified domain Name
2. store the generated unique key pair in blockchain
3. return the generated unique key pair to the user
4. end

The function IssueNewCertificate is used to check the public zone signing key (ZSK) which is published in the DNSSEC record of the domain owner with the zone signing key written in blockchain when it was first issued. If the validation is successful, then the certificate is issued either for the first time or renewal after one year. In Fig. 4, a new certificate is generated and the details are shown in it.

Algorithm 2: Function IssueNewCertificate (dnsRecordKey, DomainName)

Function IssueNewCertificate(DNSRecordKey, DomainName)

1 if dnsRecordKey matches with the key in the blockchain
  then generate a new certificate (X.509 standard)
2 propagate the new certificate to the CA network
3 return True
4 else
5 return False
6 end

The function VerifyCertificate is used for verification of the existing certificate by taking the certificate hash written in the blockchain and certificate hash of the domain owner. The validated certificate is the original certificate else it is a spoofed certificate. The tamper-proof nature of the blockchain facilitates these functions.

Algorithm 3: VerifyCertificate (certHash)

```
Function verify Certificate (certHash)
  if certHash of the domain owner matches with
  certificate public  key hash written in the blockchain
  then return True
  else
  return False
  end
```

3.1.2 Verification and Revocation of Issued Certificates

Decentralized public key infrastructure (DPKI) verification and revocation system on public blockchain networks are done with smart contracts. With this approach, we can have the certifying authorities to time stamp the issued certificates on a public blockchain and a smart contract can be used to verify the authenticity of the certificates. Also, the smart contract can auto-renew certificates after verifying the authenticity of the server owner. A proof of authority (PoA) consensus algorithm is used for the mining and validation of transactions in the proposed approach. In PoA, the validators of the block go for staking their reputation as well as their money. In each round, a validator in the list of authenticated validator nodes becomes the leader and generates the block and sends it to other validating nodes. The validating nodes accept or reject the transaction and if the leader continues to send an invalid block, then the reputation of the leader decreases and eventually it gets excluded from the validator's list. The smart contract verifies the authenticity of the issued certificates, revokes issued certificates, and automatically renews certificates if the registered period expires. The data stored on public blockchain using the smart contract is privacy protected. Instead of storing the whole certificate, we just store the SHA256 hashed form of the certificate public key. Once a server issues the certificate to the requesting client, the client application will create a certificate hash and ping the smart contract for verification.

3.2 *Phase 2*

3.2.1 Decentralized Control of the Certifying Authorities (CA) and the Governing Body (Root CA) Using Permissionless Blockchain

The governing body or root CA implements a permissionless blockchain to maintain the integrity of the certifying process. Only CAs thoroughly audited by the governing body is allowed to be part of the permissionless blockchain network. The governing body regulates the whole network as in the traditional approach but with transparency because of the blockchain. Also, the CAs can approve and validate certificates issued by peers. A peer to peer approach in issuing certificates is proposed with some permissions in place to regulate the process and maintain balance in the network.

4 Implementation

The smart contract can be written in Python and deployed on the blockchain. The smart contract has three functions.

Get_dns_record
    Inputs: Domain name
    Output: TXT Record value for domain ownership
Check certificate
    Inputs: Certificate hash
    Output: Boolean (True | False)
Revoke certificate (Called by the CA or Root CA)
    Inputs: Certificate Hash
    Output: Boolean (True|False)
Renew certificate (Called by the server owner)
    Inputs: Certificate Hash, Server Control Key
    Output: New Certificate Hash

This blockchain network runs on proof of elapsed time (PoET) consensus algorithm. The root CA/governing body is the root node in the permissioned blockchain network. The root node designs the initial governing rules for the network and the rules change over time after taking feedback from other network participants. The certificate standards are designed by the root node. The first new node is added to the network only after a thorough audit by the root CA. Next new nodes are added to the network upon audit by the root CA/Node and then confirmation by other nodes in the network. When a request is made to issue a new certificate to the CA network, the PoET algorithm will take care of consensus and assign an issuing task to the required node in the network. Once the node authenticates the requesting body/server the certificate will be issued the requestor upon confirmation from the other CA nodes. Once confirmed by the CA nodes the certificate will be time-stamped on a public blockchain network and the public key hash will be sent to the requester. To test this implementation, a multichain blockchain on five nodes is created with Python and with postman API. The notion of multichain since we use multi-signatures in our system. The transaction data has the certificate data in the specified standard or X.509 along with the requestor information and the time stamp. The multichain is created and implemented in Python and it is visualized with postman API. The first node which is the genesis node is created in the blockchain.

5 Results and Discussion

The maximum block size which has the list of all new transactions, the DNS transactions in our context, the target time for generation of the block that is to validate all transactions in a block and mine it, the size of the chunk which can be used in off chains (the chunk size when a streamed offline data has been opted) are given below. The maximum size of the block in the blockchain is 8 MB and the target time for the generation of a block is 15 s. The chunk stream size for offline streams is 1,048,576 KB (Table 1).

In Fig. 2, the time taken to generate a new block is 1000 ms which is much smaller than a proof of work. The key size used is 1024 bits and we use secp521r1. The notion of multichain is used in the diagram since it uses multi-signatures of the blockchain nodes.

In Fig. 3, the creation of the root node which has the genesis block is done in blockchain and is viewed with postman API. The smart contract written in Python is accessed as an API by the client for certificate generation or renewal. The block which has the certificate is added to the blockchain by the mining node with PoET consensus, the very first node whose timer has expired can propose the block, provided that node has a high reputation and good stake. The ECDSA signature algorithm is used for signing the transactions by the originator of the transactions, by the miner and can be used for source authentication and non-repudiation.

In Fig. 4, the X.509 certificate is generated and is written in the blockchain in JSON format. The X.509 certificate has the time stamp of when certificate issued or renewed, the public key of the domain name owner, and the multi-signatures of m nodes.

Table 1 Parameters in the proposed system

Maximum block size	8 MB
The target time for the generation of a block	15 s
Chunk size	1,048,576 KB

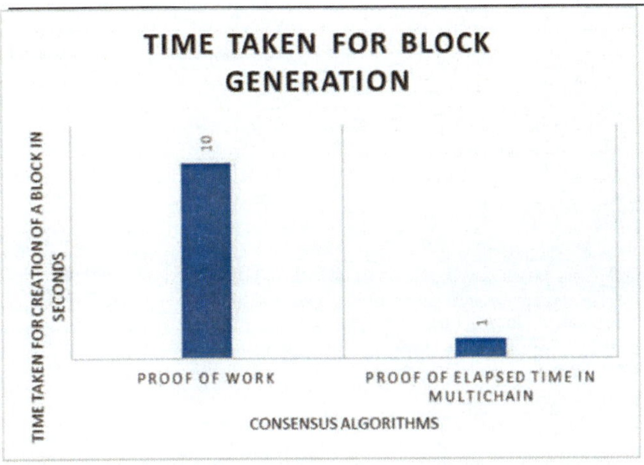

Fig. 2 Time taken for generating a block in blockchain with proof of elapsed time

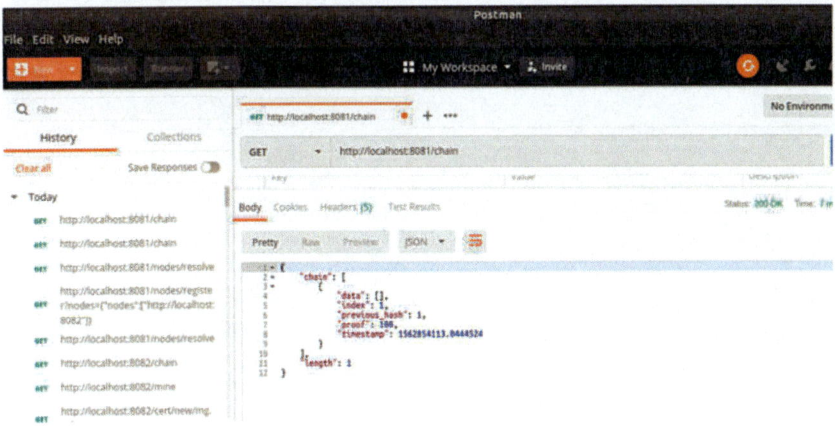

Fig. 3 Creation of the first root node

In Fig. 5, any new node can be to the network of nodes which will also take part in the mining process. The m nodes out of total n nodes are chosen randomly and every node has the current consistent copy of the blockchain with the log of certificates generated, issued, renewed, and revoked. Thus, the creation of new nodes, generation of certificates to the requesting clients is shown in the above figures. The proof of concept is shown with the simulation of nodes added to the network and the certificate is generated when the client requests it.

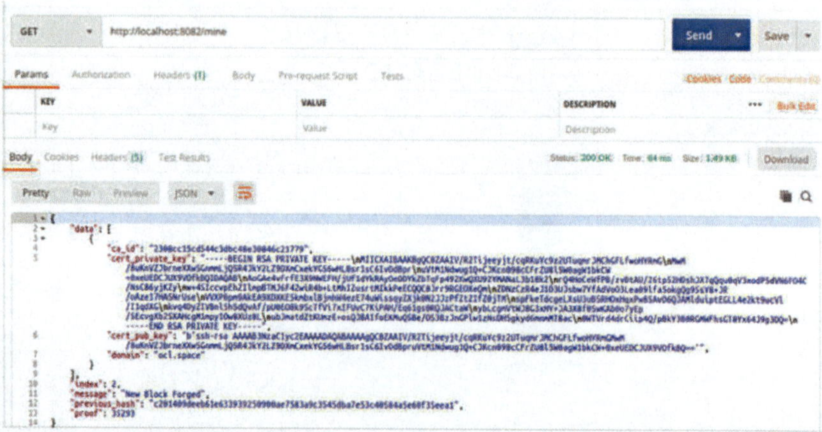

Fig. 4 Certificate generation for the client

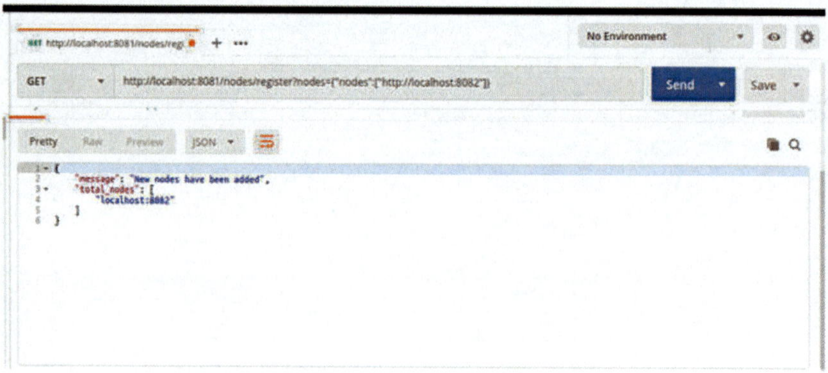

Fig. 5 New node added to the multichain

References

1. Sukhodolskiy I, Zapechnikov S (2018) A blockchain-based access control system for cloud storage. In: 2018 IEEE conference of russian young researchers in electrical and electronic engineering (EIConRus). IEEE
2. Jiang W, Li H, Xu G, Wen M, Dong G, Lin X (2018) A privacy-preserving thin-client scheme in blockchain-based PKI. In: 2018 IEEE global communications conference (GLOBECOM). IEEE, pp 1–6
3. Yin W, Wen Q, Li W, Zhang H, Jin Z (2018) An anti-quantum transaction authentication approach in blockchain. IEEE Access 6:5393–5401
4. Wang Q, Li X, Yu Y (2017) Anonymity for bitcoin from secure escrow address. IEEE Access 27(6):12336–12341
5. Karaarslan Enis, Adiguzel Eylul (2018) Blockchain based DNS and PKI solutions. IEEE Commun Stand Mag 2(3):52–57

6. Singla A, Elisa B (2018) Blockchain-based PKI solutions for IoT. In: 2018 IEEE 4th international conference on collaboration and internet computing (CIC). IEEE
7. Friebe S, Sobik I, Zitterbart M (2018) Decentid: Decentralized and privacy-preserving identity storage system using smart contracts. In: 2018 17th IEEE international conference on trust, security and privacy in computing and communications/12th IEEE international conference on big data science and engineering (TrustCom/BigDataSE). IEEE
8. Yeow K et al (2017) Decentralized consensus for edge-centric internet of things: a review, taxonomy, and research issues. IEEE Access 6:1513–1524
9. Isirova K, Oleksandr P (2018) Decentralized public key infrastructure development principles. In: 2018 IEEE 9th international conference on dependable systems, services and technologies (DESSERT). IEEE
10. Won J et al (2018) Decentralized public key infrastructure for internet-of-things. In: MILCOM 2018-2018 IEEE military communications conference (MILCOM). IEEE
11. Anada H (2018) Detailed instantiation of the decentralized multi-authority anonymous authentication scheme and tighter reduction for security. In: 2018 Sixth international symposium on computing and networking (CANDAR). IEEE
12. Maksutov AA et al (2019) Detection of blockchain transactions used in blockchain mixer of coin join type. In: 2019 IEEE conference of russian young researchers in electrical and electronic engineering (EIConRus). IEEE
13. Yao H et al (2018) Homomorphic hash and blockchain based authentication key exchange protocol for strangers. In: 2018 Sixth international conference on advanced cloud and big data (CBD). IEEE
14. Hu Y et al (2018) KeyChain: blockchain-based key distribution. In: 2018 4th International conference on big data computing and communications (BIGCOM). IEEE
15. Li D, Du R, Fu Y, Au MH (2019) Meta-key: a secure data-sharing protocol under blockchain-based decentralized storage architecture. IEEE Networking Lett 1(1):30–33
16. Aitzhan NZ, Svetinovic D (2016) Security and privacy in decentralized energy trading through multi-signatures, blockchain and anonymous messaging streams. IEEE Trans Dependable Secure Comput 15(5):840–852
17. Suma V (2019) Security and privacy mechanism using blockchain. J Ubiquitous Comput Commun Technol (UCCT) 1(01):45–54

Decentralized Privacy-Preserving Framework for Health Care Record-Keeping Over Hyperledger Fabric

Baddepaka Prasad and S. Ramachandram

Abstract Some of the exciting features of blockchain technology are immutability, distributed ledger technology, peer-to-peer network, and consensus mechanisms. Blockchain applications include IoT, supply chain, finance institutions, and e-governance. Blockchain technology is emerging in the health care applications for maintaining electronic healthcare records to improve the accuracy of diagnosis. Privacy is more important in health care systems, especially for electronic medical records. Here, privacy is defined in two ways, i.e. user privacy and data privacy. For user privacy, the anonymity of the user is very important as it defines that no one can find who the user is and who is the validator. But, user anonymity is not maintained in the existing systems. Hence, the proposed model provides a novel solution for privacy-preserving of electronic medical records. The proposed architecture implements group signatures for providing privacy-preserving to the electronic medical records. Using group signatures also avoids problems like Unforgeability, Anonymity, Traceability, and Unlinkability.

Keywords Security and privacy · Blockchain · Hyperledger · Signatures · Electronic medical records anonymity · Unforgeability · Unlikability

1 Introduction

Electronic medical records are being utilized in the medical industry and research which are termed as e-health. Records can be organized in the form of digital format so that it can be used for improving the efficiency of diagnosis. Blockchain proposed by Satoshi Nakamoto for implementing Bitcoin, the first application of blockchain [1]. Blockchain has public and private models. Public blockchains are

B. Prasad (✉) · S. Ramachandram
Osmania University, Hyderabad, India
e-mail: prasad.baddepaka@gmail.com

S. Ramachandram
e-mail: schandram@gmail.com

G. Ranganathan et al. (eds.), *Inventive Communication and Computational Technologies*, Lecture Notes in Networks and Systems 145,
https://doi.org/10.1007/978-981-15-7345-3_39

useful for cryptocurrencies like Bitcoin, Ether coins, Lite coins, and Peer coins, etc. under the platforms like Bitcoin [2], Ethereum [3], Neo [4], Nem [5] and Cardano [6]. These public blockchains are implemented in permissionless environments and private blockchains are used for smart contracts and decentralized applications. Smart contracts proposed by Szabo in 1996, but due to the lack of resources and computing power, the smart contracts were unable to reach public attention [7]. Public and private models have some list of properties like immutability with SHA-256 hash functions and every block connected with hash functions like a linked list. Some applications implementing in blockchains such as IoT, healthcare, financial institutions, supply chain, ıdentity management, record-keeping, and voting. To implement these applications some environments exist like Hyperledger Fabric [8], Quorum [9], R3 Corda [10], Tender mint [11]. The private blockchain model has some research gaps like Security and Privacy issues, Scalability problems, Interoperability, Smart contracts security, and consensus protocol issues. These research gaps play a crucial role in the private blockchain system. In this scenario, we present the security and privacy issues for private blockchain models, especially specific to healthcare applications. Electronic records are divided into four ways like electronic medical records, biomedical research, and education, remote patient monitoring, drug/pharmaceutical supply chain, [12]. Electronic medical records maintenance is very important for the analysis of perfect treatment for patients and also to find treatment methodology and also can be used by the government for further action.

2 Related Work

Since the last decade, some interesting research is done on e-health medical information with the help of blockchain. Blockchain features such as immutability and ledger technology have been implemented for electronic medical records and are used for sharing management and security and privacy issues.

The Father of blockchain is Satoshi Nakamoto and the first application started with cryptocurrency Bitcoin [1] then the journey of blockchain spread into other areas like IoT, healthcare, and financial institutions, etc. Some work is done in healthcare applications using smart contracts in blockchain. Agbo et al. [12] addressed systematic reviews on healthcare applications based on preferred reporting items for systematic reviews and meta-analysis (PRISMA) guidelines and listed some research gaps in healthcare. Mettler et al. [13] addressed issues in the drug industry. The drugs are more valuable for the patient but it has been traveling from some other country and can be subjected to change. Azaria et al. [14] proposed a new model for sharing electronic records and access controls for medical records and his proposed model discussed confidentiality, accountability, and authentication. In this model, the author maintained the records based on the Ethereum platform. But, the architecture had some problems like 51% attack, also needs huge energy for calculation of nonce and if miners are compromised, then data may be lost. This model used smart contracts to implement healthcare record maintenance—register contract, patient contract, and detailed contract. Jiang et al. [15] proposed architecture for record management and sharing for healthcare information. The model is implemented in emr-chain and PhD-chain but some problems exist when Proof of Work (POW) is used in any platform. POW can provide a consensus for Ethereum and Bitcoin platforms. Maintaining medical records for data sharing should not only perform sharing but maintain privacy, which is a vast area for research in healthcare. Most of the countries are not following any policies regarding healthcare. HIPPA [16] rules are very sophisticated for healthcare management. HIPPA rules are provided with privacy and security. Privacy needs to be provided for every user and privacy can applicable to data privacy and user privacy. Li et al. [17] proposed a novel architecture for privacy preservation of medical data. The author proposed DPS (blockchain data preservation system) for providing privacy on data through the blockchain. The system performs four major operations like submission, manipulation, verification, and query operation. In this model, miners can work for construct block but the miner has taken more human and machine requirements to perform POW. Dubovitskaya et al. [18] proposed a new prototype for providing the privacy and fine grinned access control in healthcare applications, and using this prototype can decrease the turnaround time and improve the solutions for medical care and has been implemented with Stony Brook University Hospital. Gordon et al. [19] proposed interoperability for healthcare data describing, how the data sharing among the different hospitals has to be performed. The author proposed interoperability for different domains of hospital data to exchange securely. He worked on five mechanisms like access rules, aggregation, liquidity, identity, and immutability to solve these issues based on blockchain technology. Yue et al. [20] proposed new architecture based on mobile compatibility and providing access only for authenticated users in Healthcare Data Gateway (HGD) for patients to work on access control and data sharing including security and privacy on medical records. This architecture needs a high-speed internet facility. Griggs et al. [21] proposed a new architecture for a remote patient monitoring system to secure health sharing between the user and professionals. The author developed this architecture for IoT by using smart contracts and Ethereum blockchain models. The Ethereum blockchain model can use POW for consensus protocols. Esposito et al. [22] proposed architecture for a cloud-based electronic medical records system. This architecture implemented for managing the medical records and sharing the records between different places-based clouds and the author discussed security and privacy issues based on HIPPA protocols. Hussein et al. [23] proposed two solutions for security and query optimizing techniques. The author discussed for security based on discrete transforms and query optimizing based on genetic algorithm and also provided some features like immutability, access controls, and high-speed verifying. Zhang, A et al. [24] proposed two models those were private blockchain and consortium blockchain. The author designed this architecture to the individual hospitals to maintain the private blockchain and all private hospital blockchain hashes put into consortium blockchain. Al Omar et al. [25] proposed new architecture that could be privacy-preserving for electronic medical records. The author proposed a medibchain for Pseudonymise and secure e-health record management. Zhang et al. [26] discussed some issues based on security and privacy for blockchain technology.

The author addressed some issues like anonymity, unlinkability, and said that these issues can be solved with the help of group signatures. Chaum et al. [27] proposed group signatures in 1991. The author said that by using group signatures can be used to resolve anonymity and unlinkability with standard assumptions.

In this paper, the idea of utilizing private blockchain-hyper ledger fabric for health-care record-keeping systems for electronic medical records in hospitals is introduced. The remaining of the paper is described as follows. First, in Sect. 3 background of Hyperledger Fabric is discussed. Section 4 explains this paper's contribution including the proposed framework. Section 5 summarizes—Main contributions and properties of the proposed framework. Section 6—Performance analysis. Finally, Sect. 7, gives the conclusion of the paper along with possible future work.

3 Hyperledger Fabric Workflow

Hyperledger fabric is a very good model to execute smart contracts. This model is developed for private blockchain systems. According to the model, fabric users can register to get public and private keys for smart contract execution with the help of certificates and signature concepts. Certificates received from a certified authority (CA) consist of some attributes like version number, serial number, validity date, the issuer name, and public key information (Fig. 1).

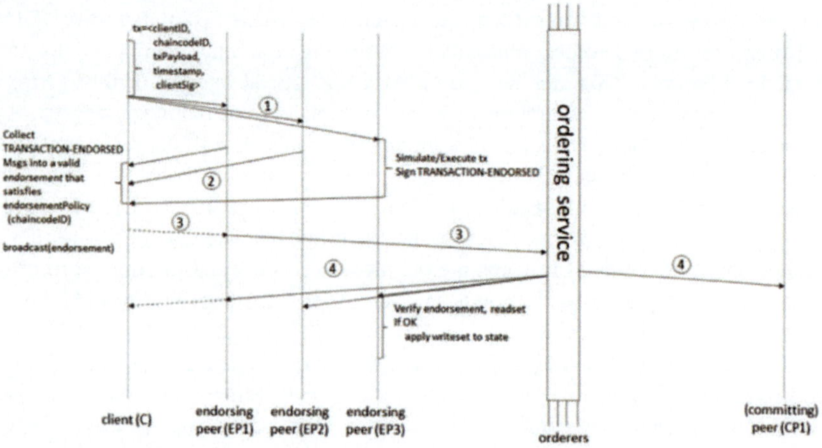

Fig. 1 Hyperledger fabric workflow [28]

Transaction flow execution:

Step 1: Propose Record/document transaction

Users can send a record transaction to multiple endorsing peers. While sending the proposal to peer nodes, the client signature of the user is attached.

Step 2: Validating and Execute proposed record/document transaction

E_0, E_1, and E_2 are endorsing peers that are ready to execute all the transactions when the proposal is received. Before executing the proposal, the peer node can validate the user signature with the help of the user public key and then return an acknowledgment to the client.

Step 3: Proposal document transaction

The client receives all the endorsers acknowledge that it has communicated to ordering service.

Step 4: Ordering services

Ordering services are responsible for receiving all the transactions and form one order for construing the block and send to the network.

Step 5: Deliver record/document Transaction

After completion of block formation, it is sent to committing peer nodes and other peer nodes. These blocks are sent to the network for verification of peer nodes. One peer node takes a responsibility to send blocks to other peer nodes by using gossip protocols.

Step 6: Verification

In this verification phase, committing nodes are ready to verify those block transactions which are based on E_0, E_1 and E_2 (endorsing peers) depending on the endorser policy. As per the Endorser Policy, every transaction must be signed by all endorser nodes (E_0, E_1 and E_2). Finally, committing node CP1 can verify all transactions and add block into the blockchain.

Step 7: Notify Transaction

In this final step, Transactions are verified by the committing nodes. Once verified successfully, a block can be added to the blockchain and notify the events to all users and peer nodes.

4 The Proposed Framework for Health Care Record-Keeping Using Hyperledger

Group signatures are very useful in providing solutions to privacy issues. In health-care systems, record management plays a crucial role but privacy issues exist while maintaining the records in hospitals and during data sharing. In this scenario, we mainly concentrate on privacy over the blockchain. Single signatures and certificates have not been perfectly providing anonymity, unlinkability, traceability, and public key and private key management. These issues are resolved using group signatures. Group signatures have three main roles to provide privacy.

- **Group Manager**: Group manager ready to generate a public key for entire group members and private keys for every group member. Finally, the group manager generates a master secrete key for identifying the fraud member in a particular group. It means if one group consists of 100 members then one public key for the corresponding group and 100 members have individual private keys and one secrete key to find a malicious member for that particular group. Finally, the group manager behaves like a certified authority.
- **Group Members**: A group member ready to sign any record or transaction by using his secret key thus his identity is concealed from any validators.
- **Validator**: Whenever validator receives record then validator can use the group public key for validating a group signature. The validator doesn't know that particular group member but the validator validates that particular record perfectly and cannot specify who is a user or group member.

Group signature scheme algorithms implementing as follows:

Syntax: Group signature scheme algorithms GS = (G.SignatureG.KeyGen, G.Validate, G.Open).

Group signature scheme algorithms implementing as follows:

Syntax: Group signature scheme algorithms GS = (G.Signature, G.KeyGen, G.Validate, G.Open).

G.Keygen(1^k, 1^N): Group key generation algorithm

G.KeyGen(1^k, 1^N): The group key generation algorithm that takes as
input: a) input the security parameter k,
b) the number of users N.
output: generate
a) group public key **gpubk**,
b) the group signing keys **$gprk_i$** and group members or users for $i \in [N]$
c) the group master key **gmsk** required to open signatures by the group manager.

G.Signature($gprk_i$, m): Group signature algorithm

The group signing algorithm takes as
input: a group member private key $gprk_i$ and message $m \in \{0, 1\}^*$
output: a group signature σ on the message.

G.Validation(σ, m, gpubk): Validation algorithm

G.Validate: The deterministic group validation algorithm takes as
input: a) a group signature,
b) a message
c) the group public key

procedure:
G.Validation(σ, m, gpubk):
G.Validation(G.Signature($gprk_i$, m), m, gpubk) = 1;
otherwise 0;
output: the signature is valid 1;
else 0;

G.Open(gmsk, σ, m): Tracing algorithm

The group opening algorithm is taken as
input: a) the group master key,
b) a signature,
c) the corresponding message
procedure:
G.Open(gmsk, σ, m):
G.Open(gmsk, G.Signature($gprk_i$, m), m) = i.
output: an identity related to σ.

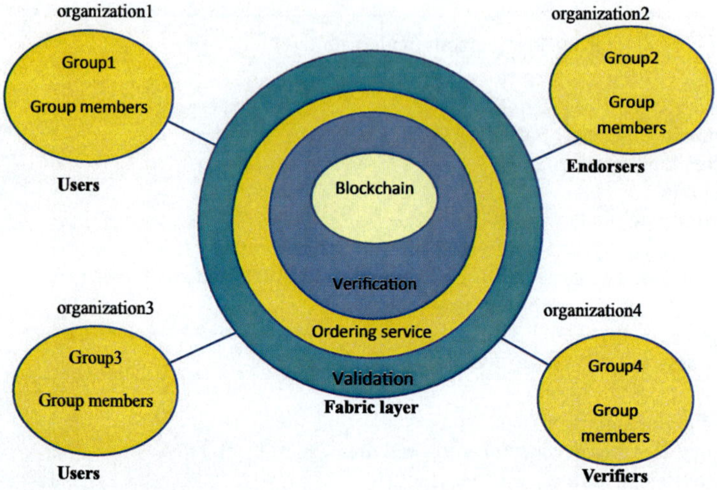

Fig. 2 Fabric group member's model

4.1 Hyperledger Fabric Group Member's Model

Figure 2. The proposed model has a total of four groups. Group1 and Group3 for users or clients and each group consist of 1 to n numbers. Group2 for endorsers or doctors for validating the patient records and Group4 for verification and verifiers or a higher authority can verify the doctor's validation after the records are put into the blockchain.

Algorithm for Validation

Input: The group member 'u' with key pair (u_{gpubk}, u_{gprk}), record proposal T_x, Endorser e_i (1<=i<=3) with endorser group member key pair (e_{gpubk}, e_{gprk}) and corresponding smart contract 'S'.
procedure:
1) 'u' signs T_x: σ=sign (T_x, u_{gprk})
2) 'u' sends σ, T_x to e_i
3) e_i checks
    a. cd1: validate (u_{gpubk}, T_x, σ)
    b. cd2: execute S and checks the format of T_x
4) if cd1 and cd2=true
    a. e_i signs σ: σ^i=sign (σ, e_{igprk})
    b. sends σ^i to user;
    else
        return fail;
output:sends σ^i to user.

Algorithm for Verification

Input: Endorser e_i (1<=i<=3) with endorsed transaction proposal σ^i, commit or CP_j (1<=j<=1) with commitor group member key pair (cp_{gpubk}, cp_{gprk}) and endorsing policy EP.
Procedure:
1) ordering services sends σ^i to CP_j
2) cp_j checks for e_i (1<=i<=3)
    a) cd1: verify (e_{igpubk}, σ^i)
    b) cd2: Endorsing policy EP
3) if cd1 and cd2 = true then
    a) marks the record as valid
    b) valid records into a block 'b' and emits notifications;
else
    marks the record transaction as invalid;
output: valid records into a block 'b'.

5 Contribution of Proposed Framework

In this paper, we proposed a novel framework to provide privacy using fabric as follows:

1. A particular group member can send a record or document proposal to the validators or endorsers. The validator can validate a particular document by using group public key but not member public key and execute that record without revealing the user identity. Endorser has to send that record to a group member.

2. A group member has received proposed records. The user sends to ordering service then sends it to committing peer nodes. Committing peer nodes to receive those transactions and verify with the help of validators group member's public key and there is no chance to reveal validator's identity.
3. If in case any group member can act as a malicious node then verify with the help of the group master key. Not only group members, but a suspicious group validator can be verified with the group master key.

Moreover, we concentrate on a list of properties in our framework.

Unforgeability: Group manager can create one public key and multiple private keys for one particular group and there is no chance to forge the signature.

Anonymity: No one can identify the user identity. In this scenario, group signatures can solve this anonymity. Members send messages with his signature then validator finds his identity by using master key otherwise validator can't reveal user identity.

Traceability: Anyone can act as a malicious node then easily find his identity by using the master key. With the help of the group manager, we can break the user anonymity to find a malicious node.

Unlinkability: Single users send multiple transactions then no one can say that these transactions belong to one user. This model can't find the user transaction linkability because of a single public key for the entire group.

6 Performance Analysis

In Hyperledger fabric provides anonymity and unlinkability by using the identity mixing model. In standard X.509 certificates, there is a problem regarding anonymity and the identity mixing technique is used to solve anonymity problems. Here, the attribute information is hidden and then the record is sent with signature and on receiving that particular record, the endorser performs validation. Endorsers can't say that this transaction belongs to a particular user. The identity mixing technique can't provide full-anonymity compared with group signatures. Every user has a public key and private key to maintain by using key management but we propose a novel solution for anonymity and unlinkability by using group signatures and only one public key to entire group members for easy key management (Tables 1 and 2).

Table 1 Performance analysis of security metrics

<table>
<tr><th rowspan="2">Security metrics</th><th colspan="3">Models</th></tr>
<tr><th>Bitcoin</th><th>Hyper ledger</th><th>Proposed model</th></tr>
<tr><td>Security</td><td>✓</td><td>✓</td><td>✓</td></tr>
<tr><td>Privacy</td><td>☒</td><td>✓</td><td>✓</td></tr>
<tr><td>Traceability</td><td>☒</td><td>✓</td><td>✓</td></tr>
<tr><td>Unforgeability</td><td>☒</td><td>✓</td><td>✓</td></tr>
<tr><td>Anonymity</td><td>☒</td><td rowspan="2">Identity mixing model</td><td rowspan="2">Group signatures</td></tr>
<tr><td>Unlinkability</td><td>☒</td></tr>
</table>

Table 2 Performance analysis on the public key and private key

Models	Public and private key management	
	Public key	Private key
Bitcoin	Every user	Every user
Hyper ledger	Every user	Every user
Proposed model	One public key for group users	Every user

7 Conclusion

Blockchain technology can modify the world what we see as most of the problems can be solved by using this technology features like immutability, consensus, distributed ledger technology, peer-to-peer network, and cryptographic functions. Blockchain technology is applied in IoT, supply chain mechanisms, healthcare, financial institutions, voting, etc. Presently, some environments exist to provide blockchain systems like Ethereum, Hyperledger, Sawtooth, and R3-Corda. The healthcare industry required research for solving some issues like record/document sharing and management of particular information without stealing the information. Many other systems cannot follow the HIPPA rules and HIPPA protocols mainly concentrating on privacy and security issues. Healthcare data can play a crucial role in the drug industry and government departments and based on this information governments can release the fund and hospitality to the patients. Sometimes intruders come into the picture and they are trying managing records to release the fund and benefits. Hence, we need perfect record sharing with solvable security and privacy issues.

To provide privacy in healthcare for electronic medical records we propose a novel solution to maintain privacy by using group signatures with the help of Hyperledger fabric. Group signatures have three main goals like group manager, group members, and validators. The group manager has generated public keys, private keys, and master keys for one particular group. By using group signatures avoid privacy problems like unforgeability, anonymity, traceability, and unlinkability.

For future work, we will implement dynamic group members for group signatures.

References

1. Nakamoto S (2019) Bitcoin: a peer-to-peer electronic cash system. Manubot
2. Bitcoin (n.d.) Retrieved from https://www.blockchain.com/explorer
3. Ethereum (n.d.) Retrieved from https://cryptocrawl.in/what-is-ethereum/Ethereum
4. Neo (n.d.) Retrieved from https://neo.org/dev
5. Nem (n.d.) Retrieved from https://nem.io/technology/
6. Cardano (n.d.) Retrieved from https://www.cardano.org/en/home/
7. Szabo N (1996) Smart contracts: building blocks for digital markets. EXTROPY J Transhumanist Thought, (16) 18(2)
8. Hyperledger Fabric (n.d.) Retrieved from https://www.hyperledger.org/projects/fabric
9. Quorum (n.d.) Retrieved from http://www.jpmorgan.com/global/Quorum
10. R3Corda (n.d.) Retrieved from https://www.r3.com/corda-platform/
11. Tendermint (n.d.) Retrieved from http://tendermint.com
12. Agbo CC, Mahmoud QH, Eklund JM (2019) Blockchain technology in healthcare: a systematic review. In: Healthcare, vol 7, no 2. Multidisciplinary Digital Publishing Institute, p 56
13. Mettler M (2016) Blockchain technology in healthcare: the revolution starts here. In: 2016 IEEE 18th international conference on e-health networking, applications and services (Healthcom). IEEE, pp 1–3
14. Azaria A, Ekblaw A, Vieira T, Lippman A (2016) Medrec: using blockchain for medical data access and permission management. In: 2016 2nd International conference on open and big data (OBD). IEEE, pp 25–30
15. Jiang S, Cao J, Wu H, Yang Y, Ma M, He J (2018) Blochie: a blockchain-based platform for healthcare information exchange. In: 2018 IEEE international conference on smart computing (smartcomp). IEEE, pp 49–56
16. HIPPA policy (n.d.) Retrieved from https://www.hhs.gov/sites/default/files/hipaa-policy-brief-hipaa-pha-final.pdf
17. Li H, Zhu L, Shen M, Gao F, Tao X, Liu S (2018) Blockchain-based data preservation system for medical data. J Med Syst 42(8):141
18. Dubovitskaya A, Xu Z, Ryu S, Schumacher M, Wang F (2017) Secure and trustable electronic medical records sharing using blockchain. In: AMIA annual symposium proceedings, vol 2017. American Medical Informatics Association, p 650
19. Gordon WJ, Catalini C (2018) Blockchain technology for healthcare: facilitating the transition to patient-driven interoperability. Comput Struct Biotechnol J 16:224–230
20. Yue X, Wang H, Jin D, Li M, Jiang W (2016) Healthcare data gateways: found healthcare intelligence on blockchain with novel privacy risk control. J Med Syst 40(10):218
21. Griggs KN, Ossipova O, Kohlios CP, Baccarini AN, Howson EA, Hayajneh T (2018) Healthcare blockchain system using smart contracts for secure automated remote patient monitoring. J Med Syst 42(7):130
22. Esposito C, De Santis A, Tortora G, Chang H, Choo KKR (2018) Blockchain: a panacea for healthcare cloud-based data security and privacy? IEEE Cloud Comput 5(1):31–37
23. Hussein AF, ArunKumar N, Ramirez-Gonzalez G, Abdulhay E, Tavares JMR, de Albuquerque VHC (2018) A medical records managing and securing blockchain based system supported by a genetic algorithm and discrete wavelet transform. Cogn Syst Res 52:1–11
24. Zhang A, Lin X (2018) Towards secure and privacy-preserving data sharing in e-health systems via consortium blockchain. J Med Syst 42(8):140

25. Al Omar A, Rahman MS, Basu A, Kiyomoto S (2017) Medibchain: a blockchain based privacy preserving platform for healthcare data. In: International conference on security, privacy and anonymity in computation, communication and storage. Springer, Cham, pp 534–543
26. Zhang R, Xue R, Liu L (2019) Security and privacy on blockchain. ACM Comput Surv (CSUR) 52(3):1–34
27. Chaum D, Van Heyst E (1991) Group signatures. In: Workshop on the theory and application of cryptographic techniques. Springer, Berlin, Heidelberg, pp 257–265
28. Hyperledger Fabric workflow (n.d.) Retrieved from https://www.skcript.com/svr/consensus-hyperledger-fabric/

Decentralized Application for Two-Factor Authentication with Smart Contracts

S. Venkata Sai Santosh, M. Kameswara Rao, P. S. G. Aruna Sri, and C. H. Sai Hemantha

Abstract Nowadays, the safety of a mobile phone cannot be assured with the password of the owner. Humans generally do not remember passwords which can be simpler to recollect and additionally use the identical password across different applications. It becomes less-secured if an intruder gains access to that particular mobile and they can easily crack the OTP sent by the officials. In this paper, we proposed a new method of securing the OTP with the help of blockchain using tokens generated by Ethereum. In this approach, the application has an authentication model using an email and password, and another one is using the Ethereum account address.

Keywords Ethereum blockchain · Decentralized · Two-factor authentication · Smart contract with truffle suite

1 Introduction

This work suggests a workaround for 2FA in this paper deprived of the need for a unified approved third party to create and allocate tokens to users once they are legitimated. Ethereum blockchain is used to enforce the proposed solution, which provides the freedom to program the logic using smart contracts for the generation and distribution of 2FA token in a trustless decentralized way. The Ethereum blockchain and OpenSSH are merged to allow the OpenSSH client for making 2FA.

Ethereum is a safe, distributed blockchain that provides a flexible and user-friendly environment for trustless, shared application development. It is a blockchain and barebone focused proof-of-work for cryptocurrency Ether. Ethereum may be viewed as a distributed computation network where everyone paying the requisite fees performs smart contracts. The integrity of implementing the smart contract depends on the

S. Venkata Sai Santosh (✉) · M. K. Rao · P. S. G. Aruna Sri · C. H. Sai Hemantha
Department of Electronics and Computer Engineering, KLEF, Vaddeswaram, Guntur, India
e-mail: santhoshsaragadam31@gmail.com

M. K. Rao
e-mail: kamesh.manchiraju@kluniversity.in

G. Ranganathan et al. (eds.), *Inventive Communication and Computational Technologies*, Lecture Notes in Networks and Systems 145,
https://doi.org/10.1007/978-981-15-7345-3_40

miners and on the underlying consensus protocol ensuring that only truthful miners are promoted. Ethereum creates smart contracts that can be designed to perform code logic based on different input criteria or occurrences along with usage of the distributed ledger's non-volatile storage. Smart contracts are mainly coded in a language named solidity and are translated into a bytecode that can be implemented on the Ethereum virtual machine after compilation. This bytecode is registered in the database, which is paired with a special identifier known as the contract key. For this contract address, we can execute multiple smart contract functions along with passing the necessary arguments as data. We have the following functionalities for developing a 2FA device. A decentralized application (DApp) is a software offering user-smart contract interaction. Using a JavaScript API for connecting with the Ethereum blockchain, Ethereum DApps usually interface users with an XML/Js Web app.

2 Related Works

The authentication solution proposed in [1] does not include any approved third party between the applicant (user) and the verifier (server). In [2], the work defines an authentication scheme using the Ethereum address. Aloul et al. [3] described a system that guarantees device verification, such as Internet banking or ATMs in a very safe way. In [4], the Ethereum foundation and issues of Ethereum tokens are addressed. Alireza et al. [5] introduced a set of authentication methods for a user utilizing a graphical password. Defigueiredo et al. [6] took a deeper look at this issue to demonstrate how two-factor authentication will provide additional protection for mobile devices. Rigney et al. [7] display a prototype of the dial-to-connect (D2C) VPN framework for the transmission of Digital Living Network Alliance (DLNA) protocols across home networks. Braz et al. [8] explored a location-based authentication method that produces authentication questions based on the positions of users monitored by smartphones. Feng et al. [9] recommend a solution for the deficiency of Das's scheme in two-factor authentication.

3 Proposed System

We incorporate decentralized application in the proposed program that allows user signs into the Web app. He gives, for example, username, password and address of his Ethereum account. If the token is right, the consumer will be authenticated. The application delivers token to the customer using his Ethereum account to submit it to the smart contract

Hash Functions The hash function is used to provide security by keeping their address in cryptography format.

Digital Signature It gives security to getting data in digital format. Every key will have a public and open key. Private key utilized for getting to data and open key utilized for exchanges.

4 System Architecture

Blockchain utilizes a mutual association (P2P), which is a network innovation expressed. Through the support of these consumers, they should be allowed to get the comparative advantage that was a focal right in comparison with the consumer deal and all such delivery, and they are the user properties of the two companies. Blockchain utilizes a mutual association (P2P), which is a network innovation expressed. Through the support of these consumers, they should be allowed to get the comparative advantage that was a focal right in comparison with the consumer deal and all such delivery, and they are the user properties of the two companies. There are a few forms of blockchains that depend on the situation and need.

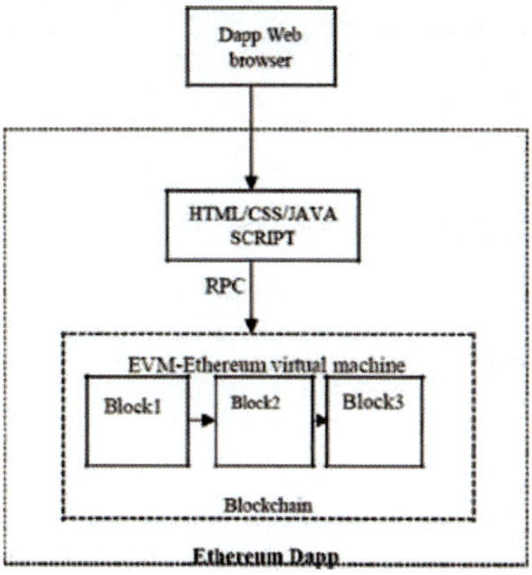

5 Ganache Overview

Ganache is software that will be used to maintain an interface between the user and a personal blockchain for Ethereum development you can use to deploy contracts, build up your applications, and run tests. It is accessible as a work area application just as an order line instrument (earlier known as the TestRPC). Ganache is accessible for Windows, Mac, and Linux.

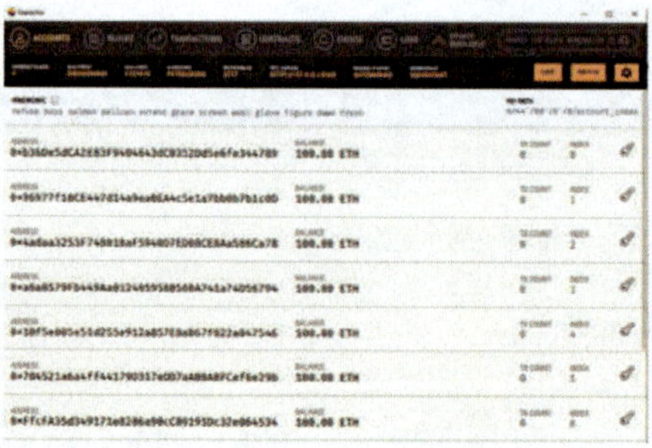

5.1 *Truffle Overview*

Truffle suite is a framework for Ethereum blockchain where we can deploy the solidity programs and run the migrations. Truffle takes care of managing your smart contracts and these smart contracts are evaluated using the browser Remix IDE. To compile the truffle project, we have to change the root of the directory. Truffle includes an integrated debugger, organization, and paired testing for quick advancement. It consists of three folders contracts, migrations, and test folder. Contracts folder consists of the smart contracts written in solidity programming language. The migrations folder consists of the scripts which are deployed on the Ethereum network. The test folder is unit tests to test the smart contracts.

5.2 *Decentralized Application (DApp)*

A decentralized application is developed to make the outfits simple to utilize participation with insightful understandings. Ethereum DApps regularly interface customers through JS and interfaces are used to maintain the relationship between the user and the application. Decentralized application interfaces close by blockchain center by methods for these methods and all these areas are used to build the projects, which licenses correspondence with close by clients. Usually, we have a suite of related concessions to basis and license consistent storing of their understanding essential state.

- Give online user interface plan for the keen agreement.
- Consider contributions from a client and store them in the blockchain.
- Collect the client information put away in the smart contract and show it to the client.

The main advantage of Ethereum blockchain is open-source, public blockchain, and many tools and frameworks that can be built on top of Ethereum.

Public blockchains are not applicable for all applications and take more time for mining and validation of the blocks [10].

5.3 *Meta Mask Extension*

Metamask is a digital money wallet which can be utilized on the Chrome, Firefox, and Brave programs. It is additionally a program expansion. This implies it works like a scaffold between typical programs and the Ethereum blockchain.

The Ethereum blockchain is where clients can construct their applications (which are called DApps) and digital forms of money. Ethereum likewise enables its clients to compose exchange rules called savvy contracts. MetaMask can be utilized to store keys for Ethereum cryptographic forms of money as it were. Along these lines, the MetaMask wallet can be utilized for putting away keys for Ether and ERC20 tokens on three diverse Internet browsers. It additionally enables clients to peruse the Ethereum blockchain from a standard program. MetaMask requires no login and does not store your private keys in any server, rather they are put away on Chrome and secret word ensured.

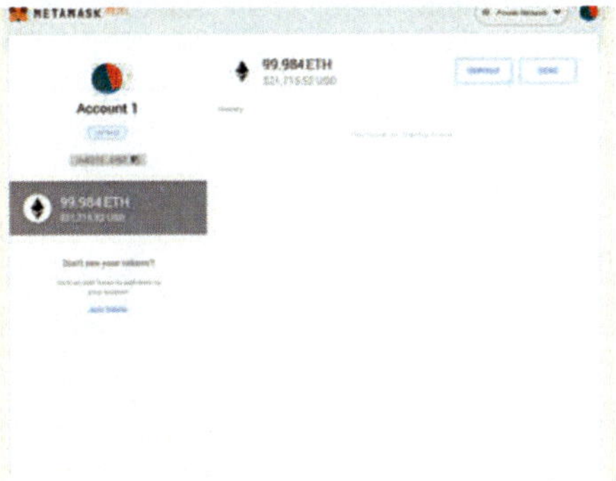

5.4 Solidity

Solidity programming is a high-level language that will be mostly used for building projects related to blockchain technology and it will provide us Ethers as a denomination so that we will use them for our transactions to develop our projects. Solidity will have some certain data types and with the help of these data types we can able to create and build our projects. For executing these programming, it will provide virtual machine for compiling and to run the program. When sending contracts, you should utilize the most recent discharged form of Solidity. This is because breaking changes just as new highlights and bug fixes are presented normally. We at present utilize a 0.x adaptation number to demonstrate this quick pace of progress

5.5 Remix IDE

Remix IDE is a browser where we can able to build our projects based on solidity programming and also we can utilize these programs for building applications based on blockchain technology.

Remix additionally bolsters testing, troubleshooting, and conveying of shrewd agreements and substantially more.

Our Remix venture with every one of its highlights is accessible at remix.ethereum.org and more data can be found in these docs. Our IDE apparatus is accessible at our GitHub storehouse.

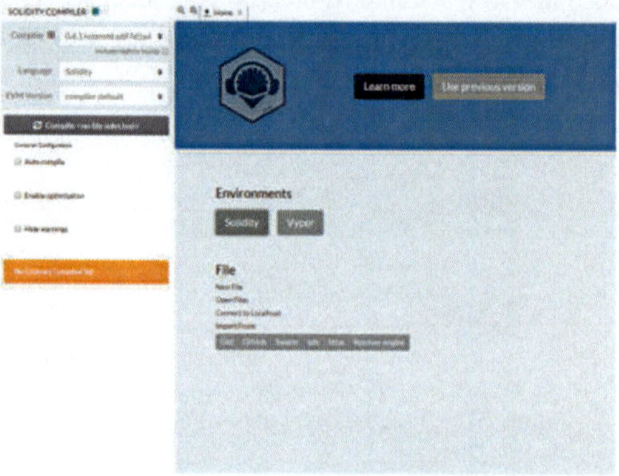

6 Implementation

In model over, we should consider the situation when 2FA is required during the login process. Just the client confirmed with the two components will be trusted by the application. Here is the way all things considered the 2FA with the blockchain may resemble:

- Client joins on the application site. He gives,
- for example, email, secret word (first validation factor) and address of his Ethereum account (e.g., 0xF00BF74C2C530).
- Client signs in with their email and secret phrase.
- Application request that he confirm with Ethereum account and gives validation token.
- The client sends confirmation token as information to the application's shrewd agreement.
- The shrewd agreement stores data that sender address 0xF00BF74C2C530 has sent given token.
- The application peruses token that was put away in brilliant agreement by an account with address 0xF00BF74C2C530. If it is a right token, at that point client is validated.

Perhaps, it sounds somewhat mind-boggling yet we may summarize it in two sentences:

- An application sends a token to the client.
- A client sends it to the smart contracts utilizing his Ethereum account.

6.1 Installation Steps

For Windows 10

Go to (https://nodejs.org/en/) to install node js.

**Note: Run command prompt as administrator for installation **

First, we need to open a command prompt and then write the following commands.

- For installation purpose, we need to type one command.
- Later that we need to install the required tools.
- We need to install python for writing programs.
- Ethereum RPC will be installed later.
- Truffle 4.1 be installed and update it if required.

(Note: Check if the truffle version is 4.1.5. If not then below steps

1. npm uninstall-g truffle.
2. npm install-g truffle@4.1.5).
3. Install the required software for establishing a connection for the application.

Install Metamask for Chrome
Setting up and deploying an application (OS: Windows)
git clone-https://github.com/venkat31/tufa-master.git
Git Reference link-https://github.com/Tufa-master.git
npm install
truffle.cmd compile//This step will create a build folder.

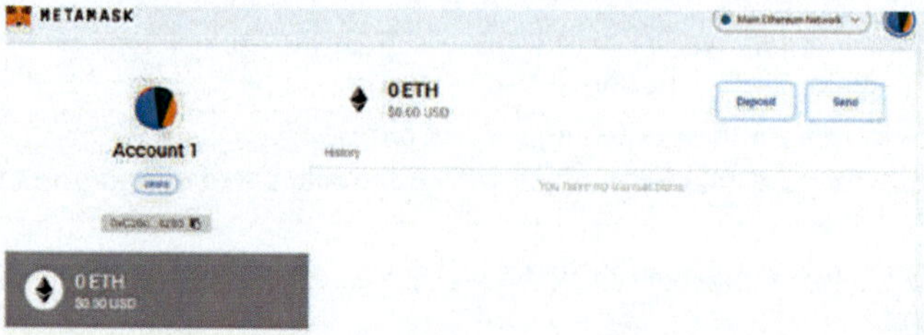

Open Terminal2/Another Command prompt
Start testrpc by command-> testrpc-m "seed words of Metamsk"
Go to a web browser (chrome)
click on metamask icon and select the network as "Localhost 8545" (As testrpc now running at localhost:8545).

7 Results

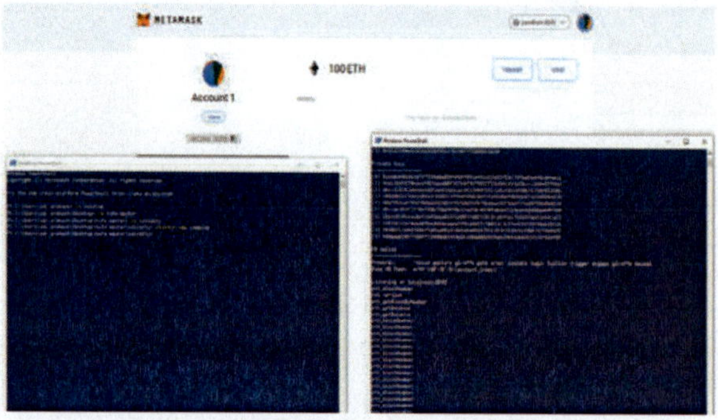

Go back to terminal 1 and execute truffle.cmd migrate

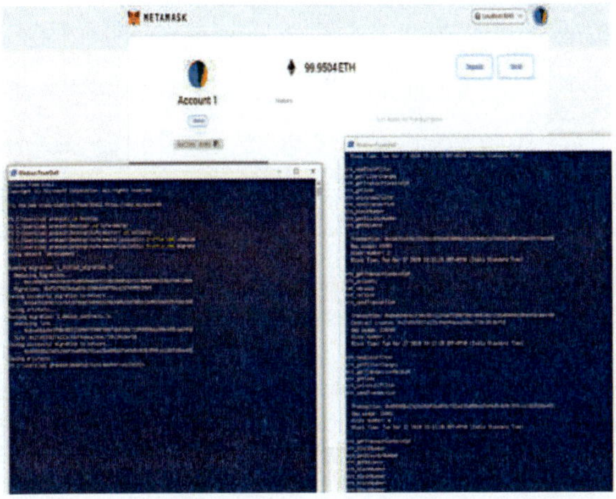

8 Conclusion

In this paper, we have implemented a new way of securing our application with the help of Ethereum blockchain technology. As we know that Ethereum is a public ledger where we can use these Ethereum for implementing and developing our projects. We are proving the security and availability and privacy of users more efficiently and effectively. Now the user is authenticated in both frontend and backend with the help of MetaMask. In MetaMask, the seed that is present are copy in command prompt and these will validate the users. Smart contract stores information of the user and it will read and authenticate the user. With the help of this system, we do not need to depend on any other third-party apps for protecting our transactions. App sends it token to the user and the user sends it to be a smart contract using the Ethereum account. Ethereum is low of cost and there is also helpful for so many developers for building their projects.

References

1. Amrutiya V, Jhamb S (2020) Trustless two-factor authentication using smart contracts in blockchains
2. https://blog.softwaremill.com/2-factor-authentication-with-smart-contracts-edd325f08b7a
3. Aloul F, Zahidi S, El-Hajj W (2009) Two factor authentication using mobile phones. In: IEEE/ACS international conference on computer systems and applications. IEEE, pp 641–644
4. (2015) The Cointelegraph. A brief history of Ethereum from Vitalik Buterin's idea to release
5. Sabzevar AP, Stavrou A (2008) Universal multi-factor authentication using graphical passwords. In: 2008 IEEE international conference on signal image technology and internet based systems

6. De Figueiredo D (2011) The case for mobile two-factor authentication. IEEE Sec Priv 9(5):81–85
7. Rigney C, Willens S, Rubens A, Simpson W (2000) Remote authentication dial in user service (RADIUS). RFC 2865
8. Braz C, Robert J-M (2006) Security and usability: the case of the user authentication methods. In: International conference of the association francophone d'interaction homme-machine
9. Huang H-F, Chang Y-F (2010) Enhancement of two-factor user authentication in wireless sensor networks. In: International conference on intelligent information hiding and multimedia signal processing
10. Amrutiya V et al (2019) Trustless two-factor authentication using smart contracts in blockchains. In: 2019 international conference on information networking (ICOIN)

A Crypto Model for Confuse-Cum-Diffuse RGB Images: A Near Zero Correlation Approach

R. Ashwin Kumar, T. Avinash, Nithya Chidambaram, and Amirtharajan Rengarajan

Abstract Image sharing has been quite common in recent days. With advancements in image sharing technology, the vulnerabilities of a cyber attack on these have also increased. Thus an effective method for encrypting these images is required. This paper proposes an Image Encryption algorithm for a color image using the Logistic Map. The correlation between the adjacent pixels is the major concern in image encryption. In this algorithm, the encryption is done using both confusion and diffusion. Statistical analysis and qualitative analysis are used for showing the feasibility and effectiveness of the algorithm.

Keywords Chaotic map · Multilayer confusion · Diffusion · Image encryption

1 Introduction

In recent days, informing sharing through various social media platforms has become quite common and we have started to use these platforms for official use too. The most common information-sharing media are digital images through various portable devices like smartphones, tablets, laptops, etc. Even though these platforms provide a faster and simpler way of information sharing, the security and privacy of this information is a matter of question [1]. Data transmitted over both wired and wireless links suffer different kinds of attacks respectively [2]. To overcome these issues image encryption algorithms are used. Chaotic maps and Attractors [3] are used for mage encryption because of highly dynamic systems [4, 5], they exhibit good properties such as pseudo-randomness which is highly sensitive to initial condition and control parameter [6], absent such information, it will exhibit random behavior. Also, it is impossible to predict the initial conditions that were provided to generate random numbers accurately.

R. Ashwin Kumar · T. Avinash · N. Chidambaram (✉) · A. Rengarajan
Department of Electronics and Communication, SASTRA Deemed University, Thanjavur 613401, Tamil Nadu, India
e-mail: cnithya@ece.sastra.edu

G. Ranganathan et al. (eds.), *Inventive Communication and Computational Technologies*, Lecture Notes in Networks and Systems 145,
https://doi.org/10.1007/978-981-15-7345-3_41

There are several methods by which image can be encrypted such as pixel permutation, image encryption based on double random-phase encoding [7], image encryption based of neural nets [8], image encryption based on chaotic maps, where efficient low dimensional chaotic maps such as Compounded Piecewise Linear Map (CPLM) [9] is used to enhance the security. But the weakness in these low dimension chaotic maps makes these algorithms vulnerable to statistical and differential attacks. In some encryption algorithms, a color image that is composed of red, green, and blue (RGB) is split into several layers and each layer is encoded by an encryption method same to that of a gray-level image. Another popular method of encrypting color image is DNA based algorithm [10, 11] where DNA encoding and DNA computing are used [12]. However, most of these algorithms cannot resist differential attack, because their ciphered images only rely on the secret key.

In this paper, chaotic logistic maps are used to generate two different unique key sets. One key is then imported to confuse the RGB pixels of a color image by sorting them according to the key. This produces an index key that can be then used to get back the original image. After the confusion process, a new key is imported for diffusion where each pixel values are changed according to key values. Hence the encrypted image is obtained.

2 Preliminary

2.1 *Logistic Chaotic Mapping*

Chaos theory is a branch of mathematics that deals with nonlinear dynamical systems. Logistic mapping is also known as pest model which determines how complex chaotic behavior can arise from very simple non-linear dynamical equations. The logistic map is determined by the Eq. (1),

$$x_{t+1} = r\, x_t(1 - x_t) \quad (1)$$

where x_t is a value between [0, 1] that represents the ratio of the existing population to the maximum possible population, and r is the branch parameter whose value lies between [0, 4]. Logistic mapping is in a chaotic state when r lies between 3.5699456 and 4. It is represented using the bifurcation diagram in Fig. 1. A bifurcation diagram shows the values visited or approached asymptotically of a system as a function of a bifurcation parameter in the system. The initial parameter is given as $r = 3.9$. Logistic mapping is in a chaotic state when 'r' lies between 3.5699456 and 4.

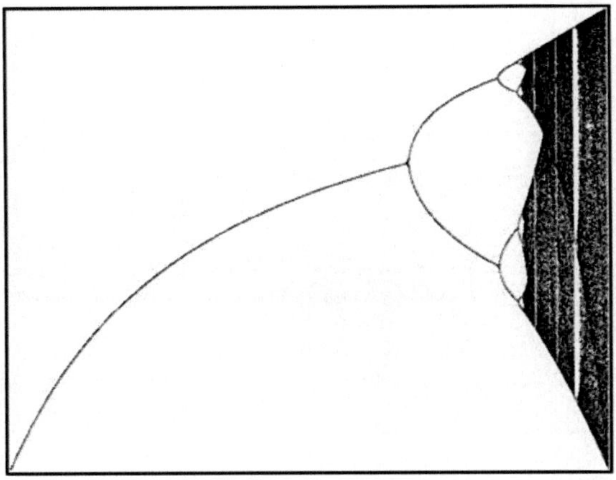

Fig. 1 Bifurcation diagram

3 Proposed Method

3.1 Key Generation

The processes of encryption and decryption are done by the use of various keys. The key defines the Quality of the encryption, so it is important to produce a key that is hard to replicate by hackers. To create such keys we use chaotic maps. The logistic map expressed in Eq. (1) is used in this proposed algorithm.

$$x[i] = (b * x[i-1] * (1 - x[i-1])),$$

where $b = 3.9, i > 0, x[0] = 0, x[1] = 0.59$ with precision of 10^{16} after decimal point.

To remove the repetition of numbers produced by the map the 1-D array is converted into a set. As per the theory of sets repetition of the same number is not allowed.

$$\text{Example}: \text{key} = [0.0844578\ 0.041245743\ 0.0844578\ 0.1540248\ \ldots\ 0.24154],$$
$$\text{SET(key)} = \{0.024154\ 0.04547\ 0.0844578\ 0.1540248\ \ldots\ 0.997458\}$$

This sorted array can again be mixed or scrambled by using only the digits after ignoring the first two or three digits. Precision is of 10^{16} decimal values are given to each value in the obtained key. There are two different keys used and each follows the same precision of 10^{16}. The keyspace becomes $10^{16} \times 10^{16} = 10^{32}$, hence making it difficult for the attackers to obtain the key. By this process, we obtain a key without repetition, and by using different values of decimal places key 1 and key 2 can be produced as seen in Fig. 2.

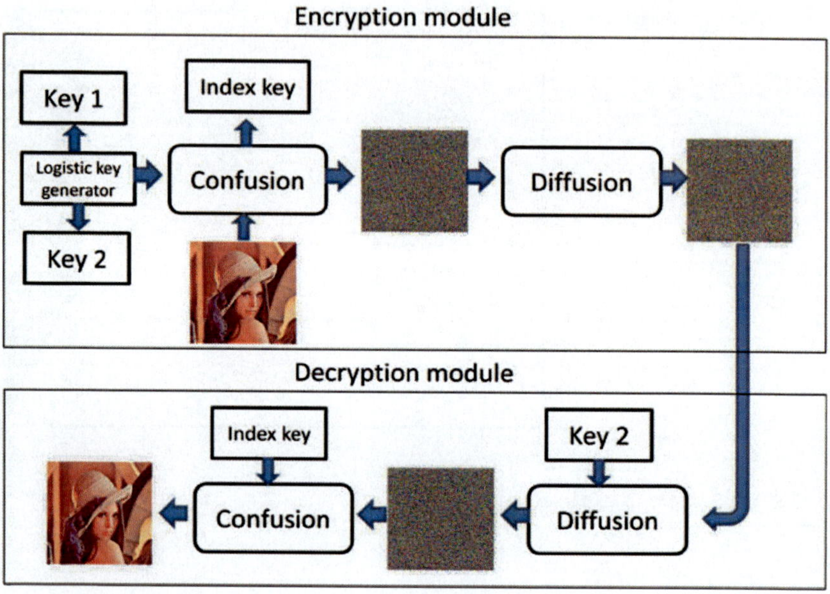

Fig. 2 Block diagram of the process of encryption and decryption

3.2 *Encryption*

Step 1 Import the keys (key 1, key 2).

Step 2 Import the original image.

Step 3 Reshape original image into a single-dimensional array OI of shape(1, R), where $R = (\text{length} * \text{breadth} * 3)$

Step 4 Create an array of dimensional $(1 * R)$ containing numbers from 0 to R − 1 in ascending order named list.

$$\text{LIST} = [[0\ 1\ 2\ \ldots R-1\ R]]$$

Step 5 Combined the three arrays key-1, OI and list into a single array of dimension $(3, R)$ in the format [[key-1[0] OI[0] list[0]], [key-1[1] OI[1] list[1]], …, [key-1[$R-1$] OI[$R-1$] list[$R-1$]]] and name it as KOL.

Step 6 Confusion is now done by sorting the array KOL which has key_1 array at its 0th index. This sorts the array key_1 but scrambles the other 2 arrays inside KOL hence performing confusion.

Step 7 Extract the array LIST from KOL, which is now used as a key to resort key_1 which contains the image data.

Step 8 Create an array final_cipher. Perform the operation final_cipher[i] = (confusion cipher[i] XOR Key-2[i]), where $i = 0$ to $(R-1)$

Step 9 Reshape the array final cipher into shape ($L * B * 3$).
Step 10 Save and transmit the image.

3.3 Decryption

Step 1 Import the Index key and key-2 as an array.
Step 2 Import the cipher image.
Step 3 Reshape Cipher final into a one-dimensional array of shape (1, R), where R = (length * breadth * 3) and name it as CI.
Step 4 Create an array of De_D.
Step 5 Perform the operation De_D[i] = (CI[i] XOR Key-2[i]), where $i = 0$ to ($R - 1$).
Step 6 Combined the arrays Index key and De_D into a single array of format [[[Index_key[0] De_d[0]], [Index_key[1] De_d[1]], ..., [Index_key[$R - 1$] De_d[$R - 1$]]] and name as ID.
step7 Now sort the Index_key array in ascending order which is present inside the array ID.
Step 8 Extract the array De_d and name it as final_Decryp.
Step 9 Reshape the array final_Decryp into shape ($L * B * 3$).
Step 10 Save the decrypted image.

4 Result and Analysis

The image encryption algorithm has been tested for several RGB images taken from the standard database. The implementation of image encryption executed using programming language: Python with the processor configuration: Intel Core i7-7500 CPU @ 2.70 GHz and RAM 8.00 GB. The test images are reshaped into 256 × 256 × 3 resolution in unit 8 format. The quality of encryption is tested using analysis of standard evaluation metrics.

4.1 Histogram Analysis

The histogram represents the distribution of the pixel intensities of an image data graphically, an image can be uniquely identified by its histogram. This makes the cryptanalyst do statistical attacks on the image easily. To avoid this histogram has to be uniformly distributed. The histogram thus obtained by cipher image will be uniformly distributed. Also, the histogram of all cipher plains and images are similar as in Table 1, thus uniqueness is destroyed.

Table 1 Histogram of the original image with its corresponding cipher for each plane

Histogram			
Image	Red plane	Green plane	Blue plane
	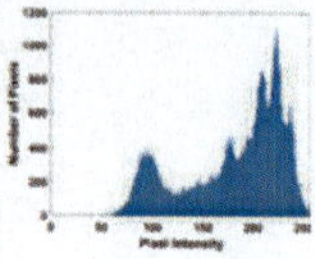	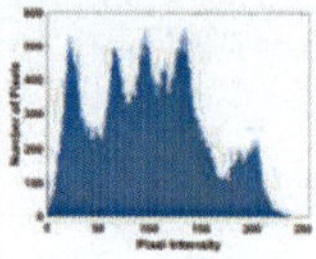	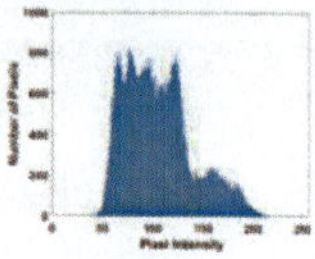
	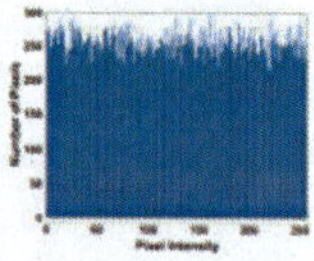	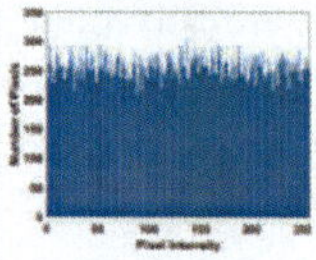	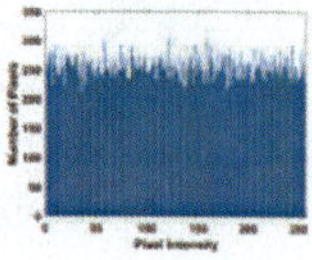

4.2 Correlation Analysis

The pixels in a digital image are highly correlated to each other. This makes it vulnerable to statistical attacks. To avoid such attacks correlation has to be made closer to zero. The encryption algorithm changes the correlation between the pixels in the digital image nearly to zero as shown in Tables 2 and 3 shows the change in correlation between original and encrypted images.

$$E(x) = \frac{1}{N} \sum_{k=1}^{k=N} x_k \tag{2}$$

$$\mathrm{VAR}(X) = \frac{1}{N} \sum_{k=1}^{N} X_k \tag{3}$$

Table 2 Correlation graphs per plane

Test image	Red plane	Green plane	Blue plane
	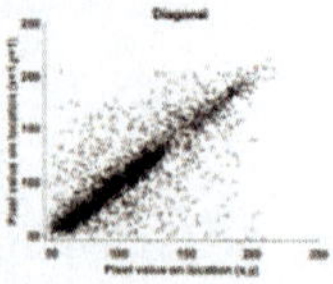	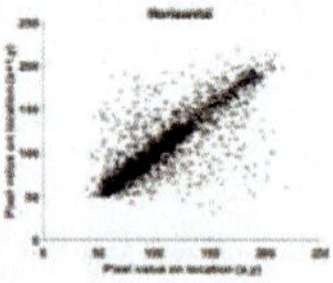	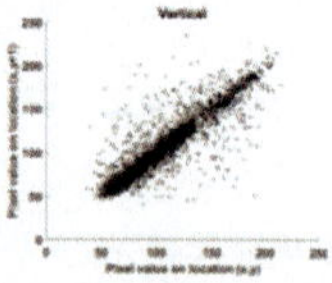
	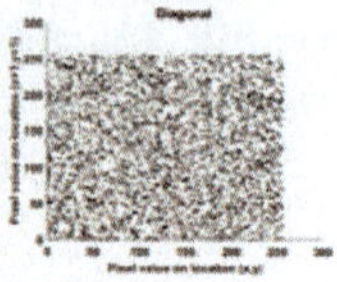	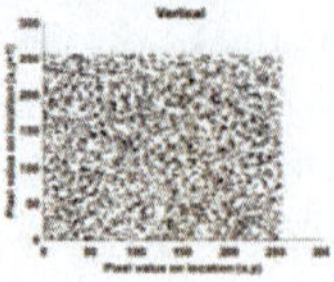	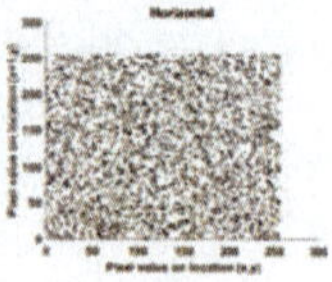

Table 3 Correlation and entropy of the original image red plane with its corresponding cipher

		Original image				Cipher image			
		Horizontal	Vertical	Diagonal	Entropy	Horizontal	Vertical	Diagonal	Entropy
Girl	R	0.9729	0.9622	0.9482	6.4200	0.0007	−0.0015	0.0037	7.9973
	G	0.9719	0.9647	0.9500	6.4457	−0.0010	0.0015	−0.0030	7.9971
	B	0.9584	0.9519	0.9377	6.3807	0.0056	−0.0038	−0.0038	7.9973
Tree	R	0.9590	0.9361	0.9159	7.2104	−0.0014	0.0013	0.0017	7.9974
	G	0.9687	0.9457	0.9318	7.4136	−0.0007	0.0018	0.0053	7.9972
	B	0.9612	0.9406	0.9265	6.9207	0.0002	0.0002	0.0027	7.9966
Couple	R	0.9493	0.9562	0.9176	6.2499	0.0018	−0.0062	0.0028	7.9974
	G	0.9308	0.9534	0.9002	5.9642	0.0008	0.0060	0.0023	7.9972
	B	0.9178	0.9442	0.8890	5.9309	−0.0026	−0.0008	−0.0025	7.9978
House	R	0.9671	0.9353	0.9126	6.4311	0.0001	−0.0054	−0.0002	7.9973
	G	0.9805	0.9474	0.9320	6.5389	0.0082	0.0005	−0.0052	7.9971
	B	0.9820	0.9749	0.9625	6.2320	0.0014	0.0049	−0.0013	7.9973
Lena	R	0.9338	0.9641	0.9074	7.2796	0.0039	−0.0026	0.0008	7.9974
	G	0.9049	0.9467	0.8802	7.6315	0.0020	−0.0081	−0.0011	7.9968
	B	0.8597	0.9061	0.8383	6.9891	−0.0088	.0043	−0.0017	7.9970
Lena [12]	R	0.9371	0.9628	0.9383	7.2933	−0.0124	−0.0001	−0.0055	7.9903
	G	0.9247	0.9513	0.9341	7.5812	−0.0038	0.0059	−0.0086	7.9890
	B	0.8741	0.9210	0.8889	7.0856	0.0075	−0.0062	0.0006	7.9893
Lena [11]	R	0.9244	0.9765	0.9366	7.2933	−0.0112	0.0026	0.0052	7.9895
	G	0.9336	0.9794	0.9368	7.5812	0.0050	0.0199	−0.0064	7.9894
	B	0.8545	0.9498	0.9046	7.0856	−0.0179	0.0120	−0.0161	7.9894

where x and X is the pixel value and N is the number of pixels

$$\mathrm{COV}(I_1, I_2) = \frac{1}{N}\sum_{k=1}^{N}(I_{1k} - E(I_1))(I_{2k} - E(I_2)) \tag{4}$$

$$r = \frac{\mathrm{cov}(I_1, I_2)}{\sqrt{\mathrm{VAR}(I_1)}\sqrt{\mathrm{VAR}(I_2)}} \tag{5}$$

where I_1 and I_2 are the horizontal and vertical pixel values respectively and $N =$ length $=$ breadth.

Where $E()$ is mean, VAR() is variance, COV() is the covariance, and r is the correlation between pixels.

4.3 Entropy Analysis

The randomness in pixel data of an image is measured using entropy. The entropy of a given image is given by

$$e = -\sum_{i=1}^{n} P(pi_i)\log(P(pi_i)) \tag{6}$$

where *P*(*pi*) is the probability of occurrence a pixel and *n* is the total number of pixels.

For an image with 256 levels of gray shades, maximum entropy of 8 bits/pixel is measured. Maximum entropy is achieved when the probability of occurrence of gray levels in equal. So a uniformly distributed histogram is required to get maximum entropy and is shown in Table 3.

4.4 Encryption Quality Analysis

The quality on which the encryption is done is an important aspect. Inspecting the image visually to check the quality of the encryption does not provide data regarding the hidden features. Hence the deviation in various aspects of the original image and encrypted image is used to check the quality of encryption. The various results are shown in Table 4.

(1) ***Maximum deviation***: It can be calculated by the histogram of the original and cipher image and then finding the absolute difference between histograms. Let '*d*' be the difference between the histogram of original and cipher pixels. Histogram difference in 0th and 255th indices is called d_0 and d_{255}. k is the index varies from 1 to 254. Using Eq. (7), maximum deviation is calculated,

Table 4 The results of the encryption quality analysis

	Maximum deviation			Deviation from the uniform histogram			Irregular deviation		
	R	B	G	R	B	G	R	B	G
Couple	8.6426e+04	8.9597e+04	9.1201e+04	0.0473	0.0492	0.0430	54352	57432	58079
Girl	8.1261e+04	81630	84437	0.0504	0.0517	0.0491	45829	51654	53224
House	73993	70391	7.9635e+04	0.0483	0.0504	0.0482	28420	31441	29321
Lena	5.1116e+04	3.6006e+04	64504	0.0481	0.0541	0.0523	19615	39946	38651
Tree	47911	4.5967e+04	6.2821e+04	0.0483	0.0504	0.0482	31925	33321	28906

$$\text{MaxDev} = \frac{d_0 + d_{255}}{2} + \sum_{k=1}^{254} d_k, \tag{7}$$

More the value of MaxDev(Maximum deviation), the higher the deviation.

(2) ***Irregular deviation***: The closeness between a uniform distribution and statistical distribution of histogram deviation is measured using irregular deviation. Let '*d*' be the absolute difference between the original image and its cipher. '*h*' indicates the histogram of '*d*'. $I = 0{:}255$ because of the 8 bit values of the pixel. Mean of the differences in histogram A_H is computed using the Eq. (8)

$$A_H = \frac{1}{256} \sum_{I=0}^{255} h_I \tag{8}$$

$$\text{IrDev} = \sum_{k=0}^{255} |h_k - A_H| \tag{9}$$

Smaller the value of IrDev(Irregular deviation), provides better quality.

(3) ***Deviation from uniform histogram***: The deviation of the obtained histogram and uniformly distributed histogram.

Since the test images are of size 256 * 256 and 8 bit (0–255) the uniformly distributed histogram will have a value of (256). H_k is the histogram of a particular pixel value.

$$\text{UniHD} = \frac{\sum_{k=0}^{255} |256 - H_k|}{256 * 256} \tag{10}$$

Lower the value of HD better is the quality of encryption.

5 Conclusion

Thus the encryption and decryption of RBG image have been successfully done by using various processes such as confusion and diffusion of the image pixels with the help of keys generated using the logistic chaotic map. The strength or validation of the process is also tested statistically by various analyses such as histogram analysis, entropy analysis, correlation analysis, and encryption quality analysis. From the results obtained from the test and analysis, it is possible to conclude that the proposed algorithm can retain the encryption standards and will able to withstand various statistical attacks. The tested images provide acceptable values of correlation, entropy, and various deviations such as maximum deviation, irregular deviation, and deviation from uniform histogram for each test, proving the efficiency of the proposed algorithm.

References

1. Xiao D, Cai H, Wang Y, Bai S (2016) High-capacity separable data hiding in encrypted image based on compressive sensing. Multimed Tools Appl 75(21):13779–13789. https://doi.org/10.1007/s11042-015-2922-9
2. El-Bendary MAM (2017) FEC merged with double security approach based on encrypted image steganography for different purpose in the presence of noise and different attacks. Multimed Tools Appl 76(24):26463–26501. https://doi.org/10.1007/s11042-016-4177-5
3. Graça DS, Rojas C, Zhong N (2017) Computing geometric Lorenz attractors with arbitrary precision. Trans Am Math Soc 370(4):2955–2970. https://doi.org/10.1090/tran/7228
4. Gopalakrishnan T, Ramakrishnan S (2017) Chaotic image encryption with hash keying as key generator. IETE J Res 63(2):172–187. https://doi.org/10.1080/03772063.2016.1251855
5. Afifi A (2019) A chaotic confusion-diffusion image encryption based on Henon map. Int J Netw Secur Its Appl 11(4):19–30. https://doi.org/10.5121/ijnsa.2019.11402
6. Ponuma R, Amutha R (2018) Compressive sensing based image compression-encryption using novel 1D-Chaotic map. Multimed Tools Appl 77(15):19209–19234. https://doi.org/10.1007/s11042-017-5378-2
7. Huang H, Yang S (2017) Colour image encryption based on logistic mapping and double random-phase encoding. IET Image Process 11(4):211–216. https://doi.org/10.1049/iet-ipr.2016.0552
8. Lin J et al (2018) An image compression-encryption algorithm based on cellular neural network and compressive sensing. In: 2018 IEEE 3rd international conference on image, vision and computing (ICIVC), pp 673–677. https://doi.org/10.1109/icivc.2018.8492882
9. Wang C, Zhang X, Zheng Z (2017) An efficient image encryption algorithm based on a novel chaotic map. Multimed Tools Appl 76(22):24251–24280. https://doi.org/10.1007/s11042-016-4102-y
10. Zhang LM, Sun KH, Liu WH, He SB (2017) A novel color image encryption scheme using fractional-order hyperchaotic system and DNA sequence operations. Chinese Phys B 26(10). https://doi.org/10.1088/1674-1056/26/10/100504
11. Wu X, Kurths J, Kan H (2018) A robust and lossless DNA encryption scheme for color images. Multimed Tools Appl 77(10):12349–12376. https://doi.org/10.1007/s11042-017-4885-5
12. Wu X, Wang K, Wang X, Kan H (2017) Lossless chaotic color image cryptosystem based on DNA encryption and entropy. Nonlinear Dyn 90(2):855–875. https://doi.org/10.1007/s11071-017-3698-4

An Efficient DFS Algorithm to Compute Least Longest Schedule Path of Software Projects

B. N. Arunakumari

Abstract The requirements of the ensuing software provided by the client are studied by software developers before the preparation of the project charter. The schedule, cost and person-days estimation are made by the developer using an activity chart. This activity chart is transformed into the graphical form using standard semiotics then manually the critical path, i.e. least long path from start to end is identified. This procedure may have several shortfalls (i) Time-consuming (ii) Manual process. In addition to this, a manual process is an error-prone approach. When the number of activities is increased, the determination of the critical path is a herculean task. In such a case, there is quintessence to develop an automated methodology that determines the critical path. This paper contains a proposed automated methodology for the determination of the critical path through a modified depth-first search (DFS) algorithm. Instead of activity chart representation, we have transformed the tabular information of dependencies and time taken by each activity into $n \times n$ matrix form and then developed an algorithm to compute critical path it takes time complexity of $O(n^2)$.

Keywords Critical path · Critical value · Schedule · Matrix element · Depth-first search algorithm · And dependency matrix

1 Introduction

Vision: To automate the determination of the software development projects' schedule.

Mission: To develop an ameliorated methodology for critical values of the project using a modified depth search method.

B. N. Arunakumari (✉)
BMS Institute of Technology and Management, Bengaluru, India
e-mail: arunakumaribn@bmsit.in

G. Ranganathan et al. (eds.), *Inventive Communication and Computational Technologies*, Lecture Notes in Networks and Systems 145,
https://doi.org/10.1007/978-981-15-7345-3_42

Objectives:

- To develop a matrix form of dependent activities of the project.
- To modify appropriately the depth-first search method.
- To determine the critical path (least long path) and value to cover all activities of the software development project.
- To optimize the determination of the critical path and value.

1.1 Motivation

The design of the activity chart is to determine the critical path (least long path). This can be identified from human skill. If the volume of activities is enhanced drastically, the human skill may not determine schedule precisely with three types of dependencies viz. the sequential, hierarchical and period dependency failing software projects as competitors may get wide publicity. Moreover, the manual process is the time-consuming and error-prone approach. The number of errors proportional to the number of activities. However, during the preparation of the project charter, the schedule may be appropriately modified as per customer developer bonhomie. As a result, sometimes though the project output software is with high precision the delay may cost heavily. On the other hand, negotiating the cost of the project with the customer, the schedule has to be linked to cost on the time base. In such a situation, there is a need to determine the schedule and accordingly negotiate the cost. Depending on the cost, the human person month are to be determined.

Moreover, the success rate of the software development project is 19% [1] and with a change in schedule and cost, the success rate may enhance to 49% [2, 3]. Therefore, there is a need to determine the schedule and cost precisely. In au-courant situation, there is no such methodology available. This paper is an attempt to develop an automated procedure for the determination of a precise schedule. This enables us to determine the cost, schedule and person-months to make a success rate to 49% this without affecting the extant software development.

1.2 Literature Survey

The project schedule is determined by designing the activity chart and then the determining critical path manually [1–3]. This is limited to fixing the project schedule. Depending on the project strategy and customer needs, if the schedule is to be modified again all the activities are manually modified and designed. The success of which is human skill dependent and herculean task. Moreover, if the activities are enhanced innumerably the determination of critical path also error-prone approach. Some of the projects are real-time projects, as the modification of the project cost, duration and person-days of project costs heavily for the developer and customer. In [4, 5], the author discussed the activity chart and critical path. Unfortunately, we could not notice the methodology to identify the critical path. Moreover, the activity taken in the illustrative example is very limited. For large projects, the activities may be voluminous. In such a case, a manual determination is of no use [6, 7]. The critical path may be determined by the DFS algorithm but if there are several dependencies viz., sequence, hierarchical and period dependency. This method does not efficiently represent a critical path. Moreover, the literature stressed mostly on the computing schedule [8–10]. The schedule per se may not be useful as the schedule needs to modified several times in the software development process. Also, travelling salesman problem (TSP) starts from one source activity must visit every activity exactly once and return to the source activity to find the shortest distance/time/cost [11, 12]. If the graph contains N activities then TSP, based solutions have $(N - 1)$ factorial possibilities. Hence, the TSP solution is infeasible to determine the critical path for in-numeral activities involved in the project. There is no automated methodology to decide the optimization factors. As a result, software developers use human skills in fixing cost, schedule and person-days. In this paper, we have modified the DFS algorithm to compute the least long schedule path to suite real-time projects. Also, the negotiation with customers out of three factors viz. cost, schedule and person-days are optimized by fixing one, two, or all the factors as per the customer requirements.

2 Proposed Methodology

Algorithm to determine the critical path and critical value (schedule, cost or person-days).

Let n be the total number of activities present in the project. These activities may be independent or dependent on some other activities per se or cumulatively. Mathematical representation is manually designed in the form of a matrix. Each row contains the values of that task.

$(i, j) \rightarrow$ position of the ith row and jth column. The value (i, j) is the numerical content. (Elements in the position (i, j). if (i, j) is blank then val $(i, j) = 0$. The critical path from (i, j) to (l, m) is denoted by $ijlm$.

The critical path is equal to $\sum_{ij}^{lm} \text{val}(i, j)$ critical path $ijlm$.

Algorithm

$i := 0, j := 0, l := 0, m := 0, k := 0$
critical_path := ijlm, critical_value = 0
for i = 1 to n do

1. If val(i + 1, i) = val (i, i) then
   critical_path := critical_path + $(ii, \overline{i+1}\ \overline{i+1})$
   critical_value := critical_value + val(ii)
   else if

$$\text{val}\sum_{k=1}^{j}(i, i+k) \geq \text{val}\sum_{j=i}^{j}(jj)$$

$$\text{critical\_path} := \text{critical\_path} + \text{val}\sum_{k=i}^{j}(ikik)$$

$$\text{critical\_value} := \text{critical\_value} + \text{val}\sum_{ki}^{j}(ik)$$

2. If val(i + k, i) = 0 for k = 1 to j and val(i + j, i) = val (ii) then

$$\text{critical\_path} := \text{critical\_path} + (iijj) - \sum_{k=1}^{j-1}(iikk)$$

critical_value

$$:= \text{critical\_value} + \text{val}(i, j) - \begin{Bmatrix} \sum_{k=i}^{j-1}(kk) \text{ if } k := i+1, \overline{j+1}) \\ \sum_{k=i}^{j-1}(ik) \text{ otherwise} \end{Bmatrix}$$

end(if)

The proposed method

- Enhances the pathetic low success rate of software development projects by at least 30%.
- Reveals when to hire or appoint expertise to optimize cost schedule person-days for the optimized benefit of the software development project.
- Automates the computation of critical path (least long path) to cover all activities of the software development project.
- Optimizes the required factor cost, schedule and person-months of a software development project.
- Resuscitate the extant activity chart by incorporating person-months and dynamism in its utility.

2.1 Case Study

1. Design a reticular network diagram of all the activities and dependency activities as shown in Fig. 1 and Table 1. Label each activity with the duration derived from process estimate activity durations. Diagonal elements should contain several days of particular activity.
2. In each row, the dependency of the rowing activity with other activities is considered. This forms a triangular matrix (see Table 2).

All possible paths for the sample software development activities and total units (duration) required to complete the activities are as follows:

$$\begin{bmatrix} \text{path1 : start} \rightarrow T_0 \rightarrow T_1 \rightarrow T_2 \rightarrow T_3 \rightarrow T_9 \rightarrow T_{10} \rightarrow T_{11} \rightarrow T_{13} \rightarrow T_{14} \rightarrow T_{15} \rightarrow \text{end} \\ \text{units : start} \rightarrow 04 \rightarrow 06 \rightarrow 30 \rightarrow 15 \rightarrow 38 \rightarrow 30 \rightarrow 33 \rightarrow 18 \rightarrow 10 \rightarrow 07 \rightarrow \text{end} \end{bmatrix}$$

Total units (duration) required to complete the activities through path 1 is 191

$$\begin{bmatrix} \text{path2 : start} \rightarrow T_0 \rightarrow T_1 \rightarrow T_2 \rightarrow T_3 \rightarrow T_{15} \rightarrow \text{end} \\ \text{units : start} \rightarrow 04 \rightarrow 06 \rightarrow 30 \rightarrow 15 \rightarrow 07 \rightarrow \text{end} \end{bmatrix}$$

Total units required to complete the activities through path 2 is 62

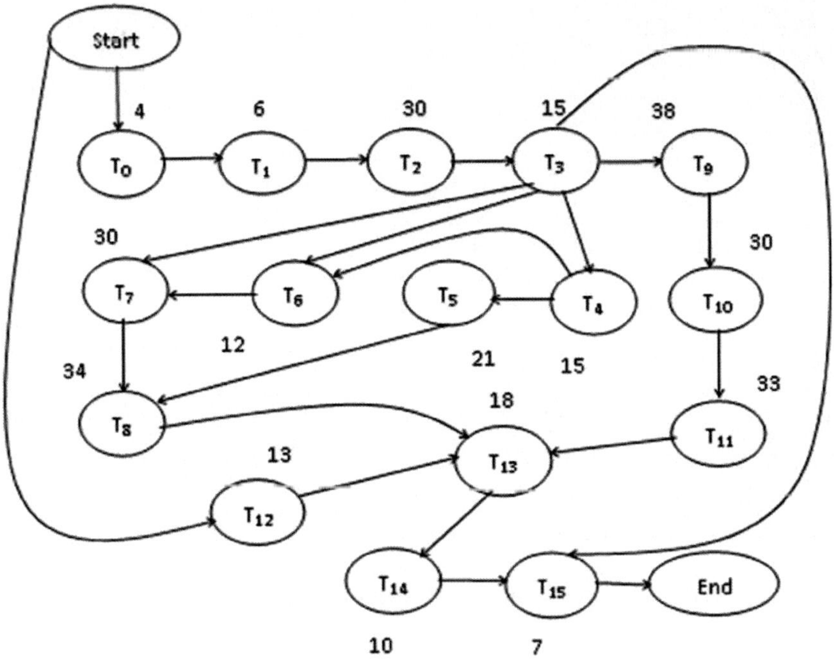

Fig. 1 Activity graph of sample software development activities

Table 1 Dependency on activities

Task	Item	Dependent on	Number of Days
T_0	Study SDLC stages	–	4
T_1	Survey literature	T_0	6
T_2	Identify reasoning for poor success rate M_0	T_1	30
T_3	Design Vision, Mission and objectives T_3–M_1	T_2, M_0	15
T_4	Organize class structures T_4–M_2	T_3, M_1	15
T_5	Identify candidate keys T_5	T_4, M_2	21
T_6	Sort table entries in definitional-referential enjambment T_6–M_3	T_4, M_2	12
T_7	Slice object methods T_7–M_4	T_3, M_1, T_6, M_3	30
T_8	Design Interrelationships T_8–M_5	T_3, M_1, T_4, M_2, T5, T_7, M_4	34
T_9	Study DDL T_9–M_6	T_3, M_1	38
T_{10}	Desiderate hiatuses of DDL T_{10}–M_7	T_9, M_6	30
T_{11}	Design desiderations in DDL T_{11}, M_8	T_9, M_6, T_{10}, M_7	33
T_{12}	Case study T_{12}–M_9	–	13
T_{13}	Write software for the design T_{13}–M_9	$T_{12}, M_9, T_{11}, M_8, T_8, M_5$	18
T_{14}	Display sample pragmatics T_{14}–M_{10}	T_{13}, M_9	10
T_{15}	Conclude	T_3, M_1	7

$$\begin{bmatrix} \text{path3 : start} \to T_0 \to T_1 \to T_2 \to T_3 \to T_4 \to T_5 \to T_8 \to T_{13} \to T_{14} \to T_{15} \to \text{end} \\ \text{units : start} \to 04 \to 06 \to 30 \to 15 \to 15 \to 21 \to 34 \to 18 \to 10 \to 7 \to \text{end} \end{bmatrix}$$

Total units required to complete the activities through path 3 is 160

$$\begin{bmatrix} \text{path4 : start} \to T_0 \to T_1 \to T_2 \to T_3 \to T_6 \to T_8 \to T_{13} \to T_{14} \to T_{15} \to \text{end} \\ \text{units : start} \to 04 \to 06 \to 30 \to 15 \to 12 \to 34 \to 18 \to 10 \to 7 \to \text{end} \end{bmatrix}$$

Total units required to complete the activities through path 4 is 136

$$\begin{bmatrix} \text{path5 : start} \to T_0 \to T_1 \to T_2 \to T_3 \to T_7 \to T_8 \to T_{13} \to T_{14} \to T_{15} \to \text{end} \\ \text{units : start} \to 04 \to 06 \to 30 \to 15 \to 30 \to 34 \to 18 \to 10 \to 7 \to \text{end} \end{bmatrix}$$

Total units required to complete the activities through path 5 is 154

$$\begin{bmatrix} \text{path6 : start} \to T_0 \to T_1 \to T_2 \to T_3 \to T_4 \to T_6 \to T_7 \to T_8 \to T_{13} \to T_{14} \to T_{15} \to \text{end} \\ \text{units : start} \to 04 \to 06 \to 30 \to 15 \to 15 \to 12 \to 30 \to 34 \to 18 \to 10 \to 7 \to \text{end} \end{bmatrix}$$

Table 2 Triangular matrix

	T_0	T_1	T_2	T_3	T_4	T_5	T_6	T_7	T_8	T_9	T_{10}	T_{11}	T_{12}	T_{13}	T_{14}	T_{15}
T_0	4															
T_1	4	6														
T_2		6	30													
T_3			30	15												
T_4				15	15											
T_5					15	21										
T_6					15		12									
T_7				15			12	30								
T_8				15	15	21		30	34							
T_9				15						38						
T_{10}										38	30					
T_{11}										38	30	33				
T_{12}													13			
T_{13}									34		30	33		18		
T_{14}														18	10	
T_{15}				15												7

Total units required to complete the activities through path 6 is 181

$$\begin{bmatrix} \text{path7 : start} \rightarrow T_{12} \rightarrow T_{13} \rightarrow T_{14} \rightarrow T_{15} \rightarrow \text{end} \\ \text{units : start} \rightarrow 13 \rightarrow 18 \rightarrow 10 \rightarrow 7 \rightarrow \text{end} \end{bmatrix}$$

Total units required to complete the activities through path 7 is 48.

Therefore, the least long schedule path for the sample software development activities (see Fig. 1) is path1 : start $\rightarrow T_0 \rightarrow T_1 \rightarrow T_2 \rightarrow T_3 \rightarrow T_9 \rightarrow T_{10} \rightarrow T_{11} \rightarrow T_{13} \rightarrow T_{14} \rightarrow T_{15} \rightarrow$ end and total units required to complete the activities through path 1 is 191.

3 Conclusion

In this paper, we have transformed the activity chart in the adjacent matrix representation to facilitate machine understanding. The modification of the depth-first search algorithm has necessitated the traversing over the dependency of activities over others has increased. The optimization of critical path and value need to be identified through traversal of matrix element $O(n^2)$.

References

1. Project Management Institute (PMI) (2012) Critical path method. In: A guide to project management body of knowledge, 5th edn. PMBOK
2. PMIs Pulse of the Profession® (2017) Transforming the high cost of low performance. In: 9th global project management survey. Project Management Institute, Inc
3. PMIs Pulse of The Profession® (2018) Success in disruptive times expanding the value delivery landscape to address the high cost of low performance. In: 10th global project management survey. Project Management Institute, Inc
4. Jalote P, Jain G (2006) Assigning tasks in a 24 h software development model. J Syst Softw
5. Jalote P (2017) A concise introduction to software engineering, planning a software project. Springer, Private Ltd, India, pp 79–102. ISBN 9788132202950
6. Lammich P, Neumann R (2015) A framework for verifying depth-first search algorithms. In: ACM CPP'15: proceedings of the 2015 conference on certified programs and proofs, pp 137–146
7. Shivanand MH, Shweta B (2010) Automated methodologies for the design of flow diagrams for development and maintenance activities, E-business technology and strategy. In: CETS 2010. Communications in computer and ınformation science, vol 113. Springer, Berlin, Heidelberg
8. Saini GS, Sanjay B (2019) Novel algorithm for software planning and development. In: ACM ICAICR'19: proceedings of the third international conference on advanced informatics for computing research, pp 1–7, Article No 4
9. Nguyen N-T et al (2018) A bayesian critical path method for managing common risks in software project scheduling. In: Association for Computing Machinery ACM. ISBN 978-1-4503-6539-0/18/12
10. Azam F, Waheed F (2019) A meta-model for planning and execution activities in software project integration management. In: ACM digital library of the 9th international conference on information communication and management, pp 32–36
11. Ricciardi LA, Vasile M (2019) Solving multi-objective dynamic travelling salesman problems by relaxation. In: GECCO'19: proceedings of the genetic and evolutionary computation conference companion, July 2019, pp 1999–2007
12. Akandwanaho SM, Adewumi AO, Adebiyi AA (2014) Solving dynamic traveling salesman problem using dynamic Gaussian process regression. J Appl Math

Priority-Centered Virtual Groups and Mobile Sink for Wireless Sensor Networks

Deivanai Gurusamy, Tucha Kedir, and Endalkachew Emare

Abstract Energy optimization is a long-lasting challenge in wireless sensor networks, and in recent years, the researchers are experimenting with mobile sink. Though many existing works focused on reducing the latency with the optimal path, the data collection time, which depends on the total number of sensor nodes deployed, is still unreduced. Even if solved with cluster heads, energy expended by nodes could still be controlled extensively. Hence this paper proposes an approach priority-centered virtual group with mobile sink (PCVG-MS). The primary focus of this approach is partitioning the sensor nodes into two virtual groups within every cluster and allowing the mobile sink to travel through the optimal path. Then, the sink collects data from only one group, which has the highest priority at a particular trip, thus ensures the long life of the nodes. The proposed method is simulated, and the experimental results show a significant drop in latency and energy consumption.

Keywords Priority · Virtual active group · Virtual passive group · Mobile sink · Wireless sensor networks

1 Introduction

Wireless Sensor Networks (WSNs) consist of a large number of sensor nodes that are densely deployed, tend to fail quickly, and limited in power, memory, and computational capability. Despite these characteristics, WSNs are used almost in all fields,

D. Gurusamy (✉) · T. Kedir · E. Emare
College of Informatics, Bule Hora University, Bule Hora, Ethiopia
e-mail: deivanaiguru@gmail.com

T. Kedir
e-mail: tuchakedir@gmail.com

E. Emare
e-mail: Endalk.emare21@gmail.com

G. Ranganathan et al. (eds.), *Inventive Communication and Computational Technologies*, Lecture Notes in Networks and Systems 145,
https://doi.org/10.1007/978-981-15-7345-3_43

such as military surveillance, environment protection, healthcare, agriculture, industrial monitoring, and home applications [1]. Moreover, WSNs play a significant role in the Internet of Things (IoT) applications by which the world is evolving smart everywhere day by day. As the energy of the sensor nodes determines the lifetime of the WSN in these applications, minimizing the energy consumption in each sensor node is crucial for WSN.

The main reason for the maximum energy depletion in every sensor node is that the nodes exhaust the energy quickly in their every long-range communication with the base station, specifically the nodes that are close to the base station [2]. So, a mobile sink [3] that goes close to the nodes and collects data with short-range communication was used later. In general, the components of WSN with Mobile Elements (vehicle, animal, or people) are sensor nodes, base stations, and support nodes for performing tasks like data collection and forwarding [4]. Moreover, to achieve better results, some researchers [2] use more than one mobile base station.

Though the mobile base stations help the nodes retain their energy for a long time, there are many challenges in it, and one among them is communication latency. So, if the travel path is well planned, the latency can significantly be reduced [5]. Though the optimal path is found, if the sensor nodes are more in number, the data collection time from all the nodes is more; again leads to increased latency. To meet these issues, a good clustering algorithm which mainly contributes to the increased lifetime of the sensor nodes is required [6]. Also, the approaches to achieve a kind of duty cycling (The nodes that are not currently required can be put in sleep mode, and the nodes that are currently active need not keep the radio continuously on [7]) can be used. So, if the clustering and duty cycle are framed with efficient algorithms and data collection process is carried out by the mobile sink through the well-defined path, energy conservation would be highly appreciated, and network lifetime can be extensively prolonged.

Hence, the paper proposes Priority-Centered Virtual Groups with Mobile Sink, where nodes are clustered, and two virtual groups are formed in each cluster. Further, the proposed method identifies the optimal travel path through which the sink travels and collects data from only one virtual group, which has a higher priority. Thus, the proposed method attempts to reduce the latency and increase the life of the sensor nodes. In the rest of the paper, Sect. 2 describes the related work, Sect. 3 describes the proposed PCVG-MS energy management approach, Sect. 4 gives the performance evaluation and results, and Sect. 5 gives the conclusion of the proposed work.

2 Related Work

This section describes the current research works, which are the primary motive for the proposed approach to deal with the mobile sink in WSNs.

Rahul et al. [3] have explored various parameters like scalability, reliability, and overhead in their work called Data MULEs (Mobile Ubiquitous LAN Extensions), where mobile elements are used to gather the data from sensors which are close in range and hand over them to the access points. The authors themselves pointed out that the main drawback of the work is increased latency in the communication of the mobile elements with sensor nodes and access points. Also, the sensor nodes deplete their energy in listening to the arrival of mobile elements. However, the paper has been a motivation for many researchers to explore more from a particular viewpoint.

Ming et al. [8] have addressed the issue of reducing the tour length of the mobile elements by identifying polling points and candidate polling points. These points are connected straight, and the mobile element travels on the straight line to collect the data from the sensor in the neighbor set. Nevertheless, the polling points are many in numbers to have each sensor as a neighbor, and it increases the tour length. Miao et al. [9] have combined the above just said algorithm with Space Division Multiple Access (SDMA) technique which is proven to be NP-Hard while Ming et al. [5] have formulated the Mixed Integer Problem (MIP) to reduce the tour length of the mobile data collector.

Charalampos et al. [10] conducted an experiment on the data collection process from the multi-hop environment. In their approach, a node with more energy and closely located to the mobile element is selected as a rendezvous node, which in turn communicates with the cluster head to transfer the data to the mobile sink. However, the necessity of two representative nodes in a cluster area is not reasoned. Also, the energy depletion and failure of either of these nodes could reduce the lifetime of the network. Miao et al. [11] have also tested the multi-hop environment with polling points, but the hops are limited.

Liang et al. [12] have proposed on-demand data collection method in which after the memory of a sensor node becomes full, the request for data collection is sent to the mobile sink that maintains an FCFS (First Come First Serve) queue to collect data. In this approach, though the mobile sink is assumed to be with unlimited energy and storage capacity, it is unfeasible as the mobile sink may happen to be continuously traveled back and forth within the area in case the state of the sensor's buffer is full one after another. Liang et al. [13] have improvised the above-said data collection method with Nearest Job Next (NJN) and Nearest Job Next Combination (NJNC). These methods based on the location of sensor nodes to the mobile sink, but in these methods also it is unclear how long the mobile sink travels around the area. Nimisha and Indrajit [14] have extended both FCFS and NJN with the Earliest Deadline First scheduling scheme in which the mobile sink gives priority to the node, which meets the deadline that is a particular period. This approach applies only to time-sensitive applications.

Ha et al. [15] have developed an optimal method for gathering the data with a mobile sink, which almost does every task that includes clustering, identifying the center of each cluster, and finding the optimal travel path. Nevertheless, collecting the data from all the nodes in the cluster leads the sink to spend a long time in each cluster and increases latency. Jerew and Liang [16] have tried reducing the latency and energy consumption of sensor nodes by collecting the data only from the cluster head. However, frequent change in the cluster head increases the overhead. Sanu and Thomaskutty [17] have contributed to reducing the travel path of the mobile sink by identifying two-chord points in each circle that denotes the communication range of a set of sensor nodes. Then, the mobile sink is let through the chord path from one circle to another. Still, the approach does not assure uniform energy depletion at each node. Jerew and Bassam [18] have formulated Adjacent Tree-Bounded Hop Algorithm (AT-BHA) and Farthest Node First-Bounded Hop Algorithm (FNF-BHA) by which the mobile sink is permitted to travel to Cluster Location in the Cluster Tree, but does not ensure the shortest path.

All these algorithms target at reducing the energy consumption with the help of mobile sink but, in many approaches, the latency and overhead are the issues. So, the proposed approach aims to reduce the latency and energy wastage without introducing additional overhead to the nodes.

3 PCVG-MS (Priority-Centered Virtual Groups with Mobile Sink)

3.1 Overview

PCVG-MS is an energy management approach which mainly focuses on the efficient utilization of the available energy from the static sensor nodes. While clustering could contribute to energy management, the proposed method creates virtual groups of active and passive nodes inside the cluster. The method generates an unusual sleep duty cycle where the virtual groups switch their roles every time the mobile sink approaches the cluster. Thus, the nodes are not consistently spending their energy. The proposed approach is applicable where the real-time data is not the necessity.

The method uses a single mobile sink as a data collector. The sink is assumed to be with high energy, ample storage, high processing power, long-range transceivers. The mobile sink starts its trip from the base station periodically. Also, it is enabled to locate the sensor's area. When using a mobile sink in WSN's, keeping the latency minimum would improve the performance of the network. So, the shortest optimal path is identified for the mobile sink to travel. Moreover, the virtual group technique reduces the time taken by the mobile sink to collect the data from each cluster. Hence, clustering, path identification, virtual group generation are collectively named PCVG-MS, which reduces the latency while increasing the lifetime of the network. Unlike the sensor nodes, the mobile sink gets recharged from the base station. For that reason, all those tasks are performed by the mobile sink.

3.2 Clustering and Locating Data Collection Points

It is the first task after the sensor nodes are deployed, and the mobile sink has arrived in the area. Generating virtual groups and saving the energy of nodes with an unusual sleep duty cycle is the primary task that differs from the existing approaches, but

Table 1 K means and Christofides algorithms

k-means clustering	Christofides algorithm

Number of clusters k
Choose cluster centers α_j for $j = 1, 2, \ldots, k$ randomly
Repeat
For $i = 1$ to m do
Compute Eq. (1)
End for
For $j = 1$ to k do
Compute Eq. (3)
End for
Until Cluster assignments β_{ij} are unchanged
Return $\alpha = \{\alpha_1, \alpha_2, \ldots, \alpha_k\}$and β_{ij}
	Input: Given Graph G
	Output: Tour path
	a. Calculate the minimum spanning tree M
	b. Find the set of vertices O with an odd degree from the calculated M
	c. Let O be the set of odd degree and even number of vertices by ensuring $\sum_{v \in V} \text{degree}(v) = 2\|E\|$ where V vertex set, v vertices and E total number of edges in M
	d. Get the new graph G_1 by forming only the vertices of O
	e. Calculate minimum-weight perfect matching; it is the graph G_2 with edges without common vertices
	f. Get the multigraph MG by uniting G_2 and M
	g. Form Eulerian circuit in MG
	h. Eliminate repeated vertices

this task requires clusters. So, initially, the nodes are clustered using the k-means algorithm [15, 19], which is given on the left side of Table 1.

Given the set of sensor nodes $S = \{\text{sn}_1, \text{sn}_2, \ldots, \text{sn}_n\}$ and the value of k, the clustering algorithm k-means partitions the nodes into k clusters $C = \{c_1, c_2, \ldots, c_k\}$ and locates the center point in each cluster thus gives set of center points $\alpha = \{\alpha_1, \alpha_2, \ldots, \alpha_k\}$ in a way that all the nodes in each cluster are similar and nodes of different clusters are dissimilar. These center points are considered as data collection points of the mobile sink. The algorithm uses two strategies, the first strategy is, the center point α is kept fixed, and the total number of nodes in each cluster β is determined. The solution to the ith node sn_i can be identified by setting

$$\beta_{i,j} = 1 \text{ if } j = \arg\min \left\| \text{sn}_i - \alpha_j \right\|^2 \tag{1}$$

$\beta_{i,j}$, which is 1 if, and only if, sn_i is assigned to cluster j and $\|x\|^2$ denotes Euclidian square.

The second strategy is the number of nodes in each cluster β is kept fixed, and the center point α is determined.

$$\sum_{i=1}^{m} \beta_{i,j}(\text{Sn}_i - \alpha_j) = 0 \text{ for all } j \tag{2}$$

and it is obtained that

$$\alpha_j = \frac{\sum_i \beta_{ij} \text{sn}_i}{\sum_i \beta_{ij}} \tag{3}$$

$\sum_i \beta_{ij}$ denotes the total number of nodes assigned to cluster j.

3.3 Determining the Travel Path

Upon clustering, the data collection points derived from it are given as inputs to determine the optimal path for the mobile sink. The movement of the mobile sink follows the graph-based mobility model, similar to the pathway model. Many researchers have developed different algorithms to find the shortest tour [10, 15, 16]. Motivated by [20], in this paper, heuristic algorithm, Christofides [21] for solving the traveling salesman problem, has been experimented, and that is shown on the right side of Table 1.

Given a graph $G = (V, €)$, where V represents the vertex set, that is the set of data collection points, and € denotes the function that gives a non-negative weight between two vertices that is the weight of each edge, generally taken as the distance between the collection points. Also, the edge weight must follow the triangular inequality.

That is, for the given three vertices v_1, v_2, and v_3.

$$€(v_1 v_2) + €(v_2 v_3) \geq €(v_3 v_1) \tag{4}$$

3.4 Constituting Virtual Groups

Once the Cluster is formed, and the best path is identified to reach the collection points, the sink gets ready to constitute virtual groups inside the cluster. Generally, if all the sensors in each cluster transmit the data to the mobile sink, the time to collect the data will be increased. Also, the energy of all the nodes is decreased gradually. According to [22], the nodes that are close in the distance sense similar kinds of data. In the proposed approach, the nodes in the cluster are closely located and may sense similar data. So instead of collecting the data from all the nodes from each cluster, the mobile sink creates Virtual Passive Group (VPG) and Virtual Active Group (VAG). Also, to collect data from only VAG, a standard priority value is associated with all the nodes in VAG of each cluster.

Moreover, the members of VAG and VPG keep changing for each trip of the mobile sink. So, energy is equally used among all the nodes in the cluster along with shorter data collection time. The procedures for creating VAG and VPG are described next.

After the mobile sink reaches the cluster, its checks its trip number and the number is odd, the nodes that are with odd ID number are put in a list else the nodes that are with even ID number are listed. Then the mobile sink broadcasts the list to all the nodes in the cluster. Upon receiving the list, the nodes in the list enter into VPG,

Table 2 Virtual group generation and data collection algorithms

Constituting Virtual Groups	Data Collection

Input: n (initialized to 1)
Output: VPG, VAG
For every n^{th} trip
Find n is odd or even
For every cluster
If n=odd
Generate a list of nodes with the odd node ID's.
Else
Generate a list of nodes with even node ID's.
End If
Broadcast the list
Form VPG and VAG
End For
End For
	Input: Priority value p, trip number n
	Output: Collected data
	For every n^{th} trip
	Constitute VAG and VPG
	For every VAG
	For every node
	If p =1
	Receive data
	End If
	End for
	End for
	n=n+1
	End For

and other nodes enter into VAG. The nodes in the VPG change their priority from the default value 1 to 0 and sleep for 60 s. The nodes in the VAG are with default priority 1 and remain active. The algorithm is shown on the left side of Table 2.

3.5 Data Collection from VAG

In the first trip of the mobile sink to the cluster, VPG and VAG are formed, and the sink checks the priority value. Then it collects the data from the nodes in VAG as per the TDMA schedule created. After collecting the data, it moves to the next cluster and repeats the process, finally updates its next trip number so that next time when the sink visits, another set of nodes can enter the VAG. The data collection algorithm follows the previous algorithm, and it is given on the right side of Table 2.

3.6 The Role of Sensor Nodes in the Proposed Approach

Each sensor node is assigned with a positive ID number and a default priority value 1. They are deployed in the area, and after the clusters are formed, the sensor nodes sense the data and store it in their memory buffer. Once the buffer becomes full, the nodes enter into the sleep mode. The mobile sink starts the tour from the base station to the area, travels via the optimal path, and meets the clusters in the data collection points. Then it broadcasts the VPG list to the nodes. Next, the nodes switch to duty mode, receive the list of nodes in VPG from the mobile sink. If the node's ID is in the

Table 3 Sensor's work routine algorithm

```
Sensor's Routine Work Algorithm
Input: ID (Positive Number), Priority Value p (Initialized to 1)
Output: Extended Life Time
While (energy! =Null)
 Sense and store it in the buffer
        If buffer=full
                Enter Sleep Mode
        Else
                Keep sensing
        End if
 Receive generated list from the mobile sink
        If ID is in list
                p=0
  Delete data from the buffer
                Sleep (60)
            p=1
        Else
                Receive TDMA slot
                While (buffer! =empty)
                Transmit data
            End while
        End If
 End while
```

list, it changes its priority to 0, deletes the stored data, and sleeps for 60 s (maximum data collection time from a cluster). After 60 s, it can again start sensing the new data and repeat the process. If the node's ID is not in the list, the node waits and receives TDMA schedule from the sink, starts transmitting the data. After the transmission is over, it repeats its regular task. So, the tasks performed by the sensor nodes are less compared to existing approaches; thus, overhead is reduced. The sensor's routine work algorithm is given in Table 3.

4 Performance Evaluation

The proposed approach is simulated, and the performance is evaluated. The sensor nodes are deployed randomly in the 200 m × 200 m area. The number of nodes is 50. The initial energy of the node is 1 J, and a mobile sink with initial energy is 10 J, and it functions in the speed 5 m/s. The MAC protocol used here is 802.11 ad, and the channel type is a wireless channel. The energy used for communication is proportional to α.

$$\propto = d^{\partial} \tag{5}$$

where d is transmission distance and ∂ is attenuation parameter, either 2 or 4. The energy consumption for transmitting and receiving a bit over 1 m distance is taken as 0.1 nJ and 0.5 nJ, respectively.

4.1 Evaluation of the Size of VAG and VPG

Even though determining the number of clusters is cumbersome, it is chosen randomly based on the total number of nodes deployed. In the simulation, 50 nodes have been deployed. So the algorithm is tested by taking the different numbers of clusters randomly. That is, for the same 50 nodes, 4 clusters, 5 clusters, 6 clusters, and 7 clusters (k-means algorithm's k values are 4, 5, 6, and 7) are created, and the simulation result for the size of VAG and VPG is observed. This is done only to verify, is there sufficient nodes are available in each group. Also, to find how does the size of VAG and VPG differ if the number of clusters is increased for the same "n" number of nodes. The results are shown in Fig. 1, and it shows that almost the nodes are in equal numbers inside VAG and VPG's. Thus, it explicitly illustrates when about half of the nodes on duty, about half of the nodes sleep in the total area. Hence, the nodes can alternatively spend the energy that leads to almost equal energy savings in all nodes.

Another thing to be noted about the size of VAG is that to receive information from each cluster, at least a node must be available in each VAG. From the simulation, it has been identified that when the number of clusters is increased, one or two VAG may happen to be without nodes. However, it does not affect the results because the neighbor VAG almost gives the same data.

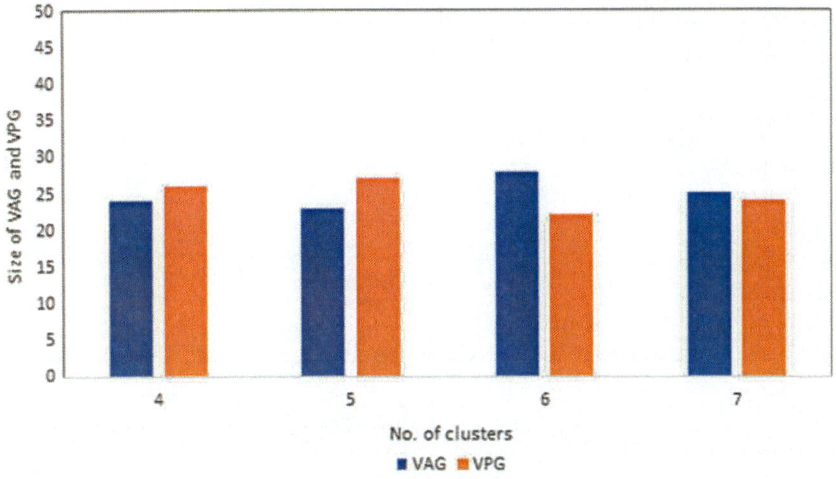

Fig. 1 Number of clusters versus total size of VAG and VPG

The total size of VAG in the entire cluster φ is given by

$$\varphi = \sum_{k=1}^{l} \sum_{x=1}^{y} a \tag{6}$$

where l denotes the number of clusters, y represents the total number of VAG and a represents the number of nodes in VAG.

The total size of VPG in the entire cluster, $\pitchfork$ is given by

$$\pitchfork = \textstyle\sum_{k=1}^{l} \sum_{p=1}^{q} b \quad (7) \tag{7}$$

where l denotes the number of clusters, q represents the total number of VPG, and b represents the number of nodes in VPG. Hence,

$$\varphi + \pitchfork = \$ \qquad (8) \tag{8}$$

where \$ denotes the total number of nodes deployed in the area.

4.2 *Evaluation of Energy Consumption*

Next, the average energy spent in VAG of all the clusters is measured, and it gives the total energy consumed by the sensors on each trip of the mobile sink. The mobile makes a trip from the base station to the collection point and collects the data from the nodes. In every trip, the total energy spent is measured by calculating the energy spent by sensor nodes in each VAG. The total energy spent ₠ is calculated as follows.

$$\text{₠} = \bar{\text{Ṛ}} + \dot{\text{T}} \quad (9) \tag{9}$$

where $\bar{\text{Ṛ}}$denotes the energy spent on receiving, and $\dot{\text{T}}$ represents the energy spent on transmission.

Energy spent in each VAG $\in$ is given by

$$\in = \textstyle\sum_{n=1}^{m} \quad \text{₠} \quad (10) \tag{10}$$

where m is the number of nodes in VAG.

So, the average energy consumed in all the clusters in one trip τ is calculated by

$$\tau = \frac{\sum_{j=1}^{p} \in}{p} \tag{11}$$

where p is the total number of VAG.

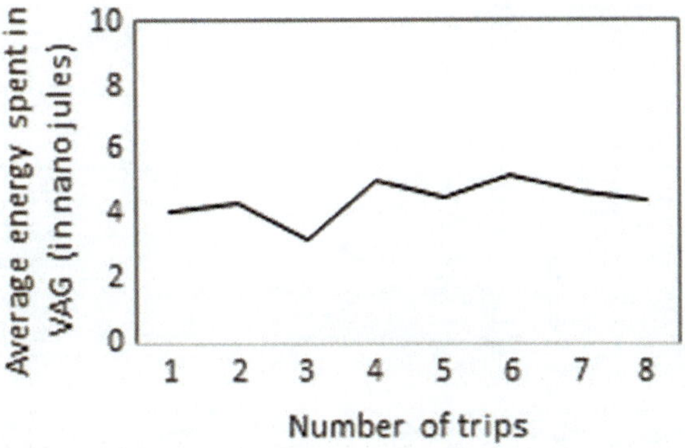

Fig. 2 Energy spent in 4 clusters

The simulation results show the average energy spent on each trip and are given in Fig. 2.

4.3 Evaluation of Latency

Latency is measured by the total time taken by the mobile sink from the base station to the sensor area and back to the base station after the data is collected. It is calculated by

$$\mathsf{U} = \Delta + \sum_{g=1}^{h} \delta + \sum_{m}^{n} \omega + \emptyset \tag{12}$$

where U represents latency, Δ denotes time taken by the mobile sink to travel from the base station to the area, h denotes the number of data collection points, δ denotes the time taken to reach each collection point, n represents the number of clusters and ω represents the data collection time and finally $\emptyset$ is the time taken from the last collection point to base station. The latency is calculated, and it is compared with the latency in each round of mobile sink in the existing approach (Optimal Method for Data Gathering) [15]. The OMDG approach does allow all the nodes in the cluster to transfer the data to the mobile sink. The mobile sink spends much time in this process. In the proposed approach, as only a few nodes transfer the data to the mobile sink, the sink quickly moves to another cluster, thereby reduces the overall time taken to reach the base station. The results are shown in Fig. 3, and it indicates that the proposed approach PCVG outperforms the existing approach OMDG in terms of latency.

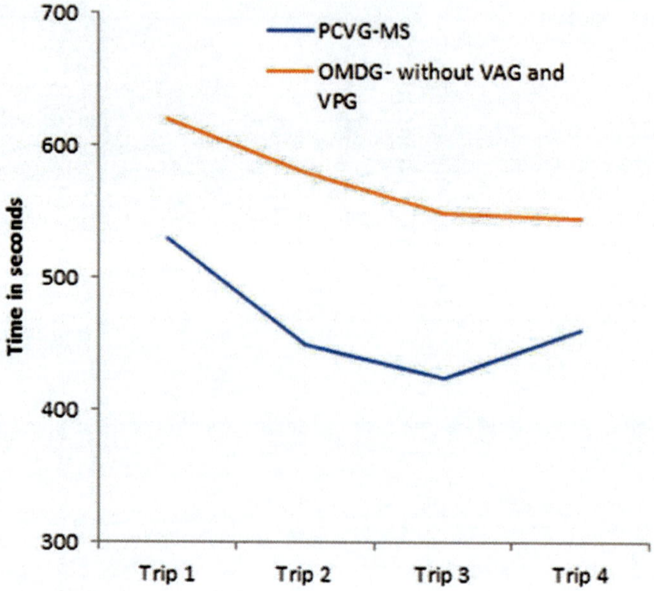

Fig. 3 Time spent in each trip by mobile sink

5 Conclusion

The paper has proposed an energy management approach as the existing approaches did not focus on the time spent on collecting data from all the nodes and the energy spent in the data collection process. In the paper, the vital part is virtual groups inside the clusters, where some of the nodes are with priority value 1, and the rest are with priority 0. The nodes with higher priority are only allowed to participate in the data collection process. Moreover, the nodes would change their priority every time the sink approaches their cluster. Along with the shortest path, this kind of data collection reduces the total time and retains the sensor's energy for a long time. The experimental results showed that the PCVG-MS approach has noticeably extended the lifetime of the network and has considerably reduced the latency in comparison to the previous approach.

References

1. Akyildiz IF, Su W, Sankarasubramaniam Y, Cayirci E (2002) Wireless sensor networks: a survey. Comput Netw 38(4):393–422
2. Gandham SR, Dawande M, Prakash R, Venkatesan S(2005) Energy efficient schemes for wireless sensor networks with multiple mobile base stations. In: Proceedings of the GLOBECOM '03. IEEE Global Telecommunications Conference, San Fransisco, USA, 1–5 December 2003

3. Rahul CS, Sumit R, Sushant J (2003) Data MULEs: modeling a three-tier architecture for sparse sensor networks. In: Proceedings of the first IEEE international workshop on sensor network protocols and applications, USA, 11–11 May 2003
4. Francesco MD, Das SK (2011) Data collection in wireless sensor networks with mobile elements: a survey. ACM Trans Sens Netw 8(1), 7:1–7:39
5. Ming M, Yuanyuan Y, Miao Z (2013) Tour planning for mobile data-gathering mechanisms in wireless sensor networks. IEEE Trans Veh Technol 62(4):1472–1483
6. Akila IS, Manisekaran SV, Venkatesan R (2017) Wireless sensor networks-insights and innovations. Intechopen, United Kingdom, pp 141–156
7. Anastasi G, Conti M, Francesco MD, Passarella A (2009) Energy conservation in wireless sensor networks: a survey. Ad Hoc Netw 7:537–568
8. Ming M, Yuanyuan Y (2008) Data gathering in wireless sensor networks with mobile collectors. In: Proceedings of 2008 IEEE international symposium on parallel and distributed processing, Miami, USA, 14–18 April 2008. IEEE
9. Miao Z, Ming M, Yuanyuan Y (2008) Mobile data gathering with space-division multiple access in wireless sensor networks. In: Proceedings of the IEEE INFOCOM 2008—the 27th conference on computer communications, Phoenix, USA, 13–18 April 2008
10. Charalampos K, Grammati P, Damianos G, Aristides M, Basilis M (2012) A rendezvous-based approach enabling energy-efficient sensory data collection with mobile sinks. IEEE Trans Parallel Distrib Syst 23(5):809–817
11. Miao Z, Yuanyuan Y (2012) Bounded relay hop mobile data gathering in wireless sensor networks. IEEE Trans Comput 61(2):265–277
12. Liang H, Yanyan Z, Jianping P, Jingdong X (2010) Evaluating on-demand data collection with mobile elements in wireless sensor networks. In: Proceedings of 2010 IEEE 72nd vehicular technology conference, Ottawa, Canada, 6–9 September 2010
13. Liang H, Zhe Y, Jianping P, Lin C, Jingdong X, Yu G (2014) Evaluating service disciplines for on-demand mobile data collection in sensor networks. IEEE Trans Mob Comput 13(4):797–810
14. Nimisha G, Indrajit B (2018) Application of mobile sink in wireless sensor networks. In: Proceedings of the 10th international conference on communication systems and networks, Bengaluru, India, 3–7 January 2018. IEEE
15. Ha I, Djuraev M, Ahn B (2017) An optimal data gathering method for mobile sinks in WSNs. Wirel Pers Commun 97:1401–1417
16. Jerew O, Liang W (2009) Prolonging network lifetime through the use of mobile base station in wireless sensor networks. In: Proceedings of MoMM2009, Malaysia, December 2009. ACM
17. Sanu T, Thomaskutty M (2018) Intelligent path discovery for a mobile sink in wireless sensor network. Procedia Comput Sci 143:749–756
18. Jerew O, Bassam NA (2019) Delay tolerance and energy saving in wireless sensor networks with a mobile base station. In: Wireless communications and mobile computing, vol 2019. Article ID 3929876
19. Alex S, Vishwanathan SVN (2010) Introduction to machine learning. Cambridge University Press, United Kingdom
20. Blaser M, Panagiotou K, Rao BVR (2012) A probabilistic analysis of Christofides algorithm, vol 7357. Springer, Heidelberg, pp 225–236
21. Christofides algorithm, personal.vu.nl › AdvancedAlgorithms › SlidesChapter2–2016, PDF
22. Manisekaran S, Venkatesan R, Deivanai G (2011) Mobile adaptive distributed clustering algorithm for wireless sensor networks. Int J Comput Appl 20(7):12–19

A Novel Mechanism for Fraud Rank Detection in Social Networks

Deepika Dasari, M. Kameswara Rao, and Nikhitha Namburu

Abstract Fraudulent behavior in the Play Store, Device's most social media app market, boosts search rankings misuse, and malicious software prevalence. Earlier work focused on malware detection executable applications and authorization investigation. The proposed Fraud rank detector discovers and uses evidence that fraudsters have left behind to detect malware and applications that are vulnerable to malware Manipulation of the hunt. Fraud Rank Detector measures review patterns and integrate uniquely identified user interactions with fraudulent and behavioral metrics gleaned from Fraud Rank Detector details the comments obtained to recognize suspect applications. Fraud Rank Detector attains more than the 95 basis points of precision in malware categorization, Fraudulent and legitimate gold standard data applications. We indicate that the Fraud Rank Detector finds the apps that oversee the search security guard's tracking technology.

Keywords Search rank · Fraud detection · Malware

1 Introduction

The financial success of Google platforms including the incentive forum they need from applications renders them perfect resources for violence and misconduct. Many dishonest programmers boost the rank and recognition of the apps on the market (e.g. by feedback, and false install credentials). Thus fraudulent programmers use mobile stores as a landing pad for the ransomware. The motive for these activities is the effect:

D. Dasari (✉) · M. Kameswara Rao · N. Namburu
Department of Electronics and Computer Engineering, KLEF, Vaddeswaram, Guntur, India
e-mail: deepikadasari985@gmail.com

M. Kameswara Rao
e-mail: kamesh.manchiraju@kluniversity.in

N. Namburu
e-mail: nikhithanamburu5@gmail.com

G. Ranganathan et al. (eds.), *Inventive Communication and Computational Technologies*, Lecture Notes in Networks and Systems 145,
https://doi.org/10.1007/978-981-15-7345-3_44

spikes in-app success turn into financial advantages and increased propagation of viruses. Deceptive programmers often use crowdsourcing platforms (e.g. Freelance writer, Fever, Good App Promotion) and recruit groups with eager employees to jointly evade taxes, trying to emulate For instance, actual, accidental actions of persons (i.e. "crowdturfing"). They label it "activity level theft" conduct. Therefore, Mobile companies are often not effective in their attempts to detect and uninstall the malware. For example, the Security guard program is used by the App store to delete ransomware. The long term the effect of online services based on the available knowledge published by their users has also produced a profitable platform to influence public opinion illegitimately. A search rank depends on the sustainability of mobile applications in industries such as Google Play.

More frequently significantly lower-ranking devices are installed and generate more income from commercials. The common belief that many good reviews help the new apps achieve greater search rankings. The resulting need for programmers to advertise their applications forcefully created an App Search Optimization (ASO) submarket. A process development approach combines application test cases and development strategies to a well-planned number of steps leading to software development construction. The research method will collaborate in the preparation of test cases, the development of test cases, the implementation of tests, and the subsequent data collection and analysis. Computer testing is a key component of product quality certainty and is also the greatest analysis of the architecture and programming of specifications. The author suggests that unpredictable app activities can differ from another framework to another. This also varies from such a specific project's atmosphere that runs on various devices. The troll might be attacking the fakeness of both there is measured by using internet-based spammer assessment tools. After the analysis has been completed, the result shows the efficacy of predicting reviews. The researchers have studied this issue of identifying hybrid schilling attacks to ranking information. The method addresses a formal question of vector-completion over the falsely inflate-symmetric vector space. The researcher shows a detailed recovery principle when the suggested solution works. They also undertake a thorough analysis of the suggested solution with artificial data and YouTube ratings qualitative review.

2 Literature Survey

Zhou and Jiang [1] Description as well as a corresponding science-based study of model communities reveals they're developing quickly so circumvent existing smartphone pro-virus software detections. Based on 4 popular network security assessment tools. Burguera et al. [2] early methods to complex device behavior analysis as a way to detect malware on a Mobile platform. The tracker is integrated into an aggregate system to collect traces from an infinite number of real targeted marketing users. Shabtai et al. [3] the recommended architecture create a network-based vulnerability scanning device that constantly tracks different features and activities collected from either the Smartphone and instead implements deep learning anomaly detection to identify the data collected as regular or irregular. Peng et al. [4] the hypothesis used selects the target of progress, complexity, and implementation, so this is an absolute requirement to understand the possible optimization algorithms and to choose the right one to get the best result. Virus Total [5] Furthermore, they connect MobiSentry for Android software that enables Mobile smartphones to collect functionality or determine if the functionality is positive or harmful. They introduce the new MobiSentry, a compact weapon framework for anti-malware and classification of smartphones. Besides common fixed functionality like privileges or connections to the Rest. PCMag et al. [6] Some of Android's main attack mechanisms against malicious programs is a risk communication system that warns the user of the privileges requested by the device when installing the item from the device, confident the user will make that's it the right decision. Greenberg [7] serves as the Android operating system's official app store, allow developers to explore and install programs that have been created and to use the Mobile Application development Kit and posted via Trawl. Google Play also acts as a multimedia platform with content for songs, newspapers, articles, photos, or Radio. Miners et al. [8] Report: This most vulnerable were applications designed to customize people's Android-based phones, as well as apps for television and gameplay. Some of the most malicious apps downloaded since 2011 in the Google Play store include Wallpaper Dragon Ball, a wallpaper app, and Finger Hockey and Subway Surfers Free Tips games. Grace et al. [9] Built an automated system called Risk Ranker to scalable evaluates the dangerous behavior of a particular app (For example, begin kernel exploitation or submit Emails). The manufacture may be used to produce a focused set of decreased requirements that once again benefit research. Ye and Akoglu [10] prevalent customers search the appropriate feedback for comparison while shopping online before buying the products. Most shops or users can write fake reviews to deceive customers to make risky purchasing decisions. Methods of misleading analysis did not take into account the relevant customer review sets and the likelihood of classification of the function weights. Sahs and Khan [11] Despite the limited resources available and limited rights granted to the user, the question of identification of such malware poses unique challenges but also presents unique opportunities in the requisite documentation added on each submission. Oberheide and Miller [12] In Play store, plan to classify both malicious and spam objectives in the search rank.

3 Search Security Gurad's Tracking Technology

Usually a guard tour program would allow office workers to set up digital checkpoints alongside a guard patrol. Checkpoints can be fully programmed by the office workers with a smart tracking device, which activates automatically when a guard enters a certain location during their patrol. No setup or installation of on-site equipment is necessary. Checkpoints can also ask for personalized questions. Similar to the asset being tested thus allowing the collection of very precise data. After entering checkpoints, guards can scan barcodes, QR codes, or even NFC tokens with their app. The computer automatically tracks when a soldier checks in at a specific location, meaning the knowledge is correctly transmitted even though the guard's computer momentarily loses service. When a guard refuses to go into the correct checkpoint, a message will be sent to back-office workers suggesting that the route has not been completed as needed. Messages may also be forwarded when a guard arrives at a checkpoint late. Following these processes, most security guard monitoring systems in real-time often introduce steps to simplify interactions between field guards and administration. Guards could use their tour program to send reports or request assistance, whereas the office staff can send intervention orders, client demands, and other critical communications. Such notifications will appear in the same device that the guard uses to search checkpoints, ensuring prompt receipt of orders.

4 Proposed Methodology

The Proposed work constructs the idea of modeling core view diagrams user-review interactions. We build PCF, which is an effective heuristic for recognizing shortly limited, core view proto-clicks established by the reviewers when the with considerably intersecting core view events across short time doors. We are using transitory dimensions of comment-time review to define suspect analysis surges obtained by apps; we show that a conman wants to post at least positive feedback to offset for an adverse review, for an application which has a score. They often classify applications with "unstable" counts for analysis, ranking, and installation, as well as apps with the ramps for asking acceptance. This work is based on then the notion that they illegitimate and fraudulent behaviors on app platforms leaving the obvious signs aside. Fraud Rank Detector achieves over 97% accuracy in the detection of malicious and the begging programs, and the 95% over the correctness the identification of malware and kindly implementations.

If instance, that heavy cost of creating legal IOS websites enables scammers to recycle the profiles by local job reviews, making them easier than normal users to evaluate popular sites. Time restrictions could cause scammers to leave feedback in the shortest possible time. Legitimate malware-affected users will record adverse comments know. Increases throughout the number of permissions needed by one edition to another, that we may consider "permit ramps", can indicate improvements to malware that are beneficial. Fraud Rank Detector surpasses Sara et AL's ransomware metrics substantially. However, ransomware also involves search fraud: while focused on malicious and harmless software, Fraud Rank Detector detected more than the 75 and bonus typical golden malware apps Fraud Rank Detector detects thousands of malicious implementations. Unlike existing solutions, the proposed system develops its research onto the finding which illegal and harmful activities on both the mobile sectors left away warning signs. That's the proposed system uncovers these nefarious works by selecting these tracks.

5 System Architecture

1. Register and Login
2. Next View Profile
3. And then Add Mobile name and OS
4. Select the mobile name and OS and Upload apps with App name, App uses, App desc (enc), App logo image, attach app
5. Add mobile booklet like Select the mobile name and OS and the attach Mobile Manuals file
6. View all uploaded the application(app) with rank and ratings and Mobile Manuals.

1. Register or Login
2. View your profile
3. Search mobile apps by keyword and review or recommend or like or give rating like 1, 2 or 3(increase rank for all)
4. Request for a secret key for download and view response
5. List all secret key permitted apps and download
6. Search top K Apps Search mobile manual by Selecting mobile name and OS and.

6 PCF Algorithm Pseudo Code

Step 1: Start:
For $d = 0$: $d <$ days. Size; d ++
Step 2: Chart PC = new ();
Best Click (PC, Days [d]);
$C = 1$; $n =$ PC ();
Step 3: $D = d + 1$; $d <$ days. Size () and $c = 1$; d ++
Best Click (PC, days [nd]);
$C =$ (PC.size () $> n$);
Step 4: Where (PC.size) $> n$); end for
All Cliques = allCliques.add (PC);
return
Step 5: BestNearClique (PC graph, Set revs)
if (PC.size () − 0)
for (root = 0; root < revs.size); root ++
Step 6: Graph candClique = (new graph);
CandClique.addNode (.get User);
CandNode = gettingMaxDensityGain (revs);
Step 7: If (candCliquedensity [{candNode}) [u])
CandClique.addNode; fi
Whereas (candNode! = zero);

Step 8: Whether (candClique. Density () > mxRho)
MaxRho = candClique);
PC = candclicks; fi ends
Step 9: Another way if (PC.size () > 0)
Do = getMaxDensityGain (revs);
If (candClique density [candNode] u) Fi
PC.addNode (candNode);
Step 10: Whereas (candNode! = null);
Return
Stop

The Database server will sign into this device using valid login details. Administrators can change privacy settings, install operating systems, and access all personal files. Management will make any changes to other accounts as well. The administrator will add the apps in this unit. If the administrator wants to add the app, the administrator must enter the title of the app, the definition of the device, the phone form, users, the name of the document, the images of the implementation, and the registry button. In the repository, the specifics will be located. View implementation in this device, when you click on the implementation, the name of the implementation, the depiction of the implementation, the portable type, consumers. Related, when admin blinks on the scientific proof for scam details will also reflect the twitter handle, mobile sort, app name, implementation ID, fraud Mac address, program name, date, and time. An admin console is a location on a database server used to hold a user name, and other information about a computer admin console allows or prohibits a user from linking to a network, another computer, or other stocks. Any system with different users requires transactions from either the user. The Web or your e-mail account is a great example of a user profile.

7 Results and Discussion

Fraud Rank Detector is extremely accurate and has a real-world effect: high accuracy. In the classification of malicious and benign applications, Fraud Rank Detector achieves over 97% precision and the over 95 million correctness in the identification of fraudulent and the harmless implementations. Google play substantially over functions of malicious software interventions against Sara et al. We also demonstrate that malware frequently includes search level cheating. Fraud Rank Detector classified more than 75 million of regular gold malware software as malicious when informed on fraudulent and benign software. Real-world impact: it demonstrates fraud and challenges. Google play identifies hundreds of fake games. We show that these requirements are suspicious: 93.3% of testers make at least. Semi-clique, 55% of all these requirements have at least 33% of the reviewer's articles of approximately 75% of all such applications produce at least 200 fraud-related terms. Using Android allowed us to explore new, deceptive assault-type review tactic where device users are the goal is to write a good brand analysis and to install and evaluate other applications. Google Play's design is based on apps that are presented as red disks. Designers, seen to add disks to phones. A designer can have programs uploaded. Players download and test software that is displayed as markers of course. A consumer can only search for an item that recently installed.

This selection of functionality leads to improved Android malware characterization and identification. For example, they merged that simple and complex technologies should obtain good accuracy in anti-malware analytics. In our view, if malware cannot be properly detected, the machine learning model must not have learned its malicious characteristics properly. Notice that every application is classified as malicious software that must have certain special properties known as malicious behavior. Hence, to identify and detect more forms of malware, it is important to collect much awesome-grained functionality which may include further malicious factors. Though Android malware detection is still a huge challenge. Scientists as well as smartphone users across the country must also join forces to contribute their recent inventions of Android malware at a shared forum such as the Contagious Forum, which since 2008 using large data sets of both the software, consumers may combine troops to protect against threats.

8 Conclusion

This paper proposed a system to look at the heaviness of the consent for applications to recognize misrepresentation and make their rank zero and further square the engineer from transferring malware applications. Also, the critical improvement in malware application recognition and the positioning of certified applications with various access authorizations. We showed the efficiency of the proposed system utilizing the dataset of App Permissions that is checked utilizing the weighting parameter. By utilizing the download tally, a noteworthy improvement in insurance is viably done when hindering the malevolent applications in the underlying phase of downloading and rating.

References

1. Zhou Y, Jiang X (2012) Dissecting android malware: characterization and evolution. In: Proceedings of 2012 IEEE symposium on security and privacy, pp 95–109
2. Burguera I, Zurutuza U, Nadjm-Tehrani S (2011) Crowdroid: behavior-based malware detection system for android. In: Proceedings of ACM SPSM, pp 15–26
3. Shabtai A, Kanonov U, Elovici Y, Glezer C, Weiss Y (2012) Anomaly: a behavioral malware detection framework for, android devices. Intel Inform Syst 38(1):161–190
4. Peng H et al (2012) Using probabilistic generative models for ranking risks of android apps. In: Proceedings of the 2012 ACM conference on computer and communications security, pp 241–252

5. Virus total-free online virus, Malware and URL scanner. [Online]. Available https://www.virustotal.com/. Last accessed on May 2015
6. Mlot S (2014, 8 Apr) Top Android app a scam, pulled from Google Play. PCMag
7. Greenberg A (2012, 23 May) Researchers say they snuck malware app past Google's 'Bouncer' android market scanner. Forbes Security. [Online]. Available http://www.forbes.com/sites/andygreenberg/2012/05/23/researchers-say-they-snuckmalware-app-past-googles-bouncer-android-market-scanner
8. Miners Z (2014, 19 Feb) Report: malware-infected android apps spike in the Google Play-store. World. Available http://www.pcworld.com/article/2099421/report-malwareinfectedandroid-apps-spike-in-the-google-play-store.html
9. Grace M, Zhou Y, Zhang Q, Zou S, Jiang X (2012) Risk ranker: scalable and accurate zero-day Android malware detection. In: Proceedings of ACM MobiSys
10. Ye J, Akoglu L (2015) Discovering opinion spammer groups by network footprints. In: Machine learning and knowledge discovery in databases. Springer, pp 267–282
11. Sahs J, Khan L (2012) A machine learning approach to Android malware detection. In: Proceedings of EISIC
12. Oberheide J, Miller C (2012) Dissecting the Android bouncer. SummerCon 2012, New York

Trusted Cooperative E-Learning Service Deployment Model in Multi-Cloud Environment

S. Udhayakumar, D. Uma Nandhini, and S. Chandrasekaran

Abstract The objective of the research work is to propose an adaptive trust model applicable to cooperative software services deployed in a cloud environment towards e-learning. The cooperation among all the participating services in terms of their compliance with the e-learning standards, risk on the unavailability of specific services, and governance in the case of disputes were not considered. The reputations of such cloud-based classrooms can be evaluated in terms of the trust between domain-specific services like registration, assessment, accounting, and results from announcement services deployed as infrastructure and platform services in a multi-cloud environment. The trust relationship is also to be ascertained between the provider and client of various virtual services in the case of emergent de-provisioning of any resource for a particular course of study.

Keywords Cloud-based e-learning · Trusted cloud computing · Game theory · Cooperative game · Coalition policy · Virtual services · Multi-cloud

---

S. Udhayakumar (✉)
Department of Computer Science and Engineering, Saveetha School of Engineering, Saveetha Institute of Medical and Technical Sciences, Chennai, Tamil Nadu, India
e-mail: mailtoudhay@gmail.com

D. Uma Nandhini
Department of Computer Science and Engineering, School of Computing, Vel Tech Rangarajan Dr. Sagunthala R&D Institute of Science and Technology, Chennai, Tamil Nadu, India
e-mail: umaudhay@gmail.com

S. Chandrasekaran
Department of Computer Science and Engineering, Bharathiyar University, Chennai, Tamil Nadu, India
e-mail: chandrasekaran_s@msn.com

Department of Tamil, Bharathiyar University, Chennai, India

G. Ranganathan et al. (eds.), *Inventive Communication and Computational Technologies*, Lecture Notes in Networks and Systems 145,
https://doi.org/10.1007/978-981-15-7345-3_45

1 Introduction

In this present era of internet computing, business and scientific software services are being deployed through major cloud providers to reach the service benefits without a huge initial investment, thereby benefiting small and medium enterprises. In such a cloud computing environment, the services are expected to preserve the private business logic of the provider and personal information of the user through their service components. Also, since these components are virtualized, control and visibility of their information in the cloud are denied, leading to privacy issues. Securing the cloud has become a challenge and exponentially harder because the digital information flow has become pervasive, data bandwidth grows exponentially, access point proliferates, the business risk multiplies and infrastructure become virtual, so consumers access to various information technology services are exposed to lack of Trust, Privacy and Security (TPS). The ability to trust in the virtualized environment where the infrastructure is not physical and static poses a major security concern in the cloud. In a public cloud, the provider hosts virtual machines (VM) running as an Infrastructure as a Service (IaaS) on the customer's premises do computation, but even then, the customer cannot protect their VMs as they don't have the control toward virtual data's. The ability to design, build and consume services delivered from a trustworthy infrastructure where providers and consumers of services can measure and attest the configuration, state, and behavior of platforms form the future of the cloud ecosystem. One such software service where trusted cooperation forms the fundamental element for collaboration is the e-learning platform. Since this e-learning service is offered worldwide by many universities and institutes, provision is to be done through a multi-cloud platform instead of a single cloud service provider. Therefore, a trustworthy e-learning service is offered to students based on their requirements like, evaluation, credit score, online evaluation, and expert guidance, etc., The cooperative nature of the e-learning model is evaluated using the age-old method of alliance formation through game theory where the stakeholders namely, the universities that offer the course, the learners, and the platform through which the course is delivered, i.e., the cloud providers. Hence the paper proposes a trust evaluation technique that is cooperative and adaptive to assure the security of information assets by providing a multi-layered policy structure to deliver contents for an e-learning solution that collaborates with multiple domains to provide a single solution to the users of the cloud ecosystem.

2 Literature Review

Trust has been studied in a varied context which specifically revolves around assurance and confidence the people, data, entities, information, or processes that will function or behave in expected ways. Trust may be defined as "the level of certainty of reliance toward the confident expectation of something or someone", thus it may be calculated as a positive value of certainty or a negative value uncertainty.

The earlier model on trust deals with an enhancement to the reputation-based trust, where the first model takes care of the feedback of the recommender that is incompatible with the initiator and is eliminated by using similarity factor, compatibility factor, and credibility factor. The second model uses two new factors namely, context and size, where the trust value changes for various contexts of job services with its associated size. The complexity of the context and size determines the behavior of the trustworthiness [1]. The implementation is done using a KVM virtual machine in java programming language [2]. The next work for the management of cloud-based services focuses on uncertainty and asset specificity in transaction frequency. It also proposes a framework for contract management, monitoring, and legacy compliance that will provide the support of contract process flow [3]. The other work focuses on the importance of trustworthy e-learning services on the mobile platform. The work proposes a radio frequency identification based e-learning data integration [4]. Further, an adaptive trust model for a general software service in a hybrid cloud environment where the adapting trust level that was evaluated for SMS cloud service deployment was proposed [5]. From a different perspective towards the learning system, emotional intelligence, and its academic performance are mined [6]. A Blockchain-based e-Health service is offered in a multi-cloud environment where malicious behavior is monitored for trustworthiness [7]. Thus, it can be inferred from the above works that trust as a means of assurance to cloud users and providers is an upcoming research area. Our model brings out the possibility of game theory to ascertaining the proof of trust.

3 Cloud-Based E-Learning Model

In any e-learning service, the cooperation of cloud providers strengthens the reputation of the organization [8]. In the proposed multi-cloud model, a cloud-based e-learning service is designed to provide the user, where the users' request is serviced by the cloud providers in a trustworthy approach by having a cooperative understanding between their partners. Earlier online learning services require the students to select the courses from a single provider or a particular University. This way of learning session has the flexibility of choosing the time of sessions and cost for each course alone. But never realized the quality of content that is delivered in terms of choice of course based on specialization and teaching methods, means one University may be specialized in delivering a quality service in one domain or area of interest using multiple teaching methods and other universities in their domain [9]. Our model tries to integrate resources from multiple universities and content providers to cooperate through Service Level Agreement (SLA) and deliver the required service in a trustworthy manner, whereby the reputation of individual trusted contents benefits through maximum provisioning. However, challenges like disagreement on service, the responsibility of data collection, and validation require

Table 1 Role and responsibilities of services

Services	Roles	Responsibilities	Players
Registration	Payment	Credit transfer	Student
Course delivery	Content access	Mail and SMS	Student and tutor
Evaluation	Assessment	Assignment	University
Course Completion	Eligibility	Examination and result	Student and University

a suitable arbiter to monitor the arbitration, negotiation, and auditing process. The selection of the arbiter is based on the highest trust value that is gained during the continuous process of service delivery with a high degree of fault-tolerant system. Moreover, a single arbiter cannot act for every servicer and therefore, the authority is entrusted based on the service that is being rendered. The initial selection would be based on the minimal reputation method that any provider may have or the one who wishes to function as a middleware agent. For example, the registration service requires the role of payment with the responsibility of credit transfer done by the student player. The course evaluation service requires the role of assessment by the University to do the responsibility of assessing the assignments submitted by the students. The course completion service requires the student to be eligible either through online attendance and other eligibility criterion set by the University to undergo the responsibility of appearing in the examination. The roles and responsibilities of each service may be depicted in Table 1.

An e-learning cloud middleware is developed as the interface between all the three stakeholders. Once registration is offered through this interface then the universities would cooperate for a feasible output to the requested service as depicted in Fig. 1.

4 Cooperative Game Theory

The concepts of game theory provide a language to formulate structure, analyze, and understand strategic scenarios. It is broadly classified into cooperative and non-cooperative game theory [10]. A Cooperative game is a game in which the player has complete freedom of pre-play communication to make joint binding agreements. These agreements may be of two kinds: to coordinate strategies or to share payoffs. The utility is that quantity whose expected gain in a risky situation is attempted to be maximized by a decision-maker all of whose choices among risky outcomes are consistent. Payoffs are in monetary terms, the utility is linear in money, interpersonal comparisons are meaningful, and monetary side payments are allowed. The N-person cooperative game theory developed by Von Neumann and Morgenstern is generally called a constant sum game whereby if there exists some constant 'c' such that at every terminal state.

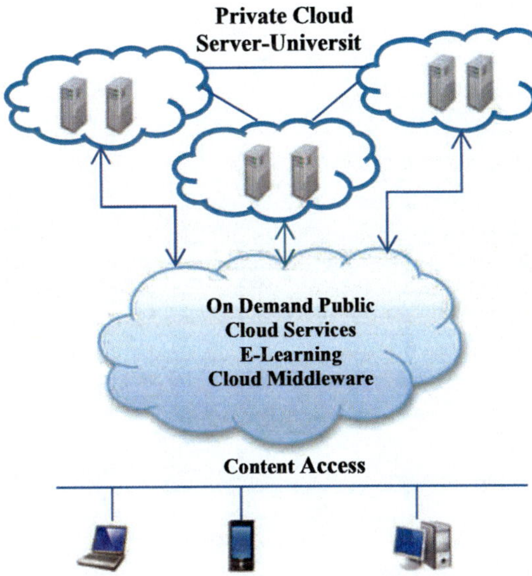

Fig. 1 Cloud-based e-learning model

$$\sum_{i=1}^{n} P_i = c \tag{1}$$

If a player is charged a fee of c/n to enter the game than the original game can be replaced by a strategically equivalent zero-sum game. The outcome of a cooperative game is a distribution of the total available payoff amongst the players rather than as a predetermined consequence of their strategic choices. The possible distribution of available payoff is called an imputation and this is managed and monitored by the arbiter of the model. In an-person cooperative game ‘Γ’ it can be represented by an imputation with a vector $x = (x_1, x_2, \ldots, x_n)$, where x_i denotes the amount received by player i. Let $N = \{1, 2, \ldots, n\}$ be a finite set of players. Each non-empty subset of N is called a coalition. The set N is referred to as a grand coalition. For each coalition S, it shall be specified with a set $v(S) \subseteq R^{|S|}$ containing $|S|$ dimensional payoff vectors feasible for coalition S.

4.1 Formation of the Game Tree

Now, the policy fixed for various trust levels, the game has to be played between the services with its availability matrix. Availability matrix for each course is given in the form of courses in the column offered by the universities in rows as given in Table 2. With this availability matrix a game tree is formed as depicted in Fig. 2.

Table 2 Availability matrix

U/A	C_1	C_2	C_3	C_4
U_1	0	1	1	0
U_2	1	0	1	1
U_3	0	1	1	0
U_4	1	0	0	1

C/N	U1 F1	U2 F2	U3 F3	U4 F4
C1	2	4	3	6
C2	5	6	4	5
C3	4	5	7	6
C4	5	8	9	8

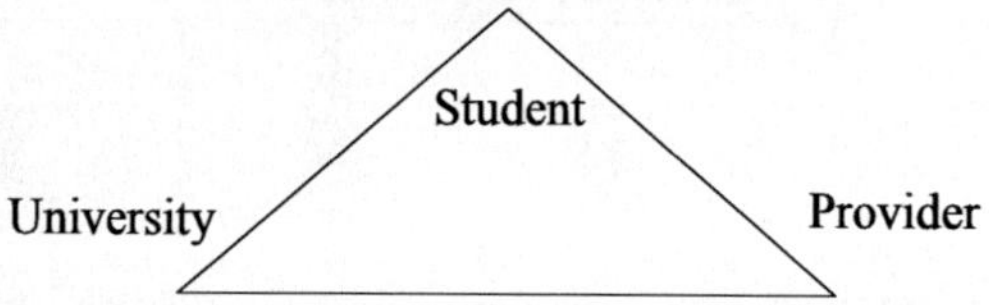

U/N	n1	n2	n3	n4
U1		8	4	
U2	5		2	6
U3		3	1	
U4	10			7

P/Y	y1	y2	y3	y4
P1	1	2	4	4
P2	1	5		2
P3		3	1	
P4	1			5

Fig. 2 Game tree for e-learning players

Based on the delivery methods the trust for teaching methodologies like SMS, mail, web, animation, and video contents can also be classified with trust values. As described in our earlier section on the basics of game theory, here N be the grand coalition and υ be the characteristics function. Then it can be defined as $N = (C, U, P)$, where, C is the No. of Course, U is the University that offers Course C, P is the service provider that hosts Universities Course.

Game policy "*p*" is how the trust that has been described and ordered in the earlier stage is gaining the value through updating by playing the game between various degrees of trust. The rules for *h*, *m*, and *l* is given in Eq. 2.

$p = \{\text{H.T.S, M.T.S, L.T.S}\}$ then

$$\upsilon = \{h : \text{H.T.S} \quad m : \text{M.T.S} \quad l : \text{L.T.S}\}$$
$$h \leftarrow H, h, \upsilon(h) > \upsilon(m)$$

$$
\begin{aligned}
&m \leftarrow M, m, \upsilon(m) > \upsilon(l) \\
&l \leftarrow L, l, \upsilon(l) \ll \upsilon(m)
\end{aligned} \tag{2}
$$

Thus, it decides on how the trust value should be gained, and where any other coalition tries to break the grand coalition to harm the grand coalition without any self-interest.

Coalition Policy Now being the trust degree established, it is up to the coalition partners to be cooperative, decide based on the SLA, which is given in Eq. 3. The agreement that needs to be imposed is through the weighted factors of each coalition. Let $t = T$ if $\upsilon(h) \equiv \upsilon(m) \equiv \upsilon(l)$, a coalition S is a subset of N then $S \subseteq N$,

$$
\begin{aligned}
&h\&\&l ==> l\&\&h \\
&m\&\&l ==> l\&\&m \\
&\neq h\&\&m
\end{aligned} \tag{3}
$$

It can be inferred from the above expression that high trusted service and medium trusted services should not cooperate, because if doing so it may suppress the existence of low trusted components for its serviceability.

Imputation Policy Since establishing trust is the core of the work, the cost-sharing that has been proposed is based on providing service to trusted agents by offering an incremental cost model which also ensures a fairness property among the players. To achieve fairness among the players we need to adopt the method of Shapley value which identifies the provider's cost based on random ordering. Now the Shapley value can be modified by ordering the service providers based on the trust they have gained by continuously cooperating with the coalition game and abiding the initial agreement to share the cost through cross monotonic method [11, 12]. Given a coalitional game (N, υ), the Shapley value of player i is given in Eq. 4.

$$
\varphi_i(N, \upsilon) = \frac{1}{|N|!} \sum\nolimits_{S \subseteq N \setminus \{i\}} |S|!(|N| - |S| - 1)![\upsilon(S \cup \{i\}) - \upsilon(s)] \tag{4}
$$

Thus, the Shapley value of a provider identifies his worth and hence the trusted provider's importance can be judged for trustworthy cloud collaboration. The trust correlation graph between players, the imputation policy, the coalition, and trust policy is given the graph in Fig. 3. The model is trained with the student's dataset based on the adaptive trust model where a supervised learning method is adopted. The probability of student registration can be done using the Bayesian model.

These cost factors are decided based on the resource usage that is being monitored by the cloud services as mentioned in Table 3.

The real-time implications of the proposed model are done using a public cloud interface, where the courses are hosted on the service providers platform. This e-learning service model analyses the students' progress of learning and the arbiter monitors the cooperative nature of the model.

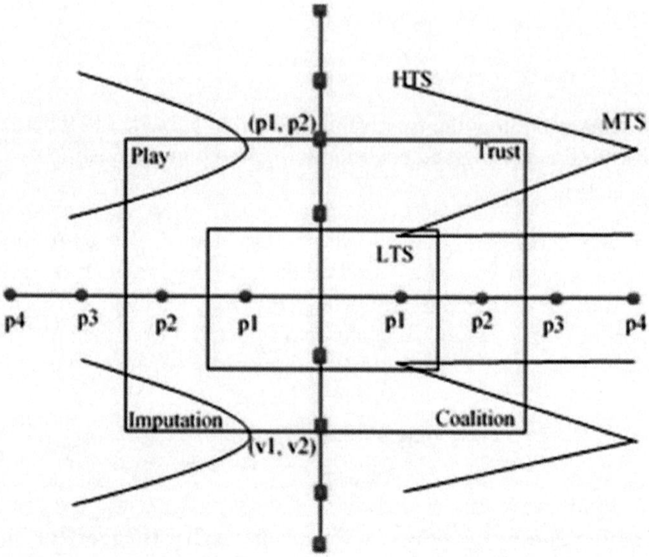

Fig. 3 Trust correlation graph

Table 3 Subject versus credit and weightage for each teaching methods

Subjects	Credit	Service providers teaching methods	Wt
Chemistry	3	Tutorial by experts	5
Physics	3	Animation	4
Computer	3	Web page tutorial	2
Maths	4	Exercise	3
Physiology	3	Presentation	6
Anatomy	2	Online examination	10
Accounting	4	Webinars	7
Multimedia	3	Email and SMS	5

5 Conclusion

Cloud is the present and future platform for e-learning services and the growth of which is phenomenal given the rise in students registering courses and taking exams online. The proposed model provides a trustworthy nature of the cloud in a multi-cloud environment for the software service of e-learning. This adaptive service model offers various features like the cost associated with subject credits, teaching method's weighted values, courses offered, and payment process. Monitoring and management policies of the model ensure that players are cooperative and adhere to game policy.

Thus, a trusted e-learning model that governs all stakeholders for cooperation can benefit the learning communities to use next-generation technology effectively.

References

1. Noor TH, Sheng QZ, Yao L, Dustdar S, Ngu AHH (2016) CloudArmor: supporting reputation-based trust management for cloud services. IEEE Trans Parallel Distrib Syst 27(2):367–380
2. Ribeiro F, Metrolho J, Costa M (2010) A bluetooth-based interactive system to improve relationships between actors in educational environments. WSEAS Trans Adv Eng Edu 7(2):33–42
3. Makhlouf R (202) Cloudy transaction costs: a dive into cloud computing economics. J Cloud Comput 9
4. Luo Z, Zhang T (2010) A mobile service platform for trustworthy e-learning service provisioning. Int J Dependable Trustworthy Inform Syst 2
5. Udhayakumar S, Latha T (2019) Trustworthy cloud federation through cooperative game using QoS assessment. In: PReMI 2019. Lecture notes in computer sciences, vol 11941. Springer Series, pp. 30–37
6. James SP, Ramasubramanian P, Angeline DMD (2020) Student learning context analysis by emotional intelligence with data mining tools. Int J Intell Eng Syst 13(2):286–298
7. Kurdi H, Alsalamah S, Alatawi A, Alfaraj S, Altoaimy L, Ahmed SH (2019) HealthyBroker: a trustworthy blockchain-based multi-cloud broker for patient-centered e-health services. Electronics 8(602):2–17
8. Badidi E (2015) Towards cooperative cloud service brokerage for SLA-driven selection of cloud services. Adv Intell Syst Comput 349:271–281
9. Caballé S, Miguel J, Xhafa F, Capuano N, Conesa J (2017) Using trustworthy web services for secure e-assessment in collaborative learning grids. Int J Web Grid Serv 13(1):49–74
10. Robinson N, Valeri L, Starkey J (2010) The cloud: understanding the security, privacy and trust challenges. Directorate-General Information Society and Media, European Commission
11. Nisan N, Rough T, Tardos E, Vazirani VV (2007) Algorithmic game theory. Cambridge University Press, pp 385–410
12. Shanmugam U, Tamilselvan L (2017) Trusted computing model with attestation to assure security for software services in a cloud environment. Int J Intell Eng Syst 10(1):144–153

Efficient Data Security Using Hybrid Cryptography on Cloud Computing

P. Chinnasamy, S. Padmavathi, R. Swathy, and S. Rakesh

Abstract Services are distributed among all servers and between the users and individuals in the cloud environment. Cloud providers have trouble guaranteeing file protection as security is the biggest issue in data handling and transfer as it can be accessed, misused and destroyed the original data form. Cloud security is a big concern in the cloud computing environment. To safeguard the cloud environment, many research works are being proposed. To overcome the security issue and achieve the CIA property (confidentiality, integrity and availability) the cryptography is used. Cryptography is the most useful technique to ensure a high level of data transfer and storage security. In traditional symmetric and asymmetric has some limitations. To solve this we are going to introducing a new hybrid technique to achieve high data security and confidentiality. In this article, we are combing ECC and Blowfish to implement a hybrid algorithm. The performance of the hybrid system is compared with the existing hybrid method and shows that the proposed method provides high security and confidentiality of patient data. The hybrid cryptography is used to defeat the inconveniences of both symmetric and asymmetric.

Keywords Blowfish · Cloud environment · CIA property · Elliptic curve · Hybrid cryptography · Security

P. Chinnasamy (✉) · S. Padmavathi · R. Swathy · S. Rakesh
Assistant Professor, Department of Information Technology, Sri Shakthi Institute of Engineering and Technology, Coimbatore 641062, India
e-mail: chinnasamyponnusamy@gmail.com

S. Padmavathi
e-mail: padukutty07@gmail.com

R. Swathy
e-mail: swathyyashodha@gmail.com

S. Rakesh
e-mail: rakeshmrk1998123@gmail.com

G. Ranganathan et al. (eds.), *Inventive Communication and Computational Technologies*, Lecture Notes in Networks and Systems 145,
https://doi.org/10.1007/978-981-15-7345-3_46

1 Introduction

Information security is the major issue in technology development, and it seems to be the most critical and necessary to maintain data privacy while transmitting through the network. More generally, cryptography is about building and analyzing protocols, which prevent private messages from being read by third parties or the public. The algorithms used for this process are known as cryptographic algorithms or ciphers (altering data from readable form to protected form) which can be categorized into two basic types based on the keys used as the symmetric key and asymmetric-key algorithm.

Symmetric encryptions, such as (DES, RC2, RC4, Blowfish, RC5, RC6, or AES) are the oldest and method of encryption in which only a single secret key is used to encrypt and decrypt data. The sender and receiver share the key, which is a major drawback because the key exchange channel can be searched by an intruder to decrypt the data [1]. To turn the secret key you need a secure channel between the sender and receiver. In comparison, asymmetric encryption, like DSA, RSA and ECC, uses two keys for plain text ciphering, both public and private. Any entity with a public key can use it to send a message but the private key is kept secret and used to decrypt the message.

A hybrid cryptosystem is a system that includes many ciphers of different kinds, each of which has its strongest strengths [2]. The sensible solution is to build a unique encryption key for secret key cryptography and afterward encode secret key with participant's public key by an asymmetric cipher. The encrypted secret key and ciphertext are then sent to the receiver.

The reason for choosing the Blowfish and Elliptic curve are as follows:

1. The data gets encrypted quickly with the help of Blowfish [3] (26 clock cycle per byte).
2. A smaller amount of memory is needed (5 KB).
3. The default key size of Blowfish is 128 bit, but based on the length of the key the Blowfish key size ranges from 32 to 448 bits.
4. The Elliptic curve is stronger than RSA. The standard 256 bit ECC key size is equal to a 3072 bit RSA key, and 10,000 times more powerful than a 2048 bit RSA key.
5. ECC requires less processing power and memory, resulting in considerably faster response times and Web server performance when it is in operation.
6. Higher safety level [4] with a smaller key size compared to other Cryptographic techniques.

The subjects covered in this paper are as follows: Introduction is discussed in Sect. 1. In Sect. 2 Related works are discussed. In Sect. 3 our proposed method with architecture explanation and flow process with methodologies and its algorithms are explained. In Sect. 4 results and analysis are explained. Ultimately, Sect. 5 addresses the conclusion and potential enhancement.

2 Related Works

Few of the hybrid cryptosystem algorithms are discussed below along with their advantages and disadvantages for secure cloud storage.

Kamara and Lauter [5] have proposed a security model that works on the public cloud, using cryptographic primitives for verifying data integrity. This paper discussed the benefits of cloud storage such as availability, reliability, efficient retrieval and data sharing, which combines recent and non-standard cryptographic primitives for secure cloud storage.

A hybrid data encryption system that would use both RSA and Blowfish was implemented in [6]. In this, they used a mathematical methodology to implement the Field Programmable Gate Array (FPGA). This strategy is very effective given its low cost and high level of protection. But key size (448 bits) is the primary issue.

Maitri and Verma [7] suggested the use of a hybrid cryptographic technique to protect cloud file storage. They used steganography with LSB by which the encryption key is covered into a picture header for key information integrity.

In [8], an innovative technique of hybrid cryptography was developed for health records. In that, they used Blowfish and enhanced RSA algorithms to improve patient data security and prevent false requests.

Wang et al. [9] introduced a new method to encrypt information and send encrypted data to another user, the user creates the public key. Decryption is done through a private key. Use of symmetric and asymmetric searchable encryption to search over encrypted data. Wang et al. have designed a model that uses security encryption techniques, and users should have prior knowledge of encrypted data.

In [10], he presented a hybrid technique (AES-RSA) for lightweight data. However, it cannot be applied to multimedia data as it provides security for lightweight data only.

In addition to Order Preserving Symmetric Encryption (OPSE), symmetric searchable encryptions were employed. System analysis has shown its usefulness in the case of a graded keyword search, but attacks, integrity and confidentiality are not relevant information. So, it might not be appropriate to provide security. Incremental encryption [11] enables data to be encrypted and exchanged with other authorized users with a different encryption key before being stored in the cloud. Agarwal and Agarwal [12] spoke about security threats in the cloud.

Dubey et al. [13] proposed to exchange data in the cloud using RSA and they have used the MD5 algorithm for data integrity. They utilized the RSA algorithm to encrypt large data files to enhance data security in the cloud.

Sarkar and Kumar [14] recommended a method for ensuring cloud data protection using hybrid encryption. This strategy would also boost data protection at a high overhead communication rate in the cloud.

Chinnasamy and Deepalakshmi [15] introduced a novel technique which produces access control as a service using multilabel (SMBACaaS). They have used an improved key generation scheme of RSA (IKGSR) for generating key and signature to achieve better confidentiality and security.

The different types of cryptographic algorithms are analyzed in [16] and are used in modern cloud storage. We gave a quick summary of various security concerns, and how we can use cryptographic methods to create stable cloud storage systems.

Singh and Kaur [17] suggested a user data encryption system before being transmitted to the cloud. AES is used to encrypt user data, and the RSA algorithm encodes the secret key. The same operation for decryption is followed, too. The hybrid strategy had been used to combat cloud DOS attacks. Similar to other methods, the only downside of this approach is more time-consuming. Akomolafe and Abodunrin [18] created a new data storage architecture using the cryptographic hybrid model. Secure data storage is obtained by using the AES, Blake2b and Schnorr Signature algorithms. The service provider is unknown about the personal encryption method to provide a high level of security because data encryption is performed on the client-side before uploading to the cloud. The method is nonetheless incompatible with multimedia files.

Karthik et al. [19] proposed the use of both symmetric key (One-Time-Pad) and Asymmetric-key algorithm (RSA) to provide strong security. The product of this approach offered better security. The time taken to encipher data is also faster than the process already in use.

Rahmani et al. [20] proposed a new method for cloud services with XaaS architecture. The authors suggested Cloud Encryption as a Service (EaaS) by which the service provider encryption security risk is reduced and client-side protection is enhanced.

From these surveys, the cloud provider is responsible for the security of client data. An asymmetric cryptosystem with hyper-elliptic curve cryptography is proposed for efficient data security, which provides secure data encryption as well as protected shield against data theft on the cloud. From the user's point of view, he believes the user has to access a high amount of cloud data in a protected way. However, the complexity of the cryptographic algorithm used, with the security concern, has not been given much importance. To resolve the complexities of the algorithm proposed earlier, the proposed model must assist directly in knowledgeable, fast and safe access to data.

The hybrid approach described above provides confidentiality only. Whereas our proposed method is novel in terms of providing features such as; enhancing client-side security through the use of hybrid cryptosystem (BLOWFISH+ECC), the efficiency of the proposed method is greatly improved in comparison with existing methods, as well as security also enhanced.

3 Proposed Method

The hybrid cryptography combines the public key cryptography with the symmetric key cryptography. The hybrid algorithms used here are Elliptic curve cryptography (public key cryptography) and Blowfish algorithm (symmetric key cryptography). Elliptical curve cryptography (public key encryption) based on the Elliptic curve theory that can be used to generate cryptographic keys that are faster, smaller and more effective. The advantage of an Elliptic curve is smaller chip size, less power consumption, increase in speed, etc. Blowfish is a freely available symmetric encryption algorithm, which is a very powerful weapon against hackers and cybercriminals used in a wide range of products including some secure email encryption devices, backup software and password. Due to the small number of rounds, Blowfish is a relatively fast block cipher (encryption tool) very powerful with a relatively simple structure. In this section, we are going to explain the basic functionalities of Blowfish and Elliptic curve cryptographic algorithms.

3.1 *Blowfish Algorithm*

Blowfish is a symmetrical block cipher that performs a Fiesta network, which consists of 16 rounds of functional decryption and iterative encryption.

The block size used is 64 bit and the size of the key can differ from any length to 448. Blowfish cipher uses 18 32 bit sub-arrays commonly known as P-boxes, and 4 32 bit replacement boxes each with 256 entries.

It consists of two stages: The first is Key Extension, and the other is data encryption. Key is converted into many sub-keys in the key expansion process, and encryption occurs in the data encryption phase across 16-round networks. Every round involves a key-dependent permutation and a substitution based on key and data (Fig. 1).

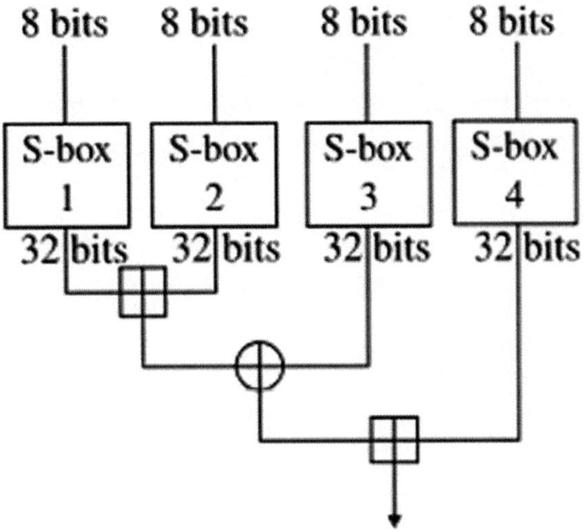

Fig. 1 The function of the Blowfish algorithm

3.2 Elliptic Curve Cryptographic Algorithm

For smaller key size, huge speed and low consumption of memory the elliptic curve cryptography (ECC) has been choosing for instantiating schemes related to the public key, digital encryption, bitcoin services and others. Those proven ECC reputations are based on its algorithmically complex, discrete problem with the logarithm (DLP). In the FP prime finite field, the Elliptic curve is about the cloud of points described in the below equation

$$y_2 = x_3 + ax + b \bmod p \tag{1}$$

where *x*, *y*, *a* and *b* are all elements within the FP. The points to be on the curve is determined by the *a* and *b* coefficients.

System Model

The cloud serves as the main storage medium where data of the patient is stored in encrypted form using the Blowfish algorithm and its key is encrypted using the public key Elliptic curve. The ciphertext of both patient's data and the Blowfish key are stored in the cloud. To decrypt the Blowfish key the private key of the Elliptic curve cryptography is used and the decrypted Blowfish key is obtained. The Blowfish algorithm uses the decrypted Blowfish key to decipher the patient's data (Fig. 2).

Here the patient's data is taken into account for storage and retrieval of the data from the cloud using the hybrid algorithms (Elliptic curve and Blowfish). The process takes place here is

1. Upload process
2. Download process.

Upload Process

If this is a script or plain text, then the path or data to be directly encrypted is defined by the client. The feature automatically generates a symmetric key called one key, based on the key size. The Blowfish is used to encipher plaintext *P* to obtain ciphertext *C*. The hidden key of Blowfish is authenticated by the Elliptic Curve cryptography process and the key that is encrypted is stored in a secure location.

Download Function

The user gets ciphertext *C* from the cloud. Elliptic curve cryptography algorithm is used for ciphertext key decryption. To obtain the plaintext *P*, the downloaded ciphertext data *C* is decrypted with the Blowfish algorithm.

Implementation

To implement the proposed method the operating system used is Windows 10 and Java 1.8 for the front end, as it is free and platform-independent. For the storage purpose, i.e. for the database SQLite is used which is the commonly used database

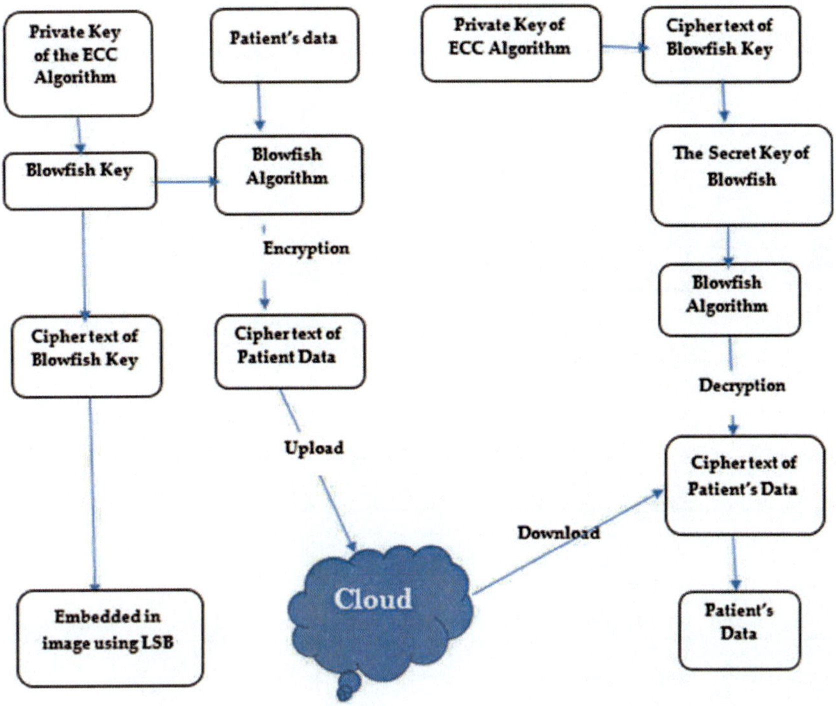

Fig. 2 The architecture of hybrid cryptography

Table 1 Key and block size settings

Technique	Keys (Bits)	Blocks (Bits)
Chinnasamy et al.	128,1024	No limit
Proposed method	128,256	No limit

as it is free and code can be available in the public domain. The IDE used here is NetBeans 8.01 which is open-source and used for Java Desktop applications and the cloud server is Apache Tomcat Server 8.0.27.0 as it is open source and implements Java server pages and Java servlets. The key settings for our proposed method are shown in Table 1.

4 Result and Analysis

In the below graph to compare the efficiency of the hybrid algorithm (Elliptic curve and Blowfish) the parameters taken into consideration are time in terms of (seconds)

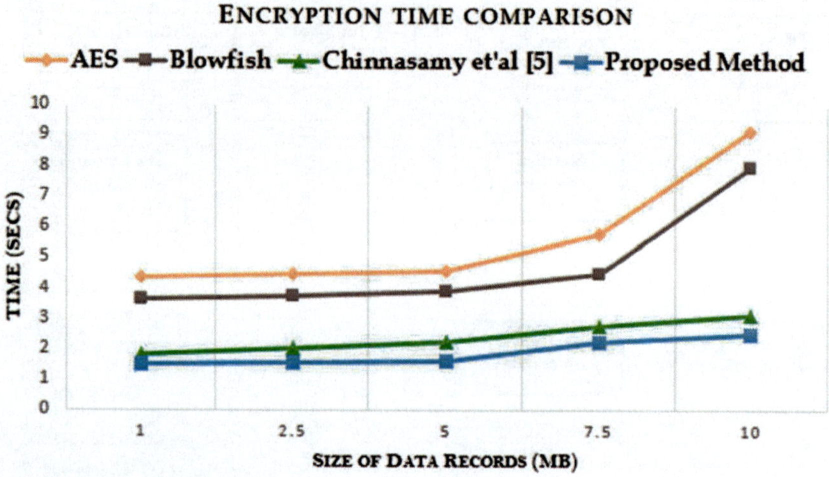

Fig. 3 Encryption time comparison

Table 2 The encryption time analysis

Data	AES	Blowfish	Chinnasamy and Deepalakshmi [8]	Proposed method
1	4.377	3.654	1.823	1.523
2.5	4.466	3.754	2.012	1.572
5	4.586	3.886	2.245	1.622
7.5	5.75	4.492	2.742	2.228
10	9.142	7.963	3.128	2.483

along *x*-axis and data size (a record) in terms of (MB) along the *y*-axis and the encryption, time is compared (Fig. 3 and Table 2).

In the below graph the algorithm (Blowfish and AES) is compared with our hybrid algorithm. Both AES and Blowfish comes under the symmetric key cryptography. Symmetric algorithms have the main advantage of faster execution and efficient for large amounts of data. By the above graph, it is evident that our hybrid algorithm is efficient than the other algorithms (Fig. 4 and Table 3).

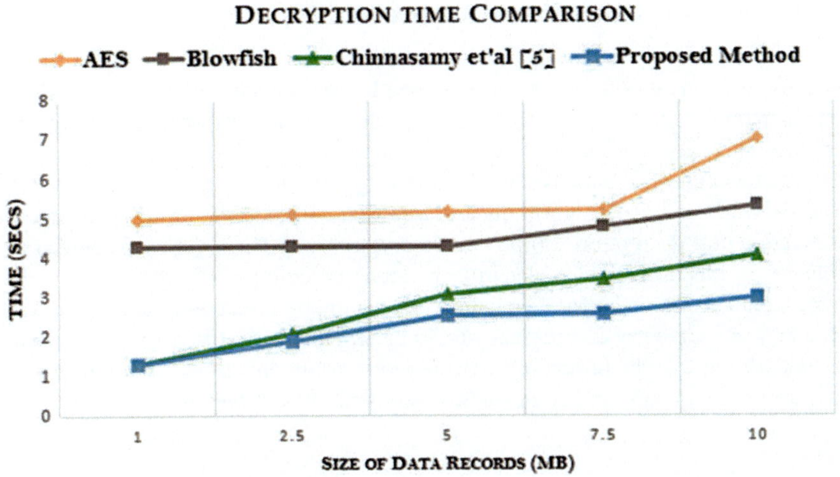

Fig. 4 Decryption time comparison

Table 3 The decryption time analysis

Data	AES	Blowfish	Chinnasamy and Deepalakshmi [8]	Proposed method
1	4.975	4.268	1.307	1.287
2.5	5.069	4.284	2.045	1.874
5	5.18	4.288	3.045	2.53
7.5	5.207	4.788	3.45	2.551
10	7.002	5.304	4.042	2.973

5 Security Analysis

5.1 Security Against Mathematical Attack

We used two separate keys for the decryption process inside the hybrid cryptosystem. This improved data and key protection even after lost one key. However, the attack cannot be carried out while the data is already in an encrypted state.

5.2 Security Against Side-Channel Attack

The security of the ECC algorithm is based on the elliptic curve discrete logarithm problem (ECDLP) is hard. ECC has many operations to compute the curve because all the operations are based on different coordinates. Also, it provides security against differential fault attacks.

6 Conclusion and Future Enhancement

The secured data storage problem is solved by introducing our proposed hybrid cryptography method. The drawbacks of the cloud are lack of greater security and privacy. This model proposed is designed and implemented in Java, incorporating the best techniques of both symmetric key (Blowfish) and asymmetric-key (ECC). The Blowfish and ECC algorithms are used for the processes of key generation, encryption and decryption. Elliptic curve cryptography (ECC) is implemented to achieve an enhanced level of security in cloud computing. ECC provides a more robust and secure model for developing and deploying a secure application in the cloud. To solve the key distribution we can incorporate with steganography method to hide the keys. In the future, to solve the key distribution we can incorporate with steganography method and compare this work with the existing hybrid method.

References

1. Al-Shabi MA (2019) A survey on symmetric and asymmetric cryptography algorithms in information security. Int J Sci Res Pub 9(3). http://dx.doi.org/10.29322/IJSRP.9.03.2019.p.8779
2. Ngwe TT, Phyo SW (2015) Hybrid cryptosystem for data security. Int J Adv Electron Comput Sci 2(6)
3. Schenier on security. https://www.schneier.com/academic/blowfish/. Last accessed 31 Oct 2017
4. Vasundhara S (2017) The advantages of elliptic curve cryptography for security. Glob J Pure Appl Math 13(9):4995–5011. ISSN 0973-1768
5. Kamara S, Lauter K (2010) Cryptographic cloud storage. Lect Notes Comput Sci 6054:136–149
6. Bansal VP, Singh S (205) A hybrid data encryption technique using RSA and blowfish for cloud computing on FPGAs. In: 2nd international conference on recent advances in engineering computational sciences (RAECS), Chandigarh, pp 1–5
7. Maitri PV, Verma A (2016) Secure file storage in cloud computing using hybrid cryptography algorithm. In: International conference on wireless communications, signal processing and networking (WiSPNET), Chennai, pp 1635–1638
8. Chinnasamy P, Deepalakshmi P (2018) Design of secure storage for health-care cloud using hybrid cryptography. In: 2nd international conference on inventive communication and computational technologies (ICICCT 2018). IEEE Xplore Compliant-Part number: CFP18BAC-ART; ISBN 978-1-5386-1974-2
9. Wang C, Cao N, Li J, Ren K, Lou W (2010) Secure ranked keyword search over encrypted cloud data. J ACM 43(3):431–473
10. Liang C, Ye N, Malekian R, Wang R (2016) The hybrid encryption algorithm of lightweight data in cloud storage. In: 2nd international symposium on agent, multi-agent systems and robotics (ISAMSR), Bangi, Malaysia, pp 160–166
11. Gansen Z, Chunming R, Jin L, Feng Z, Yong T (2010) Trusted data sharing over untrusted cloud storage providers. In: Proceedings of the 2nd IEEE international conference on cloud computing technology and science (CloudCom), pp 97–10
12. Agarwal A, Agarwal A (2011) The security risks associated with cloud computing. Int J Comput Appl Eng Sci I(CNS). ISSN 2231-4946
13. Dubey AK, Dubey AK, Namdev M, Shrivastava SS (2012) Cloud-user security based on RSA and MD5 algorithm for resource attestation and sharing in Java environment. In: CSI sixth international conference, software engineering (CONSEG)

14. Sarkar MK, Kumar S (2016) Ensuring data storage security in cloud computing based on hybrid encryption schemes. In: Fourth international conference on parallel, distributed and grid computing (PDGC), Waknaghat, pp 320–325. https://doi.org/10.1109/pdgc.2016.7913169
15. Chinnasamy P, Deepalakshmi P (2018) A scalable multilabel-based access control as a service for the cloud (SMBACaaS). Trans Emerg Telecommun Technol 29(8):e3458. https://doi.org/10.1002/ett.3458,2018
16. Yong P, Wei Z, Feng X, Zhong-hua D, Yang G, Dongqing C (2012) A secure cloud storage based on cryptographic techniques. J China Univ Posts Telecommun 19:182–189
17. Singh N, Kaur PD (2015) A hybrid approach for encrypting data on cloud to prevent DoS attacks. Int J Database Theor Appl 8(3):145–154. http://dx.doi.org/10.14257/ijdta.2015.8.3.12
18. Akomolafe OP, Abodunrin MO (2017) A hybrid cryptographic model for data storage in mobile cloud computing. Int J Comput Netw Inform Sec 6:53–60
19. Karthik, Chinnasamy, Deepalakshmi (2017) Hybrid cryptographic technique using OTP:RSA. In: 2017 IEEE international conference on intelligent techniques in control, optimization and signal processing (INCOS), Srivilliputhur, pp 1–4
20. Rahmani H, Sundararajan E, Zulkarnain Md, Ali AMZ (2013) Encryption as a service (EaaS) as a solution for cryptography in cloud. Procedia Technol 11:1202–1210

Performance Comparison of MQTT and CoAP Protocols in Different Simulation Environments

Malti Bansal and Priya

Abstract Smart healthcare is among the most significant IoT applications. The Internet of Things (IoT) revolutionizes our viable needs that include the healthcare system. This review paper presents an overview of the two extensively used application layer protocols for IoHT systems: MQTT and CoAP. However, the choice of a standard and efficacious application layer protocol out of the two is a challenging task since it requires knowledge about the area of application and its messaging requirements. Hence, it is important to introduce their characteristics comparatively along with the selection of appropriate simulator tools to test, analyze, and validate the preexisting concepts and optimize the performance of prototypes. Subsequently, it has performed relative research based on interdependent criteria and different simulation tools to have an in-depth understanding of their pros and cons. Therefore, researchers can select their relevant application layer protocol and simulation tools based on their suitability and necessity.

Keywords Application layer protocol · MQTT · CoAP · IoHT · Cooja simulator · NS-3 · OMNeT++

1 Introduction

The Internet of Healthcare Things (IoHT) is primarily a network system based on IoT which intercommunicates between a patient and facilities related to healthcare resources such as hospitals and electronic health systems for ECG or EKG (Electrocardiogram), HR (heart rate), EEG (electroencephalogram), blood sugar for diabetes, and biomedical sensors based vital signs related to patient's body. These vital signs involve pulse, oxygen content in blood, galvanic skin response, airflow (breathing), glucometer for blood sugar level, body temperature, blood pressure (BP), patient

M. Bansal (✉) · Priya
Department of Electronics and Communication Engineering, Delhi Technological University, Delhi 110042, India
e-mail: maltibansal@gmail.com

G. Ranganathan et al. (eds.), *Inventive Communication and Computational Technologies*, Lecture Notes in Networks and Systems 145,
https://doi.org/10.1007/978-981-15-7345-3_47

location and activity (accelerometer), and measure of electrical activity, i.e. electromyography [1]. This vital information is gathered through various biomedical sensors attached to the patient's body and then further sorted by mobile applications or other devices such as computers system, smartphones, specific embedded devices, etc. [2]. These devices are connected to a gateway through various messaging protocols such as AMQP, MQTT, RESTFul, COAP, HTTP, DDS, 6LowPAN, etc. Furthermore, Cloud services are used as a repository to which a Gateway connects sorts and stores big data collected continuously from the patient. On the other side, patients' information can also be hoarded in a Health Information System (HIS) with the help of e-health records which thereby allows the doctor/relative/patient to access the clinical history of patients whenever a patient visits a medical doctor (Fig. 1).

In the further section, we will focus on various application layer protocols in the IoHT system to establish the interconnection between devices like computers, mobile phones, embedded chips, and a gateway which further stores the data on server/cloud.

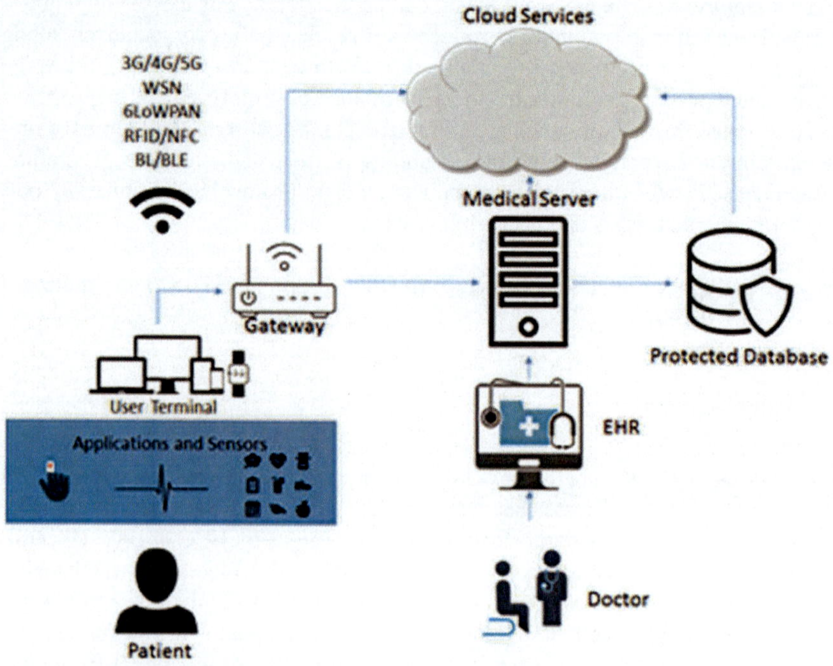

Fig. 1 IoHT-based structure [16]

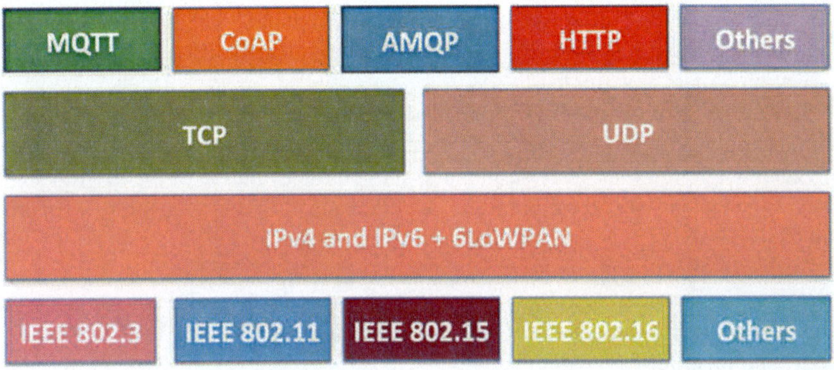

Fig. 2 Protocol stack for IoHT systems [3]

1.1 Application Layer Protocols for IoHT

IoHT can use numerous application layer protocols available depending upon the kind of requirements of the system. AMQP, DDS, MQTT, CoAP, DDS, HTTP, XMPP, SMQTT, RESTful are some of the application layer protocols generally used. AMQP and JMS are applicable when the development and designing of applications related to address are required along with speed and reliability in transactions related to business. MQTT and CoAP which are constrained network mainly applicable for gathering the data (e.g. sensor updates). Few protocols are developed for applications that require real-time text transmission over the internet and spotting online status. Few protocols such as RESTful, HTTP, and CoAP [3] are mainly developed for web applications that require communication over the internet. In further segments, the two most prominent and emerging application layer protocols for IoHT systems: MQTT and CoAP have been discussed. The protocol stack of IoHT systems contains these two protocols in the topmost layer (Fig. 2).

1.2 IoHT Simulators

Before the new IoHT products and protocols are deployed in the target environment, extensive research is carried out for its design and development which requires suitable testing and extensive assessment that leads to the requirement of a broad range of different tools and techniques. In addition to this, it is quite challenging to set up reliable and repeatable experiments that only not require real-time software and hardware but also require expertise and domain knowledge of the same specifically [4].

An IoHT simulator is generally expected to possess high precision for different frameworks consisting of disparate elements, supporting extensibility (scalability), providing less energy consumption and should be computationally efficient along with scalable enough to be able to support traditional needs like optimization and evaluation of new protocol [5].

Simulators can be divided into three categories that are applicable in IoHT research and development depending upon architecture and scope. These are:

Full-stack simulators: were designed to support the evolution of the IoT models. They provided end-to-end support to all the elements of IoT. For example, iFogSim and CloudSim.

Big Data Processing Simulators: simulators which focused on the processing aspects of big data related to IoT's various areas of applications. Their main focus is Cloud Performance. For example, IoTSim and SimIoT working on the MapReduce programming model.

Network simulators: The majority of the available simulators are network simulators. They are mainly used in research purposes which employs wireless sensor network (WSN). The most widely used network simulators are as follows: Cooja using Contiki OS, CupCarbon, NS-2, OMNeT++, NS-3, and QualNet [6]. This broad classification is shown through the block diagram (Fig. 3).

The criteria on which simulators are selected can be enumerated as Scope, cyber-attack simulation, Last update, citation count, type, language, evaluated scale, built-in IoT standards, precision, mobility, domain target, overall feasibility, etc. [6].

Application Layer protocol mainly deals with the exchange of data between devices and a gateway. To research and evaluate various characteristics and parameters of application layer protocols, we needed to use simulators that are used for

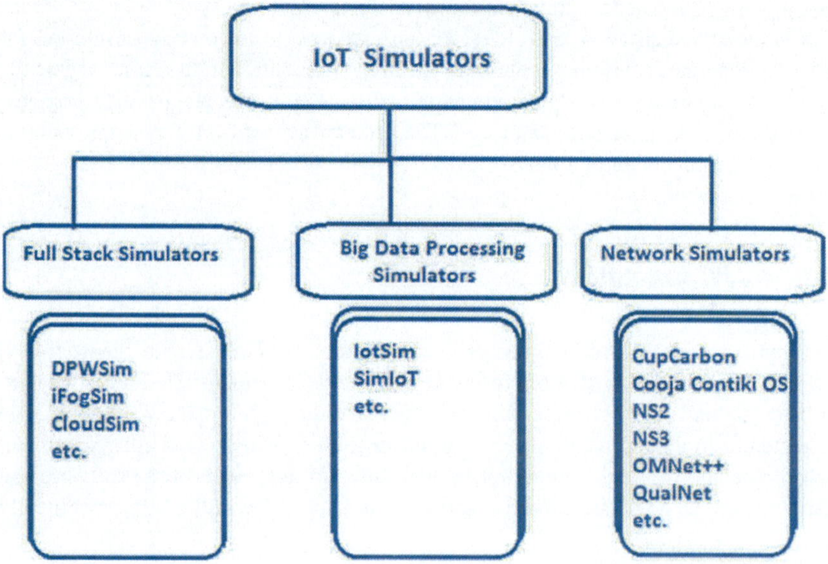

Fig. 3 Broad categories of IoT simulators

Wireless Sensor Network (WSN) since both deals with the exchange of data. The most suitable type of simulators used in this context is the network simulators. We further discuss a few extensively used Network Simulators.

Cooja simulator comes as embedded software in the Contiki OS. Contiki OS works on a hypervisor which makes it the most approved and easily accessible OS. It can be easily installed on any operating system using VM Ware. It is used to test and evaluate the performance of various IoHT nodes generally called sensors [7]. There are more than one hundred research papers accessible regarding Cooja. Researchers can recreate real-time experimental setups that consist of widely tested messaging protocols such as MQTT and CoAP which are carried out over an address 802.15.4 [6].

The OMNeT++ is a module-based, extensible, component-oriented C++ simulation library and framework which is in demand nowadays. Like Cooja, OMNeT++ was also developed for wireless sensor networks with the advantage of being operated on any platform. It has a well-established framework. Its extensibility and user-friendly interactive ability make it popular [1].

The NS-3 simulator is next in line to the NS-2 simulator which can also be used to create a practical approximation of wireless sensor networks frameworks for reproducing the lowest layer in the architecture of IoT. Despite having its similarities to OMNeT++ and assist for 6LoWPAN protocol over the address 802.15.4, NS-3 does not implement application layer protocols like MQTT and CoAP. It needs special libraries to support them [6].

2 MQTT (Message Queuing Telemetry Transport) Protocol

MQTT is among the widely and formerly used device to device or machine to machine communication protocols. It was put forward in 1999 and designed by Andy Stanford Clark (IBM) and Arlen Nipper (Arcom Control Systems Ltd). It is based on a server/client model which is lightweight specifically developed for the device to device communications in constrained environments. It's often called the publisher/subscriber model [8]. Data privacy between the different subscriber/publisher maintained with the help of security models, where privacy is facilitated by an access control solution, controlling functions, and rights for the system entities. These methods control access in the pub/sub model by matching and routing. Another way is to perform encrypted matching which allows the broker to determine whether an encrypted publication matches an encrypted subscription or not. Cryptography is used to provide access control and thus privacy in the system. The initial encryptions and final decryptions of pub/sub messages before their delivery are carried out by the broker [9].

There are three main components of MQTT based on the publisher/subscriber model (Fig. 4): (i) subscriber (ii) publisher (iii) broker. A subscriber such as gateways can request, send, and receive data messages and information regarding patients from a publisher such as embedded systems, computers, and mobile phones. The broker

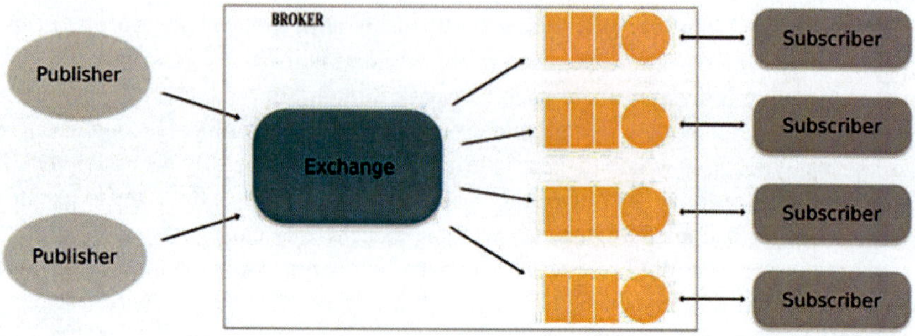

Fig. 4 MQTT architecture [17]

is the heart of the publish/subscribe model. A broker can handle several clients at a time and is responsible for receiving and filtering the messages, determining who is subscribed to each message, and sending the message to the subscriber. The broker holds the session data of all clients with persistent sessions, including subscriptions and missed messages along with authentication and authorization of clients. In short, the broker is the central hub through which every message must pass [10]. An MQTT publisher/server with a broker to send messages from a specific topic to its subscriber/clients. Eventually, the broker received data messages from publishers that are forwarded to the subscribers who are interested in the topic for receiving the data message. Every message is sent through an address commonly called a Topic [11]. To start a connection, the publisher sends a CONNECT message to the broker while it opens the network socket. The broker closes the connection if there is a time gap between these two processes. The broker uses their unique ID to identify the publisher and its current state. The clean session flag tells the broker whether the publisher is willing to establish a persistent session or not depending upon (QoS) level. The broker facilitates custom client authentication and authorization systems by user name and password and integration into backend systems. The client specifies the time frame in sec for the exchange process [10].

It is an application layer protocol that demands a predetermined header of 2 bytes along with short (binary) data message payload of max size, i.e. 256 megabytes [3]. Transport layer protocol which ensures the transparent transfer of data uses TCP (Transmission Control Protocol) and TLS/SSL cryptographic protocols that provide authentication and data encryption between the servers in MQTT, hence adding a security feature to MQTT Protocol. Thus, a connection has to be established between client and broker for the communication of data message which should be secure so that private data of the patient is not accessed by the third person. So, MQTT is called connection-oriented (Fig. 5).

Another important attribute of MQTT is its three (QoS) levels which make it authentic for the conveyance of data messages between server and clients [3]. In QoS-0, the data message is sent only for a single time without any confirmation which means it is not guaranteed whether it is delivered properly or not; QoS-1, the data message is sent over the topic to the subscriber at least one time and its delivery

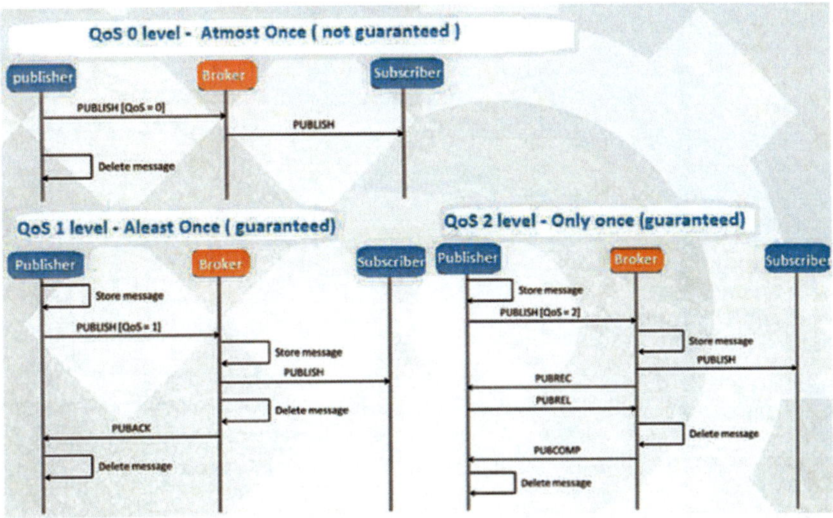

Fig. 5 Different QoS levels in MQTT

is guaranteed to the publisher and it may be possible with several identical copies; and QoS-2, only one message is assured to be delivered to subscribers only once and assured to be without any identical copies. It is guaranteed to be delivered.

MQTT protocol is generally appropriate for substantially big networks that have tiny devices and machines which are required to be supervised and observed from the server-side over the web connection. It is a very fundamental application layer protocol which only offers a few control criteria to work upon and was not developed for multiple device-to-device transfers and doesn't support multicasting [11].

3 CoAP (Constrained Application Protocol)

CoAP is another HTTP Protocol which mainly works for device-to-device communication. It was designed by the Internet Engineering Task Force CoRE (Constrained RESTful Environments) Working Group. Like MQTT, the CoAP messaging protocol is lightweight and supported by both resource/observe (a form of publishing/subscribe) and request/response, i.e. server/client model [8] (Fig. 6).

CoAP was mainly designed to interact with HTTP and the RESTful based resource-constrained IoT systems via simple proxies. In MQTT, Topics are used but in CoAP Universal Resource Identifier (URI) [12] are employed that recognizes the resources which are controlled by the server. Publishers will publish the data message to the Universal Resource Identifier and subscribers will connect to a particular resource identified and suggested by the URI. All the subscribers are notified whenever a publisher sends new data value to the URI. An entity-tag (E-Tag) is used

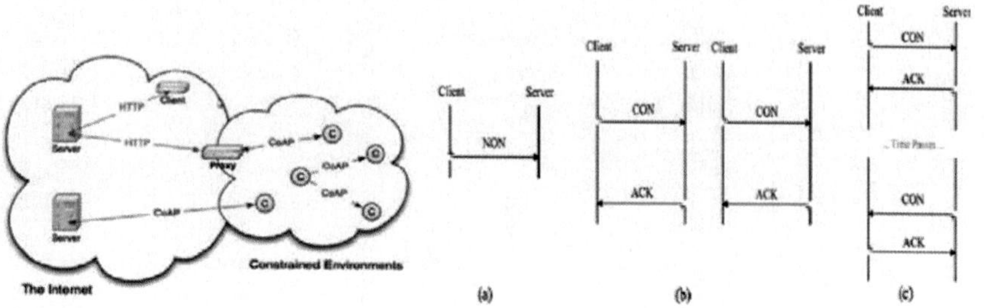

Fig. 6 **i** CoAP architecture [15], **ii** CoAP: **a** Non-conformable; **b** piggybacked response; **c** separate response [17]

as a resource local identifier for differentiating between representations of the same resource that changes with time. It is generated by the server providing the resource, which may generate it in any number of ways including a version, checksum, hash, or time [13].

It is binary which makes efficient and optimized and generally needs a predetermined header of 32 bits, i.e. 4 bytes with tiny data message payloads up to a max value which is determined by the webserver [12]. CoAP requires UDP (User Datagram Protocol) as support from transport for low latency and DTLS (Datagram Transport Layer Security) for security and privacy of data contained in communication [14]. Thus, the transfer of data messages between clients and servers is done through connectionless datagrams. Although it provides two levels of QoS by employing a 'confirmable' or 'non-confirmable' message still it is less reliable [3].

Messages in the CoAP Application Layer protocols are broadly categorized into four classes: (i) Confirmable (ii) Non-confirmable (iii) Acknowledgment (iv) Reset.

(i) Confirmable (CON): A confirmation response is required when the message is sent, i.e. the receiver should confirm the message receipt.
(ii) Non-confirmable (NON): A confirmation response is not needed when the message is dispatched. Hence there is no requirement of attesting the message receipt by the receiver.
(iii) Acknowledgment (ACK): An acknowledgment is acquired to reciprocate to the confirmable data message that verifies that it was the last reception of data message.
(iv) Reset (RST): If there is an error in the sent data message or if it is not comprehensible or if the receiver is not willing to send/receive data message to the sender, in these cases, reset message will be dispatched [15].

CoAP Application Layer Protocols provides more operations than MQTT such as support for the content which allow publisher and subscriber to develop gradually and independent of each other thereby adding up to date representations without causing any remarkable change in each other [3].

4 Comparison Between CoAP and MQTT Protocols

1. *Comparative Analysis of MQTT and CoAP based on Static Criteria*:

A comparative analysis of the most extensively used application layer protocols such as MQTT and CoAP has been presented in the tabular form. Table 1 enumerates the differences and similarities between these two emerging messaging protocols for IoHT systems which are based on the static criteria like the year, architecture, abstraction, header size, message size, cache and support proxy, reliability, standard, TCP used security, encoding and licensing models required in both MQTT and CoAP protocol.

2. *Comparative Analysis of MQTT and CoAP based on Dynamic Conditions*:

A relative analysis of MQTT and CoAP for IoHT is presented in the Tabular Form. Based on their needs and development processes, these two application layer protocols are very different from each other. Hence, they are used as per the requirement and precision of an IoHT system. Nevertheless, this differentiation is mentioned in vague lingual assessments ranging from 'Lower' to 'Higher' to provide a simpler and inclusive view of both the protocols. This comparison might vary under a few circumstances such as different IoHT modules and may show different outcomes than mentioned below (Table 2).

Table 1 Comparative analysis of MQTT and CoAP [3] based on static criteria

S. No.	Criteria	MQTT	CoAP
1.	Year	1999	2010
2.	Architecture	Client-broker	Client-server or Client-broker
3.	Abstraction	Publisher-subscriber	Publisher-subscriber Request-response
4.	Header value	2 Byte	4 Byte
5.	Message range	Tiny and not defined (of 256 MB max size)	Tiny and not defined (small to fit in single IP datagram)
6.	Standard	OASIS, Eclipse foundation	IETF, Eclipse foundation
7.	Reliability	QoS 0—At max once, QoS 1 At least once, QoS 2—Exactly once	CON (At max once) NON(At least once)
8.	Cache and support proxy	Partial	Yes
9.	Transport layer	TCP	UDP, SCTP
10.	Security	TLS/SSL	DTLS
11.	Encoding	Binary	Binary
12.	License	Open source	Open source

Table 2 Comparative analysis of MQTT and CoAP based on dynamic conditions

S. No	Dynamic conditions	MQTT	CoAP
1.	Power consumption versus resource requirements	Medium	Lower
2.	Size of message versus overhead of message	Medium	Lower
3.	Bandwidth versus latency	Medium	Lower
4.	Reliability versus interoperability	Higher	Medium
5.	Security versus provisioning	Lower	Medium
6.	Usage versus standardization	Higher	Medium

3. *Comparative analysis of simulators depending upon parameter it extracts for MQTT and CoAP*:

This section presents a comparative analysis depending upon the parameter a particular simulator can evaluate for both MQTT and CoAP protocols in IoHT systems. Based on this analysis, a researcher can find out which simulator to be used for the concerned parameters in the respective protocols. It depends upon the research paper available in the respective domains (Table 3).

Table 3 Comparative analysis of simulators depending upon parameter it extracts for MQTT and CoAP

S. No.	Simulators	MQTT	CoAP
1.	Cooja-Contiki OS	RTT delay QoS efficiency [18] Average power consumption Average duty cycle Program size	Response time Payload [19] Energy consumption [15] Average time response Energy type CPU [20] Full transaction No HandShake transaction HandShake cost RAM/ROM overhead Energy per bit transaction [21] Packet delivery ratio Bandwidth consumption [18]
2.	OMNeT++	End to end delay Payload QoS efficiency Bit error rate	–
3.	NS-3	–	–
4.	NS-2	Probabilistic content Delivery and end to end delay [22]	–

5 Conclusion and Future Scope

MQTT and CoAP; the two widely approved Application Layer protocols have been evaluated for IoHT systems. This paper presents an all-inclusive comparison between these two protocols to enumerate and compare their characteristics along with the types of simulators suitable for the performance analysis of both protocols. It also has performed an exhaustive and relative analysis which is rooted in some basic interdependent criteria to in-depth knowledge of their pros and cons. To make the relative analysis of protocols facile and uncomplicated, it has been illustrated by using a simple table to provide a quick and broad view of both the messaging protocols. Consequently, the researchers/readers can determine their relevant usage and resolve the queries in IoHT systems depending upon their necessity and aptness. Both static and dynamic parameters have been analyzed to obtain a bigger and relative overall idea of the two main application layer protocols. It also considers changeable parameters and other overheads to sustain the transmission of data messages from the device to Gateway. Furthermore, it is a rapidly growing and ever-changing scope that has a lot of potentials to develop and evolve hereafter. Therefore, it is fascinating to evaluate these protocols in the practical environment of the IoHT system.

References

1. Bansal M, Priya (2020) Application layer protocols for internet of healthcare things (IoHT), In: 4th international conference on inventive systems and control (ICISC 2020), pp 366–371
2. Riazul ISM, Kwak D, Humaun KMD, Hossain M, Kwak K (2015) The internet of things for healthcare: a comprehensive survey. IEEE Access 5:678–708
3. Imane S, Tomader M, Nabil H (2018) Comparison between CoAP and MQTT in smart healthcare and some threats. In: 2018 international symposium on advanced electrical and communication technologies (ISAECT), Rabat, Morocco, pp 1–4
4. Papadopoulos GZ, Beaudaux J, Gallais A, Noël T, Schreiner G (2013) Adding value to WSN simulation using the IoT-LAB experimental platform. In: Proceedings of 2013 IEEE 9th international conference on wireless and mobile computing, networking and communications (WiMob), Lyon, France, pp 485–490
5. Sharif M, Sadeghi-Niaraki A (2017) Ubiquitous sensor network simulation and emulation environments: a survey. J Netw Comput Appl 93:150–181
6. Chernyshev M, Baig Z, Bello O, Zeadally S (2018) Internet of things (IoT): research, simulators, and testbeds. IEEE Int Things J 5(3):1637–1647
7. Dunkels A, Gronvall B, Voigt T (2004) Contiki—Lightweight and flexible operating system for tiny networked sensors. In: Proceedings of 29th annual IEEE international conference on local computer networks, Tampa, FL, USA, pp 455–462
8. Behrisch M, Bieker L, Erdmann J, Krajzewicz D (2011) SUMO—simulation of urban mobility: an overview. In: Proceedings of SIMUL 3rd third international conference on advances in system simulation, pp 55–60
9. Onica E, Felber P, Mercier H, Riviere E, Cuza AI (2016) Confidentiality-preserving publish/subscribe: a survey, vol 49(2). ACM. University of Iasi, Romania, Institut d'Informatique, Universite de Neuchatel, Switzerland, November 2016
10. https://www.hivemq.com/blog/mqtt-essentials-part-3-client-broker-connection-establishment/

11. Opadhyay SB, Bhattacharyya A (2013) Lightweight internet protocols for web enablement of sensors using constrained gateway devices. In: 2013 international conference on computing, networking and communications (ICNC). IEEE, pp 334–340
12. Naik KP, Joshi UR (2017) Performance analysis of constrained application protocol using Cooja simulator in Contiki OS. In: 2017 international conference on intelligent computing, instrumentation and control technologies (ICICICT), Kannur, pp 547–550
13. https://tools.ietf.org/html/rfc7252
14. Shelby Z, Hartke K, Bormann C (2014) The constrained application protocol (CoAP). Tech Rep
15. Jaffey T (2014, February) MQTT and CoAP, IoT protocols. [Online]. Available https://eclipse.org/community/eclipsenewsletter/2014/february/article2.php
16. Rodrigues JJPC et al (2018) Enabling technologies for the internet of health things. IEEE Access 6:13129–13141
17. Thangavel D, Ma X, Valera A, Tan H-X, Tan CK-Y (2014) Performance evaluation of MQTT and CoAP via a common middleware. In: 2014 IEEE ninth international conference on intelligent sensors, sensor networks and information processing. IEEE, pp 1–6
18. Fournaris AP, Giannoulis S, Koulamas C (2019) Evaluating CoAP end to end security for constrained wireless sensor networks. In: 2019 10th IFIP international conference on new technologies, mobility and security (NTMS), Canary Islands, Spain, pp 1–5
19. Govindan K, Azad AP (20150 End-to-end service assurance in IoT MQTT-SN. In: 2015 12th annual IEEE consumer communications and networking conference (CCNC), Las Vegas, NV, pp 290–296
20. Ferdousi R, Helaluddin M, Akther A, Alam KM (2017) An empirical study of CoAP based service discovery methods for constrained IoT networks using Cooja simulator. In: 2017 20th international conference of computer and information technology (ICCIT), Dhaka, pp 1–6
21. Abosata NRA, Kemp AH, Razavi M (2019) Secure smart-home application based on IoT CoAP protocol. In: 2019 sixth international conference on internet of things: systems, management and security (IOTSMS), Granada, Spain, pp 13–17
22. Ludovici A, Moreno P, Calveras A (2013) TinyCoAP: a novel constrained application protocol (CoAP) implementation for embedding RESTful web services in wireless sensor networks based on tinyos. J Sens Actuator Netw 2(2):288–315

A Compact Vertex Fed Heptagon Monopole Antenna in a Wide Diamond Slot for UWB Applications

A. Priya, M. Saravanan, D. Balasubramaniam, A. Subahar, and V. Purushothaman

Abstract This paper proposes a compact ultra-wideband (UWB) monopole antenna. The antenna consists of a wide diamond-shaped slot along with a heptagonal shaped patch placed at the center. Two T-shaped stubs are attached with radiating patch along vertices of the patch element. Parametric analysis is carried on antenna dimensions to optimize antenna performances. The prototype model is fabricated and the measurement results are compared with simulated results. The proposed antenna has achieved −10 dB impedance bandwidth of 7.9 GHz (3.1–11) GHz and covers Wireless LAN (5.18–5.85) GHz/IEEE WiMAX (3.40–3.69) GHz, (3.33–3.96) GHz, radar applications (LRIR—Long-Range Imaging Radar) (9.5–10.5 GHz) and remote sensing and tracking applications (10.6–10.7) GHz. The antenna gives a measured peak gain of 6.59 dBi. The results obtained are compared with other conventional UWB antennas and proved that the antenna is compact with improved gain and hence it is more suitable for modern wireless ultra-wideband communications.

Keywords Vertex fed · Monopole antenna · Slot · Polygonal · Ultra-wideband antennas

A. Priya (✉)
B. S. Abdur Rahman Crescent Institute of Science and Technology, Chennai, India
e-mail: priyamarish@crescent.education

M. Saravanan · D. Balasubramaniam · A. Subahar · V. Purushothaman
Vel Tech Rangarajan Dr. Sagunthala R&D Institute of Science and Technology, Chennai, India
e-mail: msarawins@ieee.org

D. Balasubramaniam
e-mail: drbalasubramaniamd@veltech.edu.in

A. Subahar
e-mail: subahara@veltech.edu.in

V. Purushothaman
e-mail: purushothamanv@veltech.edu.in

G. Ranganathan et al. (eds.), *Inventive Communication and Computational Technologies*, Lecture Notes in Networks and Systems 145,
https://doi.org/10.1007/978-981-15-7345-3_48

1 Introduction

The ultra-wideband has drawn attention many years back because of the pulse type of communication which occupies narrow bandwidth. But the major problem with UWB was the low spectral efficiency because of the large spreading of signals. But in the late 90s it was proved that spectral efficiency was not considered as a major factor in military radar application wherein the spatial resolution was so important rather than spectral efficiency in such applications. Slot antennas and printed antennas have been proved to be more efficient for UWB and in military radar. A single antenna to work for different applications and also to sense and track different frequencies is the need of the time. Several antenna designs have been proposed by antenna researchers for various applications. Various techniques for bandwidth enhancement has been proposed by many researchers. [1, 2] has proposed the technique of bandwidth enhancement in UWB antennas using stubs and [3, 4] using slots. In [5], a technique that the bandwidth can also be increased in UWB antennas using tapered transitions in an inset feed type. An antenna to radiate in all directions is the interest of most of the researchers. In [6, 7] the omni-directional radiation characteristics of a CPW-fed Multiband UWB antenna was explained. The omni-directional characteristics of an antenna can be converted to unidirectional radiation when the ground plane is partial along with CPW feed is described in [8]. UWB antennas for short-range wireless communications using notch feed and by modifying the ground plane is presented in [9]. Various printed slot antennas of different shapes for various wireless applications have been presented in [1–10]. Various self-complementary structured antennas have been described in [11, 12], complementary structured antennas with near omni-directional characteristics in the UWB range is described in [12, 13]. The bandwidth enhancement technique by introducing a loop feeding structure is discussed in [14, 15]. The UWB radar system is designed for military applications [16]. In [17] a C-shaped monopole antenna designed to achieve an acceptable level of in-band return loss is explained. The performance characteristics of the UWB antenna for WBAN applications are analyzed in [18]. A UWB antenna loaded with slots for dual orthogonal polarization is explained in [19]. The basic ideas about ultra-wideband technology, its design techniques, issues, and system considerations have been clearly described in [20]. The above-mentioned antenna designs though concentrated on the basic parameters has a constructive ground plane which increases the antenna geometry. Also, the above-mentioned designs have not concentrated much on radar application. The reported antennas occupy large space and complex structures as compared with the proposed antenna. So we propose a new heptagonal patch element that is electromagnetically coupled with the diamond slot through the microstrip line feed. The antenna is suitable for the entire UWB mentioned by FCC along with WLAN/WiMAX and radar applications. The group delay is an important parameter to be considered in radar and imaging applications and that is taken into account in this design. The heptagonal antenna presented in this paper operates at ultra-wideband and is modeled using Ansoft high-frequency structure simulator. The performances of the proposed antenna are validated by fabricating

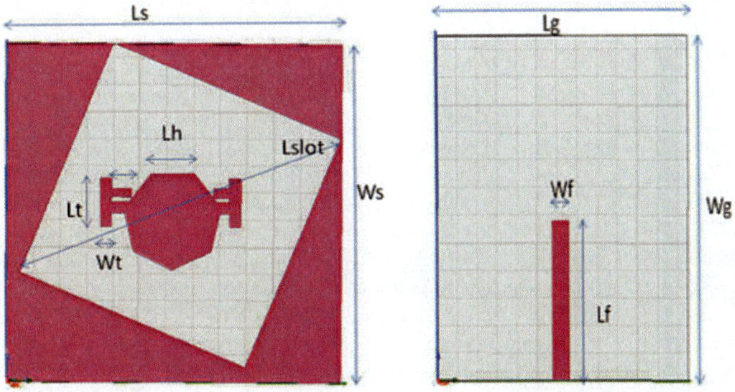

(a) Front (b) Back

Fig. 1 Simulated model

a prototype antenna on the FR4 substrate and measured using antenna test systems and are compared with simulated results. Simulation results agree well with measured results and it is more suitable for wideband applications.

2 Proposed Antenna Design Procedure

The heptagonal structure is designed in HFSS 13.0 version with FR4 substrate having 4.4 relative dielectric constant and 0.02 loss tangent. The antenna is sized about 25 $mm^3 \times 25\ mm^3 \times 1.6\ mm^3$. Figure 1 shows the proposed structure simulated using HFSS 13.0 V. The front portion is a heptagonal patch with two T strips inside a diamond-shaped slot and the back portion is a microstrip feed. The parasitic heptagonal patch is placed at the center which is surrounded by a wide diamond-shaped slot. The area of this slot is controlled by changing the dimensions of the square patch or by changing the dimensions of the diamond slot. The feed length and slot dimension are optimized to get impedance matching for maximum transmission of the signal from the microstrip feed line to the heptagonal parasitic patch. Figure 2 shows the front and back view of the prototype with its optimized dimensions being listed in Table 1.

3 Parametric Analysis

The optimized dimensions of the proposed heptagonal antenna are based on parametric analysis using HFSS. Initially, an optimized square-shaped substrate measuring 25 mm × 25 mm is chosen and a slot of various shapes like triangle, pentagon, hexagon, septagon, circular, and diamond is created on the substrate.

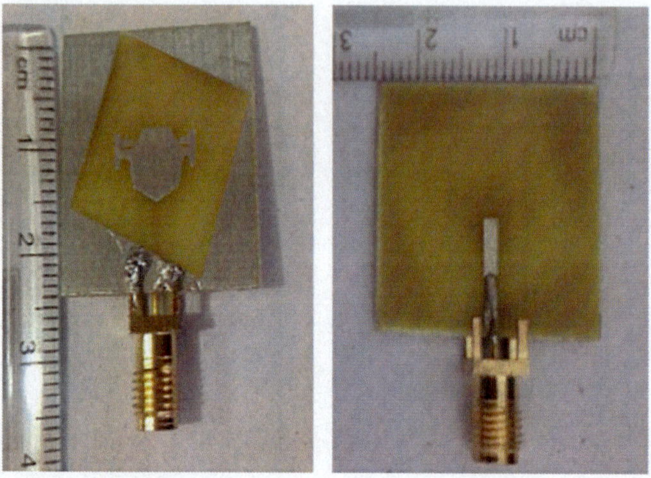

(a) Front (b) Back

Fig. 2 Hardware model

Table 1 Proposed heptagonal antenna parameters and dimensions

Optimized parameters	Dimensions (mm)
Substrate length (Ls), Substrate width (Ws), T strip length (Lt)	25, 25, 3.5
Width of the T strip (Wt), Length of one side of a heptagon (Lh), the height of the substrate (Hs)	0.8, 3.2, 1.6
Length of the diamond slot (Lslot), Length of T-arm, Length of the ground plane (Lg)	9.7, 1.5, 25
Ground plane width (Wg), Feed length (Lf), Feed width (Wf)	25, 11.5, 1.6

Under the careful parametric analysis of all the above shapes of slots, the proposed diamond-shaped slot optimized for 9.7 mm length diagonally gives excellent return loss characteristics for the full UWB band as shown in Fig. 3.

Inside the diamond slot optimally designed heptagonal shaped patch is placed. The heptagonal shape is finally arrived at and proposed as our structure after checking other patches of shape including triangular, diamond, pentagonal, hexagonal, and circular. Only the heptagonal patch satisfies the UWB band of frequency and this can be seen from Fig. 4. To the heptagonal structure, T-shaped strips are added on both the ends. First, a single T-shaped strip is added on both the sides and the characteristics are compared with the heptagonal patch without T strips. Then the second T strip is added to the heptagonal patch and the results are compared for all three conditions say without strip, with one T strip and two T strips. The proposed structure with two T strips satisfies the UWB frequency covering the wireless applications.

From Fig. 5 it can be understood that without T strips the application bands are not covered. But when one T strip is introduced the sensing band of operation is

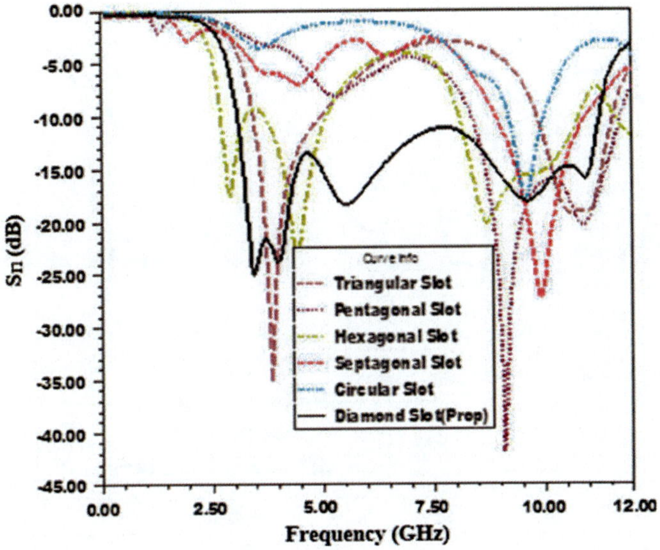

Fig. 3 Reflection coefficient for different slot configurations

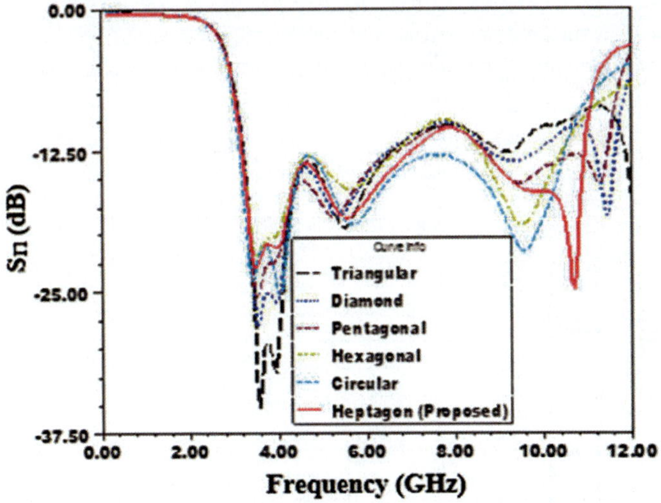

Fig. 4 Reflection coefficient for different parasitic patch configurations

achieved but the return loss was less compared to the results of the antenna without T strips. Also, there is a small frequency shift in the bands. This has been overcome by introducing another T strip which has resonated with the antenna for four bands of applications namely Wireless LAN (5.18–5.85 GHz)/WiMAX (3.40–3.69 GHz),

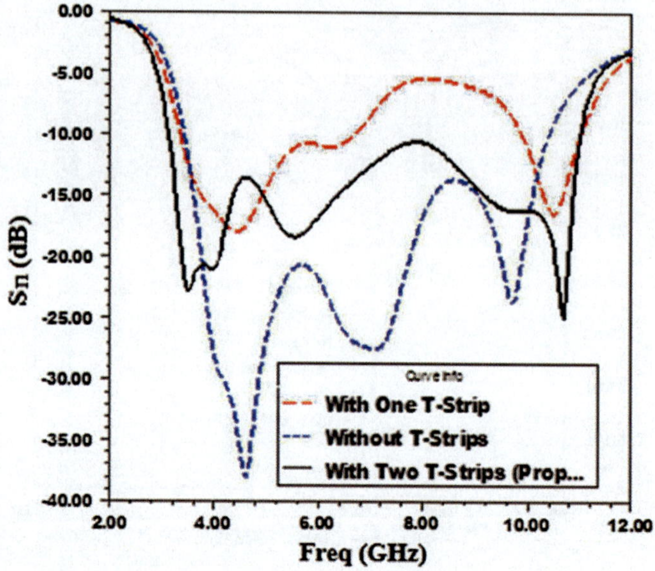

Fig. 5 Simulated reflection coefficient with and without T strip

(3.33–3.96 GHz) and remote sensing and tracking applications (10.6–10.7 GHz). The specified diamond slot is chosen slanted for the bandwidth enhancement. Figure 6

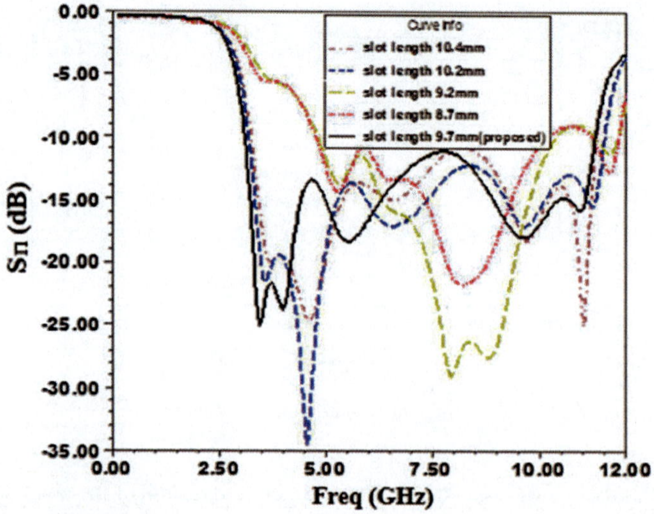

Fig. 6 Simulated reflection coefficient for various diamond slot length

shows the band achievement by optimizing the diamond slot length to 9.7 mm thus covering the LRIR band (9.5–10.5 GHz).

Figure 7 shows the return loss versus frequency characteristics with feed length varied from 10 to 12 mm and the optimized feed length for the proposed antenna is 11.5 mm. The feed exactly touches the vertex of the heptagon and the parasitic effect is well realized on the patch for all bands of frequencies. The feed dimensions are optimized to get an overall improvement in bandwidth. Figure 8 shows the reflection characteristics for various lengths and width of the ground. The optimized dimension of the ground plane for the proposed design is 25 mm × 25 mm.

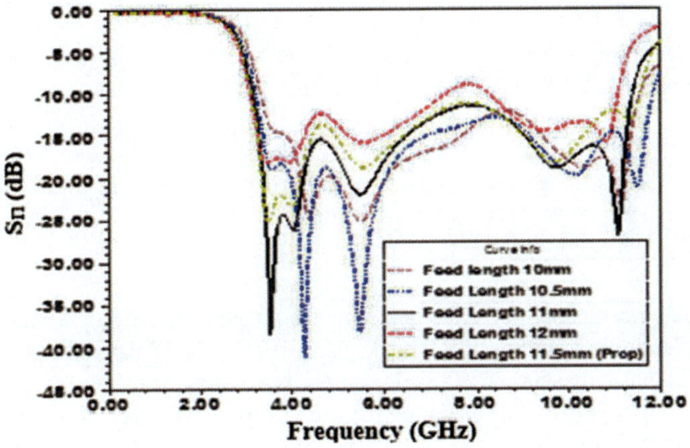

Fig. 7 Reflection coefficient for different length of the feed line

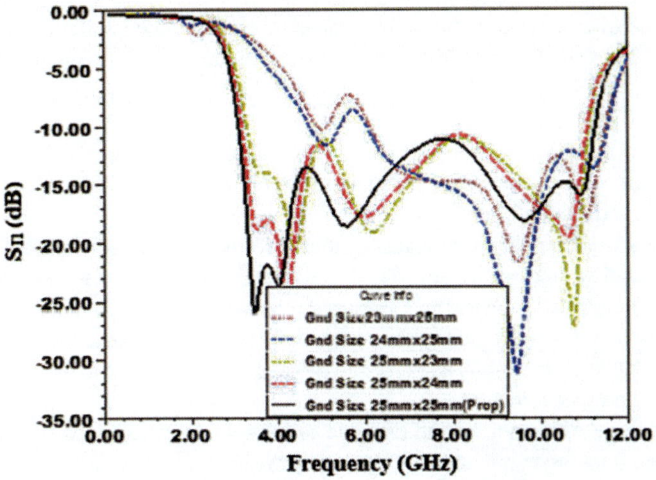

Fig. 8 Reflection coefficient for various ground size

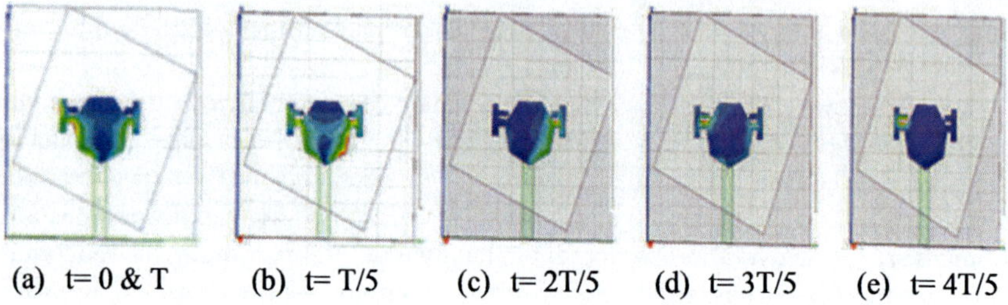

Fig. 9 Distribution of surface current at **a** 3.45 GHz, **b** 3.96 GHz, **c** 5.56 GHz, **d** 9.73 GHz, **e** 10.68 GHz

The distribution of the surface current of the heptagonal antenna is shown in Fig. 9. The concentration of maximum surface current leads to resonance at a particular frequency.

4 Result and Discussion

The parameters considered for the performance measure of the Heptagonal antenna are the reflection magnitude (dB), peak gain (dBi), group delay, and the radiation pattern. The basic measurement results for the fabricated prototype is taken using a standard VNA N9916A. The radiation pattern and gain measurements were taken in the anechoic chamber.

Figure 10 shows the reflection coefficient characteristics of the prototype both simulation and measurement. The simulated and measured result has less than −10 dB return loss in the UWB band covering the WLAN, IEEE WiMAX, LRIR, sensing, and tracking applications. The impedance bandwidth is calculated from this figure and it is found to be 114%. The radiation pattern for the heptagonal antenna is presented in Fig. 11a–e. For the first three bands, the pattern is omni-directional in H-plane but in E-plane it resembles the pattern of a dipole for the simulation as well as measurements as in Fig. 11a–c. The remaining two bands have a pattern that is nearly omni-directional in H-plane and resemble a dipole in E-plane in simulation as well as measurements as shown in Fig. 11d, e.

From Figs. 10 and 11, it is observed that there is a small deviation between the simulation and measurements due to material tolerances and fabrication inaccuracies. The losses associated with the antenna are dielectric loss due to its poor loss tangent of 0.02. For better understanding, the comparison of simulation results and measured results are listed in Table 2. The peak gain results closely resemble the same in both simulation and measurement.

The average efficiency of the antenna is 89% for all the bands for LRIR and sensing and tracking applications. This makes the antenna suitable for military radar applications. Table 3 gives the parameter comparison of the proposed heptagonal

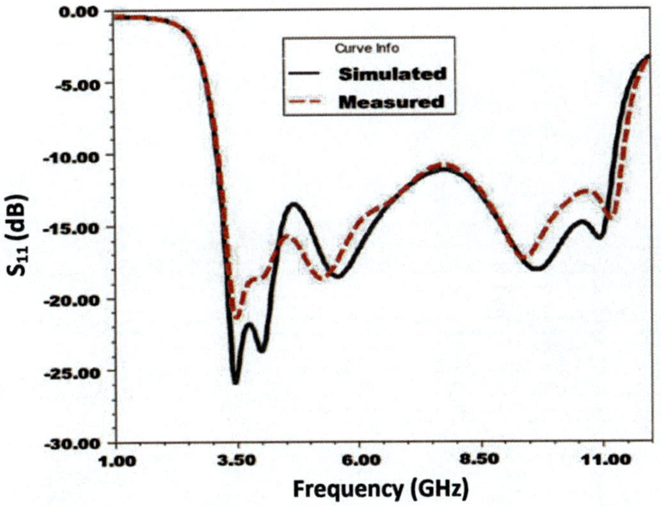

Fig. 10 The simulated and measured reflection coefficient

antenna with existing UWB designs. The proposed heptagonal antenna results are comparably high to the results of existing designs.

5 Conclusion

A new compact UWB antenna suitable for multiple wireless applications with a minimum size of 25 mm × 25 mm is simulated and validated by fabrication and measurement. The fabricated prototype operates from 3.1 to 10.6 GHz having S_{11} <-10 dB with impedance bandwidth of 114%, 6.59 dBi peak gain, and 4.41 dBi average gain. Thee result obtained implies that this antenna is suitable to work in the UWB range covering multiple wireless bands.

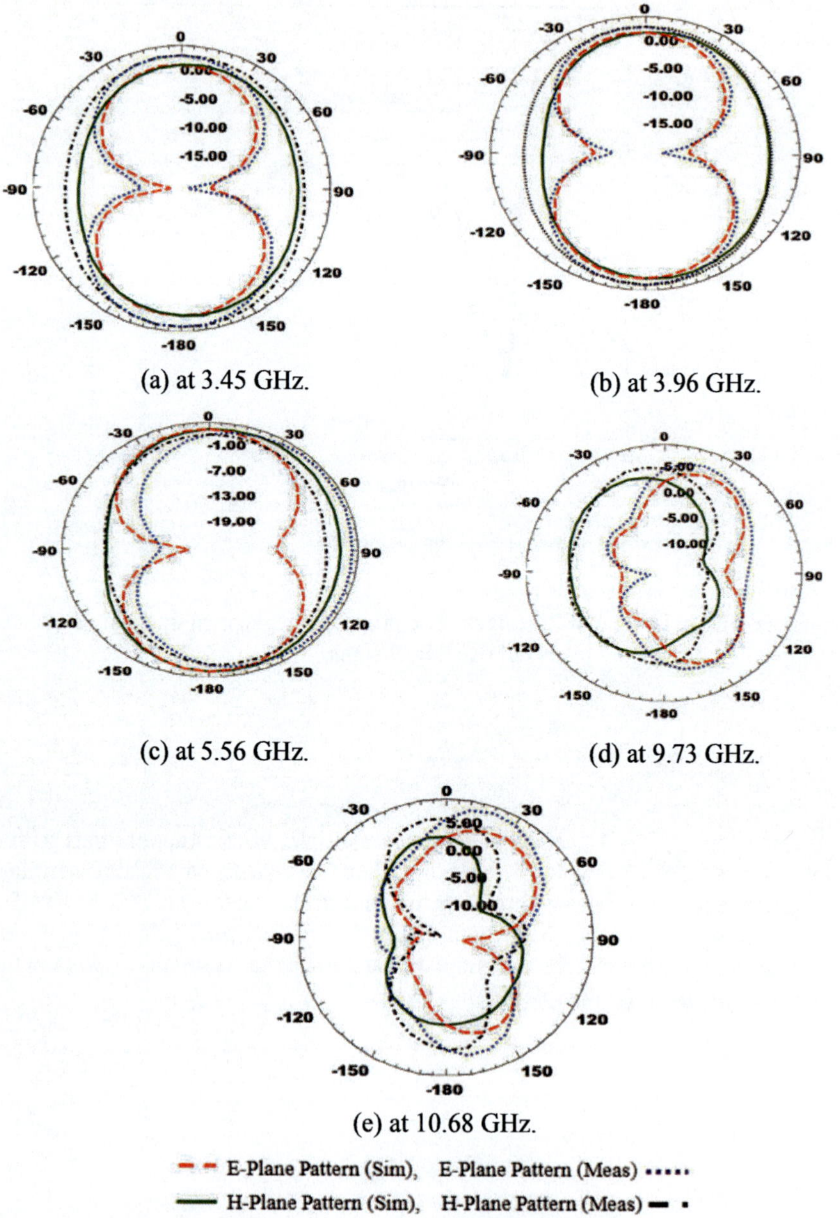

Fig. 11 Simulated and measured E-plane and H-plane radiation pattern

Table 2 Comparison of simulation results and measured results

Parameters	Simulated					Measured				
Freq (GHz)	3.45	3.96	5.56	9.73	10.68	3.48	4.01	5.24	9.37	11.16
$\|S_{11}\|$ dB	22.85	21.14	18.41	16.26	29.96	21.41	18.61	18.67	17.31	14.58
Gain (dBi)	2.39	3.01	3.81	6.24	6.59	3.20	3.19	2.41	4.66	6.48

Table 3 Comparison of various existing antennas with proposed antenna

Ref.	Resonance frequency (GHz)	Number of bands	Antenna size (mm)	Maximum gain (dBi)
[2]	2.3–14	04	50 × 50 × 1.6	6.5
[9]	3.26–10.68	03	34 × 32 × 1.6	4.83
[12]	3.1–15	04	30 × 15.5 × 1	3.9
[14]	3.1–10.75	02	10 × 30	4.8
[18]	2.8–11.4	03	64 × 64 × 7	4.41
Proposed antenna	3.1–10.6	05	25 × 25	6.8

References

1. Takemura N, Ichikawa S (2017) Broad banding of printed bell-shaped monopole antenna by using short stub for UWB applications. Prog Electromagn Res C 78:57–67. https://doi.org/10.2528/pierc17060702
2. Ellis MS, Zhao Z, Wu J, Ma K, Nie Z-P, Liu QH (2014) A planar monopole UWB antenna with improved lower end bandwidth using an L-shaped stub extended on the ground plane. Prog Electromagn Res C 52:109–114. https://doi.org/10.2528/pierc14052001
3. Ali T, Biradar RC (2017) A miniaturized Volkswagen logo UWB antenna with slotted ground structure and metamaterial for GPS, WIMAX and WLAN applications. Prog Electromagn Res C 72:29–41
4. Meena ML, Kumar M, Parmar G, Ram SM (2016) Design analysis and modeling of directional UWB antenna with elliptical slotted ground structure for applications in C-bands and X-bands. Prog Electromagn Res C 63:193–207
5. Bozdag G, Kustepeli A (2015) Wideband printed planar monopole antenna for PCS, UWB and X-band applications. Prog Electromagn Res C 60:95–103. https://doi.org/10.2528/pierc15090301
6. Joseph S, Paul B, Shanta M, Pezholil M (2015) CPW-fed compact UWB spiral antenna for multiband applications. Int J Ultra Wideband Commun Syst 3(2):85–89. https://doi.org/10.1504/ijuwbcs.2015.077111
7. Naidu PV, Kumar R (2014) Design of CPW-fed dual-band printed monopole antennas for LTE/WiMAX/WLAN and UWB applications. Prog Electromagn Res C 54:103–116
8. Edalati A, Shao W, McCollough T, McCollough W (2017) A novel cavity backed monopole antenna with UWB unidirectional radiation. Prog Electromagn Res C 72:1–13. https://doi.org/10.2528/pierc16121610
9. Susila M, Rao TR, Gupta A (2014) A novel smiley fractal antenna (SFA) design and development for UWB wireless applications. Prog Electromagn Res C 54:171–178. https://doi.org/10.2528/pierc14091803
10. Xu L, Yuan B, He S (2013) Design of novel UWB slot antenna for bluetooth and UWB applications. Prog Electromagn Res C 37:211–221. https://doi.org/10.2528/pierc13011002

11. Sayidmarie KH, Fadhel YA (2013) A planar self-complementary bow-tie antenna for UWB applications. Prog Electromagn Res C 35:253–267. https://doi.org/10.2528/pierc12103109
12. Abbosh AM, Bialkowski ME (2008) Design of ultrawideband planar monopole antennas of circular and elliptical shape. IEEE Trans Antennas Propag 56(1):17–23. https://doi.org/10.1109/tap.2007.912946
13. Siddiqui JY, Saha C, Antar YMM (2015) A novel ultrawideband (UWB) printed antenna with a dual complementary characteristic. IEEE Antennas Wirel Propag Lett 14:974–977. https://doi.org/10.1109/lawp.2014.2388272
14. Addaci R, Ferrero F, Staraj R, Fortaki T, Seetharamdoo D, Hamdiken N (2014) Simple bandwidth-enhancement technique for miniaturised low-profile UWB antenna design. Electron Lett 50(22):1564–1566. https://doi.org/10.1049/el.2014.3163
15. Kumar T, Rambabu K, Gautam AK, Kanaujia BK (2015) Design of miniaturised UWB antenna for oil pipeline imaging. Electron Lett 51(21):1626–1628. https://doi.org/10.1049/el.2015.1822
16. Allen B, Dohler M, Okon EE, Malik WQ, Brown AK, Edwards DJ (2007) Ultra-wideband antennas and propagation for communications, radar and imaging. Wiley
17. Koziel S, Bekasiewicz A (2016) A structure and simulation-driven design of compact CPW-Fed UWB antenna. IEEE Antennas Wirel Propag Lett 15:750–753. https://doi.org/10.1109/lawp.2015.2471848
18. Jeong W, Tak J, Choi J (2015) A low-profile IR-UWB antenna with ring patch for WBAN applications. IEEE Antennas Wirel Propag Lett 14:1447–1450. https://doi.org/10.1109/lawp.2015.2411263
19. Abed AT, Singh MSJ, Islam MT, Khaleel AD (2017) Dual crescent-shaped slot antenna fed by circular polarisation into dual orthogonal strip lines. IET Microwaves Antennas Propag 11(15):2129–2133. https://doi.org/10.1049/iet-map.2016.0747
20. Galvan-Tejada GM, Peyrot-Solis MA, Aguilar HJ (2015) Ultra wideband antennas: design, methodologies and performance. CRC Press, Taylor and Francis Group

Performance Investigation of Various SRAM Cells for IoT Based Wearable Biomedical Devices

J. R. Dinesh Kumar, C. Ganesh Babu, V. R. Balaji, K. Priyadharsini, and S. P. Karthi

Abstract The field of IC technology grows every day and it plays a decisive role in a day to day activity. One of the most emerging fields is biomedical. This field is getting wider and the size of the devices becomes reduced from ECG machines to portable medical devices. Nowadays wearable devices and implantable biomedical devices on the body captures the biosignals and convert it as potential signals to measure the variation and the devices which are connected to the cloud using IoT technology is helpful to digest the types of problem with less time and better accuracy. So the biopotential signal has to be shared and processed between the memory and the processor. The more sensible processor has to collect the required data from the associated memory to make an agile response. Hence this paper describes the logic and criteria on choosing the most prominent memory SRAM. The SRAM architecture in this paper is discussed based on transistors count like 5T, 6T, 8T,9T, and 10T. The no of transistors on the SRAM used to decide the level of leakage power and other junction power and also the die area it occupies. Here the 5T SRAM shows the notable variations in the delay and has a good level on leakage power and dynamic power reduction. Henceforth using the proposed SRAM in wearable biomedical devices could cause betterment in speed, agility in response, and consume less power. This SRAM can be used to access denser data with less delay time.

Keywords SRAM · Biomedical devices · Leakage power · Delay · IoT

J. R. Dinesh Kumar (✉) · V. R. Balaji · K. Priyadharsini · S. P. Karthi
Department of Electronics and Communication Engineering, Sri Krishna College of Engineering and Technology, Coimbatore, Tamil Nadu, India
e-mail: dineshkumarjr@skcet.ac.in

C. Ganesh Babu
Department of Electronics and Instrumentation Engineering, Bannari Amman Institute of Technology, Sathyamangalam, Tamil Nadu, India
e-mail: bits_babu@yahoo.co.in

G. Ranganathan et al. (eds.), *Inventive Communication and Computational Technologies*, Lecture Notes in Networks and Systems 145,
https://doi.org/10.1007/978-981-15-7345-3_49

1 Introduction

The Modern era of integrated circuits has many advantages in power reduction, low area, and lesser density packages and a minimum level of die size [1] with portable and comfortable devices at the same scale of orientation and improvement of usage of deeper technology in day to day activity and growth of the technology scale demands more no of the research area to be focused on power consumption and portable devices [2]. One of the most materialize fields is to develop light-weighted biomedical devices. The development of nano sensors [3] and various devices at micro-level leads to the measurement of physical parameters is quicker and one who can check their health parameters [1] using the more compact and highly précised medical devices. Specifically focusing on the biomedical devices which are not harmful to the human should ensure that two major criterions one is more no of devices are designed in such a way that it observes the biosignals (potential) whose magnitude is comparably minimum and frequency of operation is the low and second one consumes less amount of power [4]. The highly designed medical devices such as ECG and BP measuring devices need a replacement of battery which are easy, but the implantable devices are difficult to manage and if the devices consume more power leads to the reduction on the life of the implantable device which again causes a surgical operation to either replace or remove.

The terms energy and power are related to the supply voltages. So the vahigh-speed computariation in the supply voltages [5] could determine the level of power dissipation and consumption. The drastic changes in voltages could lead to the scaling of devices in two ways. One is producing the new architecture at top-level design and another is implementing a new routing protocol [5] or algorithm to the existing methodology. However the scaling of voltages could affect the delay between the transistors at deep sub-threshold regions. Perhaps the increase of delay between the path directly improves the leakage energy at the subthreshold region. So scaling voltage level [6, 7, 16, 17] should create a flawless model to maintain balanced energy and delay between the circuits of transistors. The center blocks of biomedical devices are designed based on Soc types. The Soc may consist of processor, ADC blocks, various sensors, and actuators along with the ports for communications and memory slot locations of on-chip and SRAM. The static ram is one of fastest type of ram which does not require any kind of refreshment like DRAM. The architecture of the cache shown in Fig. 1, One of the predominant temporary storage is cache, all the processors are using the different levels of cache and at the different locations from the processor, it is named as L_1, L_2, and L_3. The speed of operations are varied for each level of the cache [8] and L_1 is faster and smaller than other levels. L_2 could be located as a single inline memory module near to the processor. So depending on the applications the L_2 could access more no of data and the operation of find in the memory which is known as hit and the probability of not finding the data in the memory is known as a miss. The high-performance computing devices could easily categorize based on their performance in terms of how to speed the processor fetching the data from the memory [9]. Hence the most of the devices associated with telecommunication [10,

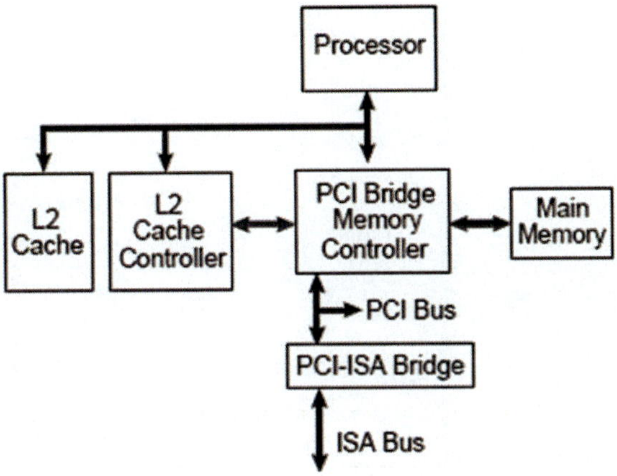

Fig. 1 Basic system blocks with cache

16] and high-speed computation need to fetch the data synchronously and process them according to the drivers. So the synchronous SRAM is used to design this caches to trigger the speed of operation which may use in wearable devices The processor is communicated with main memory and other blocks via PCI bridge memory controller and also the data can be accessed from any levels of cache like L_1, L_2, etc.

The most important features of biomedical devices which are implantable or wearable to measure the biosignals which are quite subsidiary amplitude and frequency and these signals to be stored in a block of memory arrays to access and set the reference points or reference signals [1] according to the changes in climatic and body condition. To maintain a large amount of data via memory array could create a problem on power and consumption. However, memory the cycle of any digital circuitry operation dealing with write, read, and hold data or value from the signal for a particular period [11]. Hence to perform high-speed data access the scaling of voltage is considered to be flexible and the leakage power and other power such as static, dynamic power should maintain at the minimum level [6]. Figure 2 describes the graphical view on variation in supply voltage to the IC technology and corresponding changes of the year. From the figure a mission is clear that reduction on supply voltage could cause a drastic reduction in power it may be due to stand by or hibernate mode of the device. The upcoming sections of this paper describe the related works followed by the methodology and the various pros and cons of the system.

2 Related Works

The implantable biomedical device and wearable biomedical devices could sense the potential variation from the biosignals to identify the response of the body or

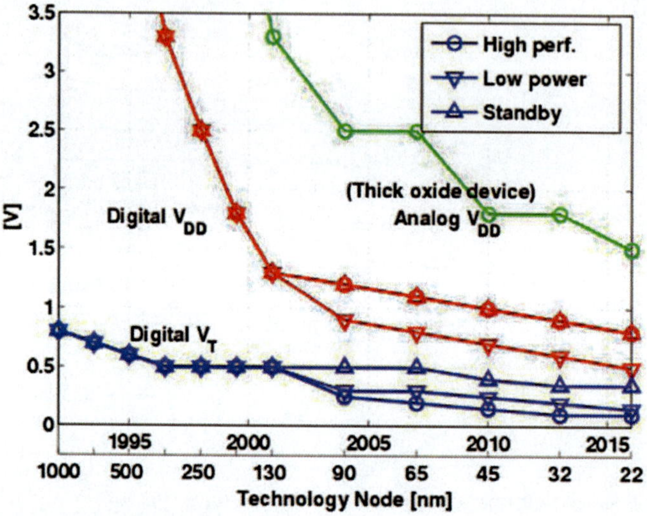

Fig. 2 Power versus scaling technology

the measurement of parameters so that it can be referenced for the other variation due to changes in body parameters [1, 11]. Hence the observed amplitude and the frequency of signals has to be stored to a memory location which may be mostly on-chip memory. To make large data storing and accessing Flash-type memory or DRAM required [12]. But for the devices associated with the biomedical application needs a faster response and high speed of data transfer between the processor and sensors associated with the device which insists on the importance of SRAM in the field of medical applications. There are more research works has been proposed on the design of SRAM and using them for different applications like IoT, communication, and wireless sensor network or even for defining high-speed computation devices in artificial intelligence. The article [13]. A 12T sub-threshold SRAM Bit-cell for Medical Device Application outlines the construction of SRAM using 12 number of CMOS based transistors and how the signal to noise margin can be reduced. According to the [13] authors description, the addition of transistors and pushing them to operate at the sub-threshold region might reduce the SNR of 45% but at the same instance the addition of transistor could lead to the increment in the die area compare to the conventional method transistor and also more no of a transistor connected with cross-connected inverter cause the additional requirement of pull up and pull down resistors logic along with the possibility of development of unwanted capacitance when making fabrication process. Interestingly the method which is having 12T has good immune power to the noise, hence the SNR is high due to the chargeable transistors N_9 and N_{10} which are connected to the WL and WL' respectively. Another article [14] "Ultra-Low Power, Process-Tolerant 10T (PT10T) SRAM with improved read/write ability for the internet of things (IoT) applications" which focus on improving the read speed of SRAM with 10 number of transistor operation. The two single lines RWL and WWL are responsible for reading and write operations. The inverter connected as cross-coupled to form the latch operation and to read or write the data of 1 or 0. The usage of 10 transistors [15] again occupies more areas when implementing and access the data of more than gigabytes as it is involved in IoT application. The MN2 and MN4 have connected to the output logic of lower inverter again which could cause more leakage power at the subthreshold region. The other pair of transistors MP1–MP2 and MN6–MN8 were connected to the supply and grounds respectively. Reduce the number of transistors without degrading the performance of SRAM such as speed, and acquiring more reliable data was getting more complex due to the sub-threshold region operation. Few more article related the way to define the read and write operation on SRAM with the different architecture and especially [12] "process corners analysis of data retention voltage (DRV) for 6, 8, and 10T SRAM Cells" define the importance of variation in the retention voltage on determining the stability of the SRAM memory array at the different regime of process corners (Fig. 3).

The process corners effects are variation starts from the IC fabrication methodology, changes in channel length modulation parameter, and changes in the scaling voltage which could cause an unpredictable response. This also makes a high impact on gate bias junction leakages, deep and sub-threshold leakages along with major reverse leakages of MOSFET, and this could vary to the different approaches on the architecture and die to die area variation or within the die width modulations.

The author [12] explains the potential changes might happen at the Data retention voltages (DRV) but the transistor count is increased which again leads to the larger die occupancy and the cross-coupled setup of two inverters connected with access transistors as shown in Fig. 4, which are controlled by the BL or BL' lines create more leakage power at the node points of coupled junctions [8]. Even though this circuit is good for analyzing the DRV based process corners, it is not suited to the wearable biomedical devices. Quite in numbers of research has been made on designing the SRAM which was a very fast response and does not require any circuit for refresh and more reliable performance on data storage with lower power consumption. To

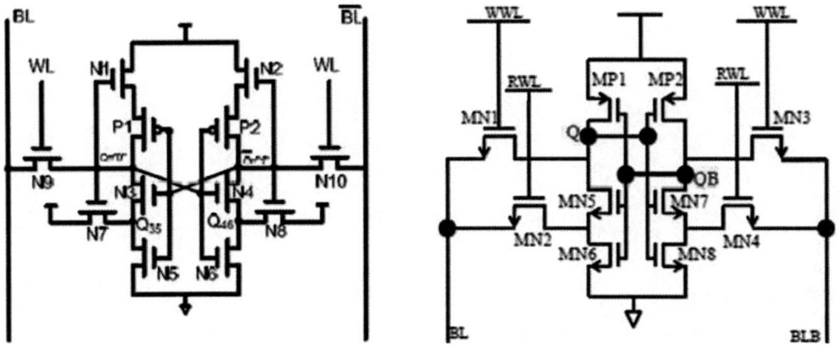

Fig. 3 **a** 12T SRAM, **b** 10T SRM

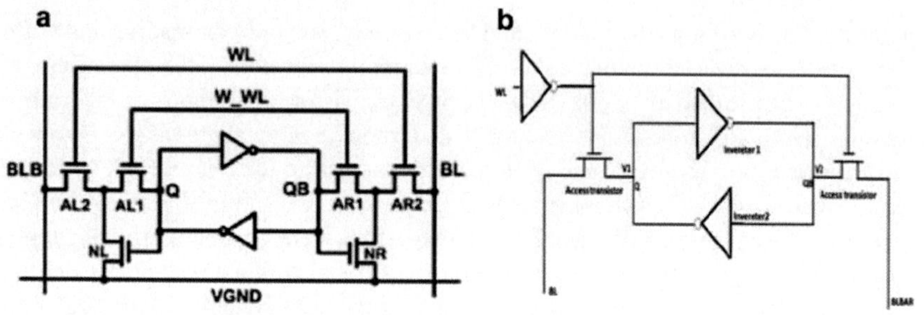

Fig. 4 **a** 9T SRAM, **b** 8T SRM with DRV

overcome the certain drawbacks on density and cost for manufacturing the SRAM for high-speed operation [3] or high computation with power reeducation at the junctions. With the rapid growth of the IC technology, the shrinking of devices could happen. Due to this, the leakage power becomes the major issue. Without giving any input signal some amount of current is derived at the output section. It is called Power discharge and the sources are mainly due to the switching activity of the transistor. Other sources are gate leakage, which is caused due to the I_{GS}, it might be less and the shrinking of devices, who makes the oxide layer to be very thin and it makes the transistor to switch on for the small variation on the input. The tunneling effect, which acts as major static power consumption along with the subthreshold leakages [1]. Another source of power reduction is charging and discharging of high parasitic capacitance

$$\mathrm{Psc} = \mathrm{Vdd} * \mathrm{IP} \tag{2.1}$$

$$E_{\mathrm{dp}} = \mathrm{Psc} * \mathrm{tsc} \tag{2.2}$$

Energy conservation

$$E = \mathrm{CL} * \mathrm{Vdd}_2 \tag{2.3}$$

$$P_{\mathrm{switch}} = E * f.\alpha \tag{2.4}$$

$$\mathrm{Isbt} = A * \mathrm{e}_n \mathrm{KT}^{q(\mathrm{VGS}-V_{\mathrm{th}})} / \left(1 - \mathrm{e}_{\mathrm{KT}}^{\frac{qV_{\mathrm{ds}}}{}}\right) \tag{2.5}$$

where,

Vdd—Supply voltage, Ip—Peak current and Tsc—Period, Isbt—Subthreshold current. The equations on power consumption and energy conservation could intent the importance of supply voltages. The variation on the Vdd the power dissipation level is reduced and the channel length modulation parameter and body effects are called secondary characteristics of MOS devices. Another parameter that affects the SRAM performance in switching activity, which is vary for the different SRAM techniques. So the proposed methodology was examined to the variation on supply voltages [4] and the power reductions related. By pushing the transistor to operate in the subthreshold region we can reduce the leakage power and then removing of stacked transistor reduce the cell count.

3 Methodology

The proposed methodology is mainly to deduce the no of transistor usage to store a single bit i.e., 1 or 0 (logic High or Low). The SRAM is a more reliable circuit and denser circuit. Usually all the memory devices are designed to perform either read or write operation i.e., extracting (recalling) the data from the on-chip memory to the processor or store the results of the computation to the on-chip memory. Like the operation states read and write, stand by (hold) operation should be performed in case of larger data has to be computed. Our proposed system is more suitable to the biomedical application on wearable and implemented type, since large no of biosignals as the potential to be processed, so that the SRAM to be defined to more optimum energy per operation on the stages of hold (stand by) condition. To perform the write and read operation corresponding control lines to be initiated. Specifically to write "0" operation the WL line should be enabled with BL to be high value (logic 1) and BL' to be a low value (logic 0) and then write "1" has to be done in vice versa. A latch operation is taken place where this operation is particularly like an SR latch circuit with variable pulses as low, in other applying signals to make rest. Similarly the read operation is done through averring of the word line and then adjusting the bit lines (BL). The Bit lines are averted accordingly the data-driven. Due to imbalance between the BL and BL' makes the circuit to get the potential difference and which could lead the operation of reading the data from the memory cell. So the transistor is placed to sense the variation on the memory cell to detect logic high or logic low value. The proposed operation of SRAM is illustrated in Table 1, where *H* and *L* represent high and low logic values corresponds to 1 and 0. Thus the value of *H* should be maintained to the value of 1.2 V of scaling voltage and logic low is maintained to 0 scaling voltage.

SRAM also implemented using feedback technology of two cross-coupled inverters connected and MOS 6 and 5 act as an access transistor for writing the

Table 1 Simplified operation of SRAM

	WL	RL	BL/BL'
Write*H*/*L*	*H*	*L*	*H*/*L*
Read*H*/*L*	*L*	*H*	*L*/*H*
Hold	*L*	*L*	*H*

data to the SRAM. Due to the cross-coupled sections the reverse leakage power is developed at the junction of inverter output. This leakage power can be reduced to 50% by using a single MOS to switch over the both access line which is operating at a deep sub-threshold region. The bit lines are pre-charged so the static noise margin value is found at the time of high value and the reading operation is taking place at this instant. Since the bit lines are pre-charged so the writing of 0 only needs assert signaling otherwise the write pin is available in logic high. The corresponding time required to change the write signal from 1 to 0 or vice versa is called to write access time. The MOS M_5 and M_6 are connected with the WL and access line. Hence the bit lines are given as BL and BL' to the two different transistors which could arise the different categories of failure like read, write, and access lines. So the probability of failure reduction in the above category will improve the functionality and robustness of the SRAM. The probability of failure could reduce by reducing the no of instances the device hit on the maximum value of VDD, i.e., scaling voltage of 1.2 or more values. One of the simplest approaches is reducing the count of transistor rather than the scaling voltage because the reduction of scaling voltage of the MOS device might cause signal integrity problem, ration of cutoff current I_{on}/I_{off} value is high along with that affects the secondary characteristics of MOS in sub-threshold regions. Hence the 5T SRAM is extricated from the 6T SRAM by replacing one MOS transistor which is connected to the access lone. However the cross-coupled logic of inverter to be used here to change the values of read and write or hold signals which may be asserted by the voltage control circuit and pre-charge circuit for bit line. To perform the read operation the 5T SRAM will undergo two different mechanisms to avoid unwanted power consumption. In the first mechanism, the read and write control signals are remains on logic high (1) value so that the voltage control circuit is connected to the NMOS of the bottom is turned off and then the voltage at the nodes of NMOS2 and NMOS1 is pulled up to the high value as following to the inverter logic. At the same instant the current flow at the bottom circuit is high which makes the read access line assert to high value making the reading operation is speed than the traditional method. The second mechanism of reading is the read logic is low which performs the inverse operation described above (Fig. 5).

The time difference between the logic high to low on the read cycle used to calculate the read access time. The value of read access time can be modified based on the fall time and rise time adjustment. The reading of logic high (1) has happened under the condition the NMOS1 and NMOS2 are turned off and on respectively which pushes the voltage to be developed at the cross-coupled junctions due to the pre-charge sections of access line, so that the voltage of the read access is maintained to a constant value and which is greater than the threshold voltage of PMOS 1 and 2. The reverse operation to be carried out for the read logic low (0) concerning the pre-charge logics. The only thing more serious concern about the SRAM design is cross-coupled connections that are similar set up to the latch circuit. This setup is combined with the inverter logic where it starts to act as oscillators, that us reclamation action takes place between the nodes of inverters. Specifically this is developed when the logic low (0) is stored and it leads to the destruction of the signal varies from the read of logic low (0) toward logic (1). To avoid this regenerative action the proper aspect

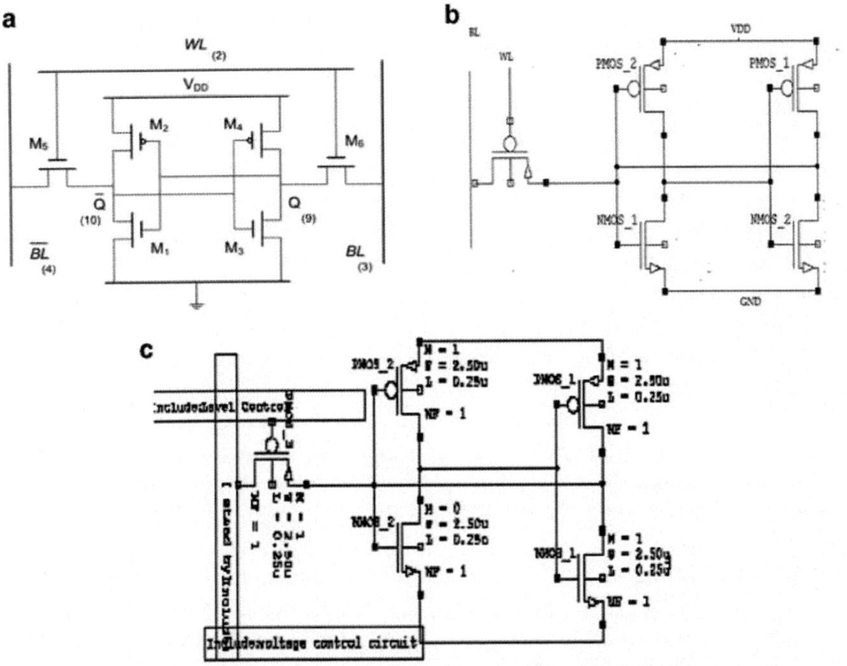

Fig. 5 a 6T SRAM, b 5T SRAM with control voltage and pre charge sections

ratio between PMOS and NMOS, and where the ratio of 1:2.5 is used and the access transistors are connected with the pull-up transistors to maintain the high ratio of pull up and also this could improve the write and read operations parallel. Likewise read and write operation the SRAM posses one more state of operation called hold mode or standby mode. Read and write operations are denied during this mode of operation. The node voltages of the inverter are trying to achieve the equilibrium condition, i.e., the potential developed at the inverter of PMOS1 and NMOS1 is slightly equal to the threshold voltage of other sections. It is due to that both read and write lines are becoming low line and the bit line is high which makes the voltage control circuit NMOS is kept high. So that the standby mode is turned on whenever the NMOS of VCC is switched to the low value which makes associated NMOS are active. The time on which the hold time is maintained is called hold access time and this delay of occurrence can be adjusted by varying the rise time and fall time value of the signals which are used for the assertion. The alternate variation of the threshold voltage at the nodes does not introduce any load to the word line. The additional circuit VCC also maintains equal load. Hence the ratio of NMOS/PMOS is not varied so the leakage power is reduced and this variation does not have to create any impact on the dynamic power. As because of the reduction in leakage power and access time on hold the overall power reduction on the standby mode is minimized. The trade-off between area and power could generate the idea on the reduction of cell count i.e., reducing the no of the transistor and reduce the switching activity of swinging from logic high (1) to logic low (0), could lead to reducing the power dissipation also the aspect ratio of the transistor (width to length) defines the power consumed and energy acquired by the electron moment on current leakage could reduce the leakage current. Thereby the power reduction can be improved in the cell reduction technique. These transistors are normalized to operate at the lower potentials to switching. The variation on the transistor ensure the high level on stability in both cases on read and write which is the basic function of the SRAM, where the proposed model of 6T SRAM could consume less power as it has less no of switching activity and the frequency of operation is defined with the charging of the capacitor. The 6T transistor-based SRAM isolates the Bl and Bl' and the cross-coupled pair of CMOS are responsible for read and write. Even though the 6T has a good speed of switching, it is not optimized to the body wearable devices since it needs the larger battery life and the processor might consume less power. Usually when we prefer 10T SRAM the lower transistors were activated for read and write following the 1 and 0 of the bit line. The stacking effect is preferred, which is a series connection of transistor at the deep sub-micron regions which changes the body effect can reduce the leakage power. So during the next cycle on read the transistor is not completely off and it stays at half of the voltage and then the prior charge circuit consumes very less power for the variation of voltages. In the case of hold operation the bl maintains at the midway of the voltage to save the data, during this stage it acts as a buffer. The drawback with the 10T technique is more no of the transistor occupies large no of the area and the switching power is high. This is also possible to perform the power reduction as a sequence of '0' and '1', '0' as writing cause the small leakage power management via partially of Bl, and during '1' of writing the remaining was done through the pre-charged point. Hence the power reduction is achieved. The area has to be compensated meanwhile the functionality of SRAM could not be affected. So its proposed that usage of the single transistor to operate for Bl and Wl access, ie the wl line is connected gate terminal of the MOS transistor which could operate at the deep threshold region, hence the subthreshold current is reduced, and It could not complete the process of fully charged and during the wl is '1' then it could conduct for the write operation and its charged up to the level of storing data and during the reverse operation the read could start to happen from the point of pre-charged value to reduce the power, kind of single transistor with two signal variation at the sub-threshold region remove the stacking effect of inverter in SRAM and case of both bl and wl are at high the hold operation is taken the SRAM will act almost as a buffer. The usage of separate transistors for pre-charging to high and low by forming separate units used in 10T SRAM can overcome by this method. Hence we suggested 5T technology where the one separated transistor was responsible for managing the power reduction. This is done by operating the M_5 transistor at the cut off sleep region. The drawback of SRAM on volatile can be compensated using the different read operations and prevents the data from corrupting. The switching activity of the SRAM could be control by adjusting the time on which the wl and Bl are varied from '0' to '1' and vice versa. The switching activity of an inverter has happened when the PMOS become open for the high input and NMOS becomes closed for the low input, like that two CMOS are cross-coupled so the switching activity of pair of CMOS are maintained. That is each CMOS was viewed as an RC network where the load capacitance is calculated for the vi variation of high to low and inverse. By controlling the variation of CMOS at switched intervals the switching activity is controlled and operating the transistor with lower potential which could control the charge and discharge operation of load Capacitance (Cl) at the subthreshold regions.

4 Results and Discussion

The SRAM with different numbers of transistors count were discussed in the previous sections are simulated in the cadence virtuoso environment and for 45 nm technology. Figure 6 shows the proposed SRAM architecture which uses 5 no of transistors for its operation such as read, write, and hold operations. The corresponding layout of the SRAM is generated to make more ASIC based applications. Similarly, Figs. 7, 8, 9 and 10 along with the layout generated for Soc and Asic based application.

Figure 11 shows the simulated waveform for the SRAM of 5, 6 and 9T. The waveforms are simulated by asserting the signal of BL, Read and write signal. During the operation of Read the process of the write line should be maintained to a lower value as shown in the graph. The strong 1/0 and week 0/1 values are stored based on the variation of the transistors turned on (NMOS on/off and PMOS off/on). The

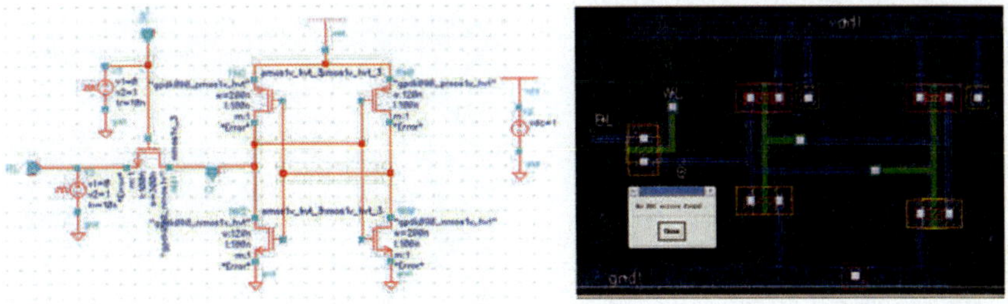

Fig. 6 No. of transistors: 5–5TSRAM and layout

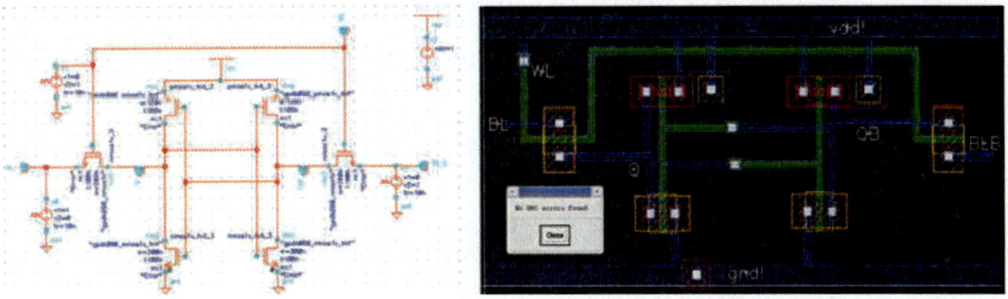

Fig. 7 No. of transistors: 6–6TSRAM and layout

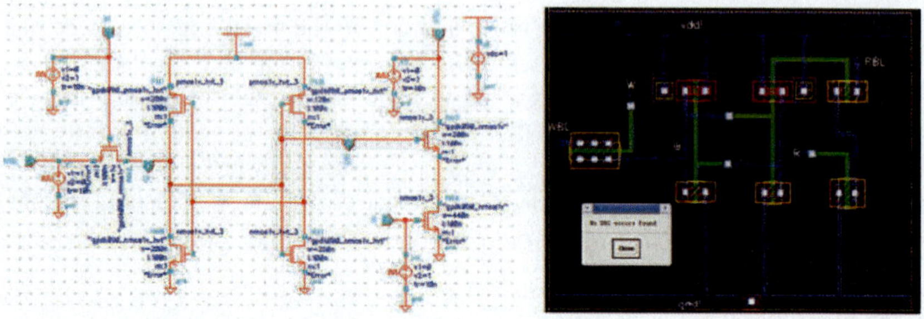

Fig. 8 No. of transistors: 7–7TSRAM and layout

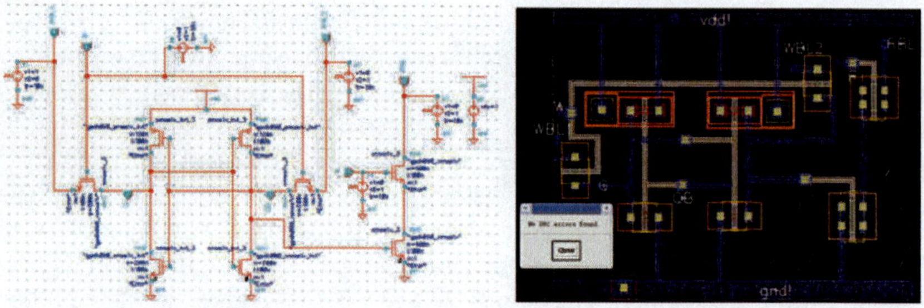

Fig. 9 No. of transistors: 8–8TSRAM and layout

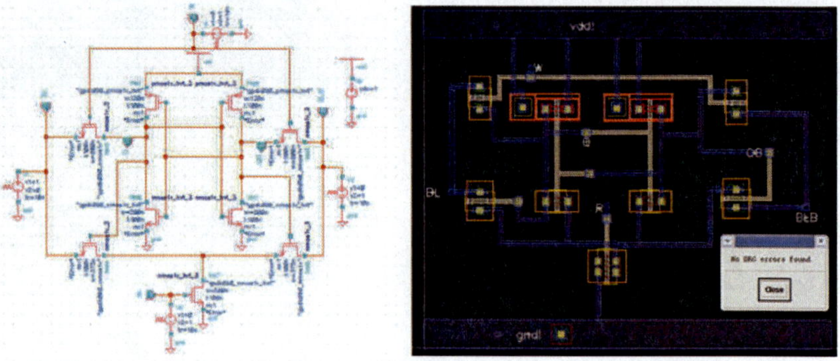

Fig. 10 No. of transistors: 9–9TSRAM and layout

simulation graphs show that to maintain the constant variation of read and write logic, the bit line BL should be kept as high, which is pre-charge logical for the read and write operation which explains in Sect. 3.

Table 2 shows the overall comparison of power and delay for the different SRAM cells like 5, 6, 7, 8, and 9T. Thus the comparison table shows the reliability of SRAM in data storage and how much time it takes to perform the read, write, and hold

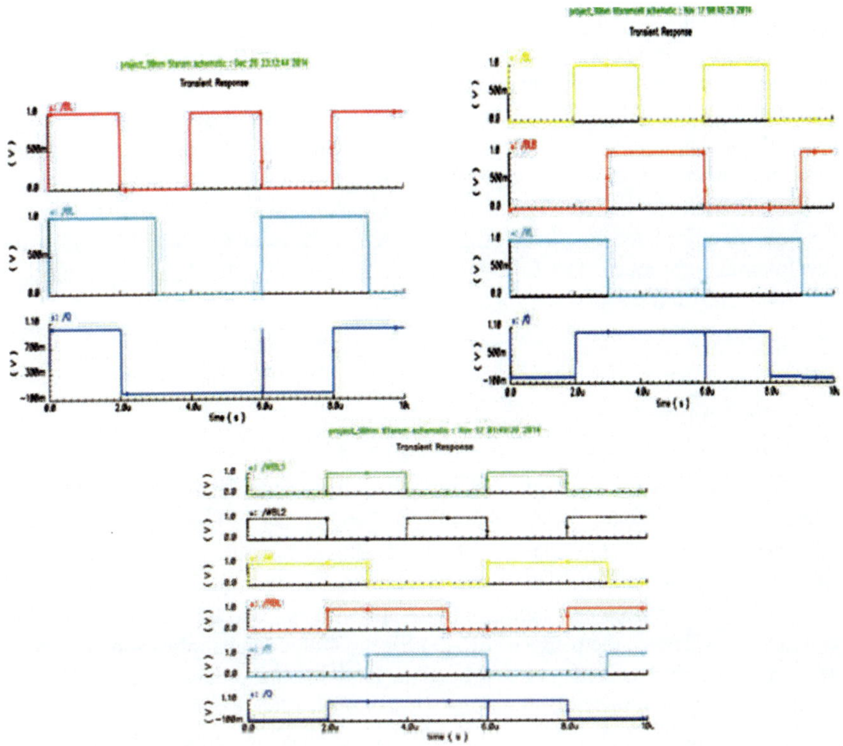

Fig. 11 Simulation graph of SRAM based on variations in read and write logic

Table 1 Overall comparison of power and delay

Different SRAM cells	Leakage power (nW)	Static power (nW)	Dynamic power (nW)	Delay (ns)
5T	3.24	8.19	0.368	2.14
6T	27.26	37.78	10.27	3.96
7T	3.45	105.34	21.20	6.34
8T	67	93.17	21.06	9.18
9T	62.64	82.89	36.68	13.98
10T	89.14	155.3	63.4	24.61

operations. Such that the 5T SRAM has the lowest leakage power 3.24 nW compare to other structures. Also other powers such as static and dynamic power are also very less. Only the delay of 5T SRAM is low due to the access time variation between the read and write operation.

The results were tabulated and it shows by using 5T SRAM we can reduce the power leakage. Hence by implementing the SRAM array using these 5T techniques we could develop a more powerful memory unit as shown in Fig. 12. Thus this memory section is fitted to the wearable devices which *I* connected to the IoT consumes more amount of power on each iteration of the process of switching happened between triggered input and the output. Henceforth the SRAM array can reduce the leakage powers of the implantable devices and produce the good power management unit in terms of the extended life span of a battery or long-running duration of portable devices. These results are shown for the single SRAM device so when we connected to the SRAM array it is mandated to give importance to fault-tolerant techniques. The fault-tolerant technique could be done on the partial reconfiguration and usage spare logics.

5 Conclusion

This paper explains about the different architectures implemented on SRAM with different no of transistors. The wearable and implanted biomedical devices are connected to the cloud (IoT) for the data processing applications. So that the importance of storing the data and access the data on the required time is more important. Also the type of processors should coordinate with the on-chip memory and the cache memory. Hence the usage of SRAM for the biomedical application is needed for retrieving the data and process the data between the processor and the chip memory. In the same instance, the memory sections of the die and on the die areas should consume lower power, specifically, the leakage power should be low enough to a reduction in the overall power. Henceforth by implementing the proposed SRAM based on 5T structure assured that it consumes very less leakage power of 3.24 nW and static power of 8.19 nW and the delay of 2.14 nS. In future the 5T SRAM structures could be used for the different applications in IoT and cloud-based applications. Further to the research area of memory cells the no of transistor event reduce to 4T or 3T structures. The applications are wider if the scaling of voltage is also less and consumes a lower amount of power leakages in all aspects.

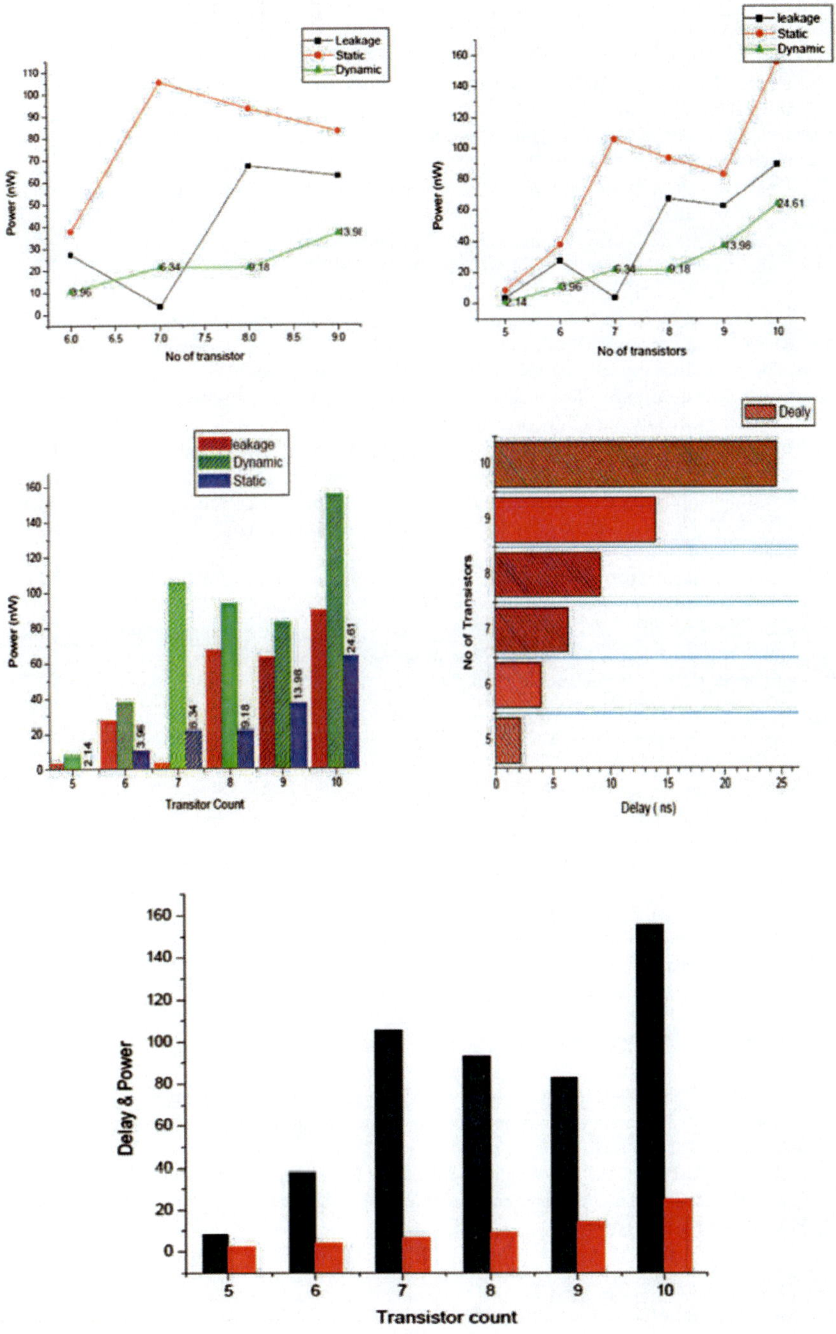

Fig. 12 Performance analysis of different SRAM

Reference

1. Kumar JRD, Babu CG, Balaji VR (2019) Analysis of effectiveness of power on refined numerical models of floating point arithmetic unit for biomedical applications. In: AIP scopus indexed proceedings of international conference on advances in materials processing and characterization, ICAMPC 2019
2. Kumar JRD et al (2019) Performance investigation multiplier for computing and control applications. JRDCS
3. Tan MLP, Lentaris G Device and circuit-level performance of carbon nanotube field-effect transistor with benchmarking against a nano-MOSFET. J Nano Scale Res Lett 21, 2005
4. Roy K, Mukhopadhyay S, Meimand-Mehmoodi H (2003) Leakage current mechanisms and leakage reduction techniques in deep sub-micron CMOS circuits. Proc IEEE
5. Malar ACJ, Kowsigan M, Krishnamoorthy N, Karthick S, Prabhu E, Venkatachalam K (2020) Multi constraints applied energy efficient routing technique based on ant colony optimization used for disaster resilient location detection in mobile ad-hoc network. J Ambient Intell Humanized Comput. https://doi.org/10.1007/s12652-020-01767-9
6. Klinefelter A, Roberts NE, Shakhsheer Y, Gonzalez P, Shrivastava A, Roy A, Craig K, Faisal M, Boley J, Oh S et al (2015) 21.3 A 6.45 μW self powered IoT SoC with integrated energy-harvesting power management and ULP asymmetric radios. In: Proceedings of the IEEE international solid-state circuits conference—(ISSCC), San Francisco, CA, USA, 22–26 February 2015
7. Prabhu E, Mangalam H, Karthick S (2016) Design of area and power efficient Radix-4 DIT FFT butterfly unit using floating point fused arithmetic. J Central South Univ 23(7):1669–16810
8. Pal S, Islam A (2016) 9-T SRAM cell for reliable ultralow-power applications and solving multibit soft-error issue. IEEE Trans Dev Mater Reliab 16:172–182
9. Pagiamtzis K, Sheikholeslami A (2006) Content-addressable memory (CAM) circuits and architectures: a tutorial and survey. IEEE J Solid State Circ 41(3):712–727
10. Balaji VR, Subramanian S A novel speech enhancement approach based on modified DCT and improved pitch synchronous analysis. Am J Appl Sci
11. Sharif KF, Islam R, Biswas SN (2018) Low power novel 10T SRAM with stabled optimized area. In: 2018 IEEE international WIE conference on electrical and computer engineering (WIECON-ECE), pp 21–24
12. Sachdeva A, Tomar VK (2020) Design of a stable low power 11-T static random access memory cell. J Circ Syst Comput 21:205–206
13. Chen H, Jun Y, Meng Z, Xiulong Z (2011) A 12T sub threshold SRAM bit-cell for medical device application. IEEE. https://doi.org/10.1109/cyberc.2011.93
14. Singh P, Vishvakarma SK (2017) Ultra-low power, process-tolerant 10T (PT10T) SRAM with improved read/write ability for internet of things (IoT) applications. J Low Power Electron Appl 7:24
15. Islam A, Hasan M (2012) Leakage characterization of 10T SRAM cell. IEEE Trans Electron Dev
16. Kumar JRD, Babu CG, Karthi SP, Soundari DV, Priyadharsini K (2020) A novel system design for intravenous infusion system monitoring for betterment of health monitoring system using ML-AI. IJRTEE 2649–2655
17. Balaji VR, Maheswaran S, Babu MR, Kowsigan M, Prabhu E, Venkatachalam K (2020) Combining statistical models using modified spectral subtraction method for embedded system. Microprocess Microsyst 73

Comparative Study of Different Beamforming Techniques for 5G: A Review

Laxmikant Shevada, Hema D. Raut, Rajeshwari Malekar, and Sumit Kumar

Abstract Due to beamforming, there will be an increase in energy of signal to the intended user and decreases in interference which is the prime requirement of mmWave communications required for high data rates and large capacity applications. Different beamforming techniques such as analog, digital, hybrid beamforming along with their associated methods have been studied and compared in this paper for searching optimum architecture in terms of energy efficiency. Antenna parameters like gain, half-power beamwidth, scattering parameters are calculated at different frequencies by various authors using tools like ADS, HFSS, MATLAB, CST, and Microwave studio have been studied. While doing this comparison, both hardware, as well as software (algorithms) aspects of the techniques has been considered.

Keywords Beamforming · Antenna · Analog beamforming · Digital beamforming · Hybrid beamforming

1 Introduction

For transmission or reception of the directional signal, beamforming is utilized in sensor arrays in which some angle experiences interference either constructive or destructive. Beamformer monitors the amplitude and phase of the signal at

L. Shevada · H. D. Raut · R. Malekar · S. Kumar (✉)
Electronics & Telecommunication Department, Symbiosis Institute of Technology, Symbiosis International (Deemed University), Pune, India
e-mail: er.sumitkumar21@gmail.com

L. Shevada
e-mail: lp.aish@gmail.com

H. D. Raut
e-mail: hdraut2083@gmail.com

R. Malekar
e-mail: rajeshwari.malekar@gmail.com

G. Ranganathan et al. (eds.), *Inventive Communication and Computational Technologies*, Lecture Notes in Networks and Systems 145,
https://doi.org/10.1007/978-981-15-7345-3_50

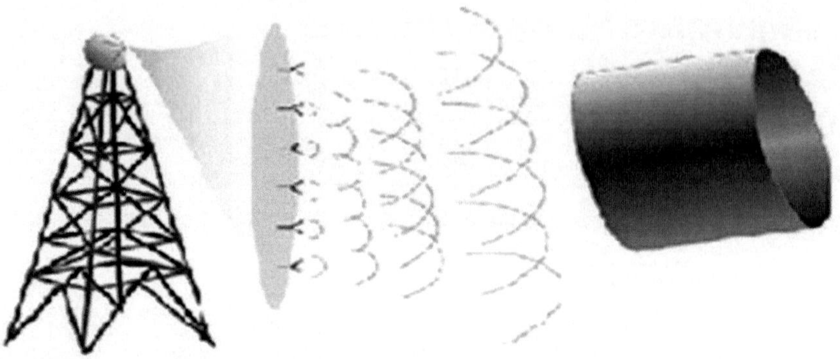

Fig. 1 Directional beam formation using an antenna array

every transmitter for changing the directivity of an array. In beamforming, multiple antennas are used to control the direction of a wave with the help of amplitude and phase of signals in an array. While doing this, space between the same signal sent from different antennas is kept as half of the wavelength. Depending upon the position of the receiver, the alignment of signal forms the constructive or destructive interference. The creation of a directional beam using an antenna array is shown in Fig. 1 [1].

In wireless networks, for enabling higher capacity and reducing co-channel interference smart antenna is the most dominant technology in which antenna centers towards the pattern of beam for the "signal of interest (SOI)" and reduces the "Signal not of interest (SNOI)". A smart antenna system divided into two types depending on the beamforming technique, the first one is switched (conventional or fixed) beam system and the second one is an adaptive (phased array) array system [2]. The first antenna system can be easily constructed by using beamforming networks, several radiating elements, and RF switches [3]. The second system requires advanced signal processing and different intelligent algorithms. Analog beamforming network can be constructed using different feeding techniques such as the Butler matrix, Blas matrix, Nolen matrix, and Rotmans Lens. In this paper we have compared Butler matrix, Blas matrix and Nolen matrix beamforming techniques. Digital beamforming can be classified on the basis of the arrival of the angle from the transmitter so, when the angle of arrival from the transmitter does not change with time, optimum array weights need not be adjusted then it is called as fixed digital beamforming whereas if optimum array weight needs to be adjusted as the angle of arrival changes with time then it is called as adaptive digital beamforming. Algorithms for these digital beamforming techniques have been presented in the literature by different authors. The huge demand requirement of the wireless communication system mainly depends upon spectral efficiency and bandwidth. Currently, the frequency range for wireless technology is from 0.3 to 3 GHz frequency band [4]. System bandwidth can be explored as "physical layer technology has already touched the Shannon capacity" [5]. Therefore, radio paradigms based fifth-generation technology relies on consideration of high-frequency mmWave frequency band ranging from 3 to 300 GHz. Multiple Input Multiple Output (MIMO) technology along with multiple antennas at Tx and Rx is considered as a promising method to enhances system efficiency [6]. While addressing these issues, researchers came up with the hybrid beamforming (analog plus digital beamforming) techniques that utilize maximum bandwidth, reduce interference, and increases the data rate of the system.

In beamforming, most of the signals generated from antenna array steers in the desired angular direction with the scale factor which is weighted in the transmitter section. Whereas to maximize the signal to noise ratio (SNR) at the receiver side, signals are combined coherently with different scale factors. The gain calculated at the receiver from signal to noise ratio is known as beamforming gain. In hybrid beamforming, the beamforming gain as a result of a variation in the slope of the error probability is called diversity gain. High propagation losses encountered in mmWave bands for 5G can be eliminated by using a hybrid beamforming technique along with MIMO. With these losses diffraction, blockages and low penetration problems can also be avoided using the compact structure of hybrid beamforming. Capacity enhancement issue can be solved by using hybrid beamforming and employing massive MIMO in the form of a heterogeneous network (HetNet) [7].

The direction of beam arrival based beamforming technique is used in a wideband code division multiple access (WCDMA) whereas, in the 2G elementary beamforming with antenna array concept is employed. In the case of 3G evolution, a precoding based MIMO beamforming technique is utilized. Fixed and adaptive beamforming technique is employed in 4G communication. The hybrid beamforming technique with a massive MIMO concept is going to be deployed in mmWave communication.

This paper is presented in the following way: In Sect. 2, literature review of different research papers has been carried out, Sect. 3 consists of a comparative discussion of various beamforming techniques such as analog, digital, and hybrid beamforming, and finally, the paper is concluded in Sect. 4 based on our observations and findings.

2 Literature Review

The feeding network is important in beamforming antenna array which allocates the required angle and the magnitude of signal to every antenna element. There are mainly three topologies for implementing feeding networks as which is series feeding, parallel feeding, and matrix feeding. Out of these, "matrix feeding is a multiple input multiple output network" which is suitable for mmWave communications [8]. 4 × 4 Butler matrix [9] is used as a beamforming network that produces 4 narrow steerable beams when the designed antenna array operates at 2.35 GHz and its simulated results on ADS and CST Microwave studio is presented in [9]. The Beamwidth and gain are calculated for 4 different ports each having bandwidth more than 100 MHz. It is formed by using phase shifters, couplers, and crossovers. The phase difference across ports is constant in the output. In [10], Hassanien et al. introduced the Rotman Lens beamforming technique in which a steerable system operating at frequency from 25 to 30 GHz is designed and covers a scanning angle from −45 to 45 degree. This system provides better outcomes on the basis of scattering parameters and beam steering characteristics. Rotman Lens provides an extensive scanning angle, low cost, and it is easy to implement [11]. It has true time delay characteristics means "it steers the beam which is independent of operating frequency" and has high bandwidth [12]. 3 × 3 Nolen matrix [8] consists of coupling ratios and phase shifters which provide arbitrary phase difference. Blas matrix [13] can be designed using load terminators, phase shifters, and couplers but has more power loss that becomes a challenging issue in the Blas matrix. Nolen matrix was developed by cutting partially Blas "matrix along the diagonal line and replacing diagonal coupler by transmission line" which solved the problem of power loss in Blas matrix and hence reduces the number of components.

In Phase shifters, the beams are steered to SOI in the case of traditional antenna arrays. But in a modern antenna array, smart antenna do the beamforming known as a digital-beamformed array which is based on algorithmic logic. According to this logic, the beam is allowed to steer towards SOI and null towards SNOI. "When the angle of arrival from the transmitter does not change with time, optimum array weights need not be adjusted is called fixed digital beamforming whereas optimum array weights need to adjust when the angle of arrival changes with time" is known as adaptive digital beamforming. Both fixed and adaptive digital beamforming algorithm performance is compared on eight elements linear patch array antenna operating at 2.4 GHz frequency for LTE applications and results are simulated on MATLAB and HFSS software [14]. Matrix Inversion algorithm applied to fixed beamforming works on linear algebra method [15] for computing complex weights. Least Mean Square (LMS) method is the simplest algorithm applied to compute adaptive weights in real-time which works on gradient descent method to calculate weights. This algorithm is not suitable for a highly noisy environment [16]. To overcome this drawback, weight in LMS has to compute recursively using the recursive LMS algorithm.

Motivation for hybrid beamforming is discussed in [6]. As compared to conventional digital beamforming, hybrid beamforming requires less hardware and consumes less power. For lowering energy consumption in mmWave systems, hybrid (analog and digital) structures have been proposed in [17]. "A Survey on Hybrid Beamforming Techniques in 5G: Architecture and System Model Perspectives," was done by Ahmad et al. in [7]. In pursuit of a suitable antenna for 5G communication, a survey of performance enhancement technique carried out in [18]. It has been observed that high gain, cost-effective, and miniaturized antipodal Vivaldi antenna [19, 20] will be a potential candidate for high-frequency mmWave communication whereas triangular patch antenna array is suitable for C-band application [21].

Table 1 Comparison of Butler matrix, Blas matrix, and Nolen matrix [8, 9]

Parameters	Butler matrix	Blas matrix	Nolen matrix
Construction	Hybrid couplers, crossover, phase shifter	Couplers, phase shifters, load terminations	Couplers, phase shifters
Degree of freedom	Less due to constant phase difference	Moderate due to unique phase difference	High due to arbitrary phase difference
Efficiency	High as integration is easy	Less as signal flows into terminating load	Moderate as limited signal flows to terminating load
Power loss	Low due to simplicity	High due to complex structure	Moderate as it is designed by cutting Blas matrix along the diagonal line
Design	Simple	Moderate	Complex

3 Review of Survey Papers on Beamforming in 5G

3.1 Analog Beamforming

The process of beamforming in which amplitude or phase variation is applied to an analog signal at the transmitter end and the signal received from different antennas are summed up before applied to ADC in the receiver end is called as analog beamforming. Analog beam former consists of transmitter modules that are used to control amplitude and phase of the transmitted signal of each antenna element.

This type of beamforming technique can be achieved by using the Butler matrix, Nolen matrix, or Blas matrix feeding techniques which have been explained by different authors compared in Table 1 based on certain parameters.

3.2 Digital Beamforming

Before DAC conversion at the transmitter end, amplitude, or phase variation is applied to digital signal is known as digital beamforming. This beamforming implies weighting these digital signals such that when added together forms the desired beam. To achieve digital beamforming, different algorithms like matrix inversion (MI) algorithm, least mean square (LMS) algorithm, and recursive least mean square (RLMS) algorithm have been studied and compared in Table 2.

Table 2 Comparison of MI, LMS, and RLMS Algorithms [14]

Parameters	Matrix inversion	Least mean square	Recursive least mean square
Logic	Linear Algebra	Gradient descent method	Gradient descent method is recursive fashion
Methodology	Matrix	Eigenvalues	Eigenvalues
Memory	Not required	Required	Required
Complexity	High due to matrix inversion	Low as no matrix is involved	Moderate as recursiveness is involved due to low convergence of LMS
Application	Fixed beamforming	Adaptive beamforming	Adaptive beamforming in a noisy environment
Quality of Service	Poor due to more interference	Moderate as it is better than MI but not than RLMS	Good as it is applicable in a noisy environment

3.3 Hybrid Beamforming

Analog beamforming suffers from the inter-user interface and fewer accuracy problems whereas digital beamforming is complex and costly. Considering these disadvantages and advantages like simplicity of analog and a high degree of freedom of digital beamforming techniques, researchers came up with a solution of hybrid beamforming that is a combination of analog and digital beamforming which can fulfill the growing energy efficiency and spectrum efficiency requirement of mmWave communication necessary for 5G communications.

Hybrid beamforming is classified into different categories based on architecture, resource management, and application areas and found out as it is cost-effective solution for 5G communication. Dimensionality reduction can be possible in mmWave communication using massive MIMO hybrid beamforming.

4 Conclusion

After a comparative study of different feeding techniques for analog beamforming, it has been observed that Butler Matrix is the potential technique to implement corresponding beamforming as it provides high directivity compared to Blas and Nolen matrix. Complex weights can be calculated using the matrix inversion algorithm in case of digital fixed beamforming whereas Recursive Least Mean Square Algorithm provides good convergence and better results for digital adaptive beamforming over the noisy environments. Hybrid beamforming enables mmWave massive MIMO communications that open the door for 5G communications.

References

1. https://www.rcrwireless.com/20180912/5g/5g-nr-massive-mimo-and-beamforming-what-does-it-mean-and-how-can-i-measure-it-in-the-field
2. Fernandes M, Bhandare A, Dessai C, Virani H (2013) A wideband switched beam patch antenna array for LTE and Wi-Fi. In: 2013 annual IEEE India conference (INDICON). IEEE, pp 1–6
3. Chang YJ, Hwang RB (2001) Switched beam system for low-tier wireless communication systems. In: APMC 2001. 2001 Asia-Pacific microwave conference (Cat. No. 01TH8577), vol 2. IEEE, pp 946–949
4. Andrews JG, Buzzi S, Choi W, Hanly SV, Lozano A, Soong ACK, Zhang JC (2014) What will 5G be? IEEE J Sel Areas Commun 32(6):1065–1082
5. Bangerter B, Talwar S, Arefi R, Stewart K (2014) Networks and devices for the 5G Era. IEEE Commun Mag 52(2):90–96
6. Molisch AF, Ratnam VV, Han S, Li Z, Nguyen SLH, Li L, Haneda K (2017) Hybrid beamforming for massive MIMO: a survey. IEEE Commun Mag 55(9):134–141
7. Ahmed I, Khammari H, Shahid A, Musa A, Kim KS, Poorter ED, Moerman I (2018) A survey on hybrid beamforming techniques in 5G: architecture and system model perspectives. IEEE Commun Surv Tutorials 20(4):3060–3097
8. Ren H, Zhang H, Jin Y, Yixin G, Arigong B (2019) A novel 2-D 3 × 3 Nolen matrix for 2-D beamforming applications. IEEE Trans Microwave Theor Tech 67(11):4622–4631
9. Zulkifli FY, Chasanah N, Rahardjo ET (2015) Design of Butler matrix integrated with antenna array for beam forming. In: 2015 international symposium on antennas and propagation (ISAP). IEEE, pp 1–4
10. Hassanien MA et al (2019) Wideband Rotman lens beamforming technique for 5G wireless applications. IEEE
11. Rotman W, Turner R (1963) Wide-angle microwave lens for line source applications. IEEE Trans Antennas Propag 11(6):623–632
12. Lambrecht A, Beer S, Zwick T (2010) True-time-delay beamforming with a Rotman-lens for ultrawideband antenna systems. IEEE Trans Antennas Propag 58(10):3189–3195
13. Blass J (1960) Multidirectional antenna—a new approach to stacked beams. In: IRE international conference record, vol 8(Part 1), pp 48–50. Nolen J (1965) Synthesis of multiple beam networks for arbitrary illuminations. Bendix Corporation, South Bend, IN, USA
14. Awan AA, Khattak S, Malik AN (2017) Performance comparisons of fixed and adaptive beamforming techniques for 4G smart antennas. In: 2017 international conference on communication, computing and digital systems (C-CODE). IEEE, pp 17–20
15. Moreira A (2013) Digital beamforming: a paradigm shift for space borne SAR. In: 14th international radar symposium (IRS), 19–21 June 2013
16. Alonso L, Ver Hoeye S, Fernández M, Vázquez C, Camblor R, Hotopan G, Hadarig A, Las-Heras F (2015) Millimetre wave textile integrated waveguide beamforming antenna for radar applications. In: Global symposium on millimeter-waves (GSMM). IEEE, pp 1–3
17. Roh W, Seol J-Y, Park J, Lee B, Lee J, Kim Y, Cho J, Cheun K, Aryanfar F (2014) Millimeter-wave beamforming as an enabling technology for 5G cellular communications: theoretical feasibility and prototype results. IEEE Commun Mag 52(2):106–113
18. Dixit AS, Kumar S (2020) A survey of performance enhancement techniques of antipodal Vivaldi antenna. IEEE Access 8:45774–45796
19. Dixit AS, Kumar S (2020) The enhanced gain and cost-effective antipodal Vivaldi antenna for 5G communication applications. Microwave Optical Tech Lett 62(6):2365–2374
20. Dixit AS, Kumar S (2020) A miniaturized antipodal Vivaldi antenna for 5G communication applications. In: 2020 7th international conference on signal processing and integrated networks (SPIN). IEEE, pp 800–803
21. Bhadoria B, Kumar S (2018) A novel omnidirectional triangular patch antenna array using Dolph Chebyshev current distribution for C-band applications. Progress Electromag Res 71:75–84

Vulnerability Analysis of FPGA Through Side-Channel Attacks in Cloud

S. Harini and Aswathy Ravikumar

Abstract Field programming gate array is a semiconductor device used for programming tasks in parallel with greater efficiency and performance that the general core processors present in our desktop. Larger tasks can be parallelized at a much faster rate. The need for more high-performance computing is increasing, especially in the field of cloud after the patches for effects like "Specter" and "Meltdown" have reasonably slowed down the processors in turn affecting the entire cloud, the need for FPGA accelerators have increased [1]. But there are kinds of security attacks on which FPGA makes it vulnerable and harm clouds. In this paper, the various vulnerabilities of FPGAs try to analyze the kind of attacks on FPGA.

Keywords Side-channel attack · FPGA · AES algorithm · Physical attack · Comparison of algorithm

1 Introduction

The rapid growth of big data has made it a necessity to securely exchange the data between locations, and this has led to the implementation of special hardware and encryption schemes for secure, confidential data transfer. Field programmable gate arrays (FPGA) which are easily reprogrammable has been a breakthrough in this field as it provides a solution for the hardware setup for the implementation of the secure encryption schemes, and many works have shown the improvement in performance and flexibility of the digital application using FPGA. But to make sure hardware is secure, we need to find the kind of attacks implemented on it so that we can have a solution for every attack and can be made more secure and less vulnerable when

S. Harini (✉) · A. Ravikumar (✉)
School of Computer Science and Engineering, VIT, Chennai 600127, India
e-mail: harini.s@vit.ac.in

A. Ravikumar
e-mail: aswathy.ravikumar2019@vitstudent.ac.in

G. Ranganathan et al. (eds.), *Inventive Communication and Computational Technologies*, Lecture Notes in Networks and Systems 145,
https://doi.org/10.1007/978-981-15-7345-3_51

implemented in the cloud. The main emphasis is on security primitive implementation and its detailed analysis of the FPGA synthesis flow vulnerabilities and its application to the cryptographic algorithms.

The cryptographic algorithm considered here is AES [2], and the hardware-based implementation of this algorithm is widely used because of the highly secure and easy implementation compared to the software setup of AES. Later works have been developed [3] for FPGA implementation of the block ciphers. The AED 128-bit algorithm [4] with 21 cycles is implemented with modification of hardware architecture with reduced complexity. The lookup table method is used for the implementation of the round function in AES using an FPGA encryption module [5].

2 Literature Survey

Due to the widespread use of FPGAs in many critical application domains, their security aspect is a major concern. The recent developments have led to the FPGA cloud with remote access to the users to the reconfigurable fabric to implement custom accelerators. This access can expose new security vulnerabilities in the entire system through malicious use of the FPGA fabric.

Types of attack discovered that lead to the malicious practice

1. Side-Channel attack:
    a. Power analysis attack
    b. Remote side-channel attack (without physical access)
2. Physical Attack: SRAM FPGAs, Flash FPGAs
3. Black Box attack.

Physical Attack: The main idea of the physical attacks launched is to get the information of the cryptographic algorithm and to obtain the secret key used in the algorithm by investigating the points inside the chips. The main target points of these attacks are the different parts other than the normal input-output pins of the FPGA.

SRAM FPGA

FPGA in SRAM is mainly used for the configuration logic cells data for the static memory modules which are an array of latches. Static RAM is volatile memory storage, and so every time, the power source is cut offf the data is lost, and this leads to the need of programming the FPGA during the booting process. Both SRAM and SRAM FPGA have a similar structure for the memory cell of the first and the internal structure of the later, and this leads to making the attacks at the settings. It is believed that SRAM being static will lead to the loss of data when the power supply is lost, but it is not always true because it depends on the semiconductors. The main physical variations are made by the effects like ionic contamination, electromigration, and hot carries (Fig. 1).

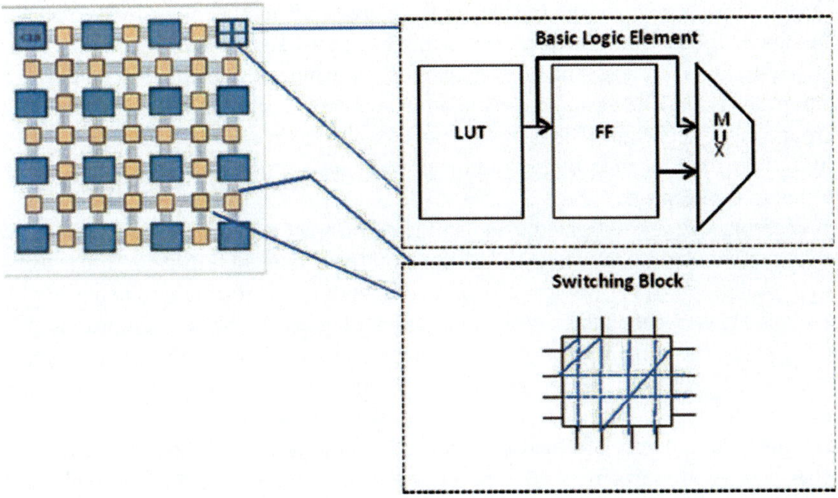

Fig. 1 SRAM FPGA [6]

Flash FPGA

In Flash-based FPGA, the main point of storage is flash, and here, there is no need for any static memory unit, and this has the main advantage over SRAM in power utilization and radiation effects. Analysis of Flash FPGA can be done in a vacuum chamber using a power supply, and later, the attacker can detect the emissions and create a display of it using an electron microscope. The analysis of such attacks is not yet finalized by experts. The other important point of attack is the flash memory unit which has a high programming time and creates the erasing of the content difficult.

FPGA in Cloud

In recent times, FPGAs are integrated into cloud data centers. They are mainly used for offloading and accelerating service-oriented tasks. These service-oriented tasks include deep learning, network encryption, web-page ranking, etc. In [7], IBM's cloud FPGA, a new technique to integrate FPGA in the cloud for scaled operation, is proposed. Leading service providers like AWS, Amazon have started integrating FPGA in its data center for accelerating task execution. In [8], custom computing through FPGA in the cloud is explained. In [9], DNN execution using FPGA on the cloud is proposed.

Attacks on FPGA in Cloud

From the literature, it could be seen that FPGA in the cloud is very vulnerable to cyber-attacks. Some of the most common attacks include black box attacks and side-channel attacks. These are explained in the following subsection.

Black Box attack: Black box attack is the method to reverse engineer a chip. The attacker makes use of a brute-force pattern in which all possible combinations are made use of and the respective outputs are saved. Then, based on the pattern obtained, the attacker will analyze with the help of Karnaugh map or other algorithms to obtain the logic in the FPGA. This is only possible in a small-sized FPGA.

Side-Channel Attack: The implementation setup allows side-channel attacks like timing behavior, electromagnetic radiation, power consumption, and differential power analysis. The analysis is mainly done to obtain access to the secret key. DPA helps in the identification of areas in the power consumption of a device that can be related to the secret key.

These attacks are mitigated using so many different ways. In [10], a symmetric re-encryption scheme to protect FPGA in a cloud environment is discussed. The technique proposed in [11] discusses secure computational infrastructure using FPGA overlays. Privacy-enhancing cloud computing for big data using FPGA is proposed in [12]. In [13], the vulnerability of the FPGA-based remote power side-channel attack is discussed. In secure FPGA as a service using various cryptographic techniques is analyzed. Based on the literature survey, it is evident that FPGAs are increasingly becoming a part of cloud data centers, and they have to be secured against cyber-attacks for security. In this paper, we have considered remote side-channel attacks because they are the most common attacks on FPGA. AES-based solutions for side-channel attacks are analyzed in this paper.

3 Proposed Work: Vulnerability Analysis of FPGA Security in Cloud Environment Using AES Encryption Techniques

For the proposed work, we have considered the vulnerability analysis of FGPA in a cloud against remote power-based side-channel attacks using AES encryption techniques. There are two steps to the analysis.

1. Simulating power analysis attack
2. Vulnerability analysis of the attack on AES encrypted FPGA in cloud.

Both these steps are explained in detail in the below sub-sections.

3.1 Simulation of Power Analysis Attack

The implementation of cryptographic devices is mainly semiconductor logic gates based using transistors. The electrons flow across the silicon substrate when current flow is applied to or withdrawn from a transistor's gate, consuming power and producing electromagnetic radiation. Simple power analysis (SPA) is a technique that involves directly interpreting power consumption measurements collected during cryptographic operations (Fig. 2).

1. Configure the environment

```
In [ ]: # we will be plotting figures right here and not in a separate window
%matplotlib inline

# generic python stuff
import matplotlib
import matplotlib.pyplot as plt
import numpy as np # make sure you use numpy-MKL build for adequate performance!
import struct
import time

# configure figure size
matplotlib.rcParams['figure.figsize'] = (15.0, 10.0)

# local packages
from desutils import *        # my DES utilities
from lracpa import *          # my LRA-CPA toolbox
from condaverdes import *     # incremental conditional averaging
```

2. Attack settings

```
In [ ]: ## Traceset, number of traces, and S-box to attack
tracesetFilename = "traces/hwdes_card8_power.npz"
sampleRange      = (0, 50) # range of smaples to attack
N                = 10000 # number of traces to attack (not more that nthe file has)
offset           = 0     # trace number to start from
evolutionStep    = 500   # step for intermediate reports
SboxNum          = 1     # S-box to attack, counting from 0

## Leakage model
## (these parameters correspond to function names in lracpa module)
averagingFunction     = roundXOR_valueForAveraging # for CPA and LRA
intermediateFunction  = roundXOR_targetVariable    # for CPA and LRA
leakageFunction       = leakageModelHW             # for CPA
basisFunctionsModel   = basisModelSingleBits       # for LRA

## Known key for ranking
knownKey = 0x8A7400A03230DA28 # the correct key
encrypt = True
```

```
In [ ]: # get the known round key
roundKeyNum = 1
if (encrypt == False):
    roundKeyNum = 16
roundKey = computeRoundKeys(knownKey, roundKeyNum)[roundKeyNum-1]
knownKeyChunk = roundKeyChunk(roundKey, SboxNum)
print "Known round key: " + format(roundKey, '#014x'),
print '[',
for i in range(8):
    print format(roundKeyChunk(roundKey, i), '#04x'),
print ']'
```

3. Load samples and data

```
In [ ]: # Readout
```

Fig. 2 Code snippet to simulate remote side-channel attacks

3.2 Implementation of AES Algorithm

In this work, we have used AES algorithm to implement a power analysis attack on FPGA

The main steps of the AES algorithm implemented:

1. Symmetric encryption in which only a single secret key is used.

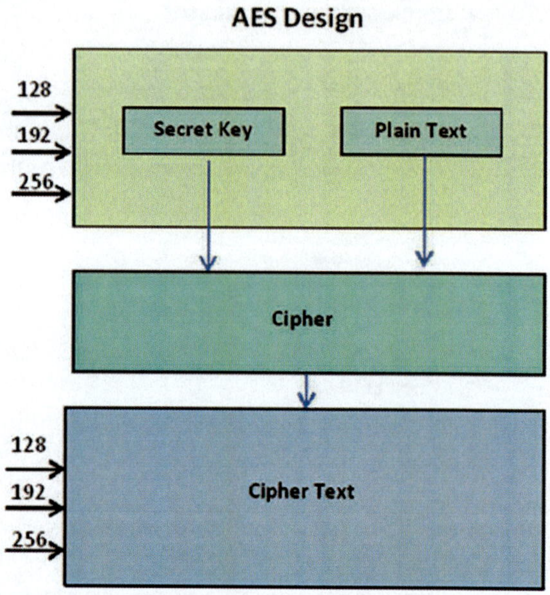

Fig. 3 AES analysis implementation [15]

2. 128-bit cipher algorithm in which the data is divided into equal parts of each fixed-length data (128 bits). The chunks are processed in AES where each round is dependent on the output of its predecessor. Large data can be encrypted using AES.
3. The strength of AES depends on the possible combinations of key permutations using the finite field analysis method.

A sequence of transformations is done on the data in AES [14] of which the first few steps are mainly making an array representation of data, and then, each round of operations is performed in order. The key size determines the number of rounds for 128-bit keys is 10 rounds, 192-bit keys are 12 rounds, and 256-bit keys are 14 rounds. The order of operations in AES is the substitution of data using a substitution table, shifts data rows, then followed by mixing columns. The last step is exclusive or (XOR) operation performed on each column using a different part of the encryption key (Fig. 3).

3.3 RSA in FPGA

1. Asymmetric key algorithm with both separate public and private keys. A public key is available to open the world, whereas the private key is possessed by the owner.
2. Public key encryption is used for exchanging data.
3. Private key encryption is used for authentication of the owner (digital signatures).

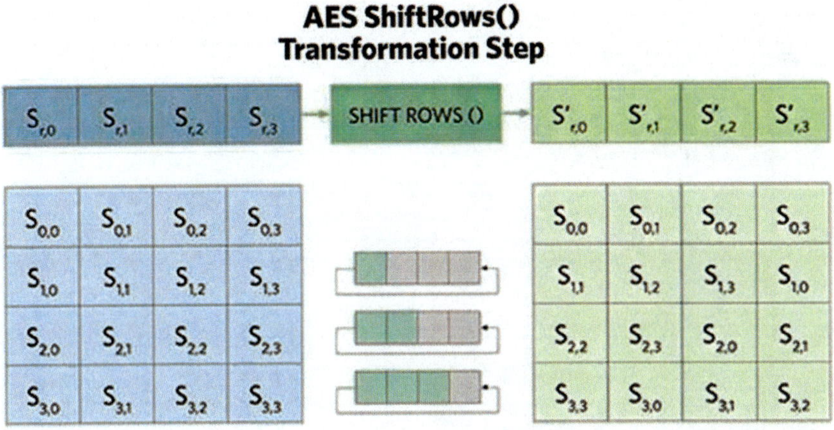

Fig. 4 AES shit row () [15]

4. It is a stream cipher algorithm. Meaning, entire data is encrypted at once, which takes more computational power. Hence, it is slow. The strength of the RSA algorithm depends on the factorization of a large number (Fig. 4).

4 Result Analysis

The power consumption is to be measured for the various cryptographic algorithms and that is represented as a trace. In the past, attacks on the power supply level required local access to the hardware. In this research, we discuss a security vulnerability in FPGAs [16] that allows a valid configuration to generate severe voltage fluctuations, which crashes the FPGA within a few microseconds.

RSA versus AES: Most recent method of cryptographic algorithms is using the brute-force attacks in which all possible combination is tried out until you get the required results, but it is time-consuming. The most secure way to encrypt a message is using enough bits in your key such that there is not enough energy in the universe to try enough candidate keys. In the AES algorithm [17], at present scenario, 128-bit key is safe, but it can be broken in the future, so we have the 256- and 512-bit versions that cannot be broken. The main idea behind breaking asymmetric algorithm is breaking the math logic behind it which can be discrete logarithm or factorization, etc. (Fig. 5).

Security of RSA is weak, and here, brute-force attack on the keys is performed by taking the factor of the modulus into primes and derives the keys used in the encryption process. This is comparatively easy compared to AES. Form this it can be concluded the "equivalent security" of RSA key length versus AES key length changes over time. There is a need to increase the RSA key size relative to the AES key size to account for technological advances.

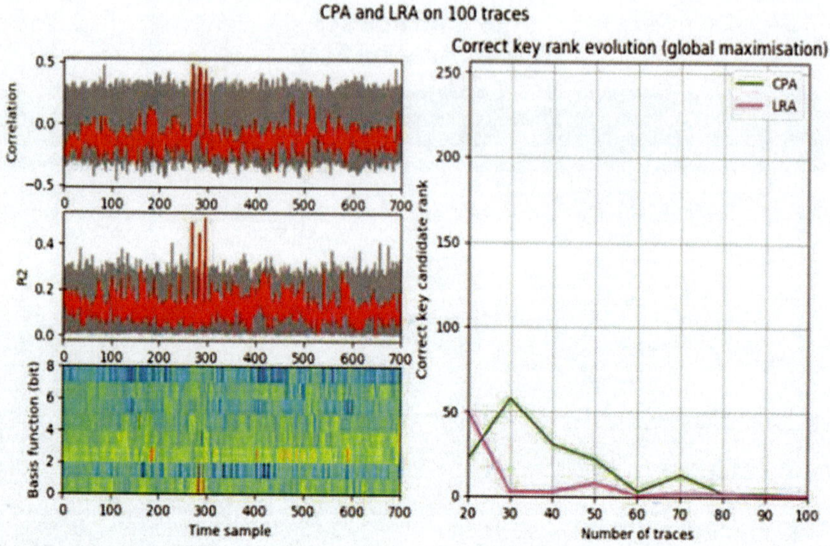

Fig. 5 Sample 100 trace

AES, DES, and Blowfish Algorithm: AES [18, 19], DES[20], and Blowfish are all symmetric key algorithms, with a single secret key for encryption and decryption. The main difficulty in symmetric encryption is the key exchange. RSA is an asymmetric key algorithm that has mainly two keys the public–private key systems, which makes the data transfer arguably safer. The algorithm and structural differences occur in AES, DES, Blowfish algorithms, but the basic functionalities are the same. Here, the AES algorithm is considered since it is more fast, standard, and more widely used and secure compared to other symmetric algorithms. The main drawback of the DES algorithm is the software version is slow, and it is the only 56-bit key which can be easily attacked Blowfish was designed for software. But it has high memory utilization and high setup time, but it uses 448-bit keys and encrypts 64 bits of data at a time.

5 Conclusions and Future Scope

The hardware and software implementation of AES [21, 22] is pretty fast and effective. Many versions are there with a 128-, 192-, or 256-bit key. The key setup time is relatively small and fast, and the memory required for AES is small. Another major advantage of the AES algorithm is the large block size in increasing the resistance to some theoretical attacks if the amount of data encrypted is large. Finding different kinds of attacks and using different algorithms find out ways to tackle the attack to make the hardware more secure and less vulnerable to attacks. Then implementing

FPGA hardware in the cloud for effective and faster usage of transfer of data after the attack analysis.

References

1. Smekal D, Ricci S, Dzurenda P, Martinasek Z (2019) Privacy-enhancing cloud computing solution for big data. In: 2019 11th international congress on ultra modern telecommunications and control systems and workshops (ICUMT). IEEE, pp 1–6
2. Daemen J, Rijmen V (1999) AES submission document on Rijndael, version 2. http://csrc.nist.gov/CryptoToolkit/aes/rijndael/Rijndael.pdf
3. Standaert FX et al (2003) A methodology to implement block ciphers in reconfigurable hardware and its application to fast and compact AES Rijndael. In: The field programmable logic array conference, Monterey, CA, pp 216–224
4. Lu CC, Tseng SY (2002) Integrated design of AES (Advanced Encryption Standard) encrypter and decrypter Application-Specific Systems. In: The IEEE International Conference On Application-Specific Systems, Architectures And Processors, 2002, pp 277–285
5. McLoone W, McCanny JV (2001) Rijndael FPGA implementation utilizing look-up tables Signal Processing Systems. In: 2001 IEEE workshop on signal processing systems, pp 349–360
6. Hoe D, Bollepalli L, Martinez C (2013) FPGA fault tolerant arithmetic logic: a case study using parallel-prefix adders. VLSI Des 1–10
7. Fahmy SA, Vipin K, Shreejith S (2015) Virtualized FPGA accelerators for efficient cloud computing. In: 2015 IEEE 7th international conference on cloud computing technology and science (CloudCom). IEEE, pp 430–435
8. Al-Aghbari AA, Elrabaa ME (2019) Cloud-based FPGA custom computing machines for streaming applications. IEEE Access 7:38009–38019
9. Chen Y, He J, Zhang X, Hao C, Chen D (2019) Cloud-DNN: an open framework for mapping DNN models to cloud FPGAs. In: Proceedings of the 2019 ACM/SIGDA international symposium on field-programmable gate arrays, pp 73–82
10. Al-Asli M, Elrabaa ME, Abu-Amara M (2018) FPGA-based symmetric re-encryption scheme to secure data processing for cloud-integrated internet of things. IEEE Internet Things J 6(1):446–457
11. Fang X, Ioannidis S, Leeser M (2019) SIFO: secure computational infrastructure using FPGA overlays. Int J Reconfigurable Comput 2019
12. Zhao M, Suh GE (2018) FPGA-based remote power side-channel attacks. In: 2018 IEEE symposium on security and privacy (SP). IEEE, pp 229–244
13. Will MA, Ko RK (2017) Secure FPGA as a service—towards secure data processing by physicalizing the cloud. In: 2017 IEEE Trustcom/BigDataSE/ICESS. IEEE, pp 449–455
14. Daemen J, Rijmen V The block cipher RIJNDAEL, NIST's AES home page. Available from: http://www.nist.gov/aes
15. https://searchsecurity.techtarget.com/definition/Advanced-Encryption-Standard
16. Standaert FX, Rouvoy G, Quisquater JJ, Legat JD (2003) Efficient implementation of Rijndael encryption in reconfigurable hardware: improvements and design tradeoffs. In: Proceedings of CHES 2003. Lecture notes in computer science, vol 2779. Springer, Berlin, Heidelberg, pp 334–350
17. Standaert FX, Rouvoy G, Quisquater JJ, Legat JD (2003) A methodology to implement block ciphers in reconfigurable hardware and its application to fast and compact AES Rijndael. In: Proceedings of FPGA 2003. ACM, pp 216–224
18. Baretto P, Rijmen V The KHAZAD legacy-level block cipher. Submission to NESSIE project. Available from: http://www.cosic.esat.kuleuven.ac.be/nessie/
19. Dandalis A et al (2000) A comparative study of performance of AES candidates using FPGA's. In: The third advanced encryption standard (AES3) candidate conference, New York, USA, 13–14 Apr 2000

20. Rouvroy G, Standaert FX, Rouvoy G, Quisquater JJ, Legat JD (2003) Design strategies and modified descriptions to optimize cipher FPGA implementations: fast and compact results for DES and TripleDES. In: Proceedings of FPL 2003. Lecture notes in computer science, vol 2778. Springer, pp 181–193. Tomov S, Dongarra J, Baboulin M (2010) Towards dense linear algebra for hybrid GPU accelerated manycore systems. Parallel Comput 36(5–6):232–240
21. Elbirt AJ et al (2000) An FPGA implementation and performance evaluation of the AES block cipher candidate algorithm finalists. In: The third advanced encryption standard (AES3) candidate conference, New York, USA, 13–14 Apr 2000
22. Gaj K, Chodowiec P (2000) Comparison of the hardware performance of the AES candidates using reconfigurable hardware. In: The third advanced encryption standard (AES3) candidate conference, New York, USA, 13–14 Apr 2000
23. Gaj LK, Chodowiec P (2003) Very compact FPGA implementation of the AES algorithm. In: Proceedings of CHES 2003. Lecture notes in computer science, vol 2779. Springer, Berlin, Heidelberg, pp 319–333

UBAPS: Inexact Unsigned Binary 5:2 Compressor Towards Power Efficient and High Speed for Three-Stage FIR Filter

M. Vaishnavi Reddy, N. Sai Pooja Reddy, and J. V. R. Ravindra

Abstract An efficient multiplier is one of the crucial modules in many of the digital circuits, entailing its prominence in many applications of VLSI such as digital signal processing, cryptography, and communications. With the advancement of technology, circuits have been enforced to encounter the power, area constraint, and propagation delay parameters. In this regard, compressors have become imperative, forming the elementary part of multipliers in reducing and associating the partial products in a parallel way. In this paper, a 5:2 compressor is proposed with abridged power consumption and propagation delay factor. Approximate compressors with synthesizable structural reductions have led to better results, subsequently higher-order multipliers, filters are constructed with simpler and higher-order compressors. Using 45 nm technology, the design has been implemented in pass transistor logic (PTL), where the speed has escalated by 98.4%, power improvement of 78.04%, compared to the other compressor design in the literature.

Keywords Compressor · Multiplier · FIR filter

1 Introduction

Multiplication being the prerequisite operation for general purpose electronic hardware devices that drives the attention of the research work, to formulate a power-efficient design. Multipliers are miniature circuits that are incorporated within the

M. Vaishnavi Reddy · N. Sai Pooja Reddy · J. V. R. Ravindra (✉)
Centre for Advanced Computing and Research Laboratory (C-ACRL), Department of Electronics and Communication Engineering, Vardhaman College of Engineering, Shamshabad, Telangana, India
e-mail: jayanthi@ieee.org

M. Vaishnavi Reddy
e-mail: vaishnavireddy165@gmail.com

N. Sai Pooja Reddy
e-mail: nadimpallypoojareddy@gmail.com

G. Ranganathan et al. (eds.), *Inventive Communication and Computational Technologies*, Lecture Notes in Networks and Systems 145,
https://doi.org/10.1007/978-981-15-7345-3_52

applications. As the technology scales down, the research in the semiconductor industry complicates the integration of circuit designs, diminishing the area of the electronic design interface. Most of the literature study on arithmetic circuits focuses on the improvement of multiplier performance.

The confronting encoding techniques for the multiplier circuits have become the key area of research. Approximate and precise algorithms have rendered their significance in demonstrating reduced error-free circuits [1] and power-efficient designs. The least probability terms in the truth table are neglected to simplify the error circuits by which the partial products are approximated to produce error-free circuits [1]. However, modern applications are error-tolerant, structured with filters, and designed as per user requirement such as implementing the notions in machine learning to overcome the adversarial attack [2]. Partial products (PP) play a crucial role in influencing the circuit speed where partial product reduction modules dominate the fraction of the silicon area.

Implementation of multipliers generally is carried out in three phases, namely, partial product generation (PPG), partial product reduction (PPR), and carry propagation; wherein, in the initial stage, the bits of multiplicand and multiplier are multiplied. This can be done in several ways depending on the input format opted by the user. The successive stages become the performance improvement areas in judging the overall functionality of the circuit. This is achieved by techniques of carry save adders (CSAs) [3], compressors of different orders, approximate multiplication techniques [4], along with the practice of buffers in the designs upgrading the driving capability. Most of the multipliers such as a bit-level pipelined multiplier, Dadda multiplier constitutes CSAs that aids the circuit in declining the latency of critical path delay [5]. Inexact computing techniques have been recognized as one of the paramount practices in redesigning the digital circuits for acquiring low complexity and area constraint results.

The hierarchy of multipliers encompasses the multiplication models of various input categories as mentioned in Fig. 1.

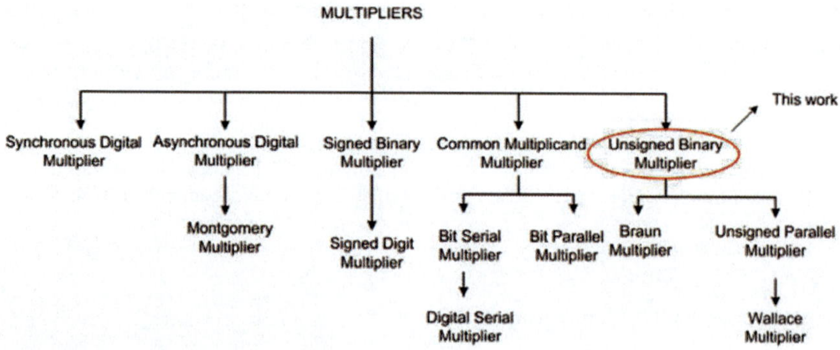

Fig. 1 Taxonomy of multipliers

Most of the multipliers such as Wallace [6], Braun are composed of compressors of unsigned binary input. The proposed compressor is an unsigned binary input circuit that has prominent use in unsigned computing networks. The other input designs like signed binary, digit serial, bit-level, digit serial shift, decimal compressor formats can also be designed with appropriate changes in the compressor logic design that can cope up with the requirement. A compressor is a multiplexer, with XOR entailed design, that counts the outputs based on the inputs of a logic circuit. Research in compressors has modulated the switch between exact and inexact modes in the logic execution period. The run-time of a system considers parameters such as error rate, error distance, and critical path delay which are conferred in the subsequent sections.

The paper continues with the literature survey in Sect. 2, followed by proposed work in Sect. 3. Simulation results are presented in Sect. 4, with an FIR filter application in Sect. 5, and concluding the paper in Sect. 6.

2 Literature Survey

Over the period, the exact compressors have been transformed into exact and inexact circuits reducing the power, area, and delay parameters. Inexact compressors are output specific circuits that are defined to produce precise outputs that can be employed to a system associated with few outputs neglecting the insignificant outputs to the circuit. Some of the research works in compressors are:

The proposed compressor in [7] has two designs, where each design result overcomes the other design in speed and power parameters. The carry output C_{o2} is made independent of input C_{i1} to limit the power dissipation. The second design is generated using a carry generation module of CMOS full adder to limit the delay propagation to 4 XOR delay. This conventional design has been implemented in 250 nm CMOS technology. The circuit is simulated for various supply voltages to overcome the scale of delay.

Carry generation modules (CGEN) have been used instead of multiplexers to generate intermediate carry outputs in [8, 9]. A CMOS full adder design consists of a group of XOR and CGEN modules, whose operation is faster than the XOR cascaded gates. The design is realized with two different XOR logics to lessen the transistor count. The overall circuit has been simulated in 90 nm technology at 1 V supply voltage.

The traditional exact 5:2 compressor composed of XOR and MUX with five primary inputs and two outputs constitutes three full adders. The compressor in [10] has been redesigned with full swing transmission gate logics which abruptly lowers the power, delay, and area parameters; buffers used in the circuit were not considered for power calculations. The setup was simulated in 32 nm technology along with layout setup. The low voltage compressor in [11], aims at improving the driving capability of the circuit by preferring the XOR-XNOR circuit along with optimal MUX in complementary CMOS design that is sturdy in voltage scaling and transistor sizing.

3 Proposed Work

The novel design of the 5:2 compressor aims at receding the critical propagation delay and power of the circuit with five inputs which are referred to as 5-bit input size. An elementary compressor which is erected with cascaded full adders is altered to a combination of XOR and multiplexer (MUX) [7] in the conventional literature designs. A compressor must abide by the input-output relationship with inputs and outputs of the same weight, which is defined as follows:

$$X_1 + X_2 + X_3 + X_4 + X_5 + C_i = \text{Sum} + 2 \times C_o + \text{Carry} \tag{1}$$

In the novel design of 5:2 compressors (UBAPS), they carry out C_{o2} is made independent of C_{i1}, which bounds the carry propagation delay of a single compressor when they are connected in large numbers as alignment series in extensive circuit designs as shown in Fig. 2. The arrangement of numerous 5:2 compressors denotes the number of compressor units precisely complexity of compressor units.

In most of the design techniques, 5:2 compressor is composed of lower-order compressors such as 3:2 compressor, 4:2 compressor [12], producing carry outputs from each compressor and propagating it to the subsequent higher-order compressor. The typical circuit designed in the literature is elucidated by modifying the intermediate logic equations of carries C_{o1}, C_{o2} [10]. The other equations are remained intact as follows:

$$\text{Sum} = X_1 \oplus X_2 \oplus X_3 \oplus X_4 \oplus X_5 \oplus C_{i1} \oplus C_{i2} \tag{2}$$

$$C_{o1} = (X_1 \oplus X_2)X_3 + \overline{(X_1 \oplus X_2)}X_1 \tag{3}$$

$$C_{o2} = (X_4 \oplus X_5)(X_1 \oplus X_2 \oplus X_3) + \overline{(X_4 \oplus X_5)}X_4 \tag{4}$$

$$\begin{aligned} \text{Carry} = {} & (X_1 \oplus X_2 \oplus X_3 \oplus X_4 \oplus X_5 \oplus C_{i1})C_{i2} \\ & + \overline{(X_1 \oplus X_2 \oplus X_3 \oplus X_4 \oplus X_5 \oplus C_{i1})}C_{i1} \end{aligned} \tag{5}$$

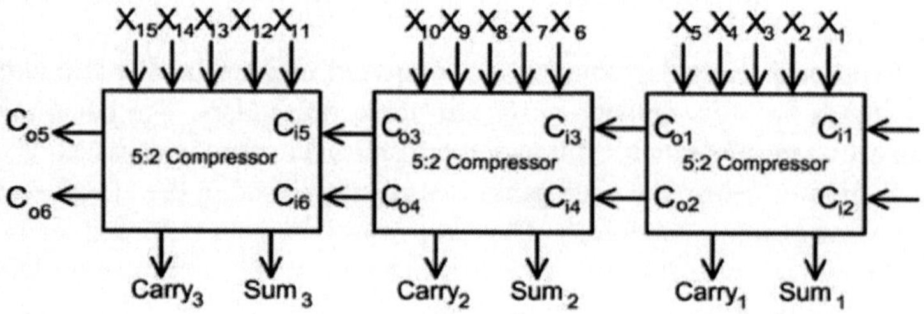

Fig. 2 System analysis of 5:2 compressor

The proposed design illustrates the power efficiency and minimal propagation delay, with the transformation of carry outputs generated from the transitional (intermediate) 3:2 compressor, into two suitable logic operations with equivalent outputs. Inquiring the fact that, as the signal propagates from one compressor to successive compressor, delay increases which signify the performance of a digital circuit. To reduce the delay of C_{o2} carry output, it is made independent of input X_5 as shown in Fig. 3, along with the PTL designs of XOR and MUX gates as shown in Fig. 4.

The logical equation of UBAPS is determined by transforming the logical expressions in [7], with the help of formulations of digital logic. The modified equations are given as:

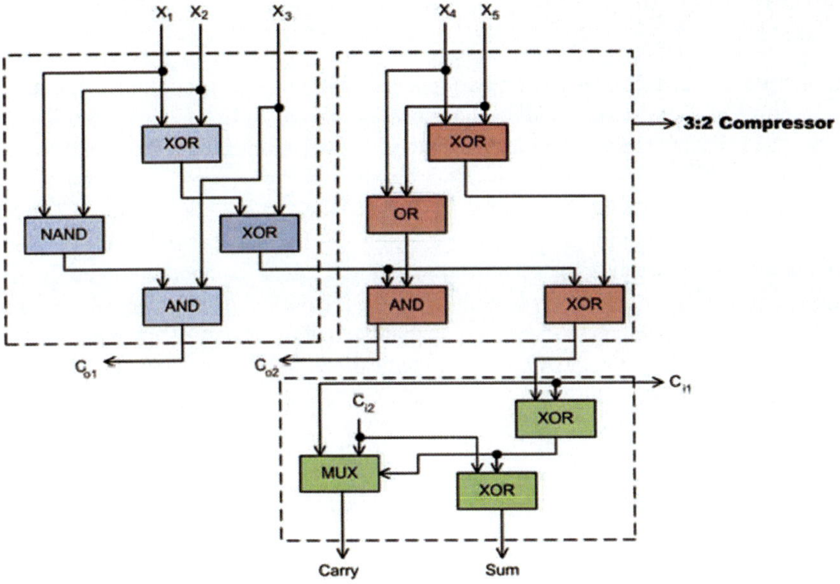

Fig. 3 Proposed 5:2 compressor design

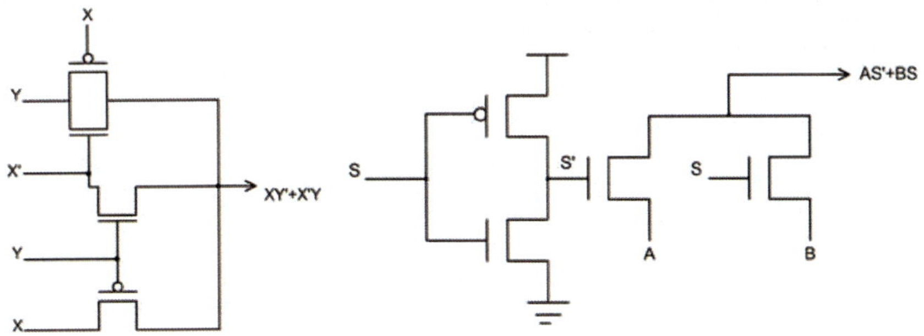

Fig. 4 Implementation of XOR and MUX logics

$$\text{Sum} = X_1 \oplus X_2 \oplus X_3 \oplus X_4 \oplus X_5 \oplus C_{i1} \oplus C_{i2} \tag{6}$$

$$C_{o1} = X_3\overline{(X_1 X_2)} \tag{7}$$

$$C_{o2} = (X_4 + X_5)(X_1 \oplus X_2 \oplus X_3) \tag{8}$$

$$\begin{aligned} \text{Carry} = {} & (X_1 \oplus X_2 \oplus X_3 \oplus X_4 \oplus X_5 \oplus C_{i1})C_{i2} \\ & + \overline{(X_1 \oplus X_2 \oplus X_3 \oplus X_4 \oplus X_5 \oplus C_{i1})}C_{i1} \end{aligned} \tag{9}$$

The derived Eq. (7) is transistor compact when compared with (2) denotes a multiplexer, whereas the UBAPS uses AND and NAND gates to generate the same output with trivial error rate. The output of the second block (2) of the compressor is shortened into a simpler one with a slender amount of change in transistor count. The novel UBAPS design handles an input bit size of 5 for a single compressor usage and is extendable by an increase in the compressor units with suitable designs as per requirement. In other cases dealing with lower bits, some of the inputs of the compressor are retained to a constant value (DC value). As mentioned, each compressor controls 5 input bit size, accordingly, the user must calculate the number of compressor units as per the specification of the circuit for computation. The significance of a compressor circuit is built up when it exhibits lower power, delay, and transistor count. This can be effectively emphasized with the help of applications like FIR filter in Sect. 5, in image processing applications like image sharpening, several multipliers integrated using compressors as transitional blocks such as signed array multiplier.

4 Simulation Results

The implemented PTL logic 1 design of compressor is simulated in the Cadence Virtuoso interface in 45 nm technology and the results are shown in Table 1.

The comparison of various compressor designs over the period, considering various parameters are illustrated in Figs. 5, 6, 7, and 8. The results are found to

Table 1 Simulation results of the proposed design compared with existing designs

S. No.	Designs	Power (uW)	Delay (ns)	PDP (fJ)	No. of transistors
1	UBAPS	0.04936	0.00269	0.0001318	58
2	D. Balobas	2.172	0.335	0.728	58
3	Amir Najafi	4.78	0.204	0.98	78
4	Ardalan Najafi [13]	4.66	0.1682	0.78	82
5	Maddisetti [14]	0.224	9.919	2.23	20

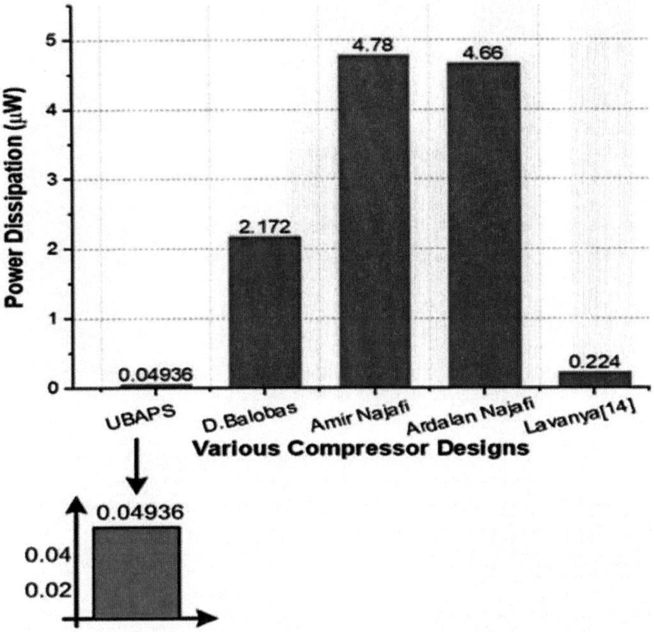

Fig. 5 Power comparisons of various compressors

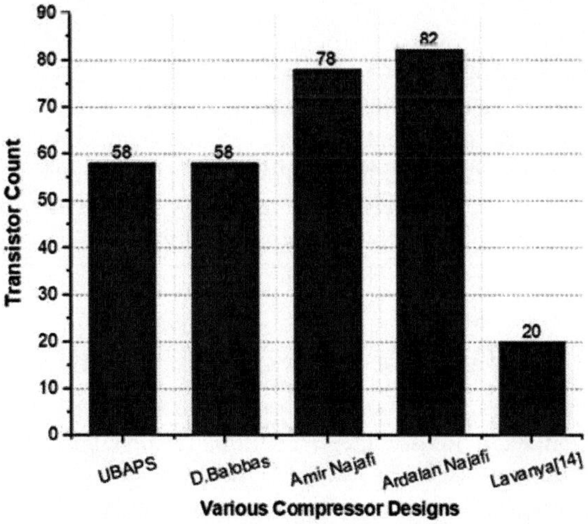

Fig. 6 Transistor count comparisons of various compressors

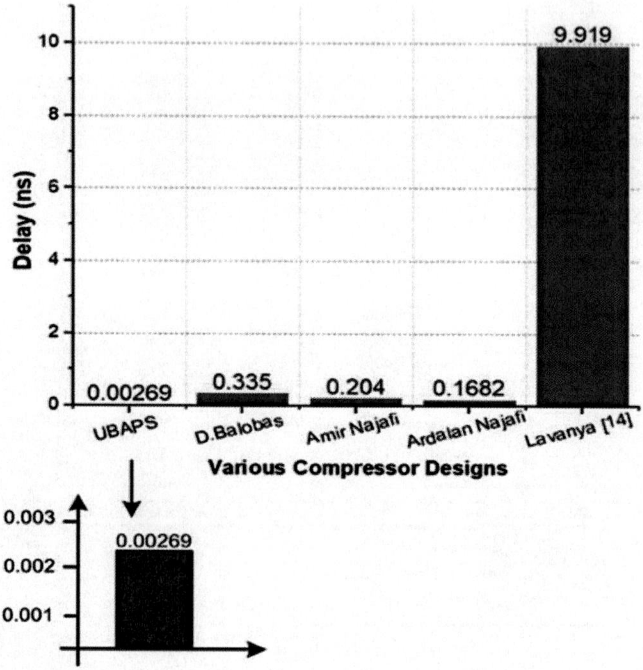

Fig. 7 Delay comparisons of various compressors

be advancing in power, delay, and power and delay product (PDP) compared with prior literature works with a significance of 78.04%, 98.6%, and 99.9%, respectively. The proposed circuit has achieved less degree of an error rate of 12.5% and 15.6% in consideration with the formulated carry outputs. The power dissipation in digital circuits can be controlled by variation in the supply voltages (Vdd). Input signal edge duration is one of the dependency factors for short circuit dissipation.

5 Designing of Three-Stage FIR Filter Using UBAPS

Digital circuits are integrated into electronic applications that are designed to drive down the power, delay, and many other parameters. Circuits such as compressors, multipliers, and accumulators are transitional circuits used in designing techniques like filters which are sequentially used in applications like digital signal processing (DSP), cryptography, and machine learning [2]. Filters are categorized as digital and analog that differ by the type of input presented. One of the prominent stable digital filters is FIR filter [14], seemingly digital filters assisting to attain enhanced performance results by performing noiseless mathematical operations in the intermediate stages of the filter with precise outputs with a diminished signal to noise

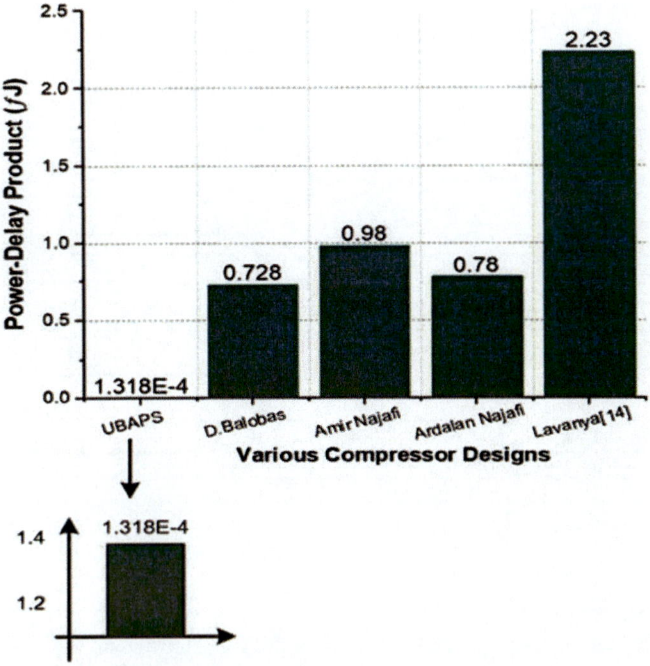

Fig. 8 PDP comparisons of various compressors

ratio (SNR). The digital filters are advantageous over analog and switched-capacitive circuits because digital filters (FIR filter) are implemented without operational amplifiers [15]. Cascading of digital inverse filters and analog filters maintain constant overall stop-band frequency response [16]. UBAPS is integrated with a low pass FIR filter that is a frequency selective network, which is used to determine performance parameters for the proposed compressor design.

The application of the third-order low pass FIR filter design with UBAPS entailed in it is shown in Fig. 9.

The design is built with a set of compressors, D flip-flops, and Adders arranged in such a manner that the outputs of compressors are propagated to flip-flops (FF), subsequently to adders with definite delay produced by FF. The performance results of the filter are obtained by implementing in Cadence Virtuoso, with power as 413.2 uW, critical path delay of 37.41 ns. The compressor in the FIR filter application has presented encouraging parameter results of the filter such as power and delay.

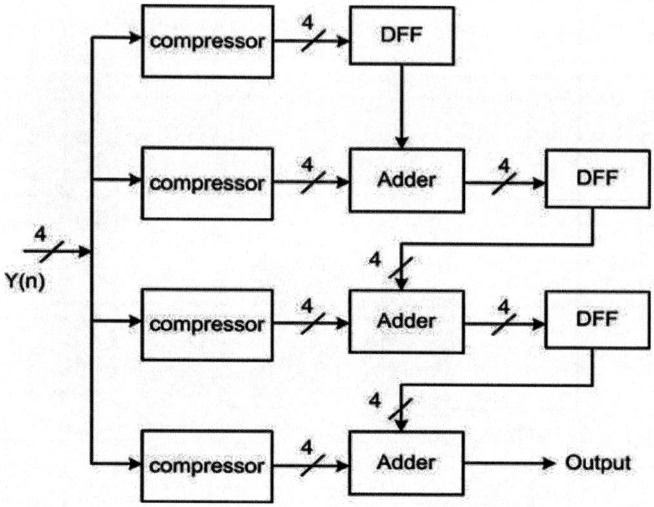

Fig. 9 FIR filter design using compressors

6 Conclusion

The proposed 5:2 compressor UBAPS demonstrated in this paper that targets the power and delay parameters. The novel compressor has been implemented in 45 nm Cadence Virtuoso technology, mainly focusing on the intermediate carry outputs. The novel Boolean expressions are devised to exercise the circuit toward power and delay factors. The simulation results have exhibited an improvement of 78.04% in power consumption and 98.4% expeditious in delay propagation. Besides, the PDP is another parameter that breaks the simulated results in the literature. The proposed work has proven its remarkable significance over the past literature work in all the parameters simulated at a supply voltage of 1 V. The compressor is transformed into an FIR filter application with boosting results. The compressor proposed is electronically feasible in applications associated with lesser power consumption and delay propagation.

References

1. Esposito D, Strollo AGM, Napoli E, De Caro D, Petra N (2018) Approximate multipliers based on new approximate compressors. IEEE Trans Circuits Syst I: Reg Pap 65(12):4169–4182
2. Maddisetti L, Ravindra JVR (2019) Low-power, high-speed adversarial attack based 4:2 compressor as full adder for multipliers in FIR digital filters. In: 2019 IEEE Nordic circuits and systems conference (NORCAS): NORCHIP and international symposium of system-on-chip (SoC), Helsinki Finland, 29–30 Oct 2019

3. Javali RA, Nayak RJ, Mhetar AM, Lakkannavar MC (2014) Design of high speed carry save adder using carry lookahead adder. In: Proceedings of international conference on circuits, communication, control and computing (I4C 2014), pp 33–36
4. Ansari MS, Jiang H, Cockburn BF, Han J (2018) Low-power approximate multipliers using encoded partial products and approximate compressors. IEEE J Emerg Sel Topics Circuits Syst 8(3):404–416
5. Anitha P, Ramanathan P (2014) A new hybrid multiplieusing Dadda and Wallace method. In: 2014 international conference on computer communication and systems (IEEE-2014), Coimbatore, India, 03–05 Jan 2014
6. Waters RS, Swartzlander EE (2010) A reduced complexity Wallace multiplier reduction. IEEE Trans Comput 59(8):1134–1137
7. Menon R, Radhakrishnan D (2006) High performance 5:2 compressor architectures. IEE Proc Circuit Devices Syst 153(5):447–452
8. Najafi A, Najafi A, Mirzakuchaki S (2014) Low-power and high-performance 5:2 compressors. In: The 22nd Iranian conference on electrical engineering (ICEE 2014), Shahid University, 20–22 May 2014, pp 33–37
9. Chang C-H, Gu J, Zhang M (2004) Ultra low-voltage low-power CMOS 4-2 and 5-2 compressors for fast arithmetic circuits. IEEE Trans Circuits Syst I: Reg Pap 51(10):1985–1997
10. Balobas D et al (2018) Low-power high-performance CMOS 5-2 compressor with 58 transistors. Electron Lett 54(5):278–280
11. Gu J, Chang C-H (2003) Ultra low voltage, low power 4-2 compressor for high speed multiplications. In: Proceedings of the 2003 international symposium on circuits and systems, 2003. ISCAS '03, 25–28 May 2003, pp 321–324
12. Edavoor PJ, Raveendran S, Rahulkar AD (2020) Approximate multiplier design using novel dual-stage 4: 2 compressors. IEEE Access (pre-print)
13. Najafi A, Timarchi S, Najafi A (2014) High-speed energy-efficient 5:2 compressor. In: MIPRO 2014, Opatija, Croatia, 26–30 May 2014, pp 80–84
14. Maddisetti L, Ravindra JVR, Performance metrics of inexact multipliers based on approximate 5:2 compressors. In: 2018 international SoC design conference (ISOCC), Daegu, Korea (South), 12–15 Nov 2018, pp 84–85
15. Jaya Kumar D, Logashanmugam E (2014) Performance analysis of FIR filter using booth multiplier. In: 2nd international conference on current trends in engineering and technology, ICCTET14, pp 414–417
16. Ahn D, Hong S (2011) A low cost analog FIR channel select filter for wireless receiver. In: 2011 IEEE radio and wireless symposium, Phoenix, AZ, USA, 16–19 Jan 2011, pp 211–214

Utilizing a Raspberry Pi for Transmitting Image using Li-Fi Transceiver

S. C. Sandeep, P. Sreenivasa Reddy, M. Shrenik, Shaista Farheen, and D. Praveen Kumar

Abstract Li-Fi technology is one of the emerging fields in the communication domain. When information is transmitted through an LED light, the flickering is faster than the human eye can perceive. The information is transmitted faster using Li-Fi when compared to Wi-Fi. In this paper, we propose a prototype where we use Raspberry Pi to retrieve images stored in it, convert it into bits, and send information using LED. This information is captured by a solar panel and is converted into an image using another Raspberry Pi. We also simulate the process of converting image into bits in the transmission side and converting received bits into an image at the receiver side using Python code. Li-Fi enables us to use a broader light spectrum when compared to the radio spectrum used by Wi-Fi which is crowded. Li-Fi has found its applications in hospitals, schools, and workplaces.

Keywords Light Fidelity (Li-Fi) · Light Emitting Diode (LED) · Raspberry Pi (RPi) · Solar Panel

---

S. C. Sandeep (✉) · P. Sreenivasa Reddy · M. Shrenik · S. Farheen · D. Praveen Kumar
Dayananda Sagar College of Engineering, Bengaluru, Karnataka, India
e-mail: sandeep-ece@dayanandasagar.edu

P. Sreenivasa Reddy
e-mail: sreenivasa450@gmail.com

M. Shrenik
e-mail: shrenik2912@gmail.com

S. Farheen
e-mail: shaistafarheen7997@gmail.com

D. Praveen Kumar
e-mail: praveenkumar2131998@gmail.com

G. Ranganathan et al. (eds.), *Inventive Communication and Computational Technologies*, Lecture Notes in Networks and Systems 145,
https://doi.org/10.1007/978-981-15-7345-3_53

1 Introduction

Light Fidelity can be regarded as light-based Wi-Fi. In Li-Fi technology, there is the utilization of light instead of radio waves for transmitting the information. The main components of a simple Li-Fi system are,

- LED transmitters: At the transmitter side, data is converted into bits, which drives the LED.
- Photodetector: At the receiver side, the photodetector receives the light signals. Received bits are converted into original data.

LEDs emit light when there is a change in the energy levels of the semiconductor diode. Depending on the type of semiconductor used and variation in the energy levels, the wavelength of the emitted light is determined. The data rate depends on the utilized LED dimensions. Micro-LEDs have the ability of handling millions of variations in light intensity. A solitary micro-LED may have data rates >10 Gbps. Since light stream travel in parallel emerging from micro-LED, data rates of Gbps can be achieved.

IEEE 802 workgroup has standardized protocols for Li-Fi. The Physical layer (PHY) and Media Access Control (MAC) layer are utilized for communication.

Data that is to be transmitted by LED is modulated into a carrier signal. Some modulation techniques used by Li-Fi systems are:

- Variable Pulse Position Modulation (VPPM): It is similar to PPM where adjusting pulse width will help us attain constant data rate and variable range.
- Color Shift Keying (CSK): The color shift keying is applied when LEDs of type RGB are utilized. Transmitted signals are encoded into color intensities emitted by red, green, and blue LEDs.
- Frequency Shift Keying (FSK): The Frequency Shift Keying can be utilized for changing frequencies of the carrier signal to represent data. For different frequencies, distinct values of 0 s and 1 s are transmitted.
- Sub-Carrier Index Modulation OFDM (SIM-OFDM): Amplitude Shift Keying (ASK) and Quadrature Amplitude Modulation (QAM) are added to amplitude/phase modulation techniques as an additional dimension. Information is carried to the receiver by using the sub-carrier index.
- On-Off Keying (OOK): The On-Off keying makes use of Manchester encoding technique for the representation of data in the form of 0 s and 1 s. The edge transition of high to low represents logic 0, whereas low to high represents 1.

Numbers are stored in the digital computers with numeric base-2 or binary rather than a decimal. Electronic circuits inside the computers are constructed by restricting to binary. The circuitry can differentiate between two voltage levels easily. Thus, images are also stored in bits or bytes (1 byte = 8 bits) inside the computer.

Color model: Digital images use the color model to create a broad hue range from a compact primary color sets. There are different color models used for images, the most commonly used are,

- Red, Green, and Blue (RGB) Model: In this model, primary colors are mixed to form different colors. The red, green, and blue channels are represented by an integer value ranging from 0 to 255, respectively, to represent a color.
- Grayscale Mode: Grayscale images consist of only gray tones of colors; there are only 256 gray colors. It is represented by an integer value ranging from 0 to 255.
- Binary images: A binary image consists of pixels that can have one of the two colors, black or white, i.e., each pixel is stored as a bit 0 or 1.

By this, the image is converted into a binary image. The reverse process is followed to convert back the binary image into an RGB image (Fig. 1).

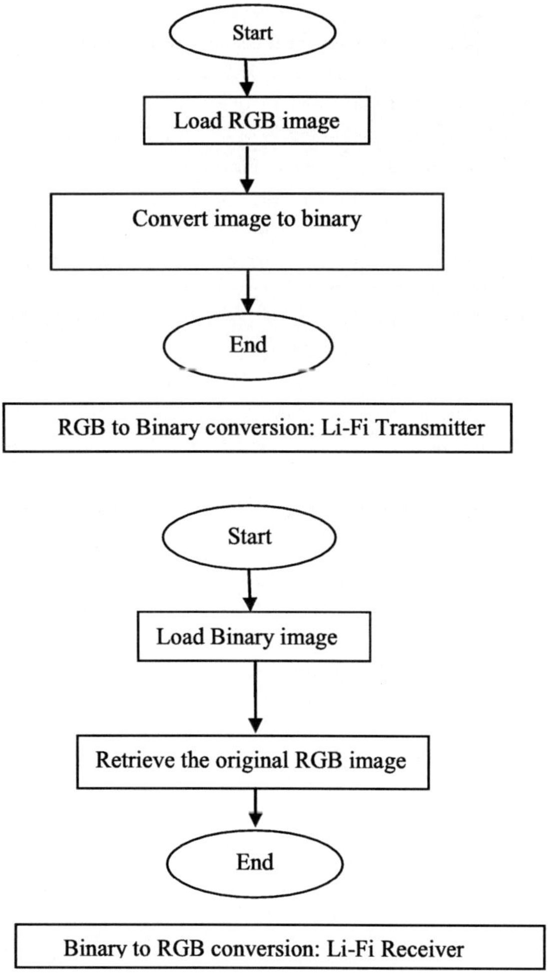

Fig. 1 Flowchart of RGB to binary conversion and binary to RGB conversion

2 Related Work

The essential and foundation data of Visible Light Communication (VLC) has been given in [1], expressing the utilization of VLC alongside its properties. One of the approaches for wireless communication is VLC. In VLC, at the transmitting side Light Emitting Diodes (LEDs) are used as transmitter and at receiving side photodiodes are used as the receiver. VLC communication properties are as follows: Separation utilizing VLC is between 1 and 100 m, which is in contrast to the Radio Wave Communication where the distance of separation in lesser, and it is essentially Line-of-Sight (LoS) communication. Some significant applications focused are location-based services which can be carried out by utilizing either photodiodes or image sensors as receivers.

The term Li-Fi or "Light Fidelity" was premier instituted in [2]; infrared as well as light in the visible range is utilized by Li-Fi technology for the conveyance of rapid information.

It has been conveyed in [3] that Li-Fi expands the concept of VLC to establish a fully connected wireless communication network. The Li-Fi cell is smaller than the Femtocell of the RF communication network but shows higher efficiency of spectrum. This can implement IoT and 5G systems. Li-Fi also provides mobility and multiple accesses. The size of IR and visible light is nearly 2.6k times the RF spectrum of 300 GHz.

It has been talked about in [4] that customarily, a VLC is a wireless communication connect consisting of an LED as a light source and a receiver such as a photodetector. The information rate relies upon the modulation technique utilized just as the lighting engineering.

White light producing commercial LEDs are fetched cost-effectively by mixing blue and yellow lights. Phosphorus color converting material converts blue light into yellow light but impedes the quickness of the frequency response (this implies that frequencies of a greater range get intensely lessened). Subsequently, the LED's frequency band is only found around 2 MHz locales. Data rates of ~1 Gbps could be got by using a blue filter at the receiver side. Because white light is delivered by blending primary hues; further developed red, green, and blue (RGB) LEDs empower information rates of $\leq$5 Gbps. A solitary micro-LED yields 8 Gbps information rate, whereas utilizing a lightning system that incorporates laser makes even 100 Gbps plausible.

Misconceptions regarding Li-Fi have been corrected in this paper, including, but not limited to:

- It is said that maybe the best-misguided judgment with regards to Li-Fi is considering it a "Line-of-Sight (LoS) technology," and also: Digital modulation plans of high order may be utilized alongside Orthogonal Frequency Division Multiplexing (OFDM), bridling accessible channel capacity.
- "Diminishing of lights not practicable": Propelled modulation techniques (like eU-OFDM) empower Li-Fi functioning near the LED's Turn-on Voltage (ToV)

implying that lights may be worked at lesser light yield levels, keeping greater information rates.

The potential effect that Li-Fi technology can have over various existing and rising industries is additionally outlined, including the noteworthy commitment it can make to existing cellular networks where mobile information traffic development will originate from more spectrum use, rather than spatial reuse.

A point-to-multipoint model for transferring videos and audio in real-time, utilizing LED lamps used in business has been proposed in [5], to get an enormous increment in transmission separation and refinement in channel space. An optical connection has been utilized, instead of a connector wire for use in cellular frameworks, which can be utilized in video conferencing, real-time video frequency monitoring, and smart traffic system: Different useful situations were brightening, as well as sending of information is in the interest of the everyday life of today.

- It has been indicated: The accomplishment of high-quality audio, video transmission with 3 m separation between transmitter and receiver is plausible utilizing basic LED lamps and enhancements can be made by including a focusing lens to center the focal point between the receiver and the sender.
- The second source is a solitary LED supplanted by LED arrays, which was fit for supporting the lighting, information communication just as monitoring.

The working model of Li-Fi has been exhibited in [6], featuring its favourable circumstances over Wi-Fi as far as two patterns being seen now: One in the expansion or up gradation of wireless services, other being increment of customer interest for such administrations, for which accessible RF spectrum is constrained. This innovation, contrasted with Wi-Fi, offers points of interest like expanded access-spectrum, reduced latency, proficiency, security, and a lot more speed. Structuring of a Li-Fi transceiver transmitting digital data has been explored, utilizing Arduino.

- In the working model illustrated, the transmitter changes discrete-time, amplitude information to visible light where an LED has been utilized, being an appropriate component for the same. To modulate LED's light intensity (result: light intensity relates to symbols transmitted), the Arduino ports were not equipped for the conveyance of perfect current measures for transmitting symbols quickly, and hence, a transistor was utilized in switch mode, making switching of a bigger current quicker.
- Utilizing a photodiode, the receiver converts approaching light into current. It is gainful to utilize a photodiode since photodiodes have a quick reaction time, spectral affectability, and an enormous radiant sensitivity zone. Since Arduino ports cannot accept voltage >5 V, the automatic gain controller was electrically structured between Arduino and photodiode to process the electrical signal, for veracious interpretation. Outcomes indicate information transfer; after guaranteeing fruitful information transfer, video frames were sent and got effectively.

It has been stated in [7] that Li-Fi uses the E.M. spectrum for data transmission and thus normally used modulation techniques for RF applications can be applied to Li-Fi with essential changes. Li-Fi has few specific modulation techniques. It is said in the paper that the broadly utilized modulation plans include On-Off Keying (OOK), Pulse Position Modulation (PPM), and Pulse Amplitude Modulation (PAM). In OOK modulation, the presence of carrier for an explicit span speaks for binary 1, and the absence of carrier represents binary 0, and an information pace of 100 Mbps can be accomplished by utilizing OOK modulation. PPM is more power productive but has lower spectral proficiency. Optical Spatial Modulation (OSM) is power productive and bandwidth-efficient. Carrier-less Amplitude and Phase modulation (CAP) uses two orthogonal signals for transmission using spectrum efficiently. Some more sophisticated schemes have varying durations of data transfer to convey additional information. These include Multi-Carrier Modulation (MCM) schemes for rapid optical wireless communication, and Li-Fi specific modulation schemes, the latter incorporating techniques to combine various shades of light with the goal that the yield information can be conveyed by the shading itself and thus the power of the yield can be close to steady.

The basic transmission and reception mechanism involved in transmitting images over Li-Fi have been illustrated in [8]. This mechanism is based on the simple concept that communication by light or VLC uses visible light from LEDs for data transmission. The light emitted by the LED is driven and modulated by a mini-computer. At the receiver side, the data arrived is demodulated by a photodetector, which is normally associated with a comparative processing device for the last recouping of the data. An application based on the On-Off Keying technique is used for transmitting an image using Raspberry Pi as a mini-computer and a module consisting of LED and photodetectors. It is concluded that the recovery of the signal depends on:

- The quantity and type of receiver (photoresistor, photodiode, or a reverse-biased LED).
- Encoding methods utilized (return-to-zero, non-return-to-zero, etc.).
- Modulation plan (OOK, NPM, VPPM, PPM, and OFDM) synchronization and distance between LED and light detector as well as the shape and the wavelengths of LEDs employed.

Communication using Li-Fi has been demonstrated in [9], by transmitting two sorts of information: Text and audio messages. Audio signals when transmitted under different topologies—SISO, MISO, MIMO, etc., a difference in characteristics was observed concerning the information transmitted. The sending of text between two users was achieved utilizing an arrangement of Arduino boards, LED, and silicon photodiode, and a conclusion was made that data transmission using IR LED has a maximum range of 2 cm, whereas LED has a maximum range of 22 cm.

The guideline of Visible Light Communication has been exhibited in [10] and it explains the attainability of using Python as the language for coding and SPI for the movement of data. It records sensible electronic components to process bitwise information signals. Python codes have been used for the transmitter and the receiver modules, and encoding/decoding schemes have been suggested to decrease the measure of data processed, transmitted, and obtained at the receiver. The prototype, made after sufficient study of datasheets of the components (Raspberry Pi 2B,

Shiji LED APA102C (DotStar), TI 74HCT125 Quad Level Shifter, etc.), shows it is suitable to use addressable LEDs as a transmitter, and they are not reasonable outside of a vigorously obliged condition.

The image processing method in [11] explains the need for converting the image to binary. Binary converted images are used as a tool for segmentation, examination, compression, and separation. For the Li-Fi transceiver system, the RGB image is converted into binary data and transmitted through light as a medium. The original image is retrieved at the receiver by converting a binary image to an RGB image by using the algorithm mentioned in the paper.

Light Emitting Diodes and their properties have been mentioned in [12] alongside their lightening capabilities, highlighting their significance in the transmission of information. It is stated that this paves a way for communication using Li-Fi technology, which was otherwise only communication utilizing radio frequencies. In this paper, the downlink was implemented using visible light and uplink using infrared LEDs.

The use of computer communication networking protocols has been portrayed in this paper for data integrity, for the detection of lost signals and to ensure correct transmission and reception of data. The paper demonstrates how visible light can be used to transfer data in the form of text from one computer to another. These computers have been used as the end devices, and to interface software to the hardware.

3 Proposed Work

3.1 Objective

This project intends to structure and implement a transceiver prototype using Raspberry Pi for transmitting and receiving an image that uses Li-Fi technology, also known as Light Fidelity. The objectives of this project are:

- The image that is to be transmitted, is converted into bits.
- The transmitter must be capable of transmitting data and the receiver must be capable of receiving data successfully.
- The received bits are converted back into the original image.

3.2 Methodology

The basic principle involved in this technology is that, if the LED is made ON (when a suitable voltage appears across the leads of the LED), digital HIGH, i.e., 1 will be transmitted and if LED is made OFF (when no voltage or voltage below the threshold has appeared across the leads of the LED), digital LOW, i.e., 0 will be transmitted. The LED starts flickering as a consequence of continuous variation in the current; therefore, data will be transmitted at a high rate. By varying the flickering rate of LED, the encoded data can be received in different sets of 0 s and 1 s. The data to be transmitted is given to the LED in the form of 1 s and 0 s, since it is based on the principle of ON and OFF keying. This makes the data to be transmitted at a higher rate by the transmitter, and at the receiver side, the data is received in the form of 1 s and 0 s and is converted to its original form.

3.3 *Working Principle*

- The image that is to be transmitted is retrieved from RPi storage (Fig. 2).
- Using Python code the image is converted into an array of bits consisting of 0 s and 1 s.
- The array of bits is sent through the GPIO pin of RPi to LED.
- If bit 1 is to be transmitted: LED is turned ON, if bit 0 is transmitted: LED is turned OFF.
- Information sent by the flickering of LED is received by a Solar Panel.
- The analog output of the solar panel is converted into digital readings by ADC (MCP 3008).
- The digital readings are converted into voltage levels, a threshold voltage is set. When the input is higher than the threshold voltage, it is considered as logic 1 and if it is lower than the threshold voltage, it is considered as logic 0.
- The received bits are converted back to the original image.

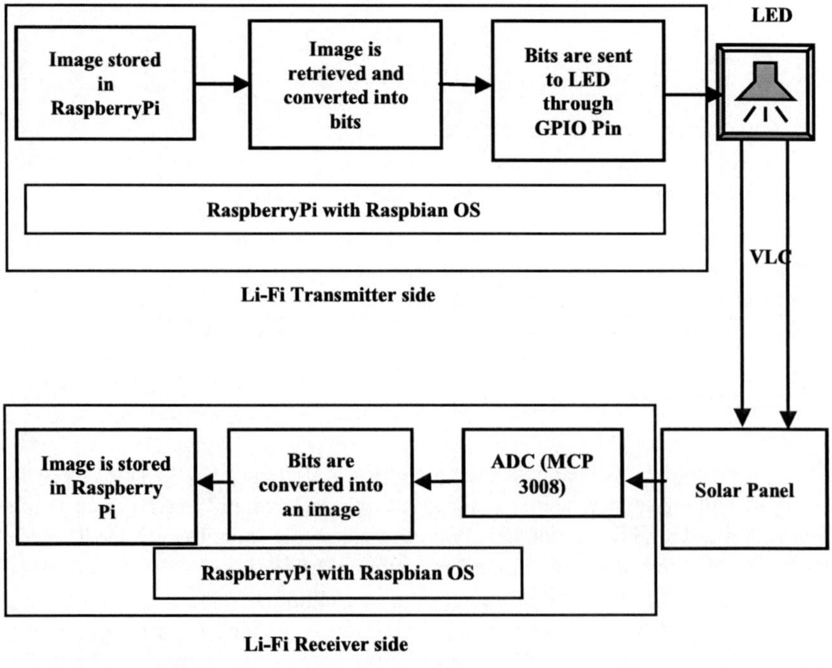

Fig. 2 Block diagram for transmission of data using Li-Fi

4 Result Analysis

4.1 Simulation Results

The image that is to be transmitted should be converted into 1 s and 0 s so that it can be easily transmitted by flickering LED. “1” can be sent as logic level high, “0” can be sent as logic level low to the LED. The received bits are used to reconstruct the image. The process of conversion and construction is done using Python code. The simulation is done on Windows 10 using Jupyter notebook IDE.

The image which is to be transmitted is retrieved from storage and converted into bits. In the simulation, the image “idct-card-mvp.png”, which is shown in Fig. 3, is converted into bits.

Figure 4 contains the code which converts the image into bits.

Algorithm for transmitter Python code

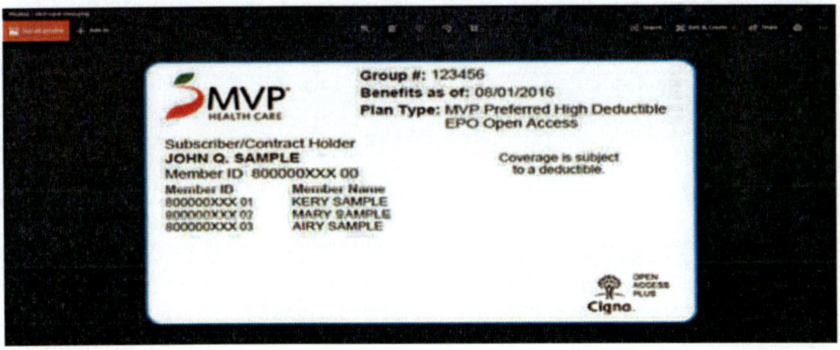

Fig. 3 The image “idct-card-mvp.png”

```
from PIL import Image
import time
```

```
buffer = bytearray()
picture = bytearray()
file = open("idct-card-mvp.png","rb")
bytes = bytearray(file.read())
for byte in bytes:

    for i in range(8):
        bit = (byte>>i) & 1

        buffer.append(bit)
```

Fig. 4 Code to convert images into bits

1. Declare buffer, picture variables as bytearray().
2. Open image that is to be transmitted as “file”.
3. Convert the file into bytearray().
4. For every byte in bytearray, convert byte into bits by right shifting bit by 1 place in each iteration and performing bitwise AND with “1”.
5. Append extracted bit to buffer.

The bits are appended to buffer array, where bits are stored as shown in Fig. 5.

For simulation purposes, it is assumed that bits are transmitted with 0% error.

At the receiver side, the bits are received and converted into an image. The Python code to convert bits to the image is shown in Fig. 6.

Algorithm for receiver Python code

1. Received binary bits are stored in the buffer.
2. Take 8 bits at a time and store in a bits to bytes array.
3. Convert 8 bits to a respective byte value.
4. Append bytes to picture array.
5. Open an empty image file and write a picture array to it.
6. Close the image file.

The image is then stored as “relogo1.jpg” as shown in Fig. 7.

The image “relogo1.jpg” is shown in Fig. 8.

```
buffer
```

```
bytearray(b'\x01\x00\x00\x01\x00\x00\x00\x01\x00\x00\x00\x00\x01\x00\x01\x00\x
\x00\x00\x01\x00\x01\x00\x01\x01\x00\x00\x00\x00\x00\x01\x00\x01\x00\x00\x00\x
\x01\x00\x00\x00\x00\x00\x00\x00\x00\x00\x00\x00\x00\x00\x00\x00\x00\x00\x00\x
\x01\x01\x00\x00\x00\x00\x01\x00\x00\x01\x00\x00\x01\x00\x00\x00\x00\x01\x00\x
\x01\x00\x00\x01\x00\x01\x00\x00\x00\x00\x00\x00\x00\x00\x00\x00\x00\x00\x00\x
```

Fig. 5 Image of the buffer array

Fig. 6 Code to convert bits into the image

```
picture = bytearray()
```

```
for i in range(int(len(buffer)/8)):
    bitstobytes = buffer[i*8:i*8+8]

    byte = 0
    for x in range(8):
        byte = byte + pow(2*bitstobytes[x],x)*bitstobytes[x]

    picture.append(byte)
#write to file and display result
refile = open("relogo1.jpg", "wb")
refile.write(picture)
refile.close()
```

Fig. 7 Image "relogo1.jpg" stored in the location

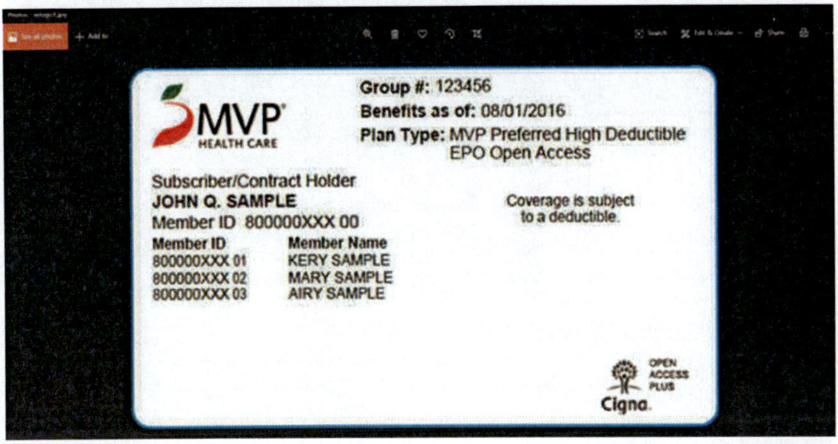

Fig. 8 Image "relogo1.jpg"

4.2 *Coverage Region*

Transmission of HIGH bits (LED ON): In this, the experiment is set up in such a way that the LED is turned on continuously by transmitting logic bit 1. The transmitting LED is kept at one end (say at distance 0 cm) and the receiving solar panel is moved along the scale.

Fig. 9 6 cm—2.82 V

Fig. 10 13 cm—1.89 V

Table 1 With ambient light ON, LED is switched ON, distance is varied between LED and solar panel, voltage reading is observed

S. No.	Distance (cm)	Voltage (V)
1	2	3.52
2	4	3.29
3	6	2.82
4	8	2.51
5	10	2.21
6	13	1.89
7	16	1.71
8	20	1.54
9	25	1.42
10	65	1

Figures 9 and 10 show the experimental setup to obtain voltages at different distances from the transmitting LED.

The voltage across the Solar Panel at different distances from the LED is tabulated at constant Luminous Flux of 90 lm/W, as shown in Table 1.

Figure 11 shows that there is a gradual decrease in the O/P voltage of the solar panel with varying distance from the panel. It is also seen after crossing a distance of 25 cm (out of the LED range), the voltage decreases to the levels of ambient lighting and natural lighting conditions.

4.3 *Data Rate*

When an array of known bits or the bits obtained from an image are transmitted with/without delay, the time till these are received is observed.

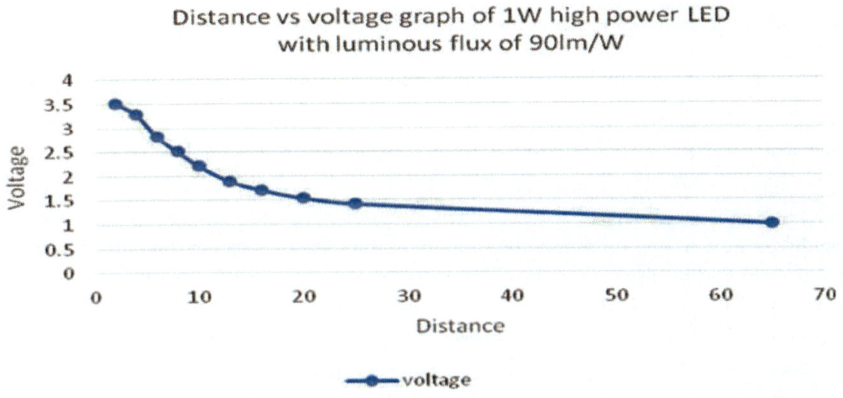

Fig. 11 Graph for the readings in Table 1 is plotted with distance taken along the X-axis and voltage along the Y-axis at a constant luminous flux of 90 lm/W

Data rates vary based on the size of the images/length of bits and are calculated for the total transmission time during which bits get transmitted.

Based on the number of bits transmitted per second, the data rate is obtained per second.

If the number of bits $= x$ and y is the number of seconds required for x bits to get transmitted, then data rate can be calculated as:

$$\text{Data rate} = \frac{x}{y} \text{ bits per second (bps)}$$

Or

$$\text{Data rate} = \frac{x}{y \times 1000} \text{kilobits per second (kbps)} \tag{1}$$

Example A black and white image consisting of 18,904 bits is taken and transmitted with delay of 0.01 s. It took 191.22 s for the image to get transmitted.

As per Eq. (1), the data rate can be calculated as follows:

$$\text{Data rate} = \frac{18{,}904}{191.22 \times 1000} \text{ kbps}$$
$$\text{Data rate} = 0.09 \text{ kbps}$$

4.4 Reception of Known Bits

An array of known bits is transmitted. At the receiver, the information is received by a Solar Panel that is connected to the MCP 3008. MCP 3008 is an 8 channel 10 bit Analog-to-Digital Converter (ADC). The analog signal received from the Solar Panel is converted into digital values. The digital values are converted into voltage levels. A threshold voltage is set. When the received voltage level is above the threshold voltage, it is considered as bit 1, when it is below the threshold voltage it is considered as bit 0.

Each bit is transmitted with a delay of 0.01 s. The Python code at the transmitter and receiver is run manually at the same time. The receiver was able to receive all required bits along with other bits which were the reading from ambient light when the transmitter is not transmitting.

The receiver and transmitter must be synchronized in such a way that the receiver must be able to receive only required bits so that bit error rate (BER) can be calculated.

4.5 Transmission of Multiple Images

The complete process of transmission of an image involves the use of, 1—Transmitter RPi, 1—LED, 1—Solar Panel, 1—MCP3008 IC (ADC), and 1—Receiver RPi. As the RPi is capable of running multiple programs at a time, multiple images can be transmitted using different LEDs driven by the GPIO pins of the same transmitter RPi. The whole setup is isolated from each other to avoid any form of interference. As the conversion of analog-to-digital voltages involves working of MCP3008 and RPi using Hardware SPI as a pair for one single image, the no. of RPi on the receiver side must be same as the no. of images transmitted at a time. Thus, the idea of transmitting multiple images at a time is possible, but at the cost of increased hardware components on the receiver side.

Thus, in this proposed model, we are limiting ourselves by transmitting only one image at a time, although multiple images can be transmitted consecutively using a single transmitter and a single receiver, by defining some delay between the images and optimizing the Python code on both the transmitter side and the receiver side to serve the purpose.

4.6 Size of Image and Time Constraint

In the initial testing, when the array of known bits was transmitted, a maximum of 0.01 s delay could be achieved between 2 bits to receive it unaltered. Thus, enabling us to transmit at a data rate of a maximum 100 bps = 0.1 kbps. For the experimental purpose, we are restricting the maximum time of transmission to 5 min (300 s).

$$\text{Max time of transmission} = 5\,\text{min}\ (300\,\text{s})$$
$$\text{Data rate} = 100\,\text{bps}$$

$$\begin{aligned}\text{Maximum number of bits transmitted} &= \text{maximum time of transmission} * \text{data rate}\\ &= 300 * 100 = 30,000\,\text{bits}\ (3.75\,\text{kB in size})\end{aligned} \quad (2)$$

Thus, any image less than 3.75 kB size can be considered for transmission using the proposed model. Although images of any size are possible to transmit, the time of transmission will increase with the size linearly making it not suitable for the demonstration.

5 Conclusion

- The voltage across the solar panel reduces gradually with an increase in distance from the LED. By this, the range of the LED is known to be 25 cm.
- Beyond the range of 25 cm, the O/P voltage was corresponding to the ambient lighting conditions.
- The receiver and transmitter must be synchronized in such a way that the receiver must be able to receive only required bits so that a bit error rate (BER) can be calculated.

References

1. Haruyama S (2013) Visible light communication using sustainable LED lights. In: 2013 proceedings of ITU kaleidoscope: building sustainable communities, Kyoto, pp 1–6
2. Haas H (2011) Wireless data from every light bulb. TED Global
3. Haas H, Yin L, Wang Y, Chen C (2016) What is Li-Fi? IEEE J Light Technol 34(6):1533–1544
4. Haas H (2017) LiFi is a paradigm-shifting 5G technology. Rev Phys 3. https://doi.org/10.1016/j.revip.2017.10.001
5. He Y, Ding L, Gong Y, Wang Y (2013) Real-time audio & video transmission system based on visible light communication. Opt Photonics J 03:153–157. https://doi.org/10.4236/opj.2013.32b037
6. Goswami P, Shukla M (2017) Design of a Li-Fi transceiver. Wirel Eng Technol 08:71–86. https://doi.org/10.4236/wet.2017.84006
7. Nivetha S et al (2019) Performance evaluation of modulation techniques in Li-Fi
8. Sandoval-Reyes S (2018) Transmission and reception of images via visible light. Res Comput Sci 147:193–202. https://doi.org/10.13053/rcs-147-12-18
9. Vinnarasi A, Aarthy ST (2017) Transmission of data, audio signals and text using Li-Fi
10. Fergusson PB (2016) Light fidelity (Li-Fi) prototype with Raspberry Pi
11. Atmaja R, Murti M, Halomoan J, Suratman FY (2016) An image processing method to convert RGB image into binary. Indones J Electr Eng Comput Sci 3:377–382. http://doi.org/10.11591/ijeecs.v3.i2.pp377-382

12. Salian PP, Prabhu S, Amin P, Naik SK, Parashuram MK (2013) Visible light communication. In: Proceedings of the 2013 Texas Instruments India Educators' conference (TIIEC '13). IEEE Computer Society, USA, pp 379–383. https://doi.org/10.1109/TIIEC.2013.74

Penalty Based Backend Path Management

Deepanshi Sengar and Anoop Kr. Patel

Abstract Ever since the technologies like big data and data mining have been introduced, keeping the data that is required to get the job done continuously available has become a requirement for almost all customer segments ranging from small business to datacenters. In contrast to connection failures in message networks—which generally could cope up by re-establishing the connection in storage networks, a network resource failure is more likely to cause a system crash. Achieving a very high level of availability can be arduous and very expensive, and it also requires that redundancy be established at multiple levels: Redundancy of storage, regular backups to help in server recovery, clustering similar servers, and redundancy of the physical path components. Load balancing, which is the distribution of reading/write (I/O) requests and fault tolerance, to surcharge the performance between the server and the storage device(s), is very important in a high load environment or settings where having consistent service level is very critical. Therefore, a system needs to have a good MPIO (multipath I/O) policy. Without an MPIO policy, a server sending I/O requests in a multiple path setup may operate such that there is a very heavy workload on some paths while other paths may be underutilized, without considering the performance of paths. The current approaches use scheduling algorithms like Round-Robin, first come first serve (FCFS), shortest seek time first (SSTF), etc., for scheduling I/O requests on paths. In such an environment, where the load on paths is not consistent or changes very frequently, Round-Robin scheduling does not work well [1]. The existing algorithms serve the I/O requests faster but not efficiently. This motivated us to study and analyze a method that can help us serve the I/O requests faster and efficiently. Our purpose is to introduce the user to a penalty based I/O path selection policy which is inspired by ant colony optimization (ACO). The ant colony optimization (ACO) is a very powerful approach for solving computational problems. Here, we will be considering improving the I/O servicing efficiency with

D. Sengar (✉) · A. Kr. Patel
Department of Computer Engineering, National Institute of Technology, Kurukshetra, India
e-mail: deepanshisengar@gmail.com

G. Ranganathan et al. (eds.), *Inventive Communication and Computational Technologies*, Lecture Notes in Networks and Systems 145,
https://doi.org/10.1007/978-981-15-7345-3_54

the help of the path selection policy. This policy will allow us to distinguish between available paths based on their past performance hence, improving the overall system performance.

Keywords Backend · Path management · I/O scheduling algorithms · Storage resources · Multipath

1 Introduction

An I/O request is a read/write request to the storage resource. Many computer systems include more than just one storage resource (devices). A path can be viewed as a queue where the I/O requests are stored one after the other. Pathing is an approach used to address the needs of storage networks by changing the way of organizing and managing multiple communication paths. The amount of use of storage resources has increased significantly. Thereby, increasing the requirement for the requests to be served at the fastest which also improves the overall performance of the system to a great extent.

MPIO facilitates the coherent potential for sending I/O over multiple paths connecting the server to storage. The various components of these redundant hardware paths are cables, host bus adapters (HBAs), switches, storage controllers, and possibly power. In some computer systems, only a single communication path is operational at any given time, whereas, in some systems, multiple paths are operational to enhance system performance. MPIO policies coherently manage these redundant connections so that I/O requests can be rerouted if a component along one path fails.

Ueda et al. [1] provided a request-based device-mapper multipath solution that focused on I/O merging for improving multipathing performance. Goggin [2] described an approach that could support any block device and was easy to implement but has the drawback that it does not have access to detailed error information. Eager et al. [3] described random, threshold, and shortest load sharing policies. Wang and Guan [4] gave weightage to the dynamic aspects of a system like CPU utilization, memory usage, etc., with the degree of influence of each factor. Qin et al. [5] proposed a scheduling algorithm that improves the availability of heterogeneous systems and improves performance of return time of task I/O's. Thomasian et al. [6] proposed a new disk I/O scheduling policy which uses look-ahead to improve performance. Prabhakar et al. [7] describes an I/O scheduling algorithm that is aware of disk-cache and parallelism in large computer systems having multiple clients that can send I/O requests concurrently. The available configuration options for MPIO includes failover, failback, Round-Robin, Least Queue Depth, Weighted Path and Least Blocks disk I/O scheduling [8]. Therefore, based on the above literature survey, the challenge [9] that is found in conventional MPIO approaches is, if a path has an I/O which is stuck and the outstanding command buffer (OCB) is reached for that path. The stuck I/O and the I/O's in the queue after it might just starve which decreases the system's performance. So, we must find a method that can efficiently explore the available paths before scheduling I/O requests. In short, a policy that might categorize the multiple paths based on their performance. So, we can know in advance if a path is slow (not responding or I/O stuck due to any reason) and prevent the system's performance from degrading. The scope of our approach is proper load balancing among all paths which will lead to a great improvement in system performance.

Our paper is distributed in five sections: Sect. 1 contains introduction; Sect. 2 contains the problem statement, Sect. 3 shows our proposed algorithm; Sect. 4 shows the experimental result and analysis, and finally, we conclude the paper in Sect. 5.

2 Problem Statement

The performance of storage devices can be measured by I/O operations per second (IOPS). Traditionally, the I/O requests are distributed to the paths in a Round-Robin fashion, without considering the performance of the path. If a path fails, then any I/O request on that path might be stuck. Also, requests in the queue after the current stuck I/O also get stuck. This might lead to I/O starvation and increases the latency for the other I/O in the queue. This, in turn, degrades the performance of the system. We can easily understand the problem with the example given in Fig. 1.

As seen in Fig. 1, we assume that the buffer capacity of each queue is of 5 tasks. Round-Robin policy is a simple and efficient algorithm when everything works fine.

Fig. 1 Distribution of I/O using Round-Robin policy

But like in our example, if IO1 gets stuck on path 1 then, IO4, IO7, IO10, and IO13 also get stuck in the queue. The other incoming I/O, say IO14 (on path 2) and IO15 (on path 3) arrived after IO4 but were processed before it. The latency of the waiting requests on path 1 also increases. So, from the example, it is evident that we need a policy that can monitor the path performance, and on the failure of a path, can react instantly without waiting for the queue limit to be reached.

3 Proposed Method

The early techniques for backend path management conventionally used FCFS or SSTF for balancing the activities over multiple paths [10]. Later, a simple Round-Robin was also started to use for bringing improvement in system performance [11, 12]. However, studies have proved that Round-Robin did not improve system performance and that many improvements could be made to the conventional Round-Robin approach [9]. The conventional policies did not take into account the performance characteristics of each path. Thus, there is a need for a policy that will help us to define whether a path is good or bad based on its past performance.

We propose an I/O path selection policy that makes use of the assigned penalty and is inspired by the ant colony optimization (ACO). This approach is a combination of Round-Robin and feedback. In case an I/O gets stuck on a path based on early feedback, our policy reacts instantly so new I/O does not get scheduled on that path. On failure of a path, the tasks in the queue will wait or rescheduled based on the recovery and monitoring process.

Latency is how fast a single I/O is handled. It is very important to understand that although a disk may handle individual I/O with an average latency of 10 ms (100 IOPS), and the actual latency as perceived by the application may be higher as some I/O must wait in line (say 15 ms or 20 ms). Waiting in queue increases the latency for that I/O to be handled, which, in turn, increases the penalty of the path. Here, we take,

$$\text{Penalty}\big[\text{path\_Index}\big] = \text{I/O Latency} + \text{int}\big(\text{Penalty}\big[\text{path\_Index}\big]/k\big) \tag{1}$$

we calculate the penalty of the paths and distribute the next I/O based on that penalty. Where 'k' is some constant used for last 'x' samples.

For our proposed method, the parameters are defined as below:

- No_of_Path—is a variable that stores the number of paths available.
- IOCounter—is a variable that keeps track of the number of I/O served till now and is initialized to '0'.
- No_of_I/O—is a variable that stores the number of I/O to be served.
- Global_penalty [No_of_Path]—is an array that stores the penalty of each path.
- Temp_Penalty [No_of_Path]—is an array that is used for finding the minimum penalty among the paths without using the minimum function from the library.

That is, by reducing 1 from each healthy path penalty. So whichever path has the least penalty its penalty will be updated to 0 first, and then will be selected for serving the next I/O. It is initialized to '0'.

- Path_Index—is a variable that stores the name of the selected path on which the next I/O will be sent.
- BanishThreshold—is the threshold penalty for each path beyond which a path will not be considered for serving the I/O (the path is unhealthy).
- UnbanishThreshold—a value slightly less than BanishThreshold, to handle the spike of I/O on the path. That is if an unhealthy path becomes healthy, and its penalty starts with 0, then all the upcoming I/O will be distributed to that path only, because of a very low penalty on that path, which means, there will be a spike on that path. Slowly it gets adjusted with other paths.
- I/O_Per_Path [No_of_Path]—is an array that keeps the count of the number of I/O sent on each path.

Figure 2 depicts the flowchart of our proposed approach in which we first initialize all the parameters. The parameters like No_of_Path, No_of_I/O, Global_penalty[No_of_Path], BanishThreshold, UnbanishThreshold can be initialized differently for each simulation. Whereas the variables like IOCounter, Temp_Penalty[No_of_Path], I/O_Per_Path[No_of_Path] will always be initialized to '0'. The parameter Path_Index will be initialized to '−1'.

After initializing the parameters, we start pumping the I/O on the path selected using the path selection algorithm (Fig. 3) and inject simulated behavior on that path. We then monitor the degraded paths (if any) in the system. If the path becomes healthy, we assign UnbanishThreshold value to the path. After which we calculate and update the Global_Penalty for the selected path (1). The above process is repeated for all the I/O requests in the system.

Figure 3 represents the flowchart for the path selection algorithm. This is the algorithm by which a path is chosen among all available paths based on its Global_Penalty and Temp_Penalty. It first checks whether the Global_Penalty[Path_Index] is less than BanishThreshold, i.e., if the path is healthy, then it can be considered for scheduling the I/O request. If a path is healthy, we check if the Temp_Penalty[Path_Index] <= 0, i.e., if the path has minimum penalty then, we choose that path for scheduling the I/O else we just decrease the Temp_Penalty[Path_Index] by one and continue the process until a path is selected. However, if the path is unhealthy, we simply move to the other path and repeat the process.

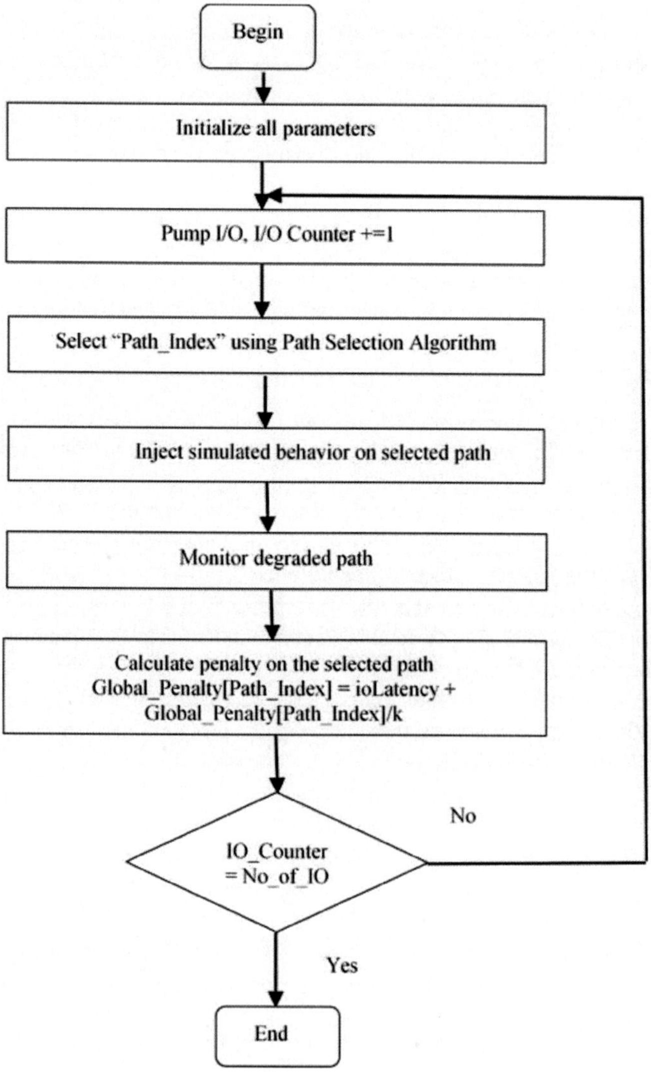

Fig. 2 Flowchart of the proposed solution

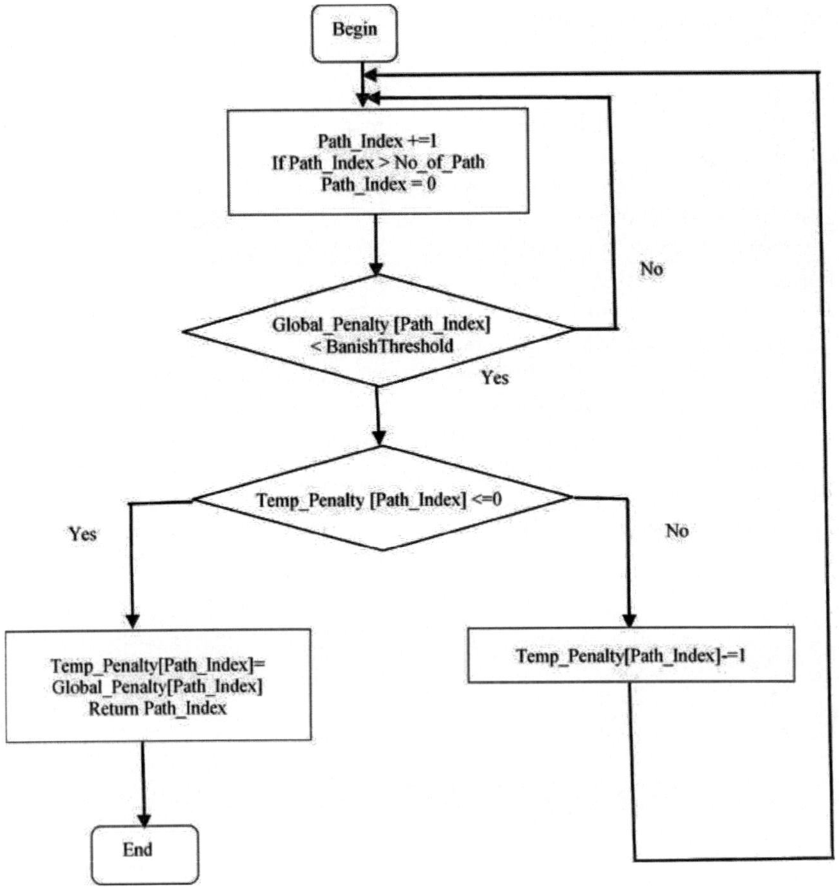

Fig. 3 Flowchart of the path selection algorithm

4 Simulation Results

4.1 Experimental Parameters

We have shown the experimental results for the proposed solution. Simulation and implementation of the proposed method and comparison with Round-Robin have been done in Python 3.14 running at Intel Cores i7-8850H CPU @ 2.60 GHz with 16 GB RAM.

For our simulation, the parameters were initialized as stated below:

No_of_Path = 4, IOCounter = 0, No_of_I/O = 2500, Path_Index = −1, Global_penalty [4] = [10, 20, 30, 80], Temp_Penalty [4] = [0, 0, 0, 0], BanishThreshold = 200, UnbanishThreshold = 150, I/O_Per_Path [4] = [0, 0, 0, 0].

4.2 Experimental Cases

The simulation was performed for the following test cases:

When one path is misbehaving: Here, we made path '2' to misbehave. For a path to misbehave the I/O latency for that path should be high. So, we manually provide high latency to path '2'. As the penalty increases on the path, no. of I/O on that path decreases. When it crossed the threshold, no I/O is scheduled on that path (Fig. 4).

When only one path is good: In this case, the paths do not misbehave at the same time. Except for one path, we made all path degraded after some interval. The result shows that the number of I/O on the healthy paths increases, and in the end, all the I/O flows on the only healthy path (Fig. 5).

When a path is degraded first and then upgraded: The path misbehaved (had very high penalty) initially and after some time it behaved fine. So, we first provide a high penalty, then UnbanishThreshold, and then low penalty (Fig. 6).

When all the paths toggle: In this experiment, we give high penalty then a low penalty to all paths, and the result shows that the number of I/O on the paths also decreases and increases (Fig. 7).

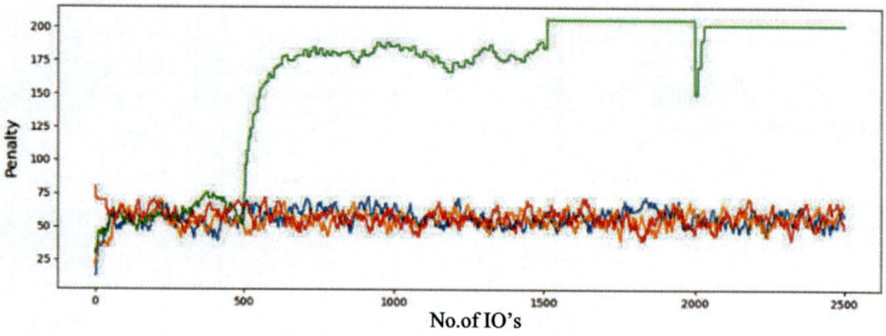

(a) Graph representing the penalty on the paths

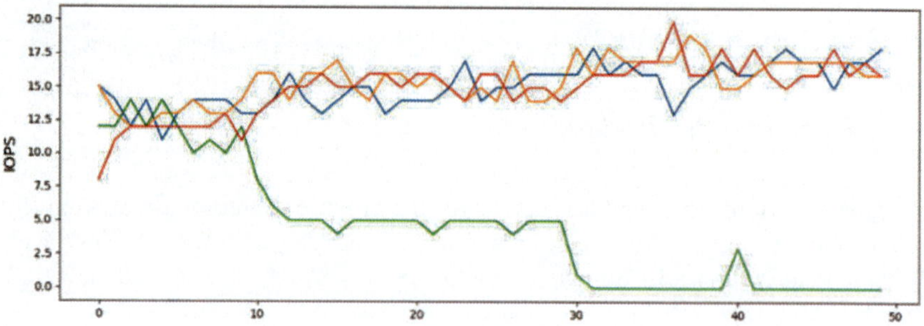

(b) Graph representing the IO's distribution on each path

Fig. 4 Performance graphs when one path misbehaves

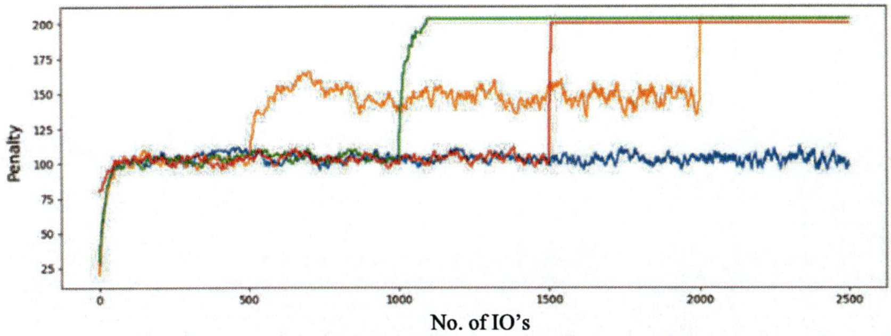

(a) Graph representing the penalty on the paths

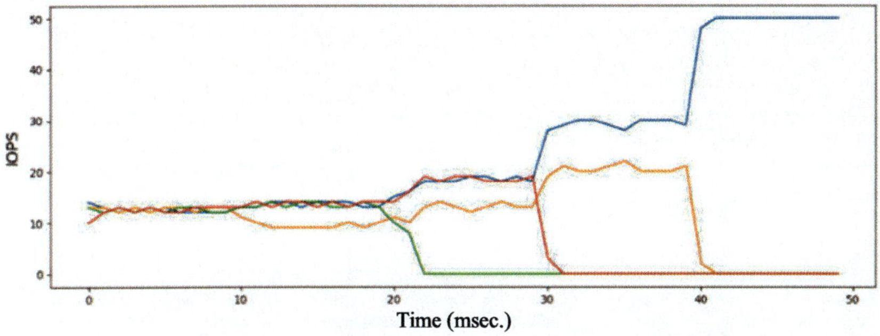

(b) Graph representing the IO's distribution on each path

Fig. 5 Performance graphs when only one path works fine

4.3 Comparison of Round-Robin with Our Proposed Approach

In this section, we have compared the Round-Robin MPIO with our proposed method. After comparison, we found that by using our approach there is proper load balancing among all paths but in the Round-Robin, some paths are heavily loaded and some paths are underutilized (Figs. 8 and 9).

In Table 1, for completing all the tasks, by using Round-Robin, it will take 9660 ms (Path 3) to complete, and as per our proposed method, it will take only 9489 ms (Path 2). We take maximum time from each approach because a schedule is complete when all the tasks have been executed, i.e., when the last task finishes execution. Utilization% columns clearly show that there is very little difference between the paths using our proposed method. It means paths are properly balanced and utilized. Improvement in completing all the tasks is also significant.

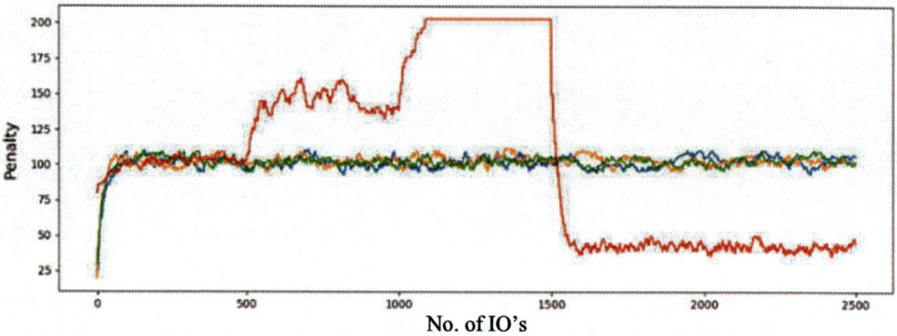

(a) Graph representing the penalty on the paths

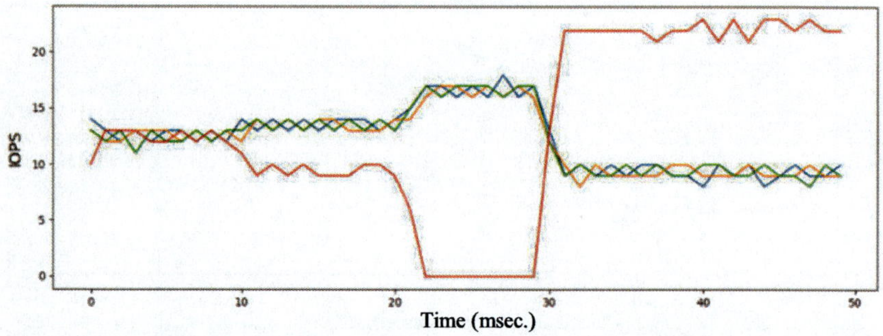

(b) Graph representing the IO's distribution on each path

Fig. 6 Performance graphs when one path toggles

5 Conclusion

The multipath I/O not only helps to provide a means for high availability (fault tolerance) but also improves the efficiency and performance of the system. Path selection problem where some path might not be working fine due to some run-time discrepancy is a real-time problem. In this paper, we have proposed a penalty based backend patch management policy. The proposed policy focused on eliminating situations that might lead to I/O starvation. Implementing such a policy in systems will not only make sure that there are multiple paths from the server to storage resources but also ensure that I/O is sent down the paths which are functioning

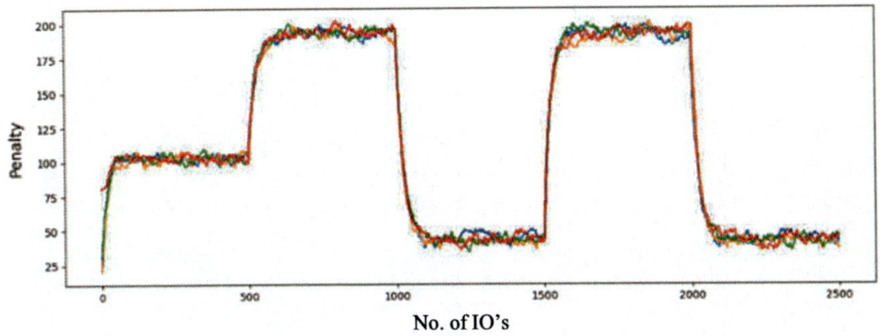

(a) Graph representing the penalty on the paths

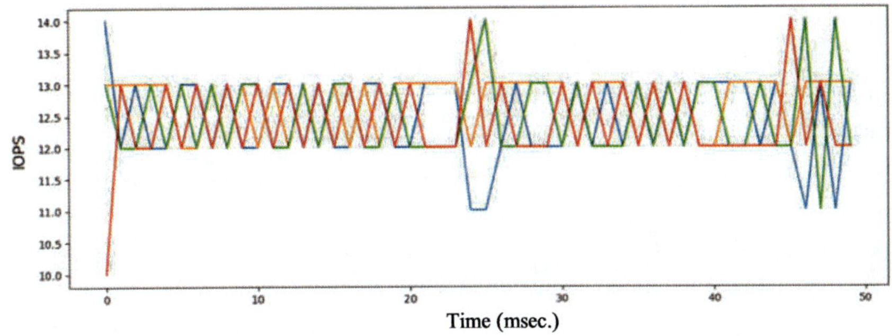

(b) Graph representing the IO's distribution on each path

Fig. 7 Performance graphs when all paths toggle

Fig. 8 Load balancing using our proposed method

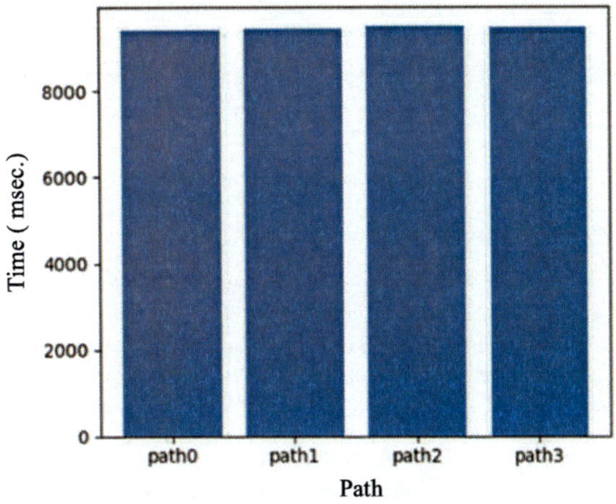

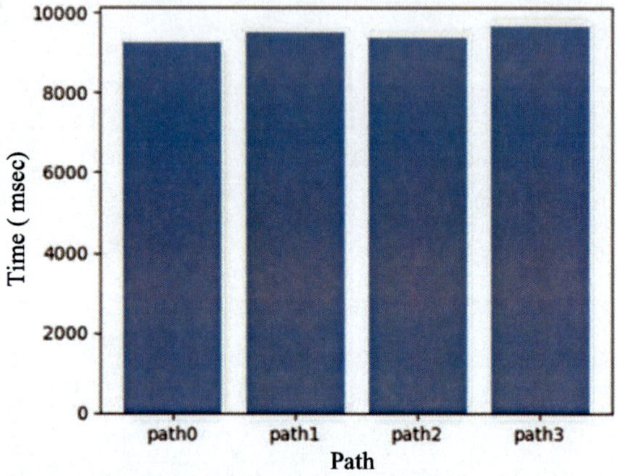

Fig. 9 Load balancing using Round-Robin

Table 1 Comparison between Round-Robin and our proposed policy

Paths	Time (in ms)		Utilization % (corresponding with the max. time of the path)	
	Round-Robin	Proposed	Round-Robin	Proposed
Path 0	9254	9410	95.7	99.1
Path 1	9511	9440	98.4	99.4
Path 2	9374	9489	97.0	100
Path 3	9660	9460	100	99.6

properly. The occurrence of faults on a path is environment-dependent, and our policy does not prevent the occurrence of the same or other faults on the path. The simulation shows that our proposed method gives better results as compared to Round-Robin multipath I/O. In the future, the work could be extended to implementation with deep reinforcement learning and machine learning.

References

1. Ueda K, Nomura J, Christie M (2007) Request-based device-mapper multipath and dynamic load balancing. In: Linux symposium, vol 2, pp 235–243
2. Goggin E (2005) Linux multipathing. In: Proceedings of the Linux symposium, p 147
3. Eager DL, Lazowska ED, Zahorjan J (1986) Adaptive load sharing in homogeneous distributed systems. IEEE Trans Softw Eng 12(5):662–675
4. Wang M, Guan J (2017) An adaptive dynamic feedback load balancing algorithm based on QoS in distributed file system. J Commun Inf Netw 2:30–40
5. Qin X, Xie T (2008) An availability-aware task scheduling strategy for heterogeneous systems. IEEE Trans Comput 57(2):188–199
6. Thomasian A, Liu C (2002) Some new disk scheduling policies and their performance. ACM SIGMETRICS Perform Eval Rev 30:266–267

7. Prabhakar R, Kandemir M, Jung M (2017) Disk-cache and parallelism aware I/O scheduling to improve storage system performance. In: IEEE 27th international symposium on parallel and distributed processing, pp 357–368
8. https://www.itprotoday.com/cloud-computing/microsoft-multipath-io-iscsi
9. Anderson M, Mansfield P (2003) SCSI mid-level multipath. In: Proceedings of the Linux symposium, p 23
10. Micha H (1980) Disk scheduling: FCFS vs. SSTF revisited. Commun ACM 23(11):645–653
11. https://support.microsoft.com/sq-al/help/2146446/on-microsoft-windows-server-2003-i-o-may-fail-when-path-failover-occur
12. https://docs.vmware.com/en/VMwarevSphere/6.7/com.vmware.vsphere.storage.doc/GUID-B7AD0CA0-CBE2-4DB4-A22C-AD323226A257.html

Multi-band Microstrip Antenna for Wireless Local Area Network

Aditya Sarin, Deveshi Thanawala, Jessica Sadavarte, and Tazeen Shaikh

Abstract This paper focuses on the designing and analysis of a rectangular microstrip patch antenna for wireless local area network (WLAN) applications operating at 2.4, 4.7, 6, 7.0, and 8.7 GHz. The antenna is designed using flame retardant (FR4) epoxy dielectric substrate with relative permittivity $\varepsilon_r = 4.4$ and thickness $h = 1.60$ mm having gain 3.12 dB. The designed antenna is simulated using high-frequency structure simulator (HFSS) software, which works on the principle of an adaptive mesh and compliant solver for a finite element method model.

Keywords Rectangular microstrip patch antenna · Microstrip line · Feed · Multi-band · HFSS

1 Introduction

The advancement of technology over the years has seen wireless technology playing a pivotal role in the communication domain. The inherent convenience of wireless technology has given it the preference over the subsequent wired technology used. However, properties of antennas that form the basis of wireless technology are pivotal when it comes to how efficient the network is.

Antennas work on electromagnetic waves among which the frequency range for communication applications is the radio frequency (RF), ranging from 20 kHz to

A. Sarin (✉) · D. Thanawala · J. Sadavarte · T. Shaikh
Mukesh Patel School of Technology, Management and Engineering, Mumbai, India
e-mail: aditya28sarin@gmail.com

D. Thanawala
e-mail: deveshi.m.thanawala@gmail.com

J. Sadavarte
e-mail: svjessica21@gmail.com

T. Shaikh
e-mail: tazeen.shaikh@nmims.edu

G. Ranganathan et al. (eds.), *Inventive Communication and Computational Technologies*, Lecture Notes in Networks and Systems 145,
https://doi.org/10.1007/978-981-15-7345-3_55

300 GHz. The major purpose of using an antenna is to radiate radio frequency waves in a certain direction and maintaining proper efficiency.

A microstrip antenna is preferred in wireless communication because it is capable of dual and triple frequency operations as well as it follows a planar configuration which can be easily attached to the host surface. An important advantage for a microstrip antenna lies in its lightweight and low fabrication cost, hence it can be manufactured in large quantities [1].

Microstrip antennas mainly comprise of ground, substrate, and a patch. The feeding technique (supply) of microstrip antenna can use mechanisms like microstrip line, coaxial, aperture coupling, and proximity coupling. A microstrip feed line is used and a conducting strip is directly connected to the edge of the microstrip patch which maintains the planar structure of the antenna as it can be etched on the same substrate.

IEEE standards developed for local area networks (LAN) are 802.11 and 802.11x family of specification developed for WLAN. According to the operating frequency of 2.4 GHz, the IEEE standard of 802.11b which is an extension of 802.11, providing transmission of 11 Mbps in the selected 2.4 GHz band [2]. WLAN can be built using several network protocols having an advantage of improved network security and supporting a large number of devices and as the connection is wireless, there is no issue relating to the length of the wire, as in wired LAN [3].

Such advantages of the microstrip patch antenna along with some disadvantages have provided it reasonable contendorship for the high speed wireless local area networks (WLAN) as well as in several different varieties of uses such as RFID and PCS [4].

2 Antenna Design

Figure 1 shows the geometry of the proposed microstrip patch antenna where the patch and ground are separated by an FR4 dielectric substrate having a permittivity (ε) of 4.4.

The following equations are required for calculating the dimensions of width, effective dielectric constant, effective length, length extension ΔL, and actual length of the patch.

The patch of width (W) is labeled as P1 having a value of 38.036 mm.

$$W = \frac{c}{2f_0\sqrt{\frac{(\varepsilon_r+1)}{2}}} \tag{1}$$

The value of width (W) of the patch is calculated by substituting the value of relative permittivity (ε_r).

The value obtained from the above equation for ε_{eff} is 4.0858

Fig. 1 Antenna geometry

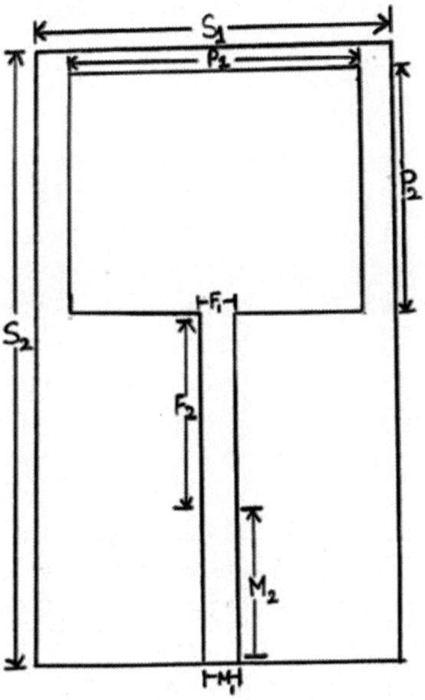

$$\varepsilon_{\mathrm{eff}} = \frac{(\varepsilon_{\mathrm{r}} + 1)}{2} + \frac{(\varepsilon_{\mathrm{r}} - 1)}{2}\left[1 + 12\frac{h}{W}\right]^{-\frac{1}{2}} \tag{2}$$

The value of effective permittivity (c_{eff}) is calculated by substituting the relative permittivity (ε_{r}), the width of the patch (W), and height (h) of the patch from the ground.

The value obtained from the above equation for L_{eff} is 30.92 mm.

$$L_{\mathrm{eff}} = \frac{c}{2f_0\sqrt{\varepsilon_{\mathrm{eff}}}} \tag{3}$$

Generally, the electromagnetic fields at the edges of the patch experience fringing effects, and such effects lead to a greater effective length than the actual length [5].

The value obtained from the above equation for L_{eff} is 30.92 mm.

$$\Delta L = 0.412h\frac{(\varepsilon_{\mathrm{eff}} + 0.3)\left(\frac{W}{h} + 0.264\right)}{(\varepsilon_{\mathrm{eff}} - 0.258)\left(\frac{W}{h} + 0.8\right)} \tag{4}$$

The value obtained from the above equation for ΔL is 0.7388 mm.

$$L = L_{\mathrm{eff}} - 2\Delta L \tag{5}$$

The length of the patch obtained from the above equation which is labeled as P2 in the diagram is having a value of 29.4424 mm.

Where f_0 is the resonant frequency
W is the width of the Patch
L is the length of the Patch
h is the height
ε_r is the relative permittivity of the dielectric substrate
c is the speed of light: 3×10^8 [1].

The antenna is simulated using HFSS operating at a frequency of 2.4 GHz and the measurements are as stated in Fig. 1. The rectangular patch has width P1 and length P2 where P2 being the shorter among the two forms the radiating edges of the microstrip patch and P1 being the non-radiating edges of the patch.

The ground layer of the antenna has width G1 and length G2. The frequency, impedance, bandwidth, and radiation beam-widths vary in nature due to changes in the dimensions of the ground plane [6]. The substrate has a width S1 and length S2.

The feeding mechanism is a microstrip line feed. The strip used is smaller in width when compared to the rectangular patch [7]. The feed line has width F1 and length F2. Microstrip has width M1 and length M2. The radiation box has a width R1 and length R2 along with height H1, such a box is created to avoid any radiation to be reflected.

To increase the capacitive behavior of the patch over a wide frequency range, the feed line is connected at a position along the edge of the model. The substrate material used in microstrip antenna designs is usually of ceramic, ferromagnetic, synthetic, or composite type.

Our model has used an FR-4 epoxy substrate which provides great flexibility for thinner substrate thickness over high frequencies. Also, FR-4 is abundantly available and is cheap. This substrate bears a tangential loss of approximately 0.02.

FR-4 can function well as an insulator and is advantageous due to the appropriate strength-to-weight ratio. The dielectric constant of 4.25–4.55 is maintained with a relative permittivity of 4.4. It can enhance the antenna radiation capability by aid in producing time-varying magnetic and electric fields.

However, FR-4 also has certain disadvantages such as due to its high tangential loss and inaccuracy of the relative permittivity can lead to unwanted complexity in the antenna by shifting the operating frequency also affecting axial ratio and gain. Hence, such antennas using FR-4 should be used in the frequency range of 2–10 GHz only.

Table 1 states the design parameters for the rectangular microstrip patch antenna.

Table 1 records the measurements of the specifications in mm for each part of the antenna which was calculated.

Figure 2 shows the simulated model of the rectangular microstrip patch antenna in HFSS.

Table 1 Design parameters for the antenna

Object	Length (mm)	Width (mm)	Height (mm)
Ground	G2 = 71.49	G1 = 47.63	–
Substrate	S2 = 71.49	S1 = 47.63	H = 1.6
Patch	P2 = 29.44	P1 = 38.036	–
Feed line	F2 = 17.45	F1 = 1.1	–
Microstrip	M2 = 20.3	M1 = 1.2	–
Radiating box	R2 = 220	R1 = 200	H1 = 50

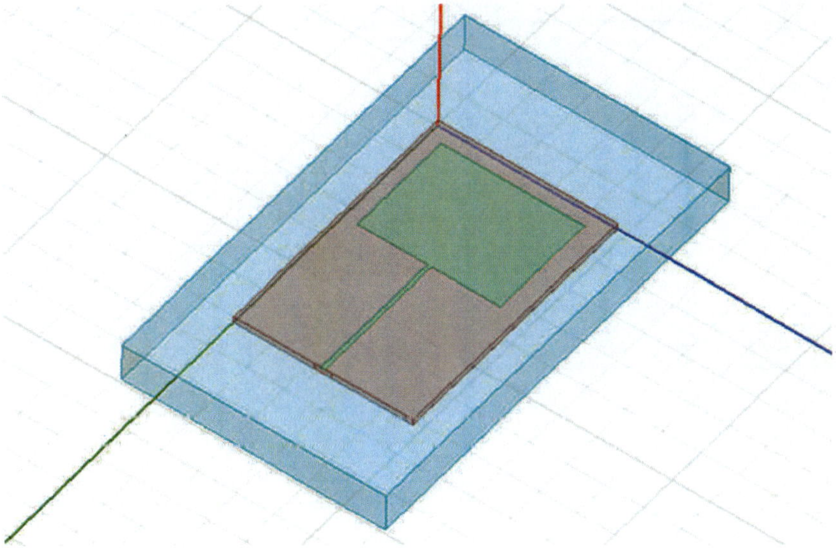

Fig. 2 Simulated design in HFSS

3 Results and Discussion

The antenna is designed and simulated in a high-frequency structure simulator (HFSS) which is a commercial tool for antenna and complex radio frequency electronic circuit design. The designed antenna gives a VSWR of 1.2463 as illustrated in Fig. 3. VSWR characterizes the impedance mismatch wherein some of the energy is reflected in the direction of the source and the remaining waves get transmitted to the antenna [8].

The return loss (S11) describes the wasted or lost power in the signal denoted as the ratio of power returning after reflection to the total power being reflected by any discontinuity in a transmission line.

The return loss obtained for our antenna at an operating frequency of 2.4 GHz is −19.1996 as shown in Fig. 4.

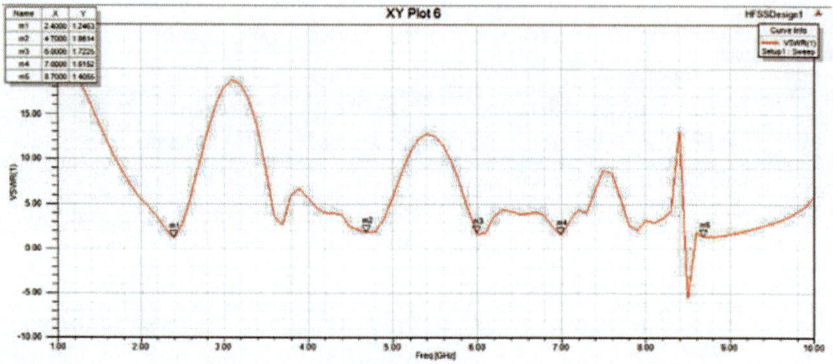

Fig. 3 Graph of VSWR versus frequency. VSWR at 2.4 GHz = 1.2463. VSWR at 4.7 GHz = 1.8614. VSWR at 6 GHz = 1.7225. VSWR at 7.0 GHz = 1.6152. VSWR at 8.7 GHz = 1.4055

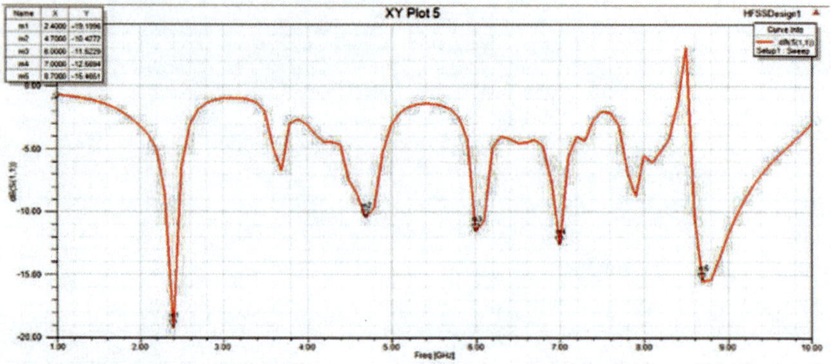

Fig. 4 Return loss (S11) obtained from simulation. Return loss at 2.4 GHz = −19.1996. Return loss at 4.7 GHz = −10.4272. Return loss at 6 GHz = −11.5229 dB. Return loss at 7.0 GHz = −12.5694. Return loss at 8.7 GHz = −15.4651 dB

In wireless communications, a viable return loss of −10 dB is expected. Hence, the higher the power ratio of P_{in}/P_{out} better the load and line are matched together [9].

The frequency marked as m1 represents 2.4 GHz with a return loss of −19.19 dB, lower limit frequency of 1.7 GHz, and upper limit frequency of 2.7 GHz thereby a total bandwidth of 1 GHz.

The frequency marked as m2 represents 4.7 GHz with a return loss of −10.42 dB, lower limit frequency of 3.5 GHz, and upper limit frequency of 5.8 GHz thereby a total bandwidth of 2.3 GHz.

The frequency marked as m3 represents 7 GHz with a return loss of −12.56 dB and a total bandwidth of 2.5 GHz (Fig. 5).

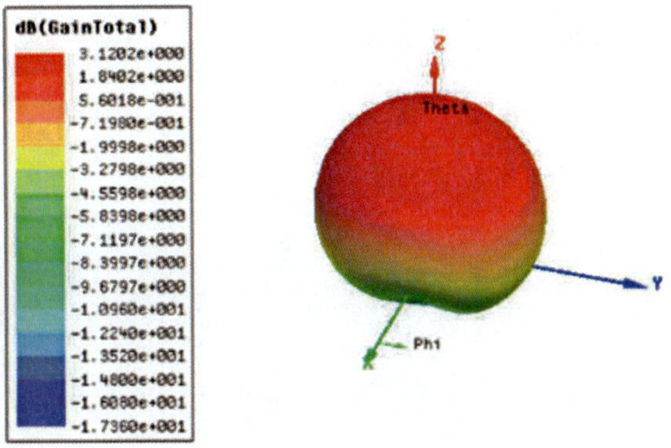

Fig. 5 3D radiation pattern for total gain (dB). Total gain = 3.12 dB

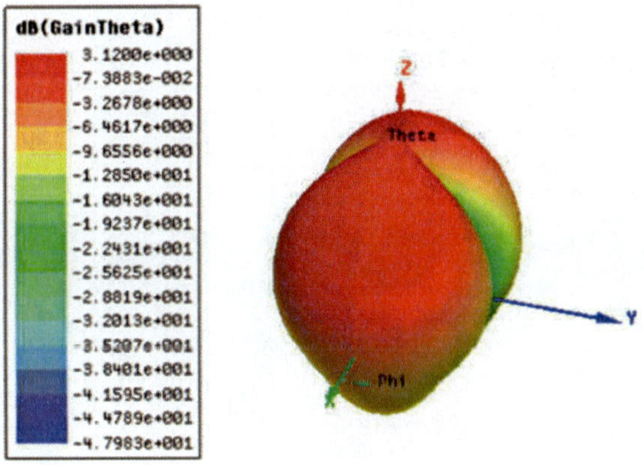

Fig. 6 3D radiation pattern for gain theta (dB). Gain theta = 3.12 dB

The bit-rate required by the antenna for wireless applications would largely depend on the bandwidth of the antenna, more precisely the bandwidth efficiency attained by the antenna (Fig. 6).

Bandwidth is largely related to the thickness of the substrate layer in the microstrip antenna. The larger the thickness of the substrate, the more bandwidth is attained but at a cost. The increase in thickness leads to low gain attainment also affected by the low permittivity (Figs. 7 and 8).

The availability of sidelobe in the spectrum of the antenna can affect the gain as well as the directivity of the antenna. However, there are techniques to eliminate these sidelobes, such as to vary the spacing in the antenna, use window functions by using FIR filters to reduce the sidelobe. Also, the amplitude and phase of each antenna are to be varied so that we can eliminate these sidelobes (Figs. 9, 10 and 11).

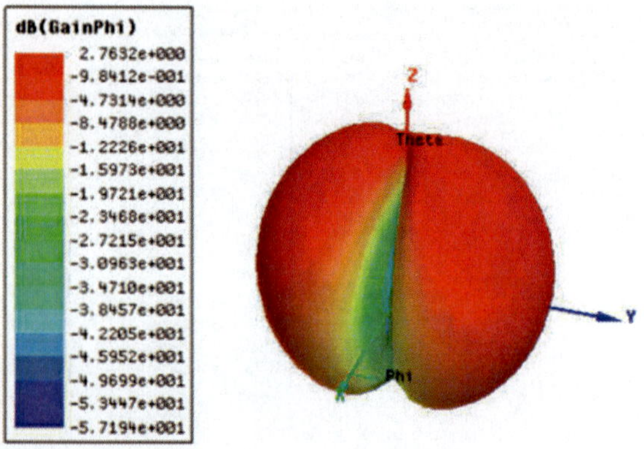

Fig. 7 3D radiation pattern for gain phi (dB). Gain phi = 2.7632 dB

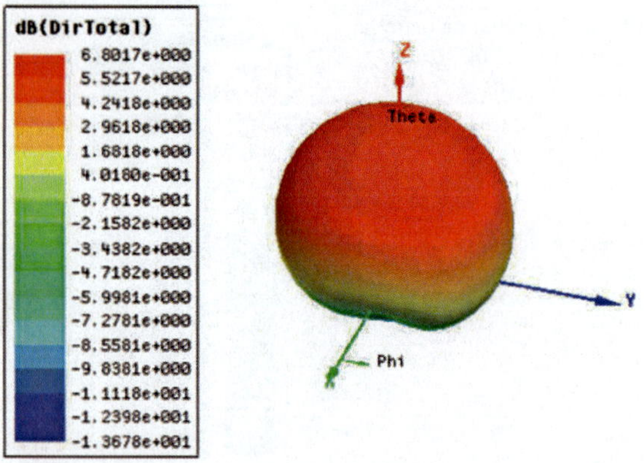

Fig. 8 3D radiation pattern for directivity (dB). Total directivity = 6.8017 dB

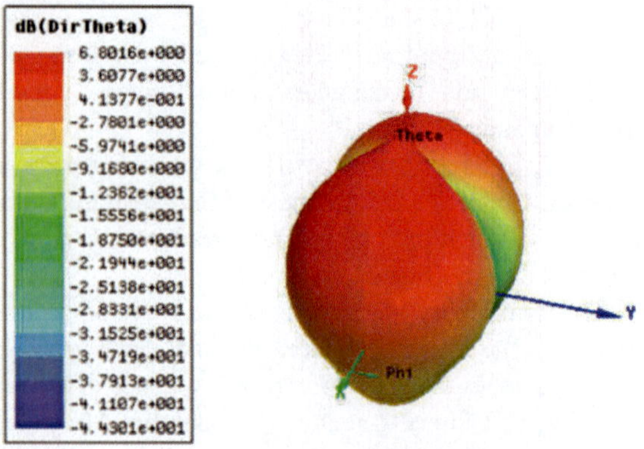

Fig. 9 3D radiation pattern for directivity theta (dB). Directivity theta = 6.8016 dB

Fig. 10 3D radiation pattern for directivity phi (dB). Directivity phi = 6.4448 dB

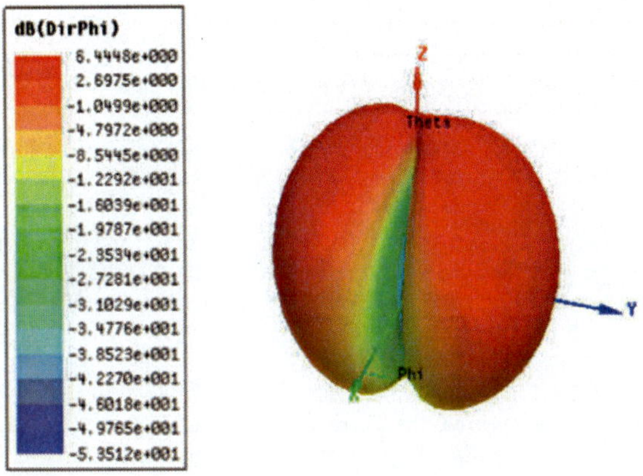

Fig. 11 3D radiation pattern for axial ratio. Axial ratio = 1.7669

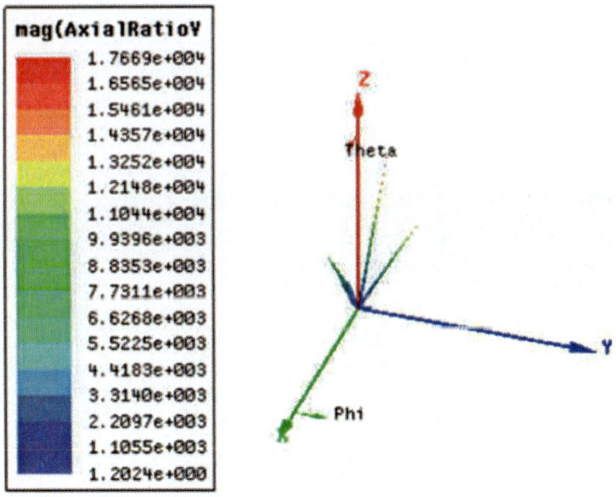

4 Conclusion

The rectangular shape of the patch eliminates feed radiation. The microstrip line feed chosen provides a planar structure and eliminates cross-polarization. The designed antenna has a sufficient bandwidth of 20 MHz and a return loss of −19.1996 dB, −10.4272 dB, −11.5229 dB, 12.5694 dB, −15.4651 dB at 2.4 GHz, 4.7 GHz, 6 GHz, 7.0 GHz, and 8.7 GHz, respectively. The value of VSWR achieved is 1.2463, 1.8614, 1.7225, 1.6152, 1.4055 at 2.4 GHz, 4.7 GHz, 6 GHz, 7.0 GHz, and 8.7 GHz, respectively, thus allowing maximum power to transmit to the antenna. The antenna achieves a total gain of 3.12 dB. A high gain makes the antenna directional with a narrow beam-width to be observed. The total directivity of the antenna is 6.8016 dB which allows the antenna to radiate a great amount of power in a specific direction thereby allowing increased performance. The axial ratio of the antenna is 1.7669 which indicates circular polarization.

References

1. Constantine A, Balanis CA (1997) Antenna theory: analysis and design, 2nd edn. Wiley, New York
2. Beal V 802.11 IEEE wireless LAN standards
3. Mitchell B (2019) Wireless local area networking explained. https://www.lifewire.com/wlan-816565
4. Balur NJ, Kulkarni SA (2012) Design of multiband microstrip antenna. Int J Adv Res Comput Sci Electron Eng (IJARCSEE) 1(10). ISSN: 2277-9043
5. Pranathi GV, Rani D, Satyanarayana GV, Rao GT (2015) Patch antenna parameters variation with ground plane dimensions. Int J Adv Res Electr Electron Instr Eng 4(8). ISSN (PRINT): 2320-3765 ISSN (ONLINE): 2278-8875
6. Mitha TH, Pour MZA (2017) Effect of ground plane size on a rectangular microstrip patch antenna based on the TE modes supported by perfect magnetic conducting surface. In: 2017 IEEE international symposium on antennas and propagation & USNC/URSI national radio science meeting, 9–14 Jul 2017
7. Balanis CA (2005) Antenna theory: analysis design, 3rd edn. Wiley, New York. ISBN 0-471-66782-X
8. Roshan P, Leary J 802.11 wireless LAN fundamentals
9. Chapter 2. Microstrip patch antenna parameters and experimental setup (simulation, fabrication and measurement)

Power Optimization by Detection and Monitoring of Sensor Event in Smart Home

A. Ganesan, K. Sujatha, N. P. G. Bhavani, V. Srividhya, and Su-Qun Cao

Abstract Sensor event monitoring and detection facilitate to recognize the activities of the human lives in a smart home daily. The major aim of this technology is to attain optimal performance in activity recognition. The symbol of the daily activity recognition involves the usage of feature inputs which have a noteworthy consequence on the output. However, commonly used representations of features dependent on daily activity have limited performance on the recognition activity. Based on the dynamic nature of the sensor events caused by daily activities, this paper presents a statistical processing approach for time series of sensor events. First time, the statistical values are extracted from sensor events dependent on time series. Subsequently, different categories of statistic formulae are proposed to solve daily activity features. To evaluate the proposed approach, two distinct datasets are adopted for activity recognition based on artificial neural network (ANN), fuzzy logic (FL), and hybrid neuro-fuzzy logic (HNFL). The experimental results reveal that the proposed HNFL approach provides power optimization.

Keywords Smart home · Sensor event · Hybrid neuro-fuzzy logic · Artificial neural networks · Fuzzy logic

A. Ganesan
Department of EEE, RRASE Engineering College, Chennai, India

K. Sujatha (✉)
Department of EEE, Dr. MGR Educational & Research Institute, Chennai, India
e-mail: drksujatha23@gmail.com

N. P. G. Bhavani · V. Srividhya
Department of EEE, Meenakshi College of Engineering, Chennai, India

S.-Q. Cao
Faculty of Electronic Information Engineering, Huaiyin Institute of Technology, Huaian City, Jiangsu Prov., China

G. Ranganathan et al. (eds.), *Inventive Communication and Computational Technologies*, Lecture Notes in Networks and Systems 145,
https://doi.org/10.1007/978-981-15-7345-3_56

1 Introduction

The way of living day by day in this world gets complicated. People are busy with their works to keep up their living standards. The lifestyle is entirely automated in such a way that where the machine replaces man as discussed by Alshehri AA, et al [5]. Mateen A, et al. [6] state that the man started re-creating his world in terms of technical advancement, modification of environmental infrastructure, agricultural revolutions, there arises a necessity to monitor the events daily using some sensing devices. These innovations created by man have not only changed and affected the world but also the lifestyle of himself has been affected and changed which leads to innovation of new devices that are smart and automated using sensors as rightly pointed out by Samanta A, et al., [7]. Statistical analysis as reported by Frezzetti A., et al., [8] indicate that there are various kinds of new sensors that are being used for automation of industrial, commercial, and residential buildings and all these have been discovered in the last decade.

The initial idea of research work started after understanding the real-time advancements. Rajaput BV, et al [9] has mentioned the technology advancements is increased not only owing to modernization and cope up with the technology. Akhtar F, et al [10] states that, to elevate the standards in terms of technology, a new method of sensor integration has been designed to discover the priority-based service.

Ye F, et al [11], have identified that the sensor power supply network is widely used in health and the environment monitoring. These include a variety of process contacts to track low, remote locations. Most likely, sensors and nodes are responsible for the four main functions of such networks: data collection, broadcasting, data retrieval, and network psychology. Most importantly, they control their own lives and performance, including memory usage, central processing unit (CPU) power, to use their resources effectively.

In addition to collecting energy, one of the main challenges is to reduce energy consumption in sensor networks lifetime. Components of the sensor are integrated into a single node or a plurality of plates and packaging a few cubic inches. The wireless sensor network of information shares the sharing of thousands of nodes communicating over a coordinated wireless channel.

Users can get information about their interest in a base station or wireless sensor network node. The wireless sensor network acts as an interface between the base station user networks. Wireless sensor networks can be thought of as a sensor that can connect to a Web network, enabling the ability to share information globally.

Compared to other systems, three common design challenges can significantly affect the entire network connectivity and productivity.

- Control protocols use network protocols to reduce data packets
- Improved spatial selection of nodes in the right place
- Node/source targeting from source node. Nodes network by performing a routing algorithm for efficiently transmitting data.

The increase in average lifespan and health costs in many developed nations are catalysts to innovation in health care. These factors along with the advancement in miniaturization of electronic devices, sensing, battery, and wireless communication technologies have led to the development of wireless sensor networks (WSNs) integrated with IoT. WSNs consist of smart miniaturized devices (sensors) that can sense, process, and store the data in the cloud using IoT, which in turn helps us to have a strong database about each home appliance and the devices, that in turn used to communicate through the proper telemedicine architecture [1] based on the high priority [2] and on-demand [3] of the emergency data. They are designed in such a way that they can be worn or implanted [4]; it monitors the signal and transmits these to specialized central servers, without much interference with the other electrical appliances. One of the most promising approaches in building the wearable home automation and monitoring systems utilizes emerging wireless area networks (WANs) integrated with IoT.

2 Contribution of This Work

TThe contribution of this work is to reduce the power consumption in wireless sensor networks (WSNs) for home automation systems that have generated tremendous efforts in recent years. In any case, in a large portion of these investigations, sensor information handling assignments, for example, health monitoring of home appliances, power optimization, making and emergency reaction, the message is sent by the remote server. Launched and handed over large volumes of data sensors use many more communication resources, bringing remote server issues and delaying the decision time notification time. This work uses the hybrid technique (Direct Power management method for base station and Self-Executing Path Resource Allocation method for server section) to reduce the utilization of power consumption based on the embedded systems. In this work, the ARM Cortex-A72 controller-based prototype is implemented with a smart gateway. Gateway is an interconnection and service management platform, especially in the WSN home automation system.

3 The Motivation for Developing a Smart Home

The motivation behind the development of monitoring sensor events in a smart home is as follows:

- The main motivation for this proposed research work is to find out the optimal solution with the help of a newly developed algorithm. The proposed work motivates to improve the performance of execution time as compared with existing and also reduces the power utilization and end–end delay response.

- An electronic gadget is a convenient way to exchange information about home appliances and between different service providers. The home automation contains the confidential information required to protect the smart home with privacy and prevent unauthorized access.
- The proposed work can be used by multiple home appliances and a huge amount of data will be shared between the various entities of the system. So, this prompted us to propose such a bandwidth-efficient solution that would reduce bandwidth requirements and increase data transfer rates in building a smart home.

4 Objectives to Construct a Smart Home

The major objectives are as follows:

- In this research work, mainly reduces the power consumption in wireless sensor network using embedded system. In this system, the unused sensor nodes will be automatically disabled to optimize power consumption. Here, the development of smart home by monitoring the home appliances is used with various sensor nodes, like air conditioner (AC), door locking system (DLC), audiovisual (AV) systems, lighting, washing machines, dishwashers, electric chimneys, liquefied petroleum gas (LPG) regulators, etc.
- The main concern of this research work to reduce power consumption and power management for a total of WSN node in such a way as to increase the lifetime.

5 Research Background

Some researchers state that the wireless sensor network (WSN) to discuss the range of power sources at different sensor nodes and greatly affect the longterm operation of the network. Therefore, the current research objectives are to design worm energy storage algorithms to improve the survival time of the network. In this paper, introduces and enhances the technology based on the artificial bee colony (ABC) method in the network simulation and synthesis. The ABC algorithm can improve the internal dynamics of the sensor's clusters of roads and terminals.

Zuhairy et al. discuss a large number of applications for cluster functionality in wireless sensor networks. Multiple compilation approaches are difficult to do, but ultimately there are unbalanced asymmetric clusters and network overload. Two of these suggested energy-efficient balanced network load balancing method. In this account, the parameters in the cluster between the remaining clusters are taken as energy, low within the cluster, low between clusters, and communication distance.

Al-Aubidy et al. discussed the home monitoring and guiding system using WSN technology. The embedded microcontroller is used to collect the data from sensors and find out the condition of the home appliances. Accordingly, the residents of the house can use the Internet to obtain the details to their respective mobile, online through the cloud service. Such systems must handle the ability of a central server to generate real-time signals from sensors and transmit measurement signals over the Internet.

SudipMisra et al. discuss power management techniques at various levels. The main purpose of WSN is to use these sensor network devices to greatly limit battery life and prevent connection degradation through aggressive energy management and it is impossible to replace batteries with tens of thousands of sensors. So we need to follow the energy conservation technique to save energy.

6 Research Gap in Smart Home Development

- All the nodes are there in an active state, so it consumes more power.
- Reducing the number of relay nodes placed on the WSNs maintains global connectivity over a limited transmission range.

WSN has been used in many emerging applications with its latest technology. But this network is vulnerable to various exposures, reducing the effectiveness of the network. It provides a thorough education of existing computer work with connected network security to refer to an existing data communications document and proposes all the problems of the proposed system to read all documents collectively.

7 Proposed Method

Wireless sensor networks (WSNs) are crucial systems because they are widely used in a variety of applications, such as hospitality, military, industrial processes, and wireless sensor networks (WSN). Data is sent via, a wireless sensor network node. To build these networks, the nodes capture information about the environment and communicate with the end-users. Each node of the WSN has a microprocessor. Memory type, RF threshold is the various sensors and aftershocks, electricity (EN, batteries, solar) cells. Dense deployment in the WSN environment often characterizes us. Control resources, processing power, storage, and most importantly, energy equipment components. This is why batteries are usually powered by them. Due to the rechargeable battery in the sensor network, it is sometimes impossible to assign a position body area network (BAN) to a body-like area. Despite the quality performance, the quality of the service depends on many conditions. Certainly, an essential role is played by the sensors in the management of power consumption using the core performance scale in a lifetime network. In contrast, the purpose of a wireless sensor network (WSN) is for users to access information of interest from data collected by spatially distributed sensors. In most applications, users only need to summarize some of the features for this distributed data. An example of a temperature sensor network is a specific trigger in the case of an average temperature alarm network

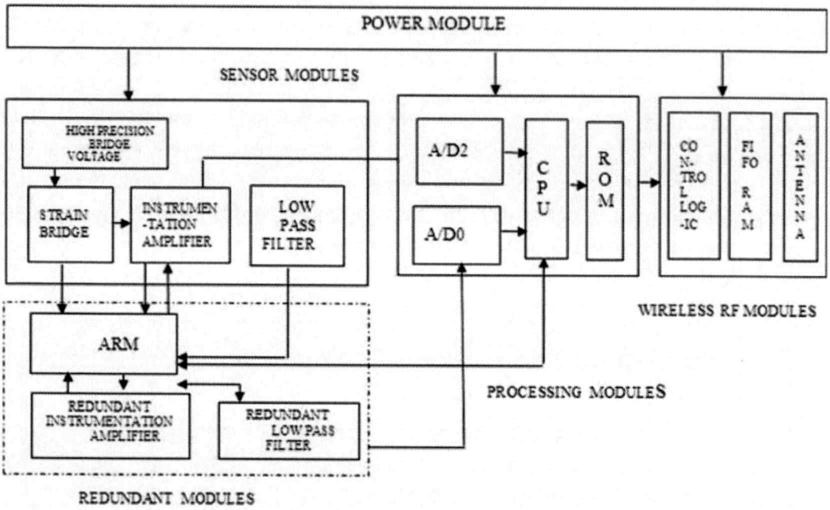

Fig. 1 Proposed block diagram for hardware model in a WSN system

or event location. All data related to the central collector node has been communicated in the end-to-end information flow model. At the same time, despite serious limitations on energy, memory, and bandwidth, worms are a very capable solution where cleaning and latency are limited. An alternative is the network calculation of the solution.

The main goal of clustering integration is to combine different clustering decisions and achieve the best accuracy for any individual cluster. Examples of known actors' methods, there is a feature-based approach that shows that clustering data (i.e., cluster labels) changes the cluster ensemble problem. The direct approach is to find the final partition through the base clustering results relabeling. The schematic representation is shown in Fig. 1.

The reliability of wireless sensor networks (WSN) is a very important application for its success in the engineering field. In normal research, when there is a node failure, this is usually discarded and rebuilt to ensure the normal functioning of the wireless sensor network. Using the appropriate WSN reconfiguration mode, although the sensor network can be adjusted, this can result in additional maintenance costs, sometimes reducing the operation of more networks. There is a re-organization in the sensor network that cannot, of course, reduce the efficiency of the entire wireless sensor network. One way to create a system is to ensure reliable and low-cost operation of WSNs wireless sensor nozzle with self-healing capability based on rechargeable hardware using the ARM controller.

8 Results and Discussion

The performance analysis of energy consumption is shown in Fig. 2. It is the proportion of the standard measure of a control message that is dealt with by the node and the ratio of information packets that is gotten by the sinks. There are many routing plans accessible every one with an unmistakable procedure and if a network is critical, it is required to have more significant measures of trade of control representing messages to be equipped for finding and making more routes. Each area in the progressive multicast information system reports to the sink along the multicast chain of its importance and any information in the middle of its locale can total from ordinary details as in Table 1.

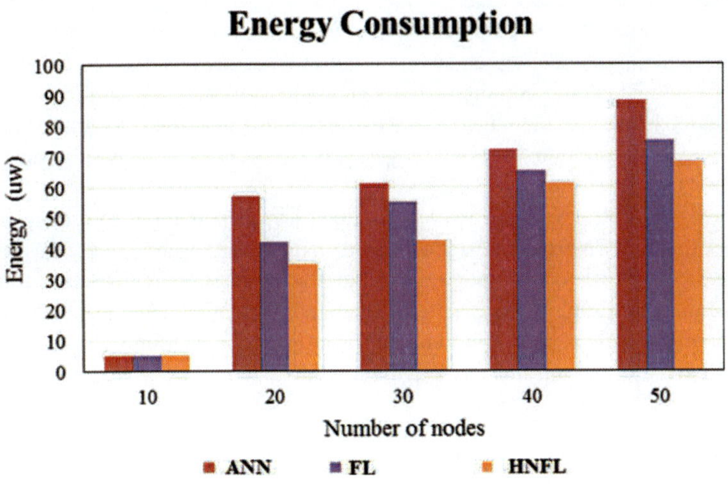

Fig. 2 Evaluation of energy consumption

Table 1 Energy consumption

Number of nodes	Power consumption by using ANN (μW)	Power consumption by using FL (μW)	Power consumption by using HNFL (μW)
10	0	0	0
20	57	42	35
30	61	55	42.3
40	72	65	61
50	88	75	68

9 Conclusion

An embedded system is a specialized computer that is constrained to resource-based sensing and control of the environment. Embedded systems usually consist of hardware and software. The most used hardware materials are processors, peripheral communication devices, actuators, sensors, power supplies, and memory storage. The application-specific algorithms, device drivers, and operating systems are typically used in the software section. Normally need a standard protocol to communicate the particular type of embedded system, for example, nodes in sensor networks are the specialized embedded systems. Sensor nodes with wireless communication capabilities can form wireless sensor networks (WSN).

Normally, two types of wireless networks are used, such as personal area networks (PAN) and wireless sensor networks (WSN). The WSN can contain hundreds or even thousands of sensor nodes. Personal area networks usually require measurement and minimization devices are implemented in small numbers. PAN devices are designed for Wi-Fi and Bluetooth common-use technologies and standard protocols, for use such as Web browsing, file transfer application, audio, and video streaming application. Today's research on WSN focuses to generate large-scale network systems of electricity using very specialized algorithms.

Therefore, in this research work, introduce a hybrid method to reduce power consumption is WSN used in residents based on the sensor event monitoring system. To reduce the power consumption in wireless sensor networks (WSNs) for home automation systems has generated tremendous efforts in recent years. In any case, in a large portion of these investigations, sensor information handling assignments, for example, health decision making and emergency reaction message are sent by the remote server.

References

1. Al-Aubidy K, Mutairi AW, Derbas A (2017) Real-time healthcare monitoring system using wireless sensor network. Int J Digit Signals Smart Syst 1(1):26–42
2. Efrem CN, Panagopoulos AD (2018) Joint power allocation and transmission time selection in duty-cycled wireless sensor networks. In: Global information infrastructure and networking symposium (GIIS), Thessaloniki, Greece, pp 1–5
3. Alagoz F, Ozger M, Akan OB (2018) Clustering in multi-channel cognitive radio ad hoc and sensor networks. IEEE Commun Mag 56(4):156–162
4. Lin C, Zhouy Y, Daiz H, Dengx J, Wu G (2018) MPF: prolonging network lifetime of wireless rechargeable sensor networks by mixing partial charge and full charge. In: Fifteenth annual IEEE international conference on sensing, communication and networking (SECON), Hong Kong, pp 1–9
5. Alshehri AA, Martins CH, Akyildiz IF (2018) Wireless FracBot (sensor) nodes: performance evaluation of inductively coupled near field communication (NFC). In: IEEE sensor application symposium (SAS), Seoul, pp 1–6

6. Mateen A, Sehar M, Abbas K, Akbar MA (2017) Comparative analysis of wireless sensor networks with wireless multimedia sensor networks. In: IEEE international conference on power, control, signals and instrumentation engineering (ICPCSI), Chennai, pp 80–83
7. Samanta A, Misr S, Bera S (2015) Link-quality-aware resource allocation with load balance in wireless body area networks. IEEE Syst J 12(1):74–81
8. Frezzetti A, Manfredi S (2015) A two-layer controller scheme for efficient signal reconstruction and lifetime elongation in wireless sensor networks. IEEE Trans Commun 16(7):2172–2179
9. Rajaput BV, Aparna P (2016) Efficient coverage and secured data transfer with load balance and connectivity preservation in wireless sensor networks. In: IEEE international conference on recent trends in electronics information and communication technology (RTEICT), Bangalore, pp 508–512
10. Akhtar F, Rehmani MH (2017) Energy harvesting for self-sustainable wireless body area networks. IT Prof 19(2):32–40
11. Ye F, Hu RQ, Qian Y (2018) Smart service-aware wireless mixed-area networks. IEEE Trans Commun 33(1):84–91

Multi-keyword Search for Multiple Data Owners Over Encrypted CloudData

Jonnakuti Lakshmi Thirusha, Yalamanchili Kavya Nandini, and P. S. G. Aruna Sri

Abstract Cloud computing is a developing IT framework model that composes a tremendous asset of figuring, stockpiling, applications controlling and arranging, and getting to online applications with high productivity and insignificant overhead. Cloud information proprietors pick to source reports in a scrambled kind for security preservation purposes. The advancement of efficient and dependable figure content pursuit procedures is fundamental. This will make it significantly harder to plan ciphertext search which conspires that can give conservative and proficient online information recuperation on the mammoth volume of scrambled data. The proposed progressive methodology structures bunches of reports are dependent on the base edge of pertinence and in this way partition the subsequent groups into sub-bunches until the farthest point on the most extreme. The proposed technique has a preferred position over the customary strategy in the security rank and significance of the recovered records. So, we implemented an email authentication, and the trapdoor and the secret key can be generated. Multiple data owners can search for multiple keywords on the cloud server for retrieving data.

Keywords Cloud computing · Ciphertext search · Ranked search · Multi-keyword search · Authentication

1 Introduction

Cloud computing is that long unbelievable vision of registering as a utility, any place cloud clients remotely store their data into the cloud hence on relish the on-request excellent applications and administrations from a common pool of configurable figuring assets . Its pleasant adaptability and monetary investment funds square

J. L. Thirusha (✉) · Y. K. Nandini · P. S. G. Aruna Sri
Department of Electronics and Computer Engineering, KLEF, Vaddeswaram, Guntur, India
e-mail: thirushajonnakuti7@gmail.com

P. S. G. Aruna Sri
e-mail: arunasri_2012@kluniversity.in

G. Ranganathan et al. (eds.), *Inventive Communication and Computational Technologies*, Lecture Notes in Networks and Systems 145,
https://doi.org/10.1007/978-981-15-7345-3_57

measure persuading every individual and undertaking to source their local propelled data of the executive's framework into the cloud. To defend the security of data and contradict unsought gets to inside the cloud and on the far side it, delicate data, for instance, messages, individual wellbeing records, photo collections, charge reports, etc., might find a good place by data house proprietors before redistributing to the modern open cloud, this is that as it may, obsoletes the ordinary data use administration upheld plaintext catchphrase search. The immaterial goals of downloading all the data and unscrambling locally are unmistakably unreasonable, on account of a large amount of data measure cost in cloud-scale frameworks, pictures conjointly contain supportive and fundamental information, and in this way, arranged framework conjointly gives picture labeling in MRSE subject. Besides, aside from disposing of the local stockpiling the executives, putting away data into the cloud does not fill any need, except if they will be essentially looked and used. Thus, investigating protection safeguarding and compelling pursuit administration over scrambled cloud data is of decent significance. Thinking about the most likely tremendous assortment of on-request data clients and a huge amount of re-appropriated data archives inside the cloud, this drawback is particularly troublesome because it is uncommonly irksome to satisfy conjointly the necessities of execution, framework convenience, and quantify ability. Record positioning is accommodated speedy inquiry, anyway the needs of all the data archives are solid and same, so the cloud administration provider and outsider stay uninformed of the vital reports, in this manner, keeping up the security of data. Hierarchic pursuit additionally can richly kill, hold, and organize traffic by causing back exclusively the chief pertinent data that is extremely interesting inside the "pay-as-you-use" cloud worldview.

2 Literature Survey

2.1 Privacy-Preserving Multi-keyword Fuzzy Search Over Encrypted Data in the Cloud

A captivating technique for the viable use of encoded data redistributed to the cloud may be allowing the keyword search straightforward over scrambled data. Existing structures deliver genuine watchword pursuit that does not survive a catchphrase writing frame malfunction or a single catchphrase fluffy investigation errors at a bound point. In this paper, we will generally propose a one-size-fits-all multi-catchphrase fluffy inquiry topic which is the delicate strategy in the field. The anticipated subject performs fluffy teamwork via algorithmic design, as opposed to extending the document record. This also disposes of the need for a predefined vocabulary and embraces various watchwords effectively. To the best of our data, this can be the essential multi-catchphrase fluffy inquiry over encoded data.

2.2 *Secure Ranked Multi-keyword Search for Multiple Data Owners in Cloud Computing*

With the presence of appropriated figuring, it ends up being coherently well known for information house owners to source their information to open cloud servers while permitting information customers to recuperate this information. For assurance considerations, secure interests over mixed cloud information convinced various asks about under the one-owner model. In any case, most cloud servers in apply do not simply serve one owner; rather, various house proprietors were urged to share the focal points that cloud servers have brought. Right now, we will for the most part propose designs in a very multi-proprietor model to impact safe stratified multi-watchword searches. We should generally effectively set up a completely one-size-fits-all safe interest show to switch cloud servers and lead secure requests while not understanding the particular information of every watchword and trapdoors. To rank the inquiry things and secure the assurance of affiliation scores among watchwords and records, we will for the most part propose a one-size-fits-all added substance request and protection.

3 Proposed Methodology

Design of vector space is used, and each record is envisioned by a vector, which derives each chronicle and is consistently seen as some degree in an exceedingly high-dimensional space. As a result of the relationship between completely unforeseen chronicles, all of the records are normally segregated into various characterizations. At the end of the day, the focuses whose separation square measure short inside the high-dimensional space are frequently characterized into a chosen classification. Contrasting all the documents inside the dataset, the sum of client-focused reports is very small. Due to the narrow reach of the required records, a class of options will progressively be partitioned into several sub-classifications. Rather than manipulate the traditional theory of grouping search, a backtracking algorithmic system is made to look at the objective records. The cloud server will first inquire about the classes and sub-classify the desired base. From the point, the cloud server can select from the base which needed sub-classification of the appropriate k records. The value of k is selected by the client and forwarded to the cloud server. We suggested an authentication-based search to generate the trapdoor and secret key of the data provided to the client. We plan a test technique to improve the rank of privacy. In the pursuit part, the cloud server can first figure the connectedness score among questions and bunch focuses on the essential level to pick the nearest group. The case that the smallest number of required documents is set by the client, the cloud server will return to the parent group of the smallest group, and the sibling groups of the smallest group can be viewed.

This technique will be iterated till the number of wanted records is glad or the premise is come to because of the exceptional hunt strategies, the rankings of records among their list items square measure entirely unexpected with the rankings got from old arrangement search. Protection rank is improved in this way. Some have been handed over a slice of the higher than job. For extra improvement, we tend to conjointly build an undeniable tree structure upon the gradable cluster strategy to check the honesty of the inquiry which leads to this paper. This real tree structure in the principle exploits the Merkle hash tree and cryptanalytic mark. Each report furthermore is because of the agent of the archive (Figs. 1 and 2).

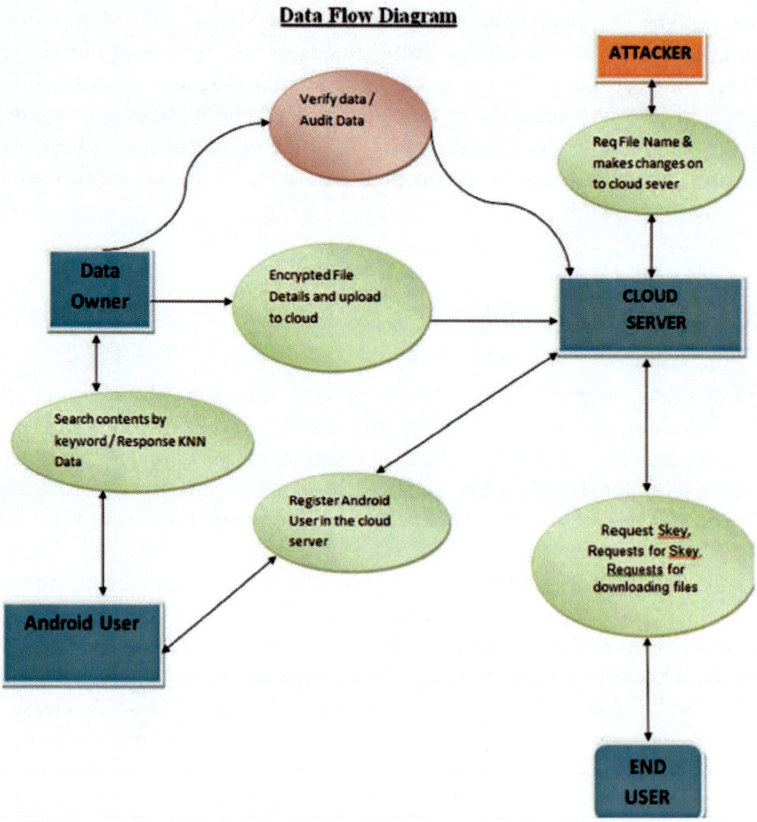

Fig. 1 Data flow diagram for multi-keyword search

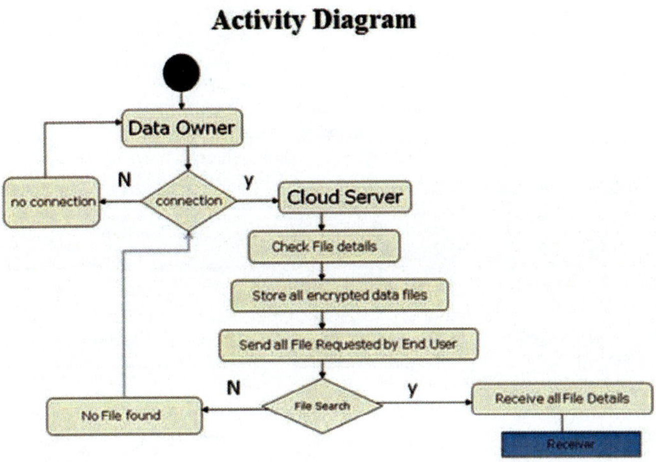

Fig. 2 Diagram for flow from one activity to another activity

4 Architecture and Algorithm

MRSE architecture is where the data owner generates the encrypted index based on the dictionary, random numbers, and secret key, the data user submits a question to the cloud server to get desired documents, and the cloud server returns the target documents to the data user (Figs. 3 and 4).

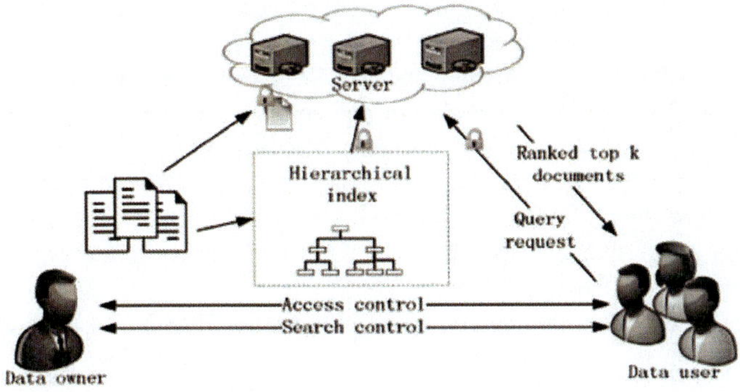

Fig. 3 Architecture of ciphertext search

Fig. 4 Algorithm for index algorithm

```
Algorithm Index
input the secret key {sk,k} and data set D
extract a dictionary Dw from D
for each document in D
      construct a document vector dv_i
invoke QHC to build hierarchical index{DC,CCV,DV}
for every element V in DV and CCV
      extend its dimension from n to (n+u+1)
      if the i^th of S is 0
          V'_i=V''_i=V_i
      else
          V'_i is set to a random number
          V''_i=V_i-V'_i
encrypt index as {M_1^T V', M_2^T V'',DC}
```

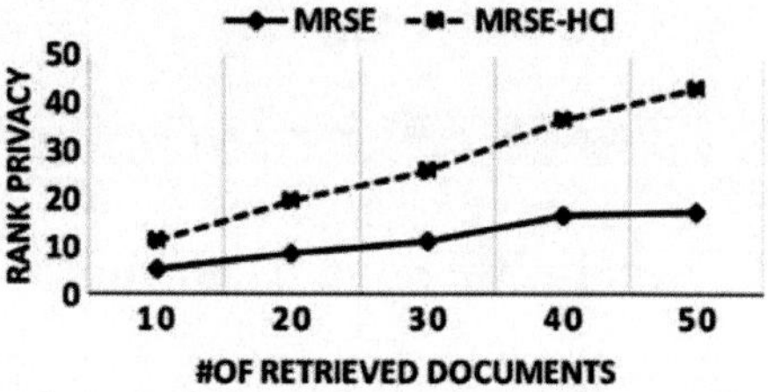

Fig. 5 Rank privacy

5 Results and Discussion

A multi-keyword search mechanism is performed by generating the trapdoor and secret key for our email address. The architecture used in this paper is MRSE architecture, where the documents and keywords are encrypted and stored in a key type. As a result, the data user, data owner, and the cloud server are the three main entities. To access the data, the data user should register first, and the user will obtain the login credentials after the user's registration. The user will get the owner's permission to view or update the data. After receiving permission from the cloud provider, the owner must grant the user permission. Using the trapdoor and the secret key provided by email, the user can access the data. The number of keywords searched in the data is given a rank.

In Fig. 5, it describes the rank of privacy between MRSE-HCI and MRSE.

6 Conclusion

We explored ciphertext search inside the circumstance of distributed storage, we will in general investigate the matter of keeping up the etymology connection between very surprising plain reports over the associated encoded records and give the structure system to fortify the presence of the semantic hunt. We will in general furthermore propose the configuration of MRSE-HCI to adjust to the needs of information blast, online data recovery, and etymology search. At the same time, an obvious device is also intended to ensure that those listed things are accurate and fulfilled. What is more, the inquiry frequency and safety can usually be dissected under two across the board risk models. Partner in the nursing exploratory stage is working to direct the strength, accuracy, and safety of the rank of the search. The trial result shows that the arranged model does not undo the multi-catchphrase search downside placed exclusively appropriately, but it also enhances the search power of associate nursing, rank.

References

1. Cao N, Wang C, Li M, Ren K, Lou W (2011) Privacy-preserving multi-keyword ranked search over encrypted cloud data. In: Proceedings of IEEE INFOCOM, 829–837
2. Vaquero LM, Rodero-Merino L, Caceres J, Lindner M (2009) A break in the clouds: towards a cloud definition. ACM SIGCOMM Comput Commun Rev 39(1):50–55
3. Cao N, Yu S, Yang Z, Lou W, Hou Y (2012) LT codes-based secure and reliable cloud storage service. In: Proceedings of IEEE INFOCOM, 693–701
4. Kamara S, Lauter K (2010) Cryptographic cloud storage. In: Proceedings of the 14th international conference on financial cryptograpy and data security
5. Singhal A (2001) Modern information retrieval: a brief overview. IEEE Data Eng Bull 24(4):35–43
6. Witten IH, Moffat A, Bell TC (1999) Managing gigabytes: compressing and indexing documents and images. Morgan Kaufmann Publishing, San Francisco, CA
7. Song D, Wagner D, Perrig A (2000) Practical techniques for searches on encrypted data. In: Proceedings of the IEEE symposium on security and privacy
8. Goh E-J (2003) Secure indexes. Cryptology. ePrint Arch http://eprint.iacr.org/2003/216
9. Chang Y-C, Mitzenmacher M (2005) Privacy preserving keyword searches on remote encrypted data. In: Proceedings of the third international conference on applied cryptography and network security
10. Curtmola R, Garay JA, Kamara S, Ostrovsky R (2006) Searchable symmetric encryption: improved definitions and efficient constructions. In: Proceedings of the 13th conference on computer and communications security (CCS'06)

Analytical Modeling of Real-World Social Network Parameters

Monika and Veenu Mangat

Abstract A social network is an interactive platform for massive communication. While analyzing a social network, its pictorial representation as a graph is required to explore and visualize it. Graph theory concepts are applied to social networks which are helpful in numerical formulations also. Social network graph theory is applicable to formulate and study various statistical parameters that can help in analyzing the social patterns of these networks. Various Python implementation commands from importing data to the computation of data for network statistics like the clustering coefficient and the degree distribution of nodes are described in this paper. These network statistics are significant as they can be used to unveil differences in the pattern of interconnections in a social network.

Keywords Social network analysis · Visualization · Networkx

1 Introduction

Social networks are considered as the backbone for sharing and communicating information among a group of various individuals. Communication between them can be represented with the help of a graph. A graph is defined as a set of nodes and edges connecting these nodes [1]. Mathematically, it can be depicted as $G = (V, E)$ where V is the set of vertices and E is the set of links between them. While analyzing a social network, the identification of how nodes are distributed across networks is an essential component. To conduct social network analysis, visualization of the network plays a key role. The main requisite of social network analysis is its visualization which can be defined as the process of creating a pictorial representation of the connections between the users of a social network. It can be beneficial in various ways [2]:

Monika (✉) · V. Mangat
UIET, Panjab University, Chandigarh, India
e-mail: monikahsp@gmail.com

G. Ranganathan et al. (eds.), *Inventive Communication and Computational Technologies*, Lecture Notes in Networks and Systems 145,
https://doi.org/10.1007/978-981-15-7345-3_58

- It helps to identify the pattern of social connection
- Results can be communicated by researchers in a more effective way
- It helps in investigating information dispersion pattern.

Social network analysis and visualization have provided researchers a perceptive view of networks. Graphs showing nodes and their interactions can be used as a framework to analyze various types of relations and processes in biological as well as information systems [3]. The fundamental attribute that describes a network is the node degree that is defined as the number of interactions it has with several other nodes in a network. It is significant due to several reasons: The influential node can be detected as the one which is having more links to other nodes. A node is considered as an information carrier if it is possessing a higher degree value. Most of the parameters that are necessary for the visualization of a network are based on the degree of nodes. The main contribution of this paper shows how an in-depth analysis of various statistical parameters can provide a better understanding of the interaction among individuals in a particular social network.

2 Dataset and Parameters Used

While analyzing a particular network, it is very important to visualize it. Two real-world datasets namely Zachary's karate club and American college football club [4] are used for this study. These datasets are undirected and unweighted, and each vertex is connected to some other vertices. This information is depicted in Fig. 1 for verification. We have used the .gml format of these datasets for analysis. In 1970s, Wayne Zachary examined interactions among the members of the karate club. It gets divided into two groups after some conflict between the instructor and the president of the group. This is the main reason which leads to the vast usage of this dataset

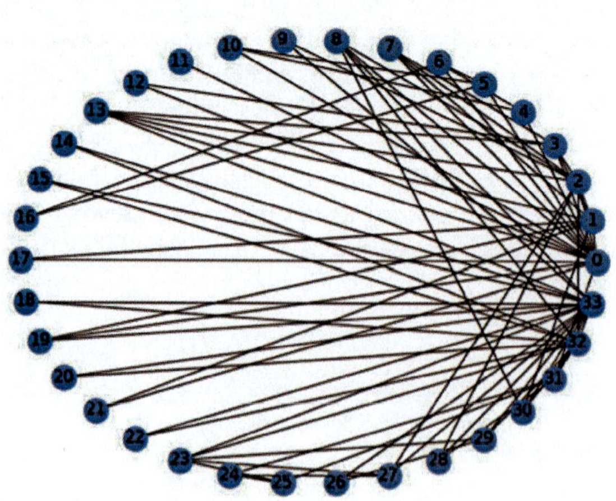

Fig. 1 Graphical representation of karate dataset

in finding multiple groups and further communities in a social network. American college football club contains the data of football teams and games between them for the season 2000. The vertices and links between them indicate to which division they belong. These datasets have been used widely for study in various types of community detection techniques [5]. So, we have chosen these datasets for analysis and visualization using the networkx package of Python. Analysis of these networks has been done based on parameters that are defined below:

No. of nodes: It represents the count of individuals represented as vertices in a network [6].

No. of edges: It represents the interactions between all the individuals [6].

Average degree: It is defined as the average number of links per node in a graph. The prerequisite for this computation is the evaluation of the degree of all nodes present in a graph. The degree of a particular vertex of a graph is the count of links occurring to it [7]. Among these values, the largest one is termed as the degree of the graph.

Density: It describes how connected a graph is. Its value can vary from 0 to 1. It depicts whether a network is sparse or dense. The density of a graph is a measure of how many ties between nodes exist compared to how many ties between nodes are possible [8].

Diameter: The diameter is an important aspect of a network. While considering the case of social network diameter indicates, in the worst case, how fast information is transferred to each individual in the network. It is computed using Eq. (1) as follows [9]:

$$\Delta(G) = \max_{x,y \in V} d(x, y) \tag{1}$$

where

$\Delta(G)$ denotes the diameter
V is the set of nodes
$d(x, y)$ is defined as the length of the longest shortest path from node x to y.

Clustering coefficient: It is a measure that depicts how close the neighboring vertices of a node are, so that they can form a complete subgraph. For a vertex i, clustering coefficient can be evaluated using Eq. (2) as follows [10]:

$$c_i = 2 * \frac{L_i}{k_i(k_i - 1)} \tag{2}$$

where

L_i represents the total number of edges existing among neighbors of vertex i
k_i represents the degree of vertex i.

Mathematically, c_i should be in the range [0, 1]. If $c_i = 0$, then no neighbor of vertex i is connected, whereas $c_i = 1$ shows all the neighboring nodes of vertex i are connected.

Degree distribution: In real-world networks, it becomes important to analyze whether all nodes possess the same degree of connections among them or is there any variation in it. There can be a case of a network having a larger number of nodes showing few connections and others having more connections. This parameter quantitatively shows variations in the number of links that various nodes of a network can have [11].

3 Implementation and Results

The first step is to import the networkx package [12] and import matplotlib function to visualize the network as a graph. Before making use of various functionalities of the networkx package, the dataset should be linked. Next, we have to check the dataset format. We are using Zachary's Karate dataset which is in .gml format.

Graphical view of the dataset along with statistical information can be shown using the following function in network package of python:

```
G = nx.read_gml('karate.gml' , label='id')
print (nx.info(G))
print (nx.is_directed(G))
nx.draw_circular(G,with_labels='true')
plt.show()
```

Figure 2 shows the graphical view of American football club dataset. It shows how various nodes are connected.

Statistical information of karate club dataset is also shown in (Fig. 3).

We can also print the degree of each node using the following function which can further be used for getting degree distribution of nodes.

```
print(nx.degree(G))
```

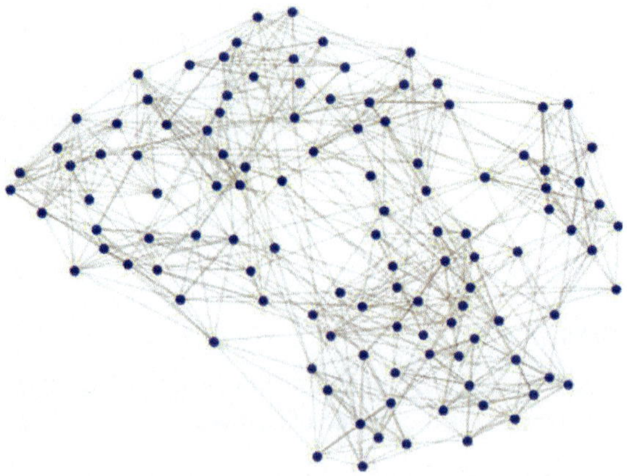

Fig. 2 Graphical view of American football club dataset

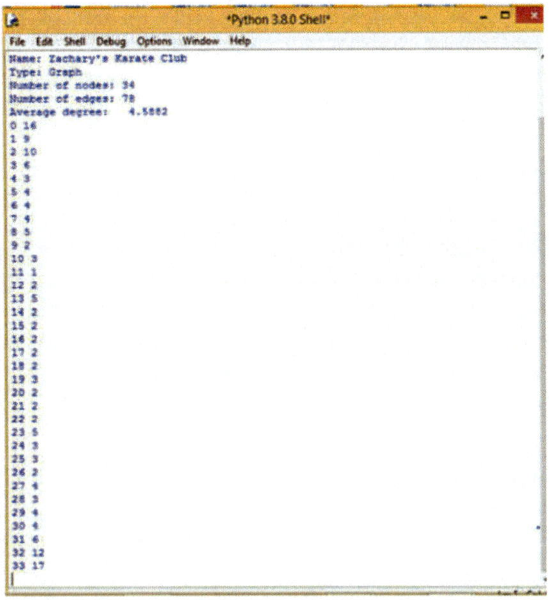

Fig. 3 Statistical information of karate club dataset

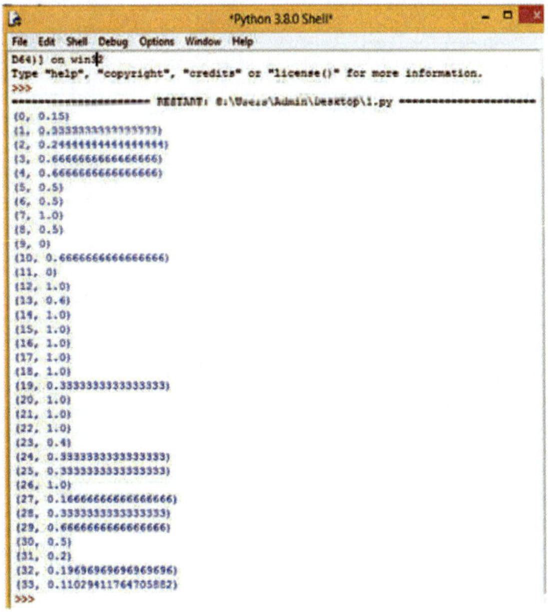

Fig. 4 Clustering coefficient of karate club dataset

To analyze whether a given social network is dense or sparse in terms of interactions among nodes and further to find the longest shortest path in a particular social network, following networkx functionality can be used:

```
print (nx.diameter(G))
print (nx.density(G))
```

For degree distribution, we can use the following function:

```
def plot_deg_dist(G)
```

The above function will plot the graph showing the count of occurrence of each degree. Most of the real-world networks follow power-law degree distribution which implies that there are very few nodes that have a very high degree and more nodes that have a very little degree. But it can be observed from Fig. 5 that the plotted line is not perfectly straight here which means this real-world network did not exactly follow power-law degree distribution. In Fig. 6, degree is not distributed uniformly across

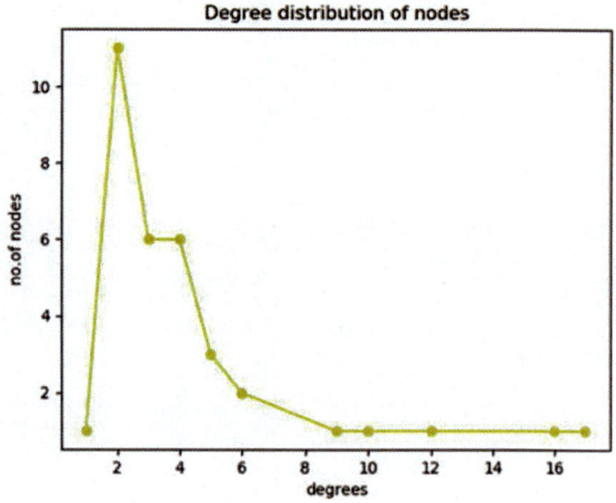

Fig. 5 Degree distribution of karate club network

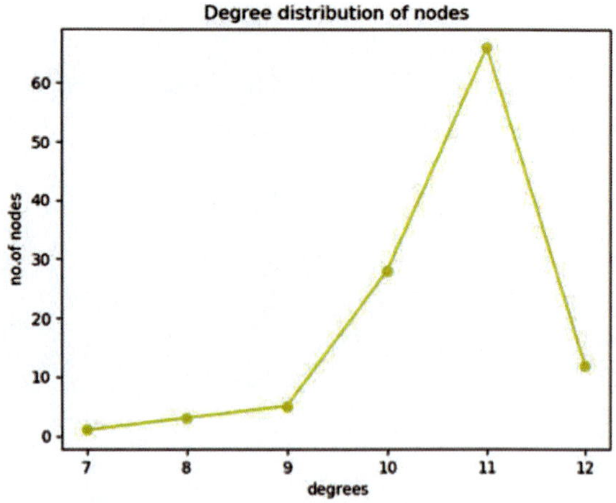

Fig. 6 Degree distribution of American football club network

Table 1 Parameter evaluation for datasets

Parameters used	Datasets used	
	Zachary's karate club	American football games club
No. of nodes	34	115
No. of edges	78	613
Average degree	4.58	10.66
Diameter	5	4
Density	0.13	0.09
Average clustering coefficient	0.57	0.40

the nodes of the American football club network. Nodes with a very high degree can also be considered as an information transport hub. The clustering coefficient of each node is calculated as shown in Fig. 4, and after getting this information, average clustering coefficient can be evaluated using networkx package of Python with the help of the following command:

```
print (nx.clustering(G))
print (nx.average_clustering(G))
```

The above-discussed parameters after evaluation are given in Table 1. Even though the American football club dataset is having a large number of nodes followed by more number of links as compared to Zachary's karate club dataset but the value of diameter, i.e., longest, shortest path is more in karate club dataset. It shows that the rate of information transmission among various nodes is higher in the American football club dataset. The average degree value of the American football club also justifies it as an average number of edges per node which is much higher than Zachary's karate club network. Figure 7 depicts an analytical view of the above statistical values. It also shows how the clustering coefficient value is related to the density parameter. Higher the value of the clustering coefficient, denser the graph will be.

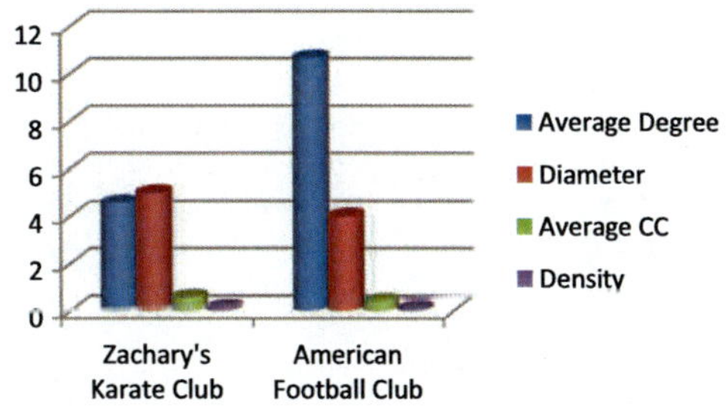

Fig. 7 Analytical view of evaluation parameters

4 Conclusion and Future Scope

This article provides a more precise statistical analysis and visualization of social network data showing interaction among various nodes of real-life datasets. Degree distribution for entire network nodes is analyzed which can be helpful in many domains of social network analysis such as community detection and rumor detection. For information about specific parameters, data is transfigured into graphs and analyzed using the networkx package of Python. Soon, we can expect to witness the development of more such packages that can help in visualizing large as well as dynamic social networks in a more refined way.

References

1. Farrang M, Nasr M (2017) A proposed algorithm to detect the largest community based on depth level. Int J Adv Netw Appl 9(2):3362–3375
2. Pfeffer J, Freeman L (2019) Social network visualization, methods of. In: Meyers RA (ed) Encyclopedia of complexity and systems science. Springer, New York, NY
3. Malathi A, Radha D (2016) Analysis and visualization of social media networks. In: International conference on computation system and information technology for sustainable solutions. IEEE, India, pp 58–63
4. http://www-personal.umich.edu/~mejn/netdata/
5. Azaouzi M, Rhouma D, Romdhane L (2019) Community detection in large-scale networks: state-of-the-art and future directions. Soc Netw Anal Min 9(23)
6. Camacho D, LLedot Á, Orgaz G, Pardo A, Cambria E The four dimensions of social network analysis: an overview of research methods, applications and software tools. https://arxiv.org/abs/2002.09485
7. Ayyappan G, Nalini C, Kumaravel A (2017) A study on SNA: measure average degree and average weighted degree of knowledge diffusion in GEPHI. Indian J Comput Sci Eng 7(6):230–237
8. Chojnacki S, Ciesielski K, Klopotek M (2010) Node degree distribution in affiliation graphs for social network density modeling. In: International conference on social informatics, Austria, pp 51–61
9. Takes F, Kosters W (2011) Determining the diameter of small world networks. In: Proceedings of 20th international ACM conference on information and knowledge management, UK, pp 1191–1196
10. Said A, Abbasi R, Maqbool O, Daud A (2018) CC-GA: a clustering coefficient based genetic algorithm for detecting communities in social networks. J Appl Soft Comput 63:59–70
11. Muller E, Peres R (2019) The effect of social networks structure on innovation performance: a review and directions for research. J Res Mark 36:3–19
12. https://pypi.org/project/networkx/

Performance Analysis of Speck Cipher Using Different Adder Architectures

Kalyani Palutla, Nagaraju Yesho, and K. Manjunathachari

Abstract This paper explores the performance of speck cipher based on different adder structures and is implemented incorporating two architectures namely folded (round based) and unfolded. Folded architecture reduces huge areas at the cost of an increment in delay which can be negligible when the area is primary metric. VLSI implementations show that Kogge–Stone tree has delay improvement in the range of (6.31–29.28)% at the cost of an increment in the area by (22.03–30.44)% when compared with the Sklansky tree, whereas Han-Carlson tree when compared with Sklansky tree, achieves an improvement in the delay for about (2.24–22.16)% at the cost of an increment in the area less than 5%.

Keywords Modulo 2^n adders · Speck cipher · Key expansion · Round function · Folded architecture · Unfolded architecture

1 Introduction

Internet of Things (IoT) with RFID tags, sensors and people, is an aisle for the next wave of innovations. Many of the devices are resource-constrained and not all of them are suitable enough to carry out conventional cryptographic algorithms. Hence, there is a requirement of flexible secure block ciphers for devices that not only include micro-controllers but also ASICs and FPGAs.

Recently, the U.S. National Security Agency (NSA) developed SIMON and SPECK families of lightweight block ciphers as an aid for securing applications

K. Palutla (✉) · K. Manjunathachari
GITAM deemed to be University, Hyderabad, Telangana, India
e-mail: kalyanipalutla1234@gmail.com

K. Manjunathachari
e-mail: Manjunathachari.kamsali@gitam.edu

N. Yesho
CDAC, Hyderabad, Telangana, India
e-mail: ynagaraju@cdac.in

G. Ranganathan et al. (eds.), *Inventive Communication and Computational Technologies*, Lecture Notes in Networks and Systems 145,
https://doi.org/10.1007/978-981-15-7345-3_59

in a very constrained environment where AES may not be suitable [1]. Several ASIC implementations of SIMON and SPECK have been reported in the literature. [2–8]. Speck has been optimized for its performance explicitly in software implementations. It is analyzed based on block size, key size, number of rounds and also structures employed to implement them. Speck is one of the lightweight block ciphers that support a range of block sizes from 32 to 128 bits utilizing allowable keys ranging from 64 to 256 bits. The number of rounds is dependent on the block and key sizes. The round function and key scheduling of this cipher are presented in [7]. Plain text is divided into two halves and in each round, it undergoes operations such as bitwise AND, bitwise XOR, modulo 2^n addition left and right circular shift of bits with a specified amount of rotation, respectively.

Unlike DESL, known as DES light-weighted utilizes an S-box in its round function thereby increases hardware implementation. Speck eliminates the use of S-boxes. The non-linearity in Speck cipher is due to modulo 2^n adder favours software performance over hardware and also one of the components in the worst-case delay path. This indicates that the speed of Speck cipher relies on modulo 2^n addition. Various adders are presented in [8] out of which Kogge–Stone, Sklansky and Han-Carlson tree-based adders [9–11] are chosen for the implementation of Speck family of ciphers. Implementation of Speck algorithm was presented in [8] is based on Sklansky tree. This results in a huge delay as this adder has an unbounded fan-out resulting in poor performance for large block sizes. The novelty of the modulo adder based on Kogge–Stone tree has an improvement in the delay at the cost of an increased area while Han-Carlson provides the best compromise between area and delay [12]. In this paper, ASIC implementation of SPECK family of ciphers is implemented in both folded and unfolded architectures utilizing the above-mentioned adders.

The rest of the paper is organized as follows: In Sect. 2, background required for the implementation of Speck cipher family is presented. Section 3 presents folded and unfolded architectures of Speck cipher. Comparisons of ciphers based on different adder structures are discussed in Sect. 4 followed by conclusion in Sect. 5.

2 Prior Knowledge

2.1 Speck Cipher

The parameters of SPECK cipher family are shown in Table 1. It is usually referred to as SPECK 2*n*/*mn*, where n is the word-size and m is the number of key words. For instance, SPECK 64/96 can be grasped as SPECK block cipher encrypts 64-bit plain text (i.e. two n-words) with key size of 96-bit which uniformly splits into three keywords (i.e. m), utilizing constants $\alpha = 8$ and $\beta = 3$ and resulting in a cipher-text after 27th round. The constants α and β differ only in the case of 32/64 block cipher (7, 2) while they remain the same for other block/key sizes (8, 3), respectively.

Table 1 Parameters of SPECK family [8]

Block size $2n$	Key size mn	Word-size n	Number of key words m	Rounds r	Constant α	Constant β
32	64	16	4	22	7	2
48	72 96	24	3 4	22 23	8	3
64	96 128	32	3 4	26 27	8	3
96	96 144	48	2 3	28 29	8	3
128	128 192 256	64	2 3 4	32 33 34	8	3

2.1.1 Algorithm for Speck Cipher

Input: Plain-text of $2n$-bit as $P = \{x^0[u], x^0[l]\}$ each n-bit size and a master key 'K' of mn-bit is split into 'm' words as $K = \{l[m\text{-}2],\ldots,l[1],l[0],k[0]\}$.

Output: Cipher-text, $C = \{x^r[u], x^r[l]\}$

For $i \leftarrow 0$ **to** r**-2 do** // key-expansion

$l[i+m\text{-}1] \leftarrow (\ k[i] + (l[i] >>> \alpha)\ )\oplus i$

$k[i+1] \leftarrow ((k[i] <<< \beta)) \oplus l[i+m\text{-}1]$

End For

For $i \leftarrow 0$ **to** r**-1 do** // Encryption

$x^{i+1}[u] \leftarrow ((x^i[u] >>> \alpha) + x^i[l]\ ) \oplus\ k[i])$

$x^{i+1}[l] \leftarrow ((x^i[l] <<< \beta\ ) \oplus\ x^{i+1}[u]$

End For

Return $C = \{x^r[u], x^r[l]\}$

2.1.2 Round Function

The architecture of round function is shown in Fig. 1. The line in red indicates the worst-case path delay. It involves modulo adder which is the crucial component affecting the worst-case delay path. High-speed modulo adders enhance the speed of the Speck cipher.

The plain text is divided into two halves namely $x^0[u]$ and $x^0[l]$ each of size n-bits. For a particular value of i, the inputs of the round function are x^i[u] and $x^i[l]$ and $k[i]$ is the corresponding key. The upper bits are right-shifted by amount α, the resultant bits are given as one of the inputs to the modulo adder. The lower bits are given as the second input to the modulo adder. The resultant so obtained are xor-ed with the

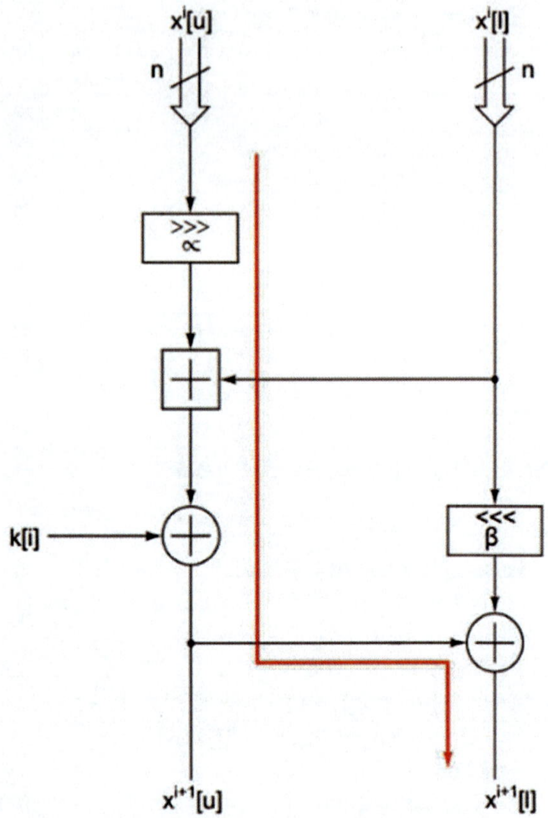

Fig. 1 Architecture of round function

respective key and the output is the most significant *n*-bits obtained after *i*th round, $x[u]$. The lower half is obtained by shifting its bits to the left by an amount β and are xored with the most significant bits obtained in the last round. The outputs so obtained are $x^{i+1}[u]$ and $x^{i+1}[l]$, respectively.

Here Bitwise XOR is denoted by $\oplus$

Modulo 2^n addition is denoted by +

Left and right circular shift operations are denoted by <<< and >>>.

2.1.3 Key Scheduling

The speck key scheduling takes a master key, it generates keys required for the rest of the rounds i.e. $(k[0], k[1], k[2], \ldots, k[r-1])$. Interestingly, it uses a round function to generate round key $k[i]$ as shown in Fig. 2. Assuming '*K*' to be the master key and the number of keywords is extracted as $(k[0], l[0], \ldots, l[m-2])$ where '*m*' can

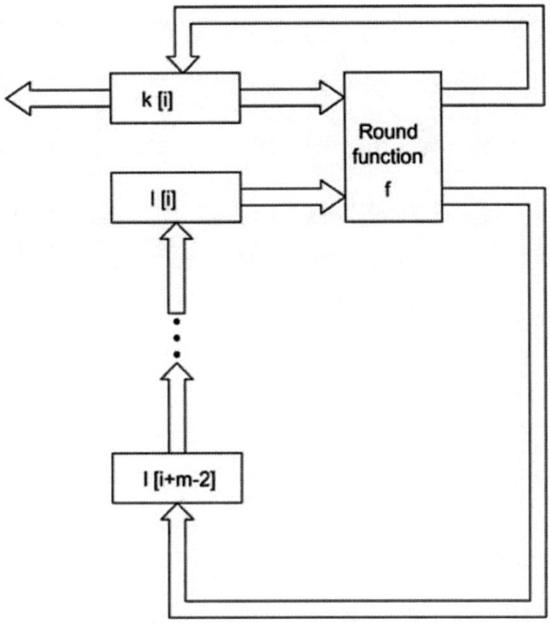

Fig. 2 Key expansion using round function

be 2, 3 or 4 as mentioned above in Table 1. In this case, the value of i ranges from 0 to $r - 2$.

2.2 *Modulo 2^n Adders*

The detailed description of the functioning of different modulo 2^n adders, their logic depth, and hardware requirements in terms of grey and black cells, fan-out are presented in [12]. In this paper, we will briefly describe the adders which are chosen for the implementation of speck cipher.

2.2.1 Kogge–Stone

The fan-out of this adder is low at each stage which results in reduced delay at the cost of an increased area. Its logical depth is similar to that of Sklansky (SK) and has a bounded fan-out of 2 at each level. Wiring congestion is one of the major problems associated with this adder. However, when compared to Sklansky and Han-Carlson (HC) there is an improvement in the delay at the cost of an increased area. It is easy to construct because of its symmetric structure as shown in Fig. 3.

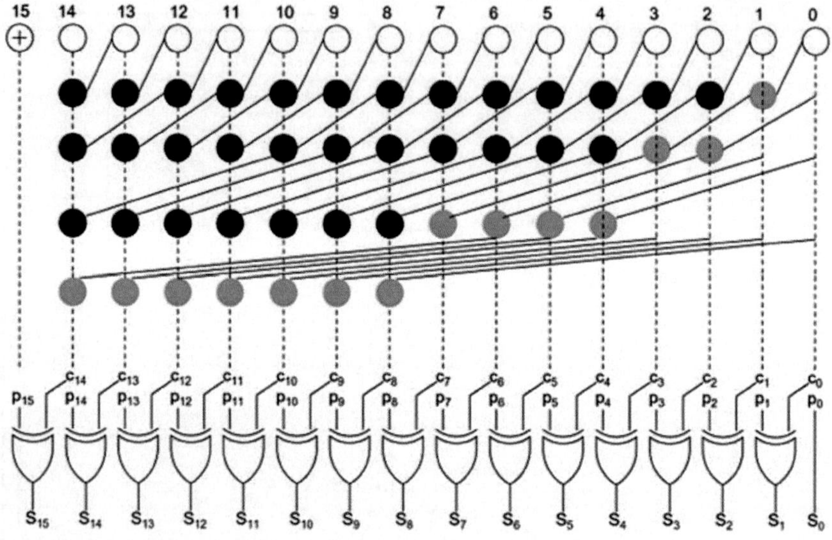

Fig. 3 Modulo adder based on Kogge–Stone adder [12]

2.2.2 Sklansky

The logic depth of this adder is optimal, i.e. $\log_2 n$. Also, wiring congestion is less when compared with Kogge–Stone (KS) as shown in Fig. 4. The longest span of wire

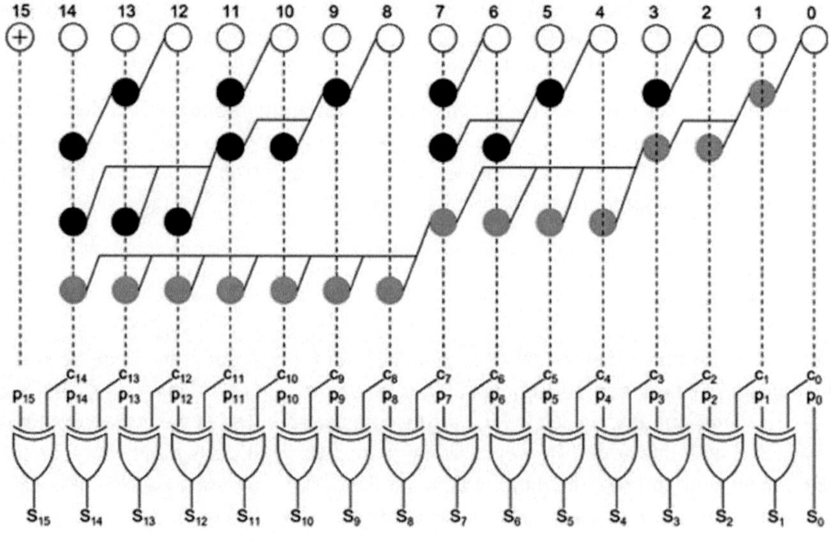

Fig. 4 Modulo adder based on Sklansky adder [12]

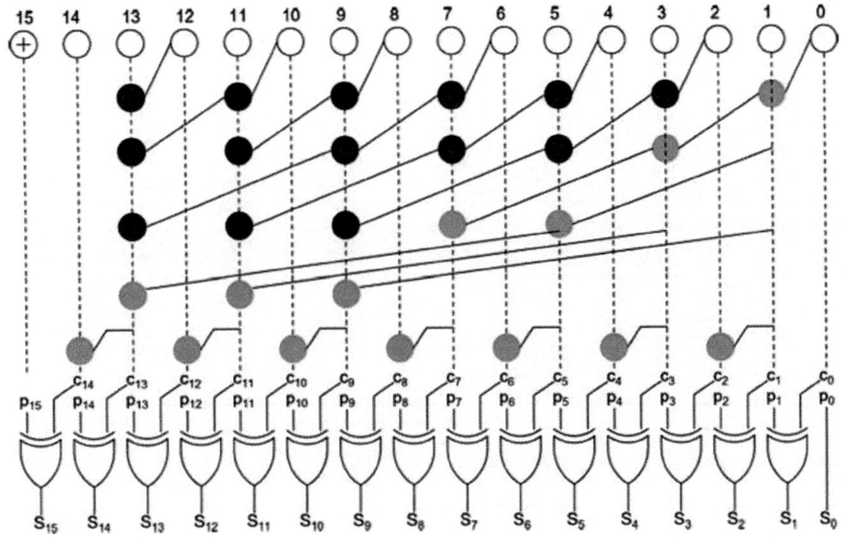

Fig. 5 Modulo adder based on Han-Carlson adder [12]

extends from ($\frac{n}{2} - 1$) to other nodes. The fan-out drastically increases when moved from one level to another which results in huge delay leading to inefficiency.

2.2.3 Han-Carlson

In comparison with Kogge–Stone the hardware complexity of Han-Carlson adder is reduced at the cost of increased logical depth. Han-Carlson adder gives a better area-delay compromise performance when compared with other prefix adders. It is shown in Fig. 5.

3 Architectures for Implementing Speck Cipher

3.1 Folded Architecture

The folded architecture is shown in Fig. 6, where round function is utilized only once. Upon the start signal '*s*' the inputs of the initial round function are plain text and key, respectively. The round function is a combinational block where the actual processing takes place and the first intermediate cipher is generated and stored in the registers. This output will be an input to the next round function. The iteration depends on the rounds for which the particular cipher is defined. The width of the

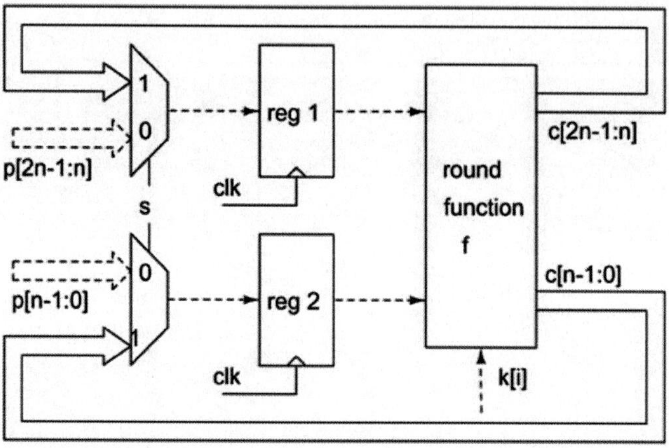

Fig. 6 Architecture of folded implementation

data path inside the round function takes the block size of the cipher while the data path width of the key scheduling will be size of the selected key.

3.2 Unfolded Architecture

It involves the single combinational block of a repetitive architecture and unrolls them for about '*r*' rounds to obtain the cipher-text. This results in a huge increase in area because of the implementation of round function and key expansion for about '*r*' rounds. Each round is interconnected with the appropriate key which is obtained from the unrolled key scheduled just before its commencement. Keys are obtained by hardware presented in Sect. 2. As shown in Fig. 7, the plain text and key are input to the first block while the remaining blocks are cascaded along with the respective keys.

4 Results and Comparison

The family of speck cipher ranging from 32/64 to 128/192 is implemented using folded and unfolded architectures. We consider cipher with modulo 2^n adder based on Kogge–stone tree, implemented in folded architecture as design D1, cipher involving adder based on Han-Carlson tree, as design D2. Design D3 is the cipher using adder based on Sklansky tree implemented in folded architecture are presented in Table 2. On the other hand, ciphers implemented using modulo 2^n adders based on

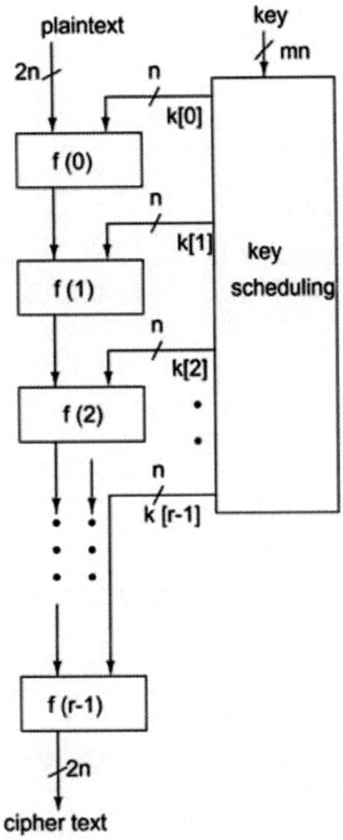

Fig. 7 Architecture of unfolded implementation

Table 2 Synthesis results of folded SPECK cipher using different adder architectures

Blocksize/keysize	D1		D2		D3	
	Area (μm^2)	Delay (ns)	Area (μm^2)	Delay (ns)	Area (μm^2)	Delay (ns)
32/64	2110.7	158.02	1645.59	172.54	1645.59	168.67
48/72	3617.5	171.09	2769.22	187.79	2632.40	199.89
48/96	3620.8	182.91	2771.60	199.36	2634.65	218.08
64/96	5169.31	211.64	3823.69	228.8	3823.69	256.82
64/128	5172.66	227.79	3826.07	246.21	3826.07	276.50
96/96	8671.53	239.62	6224.03	263.95	6031.53	338.8
96/144	8676.51	256.47	6227.48	282.31	6034.86	352.17
128/128	12,220.90	293.92	8665.34	315.04	8668.93	402.93
128/192	12,225.88	312.18	8669.06	334.68	8672.48	415.27

Kogge–Stone, Han-Carlson and Sklansky implemented using unfolded architecture are considered as D4, D5 and D6 presented in Table 3.

Table 3 Synthesis results of unfolded SPECK cipher using different adder architectures

Blocksize/keysize	D4		D5		D6	
	Normalized area (μm²)	Delay (ns)	Normalized area (μm²)	Delay (ns)	Normalized area (μm²)	Delay (ns)
32/64	0.464	145.09	0.362	158.47	0.362	155
48/72	0.795	157.09	0.609	172.42	0.579	183.71
48/96	0.832	164.23	0.637	180.26	0.605	192.06
64/96	1.344	194.32	0.994	210.15	0.994	235.99
64/128	1.396	201.79	1.033	218.24	1.033	245.07
96/96	2.428	220.10	1.742	242.58	1.688	310.10
96/144	2.516	227.96	1.805	251.24	1.750	322.21
128/128	3.910	269.90	2.772	289.39	2.774	358.81
128/192	4.034	275.58	2.862	298.43	2.861	370.02

4.1 Standard Cell Implementation

The speck cipher family is modelled in Verilog HDL for 32/64 to 128/192 in both unfolded and folded architectures. They are mapped to standard cells of tsmc 500 *nm* technology using cadence RTL compiler 14.10. Functionality of each cipher is verified using test vectors for SPECK cipher presented in [7]. The results based on Kogge–Stone, Han-Carlson, Sklansky adders are presented in Tables 2 and 3 respectively. The units of area are square microns and delay is in terms of nano-seconds.

From Table 2, it can be inferred that there is an improvement in the delay in D1 compared with D3 for about (6.31–29.82)% at the cost of an increment in the area by (22–30.04)%. D2 almost occupies the same area as that of D3 but there is an improvement in the delay for about (2.24–22.16)% at the cost of least increment in the area, i.e. less than 5%. D1 is being compared with D2 and it is concluded that there is an improvement in the delay for about (6.7–9.21)% at the cost of an increased area of (22.03–29.04)%.

For our convenience, since the area is huge it is normalized to some constant 10^5. From Table 3, it can be concluded that though the logic depth of KS and SK are same, D4 takes over the improvement in the delay over D6 for about (6.43–25.52)% at the cost of an increased area of (22–30.44)%. Hence it can be concluded that D4 is preferable in high-speed lightweight cryptography. Also, D5 is better than D6 in terms of delay for about (2.18–22.02)% at the cost of an improvement in the area for less than 5%. Lastly, on comparing D4 with D5 there is an improvement in the delay by (6.73–9.26)% at the cost of an increment in the area by (22.03–29.09)%.

Comparison of area and delay of ciphers incorporating different adders implemented in folded architecture is presented in Figs. 8 and 9.

Also, area and delay of ciphers involving different adders implemented in unfolded architectures are compared, presented in Figs. 10 and 11. Few applications where the

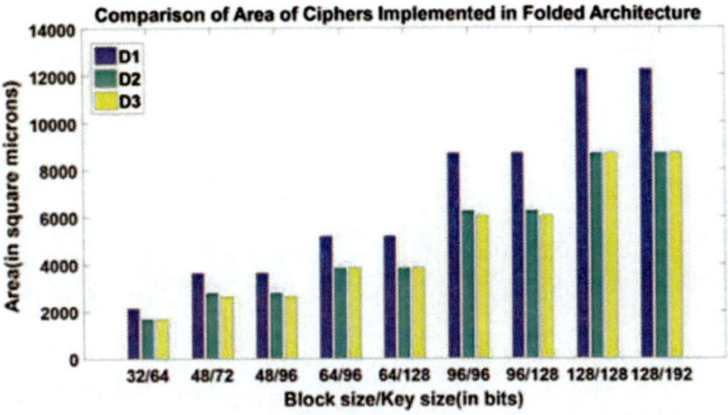

Fig. 8 Area versus block size/key size (folded architecture)

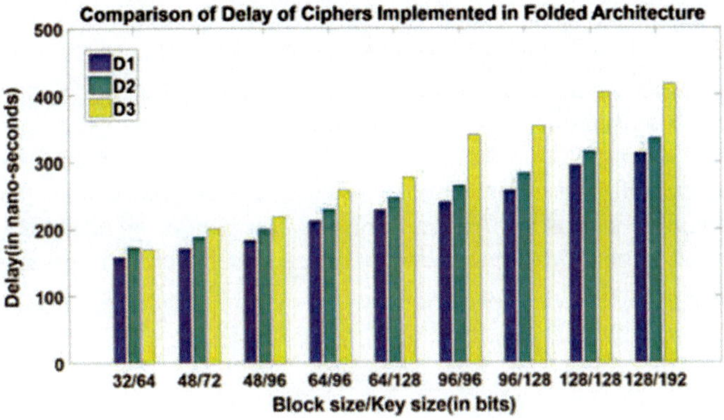

Fig. 9 Delay versus block size/key size (folded architecture)

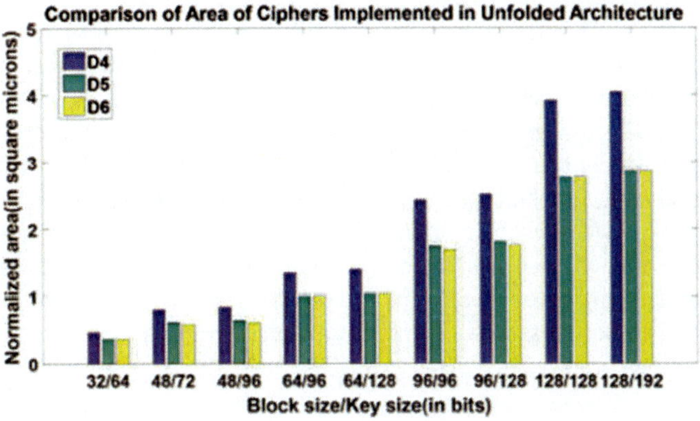

Fig. 10 Normalized area versus block size/key size (unfolded architecture)

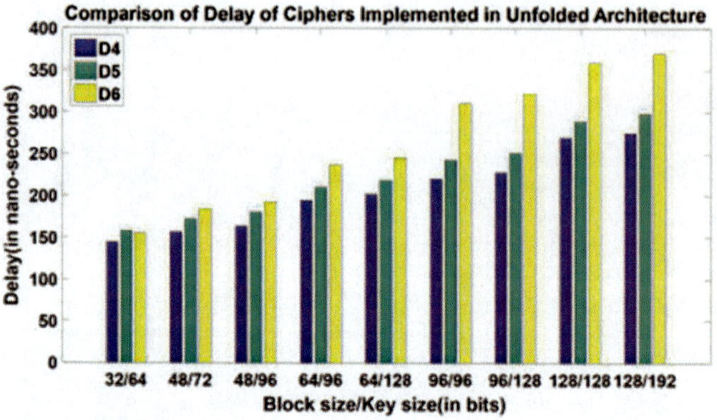

Fig. 11 Delay versus block size/key size (unfolded architecture)

delay is the primary concern, cipher implemented based on Kogge–Stone tree is efficient. Applications, where area and delay are of importance then cipher implemented based on Han-Carlson tree, outperforms both Kogge–Stone and Sklansky.

It is observed from Figs. 8, 9, 10 and 11 that delay for ciphers implemented using both folded and unfolded architecture are the same. The difference lies in the area, i.e. Hardware requirements of ciphers implemented using folded architectures are huge when compared to unfolded architectures. Hence folded architectures are preferred for resource-constrained devices.

5 Conclusion

This paper aims to implement SPECK family of ciphers in both folded and unfolded architectures. The architectures are implemented on the grounds of Kogge–Stone, Han-Carlson and Sklansky based adders. It is observed that the fastest encryption can be achieved by invoking adder based on the Kogge–Stone tree while adder based on Han-Carlson gives a better compromise between area and delay. Implementation results conclude that requirements of delay of folded and unfolded architecture are almost the same but in case of folded architecture hardware requirements can be reduced greatly. For IoT applications, when speed is primary concern ciphers implemented on the grounds of Kogge–Stone are significant whereas, for area constrained environment cipher based on Sklansky is the best choice. In terms of AT and AT^2, cipher designed using Han-Carlson is an appropriate choice. Efficient implementations of key schedulers for SIMON and SPECK will be analyzed in the future.

References

1. Beaulieu R et al (2015) SIMON and SPECK: block ciphers for Internet of Things. In: NIST lightweight cryptography workshop, 20–21 July 2015
2. Maene P, Verbauwhede V (2015) Single-cycle implementations of block ciphers. In: Proceedings of 4th international workshop on lightweight cryptography for security and privacy, vol 9542, pp 131–147
3. Beaulieu R, Shors D, Smith J, Treatman-Clark S, Weeks B, Wingers L (2015) SIMON and SPECK: block ciphers for the internet of things. In: Proceedings of NIST lightweight cryptography workshop, pp 1–15
4. Yang G, Zhu B, Suder V, Aagaard MD, Gong G (2015) The Simeck family of lightweight block ciphers. In: Proceedings of CHES, LNCS, Springer, Berlin, vol 9293, pp 307–329
5. Shah Verdi A, Taha M, Eisenbarth T (2015) Silent SIMON: a threshold implementation under 100 slices. In: IEEE international symposium on hardware oriented security and trust (HOST), Washington, DC, USA, pp 1–6
6. Wan T, Salman E (2018) Ultra-low power SIMON core for lightweight encryption. In: IEEE international symposium on circuits and systems (ISCAS), Florence, Italy, pp 1–5
7. Beaulieu R et al (2013) The SIMON and SPECK families of lightweight block ciphers. Cryptology ePrint Archive Report 2013/404; San Francisco, USA, 19 July 2013
8. Rashidi et al (2019) High-throughput and flexible ASIC implementations of SIMON and SPECK lightweight block ciphers. Proc Int J Circ Theor Appl 47:1254–1268
9. Ladner RE, Fischer MJ (1980) Parallel prefix computation. J ACM 27:831–838
10. Kogge PM, Stone HS (1973) A parallel algorithm for the efficient solution of a general class of recurrence equations. IEEE Trans Comput C-22:786–793
11. Han T, Carlson DA (1987) Fast area-efficient VLSI adders. In: Proceedings of 8th symposium on computer arithmetic, pp 49–56
12. Kalyani P et al (2020) Implementation of efficient modulo 2^n adders for cryptographic applications C. In: 4th IEEE international conference on trends in electronics and informatics, India

Evaluating Heterogeneous Ensembles with Boosting Meta-Learner

Seba Susan, Aishwary Kumar, and Anmol Jain

Abstract In this paper, heterogeneous ensemble of classifiers is evaluated and the outputs are integrated by a boosting meta-learner. Both ADABOOST and XGBOOST are tried for the meta-learning stage, and XGBOOST performed best. The heterogeneous ensemble consists of a diverse set of base classifiers—k-nearest neighbors, logistic regression, Support Vector Machines (linear, Gaussian kernels), Random forest of decision trees, and Naïve Bayes classifier. Smaller ensembles are also hierarchically formed by removing the weak learner in every stage. The Entropy of base classifier predictions is computed to identify the presence of weak learners. The predictions of the base classifiers are learnt by the boosting meta-learner using a 9:1 split of the training data, where 9 parts are used for training the base classifiers and 1 part for obtaining the ensemble predictions and training the meta-learner. A 10-fold cross-validation is introduced to avoid bias. Experimental results show higher scores on evaluating the Human Action Recognition (HAR) smartphone dataset using our ensemble model as compared to the other state-of-the-art models.

Keywords Meta-learning · Boosting · XGBOOST · Heterogeneous ensemble

1 Introduction

Recognizing human actions from mobile or smartphone information is evolving research with significant outcomes. Mobile phones contain the accelerometer sensor that tracks motion over time. Potential applications include tracking the motion of the elderly [1], detecting suspicious motion in crowded places [2], and analysis of sports activities [3]. The limitations of most of the current datasets based on smartphone sensors are that they are specific and relatively smaller in size. Hence complex models like deep neural networks do not yield good performance. Using multiple classifiers

S. Susan (✉) · A. Kumar · A. Jain
Department of Information Technology, Delhi Technological University (DTU), Bawana Road, Delhi 110042, India
e-mail: seba_406@yahoo.in

G. Ranganathan et al. (eds.), *Inventive Communication and Computational Technologies*, Lecture Notes in Networks and Systems 145,
https://doi.org/10.1007/978-981-15-7345-3_60

is a reliable technique of performance enhancement in machine learning. For smaller datasets, stacking combinations of classifiers is more useful rather than relying on a single model [4]. It was observed in [5] that heterogenous models give an improved performance than homogenous models for small datasets. Integrating the outputs of multiple classifiers is achieved by bagging, boosting, or stacking [6]. Various strategies have been proposed to integrate the outputs of multiple classifiers in a classifier ensemble, ranging from majority voting to advanced meta-learners [7]. In the meta-learning scheme, the predictions of the base classifiers in the ensemble are learnt by a meta-learner classifier in a non-linear manner. A combination of stacking and boosting is investigated in this work. Stacking of heterogenous classifiers is implemented for the formation of a diverse ensemble whose predictions are learnt by a boosting meta-learner. An Entropy score is assigned to the ensemble based on the feature weights computed by the boosting algorithm that serves to measure the efficiency of the ensemble. The organization of this paper is as follows: a review of meta-learners and base learners in a stacked ensemble classifier model is given in Sect. 2, the proposed heterogeneous ensemble model based on boosting meta-learner is presented in Sect. 3, the results are analyzed in Sect. 4 and the final conclusions are drawn in Sect. 5.

2 Ensemble Learning—A Review of Basic Concepts

2.1 Meta-Learning with Stacked Ensemble of Classifiers

A meta-learner is an additional classification layer in an ensemble model for fine-tuning the base classifier predictions. The term was introduced by Wolpert in [8], though several indicative works pre-exist [9]. Various learning strategies for stacked classifier ensemble were investigated by Chan and Stolfo in [10] to improve the performance of the ensemble as compared to that of the base classifiers. Binary classifier predictions and fusion of attributes with base classifier predictions were found to be successful in this regard. The precursor of meta-learning is classifier ensembles in which a pre-defined number of trained classifiers individually predict the class label of the test sample and the decisions are compiled, usually, by majority voting [7]. Majority voting finds its genesis in the history of social sciences, wherein decisions are taken by social committees by a popular vote among the committee members [11]. A variety of meta-learners have been tested and tried over the years. Table 1 summarizes the base and meta-classifiers in a few of these works.

Approximate Ranking Tree Forests are proposed as the meta-learner in [12], which involves comparing base classifiers with each other to generate meta-features. Decision trees are observed to be used by many researchers as the base classifier. In [13], a meta-learning approach is proposed with the goal of interpreting the hidden structures (having H neurons) in Convolutional Neural Networks (CNN). Here, a $H \times N$ feature matrix of cluster IDs is fed to the meta-learner (Decision Tree) along with

Table 1 Examples of a few distinctive ensemble models and Meta-classifier combinations in literature

Method	List of base classifiers	Meta-learner
Aydin (2018) [1]	LDA, SVM, neural networks	Fuzzy Integral
Rodriguez et al. [26]	Rotation forest: Decision trees (with PCA)	Majority voting
Nguyen et al. (2018) [27]	LDA, Naïve Bayes, k-Nearest Neighbors	Fuzzy IF-THEN rules
Alexandropoulos et al. (2019) [28]	Decision trees	ACO-stacking
Melville and Mooney (2004) [29]	Decision trees (with artificial data)	Meta-learning algorithm 'ACTIVE-decorate'

N training labels for training. The test vector comprising of cluster IDs is classified by the meta-learner into the actual label of the test sample.

2.2 *Learning with Heterogenous Ensemble of Classifiers*

Most of the ensemble methods involve homogeneous classifiers due to ease of programming and parameter selection [14]. The ensemble output is compiled in some form of average error performance across the classifiers in the ensemble. The common ensemble methods that involve homogeneous classifiers are bagging, boosting, and random forest, depending on the integration technique [15]. However, lack of diversity in classifier predictions due to similar type of errors in predictions limits the performance of homogeneous ensembles. Classifiers such as neural networks are subject to local entrapping of solutions due to the derivative optimization process [16]. The network is hence, not sensitive to the variety present in the dataset. Corrective procedures do exist, such as, Network Architecture Search (NAS) for optimal parameters [17] and dynamic neural networks that evolve as training progresses [18]. Locality of solutions and lack of variety is the motivation behind using a heterogeneous mix of classifiers in an ensemble. For the homogenous classifiers in an ensemble, variety can be induced by carefully sampling the training data and the feature subspace for custom-training of each member in the ensemble [19]. Lack of variety in predictions promoted research in heterogenous ensembles where a variety of popular classifiers like logistic regression, k-Nearest Neighbors, Naïve Bayes, Support Vector Machines, and decision trees are trained on the same data. The outputs of individual classifiers are integrated by committee voting or weighted average [20]. In a rather unconventional approach in [1], the predictions of the base learners, comprising of Linear Discriminant Analysis classifier (LDA), Support Vector Machines (SVM), and neural networks, are combined using the fuzzy integral. Prior to the integration, the confidence-scores of the base learners are optimized using cuckoo search. The idea is to efficiently utilize the diversity of the classifiers in the ensemble.

3 Proposed Meta-Learning for Heterogeneous Ensemble

3.1 The Problem Statement

The focus in this research is on heterogeneous ensemble. Its characteristic is the diversity of the base classifiers that are the constituents of the ensemble. Due to the diversity, the predictions made by the base classifiers will have a variety, which if integrated by a carefully planned meta-learning classification stage, would yield high accuracies of prediction. All the base classifiers are trained on the same data. The meta-learner for most heterogeneous ensembles in literature is majority voting. The maximum vote amongst the base classifier predictions indicates the class label of the test sample. In case of an unclear majority, an ambiguity exists, that is primarily ignored by most researchers. In our work, this problem is addressed by introducing a boosting classifier in the meta-learning stage that learns the entire set of predictions made by the base classifiers. Boosting algorithms are by themselves, based on ensembles of decision trees. These are known to concentrate on ambiguous data points for decision-making. This addresses the issue of those predictions that do not form a clear majority and are difficult to summarize. The subset of training data used in the meta-learning stage is different from that used for training the base classifiers.

3.2 Proposed Meta-Learning for Heterogeneous Ensemble

In our proposed method, the boosting algorithm is employed as the meta-learner in the second stage of classification where it learns the predictions from the base learners in the first stage. The meta-classifier integrates the predictions of all the base classifiers in a decision-making module. The base learners are a heterogeneous mix of classifiers that are trained on the same training data. Ensembles of six, five, four, and three classifiers are tested, with the heterogenous classifier models being-k-nearest neighbors, logistic regression, Support Vector Machines (linear, Gaussian kernels), Random forest of decision trees and Naïve Bayes classifier. The diversity of the base classifiers aids in unbiased prediction. The training set is divided into 10 parts out of which 9 parts are used for training the base classifiers and 1 part out of 10 is used for training the meta-learner. Ten-fold cross-validation is used to avoid bias. The algorithm for our meta-learning model incorporating boosting in the latter stage is given below.

Algorithm Meta-learning by boosting for heterogeneous ensemble

Input: *Training set_9parts, Training set_1part, Test set with unlabeled data*
Parameter: Base-learner specific
Output: Class label of *Test* sample
1: For each base-learner C*i* say, Support Vector Machines
2: perform learning on *Training set_9parts*
3: obtain predicted class label C*i*(***T***), for ***T***= *Training set_1part*
4: From all the *n* base-learners, extract:
    Meta-features={C*1*(***T***), C*2*(***T***),....C*n*(***T***)}
5: Learn the Meta-features using boosting based meta-learner
6: Extract the Meta-features from *Test* sample and give as input to the trained meta-learner for predicting the class label of *Test*
7: **return** solution

The choice of the boosting algorithm is decided next. Freund et al. (1999) proposed ADABOOST [21] as a novel boosting ensemble technique that combines the outputs of a number of weak learners or estimators that constitute the boosting ensemble. Since the method intrinsically assigns higher weights to noisy samples that are least correlated with the output label, it generally yields high accuracies and is known to be a competitor to ensemble models. In [22], a new meta-learning algorithm was devised on ADABOOST principles called meta-boosting that combines the outputs of weak learners by applying a strong learner on them. XGBOOST (Chen and Guestrin, 2016) [23] stands for eXtreme Gradient Boost that enumerates splitting points in a tree in a greedy manner. It starts with a leaf and enumerates the gradient of the dynamically changing loss function and adds branches in an additive manner. While in ADABOOST, the focus is on the misclassified samples that are assigned higher weight, in XGBOOST, the focus is on the gradient with faster optimization algorithms. XGBOOST employs L1 and L2 regularization that prevents overfitting. Since the conventional approach to integrate ensemble decision is by majority voting [20], we explore the use of boosting as a meta-learner that focuses on the difficult-to-learn-examples in the base classifier predictions. Both ADABOOST and XGBOOST are applied, within our heterogeneous ensemble framework, as the meta-learner that receives the meta-features or predictions of individual base classifiers. In our case, the difficult-to-learn-examples in the meta-features are the prediction vectors that indicate no clear majority amongst the base classifiers in the ensemble. We do not partition the dataset into smaller subsets for training the base classifiers, following the observation by Chan and Stolfo (1995) [24], that partitioning smaller datasets may have an overall negative impact on the learning performance if do not have remedies at hand to reduce the bias that ensues because of the partitioning. XGBOOST which is the gradient boosting algorithm has won several Kaggle competitions due to its unique features such as scalability and fast execution, its tendency to form deeper trees with less variance, and computing similarities with data points in an adaptive neighborhood. To identify the presence of weak learners in an ensemble, an Entropy grade is now computed, as

$$H = -\sum_{p} p \log p \tag{1}$$

The Entropy measures the randomness among a set of points [3] and is computed here from the normalized feature weights (=*p*) assigned by the boosting algorithm; the features being the base classifier predictions and weights being the frequencies of occurrence of the features in all the tree splits. A higher Entropy indicates more equal importance among the features which are the predictions of the base classifiers. A lower Entropy indicates the presence of one or more weak learners whose predictions are not contributive to the final decision taken. The ensemble with a high normalized Entropy (Entropy/Maximum value of Entropy) is considered the optimal choice of an ensemble for the given dataset.

The functional block diagram of the proposed model is shown in Fig. 1. The six heterogeneous classifiers are shown along with the meta-learner. The overall process flow for the learning model shown in Fig. 1, and summarized in the following steps:

Step 1: Split the training set in a 9:1 ratio (with ten-fold cross-validation) labeled *Training set_9parts* and *Training set_1part.*

Step 2: Train the base classifiers with *Training set_9parts* (shown by blue (solid line) arrows in Fig. 1).

Step 3: Train the boosting meta-learner on the predictions made by the trained base classifiers on the data *Training set_1part* (shown by red dotted arrows in Fig. 1).

Step 4: In the testing phase, apply the predictions made by the trained base classifiers on the test data as input to the trained boosting meta-learner (red dotted arrows only).

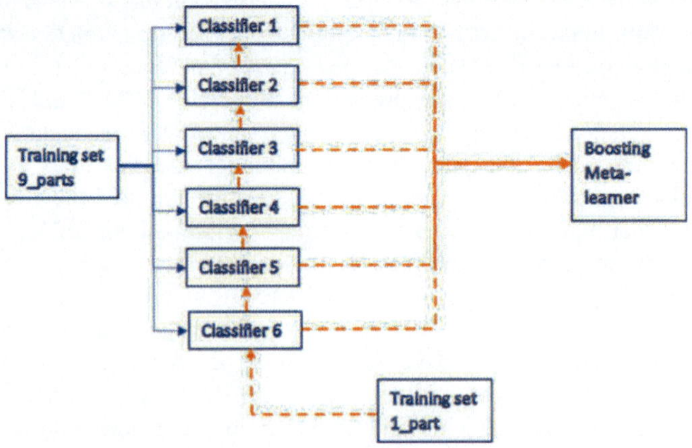

Fig. 1 Functional block diagram of the proposed model

4 Experimental Results and Discussions

The experiments are conducted in Python 3.7 version software on an Intel Pentium processor. The benchmark dataset on Human Activity Recognition (HAR) based on smartphone data [25] is used for the experiments. It is segregated at the source into 7352 samples (from 21 subjects) and 2947 samples (from 9 subjects) for training and testing, respectively. There are 561 features in all, representing 2.56 s of human activity. The activities themselves are divided into six categories—*walking, upstairs, downstairs, sitting, standing, lying.*

4.1 Results of Meta-Learning by Boosting for Heterogeneous Ensemble

The implementation of our meta-learning model is done as per the guidelines in Sect. 3. Experiments are conducted, in hierarchical order, for ensembles of 6, 5, 4 and 3 heterogenous classifiers with the heterogenous models being—k-nearest neighbors ($k = 8$), logistic regression, support vector machine (linear, Gaussian kernels), Random forest of decision trees (300 trees) and Naïve Bayes classifiers. The compositions of the heterogeneous ensembles are shown in Tables 2 and 3 respectively. The worst performer is removed at each stage to form the smaller ensemble. Two different boosting meta-learners-ADABOOST, XGBOOST are investigated for our scheme. A Grid Search strategy is used for tuning the hyperparameters (learning rate, number of estimators, etc.) of the boosting algorithms. A stratified 10-fold cross-validation scheme is used for the training data in our ensemble learning. As per this scheme, the training data is split into ten parts, nine parts are given as aggregate input to the

Table 2 Our Meta-learning strategy for 6- and 5-classifier ensembles (Test accuracy in %) with comparison to majority voting scheme

Type of classifier	Ensemble # (*5 heterogenous models*)	Ensemble # (*5 heterogenous models*)	Ensemble # (*6 heterogenous models*)
Base classifiers	*Naïve Bayes—77.02% *Logistic Regression—96.19% *SVM (Gaussian)—94.02% *Random forest (100 trees)—90.19% *k-NN (k = 8)—90.73%	*SVM (linear)—96.40% *Logistic Regression—96.19% *SVM (Gaussian)—94.02% *Random forest (300 trees)—92.63% *k-NN (k = 8)—90.73%	*Naïve Bayes—77.02% *SVM (linear)—96.40% *Logistic Regression—96.19% *SVM (Gaussian)—94.02% *Random forest (300 trees)—92.63% *k-NN (k = 8)—90.73%
Meta-learner	*Majority voting*—95.12%	*ADABOOST*—**96.64%** XGBOOST—96.64%	*ADABOOST*—**96.26%** XGBOOST—96.77%

Table 3 Our Meta-learning strategy for 4- and 3- classifier ensembles (Test accuracy in %)

Type of classifier	Ensemble # (4 *heterogenous models*)	Ensemble # (3 *heterogenous models*)
Base classifiers	*SVM (linear)—96.40% *Logistic Regression—96.19% *SVM (Gaussian)—94.02% *Random forest (300 trees)—92.63%	*SVM (linear)—96.40% *Logistic Regression—96.19% *SVM (Gaussian)—94.02%
Meta-learner	*XGBOOST*—96.40% ADABOOST—96.40%	*XGBOOST*—96.36% ADABOOST—96.33%

ensemble of classifiers. The tenth part is used to evaluate the predictions of the base classifiers that are given as input to the meta-learner. The results are shown in Table 2 for six and five ensemble combinations for both boosting and the majority voting scheme. The results in Table 2 indicate that XGBOOST (*base-estimator* = *Random forest, number of estimators* = 300) gives the best results, much higher than majority voting.

Among the base learners, SVM with linear kernel gives the individual best accuracy followed by logistic regression. The Naïve Bayes classifier proves to be a weak learner in Table 2 with an accuracy of 77.02%. The weak learner affects the results of ADABOOST for which the best results are observed for the 5-classifier ensemble. The gradient boosting scheme is found to override the weak learner as observed from the highest results of 96.77% in Table 2 for the six-classifier ensemble. A similar observation is made for the results 4- and 3-classifier ensembles in Table 3.

Overall, the best accuracy is observed for the XGBOOST meta-learner for the 6-classifier ensemble. The Entropies of various ensembles are computed in Table 4 from the relative feature importance plots shown in Fig. 2. The feature weights are shown normalized so that they form a complete probability distribution (sum of all p is 1).

From Table 4, a higher normalized Entropy of features is observed for XGBOOST as compared to that of ADABOOST. The heterogeneous ensemble with the highest Entropy (=0.978 *for 6-classifier ensemble*), in the training phase, is determined to be most suitable for the given dataset. Overall, gradient boosting has proved to be the best meta-learner for our ensemble. It is generally observed that due to meta-learning there is improvement over the performance of all the individual classifiers in the ensemble.

Table 4 Entropy of ensemble (Highlighted values indicate the normalized Entropies)

Number of base classifiers in ensemble = n	3	4	5	6
Max Entropy = *log(n)*	1.0986	1.3863	1.6094	1.6094
Entropy—ADABOOST	0.7430	0.8741	1.1143	1.4847
(*Entropy*—ADABOOST/*Max Entropy*)	**0.676**	**0.6305**	**0.692**	**0.9225**
Entropy—XGBOOST	1.0954	1.1571	1.1143	1.5740
(*Entropy*—XGBOOST/*Max Entropy*)	**0.997**	**0.8347**	**0.6924**	**0.978**

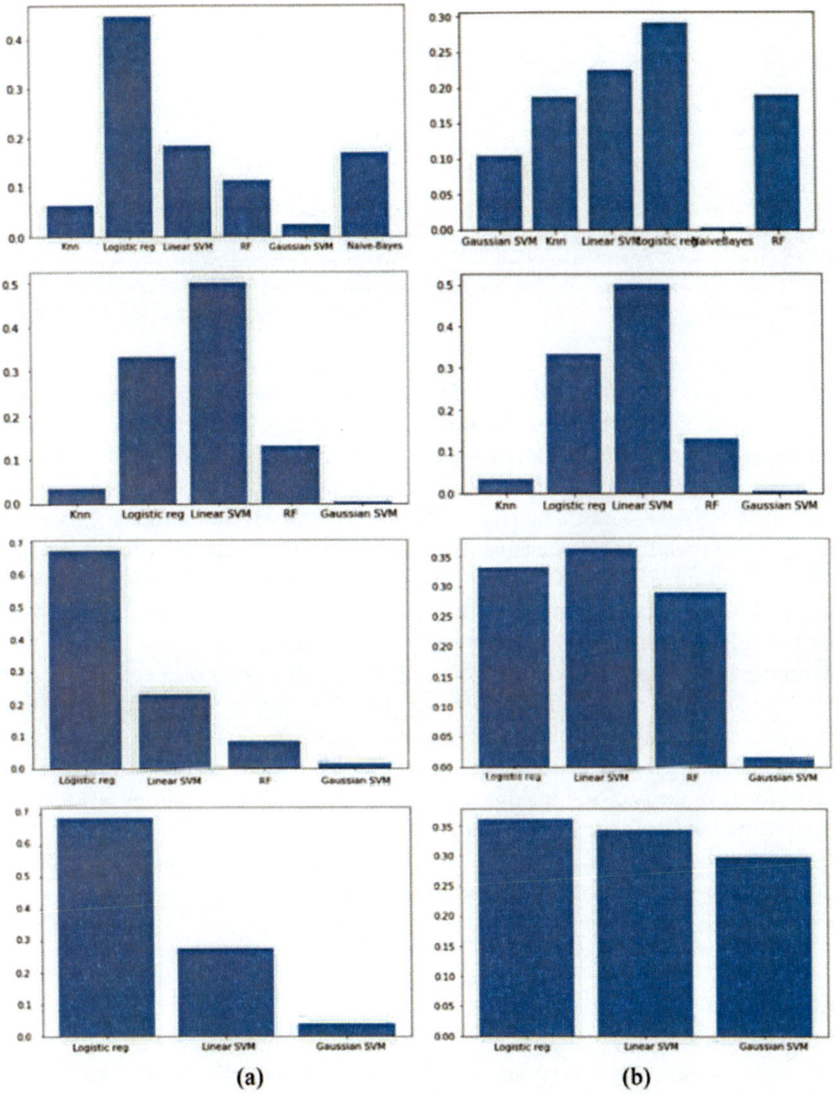

Fig. 2 The relative feature importance plot for base classifiers in the ensemble **a** ADABOOST meta-learner **b** XGBOOST meta-learner [*top to bottom: 6-, 5-, 4-, 3-classifier ensembles*]

4.2 Comparison to the State-of-the-Art Techniques

Comparison of our results to some state-of-the-art methods listed in Table 5, highlights the efficiency of our approach from the higher accuracies achieved by gradient boosting (=96.77% for XGBOOST as meta-learner for the 6-classifier ensemble).

Table 5 A comparison of the recent state-of-the-art results: test accuracy (in %)

Recent state-of-the-art results for smartphone HAR	Accuracy (%)
Probabilistic-First-Take-It-All (PFTA) [30] (Ye et al. 2018)	92
Deep ConvNet [31] (Jiang and Yin 2015)	95.18
CNN + Attention model [32] (Wang et al. 2019)	93.4
Deep Neural Net [33] (Ronao et al. 2016)	94.61
Proposed 6-classifier heterogeneous ensemble with XGBOOST meta-learner	96.77

5 Conclusion

A heterogenous ensemble of diverse classifier models with a boosting meta-learner is proposed in this work for recognizing human activities from smartphone data. In this paper, the boosting meta-learner is employed to take advantage of the diversity of predictions made by the ensemble. A heterogenous mix of classifiers namely—k-nearest neighbors, logistic regression, Support Vector Machines (linear, Gaussian kernels), Random forest of decision trees, and Naïve Bayes classifier forms the six-classifier ensemble. XGBOOST, that is a tree-based gradient boosting algorithm, is the meta-learner that performs the final decision regarding the test label based on the individual predictions of the ensemble classifiers. The ADABOOST boosting algorithm is also used for comparison. Higher classification scores especially with XGBOOST validate the efficiency of our meta-learning model despite the presence of a weak learner in the ensemble. Entropy of the feature weights assigned during boosting is computed for grading the ensemble based on participation of the constituents in the decision-making. A high Entropy coupled with high accuracy is observed for the six-classifier ensemble, in the case of XGBOOST, as compared to the smaller classifier ensembles. Application of our ensemble on large datasets with federated learning forms the future scope of our work.

References

1. Aydin I (2018) Fuzzy integral and cuckoo search based classifier fusion for human action recognition. Adv Electr Comput Eng 18(1):3–11
2. Susan S, Hanmandlu Ma (2015) Unsupervised detection of nonlinearity in motion using weighted average of non-extensive entropies. SIViP 9(3):511–525
3. Susan S, Chaurawat S, Nishad V, Sharma M, Sahay S (2016) Speed and trajectory based sports event categorization from videos. In: 2016 international conference on signal processing and communication (ICSC). IEEE, pp 496–501
4. Yu C, Skillicorn DB (2001) Parallelizing boosting and bagging. Queen's University, Kingston, Canada, Tech. Rep
5. Merkwirth C, Wichard J, Ogorzałek MJ (2009) Ensemble modeling for bio-medical applications. In: Modelling dynamics in processes and systems. Springer, Berlin, Heidelberg, pp 119–135

6. Matan O (1996) On voting ensembles of classifiers. In: Proceedings of AAAI-96 workshop on integrating multiple learned models, pp 84–88
7. Sagi O, Rokach L (2018) Ensemble learning: a survey. Wiley Interdiscip Rev Data Min Knowl Discov 8(4):e1249
8. Wolpert David H (1992) Stacked generalization. Neural Netw 5(2):241–259
9. Mitchell TM (1980) The need for biases in learning generalizations. Department of Computer Science, Laboratory for Computer Science Research, Rutgers University, New Jersey
10. Chan PK, Stolfo SJ (1993) Experiments on multistrategy learning by meta-learning. In: Proceedings of the second international conference on information and knowledge management. ACM, pp 314–323
11. Brams SJ, Fishburn PC (2002) Voting procedures. Handb Soc Choice Welfare 1:173–236
12. Sun Q, Pfahringer B (2013) Pairwise meta-rules for better meta-learning-based algorithm ranking. Mach Learn 93(1):141–161
13. Liu X, Wang X, Matwin S (2018) Interpretable deep convolutional neural networks via meta-learning. In: 2018 international joint conference on neural networks (IJCNN). IEEE, pp 1–9
14. Wang Y, Wang D, Geng N, Wang Y, Yin Y, Jin Y (2019) Stacking-based ensemble learning of decision trees for interpretable prostate cancer detection. Appl Soft Comput 77:188–204
15. Seni G, Elder JF (2010) Ensemble methods in data mining: improving accuracy through combining predictions. Synth Lect Data Min Knowl Discov 2(1):1–126
16. Susan S, Rohit R, Udyant T, Shivang R, Pranav A (2019) Neural net optimization by weight-entropy monitoring. In: Computational intelligence: theories, applications and future directions. Springer, Singapore, vol II, pp 201–213
17. Verma M, Sinha P, Goyal K, Verma A, Susan S (2019) A novel framework for neural architecture search in the Hill Climbing Domain. In: 2019 IEEE second international conference on artificial intelligence and knowledge engineering (AIKE). IEEE, pp 1–8
18. Susan S, Dwivedi M (2014) Dynamic growth of hidden-layer neurons using the non-extensive entropy. In: 2014 fourth international conference on communication systems and network technologies. IEEE, pp 491–495
19. Queipo NV, Nava E (2019) A gradient boosting approach with diversity promoting measures for the ensemble of surrogates in engineering. Struct Multi Optim 60(4):1289–1311
20. Prodromidis AL, Stolfo SJ (1999) A comparative evaluation of meta-learning strategies over large and distributed data sets. In: Workshop on meta-learning, sixteenth international conference on machine learning, pp 18–27
21. Freund Y, Schapire R, Abe N (1999) A short introduction to boosting. J Jap Soc Artif Intell 14(771–780):1612
22. Liu X, Wang X, Japkowicz N, Matwin S (2013) An ensemble method based on adaboost and meta-learning. In: Canadian conference on artificial intelligence. Springer, Berlin, Heidelberg, pp 278–285
23. Chen T, Guestrin C (2016) Xgboost: a scalable tree boosting system. In: Proceedings of the 22nd acm sigkdd international conference on knowledge discovery and data mining. ACM, pp 785–794
24. Chan PK, Stolfo SJ (1995) A comparative evaluation of voting and meta-learning on partitioned data. In: Machine learning proceedings 1995. Morgan Kaufmann, pp 90–98
25. Anguita D, Ghio A, Oneto L, Parra X, Reyes-Ortiz JL (2012) Human activity recognition on smartphones using a multiclass hardware-friendly support vector machine. In: International workshop on ambient assisted living. Springer, Berlin, Heidelberg, pp 216–223
26. Rodriguez JJ, Kuncheva LI, Alonso CJ (2006) Rotation forest: a new classifier ensemble method. IEEE Trans Pattern Anal Mach Intell 28(10):1619–1630
27. Nguyen TT, Nguyen MP, Pham XC, Liew AW-C (2018) Heterogeneous classifier ensemble with fuzzy rule-based meta learner. Inf Sci 422:144–160 (2018)
28. Alexandropoulos S-AN, Aridas CK, Kotsiantis SB, Vrahatis MN (2019) Stacking strong ensembles of classifiers. In: IFIP international conference on artificial intelligence applications and innovations. Springer, Cham, pp 545–556

29. Melville P, Mooney RJ (2004) Diverse ensembles for active learning. In: Proceedings of the twenty-first international conference on machine learning. ACM, p 74
30. Ye J, Qi G, Zhuang N, Hao Hu, Hua KA (2018) Learning compact features for human activity recognition via probabilistic first-take-all. IEEE Trans Pattern Anal Mach Intell (2018)
31. Jiang W, Yin Z (2015) Human activity recognition using wearable sensors by deep convolutional neural networks. In: Proceedings of the 23rd ACM international conference on Multimedia. ACM, pp 1307–1310
32. Wang K, He J, Zhang L (2019) Attention-based convolutional neural network for weakly labeled human activities recognition with wearable sensors. IEEE Sens J
33. Ronao CA, Cho S-B (2016) Human activity recognition with smartphone sensors using deep learning neural networks. Expert Syst Appl 59:235–244

Radial Basis Function Neural Network Based Speech Enhancement System Using SLANTLET Transform Through Hybrid Vector Wiener Filter

V. R. Balaji, J. Sathiya Priya, J. R. Dinesh Kumar, and S. P. Karthi

Abstract In real environment, humans have the difficulty to recognize the voice signal in noisy situations. Recently, many speech enhancement methods, based on transforms, provide substantial success over the conventional method. In this paper, deep learning along with conventional transforms is introduced for enhancing the speech signal. The deep learning approach is used to train the speech frames for classifying the signal either as voiced or unvoiced, followed by pitch synchronous analysis. A proper windowing technique is used along with the hybrid vector wiener filter. However existing algorithms purely depends on supervised learning which tries to reduce the Mean Square Error. The Minimum mean square error can be reduced by means of comparing the output signal with an appropriate clean speech signal. Processing delay and reliable architecture are the best characteristics of deep learning speech enhancement algorithms. Radial basis function reduces the noise coefficient which in turn increases the speech quality.

Keywords Speech enhancement · Deep learning · RBFNN · Slantlet transform

1 Introduction

In day to day life, the speech signal gets added with the noisy signal, which is inevitable. This unwanted noise is called as background noise which comes from various sound sources like speakers, environment noise, etc. This will lead to the degradation of both speech intelligibility and quality. Understanding noisy speech is tough for normal persons but it will be challenging for hearing-impaired listeners. This leads to loss of effective communication among people. Also, devices which are purely based on speech such as speech recognition devices fail poorly, due to adverse noise conditions. So, speech is one of the attracted research areas in the past

V. R. Balaji (✉) · J. Sathiya Priya · J. R. Dinesh Kumar · S. P. Karthi
Department of Electronics & Communication Engineering, Sri Krishna College of Engineering & Technology, Coimbatore, Tamil Nadu, India
e-mail: balajivr@skcet.ac.in

G. Ranganathan et al. (eds.), *Inventive Communication and Computational Technologies*, Lecture Notes in Networks and Systems 145,
https://doi.org/10.1007/978-981-15-7345-3_61

decade. In recent years, enhancement techniques gradually shifted from conventional methods to newer methods such as deep learning. In Deep learning techniques, the adaptiveness is higher, when compared to signal processing methods. The adaptive nature comes from deep learning by means of training a specified algorithm for certain duration and iterations can be done till it gets converged with the target. [1]. In recent studies, various deep learning algorithms have been used for speech enhancement methods. Promised performance improvements can be achieved by a deep learning approach over the signal processing method. The supervised task involved in deep learning, trains the input signals so that it is easier to process the signals. For enhancing the signal, nonlinear mapping of the clean and noisy signal is performed. The ideal ratio mask concept was also used for speech enhancement for extracting the enhanced speech from noisy speech.

A phase-sensitive mask approach is used to provide good signal to distortion ratio value by finding the phase difference to obtain better performance. In some cases, various ratio mask is used in deep learning for speech enhancement. Many approaches are used to reduce the MSE between the output and training target. If the value of the MSE signal is zero, the iteration will get over and the target will be reached. But the value of MSE may not be zero since the residual is high based on the SNR value. If the SNR value is low, the MSE value will be high. So MSE value cannot be taken as a criterion for enhancing the speech quality. The other possibility to enhance speech is Short Time Objective Intelligibility measure (STOI) which is based on the subject's observation. This metric shows accuracy in terms of measuring intelligibility. Perceptual metrics such as perceptual evaluation of speech quality and perceptual evaluation methods are also used for speech enhancement. These metrics also can be combined with Reinforcement Learning (RL) for enhancing speech. [2]. This research work focuses on slantlet transform combined with deep learning for speech enhancement . The problem in processing speech is differentiated silence sounds and speech signals. Slantlet transform does the processing by taking the speech signals as frames and deep learning is used in depth into the signal for analyzing the coefficients. To further improve efficiency, a distinct filter is used to reduce the noise. In this transform, the speech frames are overlapped between ranges of 50% and 75%. SLT minimizes artifacts even though the speech signals get overlapped. So it is combined with pitch synchronous analysis for speech enhancement [3]. In the deep learning technique, large data sets are used for training with specified activation function for processing and restoration. The pitch synchronous analysis combined with the Maximum Alignment technique also helps in selecting the proper window. The Hanning window filters signal coefficients for further processing.

2 Related Works

2.1 *Slantlet Transform*

Slantlet Transform (SLT) is similar to DWT. But the advantage is better time localization available in this transform. DWT is constructed as a tree structure by having iterated banks but Slantlet transform will have a parallel structure with parallel branches. The input speech data is applied to filter structures as shown and then it is down-sampled using the factor of four to obtain coefficients. The coefficients are the then threshold by a selective factor. To reconstruct the data, the inverse transform has to be applied based on a threshold value. SLT has a higher capacity to secure the speech data and this enhances the value. So SLT based transform is suitable choice for enhancing the speech signals. For reducing the noise further, hybrid vector wiener filtering is used.

2.2 *Slantlet Transform Based Speech Enhancement*

The proposed method is illustrated in Fig. 1. The degraded signal is segmented into speech frames. Then filtering is done with the help of a suitable filter. With the help of speech frames voiced or unvoiced decision is made for further processing of the signal [4]. Based on this decision, the time shift will be either a fixed one or shifted by one type period. One pitch period will be shifted if the frame is a voiced signal or it will be retained as such. Here the window has to be adapted based on the speech properties. The window should be flexible and must not be fixed [5] (Fig. 2).

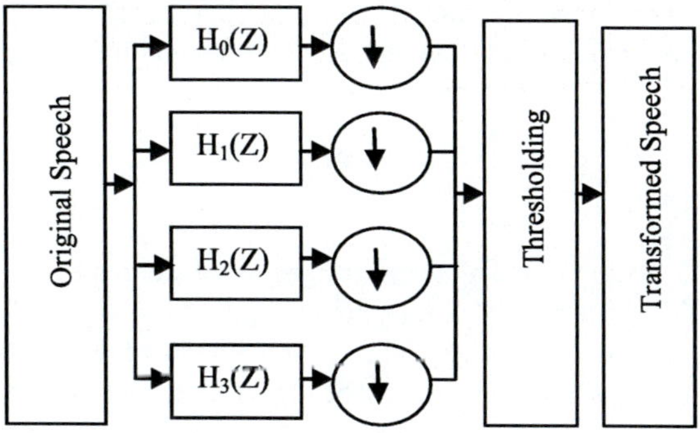

Fig. 1 Two-level SLT based data transformation

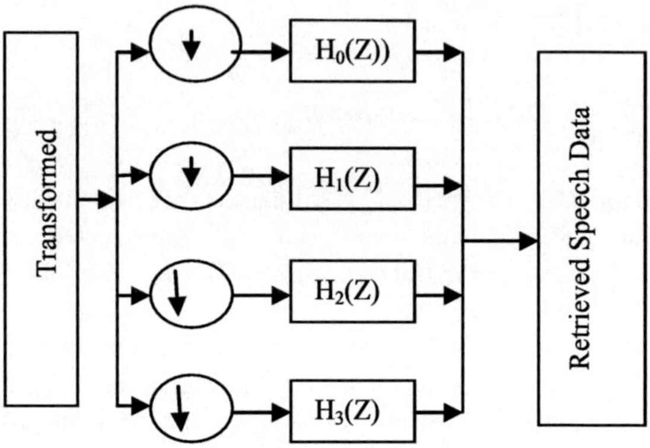

Fig. 2 Two-level SLT based reconstruction scheme

2.3 Pitch Synchronisation

For the implementation of SLT, the extraction of pitch period should be done. But, most of the algorithms can be able to obtain the pitch period only in clean situations. So the noise reduction has to be done initially by filtering Wiener filtered speech $\hat{\xi}_{m,k}$ can be given by:

$$\hat{S}_{m,k} = \frac{\hat{\xi}_{m,k}}{\hat{\xi}_{m,k} + 1} Y_{m,k} \tag{1}$$

where SNR $\hat{\xi}_{m,k}$ is the estimated apriori signal. Once the noise reduction is done, enhanced speech can be calculated by performing an inverse slantlet transform. One of the simple methods is the autocorrelation method due to robustness in a noisy environment. For performing time shifting, the extraction of the pitch period is done followed by clipping. The level of clipping is done by taking the samples from the speech frame. The autocorrelation function of the output signal $\hat{s}(n)$ is:

$$R(n) = \sum_{m=0}^{N-m-1} \hat{s}(m)\hat{s}(n+m) \tag{2}$$

The presence of a voiced signal can be found by the presence of peak magnitude. The absence of peak magnitude implies the absence of a voiced signal [6]. The pitch period determines the length of the window followed by window analysis.

The window length is to be higher for longer pitch period. The weighting function has to be calculated based on the current and the previous frames in real time. The weighting function is to be optimized for getting the enhanced speech signal [7]. The maximum alignment technique is used further to improve the pitch synchronization. Based on the SLT coefficients, the window shift is to be realigned and the signal can be represented using different frequencies and amplitudes. The coefficients are reordered with the help of basis functions [8].

3 Methodology

3.1 Deep Learning

Machine learning is one of the emerging fields and Deep learning is one among them. It purely depends on algorithms that can be trained for certain iterations. The set of neurons is used for this training purpose. The input signal is sent to one set of neurons and the output can be obtained from the other set. The input signal gets modified in the second layer whereas the first layer receives the actual input. Based on the application, the number of layers between input and output gets varied and parallel processing is done to get the exact output. The accuracy of the output depends on the number of iteration. It is used in applications such as speech processing, speech recognition, and various recognition techniques [9]. Various architectures used in deep learning are Generative deep architectures, Discriminative deep architectures and Hybrid deep architectures. Among these complex deep learning architectures, the most commonly used are Deep feed-forward networks, Convolution networks, and Recurrent Networks [10].The main aim of an RBFNN is used for approximation of a function. In a classifier, $y = f * (x)$ maps an input x to a category y. The mapping is done using $y = f\ (x;\ \theta)$. It updates the parameter value θ which gives the best approximation [11]. For analyzing the speech signal 15 ms frame is used. Based on voiced speech or unvoiced speech, the speech frame can be taken either as long or short frame [12] (Figs. 3 and 4).

3.2 Radial Basis Forward Neural Network

RBFNN uses radial basis functions as an activation function for an artificial neural network. The RBFNN structure has 3 layers that uses feed forward technique.

1. Input Layer: Input signal is given which is linear.
2. Second layer: This layer is Non-linear and uses Gaussian functions.
3. Final layer: Gaussian outputs are combined.

During training, the tap weights are updated in the second layer and third layer. RBFNN is best suited for optimization. It can be done with the help of 5 parameters such as

1. Weights in the Second and Final layer.
2. Activation function.
3. Center of activation functions.

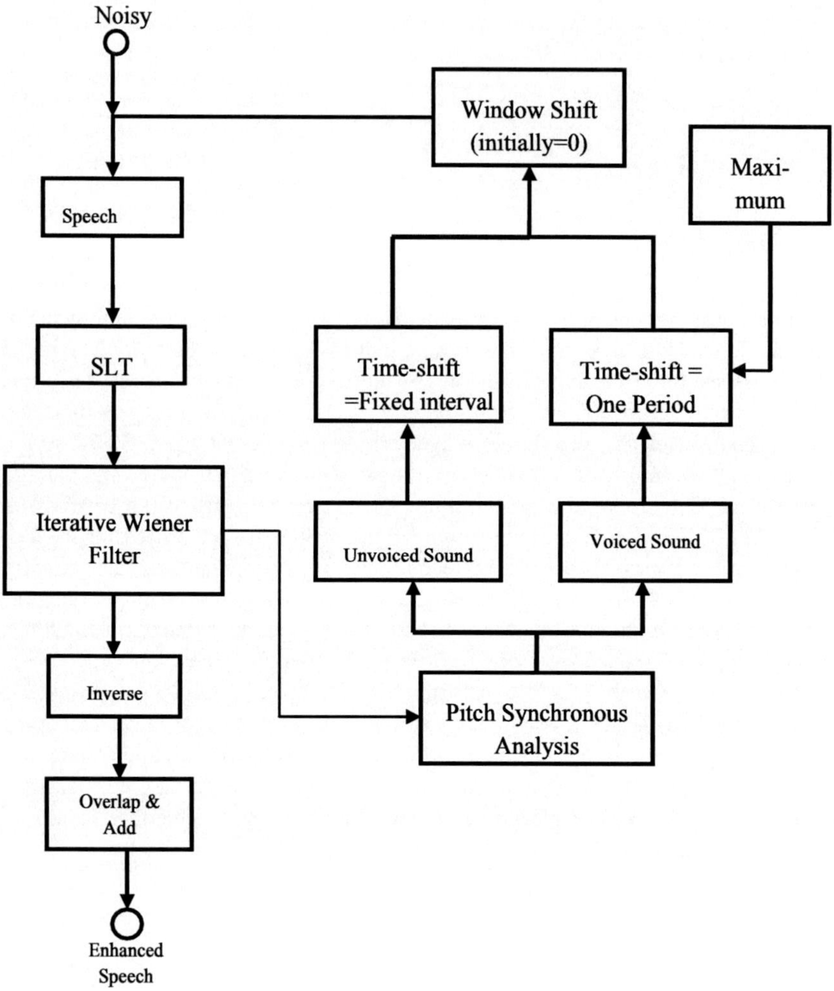

Fig. 3 Proposed block diagram for speech enhancement

4. Distribution of the center of activation functions.
5. The number of hidden neurons.

The network training starts by determining the weights between the second layer and final layer. This algorithm has many advantages when compared to conventional algorithms to overcome several issues. Since the training duration in this method is reduced and has the feature of generalization capacity it is applicable for real-time cases [13]. The activation functions will have similar characteristics based on center and distribution functions. The Gaussian coordinates are determined by means of the K-means clustering algorithm. The approximation of a function can be done with the

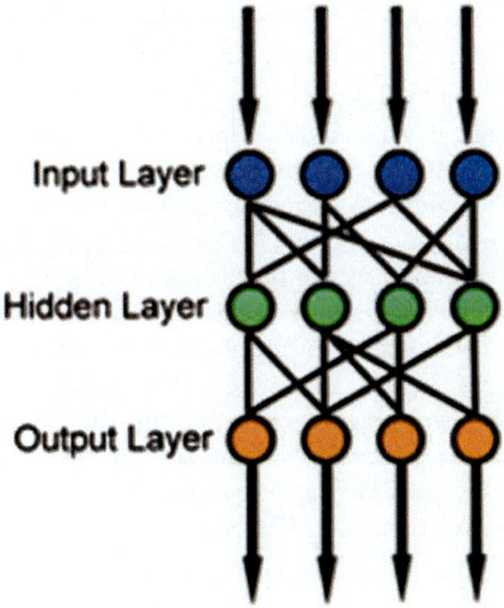

Fig. 4 Feed-forward neural network

help of the hidden layer [14]. The accuracy can be achieved by adding more number of hidden layers. RBFNN is popularly used in signal processing applications even though it has low classification speed. The optimization is done by means of faster training.

3.3 Window Function

For manipulating signals, the window function is needed for analysis of certain duration in a continuous signal. In most signal processing applications, a rectangular window is used for analysis. But the problem in using the rectangular window is the spectral leakage effect. If two signals have the same frequency, the overlapping of spectral leakage will happen. But if there are two signals with varied frequencies, one signal will obstruct the other signal. While selecting the window, side lobes play an important role [5]. A rectangular window with strong side lobes will be selected. Slantlet transform has some disadvantages when using a rectangular window because of having discontinuities at the endpoints. Due to discontinuities, the coefficient's value may get changed. This leads to finding a new windowing technique for speech application. For audio coding applications, a sine window is used to provide better attenuation. Similarly, a rectangular window is used when the signal is having high strength due to the narrow main-lobe. So while selecting the window, there should be a mutual relationship between spectral resolution and leakage effect. Hann window is very popular since it can avoid spectral leakage.

3.4 Hybrid Vector Wiener Filter

An optimal filter is needed to handle the real-time coefficients in Mean square error logic. It is based on a priori SNR for wiener filter implementation. Various ways can be used to calculate the value. Decision directed approach is one way to calculate the values. With respect to multiple noisy signals, observations are done over a continuous signal [15]. There is a need to recover the original speech signal from the continuous speech frames [16]. It may also lead to inaccuracy if the algorithms are not able to separate the discrete signals. In that case, the signal has to be modeled as a continuous time signal. The extracted samples are used for calculating minimum Mean square error [17].This wiener filter helps in noise reduction and also to reduce the Minimum Square Error of the speech signal. The extracted signal which is noise free is processed by Time Domain Pitch Synchronous Overlap-Add method. With the help of pitch synchronization, the processing is done by dividing the speech into voiced or unvoiced based on the pitch period [3].

4 Results and Discussion

The proposed system is evaluated using objective measures such as STOI (Short Time Objective Intelligibility), Segmental SNR (SegSNR), Perceptual Evaluation of Speech Quality (PESQ) [18]. Ten different segments of speeches (half females and half males), are randomly chosen from the TIMIT database. They are resampled at 8 kHz and corrupted by three additive noise types including white noise, fan noise and car noise. The total speech duration of all these test speech segments is 313.998 s including the silence period. The speech segments may contain voiced and silence period. It is assumed 50% are voiced speech [19].The proposed RBFNN based technique along with SLT transform was evaluated with the above-said parameters for two different windows hanning and rectangular [4]. The three different noise types are evaluated with the two windows and rectangular window produced better results under various SNR than hann window, since it has stronger side lobes. Since SegSNR is similar to the opinion score of subjects, it is used to determine the enhanced speech values. The Segmental SNR is calculated for various noise types and it is trained with RBFNN. The values are compared before and after the training with the network under different SNR values. The value of segmental SNR is based on input SNR values [8]. It will predict the quality of enhanced speech since each speech frame is fixed to a threshold level. An objective measurement tool that is used to measure the quality of speech is PESQ which is described in the ITU-T recommendation. It is purely based on listening tests by subjects. The subjects are made to observe both the signals, processed and noisy speech frames. Based on the observation, they provide the score as high, low or medium [15]. By taking the average of that score the mean opinion score can be calculated. So the two parameters are used to evaluate the speech intelligibility [20]. SegSNR is an accurate evaluation of speech quality, whereas the PESQ provides accurate values based on speech distortion. PESQ is a better measure when compared to conventional methods since it is highly reliable. Before verifying the SLT based speech enhancement method, the window functions are to be compared. The window length is predefined as 32 ms. SegSNR results are shown in Fig. 5. It clearly shows that the rectangular window is better than Hann window under various SNR.

To illustrate the benefits of the proposed method, the segmental SNR results are compared before and after training with the help of Radial basis function neural network. The results are better after RBFNN training. Based on the noise type, there is a slight variation in the values of Seg SNR. Table 1 lists the comparison of ΔSegSNR results. Various noise types are taken for comparison and the input SNR is 0, 5, 10 and 15 [9]. The results indicate the proposed method gives better Δ SegSNR after training with RBFNN (Figs. 6 and 7).

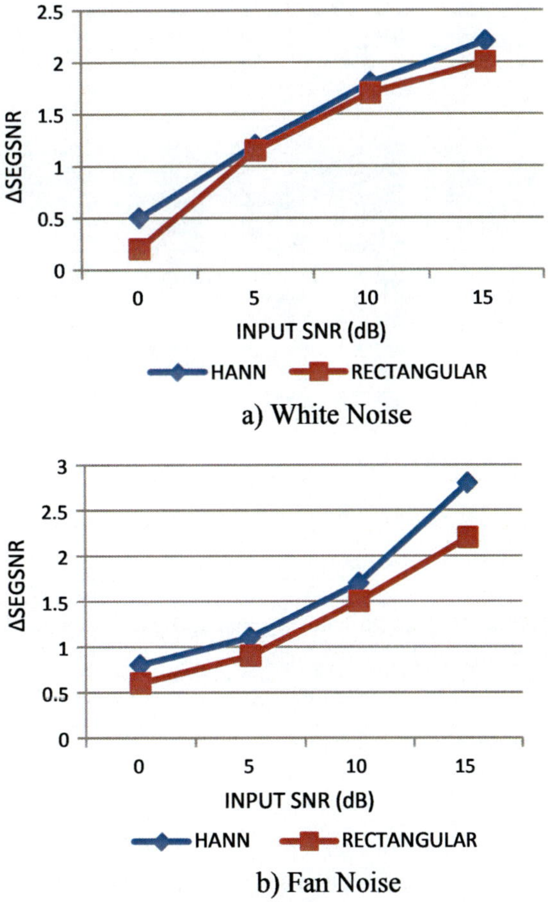

Fig. 5 Comparison of SegSNR results for various noise sources

Table 1 Δ SegSNR results comparison

Noise type	SNR (dB)	SLT using RBFNN—Before training	SLT using RBFNN—after training
White noise	0	6.23	6.01
	5	5.33	5.13
	10	4.92	4.63
	15	3.67	3.28
Fan noise	0	9.67	9.41
	5	9.33	9.13
	10	8.92	8.61
	15	8.22	8.01
Car noise	0	12.66	12.36
	5	12.01	11.99
	10	11.63	11.12
	15	10.22	9.83

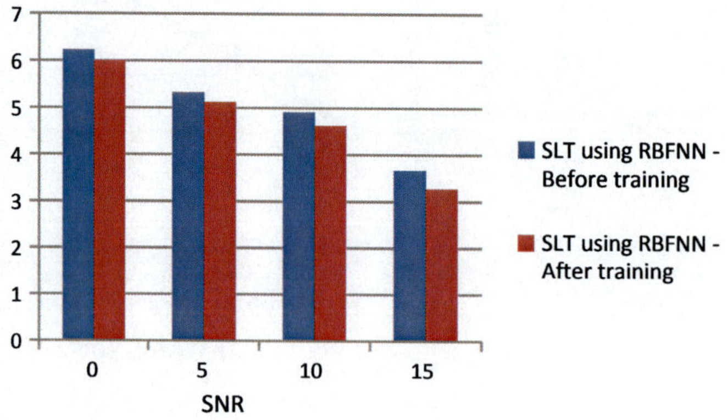

Fig. 6 Comparison of ΔSEGSNR results for white noise

Table 2 shows the comparison of PESQ and STOI measures for white noise and car noise. It also shows that RBFNN provides better results in terms of PESQ and STOI scores (Figs. 8 and 9).

5 Conclusion

This research paper focuses on using RBFNN for improving the speech intelligibility of the signal. Conventional signal processing methods provide speech enhancement with the help of transforms. Here the Slantlet transform is used along with RBFNN

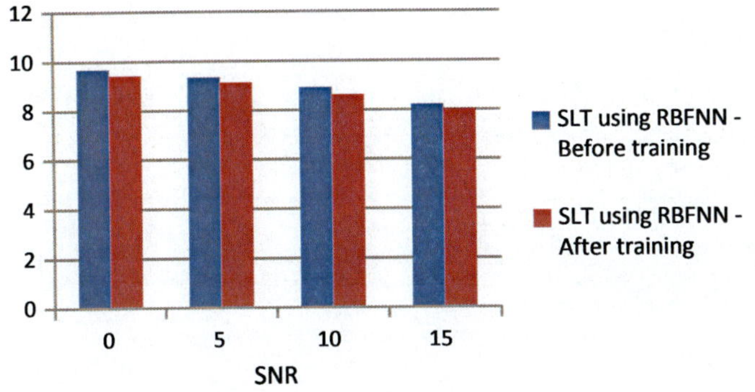

Fig. 7 Comparison of ΔSEGSNR results for fan noise

Table 2 Comparison of PESQ and STOI score for white noise and car noise

SNR	Before DNN training		After DNN training		Before DNN Training		After DNN training	
	White noise				Car noise			
	PESQ	STOI	PESQ	STOI	PESQ	STOI	PESQ	STOI
20	3.452	0.988	3.052	0.962	4.034	0.963	3.932	0.982
15	2.879	0.921	2.341	0.9	3.634	0.912	3.42	0.943
10	2.397	0.897	2.032	0.812	2.317	0.845	2.124	0.821
5	1.984	0.765	1.782	0.721	1.764	0.723	1.475	0.711
0	1.659	0.526	1.524	0.498	1.329	0.513	1.299	0.467

Fig. 8 Comparison of PESQ score

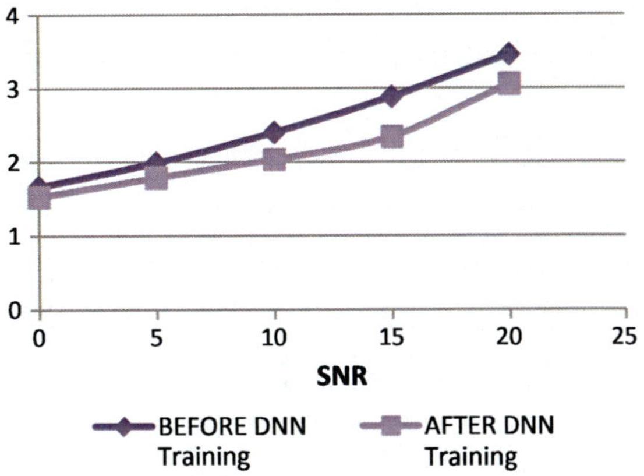

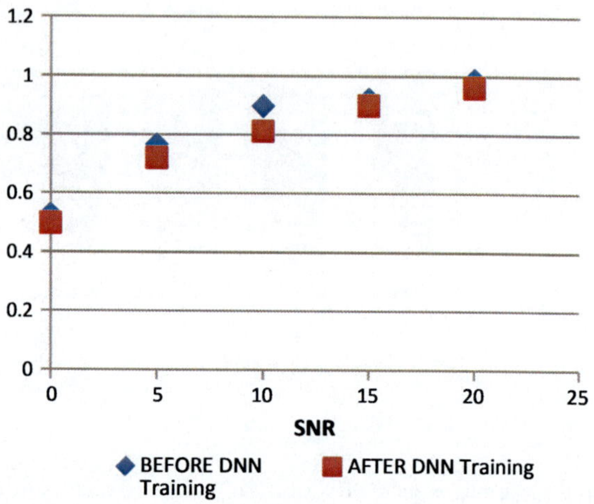

Fig. 9 Comparison of STOI score

which helps to improve the accuracy since it performs training till the convergence is obtained. In this method, the signal taken from the TIMIT database is split into overlapping frames which produces the coefficients. The coefficients are then given to a filter and trained using the algorithm for optimum results. Once the training is done, it is combined with pitch synchronous analysis to detect the silence period. Based on the pitch period, the amount of shift can be decided. The performance measure shows that radial basis function reduces the noise coefficients which in turn increases the speech quality. In the future, the activation function can be adjusted such that the training can be done in less time so that the convergence rate can be faster.

References

1. Le Roux J, Vincent E, Erdogan H (2016) Learning-based approaches to speech enhancement and separation. In: INTERSPEECH tutorials
2. Williamson DS, Wang Y, Wang DL (2016) Complex ratio masking for monaural speech separation. IEEE/ACM Trans Audio Speech Lang Process 24:483–492
3. Balaji VR (2018) A comparison of compression sensing algorithm and DUET algorithm for advanced DCT based speech enhancement system for vehicular noise. Int J Pure Appl Mathematics 119(12):1385–1394
4. Mirsamadi S, Tashev I (2016, May) A causal speech enhancement approach combining data-driven learning and suppression rule estimation. In: Proceedings of inter speech
5. Li Y, Kang S (2017) Deep neural network-based linear predictive parameter estimations for speech enhancement. IET Signal Process 11(4):469–476
6. Zhang X, Wang Z-Q, Wang DL (2017) A speech enhancement Algo Rithm by iterating single- and multi-microphone processing and its application to robust ASR. In: IEEE international conference on acoustics, speech and signal processing, pp 276–280
7. Erdogan H et al (2016) Wide residual blstm network with discriminative speaker adaptation for robust speech recognition. In: CHiME-4 workshop

8. Barker J, Marxer R, Vincent E, Watanabe S (2017) The third "CHiME" speech separation and recognition challenge: analysis and outcomes. Comput Speech Lang 46:605–626
9. Kounovsky T, Malek J (2016) Single channel speech enhancement using convolutional neural network. In: IEEE international workshop of electronics, control, measurement, signals and their application to mechatronics (ECMSM), pp 1–5, 666–676
10. Koizumi Y, Niwa K, Hioka Y, Kobayashi K, Haneda Y (2017) DNN-based source enhancement self-optimized by reinforcement learning using sound quality measurements. In IEEE international conference on acoustics, speech and signal processing (ICASSP), pp 81–85
11. Balaji VR, Sathiya Priya J (2018) Enhancement of speech signal modified binary mask based algorithm for vehicular noise. J Adv Res Dyn Control Syst 10(12-Special Issue)
12. Nugraha AA, Liutkus A, Vincent E (2016) Multichannel audio source Separation with deep neural networks. IEEE/ACM Trans Audio Speech Lang Process 24(9):1652–1664
13. Kolbaek M, Tan ZH, Jensen J (2017) Speech intelligibility potential of general and specialized deep neural network based speech enhancement systems. In: IEEE/ACM Trans Audio Speech Lang Process 25(1)
14. Balaji VR, Maheswaran S, Babu MR, Kowsigan M, Prabhu Venkatachalam K (2020) Combining statistical models using modified spectral subtraction method for embedded system. Microprocessors and Micro Systems, Elsevier
15. Dinesh Kumar JR, Ganesh Babu C, Balaji VR (2019) Analysis of effectiveness of power on refined numerical models of floating point arithmetic unit for biomedical applications. In: AIP scopus indexed proceedings of international conference on advances in materials processing and characterization. ICAMPC 2019
16. Balaji VR, Sathiya Priya J, Dinesh Kumar JR (2019) FPGA implementation of image acquisition in marine environment. Int J Oceans Oceanography 13(2):293–300. ISSN 0973-2667
17. Weninger F, Erdogan H, Watanabe S, Vincent E, Le Roux J, Hershey JR, Schuller B (2015) Speech enhancement with LSTM recurrent neural networks and its application to noise-robust ASR. In: International conference on latent variable analysis and signal separation, pp 91
18. Barker J, Marxer R, Vincent E, Watanabe S (2015) The third "CHiME" speech separation and recognition challenge: dataset, task and baselines. In:IEEE workshop on automatic speech recognition and understanding, pp 504–511
19. Goehring T, Bolner F, Monaghan J, Dijk B, Zarowski A, Bleeck S (2017) Speech enhancement based on neural networks improves speech intelligibility in noise for cochlear implant users. Hear Res 344:183–194
20. Gannot S, Vincent E, Markovich-Golan S, Ozerov A (2017) A consolidated perspective on multi-microphone speech enhancement and source separation. IEEE/ACM Trans Audio Speech Lang Process 25:692–730

Detection and Deactivation of Application Layer-Based DDoS Attack from Private Tor Network

Yogita Deepak Mane and Uday Pandit Khot

Abstract The Botnet is the network of personal computers with malicious software, connected to the Internet and controlled by BotMaster. It is considered to be the most catastrophic cyber-security threat such as phishing, spamming, DDoS, or any kind of malware. In this work, our contribution is to provide a secure network service against DDoS attacks for Tor users. The detection and deactivation of Application Layer Based DDoS attack known as "Hammer.py Bot" is explained in detail. Application layer based (L7) attacks are serious yet challenging to detect. The detection framework is based on Delta time ($T\Delta$), which is the time between two threads/requests in milliseconds. When attack "Hammer.py" does exist in the target machine, it will start sending multiple requests to victim machine, the machine became busy and stops performing its normal activities. The proposed technique leads to detect the bot as soon as it exists in the system. The average true positive rate (TPr) and accuracy of the proposed technique is 95.92%, and the average false negative rate (FNr) and error rate is 4.08%.

Keywords Botnet · Bot · PTN · Trusted middle node · DDoS · Application layer based attack · Delta time · TPr · FPr · Accuracy · Error rate

Y. D. Mane (✉)
Department of Computer Engineering, St. Francis Institute of Technology, Borivali, Mumbai, India
e-mail: yogita.ydmane@gmail.com

U. P. Khot
Department of Electronics and Telecommunication, St. Francis Institute of Technology, Borivali, Mumbai, India
e-mail: udaypandit@rediffmail.com

G. Ranganathan et al. (eds.), *Inventive Communication and Computational Technologies*, Lecture Notes in Networks and Systems 145,
https://doi.org/10.1007/978-981-15-7345-3_62

1 Introduction

The term Botnet is a crew of unprotected machines connected in a network and controlled by BotHearder (BotMaster). The BotHearder performs illegal operations through Botnet such as DDoS attack, credential frauds, keystroke attacks and many more. There are lot of tools and research available about the Bots, and how to detect the Bot activities [1, 2]. The five major existing techniques to detect the Bot activities are Signature based, Anomaly based, DNS based, Network based and Mining Based [1]. Lately, to make the detection framework more difficult the BotHearder moved his activity from normal Web browser to low latency anonymous communication such as Tor browser. Tor browser is nothing but "The Onion Router", which sends the data from one end to another through three different relays/nodes (namely Entry, Middle and Exit Node) in concealed manner. Therefore it becomes very difficult to track the actual source or the attacker. The Botnet detection tools which are available at present are became ineffective against Tor based Bots [3, 4].

The purpose of this research work is to identify Application Layer based Bot activity under low latency anonymous communication (Tor).To detect the Bot under low latency anonymous communication, a simulation environment [5–7] has been developed which is nothing but Private Tor Network. This provides full control for doing practical experiments. Doing experiments on Tor browser is not advisable because it is highly secure. The simulation environment is developed in Java and named as Private Tor Network (PTN) with trusted Middle node, as Middle node is acting as a bridge between Entry and Exit node. The trusted Middle node is identified by calculating trust value based on free CPU, Free RAM and number of cores/processors available in the system/Nodes as shown in Eq. 1 in Sect. 3. The Difference between PTN and Live Tor Network (LTN) is that PTN is not making use of Diffie Hallman (DH) algorithm because this algorithm is use to exchange the data over public network and the proposed system is working under Local Area Network in a Computer Laboratory. AES-128 bit and RSA-1024 bit algorithms are used to create a PTN to send the data between client and server in a concealed manner. The Security breach of the PTN is measured in two ways. The first breach is transferred of data in concealed manner from source to destination and second by making the highly secure Middle node. The testing is done in computer laboratory with 30 nodes. The communication between all nodes is done through socket programming.

Further paper is divided into 6 sections as Problem statement, Sect. 3 is Circuit construction and Data Transfer in Private Tor Network with Trusted Middle Node, Sect. 4 describes Related work, Sect. 5 gives detail explanation of Working of Hammer.py Bot, and Sect. 6 is Detection of hammer.py Bot followed by Conclusion.

2 Problem Statement

The focus of research work is on Detection and Deactivation of Application Layer (L7) based DDoS attack from Private Tor Network. Application layer based (L7) attacks are serious yet challenging to detect because of their secretiveness and appearing to request legitimacy. Earlier, a research on detection and deactivation of "Tor'sHammer" Bot is published in Paper [6] along with implementation details of Private Tor Network. This paper gives a full insight view of detection and deactivation of "Hammer.py" Bot. This Bot is available publically for education purpose. In the research work, the Private Tor network has been implemented with a trusted Middle Node to do the experimental setup. The performance analysis and comparative analysis is done with the help of Hammer.py and Tor'sHammer Bot detection activity.

3 Circuit Construction and Data Transfer in Private Tor Network with Trusted Middle Node

Circuit construction and Data Transfer is the essential part of Private Tor Network (PTN). As shown in Fig. 1, the first step to create a Private Tor Network under LAN infrastructure by creating a Client socket and Server socket with all available nodes. Once the socket connection gets established, the Server socket will start listening to the ports and IP addresses that are available in the network for connection, while the client socket starts with the connection with the server socket as part of connection

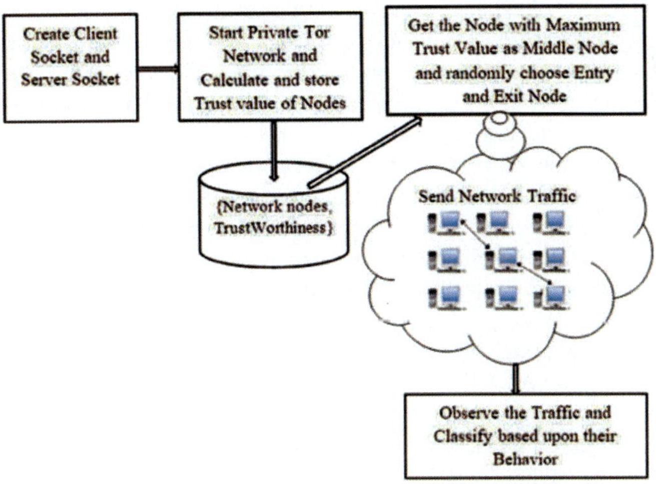

Fig. 1 Circuit construction and data transfer in PTN with trusted middle node

establishment. The next step is to start the Private Tor Network and to calculate the Trust factor of all active nodes. After that stores the calculated trust values into a database (D1: {Network Nodes, Trustworthiness}) for further analysis. The Trust value of all available IP addresses are calculated based on free physical RAM, free CPU usage and number of logical processors available by using Eq. 1.

$$\text{Trust} = (\text{physicalfreeMemorySize} + ((100 - \text{cpuPercent}) * \text{cpuCount})) \quad (1)$$

From stored trust values in D1, the system finds and assigns Middle Node with maximum trust value. After that it will randomly choose the Entry and Exit Node from a database D1 based upon quick response to ARP—a command. Once the connection gets established, both client socket and server socket start with sending and receiving the data in a network. At that time it starts observing the network traffic and classify it based upon their behavior which is shown in Fig. 1.

4 Related Work

There are 3 types of DDoS attacks presents based on their features such as Volume based attack, Protocol based attack, and Application based attacks. The purpose of volume based attack is to saturate the network bandwidth by performing UDP flood, ICMP flood and other spoofed packet floods and it is measured in bits per seconds. The server resources (such as CPU, RAM) are consumed by protocol based DDoS attack by using SYN flood, Ping of Death attack. Whereas the purpose of the Application layer attack is to crash the server by using either low rate or slow rate attacks, HTTP GET/POST floods and it is measured in requests per seconds. Application based (L7) attacks are serious yet challenging to detect because of their secretiveness and appearing to request legitimacy as it does not saturate the network bandwidth as well as not consuming system resources. The proposed system is used to detect Application layer (L7) based attacks. For experimental purpose, example of two Application Layer based attacks is taken, they are Tor'sHammer and Hammer.py. Tor'sHammer is slow rate HTTP POST attack whereas Hammer.py is HTTP GET attack. These bots are considered for research purpose as it is mainly build for educational purpose. In Paper [6] successfully detected and deactivated "Tor'sHammer" Bot based on Delta time value of two successive threads. The same methodology is used to detect and deactivate "hammer.py" bot. It performs DDoS attack and makes the victim machine busy. As a side effect the victim machine stops responding to the other legitimate requests.

Paper [5] gives a brief idea about the security to Darknet where authors have described different types of attacks under Tor network. The taxonomy represented in paper can be used to understand the functioning of cyber-attacks.

In paper [6] authors describes the methodology to detect and deactivate "Tor'sHammer" DDoS attack from Private Tor Network. Delta time between two threads is used to determine the presence of Bot.

The paper [7] focuses on study of several DDoS attack with bandwidth utilization property. This paper includes the effect of bandwidth based DoS attack in terms of cost and performance of the Tor network. For the research Shadow simulation environment is used to generate Private Tor Network.

Predication of user behavior in Tor network is presented using deep recurring neural network concept in paper [8].

5 Working of Bot: Hammer.py

Hammer.py [9] is Python based code used to perform DDoS attack. The application of this script is to perform load testing and it is available for research purpose. The load testing can be in form of either URL Flooding/HTTP POST Flooding or TCP Flooding. The default number of attacking threads for Hammer.py is 135. Hammer.py script will not allow creating more than 1800 queue items or Threads or requests to avoid memory crash. These threads can run for indefinite time and waits only if the target stops responding. The purpose of Hammer.py Bot is to gain experience with Socket programming by creating a DDoS attack. The Bot coordinates with victim machine by using BotMaster. BotMaster remotely controls all the activities of Botnet. As a result Bot attack the Target Server at the specified time with only two main parameters as target machines IP address and port number as shown in Eq. 2. Once you fire the command given in Eq. 2, it will start sending number of threads on specified target IP.

C:\Work\project\hammer-master > python hammer.py -s TargetIP -p Port number (2)

Command to do the attack on 192.168.2.118 IP address with 5000 as target port number is shown in Eq. 3. Once the command gets executed the Bot will start with his activity in terms of sending with infinite threads to target machine. And only source can stop sending the requests to destination by pressing Ctrl-C command from source machine.

C:\Work\project\hammer-master > python hammer.py–s 192.168.2.118 -p 5000 (3)

The output to the destination/Victim machine will be as given in below print statement written in Bash code [10]. As an output to source/server it gives timestamp of attack and Message as "Packet sent: hammering" as shown in Fig. 2.

(("\033[92m",time.ctime(time.time()),"\033[0m \033[94m <--packet sent! hammering--> \033[0m").

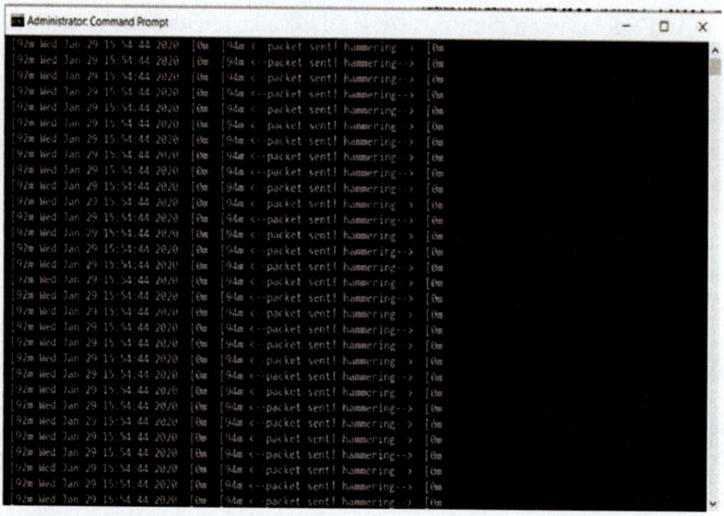

Fig. 2 Hammer.py attack

6 Analysis of Hammer.py Bot Detection

When the source performs the Hammer.py Attack on Victim Machine, Delta time of Victim machine remains zero as it only receives command from Attacker/Source and not receiving any acknowledgment from source to forward data to destination, so the request cannot complete with 3-way handshaking. As a side effect, Victim machine became busy for long time and not able to provide services to the legitimate nodes. The Delta time of each thread is less than 10 ms for more than 95% of the Total number of Bot Threads (TBT) based on the observed Delta time values after Hammer.py attack. To detect the malicious activity, the traffic with Application Layer based Bot (Hammer.py) has been manually analyzed for more than 100 observations. And in paper the analysis of three observations are given. The $T\Delta$ observations for example 1 are shown graphically in Fig. 3 with 49 attacking threads where

Fig. 3 Example 1: delta time observation of victim and attacker machine

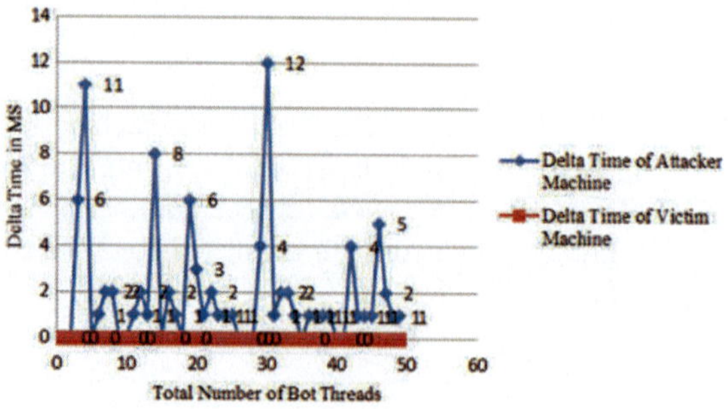

Table 1 Delta time of malicious activity

Delta time (ms)	Observed number of values of Delta time of malicious activity for example 1,2 and 3 respectively		
	TBT = 49	TBT = 135	TBT = 135
$T\Delta < 10$	47	128	132
$10 \leq T\Delta < 25$	2	7	3
$T\Delta \geq 25$	0	0	0

Victim machine Delta time is zero for all the threads and attacker machine delta time varies based on threads sent to victim machine. Total Bot threads for Example 2 and Example 3 is 135 as illustrated in Table 1.

$$T\Delta = \text{Current time} - \text{Last time} \quad (4)$$

Table 1 gives an idea about why to choose 10 ms as a Delta time. After Hammer.py attack the analysis is perform to chose threshold value of Delta time for attack detection ($T\Delta < 10$ ms, 10 ms $\leq T\Delta < 25$ ms and $T\Delta \geq 25$ ms). And reached to the conclusion that there are more number of threads (Avg 95.92%) lies in the range of $T\Delta < 10$ ms, so for the proposed system Delta time ($T\Delta$) of malicious activity is consider as less than 10 ms. Once the bot activity is detected the next task is to deactivate the bot activity by closing the socket connection.

7 Detection of Hammer.py Attack

The Bot is getting detected as soon as delta time hits to less than 10 ms this will help starting preventive action at the earliest. The Delta time value is evaluated manually by seeing the results of Delta time of attacking traffic and Delta time with normal traffic. Detail algorithm is explained in our contribution in paper [6]. As shown in Fig. 4; the first phase is Passive network Traffic monitoring which observes the incoming traffic. In Preprocessing, Delta time ($T\Delta$) for each process is calculated in milliseconds. At that instance, calculated Delta time is processed to next phase to identify the pattern of traffic then that pattern is used to do the traffic classification based on two condition; if ($T\Delta \geq 100$) then it is a normal traffic and if ($T\Delta < 10$) then Bot is detected [6]. Once the bot is detected; next to deactivate bot activities, close the socket.

Figure 5 shows the result of "hammer.py" Bot detection from Private Tor Network. Hammer.py is sending multiple threads to specified node along with port number in order to break the service between Source/Server to Destination as shown in Eq. 3. Once the bot gets initiated; for each thread or request system will check the ΔT and if ΔT is less than 10 ms then it gives an output message as "*Bot activity detected and it seems like Bot. Closing Connections*". The analysis of Bot activity includes

Fig. 4 Tor'sHammer detection and deactivation

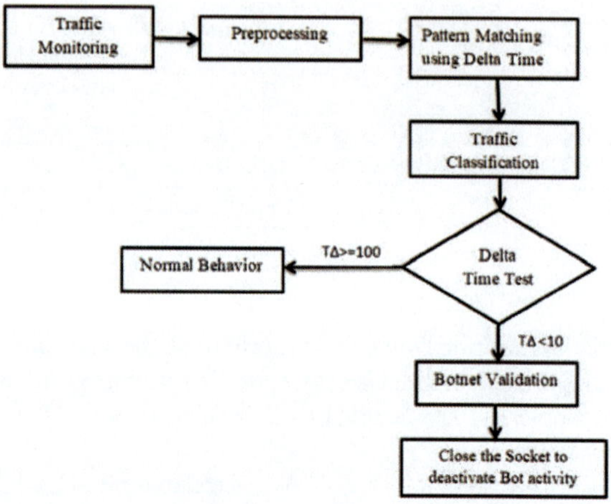

system timestamp, Delta time of Victim Machine and Delta Time of Source Machine as shown in Fig. 5 for the example 1 with 49 attacking threads.

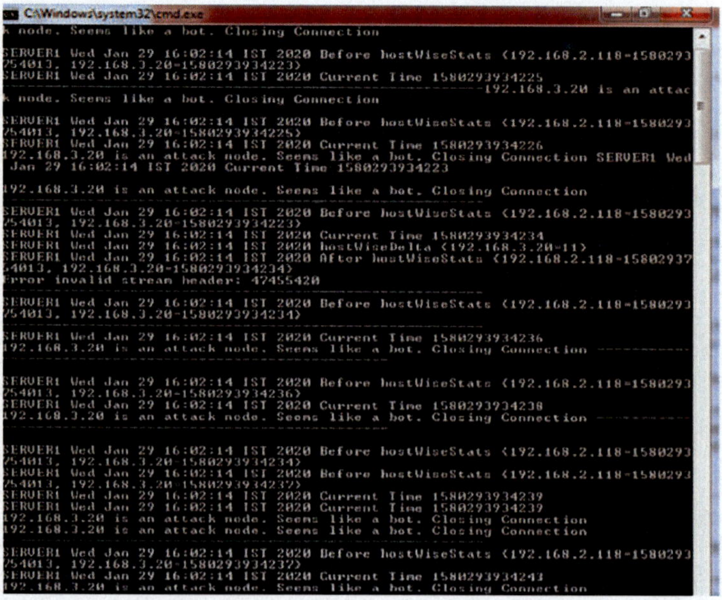

Fig. 5 Detection of Hammer.py Bot activity

Table 2 Performance analysis

Name of Bot	Observation	TBT	Performance parameters			
			TPr (%)	FNr (%)	Accuracy (%)	Error rate (%)
Hammer.py	Observation 1	49	95.92	4.08	95.91	4.09
	Observation 2	135	94.81	5.19	94.81	5.19
	Observation 3	135	97.04	2.96	97.03	2.97
	Average performance ration ratio		95.92	4.08	95.92	4.08
Tor'sHammer [6]	Observation 1	49	93.88	6.12	93.88	6.12
	Observation 2	135	96.30	3.70	96.30	3.70
	Observation 3	135	94.07	5.93	94.07	5.93
	Average performance ration ratio		94.75	5.25	94.75	5.25

7.1 Performance Analysis

Performance analysis is evaluated by False Negative ratio (FNr) and True Positive Ratio (TPr). FNr is used to determine ratio of Number of Bot threads not identified (FN) from the Total number of Bot Threads (TBT) = TP + TN as specified in Eq. 5.

$$\text{FNr} = (\text{FN}/\text{TBT}) * 100 \tag{5}$$

TPr is used to define ratio of Number of Bot threads correctly identified (TP) from the total number of bot threads as specified in Eq. 6.

$$\text{TPr} = (\text{TP}/\text{TBT}) * 100 \tag{6}$$

As shown in Table 2 for Hammer.py Bot out of 49 bot requests the incorrectly identified bot threads are 2 as the value of Delta time is more than 10 and the condition to detect bot is ($T\Delta < 10$) so the FNr in percentage is ((2/49) * 100) = 4.08%. And the remaining bot requests (47) are correctly identified as the value of Delta time is less than 10. So the TPr in percentage is ((47/49) * 100) = 95.92%. Accuracy for Hammer.py with 49 Total Bot threads is calculated as ((47 + 0)/49) = 95.91% as shown in Eq. 7. Where TN is Number of Bot threads not identified when there are no attacking threads.

$$\text{Accuracy} = ((\text{TP} + \text{TN})/\text{TBT}) * 100 \tag{7}$$

Error Rate is measure as 1-accuracy which is calculated as shown in Eq. 8. For Hammer.py the Error Rate is 4.09%.

$$\text{Error Rate} = 100 - \text{Accuracy} \tag{8}$$

Table 3 Comparison of different detection methods

Methodology used	TPr	FNr
Normalized cumulative amplitude spectrums (NCAS)	88	12
Kalman filtering	89.6	10.4
Multifractal approach	91	9
Delta time value. (Research contribution)	95.33	4.67

Similarly performance analysis for all observed values of Hammer.py attack is calculated. Performance analysis of Tor'sHammer Bot is calculated and explained in Paper [6]. The Table 2 shows the performance analysis with respect to TPr, FNr, Accuracy and Error Rate for observation 1, 2, and 3. So, average TPr of Hammer.py and Tor'sHammer Bot is 95.33%, FNr = 4.67%, Accuracy = 95.34% and Error Rate = 5.25% for ($T\Delta < 10$ ms). In order to make an objective and fair comparison analysis, three methods that have been tested in real network are chosen and comparison is shown in Table 3.

For NCAS [11] method, TPr is 88% and FNr is 12%; by using Kalman filtering [12], the TPr is 89.6% and FNr is 10.4%; by Multifractal [13] approach, TPr is 91% and FNr is 9%. And with the help of Delta time, the TPr is 95.33 and FNr is 4.67.

8 Conclusion

This research work performs detection and deactivation of Application Layer based DDoS attacks from Private Tor Network. The system is proposed to detect and deactivate DDoS attack named "Hammer.py" bot. Our contribution toward the research proposal is to create a Private Tor Network under LAN Infrastructure, to implement Trustworthy Middle Node in Private Tor Network and to design Botnet detection system based on "Passive Network Monitoring". To meet the required expectation, a Private Tor Network is created with a trustworthy Middle Node as a simulation environment. Trustworthy Middle node is identified by Free RAM, Free CPU and number of logical processors available for the system. An analysis of traffic analysis is done manually based on a threshold value ($T\Delta$) to detect Hammer.py and Tor'sHammer bot with $\alpha_{\text{normal}} >= 100$ ms and $0 \geq \alpha_{\text{attack}} \leq 10$ ms. The advantage of the proposed system is that it detects the bot the instant it exists in the system. The Average TPr and Accuracy of the proposed technique is 95.92%; the Average FNr and Error Rate is 4.08%. The future scope of this research is to work on Volume based and protocol based DDoS attacks.

References

1. Mane Y, Devadkar K (2014) Analysis of Botnet Detection. Int J Emerg Trends Eng Dev 3(4)
2. Mane Y (2017) Detect and deactivate Zeus Bot. In: Proceedings of IEEE conference on ICCCNT 2017, Delhi, India. https://doi.org/10.1109/icccnt.2017.8203918
3. Mane Y, Khot U (2018) Botnet detection in low latency anonymous communication network: a branch of knowledge. In: Proceedings of IEEE conference on ICSCET, pp 1–6. https://doi.org/10.1109/icscet.2018.8537269
4. Casenove M, Miraglia A (2014) Botnet over Tor: the illusion of hiding. In: Proceedings of IEEE conference on cyber conflict, pp 273–282
5. Cambiaso E, Vaccar I, Patti L, Aiello M (2019) Darknet security: a categorization of attacks to the tor network. In: CEUR workshop proceedings, vol 2315
6. Mane Y, Khot U (2019) An efficient technique to detect slow rate DDoS attack from private tor network. In: Int J Electron Secur Dig Forensics. Accepted December 2019
7. Jansen R, Vaidya T, Sherr M (2019) Point break: a study of bandwidth denial-of-service attacks against Tor. In: 28th USENIX security symposium (USENIX Security 19), pp 1823–1840
8. Ishitaki T, Obukata R, Oda T, Barolli L () Application of deep recurrent neural networks for prediction of user behavior in Tor networks. In: 31st international conference on advanced information networking and applications workshops
9. (WAINA) (2017) Taipei, pp 238–243. https://doi.org/10.1109/waina.2017.63
10. Körner S (2016) hammer.py. Accessed 3 Mar 2016. https://github.com/cyweb/hammer/blob/master/hammer.py
11. McIntyre R (2019) public-bash-scripts. Accessed September 19. https://github.com/ryanoasis/public-bash-scripts/blob/master/unix-color-codes.sh
12. Zhijun W, Meng Y (2008) Detection of LDDoS attack based on Kalman filtering. ACTA Electronica Sinica 36(8):1590–1594
13. Wu Z, Zhang L, Yue M (2015) Low rate Dos attack detection based on network multifractal. In: IEEE TransDependable Secure Comput 13(5):559–567. https://doi.org/10.1109/TDSC.2015.2443807

A Critical Survey on Fractal Wearable Antennas with Enhanced Gain and Bandwidth for WBAN

Sandhya Mallavarapu and Anjaneyulu Lokam

Abstract Due to the increasing demand for wearable technology in wireless communication, the consumer demand for multiple operational bands, low cost, increased gain, wideband, and electrically low volume antenna raises gradually. On the other hand, the wearable antenna suffers from the effects of bending, crumpling, and operable electromagnetic exposure limits as it is conformal to the human body. These difficulties can be justified by the procedure of creating fractal-like structures on flexible antennas for wearable applications at various wireless standards. In this survey first, a comparison study has been made of different fractal geometry antennas on wearable fabric substrates, the techniques to improve the bandwidth and gain of such antennas, which follows the discussion of groundbreaking features and approaches in slackening these issues that are anticipated recently by the researchers in this field. These specified antennas can found applications in Wi-Fi, WLAN, and various wireless standards.

Keywords Flexible antenna · Fractal wearable antenna · Defected ground structures (DGS) · Wireless body area networks (WBAN) · ISM · Super wideband (SWB) antennas

1 Introduction

Body-worn antennas gain more popularity in the current trends due to their striking features and potentials in permitting low weight, flexible, cost-efficient, and handy wireless body-centric communications. Such antennas necessitate maintaining proper shape and angle once worn on various locations of the human body; consequently, it is essential to realize those were using supple materials and must be

S. Mallavarapu (✉) · A. Lokam
Electronics and Communications Engineering, National Institute of Technology, Warangal, India
e-mail: sandhyamallavarapu@student.nitw.ac.in

A. Lokam
e-mail: anjan@nitw.ac.in

G. Ranganathan et al. (eds.), *Inventive Communication and Computational Technologies*, Lecture Notes in Networks and Systems 145,
https://doi.org/10.1007/978-981-15-7345-3_63

small assembly construction. In the end, these antennas required to operate with the least deviation of antenna parameters such as bandwidth and radiation characteristics in the vicinity of the human body. All these desires making the antennas design more perplexing, particularly given their size insistence, effects of constructional and material cramps, near-human body effects, manufacturing difficulties, and correctness. Even though minor differences in severity concerning applications, most of these problems occur in body-worn environments

Notably, in the modeling of the wearable antenna, the planar patch antenna is more prevalent in the research community as it is made conformal for integration into clothing [1] and also, because of its benefits like reduced dimensions, lightweight, high operating frequency, capable of supporting multiple frequencies, dual-polarization and lower fabrication cost [2]. Despite these advantages, it suffers from disadvantages like lower gain and low power handling capacity. In the design of the wearable antenna, the conventional substrate materials like FR-4 and Rogers can be replaced with flexible substrates like silk, felt, fleece, denim, and washable cotton for wearable applications [3]. In general, the electromagnetic properties of such flexible/textile materials are not well known, and measuring them is quite impressive. There are many types of measurement techniques and analysis for determining the dielectric properties of fabric materials [4–6]. Usually, the measured dielectric constants for flexible/textile substrates are in the range of 1–2 [7].

Now, the planar patch antenna with a flexible substrate can be designed by first choosing the proper flexible substrate with the known dielectric properties and working frequency. Then the dimensions of the patch and the substrate can be estimated from the equations exist in the literature [8] with the help of MATLAB software. The primary rectangular patch antenna for wearable applications was designed using CST/HFSS software in [9]. Enlargement of various features of the wearable antenna can be achieved using fractal geometry, metamaterial, different feeding method, Defected Ground Structure (DGS) on the ground, and by cutting slots etc., In most cases, High gain and high bandwidth are needed in any communication system. To achieve all these features simultaneously in a single device, the concept of metamaterial is used along with microstrip antennas (MSA) [10–13]. In the context of wearer's comfort and dynamic movements of the human body, the size of the antenna and electromagnetic exposure also paid attention along with improved bandwidth and gain. To fulfill these needs, a wearable antenna is implemented with the fractal-like structure on metamaterial surfaces [14].

The rest of the paper is organized as follows. Section 2 describes the fractal geometry, and a comparison has been made for different fractal geometries on wearable fabric substrates. Section 3 describes the bandwidth and gain enhancement techniques from fractals to wearables. Section 4 describes the summary and future scope.

2 Fractal Geometry

In this day and age, the miniature antenna with healthier characteristics and multiple bands as operating frequencies is one of the tendencies in modern wireless communications systems [15]. The excellent technique to reduce the antenna size is the use of fractal structures in the design. The fractal shape is an uneven or fragmented ordered shape that can be further divided into small portions; every portion obtained is compact without altering the other characteristics. Fractals are a new branch of mathematics and art. Fractal antennas have many advantages such as; reduced size, excellent efficiencies, higher gain, and broad bandwidth, etc. In general, fractal geometries have self-similarity and self-governing features. Some widespread fractals can establish new and advanced antenna designs. They are Koch Curves, Sierpinski's gasket, Sierpinski's Carpet, Cantor's comb, and Mandelbrot set, etc. The main intention of these antenna geometries is to broaden design and synthesis models outside Euclidean geometry. Fractal and random fractal arrays added numerous features in the designs [16]. The utilization of cloth/fabrics as substrates for the improvement of wearable systems has experienced fast growth due to the latest contraction of wireless standards [17] and the requirement of enhanced antenna parameters such as gain, bandwidth, and efficiencies [18, 19] for the specific applications. But practically only a few iterations are possible to design the antenna with fractal geometries after that benefits start to diminish [20].

The following Table 1 summarizes some of the techniques reported in the literature to enhance the performance metrics of the wearable fractal antennas on flexible substrates.

Table 1 Comparison table of existing techniques to improve the performance of fractals

Ref.	Technique	Application	Limitations
[21]	Microstrip line fed consists of 50 Ω line and the tapered line; also, a meandered spiral is placed in the ground plane	It is integrated with the lifejackets to help in locating the human body if any accident happens	The total efficiency and gain of the antenna were reduced due to the presence of the human body
[22]	Minkowski geometry is used for the design of wearable electro-textile	WiBro (Wireless broadband) band, WLAN and for GSM 1900 applications	Dielectric properties of the polyester are not provided experimentally
[23]	Koch Fractal Slot Antenna with CPW-Feed	Appropriate for WLAN and WiMAX applications	The substrate is not flexible
[24]	A triangular Sierpinski gasket incorporated on PIFA	Suitable for Dual-band wireless applications	Fabrication inaccuracy and inconsistency in thickness of the substrate
[25]	CPW-EBG Antenna	Used for millimeter-wave applications	The proposed EBG cell fabrication is quite complicated on a fabric substrate

3 Bandwidth and Gain Enhancement Techniques for Wearable Fractal Antennas

To cope up with the present day scenario, the performance parameters of wearable fractal antennas such as gain and bandwidth should be enhanced and also contribute to the successful communication system development and their adaptation to specific needs. Various techniques are needed to improve the performance of the wearable fractal antennas. Generally, a microstrip patch antenna structure with added stages of fractal shapes is utilized in the design as it finds the required size and characteristics. Frequently metamaterials are used to embed the microstrip antennas for filling them partially underneath the substrate of patch or placing the metamaterial reflective surfaces on the upper surface of the patch. These methods greatly improve the gain, bandwidth, directivity of the microstrip patch antennas with considerable size reduction [26]. The multi-band characteristics can be obtained with the principle of the self replica of fractals [27] and some of the fabricated prototypes of the wearable antenna on flexible substrates from the references are shown in the following Fig. 1.

3.1 Bandwidth Enhancement

Some of the proposed fractal wearable antennas in the Industrial Scientific and Medical (ISM) band cannot cover the entire impedance bandwidth as per FCC and also to fulfill the bandwidth requirements of voice and data transmission in wireless communications, numerous bandwidth enhancement techniques have been used. Primarily, the factors that will diminish the bandwidth of planar wearable antennas are the shape of the radiator, the feeding method, and the substrate's dielectric constant. Fundamentally, the broadband nature of the planar antennas defends on the lower value of Q and the use of various impedance matching or introducing

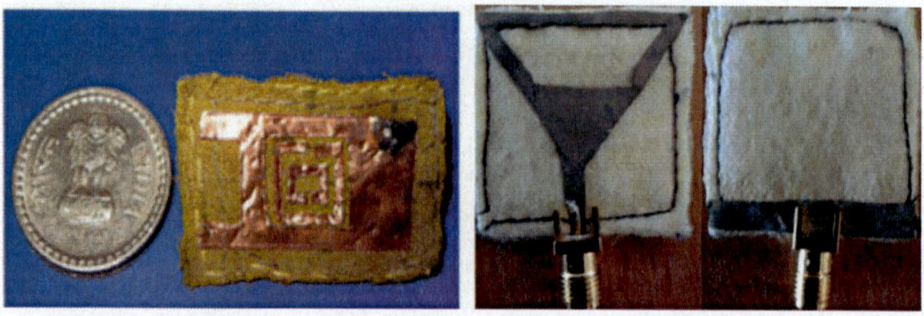

(a) (b)

Fig. 1 Fabricated prototypes from the references for wearable applications. **a** MPA with metamaterial reflective surface for enhancing gain and bandwidth [26]. **b** Dual-band textile antenna at ISM band [27]

Table 2 Bandwidth enhancement techniques of traditional patch radiators

Methodology	Techniques
Small Q	The shape of the radiator The thickness of the substrate Lower the dielectric constant Increase the losses
Impedance matching networks	Insert a matching network Add tuning elements Use slotting and notching patches
Multiple resonances	Use parasitic elements Use slotting patches Insert impedance networks Use an aperture, proximity coupling

multiple resonators. A variety of broadband approaches are exploited in [28], which are summarized in Table 2.

Further, from the literature, the bandwidth enhancement of wearable fractal antennas can be achieved by the following methods

1. Wide-slot/fractal-like slot feed-slot combinations, for example, feed gap width, and slot shape
2. Meandering slits and DGS (Defected ground structures) [14]
3. Optimization techniques at the defected ground planes [29].

Altogether, the wearable antennas are designed with fabric/textile materials and whose dielectric constants lie in the range 1–2, which is less when compared to rigid substrates such as FR4, thereby bandwidth of the wearable antennas is slightly higher than regular antennas. Additionally, considering the use of the thick substrate, the bandwidth of the wearable fractal antenna can be enhanced with the added advantage of lower bending losses when on the body. But on the other hand, the surface wave modes increase because of the thick substrate, thereby increase in mutual coupling in antenna arrays. This will result in severe degradations in impedance mismatch, large radiation loss, polarization distortion. Afterward the straightforward method to improve bandwidth is cutting slots of different shapes, thus creating a conductor loss, i.e., Lesser the Q, more the bandwidth. Later DGS is designed by etching off different shapes in the ground plane. The shielded current distribution in the ground plane depends on the shape and dimensions of the defect, which leads to a carefully regulated excitation and propagation of the EM waves through the substrate, which will enhance the bandwidth.

Among the published works, the following can be highlighted under bandwidth enhancement for the fractal shapes. A Sierpinski's square fractal slots antenna on the Rogers RO4350B substrate is fed by a new feeding type called coplanar waveguide

(CPW) [26], a Minkowski microstrip fractal antenna using method of stacking [27], and a printed wide-slot antenna with a slot of fractal shape [28] is introduced for extremely wider bandwidth.

3.2 *Gain Enhancement*

Generally, Microstrip Patch Antenna is widely used because of its low profile, which leads to Lower Bandwidth and Lower Gain. The improvement in the performance of a new wearable microstrip patch antenna through added stages of fractal geometry is necessary for its use in security, military, and Wi-Fi. There are many methods to improve the gain of such antennas, and some of them give the best results. The factors that are responsible for less gain in microstrip antennas are losses due to conductors, dielectric losses, and excitation of surface waves. Dielectric and conductor losses are unavoidable in the designing of the wearable antenna as it uses flexible substrates whose dielectric properties not known and use adhesive copper foils for metallic parts. Several methods are proposed to get the dielectric properties of the flexible materials, but in reality, its properties may not maintain the uniformity. Fundamentally, the gain of the microstrip patch antenna can be enhanced by suppressing the surface wave. This can be achieved by

1. Electromagnetic bandgap structures [30]
2. Partial removal of the substrate [31]
3. The antenna structure is backed by AMC (Artificial magnetic conductors)
4. Metamaterial reflective surfaces.
5. RIS (Reactive impedance surface).

These methods are summarized in the following Table 3.

These techniques significantly enhance the gain, bandwidth, directivity of the microstrip patch antennas with considerable size reduction. Besides these techniques to improve gain and bandwidth, there are many more other techniques such as controlling the coupling between patch and slots in the printed ground plane [35], use of metasurfaces above the patch at a proper distance such as FSS substrates and stacked fractal patches [36] and defective ground structures with new feeds etc., All these techniques are also applicable to wearable/flexible antennas just by replacing the conventional rigid substrates with the flexible substrates.

4 Conclusion

From the study, it is concluded that the wearable antenna is a significant element in the design of flexible and miniature wireless devices. Because of their low weight, flexibility, cost-efficient, and conformal characteristics, they are suitable for various

Table 3 Gain enhancement techniques of patch antennas for various applications

Ref.	Method	Gain (dB) increment	Methodology
[14]	Simple EBG structures and Fractal EBG structures	1.5 and 4 dBi Approx.	EBG structures are artificial periodic structures that allow the EM waves propagate in a particular band of frequencies for all incident angles and all polarization states, leads to the advantage of the reduction in side lobe level and higher front to back ratio
[32]	NZI metamaterial	More than 2 dBi	Near-zero index metamaterial (NZIM) can control the direction of radiation
[33, 34]	AMC	~4 dBi	The AMC structure is placed near the antenna, which doesn't alter the antenna parameters and improves the gain. This is because of the fact that there is no interaction between the antenna current and its image as it spatially distributes the antenna image current

wireless communication applications such as WLAN, Wi-Fi, etc., in a human wearable form. So to design a small-sized, low volume, economic and vast bandwidth antenna, there are many types of fractal geometries available that are used to reduce the size of the antenna and to create multi-band structures. Having the advancement, the fractal antennas consist of some limitation that makes problems in modern wireless standards. There are a lot of gain losses and numerical complexities in fractal antenna designs that cause huge boundaries in working. On the other hand, with an increasing number of iterations of Fractal, resonant frequency detunes, and antenna performance degrades. Through characterizing the fractal geometries and their results from the study, it can be accomplished that implementing the fractal geometries on flexible/wearable substrates with the technologies of metamaterial, meander line slits, defected ground structures can enhance the antenna gain and bandwidth. These are few examples of the fascinating current advancements in the field of wearable antennas, which tend to be used in future wireless and 6G eras. Therefore in the future amalgamation of various such techniques with a flexible substrate, a wearable fractal antenna with better gain and bandwidth can be designed.

References

1. Pekka Salonen HH () A novel fabric WLAN antenna for wearable applications, pp 100–103
2. Kumar G, Ray KP, (2003). Broadband microstrip antennas. Artech house
3. Hertleer C (2006) Design of textile antennas for smart clothing. In: 7th UGent Ph.D. *symposium*, pp 3–5
4. Sankaralingam S, Gupta B (2010) Determination of dielectric constant of fabric materials and their use as substrates for design and development of antennas for wearable applications. IEEE Trans Instrum Meas 59(12):3122–3130
5. Shebani NM, Mohammed AE, Khamoudi BM (2011) Design curves of micro strip ring resonator. In 12th International conference on Sciences and Techniques of Automatic control & computer engineering December. pp. 18-20
6. Resonator MR, Low PJ, Esa F, You KY, Abbas Z (2017) Estimation of dielectric constant for various standard materials using estimation of dielectric constant for various standard materials using microstrip ring resonator
7. Priya A, Kumar A, Chauhan B (2015) A review of textile and cloth fabric wearable antennas. Int J Comput Appl 116(17):1–5
8. Balanis CA (2016) Antenna theory: analysis and design. In: wiley
9. Agbor I, Biswas DK, Mahbub I (2018) A comprehensive analysis of various electro-textile materials for wearable antenna applications. In: Proceedings of 2018 texas symposium on wireless and microwave circuits and systems (*WMCS* 2018), pp 1–4
10. Lou RKM, Aribi T, Ghobadi C (2010) Improvement of characteristics of microstrip antenna using of metamaterial superstrate. In: International conference on communication engineering, pp 126–129
11. Panda AK, Sahu A (2011) An investigation of gain enhancement of microstrip antenna by using inhomogeneous triangular metamaterial. In: Proceedings of 2011 international conference on computational ıntelligence and communication networks (CICN 2011), pp 154–157
12. Kavitha K, Seyatha K (2018) Metamaterial superstrate antenna design with gain enhancement. Int J Appl Eng Res 13(24):16945–16949
13. Suraya AN et al (2019) Wearable antenna gain enhancement using reactive impedance substrate. Indones J Electr Eng Comput Sci 13(2):708–712
14. Arif A, Zubair M, Ali M, Khan MU, Mehmood MQ (2019) A compact, low-profile fractal antenna for wearable on-body WBAN applications. IEEE Antennas Wirel Propag Lett 18(5):981–985
15. Crownover RM (1995) Introduction to fractals and chaos. Jones & Bartlett Pub
16. Sangeetha M, Elizabeth B (2016) A survey on fractal wearable antennas with different substrate materials, 2(11):1–7
17. Mai MA, Osman AR, Rahim MKA, Samsuri KKNA, Zubir F (2011) Design, implementation and performance of ultra-wideband textile antenna. Prog Electromagn Res B 27:307–325
18. Abed AT, Singh MSJ, Islam MT (2018) Compact fractal antenna circularly polarised radiation for Wi-Fi and WiMAX communications. IET Microwaves Antennas Propag 12(14):2218–2224
19. Sabban A (2018) Small wearable antennas for wireless communication and medical systems. In: 2018 IEEE radio and wireless symposium (RWS), pp 161–164
20. Parmar SR, Singh H (2017) Analysis of fractal geometries and its applications in microstrip antenna. Int J Adv Res Comput Sci 8(4):194–198
21. Ahmed MI, Ahmed MF, Shaalan A-H (2017) Sar calculations of novel wearable fractal antenna on metamaterial cell for search and rescue applications. Prog Electromagn Res M 53:99–110
22. Sankaralingam S, Dhar S, Bag AK, Kundu A, Gupta B (2014) Use of Minkowski fractal geometry for the design of wearable fully fabric compact antenna. J Phys Sci 18(March):6–13
23. Krishna DD, Gopikrishna M, Anandan CK, Mohanan P and Vasudevan K (2008) CPW-fed koch fractal slot antenna for WLAN/WiMAX applications. IEEE Antennas Wirel. Propag. Lett 7:389–392
24. Soh PJ, Vandenbosch GAE, Ooi SL, Husna MRN (2011) Wearable dual-band Sierpinski fractal PIFA using conductive fabric. Electron Lett 47(6):365

25. Lin X, Seet BC, Joseph F, Li E (2018) Flexible fractal electromagnetic Bandgap for millimeter-wave wearable antennas. IEEE Antennas Wirel Propag Lett 17(7):1281–1285
26. Joshi JG, Pattnaik SS, Devi S (2012) Metamaterial embedded wearable rectangular microstrip patch antenna. Int J Antennas Propag 2012
27. Mishra SK, Shukla S, Mishra V (2015) Design of dual band textile antenna for ISM bands using fractal geometry. In 2015 International Conference on Signal Processing and Communication (ICSC), IEEE. pp. 161–165
28. Chen ZN, Chia MYW (2006) Broadband planar antennas. Hoboken^ eNJ NJ, Wiley. pp. 978–0
29. Bhatia SS, Sivia JS (2016) A novel design of wearable fractal antenna for wideband applications. In: 2016 International conference on *Advances in Human Machine* Interaction HMI 2016, pp 140–143
30. Rao N, Vishwakarma DK (2016) Gain enhancement of microstrip patch antenna using Sierpinski fractal-shaped EBG. Int J Microw Wirel Technol 8(6):915–919
31. Patil A, Suryakanth B (2015) A survey and review on gain enhancement methods of microstrip patch antenna. Int J Emerg Technol (Special Issue NCRIET-2015) 6(2):98–104
32. Suthar H, Sarkar D, Saurav K, Srivastava, KV (2015) Gain enhancement of microstrip patch antenna using near-zero index metamaterial (NZIM) lens. In: 2015 21st National conference on communications NCC 2015, pp 8–13
33. Zhong L-R, Yang G-M, Zhong Y-W (2015) Gain enhancement of bow-tie antenna using fractal wideband artificial magnetic conductor ground. Electron Lett 51(4):315–317
34. Zhang B, Yao P, Duan J (2018) Gain-enhanced antenna backed with the fractal artificial magnetic conductor. IET Microwaves Antennas Propag 12(9):1457–1460
35. Radhi AH, Nilavalan R, Wang Y, Al-Raweshidy H, Eltokhy AA, Aziz NA (2018) Mutual coupling reduction with a novel fractal electromagnetic bandgap structure. IET Microwaves Antennas Propag 13(2):134–141
36. Shi Y, Zhang W (2012) High-gain stacked Minkowski fractal patch antenna with superstrate for 60 GHz communications. Adv Intell Soft Comput 135:53–59

PERAM: Ultra Power Efficient Array Multiplier Using Reversible Logic for High-Performance MAC

E. Rishi Kiran, Swathi Vangala, and J. V. R. Ravindra

Abstract Reversible logic is an emerging technology that is helpful in diverse fields such as genetic programming, high-speed VLSI design, DNA computing and bioinformatics, quantum computing, etc. Reversible computation differs from conventional computation as it safeguards information while manipulating it. Multiplier is a crucial component of Digital Signal Processing (DSP). In general, DSP application requires low power dissipation multiplier. But in conventional computation, the multiplier dissipates more power than a reversible multiplier. In this paper, PERAM deals with reversible array multiplier. As it consists of rudimentary reversible gates like CCNOT and CNOT, analysis will be uncomplicated. The proposed design methodology is implemented and verified in cadence© virtuoso of 45 nm technology showing the improvement of 77.76% in terms of power, 71.39% in terms of power delay product. PERAM shows a great variation in the power and power delay product.

Keywords Reversible gates · CNOT gate · 1-bit full adder · Array multiplier · Multiplier accumulator unit (MAC)

E. Rishi Kiran · S. Vangala · J. V. R. Ravindra (✉)
Centre for Advanced Computing and Research Laboratory (C-ACRL), Department of Electronics and Communication Engineering, Vardhaman College of Engineering, Shamshabad, Telangana, India
e-mail: jvr.ravindra@vardhaman.org

E. Rishi Kiran
e-mail: rishikiran431@gmail.com

S. Vangala
e-mail: swathivangala8@gmail.com

G. Ranganathan et al. (eds.), *Inventive Communication and Computational Technologies*, Lecture Notes in Networks and Systems 145,
https://doi.org/10.1007/978-981-15-7345-3_64

1 Introduction

Multiplication being the prerequisite operation for general purpose electronic hardware devices that drives the attention of the research work, to formulate a power-efficient design. Multipliers are miniature circuits that are incorporated within the applications. As the technology scales down, the research in the semiconductor industry complicates the integration of circuit designs, diminishing the area of the electronic design interface. Most of the Literature study on arithmetic circuits focuses on the improvement of multiplier performance.

In VLSI designs, the nature of components is generally irreversible. Landauer proposed a principle which states that consumption of power is imminent in irreversible logic and the loss of power leads to the dissipation of energy in its computation. The dissipation energy is kTln2 Joules which is due to the removal of each distinct bit of information (k is Boltzmann's constant which is equal to 1.3807×10^{-23} joules per kelvin (J K^{-1}) and T is absolute temperature) [1]. Energy loss is a significant consideration in circuit synthesis. But according to Moore's law, after every eighteen months, the number of transistors in the design will be doubled. Therefore, the amount of energy dissipation will grow exponentially. High levels of integration have drastically minimized heat loss over the past few years. Many designs and algorithms are proposed to overcome this problem and the solution found is reversible logic.

Reversible logic holds up the system in both forward and backward directions [2] and exhibits one to one mapping [4]. In reversible logic, inputs and outputs can be restored from each other and can't eliminate information, therefore it dissipates zero heat energy. Unless energy dissipation per logic operation can be lowered below kT, the raw cost of electricity might well prove extortionate and the system might instantly overheat. Even today, the use of reversible logic operations can be a useful didactic in the design of systems that consume minimal power. This reversible logic is achieved by using reversible gates. There are two limitations in reversible logic which are taken into consideration [3].

Fan-Out is not acceptable.

Feedback is not possible in reversible logic (Fig. 1).

Multiplication is a crucial component of Digital Signal Processing (DSP) applications like Fast Fourier Transform (FFT), Digital Filters, and also in Image Processing. To reach high execution speed or to meet with the requirements in DSP applications, there are many types of multipliers being used such as Braun, Wallace, Booth, Array, Sequential and Combinational Multipliers. Efficiency of the multiplier is compared based on parameters like power, area, delay, and PDP for analysis [5]. Various methods have been adapted to attain power-efficient multiplier designs.

The paper contains a literature survey mentioned in Sect. 2. PERAM design is discussed in detail in Sect. 3. Section 4 shows up the simulation results while Sect. 5 discusses an application to demonstrate the efficiency of the proposed multiplier. Finally, Sect. 6 concludes the remarks of this research work.

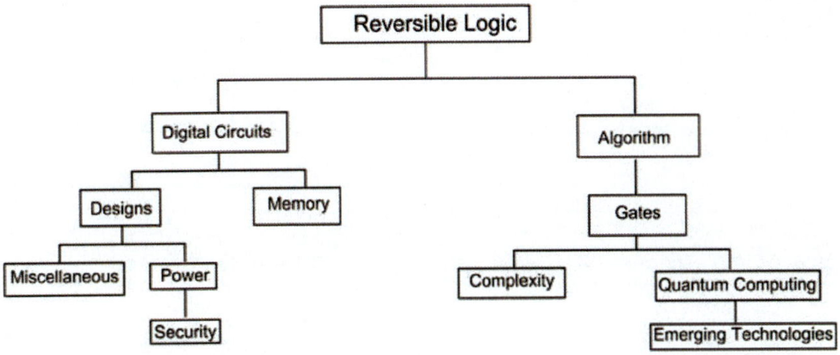

Fig. 1 Taxonomy of reversible logic

2 Literature Survey

The gate is said to be a reversible logic gate when the input is obtained from output and vice versa and the circuits with such reversible logic gates are called reversible logic circuits. There are several methods for synthesis of reversible logic gates. In [6] an array multiplier is implemented using Verilog coding. It is observed that the multiplier consumes minimal power and occupies the same area without disturbing the speed. The main drawback of this design is the complexity of the HNG gate which is used to produce partial products and full adders. So, to overcome this problem used CNOT and CCNOT gates. In [7], the design they used is a Wallace multiplier which analyses the usage of compressors to fasten the multiplication by minimizing the carry generation. It has zero garbage outputs. This [8] design contains radix-2 booth's recoding architecture of reversible multiplier which shows an improvement in quantum cost and garbage output. The area efficiency can be attained by reducing some of the reversible circuit parameters. In [9] the proposed multiplier is implemented by a partial product generation circuit and adder circuit by reducing the reversible gate parameters. In [10] the proposed design is a binary multiplier with MFA which has low complexity in hardware. This binary multiplier has a less number of garbage outputs and ancilla bits and $N \times N$ bit multiplication are also obtained. This [11] designed a quantum circuit that is more efficient than others, is obtained by computing squares of the operand. Improvement in gate count, garbage outputs, ancilla inputs, and quantum cost to zero is acquired by reducing partial product generation circuitry. This [12] established the layout of all the reversible gates using 250 nm technology and designed 4-bit Adder and 4×4 multipliers by using those gates. Thus, the simulation results give accuracy in the design. Paper [13] presents $N \times N$ multiplier of reversible logic which uses the reversible adder for addition operation and has no garbage output and no ancilla bit by binary tree methodology. [7–11, 13] mainly focuses on the quantum cost, garbage output, and constant input. But not about the power, delay, and power delay product. So, the main aim of the PERAM is to produce a power-efficient array multiplier using reversible logic.

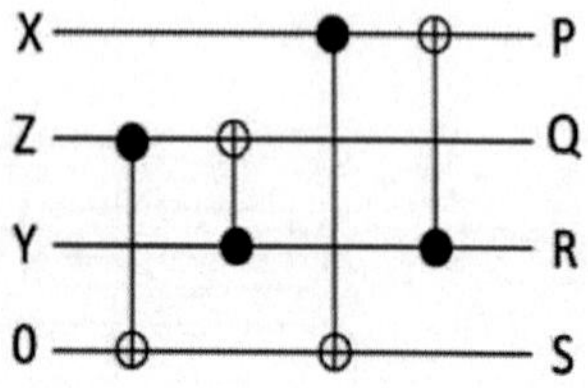

Fig. 2 Full adder

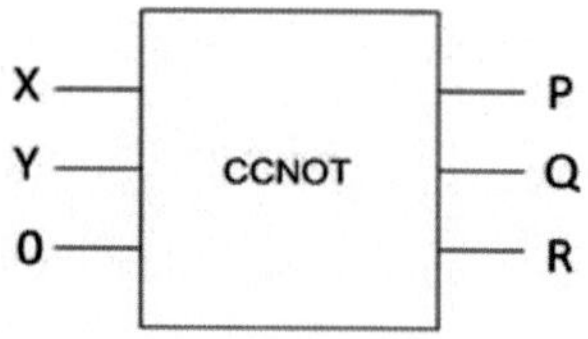

Fig. 3 CCNOT gate

3 Proposed Work

A 4 × 4 Array multiplier has been designed using CNOT and CCNOT gates. The CNOT gates are used in constructing full adder and CCNOT gate is used as the logical AND gate in the multiplier design for producing partial products for multiplication and full adders are used for the addition of those partial products.

3.1 *Reversible AND Gate and 1-Bit Full Adder*

When the input of CCNOT gate is IP = $(X, Y, 0)$ then the CCNOT is realized as AND gate and the outputs of CCNOT are $(P = X, Q = Y, R = XY)$ as shown in Fig. 2. A full adder is obtained when the inputs vectors are IP = $(X, Z, Y, 0)$ [7], and output vector of full adder OP is $(P = X, Q = \text{SUM}, R = Y, S = \text{CARRY})$ as shown in Fig. 3. The modification in the existing design provides low power and high speed through reversible gates.

$$\text{Sum} = X \oplus Y \oplus Z \quad (1)$$

$$\text{Carry} = XY + YZ + ZX \quad (2)$$

3.2 *Standard Array Multiplier*

Consider a two 4-bit numbered array multiplier as shown in Fig. 4, where $A = A0$ $A1$ $A2$ $A3$ is the multiplicand and $B = B0$ $B1$ $B2$ $B3$ is the multiplier. The output of

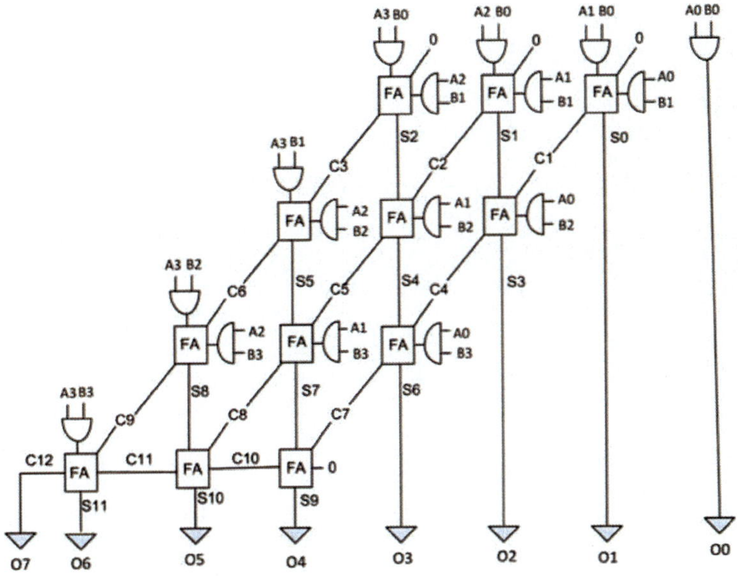

Fig. 4 Conventional standard array multiplier

the multiplier can be written as O7 O6 O5 O4 O3 O2 O1 O0. In the multiplication steps, output is acquired by multiplying the inputs and producing partial products [6]. Adding and shifting methods can be done to generate the final result. The standard array multiplier is built using a full adder cell and AND logic gates. This is an example of conventional computation.

3.3 *Proposed 4 × 4 Array Multiplier*

Multiplier performance is straight away affected by the adders in the multiplier, therefore using improved adder designs can solve problems due to power dissipation. In this paper, analysis is done to propose a power-efficient array multiplier. The PERAM as shown in Fig. 5, is the same as the standard array multiplier but in spite of using irreversible logic gates like AND, OR and XOR, reversible gates like CNOT and CCNOT gates are used for implementation. In general, $N \times N$ array multiplier operation starts from generating the partial products followed by summation of those products to get the multiplier output. This PERAM design is implemented by using CNOT and CCNOT gates. A total of 16 CCNOT gates and 12 1-bit full adders are used to perform AND and addition operation, respectively, to design the array multiplier. The gates used in this design are very basic gates with minimal power and power delay product(pdp). So that the analysis becomes easier.

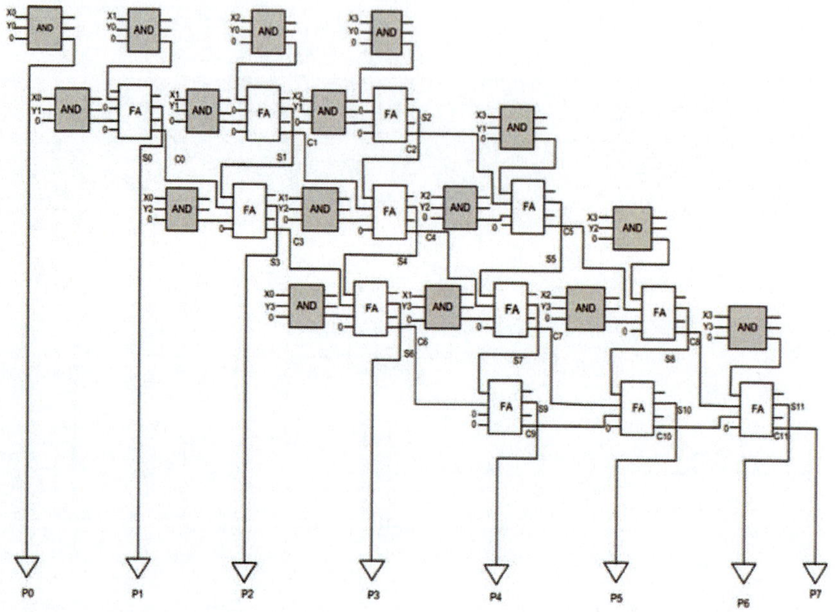

Fig. 5 PERAM-Proposed array multiplier

4 Simulation Results

Reversible multipliers are more efficient than the multipliers with irreversible gates. On comparing the existing work and proposed work, a great variation in the delay and power is observed. These power and delay are calculated by using the calculator provided by the Cadence Virtuoso tool and PDP can be obtained by multiplying power and delay of the design. The power and PDP are compared with the least values in the table to give the percentage efficiency gain. The proposed multiplier has 71.36% power delay product than the [17] design and consumes 77.26% less power compared to [18] design and the gates used in the design are very simple when compared to the gates used in the existing work (Figs. 6, 7 and 8; Table 1).

5 Designing of Multiplier-Accumulator Unit Using Proposed Multiplier

The elementary operation of the Multiplier-Accumulator Unit is multiplication. Its functional components are multiplier, adder, and an accumulator. In the present day of VLSI field, switching of signals drives large power dissipation. In the irreversible computations, the removal of distinct bit dissipates heat energy [1].

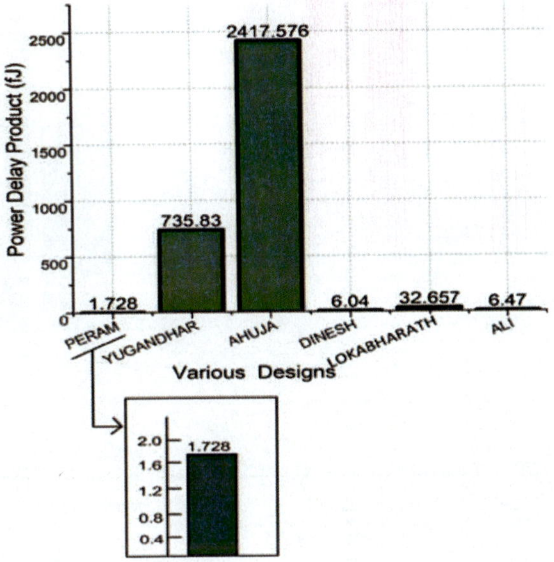

Fig. 6 PDP of various designs

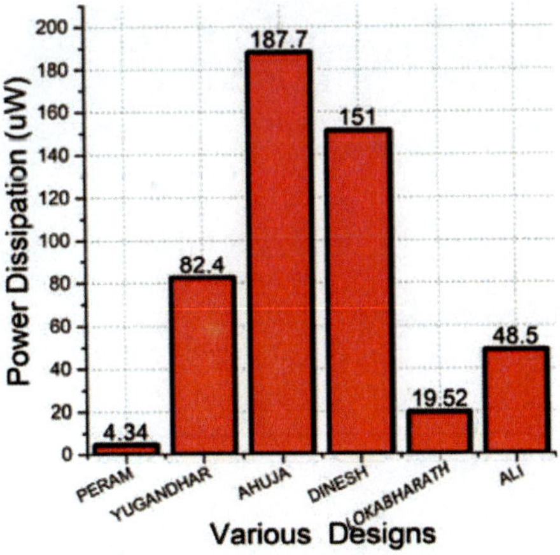

Fig. 7 Power dissipation of various designs

There are three steps involved in multiplying function, which are given as follows:

1. Generation of partial products by multiplier,
2. partial product addition by Adder,
3. final overall addition.

It performs multiplication between input bits and then adds it with the result of the previous output of the multiplier as shown in Fig. 9. The multiplier needs a significant amount of delay in the digital system. Therefore, in this paper, we have

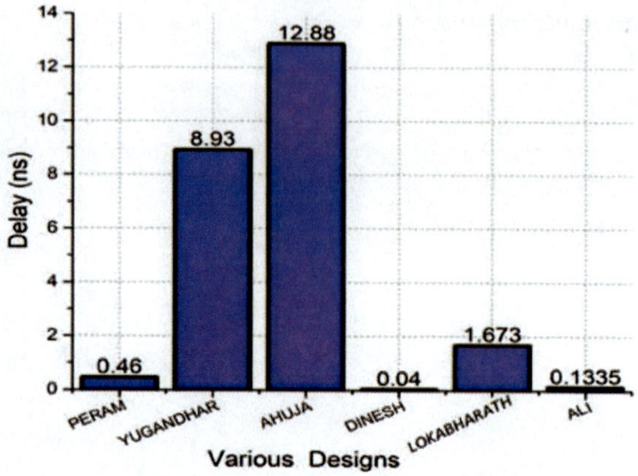

Fig. 8 Delay of various designs

Table 1 Simulation results of proposed design compared with existing designs

S. No.	Designs	Power (uW)	Delay (ns)	PDP (fJ)
1	**PERAM**	**4.34**	**0.46**	**1.728**
2	Yugandhar[15]	82.4	8.93	735.83
3	Ahuja[16]	187.7	12.88	2417.576
4	Dinesh[17]	151	0.04	6.04
5	Lokabharath[18]	19.52	1.673	32.657
6	Ali[19]	48.5	0.1335	6.47

used reversible multipliers in place of conventional multiplier. The average power of 30 uW and 403.22 fJ power delay product is obtained by using array multipliers in the MAC unit.

6 Conclusion

The proposed 4 × 4 array multiplier using reversible gates have been designed and implemented in the cadence©45 nm technology. Targeting the requirements, several designs are proposed in the literature. The comparison among the multiplier designs showed that there is a great variation in the results of PERAM. From the simulation results, it is observed that PERAM has an improvement of 77.76% in terms of power and 71.39% in terms of power delay product than the existing multipliers. The proposed multiplier can be further used for many applications to attain a power-efficient design.

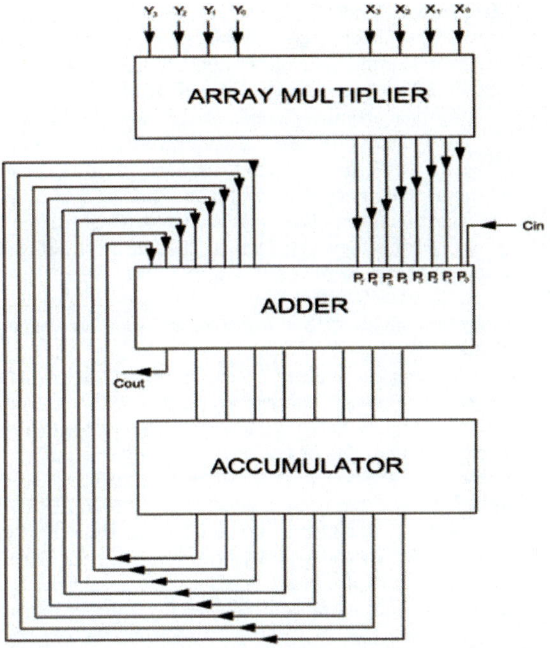

Fig. 9 Multiplier and accumulator

References

1. Landauer R (1961) Irreversibility and heat generation in the computing process. IBM J Res Dev 5(3):183–191
2. Talawar K, Hosamani P (2014) Ultra-area efficient reversible quantum radix-2 booth's recoding multiplier for low power applications. In: 2014 IEEE international conference on computational intelligence and computing research
3. Fredkin E, Toffoli T (1982) Conservative logic. Int Theor Phys 21(3–4):219–253
4. Gupta P, Agrawal A, Jha NK (2006) An algorithm for synthesis of reversible logic circuits. IEEE Trans Comput Aided Des Integr Circ Syst 25(II):2317–2330
5. Toan NV, Lee J-G (2020) FPGA-based multi-level approximate multipliers for high-performance error-resilient applications, IEEE Access 8:25481–25497
6. Radha N, Maheswari M (2018) High speed efficient multiplier design using reversible gates. In: 2018 international conference on computer communication and informatics (ICCCI-2018), 4–6 January 2018, Coimbatore, India
7. Nagamani A, Sharath Kumar S, Agrawal VK (2017) Design of garbage free reversible multiplier for low power applications. In: 2017 4th international conference on power, control and embedded systems (ICPCES)
8. Muñoz-Coreas E, Thapliyal H IEEE Trans Comput. https://doi.org/10.1109/tc.2018
9. Banerjee A, Pathak A (2010) Reversible multiplier circuit. In: Third international conference on emerging trends in engineering and technology
10. Ehsanpour M, Moallem P, Vafaei A (2010) Design of a novel reversible multiplier circuit using modified full adder. In: 2010 international conference on computer design and applications (ICCDA 2010)
11. Jayashree HV, Thapliyal H, Agrawal VK (2014) Design of dedicated reversible quantum circuitry for square computation, In: 2014 27th international conference on VLSI design and 2014 13th international conference on embedded systems

12. Mahapatro M, Panda SK, Satpathy J, Saheel M, Suresh M, Panda AK, Sukla MK (2010) Design of arithmetic circuits using reversible logic gates and power dissipation calculation. In: 2010 international symposium on electronic system design
13. Kotiyal S, Thapliyal H, Ranganathan N (2014) Circuit for reversible quantum multiplier based on binary tree optimizing ancilla and garbage bits. In: 2014 27th international conference on VLSI design and 2014 13th international conference on embedded systems
14. Wang Z, Chen S, Liu W (2015) A novel method to reduce ancilla and garbage bits of reversible quantum multipliers. In: 2015 11th international conference on natural computation (ICNC)
15. Yugandhar K, Ganesh Raja V, Tejkumar M, Siva D (2018) High performance array multiplier using reversible logic structure. In: 2018 international conference on current trends towards converging technologies (ICCTCT)
16. Ahuja M, Bajaj S (2013) Design and analysis of bypassing multiplier. In: Fifth international conference on advances in recent technologies in communication and computing (ARTCom 2013)
17. Dinesh B, Venkateshwaran V, Kavinmalar P, Kathirvelu M (2014) Comparison of regular and tree based multiplier architectures with modified booth encoding for 4 bits on layout level using 45 nm technology. In: 2014 international conference on green computing communication and electrical engineering (ICGCCEE)
18. Lokabharath Reddy N, Bassi M, Mishra RS (2016) Design of a high performance 4 bit multiplier using UT algorithm with domino logic. In: 2016 IEEE 7th annual information technology, electronics and mobile communication conference (IEMCON)
19. Albadry OA, El-Bendary MM, Amer FZ, Singy SM (2019) Design of area efficient and low power 4-bit multiplier based on full-swing gdi technique. In: 2019 international conference on innovative trends in computer engineering (ITCE 2019), Aswan, Egypt, 2–4 February 2019

Development of Social Media User Interface Portal for Maintaining Students Portfolio

Aswathi Krishnan, A. Ganesh, S. Gayathri, S. Koushik, M. Varsha Nair, and Gayathri Narayanan

Abstract In this era, social media has become a platform for education, e-commerce, job seeking and many more. This work intends to create a digital portfolio which helps freshers to upgrade their skillsets, create their resume according to the company profile and as per the company requirements, interacting with people of similar interests and giving a chance to showcase their skills on the current platform. Students will be rated according to their performances and will be awarded badges. This platform will not only help students to work on themselves but also give confidence to them while entering into the corporate world.

Keywords Front end · Technology · Unified modeling language · User interface

A. Krishnan · A. Ganesh · S. Gayathri · S. Koushik · M. Varsha Nair · G. Narayanan (✉)
Department of Electronics and Communication Engineering, Amrita Vishwa Vidyapeetham, Amritapuri, Kollam, India
e-mail: gayathrin@am.amrita.edu

A. Krishnan
e-mail: aswathikrishnan20@gmail.com

A. Ganesh
e-mail: ganeshanil964@gmail.com

S. Gayathri
e-mail: gayathrisankar798@gmail.com

S. Koushik
e-mail: koushiksivan98@gmail.com

M. Varsha Nair
e-mail: varshanairm8991@gmail.com

G. Ranganathan et al. (eds.), *Inventive Communication and Computational Technologies*, Lecture Notes in Networks and Systems 145,
https://doi.org/10.1007/978-981-15-7345-3_65

1 Introduction

To overcome the skill unidentified by the recruiter, a fully structured front-end development of the social media website is created using existing front-end technologies and frameworks. Though there are existing platforms such as LinkedIn, it has its own advantages and disadvantages [1]. Usually a social media website consists of one role for the registered user, but this design of social media consists of two roles that are allocated to the users. The roles allocated are beginner role and expert role.

Once a user registers into the website he/she will be allocated to the beginner role. He/she should enter the relevant skills which they want to improve. The user should post relevant articles and projects related to the skills that are registered earlier. The expert role users have the ability to review the article and projects that are posted by the beginner role users. Those review points will be taken an average and get added up in the progress bar which is visible in the profile page of the user. Meanwhile other users can also like/comment/share their opinion on the article/project. Once the progress bar reaches a threshold level, the beginner role user of that particular skill will be promoted to an expert role where they will be reviewing the other users' project. Likewise, a community will be developed and a platform will be created to showcase their skills.

This section follows system design wherein discuss the website's UI and the importance of it to provide a good user experience. Then the next section is UML behavioral diagram, use case diagram which describes the different actions the system performs, the interactions of things outside the system with the system itself. Following this is the UI design, the framework of our website and then the different technologies used to build the User Interface. Then, finally provide the results achieved.

2 System Design

Designing a system before building its Graphical User Interface is extremely important and plays an important role in creating a better user experience for the website [2]. Especially a social media website with lots of features needs a good system design for the better understanding of its usage. It also indicates how the system will fulfill its requirements and objectives. A user-centered Web development is essential. It refers to a design process that accommodates the needs of the user [3].

2.1 *Unified Modeling Language*

Unlike any other programming language, Unified Modeling Language can be described as a visual representation of a complex system. The language helps us

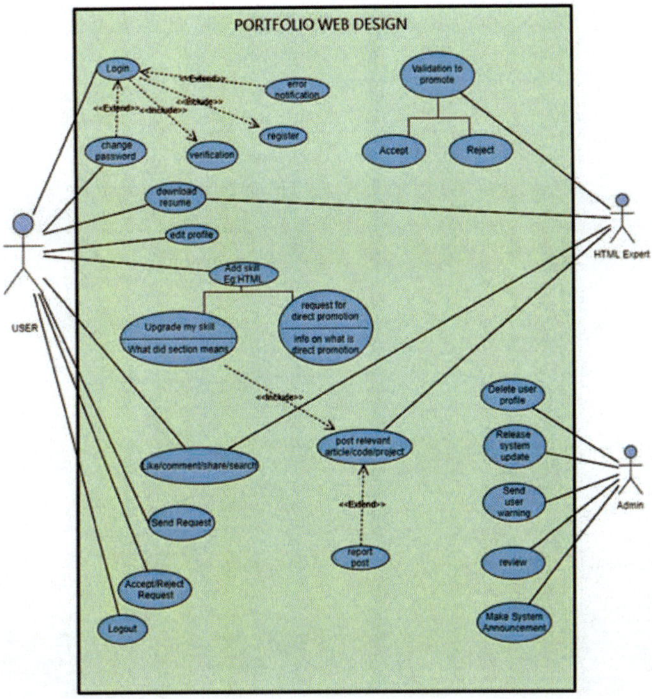

Fig. 1 Use case diagram of the system

to manage the project which is considerably large and needs a huge team [4]. The general purpose of a UML diagram is to portray the behavior and structure of the system. The system is modeled using the following diagram.

2.2 *Use Case Diagram*

It is a dynamic or behavior component of UML. Use case diagrams are imperative for visualizing the functional prerequisites of a system that will decipher into development priorities and design choices (Fig. 1).

2.3 *User Interface Design*

Graphical User Interface design serves an imperative role in designing a framework as it creates a good or bad impression about the software or a website that is built. With the use of prototyping tools like Adobe XD and Sketch, the system's GUI is designed and developed using front end technologies like HTML, CSS and JavaScript [5] (Figs. 2 and 3).

The user creates an account by giving the general profile information. During registration his skill sets are asked, which are the primary attribute of the student

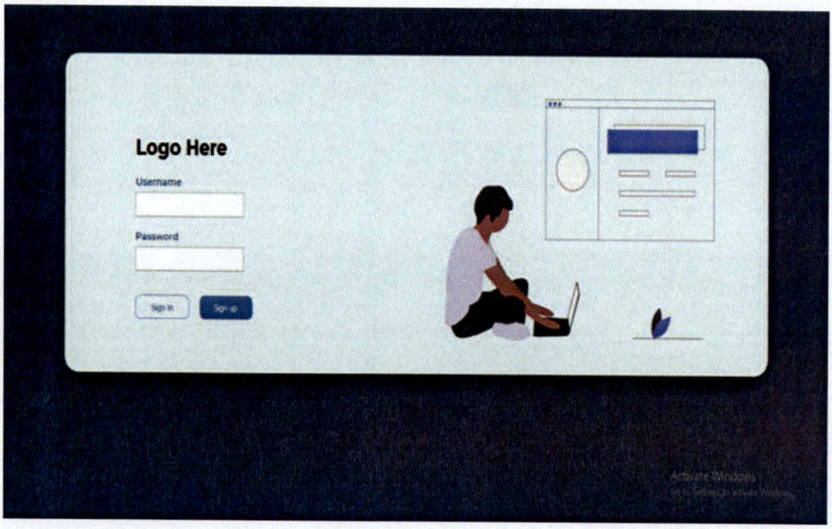

Fig. 2 Login page user interface

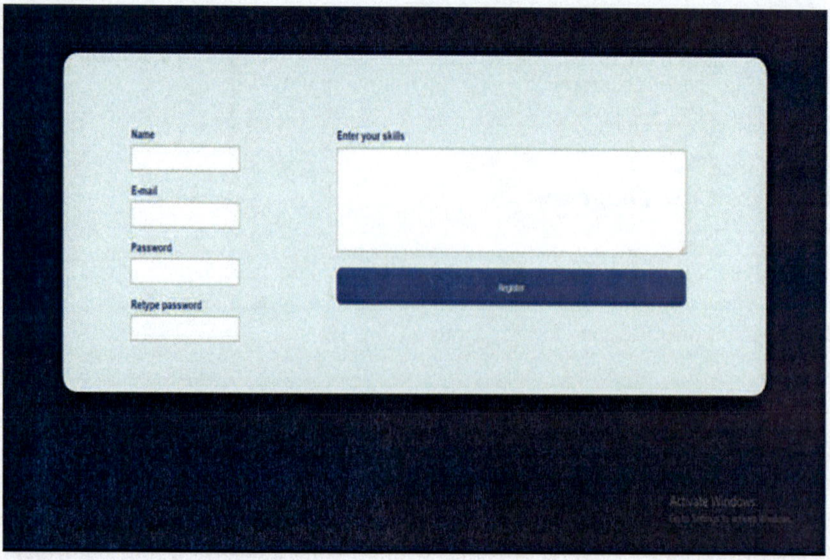

Fig. 3 Register page user interface

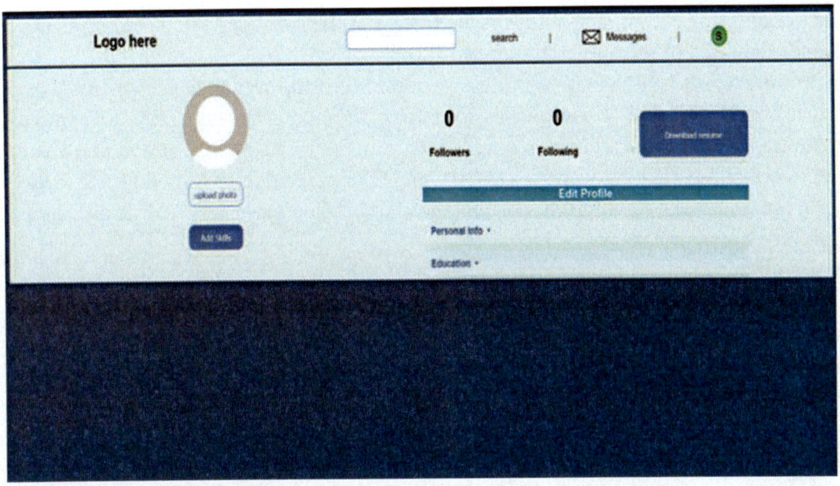

Fig. 4 Profile page user interface

or the user. As soon as he registers, he enters the dashboard which shows a skill bar showing his current skill level. In the dashboard page also ask for the student's general academic details, his projects and experiences or internship details for the resume (Fig. 4).

The skill bar is filled based on the acknowledgment that is the score given by the expert users. In case of students with equal skill bar levels the number of contributions made and how consistent the contributions occur are considered for ranking them. Thus the judgments are highly credible and the users also get expert advice and a community of similar interests.

The contributions are done in the form of posts. As soon as a user posts a code or a design it reaches the experts of that skill for review. Just like facebook when a facebook user posts his/her pictures it is visible for all his/her friends, here the user posts his/her code or design, which is visible to all experts of that skill and is reviewed by them.

HTML: Hypertext Mark-up Language (HTML) is the dialect used broadly for documents outlined to be shown in a web. It characterizes the construction of a Web page. Cascading Style Sheets (CSS) and scripting languages are used for giving appearance to Web pages such as JavaScript. HTML decides the layout and the structure of a Web page by applying various types of tags and attributes. In HTML, JavaScript is used to build interactive Web pages.

CSS: It means Cascading Style Sheets. It is introduced as a constraint for controlling the appearance of HTML pages [6]. It defines how HTML components are presented on screen, paper etc. CSS can be used to curtail a substantial amount of work. The layouts of numerous Web pages are controlled by it all at once. On CSS files external style sheets are stored .CSS may be used with any XML primarily based mark-up language and it is also independent of HTML. Developing a CSS and maintaining it is an imperative issue from a Web developer's perspective as they endure the need of thorough strategies [7].

Bootstrap: It is a framework to assist one to develop websites easier and faster. It is a framework that accomplishes to deliver balance between design and implementation which in turn enables developers to make way to better styled and easily viable websites [8]. It incorporates HTML and CSS based design templates for forms, typography, tables, navigation, buttons, modals, image carousels, etc. It also helps JavaScript plug-in.

JavaScript (JS): It is a scripting terminology, basically utilized on the Web and it is fundamental to modern Web applications [9]. The use of it is to improvise HTML pages. It assesses a system's performance on an assortment of expansive real-world websites [10]. It can generally be found encapsulated in HTML code. It ought not to be compiled as it is an interpreted language. Using JavaScript, Web pages are rendered in an associative and dynamic construct. The biggest preference to JavaScript is its capability to deliver synonymous results on all contemporary browsers.

JQuery: It is a quick, small, and feature-rich JavaScript library. The main objective of JQuery is to compose accessibility to utilize JavaScript on your website. JQuery can be used for creating a single line of code which includes a bundle of common tasks that desires numerous lines of JavaScript code.

3 Simulations and Experimental Results

A survey was conducted to know the current scenario of the fresher/students and their thoughts about the idea of a new social media platform.

Figure 5 depicts that almost all the candidates prefer to be judged based on skills and not on marks and certifications.

98.5% of the responses need a separate platform other than HackerRank or other coding platform to showcase their skills (Figs. 6 and 7).

Many online resume builders are paid and have less variety of resumes to choose from. 86.8% of the responses are in need of a good quality resume building service.

4 Conclusion

The social media website is designed in such a way that the entire system is easy to navigate and the minimal design approach made each and every component of the website to be easily readable and understandable [11]. The system makes a huge impact on the life of students as well as the job seekers. The results from the survey conducted were fully utilized while designing the system and also considering some of the current pattern and imminent trends of Social Media [12]. The data that was obtained was beneficial in terms of adding the feature to the website as well as understanding the current situation regarding the job opportunities for freshers.

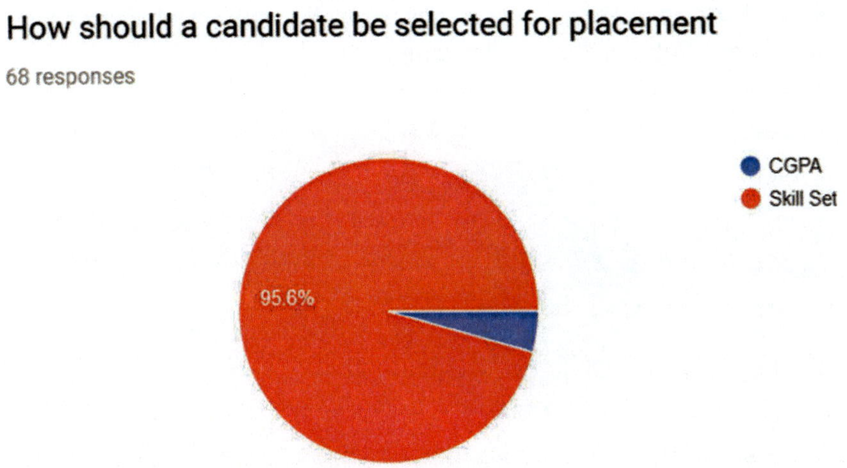

Fig. 5 Skill survey chart 1

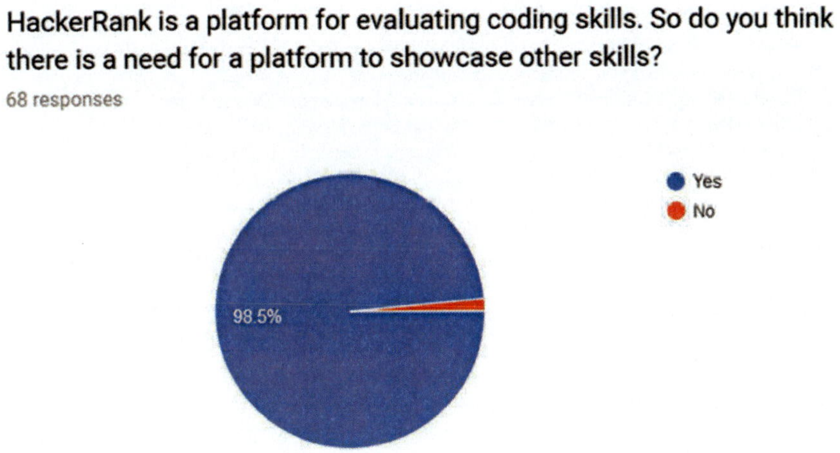

Fig. 6 Skill survey chart 2

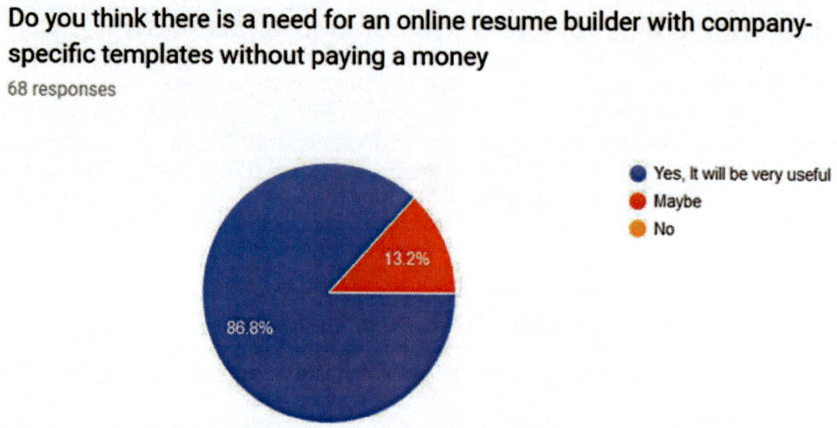

Fig. 7 Resume builder survey

References

1. Waiz S (2019, September 29) What are the disadvantages of LinkedIn? Retrieved from https://advergize.com/marketing/disadvantages-of-linkedin/
2. Design and Application of University Students Management System based on web Platform. (2016). Revista De La Facultad De Ingeniería. https://doi.org/10.21311/002.31.8.10
3. Lazar J (2001) User-centered Web development. Jones & Bartlett Learning
4. Unified Modeling Language (UML): An Introduction. (2019, April 1). Retrieved from https://www.geeksforgeeks.org/unified-modeling-language-uml-introduction/
5. https://www.academia.edu/9021632/DESIGN_AND_IMPLEMENTATION_OF_A_SOCIAL_MEDIA_BASED_WEB_APPLICATION_FOR_PROSPECTIVE_UNIVERSITY_STUDENTS_A_Case_Study_of_Covenant_University_
6. Badros GJ, Borning A, Marriott K, Stuckey P (1999, November) Constraint cascading style sheets for the web. In: Proceedings of the 12th annual ACM symposium on User interface software and technology, pp 73–82
7. Geneves P, Layaida N, Quint V (2012, April) On the analysis of cascading style sheets. In: Proceedings of the 21st international conference on World Wide Web, pp 809–818
8. Balasubramanee V, Wimalasena C, Singh R, Pierce M (2013, September) Twitter bootstrap and AngularJS: frontend frameworks to expedite science gateway development. In: 2013 IEEE international conference on cluster computing (CLUSTER). IEEE, pp 1–1
9. Jensen SH, Møller A, Thiemann P (2009, August) Type analysis for JavaScript. In: International static analysis symposium. Springer, Berlin, Heidelberg, pp 238–255
10. Chugh R, Meister JA, Jhala R, Lerner S (2009, June) Staged information flow for JavaScript. In: Proceedings of the 30th ACM SIGPLAN conference on programming language design and implementation, pp 50–62
11. Akakandelwa A, Walubita G (2017) Students' social media use and its perceived impact on their social life: a case study of the University of Zambia. Int J Multi Res 5(3):1–14
12. Baatarjav EA, Dantu R (2011, October) Current and future trends in social media. In: 2011 IEEE third international conference on privacy, security, risk and trust and 2011 IEEE third international conference on social computing. IEEE, pp 1384–1385

Spectrum Aware Dynamic Slots Computation in Wireless Cognitive Radio Sensor Networks

Veeranna Gatate and Jayashree Agarkhed

Abstract Cognitive Radio Sensor Networks (CRSNs) are resource-constrained networks that require dynamic mechanisms for data transfer. The dynamic nature of CRSN demands dynamic time slots be allocated for node communication. The Licensed users predominantly occupy the spectrum which makes unlicensed users be deprived of accessing the channel. Allocating underutilized portions of the spectrum to unlicensed users must be dynamic. In this work, Dynamic Slots Computation (DSC) based on delay and traffic parameters is implemented for dynamic spectrum allocation for unlicensed users in CRSN. Based on network parameters, a legitimate channel, possessing good quality is selected. Simulation results show that DSC algorithm significantly improves the energy efficiency, packet delivery ratio, and throughput of CRSN in comparison with other existing CRSN protocols.

Keywords Cognitive radio · Spectrum allocation · Dynamic slots

1 Introduction

Wireless Sensor Networks (WSNs) operate in Industrial Scientific and Medical (ISM) band where other communicating technologies compete for having access over the licensed spectrum. Recent research work carried out in the field of spectrum utilization shows that the majority portion of the radio spectrum is underutilized. According to the Federal Communication Commission (FCC) radio frequency spectrum allocated to wireless technologies is not efficiently utilized. Approximately 6% of spectrum is in use at a given location and at a given time when the frequency

V. Gatate (✉) · J. Agarkhed
Computer Science and Engineering Department, Poojya Doddappa Appa College of Engineering, Kalaburagi, Karnataka, India
e-mail: vcgatate@gmail.com

J. Agarkhed
e-mail: jayashreeptl@yahoo.com

G. Ranganathan et al. (eds.), *Inventive Communication and Computational Technologies*, Lecture Notes in Networks and Systems 145,
https://doi.org/10.1007/978-981-15-7345-3_66

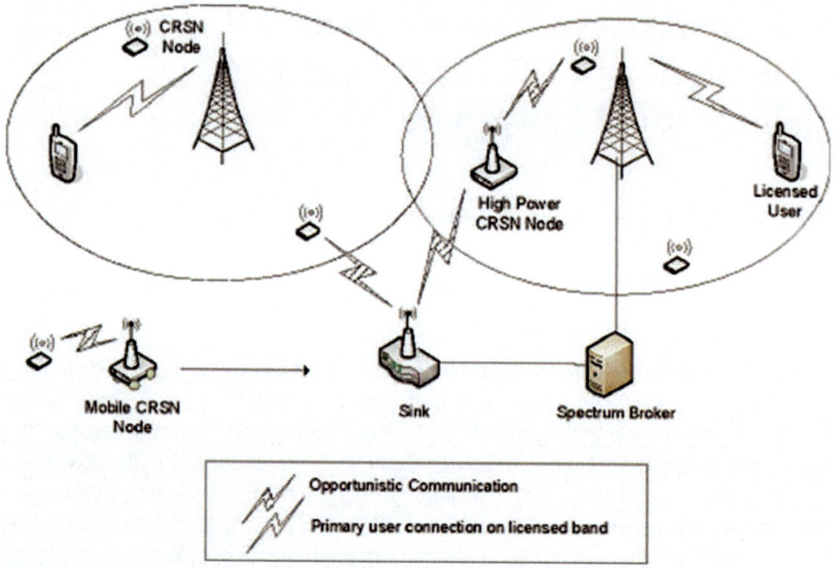

Fig. 1 Architecture of CRSN

band considered is below 3 GHz [1]. Cognitive radios have the ability to identify unutilized portions of spectrum called white spaces. These white spaces are temporal and location-independent [2]. If CR technology is imbibed in WSN, a new network paradigm can be formed called Wireless Cognitive Radio Sensor Networks (WCRSNs). İn this work, WCRSNs referred to as simply CRSNS. CRSNs are special categories of WSN in which nodes possess cognitive radio capabilities. CRSN is defined as a distributed network of Cognitive Radio (CR) wireless sensor nodes that can convert the sensed readings into an event signal and communicate collaboratively to a reliable and non-compromising Base Station (B_S) dynamically over the available spectrum bands [3]. A typical CRSN architecture is shown in Fig. 1 referred from [4]. There are two types of users namely Primary Users (PUs) or licensed users who operate on licensed spectrum band and Unlicensed users or Secondary Users (SUs) who can access the spectrum band only when a PU is not using it [5]. Therefore, SU has to access the spectrum opportunistically. If a PU has already occupied the spectrum then SU has to vacate the spectrum immediately since more priority is given for PU. The main objective of Cognitive Radio (CR) technology is to make SUs have access over the spectrum so that the spectrum is efficiently utilized. İf two or more SUs choose the same frequency band co-channel interference may occur [6]. Therefore, a promising challenge is to design mechanisms for dynamic spectrum allocation enabling SUs to occupy white spaces eliminating co-channel interference [7]. Additionally in terms of spectrum utilization cognitive radios impose additional challenge—which CR node has to be allocated with how wide a spectrum band following what frequency and for what time duration [2]. In [6], our previous work

Interference Aware Cluster Formation (IACFC) is proposed. In IACFC clusters are constructed and channels are assigned based on channel fairness and priority. Once channels are assigned static Time Division Multiple Access (TDMA) mechanism is implemented for data transfer. In terms of cognitive cycle, dynamic slots are assigned once the spectrum sensing stage is completed based on spectrum availability. In this work, IACFC is extended to have dynamic time slots for more efficient spectrum utilization. Our contributions in this work include designing a novel algorithm for dynamic slot computation and channel assignment for CRSNs.

The remaining sections of this paper are organized as follows: Sect. 2 briefly describes the related works of spectrum aware protocols with dynamic channel allocation. Section 3 describes the network model and detailed working of proposed DSC algorithm. Section 4 gives a detailed performance analysis of DSC algorithm and finally Sect. 5 concludes the paper.

2 Related Works

Research in CRSNs in terms of dynamic spectrum allocation is still in infancy. A dynamic TDMA slot reservation protocol is proposed in [8]. A dynamic frame structure is implemented which decreases its frame structure by allowing the nodes to release unused slots. Based on the traffic loads and number of mobile nodes, unassigned timeslots are controlled dynamically by dynamically changing the frame length and transmission schedule [9].

A dynamic channel selection in multichannel cognitive radio networks is proposed in [10]. A synchronous time-slotted system is considered where each timeslot has three periods—sensing the spectrum, access the channel, and data transmission. Each SU senses the spectrum to determine available channels and then selects one among available channels. Source monitors an idle channel by having countdown from k to 1 where k is the number of mini-slots. If the countdown reaches zero and still, channel remains idle then source sends a Request To Send (RTS) packet and waits for its reply. If destination replies back then Source-Destination (S-D) pair is identified for this channel and channel is reserved for S-D pair.

Optimal resource management and allocation algorithm for energy harvesting CRSNs are proposed in [11]. The list of available channels is made available at the beginning of each time slot based on the channel access probability. One channel is allocated to one senor to avoid interference. Since a channel is reserved for a node, it results in channel scarcity if the node has no data to transfer.

Novel dynamic spectrum access based on reinforcement learning is proposed in [12]. A new method of spreading spectrum density in the control channel to utilize unused portions of the spectrum is described. Authors in [2] introduce time spectrum block which indicates the time for which the cognitive radio uses the available portion of spectrum. A novel mechanism called b-SMART protocol is described for spectrum allocation.

In [13], an extension of LEACH protocol called CogLEACH in CRSN is proposed. CogLEACH protocol constructs clusters in order to conserve energy. A Cluster-Head is responsible for aggregate data collected from all non CH nodes and forward to neighbor CH. TDMA approach is implemented where CH broadcasts time slots to all non CH nodes within cluster and communication takes within allotted transmission slots. CogLEACH uses Common Control Channel (CCC) to exchange control messages before actual data transfer. If there is no data to forward then CCC starvation problem occurs which is not addressed. Also having fixed allotted transmission slots is not suitable for dynamic networks like CRSN.

In [14], authors propose a power-aware mechanism called Distributed Spectrum Aware Clustering (DSAC) in CRSN. DSAC forms clusters by selecting nodes that have a common control channel. CH rotation policy is implemented since CH consumes more energy in aggregating data from all nodes within the cluster and forwarding to BS so as to conserve energy. Power consumption is minimized by selecting clusters available within short distances. However, data transmission takes place on common channels which may lead to packet collisions.

Interference Aware Cluster Formation in Cognitive Radio Sensor Networks (IACFC) [6] is our previous work which forms clusters in CRSNs and allocates channels to clusters based on channel priority and fairness. Like CogLEACH Static TDMA schedules are constructed and forwarded to all cluster members. It considers channel occupancy and channel overlapping and attempts to allocate non-overlapping channels to clusters with minimum distance to CH.

3 Dynamic Slots Computation (DSC) Algorithm-Implementation Details

In this section, a detailed description of DSC algorithm is presented. Section 3.1 describes the network model. Section 3.2 explains the dynamic slot allocation and channel selection. Sect. 3.3 presents the flow chart of DSC algorithm implemented in CRSN.

3.1 Network Model

The network is built with a set of sensor nodes classified as $S = s_1 \ldots s_n$, base stations B_S. The ability to sense the availability and quality of channels is referred to as cognitive knowledge. All the sensor nodes in the network are built with cognitive knowledge with channel sensing Intelligence C_K to pick the available channels C within the sensing communication range for region S_R. Each node has a state timer for updating the channel state information and channel state probability. Each node maintains a routing table containing a list of neighbor nodes that are immediate one-hop neighbors. Channel sensing Intelligence helps the sensor nodes to identify the channel with high bandwidth and low latency, minimize the channel scarcity, and reduce the channel interference I among the S_R nodes. The sensor nodes are classified as primary users P_U and secondary users S_U. In the proposed protocol each S_U senses the spectrum in distributed manner and determines the list of available channels at its location. Among the list of channels based on the channel parameters, the best available channel is selected.

3.2 Computation of Dynamic Slots and Channel Selection

The communication begins by S_U node initially broadcasting sync message including timestamp which indicates the data packet transfer duration. Let N indicate the total number of SUs in the CRSN. The receiver node computes the delay by finding the difference between Sending Time (T_{send}) and Recieving Time (T_{receive}) computed as in Eq. (1)

$$\text{Delay} = T_{\text{recieve}} - T_{\text{send}} \tag{1}$$

All the nodes compute the delay. Initially, some slots are occupied by SUs termed as slot occupancy. So Initial slots are estimated by computing average slot occupancy for all the neighbor nodes N.

$$\text{Initial\_ Slot} = \text{Slot\_ occupancy}/N \tag{2}$$

Average Communication Delay is estimated for all immediate neighbors as

$$\text{Comm\_ Delay}_{\text{AVERAGE}} = \text{for all nodes} \sum \text{Delay}/N \tag{3}$$

By taking the difference of Eqs. (1) and (3) Delay Difference is computed as

$$\text{Delay}_{\text{DIFFERENCE}} = \text{Delay} - \text{Comm\_ Delay}_{\text{AVERAGE}} \tag{4}$$

For all the Delay Differences, Variance is computed as

$$\text{Delay}_{\text{VARIANCE}} = \text{for all nodes} \left(\text{Delay}_{\text{DIFFERENCE}}/N\right) \tag{5}$$

Standard Deviation (S_D) is computed by

$$S_D = \sqrt{}\text{Delay}_{\text{VARIANCE}} \tag{6}$$

The lower and upper bounds of Delay are computed by merging S_D and Average Delay termed as Latency. A threshold value for latency is determined depending on the number of Users and relative distance of nodes to BS. If the estimated latency is higher than threshold, increase the sleep time by twice. Else reduce the sleep time by half. Sleep time variation depends on the transmission period of data and is chosen dynamically in a random fashion. Introducing sleep time will save node energy as CRSNs are battery constrained networks. Initial Slots computed by Eq. (2) are now updated by checking the estimated latency with threshold.

Once the slots are updated all S_U nodes announce their updated slots by transmitting new sync messages. The channel parameters considered are data rate, traffic load, Jitter and buffer occupancy. The data rate is computed as sum of S_D and $\text{Comm\_Delay}_{\text{AVERAGE}}$. If B_W indicates the channel bandwidth then traffic load is computed as

$$\text{Traffic}_{\text{LOAD}} = \text{Recieved\_ packet\_ count}/B_W \tag{7}$$

select channel having maximum traffic load and data rate, with minimum Jitter. S_U has complete information on updated slots and best available channel. The updated slots are periodically updated at BS. Using newly updated slots and best available vacant channel, communication is initiated by constructing a tree topology.

3.3 Flow Chart of Proposed DSC Protocol

In this section, a flow chart is designed for the proposed protocol. Figure 2 shows the detailed working of DSC protocol.

4 Results and Discussions

In this work, Network Simulator NS-2 is used and implement the proposed DSC protocol and compare our simulation results with existing spectrum aware protocols in CRSN like CogLEACH [13], DSAC [14] and with our previous work IACFC [6].

4.1 Average Energy Consumption

SU nodes in a CRSN significantly spend their energy in sensing the surrounding environment, data transfer, and to check the presence of PU activity. Since CRSNs are devices with limited power, energy consumption will significantly increase with static TDMA schedules. In CogLEACH and DSAC energy consumption is high because of static TDMA. Even though the nodes have data to transfer they have to

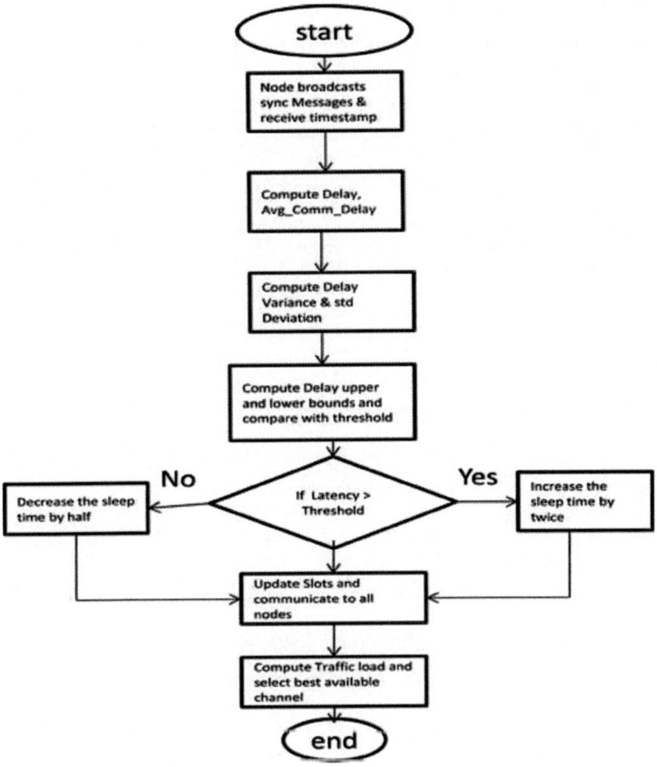

Fig. 2 Flow chart of DSC protocol

wait for their turn so energy consumption is high. İn IACFC clustering process is used to minimize energy consumption, but due to static TDMA energy consumption is not minimum. In DSC algorithm, since sleep times are assigned dynamically, nodes save a significant amount of energy. Therefore, nodes in DSC spend minimum energy. Figure 3 shows the energy consumption comparison.

4.2 *Packet Delivery Ratio (PDR) and Packets Dropped*

PDR is defined as ratio of total number of packets received to the total number of packets sent from the source.

$$\text{PDR} = \text{Total packets recieved/Total packets sent} \tag{8}$$

Higher values of PDR indicate more efficient is the protocol. In CogLEACH and DSAC, more packets are dropped due to static TDMA. İn IACFC clustering process

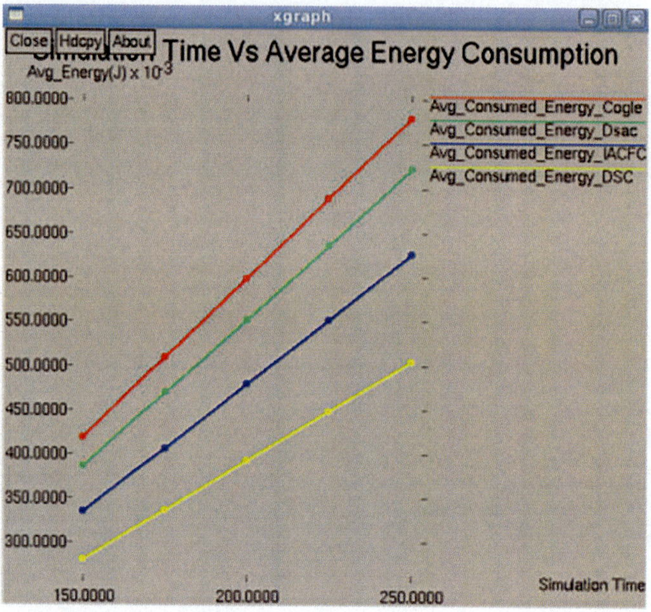

Fig. 3 Simulation time versus average energy consumption

will increase PDR as nodes have to forward data packets only to CH. But CRSN is dynamic in nature where PU activity is not predictable hence static timeslots will reduce PDR. In DSC since there is dynamic slot computation, SUs forward their data dynamically once the slot is available. Nodes need not wait for their turn as in static TDMA used in CogLEACH, DSAC, and IACFC.

Packets dropped are defined as the difference between the total packets sent by source node and the total packets received by the destination.

$$\text{Packets dropped} = \text{Total packets sent} - \text{Total packets recieved} \tag{9}$$

As PDR increases the packets dropped decreases. So lesser the packets dropped, more efficient is the protocol. Figures 4 and 5 depict the PDR and dropped packets comparison with other CRSN protocols.

Since CogLEACH and IACFC implement static TDMA, DSAC implements same channel for data and control packets there is packet drop in these protocols. In static TDMA data packets have to wait for their turn even if the nodes have data to send. As nodes have limited buffer capacity, nodes cannot wait for their turn as CRSN is dynamic in nature. In DSAC packets are dropped because same channel is used for control and data transfer. To avoid these conditions DSC introduces dynamic slots so packet drop is significantly reduced.

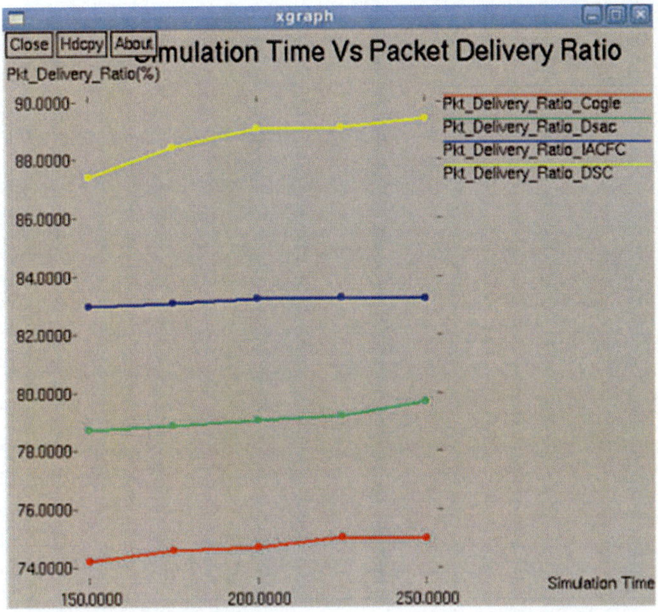

Fig. 4 Simulation time versus PDR

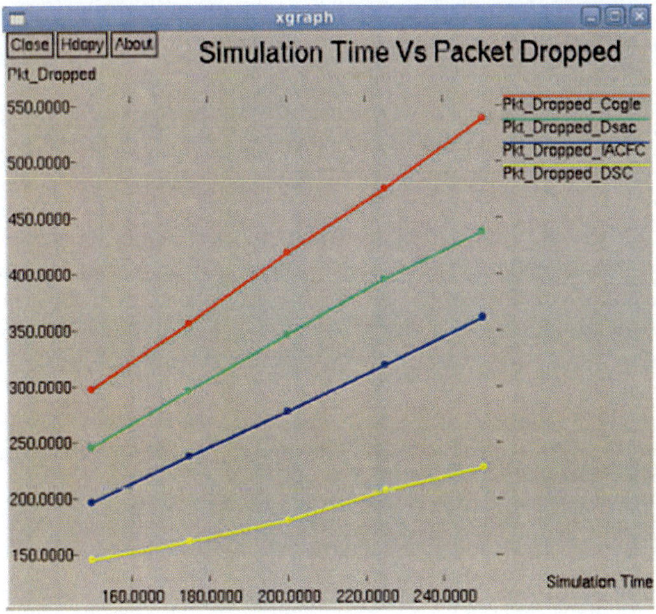

Fig. 5 Simulation time versus packets dropped

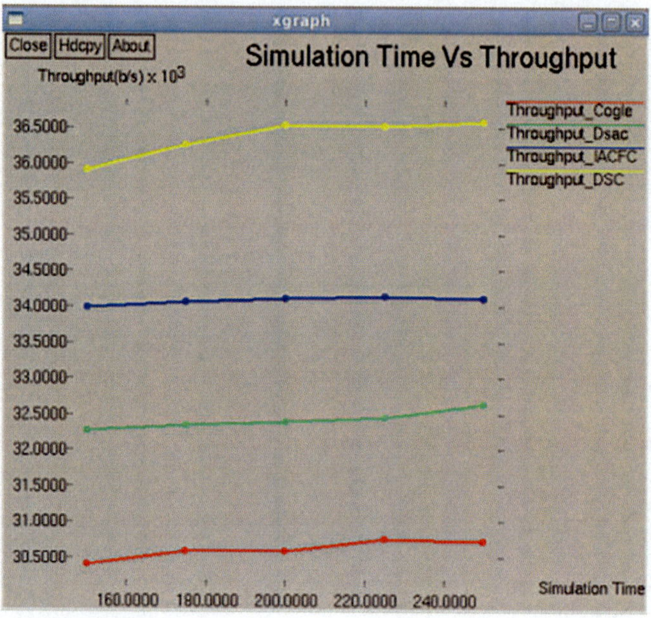

Fig. 6 Simulation time versus throughput

4.3 *Throughput*

Throughput is defined as the total number of bytes received in unit time. Higher the throughput measure, the more efficient is the protocol. As can be observed from Figs. 4 and 5 the PDR and dropped packets is optimal for DSC as compared to other protocols due to the implementation of dynamic slots. The delay-based slot computation described in Sect. 3.2 results in more packets being delivered to the destination in DSC algorithm. Also among the available channels, based on traffic load estimations best channel is selected. This shows better throughput for DSC in comparison with DSAC, CogLEACH, and IACFC. Figure 6 shows the throughput comparison.

5 Conclusion and Future Scope

Due to inherent complications of CRSNs availing dynamic spectrum allocation for Secondary Users is a challenging task. In this paper, a novel mechanism for computing dynamic slots in CRSNs is presented. As SU nodes have to constantly sense for vacant channels in CRSNs introducing sleep times for SU nodes aid in

Table 1 Performance variations of existing protocols

Protocol	Methodology	Performance variations
CogLEACH [13]	CCC and Static TDMA are used. When nodes have some data to send, using static TDMA nodes should wait for their turn. CCC starvation problem may occur when nodes have no data to send	Due to static TDMA, energy of node may get depleted. Since CRSN is dynamic, data transmission based on static TDMA leads to more energy consumption. CCC starvation problem is not addressed
DSAC [14]	Beacon frames are used to identify vacant channels. Same channel is used for control message exchange and data transfer	Nodes spend significant amount of energy in exchanging beacon frames. Same common channel for beacons and data packet transmission will lead to packet collisions leading to packet retransmissions and more energy consumption
IACFC [6]	Static TDMA as in CogLEACH is implemented. Implements fairness-based channel allocation	Data transmission based on static TDMA will result in more energy consumption and have poor performance

reducing energy consumption significantly. By introducing dynamic slot computation and channel selection based on delay and traffic estimations, proposed DSC protocol performs significantly better in terms of energy consumption, throughput, and packet delivery ratio. In the future an attempt to introduce countermeasures for security threats on CRSNs will be given primary importance along with testing DSC protocol considering other QoS parameters. Table 1 describes the performance variations between DSAC [14], CogLEACH [13] and IACFC [6].

References

1. Idoudi H, Daimi K, Saed M (2014) Security challenges in cognitive radio networks. In: Proceedings of the World Congress on Engineering, vol 1
2. Yuan Y, Bahl P, Chandra R, Moscibroda T, Wu Y (2007) Allocating dynamic time-spectrum blocks in cognitive radio networks. In: Proceedings of 8th ACM symposium on Mobile ad hoc networking and computing, New York, pp 130–139
3. Joshi GP, Nam SY, Kim SW (2013) Cognitive radio wireless sensor networks: applications, challenges and research trends. Sensors 13(9):11196–11228
4. Wireless sensor networks: a survey on recent developments and potential synergies—scientific figure on ResearchGate. Available from https://www.researchgate.net/figure/Topology-of-a-typical-cognitive-radio-sensor-network-CRSN_fig16_258165429
5. Kim H, Shin KG (2006) Adaptive MAC-layer sensing of spectrum availability in cognitive radio networks. University of Michigan, Tech. Rep. CSE-TR-518-06
6. Agarkhed J, Gatate V (2020) Interference aware cluster formation in cognitive radio sensor networks. In: Bindhu V, Chen J, Tavares J (eds) International conference on communication, computing and electronics systems. Lecture notes in electrical engineering, vol 637. Springer, Singapore

7. Cesana Matteo, Cuomo Francesca, Ekici Eylem (2011) Routing in cognitive radio networks: challenges and solutions. Ad Hoc Netw 9(3):228–248
8. Kamruzzaman S, Alam MS (2010) Dynamic TDMA slot reservation protocol for cognitive radio ad hoc networks. 46:142–147. https://doi.org/10.1109/iccitechn.2010.5723844
9. Ren J et al (2016) Dynamic channel access to improve energy efficiency in cognitive radio sensor networks. IEEE Trans Wirel Commun 15(5):3143–3156
10. Hou F, Huang J (2010) Dynamic channel selection in cognitive radio network with channel heterogeneity. In: 2010 IEEE global telecommunications conference GLOBECOM 2010. IEEE
11. Zhang D et al (2016) Utility-optimal resource management and allocation algorithm for energy harvesting cognitive radio sensor networks. IEEE J Sel Areas Commun 34(12):3552–3565
12. Lin Y et al (2016) A novel dynamic spectrum access framework based on reinforcement learning for cognitive radio sensor networks. Sensors 16(10):1675
13. Eletreby RM, Elsayed HM, Khairy MM (2014) CogLEACH: a spectrum aware clustering protocol for cognitive radio sensor networks. In: Proceedings of IEEE CROWNCOM 2014, pp 179–184
14. Zhang H et al (2011) Distributed spectrum-aware clustering in cognitive radio sensor networks. In: 2011 IEEE global telecommunications conference-GLOBECOM 2011. IEEE

High Gain Wideband Antennas for 5G Applications: A Review

Hema D. Raut, Laxmikant Shevada, Rajeshwari Malekar, and Sumit Kumar

Abstract The generation today is heading toward 5G technology and hence lot of research work is carried out in the field of antenna design in terms of compact antenna designs, bandwidth and gain improvement, high radiation efficiency. In this paper the 5G antenna designs mainly in three frequency bands that include 3–8 GHz, 22–36 GHz and above 50 GHz are discussed. The parametric study includes comparison of gain, return loss, radiation efficiency, bandwidth and axial ratio. All papers that are reviewed here have performed simulations on HFSS and CST Microwave studio. After comparative study of single element antennas it is observed that low-RCS circularly polarized antenna provides a gain of 9.3 dBi and wideband of 19.64%. Similarly, it is also observed that a Ka-band antenna array provides a gain of 19.88 dBi and a %bandwidth of 24.4%. While in the case of the multiple input multiple output antennas a metamaterial-based 2 element antenna provides %bandwidth of 22.3% and isolation of 29 dB.

Keywords Multiple input multiple output (MIMO) antenna · Dielectric resonator antenna · Metamaterial-based antenna · Phased array antenna · Mm-wave applications

H. D. Raut · L. Shevada · R. Malekar · S. Kumar (✉)
Electronics & Telecommunication Department, Symbiosis Institute of Technology, Symbiosis International (Deemed University), Pune, India
e-mail: er.sumitkumar21@gmail.com

H. D. Raut
e-mail: hdraut2083@gmail.com

L. Shevada
e-mail: lp.aish@gmail.com

R. Malekar
e-mail: rajeshwari.malekar@gmail.com

G. Ranganathan et al. (eds.), *Inventive Communication and Computational Technologies*, Lecture Notes in Networks and Systems 145,
https://doi.org/10.1007/978-981-15-7345-3_67

1 Introduction

The current 4G/long term evolution (LTE) technology suffers from various problems like latency, frequency spectrum, data bandwidth, etc. The emerging 5G technology aims to respond to these technological advancements, providing an increase in capacity, coverage, connectivity, energy efficiency and reduced cost when compared with the current 4G. To overcome above problems and to enhance wide range of 5G applications, an improved mobile system with high data rate is required and this is possible with 5G technology. The mm-Wave 5G network provides wider bandwidth, improved coverage, enhanced spectral efficiency, reduced power consumption. The various wideband antenna structures include, low-profile wideband antenna, Compact wideband circularly polarized antenna array, Wideband MIMO antennas, metamaterial-based antennas.

To enhance the impedance bandwidth multi-resonant structures are designed. Within the operation band, it is possible to obtain a stable radiation pattern with nearly constant gain [1]. In mm-wave range better radiation efficiency is obtained using dielectric resonators (DRA). MIMO antennas with high isolation are needed. By hindering the displacement current between the elements of antenna the isolation improvement can be achieved in MIMO dielectric resonator antennas. Isolation can also be improved using various techniques available in the literature such as orthogonal modes hybrid feeding mechanism, parasitic structures including metallic strips and frequency selective surfaces. Isolation can also be enhanced using DRA by reprinting metal strip on the top [2]. Use of high-gain antennas at the base stations and mobile terminals helps to reduce path loss in a communication system. The design of high-gain antennas includes phased array schemes, stacked patch antennas, electromagnetic band gap (EBG) resonator antennas, zero index metamaterials [3]. The use of directional and high-gain antennas with steerable beams can compensate for path loss and transmission loss. Antennas with high gain can be designed by phased array schemes, stacked patch antennas, EBG resonator antennas, zero index metamaterials.

The metamaterial-based antipodal vivaldi antenna (AVA) provides gain enhancement by minimizing the radiations in undesired directions [4]. The design is complex but by reducing the lower cut-off frequency it is capable of exhibiting wideband and reduced return losses. In this survey paper, the AVA design and performance is enhanced by introducing unit cells based on metamaterials [5–7]. The wideband high-gain array antenna systems with desired matching, stable radiation pattern, high beam scanning capability with reduced sidelobe level (SLL) and the pointed radiating beam is achieved [8]. The various metamaterial-based techniques have been meticulously explored to amplify the performance of antenna that includes high gain [9, 10], multi-band structures [11, 12], and beam steering [13, 14]. A worthless method that is suitable for mm-wave communication is presented to realize a high-gain planar bowtie antenna characterized by dual beam that is symmetric in nature [15]. Antipodal vivaldi array antenna with corrugated structure is designed and it provides a maximum gain of 13.2 dB in the frequency range of 24.04–40.85 GHz

[16]. The survey paper in Sect. 2 is divided into 3 sub-sections, in the first subsection single element antenna designs are studied, in the second parametric study of array antennas are summarized and finally MIMO antenna structures are studied. The feeding methods used in the surveyed papers are discussed in Sect. 3 and finally the literature survey is concluded in Sect. 4.

2 Review of Survey Papers on High-Gain Wideband Antennas

2.1 Single Element Antenna

Antenna for mm-wave application is presented and it comprises of a magnetoelectric dipole antenna which is fed by a substrate integrated waveguide microstrip. The design provides a gain of 9.2 dBi with a bandwidth of above 50% (57–71 GHz). The observed return loss at 65 GHz is −15 dB. The antenna design is compact with dimensions as 4.3 × 5.1 × 0.787 mm^3 and is a suitable candidate for mm-wave because of low back-radiation level, wider bandwidth. Compared to other reported single layer antennas the size is relatively large [17].

The antenna designed is a wideband low-profile antenna suitable for 5G wireless applications and it provides a 6.2 dBi gain with 58.3% (2.84–5.17 GHz) bandwidth. The bandwidth is enhanced by combining four resonant modes with different operating frequencies into a single framework. Impedance matching of this multi-resonant structure is enhanced by using folded walls. FR4 substrate of 4.5 mm thickness is used and the dimensions of fabricated design is 63 × 51 × 4.5 mm^3. The radiation pattern is unidirectional with measured average radiation efficiency of 64% (2.8 to 5.2 GHz). The observed return loss is −22 dB at 3 GHz and the antenna pattern exhibits low back lobe level and low cross polarization in both the planes [1].

The antenna gain is improved by placing a frequency selective surface (FSS) with high reflection magnitude over a patch antenna with a dense dielectric (DD). For further gain enhancement of an antenna a superstrate layer which in turn act as a len is deposited over the dense dielectric patch. As a superstrate layer a set of highly reflective unit cell is introduced. The air gap between the FSS layer with an identical set of FSS unit cells and the square DD patch enriches the gain and impedance matching and this in turn makes the design apt for mm-wave applications in 5G. The patch dimension is 32 × 32 × 0.508 mm^3 and it consists of two substrates of same height. A 50 Ω microstrip line feeds the patch. The parametric investigation of the designed antenna results in a gain of 17.78 dBi at 28 GHz with bandwidth of 9% and 90% radiation efficiency. The observed return loss at 29.75 GHz is −26 dB [18].

Orthogonal pattern diversity of a compact shared-aperture antenna is proposed in this reference. In the shared-aperture topology the dual-polarized nature is productive and also the dual-polarized zero index metamaterial (DPZIM) unit cell is combined with the radiating aperture. The designed antenna has a dimension of 31 × 31 ×

0.508 mm^3. The observed parametric analysis includes gain of 9 dBi, the 1 dB Gain Bandwidth of 35%, mutual coupling of −22 dB, pattern diversity of 0° and 90°. Reduced mutual coupling results in a better radiation pattern with larger beam integrity, also the design provides a front to back ratio of above 15 dB. The observed return loss is 25 dB at 28 GHz. The antenna design provides a wideband and is a good candidate for 5G base stations [19].

Based on the concept of metasurface a circularly polarized broadband antenna is designed to provide high gain. The proposed low-profile antenna is characterized by low-radar-cross-section (RCS). By means of passive cancelation a wideband monostatic RCS reduction is accomplished via proper arrangement of three types of artificial magnetic conductors surrounding the antenna. The patch is of dimension 64 × 64 × 3.5 mm^3 and the design provides a gain of 9.3 dBi with aperture efficiency of 31.2% and return loss bandwidth of 19.64% (5.1–6.2 GHz). The 3-dB axial ratio bandwidth is about 14.9% (5.37–6.25 GHz) and the return loss obtained is −22.5 dB. Since the design is based on polarization conversion metasurface, it provides high gain and enhanced axial ratio bandwidth and serves the purpose of radar system, communication and sensor systems [20].

The dielectric resonator antennas (DRA) provides low conductor losses and has high radiation efficiency. The designed antenna is composed of a 1 × 4 negative refractive index metamaterial (NRIM) and a DRA which is double fed thereby providing beam tilting. The unit cell of NRIM which is based on fractal cross ring resonator design enables change in the maximum beam direction of the DRA. Due to its symmetric structure the NRIM provides dual polarization and gives a stable negative refractive index within the frequency range of 5–5.5 GHz. The patch dimension is 100 × 50 × 43.5 mm^3 and the designed antenna structure provides return loss − 30 dB at 5.45 GHz with gain below 9 dB and side lobe level of −8 dB. A maximum beam tilting is provided in the xoz-plane is 38° [21]. Antipodal vivaldi antenna (AVA) design is fabricated on a FR4 substrate. The simulation is done on HFSS and the simulated results are verified with the measured results. Two metamaterial-based AVA are designed. The first metamaterial-based AVA design is without corrugation structures whereas the second metamaterial-based AVA design incorporates the corrugation structures. The second AVA design provides gain enhancement from 7.75 to 9.53 dB within the frequency band of 24.8 to 34.52 GHz with a %bandwidth of 35.95% and is a suitable candidate for 5G applications [22].

The Table 1 indicates that in an attempt to improve gain one has to compromise with the bandwidth. The antenna designed in reference [17] provides a design capable of providing a gain above 9 dBi with a percent bandwidth of 50%. The metamaterial-based antenna designed in reference [20] provides a gain of 9.3 dBi but with a merely 19.64% of bandwidth.

Table 1 Comparison table for single element antennas

S. No.	Antenna type	Platform used	Frequency range (GHz)	Gain (dBi)	Bandwidth (%)	Return loss (dB)	Ref. No.
1	Low-RCS CP antenna	Not given	5.4	9.3	19.64	−22.5	20
2	Wideband antenna	HFSS	2.84–5.17	6.2	58.3	−22	1
3	FSS based dense dielectric (DD) patch	CST microwave studio	28	17.78	9	−26	18
4	Magnetoelectric dipole antenna	HFSS	57–71	9.2	50	−15	17
5	Metamaterial-based shared-aperture antenna	HFSS	27–30	9	35	−25	19
6	Metamaterial-based DRA	CST microwave studio	5–5.5	8	12.84	−30	21
7	Metamaterial-based AVA	HFSS	24.8–34.52	9.53	35.95	−43	22

2.2 Array Antennas

To support wireless communication system in E-band (77–86. 5 GHz) a broadband antenna which is planar and cost effective is designed. The antenna design is based on single layer substrate integrated waveguide (SIW) method and is fabricated using single layer printed circuit board which is low cost. On the broad surface of a SIW a pentagonal slot is etched that is open-ended and inside the ring slot a short circuited metallic via is introduced. For the purpose of demonstration, a 2×8 array is designed and fed by a multiway SIW power divider along with phase shifter and SIW-WR10 conversion. The antenna has a size of $38.9 \times 38.7 \times 0.508$ mm^3. The observed return loss at 82.6 GHz is −35 dB. The parametric analysis exhibits return loss %bandwidth of 11.62% (77 to 86.5 GHz) and gain of 15 dBi at 84.75 GHz [23].

Wideband array antennas designed that are capable of providing circular polarization in mm-wave frequency band includes grid antennas, helical antenna, patch antenna and open loop antennas and are fabricated using many layers of glass ceramics. The fabrication is difficult and costly. In this paper, two 4×4 array antennas are designed one provides linear polarization whereas the other design provides circular polarization. A circularly polarized wideband array antenna is designed using coplanar waveguide feeding method. Both the antenna designs provide high gain and wideband and are manufactured on a double-sided single layer printed circuit board. Antenna array with linear polarization provides a gain of about 15.2 dBi and impedance bandwidth of 25.5% whereas a circularly polarized array design provides slightly lower gain of about 14.5 dBi with impedance bandwidth of 17.8%. The 3-dB axial ratio %bandwidth is about 15.6%. The antenna design supports low radiation losses [24].

In 5G due to complex design of parallel feed network, the coupled-feed-type planar microstrip antennas is rarely designed. The direct feed based planar arrays are widely designed in 5G because of their ease of fabrication and simple design, but they suffer from narrow bandwidth. The bandwidth can be enhanced using various techniques that include substrate integrated waveguides, traveling wave antennas. In this paper, proximity-coupled 6 × 5 planar array is designed and is fed by a difficult to design coupled-feed structure. The array antenna designed provides a %bandwidth of 9.8% (26.04–28.78 GHz) with gain of 21 dBi (27.5–28.5 GHz) and is a suitable candidate for 5G applications. The observed return loss at 27.2 GHz is approximately −28 dB [25].

Four modes of the designed planar phased array antenna are combined so as to obtain wider bandwidth with wide scan angle. Instead of using a balun structure to feed the array elements, a coaxial-to-differential-strip line transition is used to differentially feed them. The antenna array designed supports the frequency range of 25–33 GHz. The design provides a gain of 5 dBi, coverage efficiency of 50%, scan angle of over 140° and impedance bandwidth of 8 GHz. The antenna is a deserving candidate for 5G mobile terminals. The observed return loss at 32 GHz is −23 dB with return loss %bandwidth of 23.72% (26–33 GHz) [26].

The phased array antenna covers the frequency range of 24 to 31 GHz (Ka-band). The design provides high gain, wide bandwidth and efficient radiation pattern because of its novel design that is based on non-alignment of antenna patches so as provide symmetric radiation pattern and introduction of parasitic patches in the design so as to enhance the bandwidth and gain. The feeding method used is inset-feed that is simple and easy to fabricate. Linear array consist of 16 elements that are fed by 16 feed ports excited by input signal of equal amplitudes and different phases and this further excites the 16 other parasitic elements. The patch dimensions are 17.45 × 99.2 × 0.254 mm^3. Parametric analysis of the designed antenna includes, gain of 19.88 dBi, radiation efficiency of 86%, SLL is 13.4 dB, return loss of about −22 dB at 29 GHz with feedback network, the scan angle is about 108° and the 3 dB beamwidth is 5°. The design provides a %bandwidth of 24.4% (24.35–31.13 GHz) [8].

The paper deal with a 2 × 2 patch antenna array based on metasurface that consists of a frame of 4 × 4 periodic metal plates. The design provides circular polarization. A sequentially phased feeding method is used and is integrated with metasurface to obtain wide bandwidth. The dimension of the fabricated array is 64 × 64 × 2.3368 mm^3 at 5.9 GHz. The prototype provides return loss bandwidth of 55.6% with 3 dB AR bandwidth of 41.67%. The parametric analysis also exhibits a 3 dB gain %bandwidth of 37.3% and a peak gain (dBic) of 12.08. The radiation efficiency is above 88%. A return loss of −30 dB is obtained at 4.5 GHz. Thus, the designed compact wideband antenna array has all the required characteristics essential for wideband wireless communication systems [3].

In 5G technology the integration of antennas operating at low frequencies with mm-wave antennas is an emerging field with lots of nuts. The combination of 5G (mm-wave) array and 4G antennas suffers from limited bandwidth, radiation patterns that are broadside and directional radiation pattern rather than beam steering. In addition, the design of multi-layer antenna is comparatively difficult and complex and also its fabrication is tedious. The antenna design in the paper uses anti-reflective layer. The high frequency array antenna is placed behind the low frequency antenna and this low frequency antenna is made transparent using grating strips between them. This design provides the end-fire radiation pattern. The printed inverted F antenna (PIFA) with frequencies (740–960 MHz) and (1.7–2.2 GHz) that indicates dual band is integrated with a four-element extremely high frequency antenna array with end-fire radiation pattern. The array antenna has a dimension of 70 × 120 × 0.764 mm^3 and operates in frequency range of 22–31 GHz with scan angle of 50°. The mm-wave antenna array at 28 GHz exhibits a gain of 9.5 dBi. The observed return loss at 22.5 GHz is about −22 dB with %bandwidth of 37.75% (21–31 GHz). The low frequency antenna covers two frequency bands namely 740–960 MHz and 1.7–2.2 GHz with acceptable gain [27].

The antenna gain can be improved by reducing radiations in the undesired directions and moving toward metamaterial-based antenna designs because of their unique properties. Table 2 reveals that at mm-wave frequency (22 GHz above) a maximum gain of 19.88 dBi is obtained using a electronically scanned array antenna with a bandwidth of 24.4%. The circularly polarized antenna array design discussed in reference [24] and [3] provides a 3-dB axial ratio %bandwidth of 15.6% and 41.67% respectively. In mm-wave frequency range, reference [23] provides comparatively excellent return loss with overall gain of 15 dBi but the %bandwidth is comparatively low and can be improved further.

Table 2 Comparison table for antenna arrays

S. No.	Antenna type	Platform used	Frequency range (GHz)	Gain (dBi)	Bandwidth (%)	Return loss (dB)	Ref. No.
1	Wideband CP	HFSS	60	14.5	19.64	−22	24
2	E-band 2 × 8 array	HFSS	77–86.5	15	11.62	−35	23
3	2 × 2 array (CP)	HFSS	4.75–7.25	12.08	55.6	−30	3
4	6 × 8 proximity-coupled array	CST microwave studio	27.5–28.5	21	9.8	−28	25
5	Quad mode phased array	CST microwave studio	25–33	5	23.72	−23	26
6	Ka-band	HFSS	24–31	19.88	24.4	−22	8
7	mm-wave beam steering	CST microwave studio	22–31	9.5	Not given	−22	27

2.3 Multiple Input Multiple Output (MIMO) Antennas

The MIMO antenna design enhances the channel capacity greatly but suffers from the problem of mutual coupling. The coupling can be reduced by using electromagnetic band gap (EBG) structures, incorporating neutralization line in MIMO antenna design, defected ground structures (DGS) and metamaterial-based superstrates. In this paper to minimize mutual coupling in a two-element MIMO antenna system a two-layer metamaterial-based superstrate is proposed. The patch dimensions are 28 × 15.5 × 3 mm^3. The designed antenna structure covers the frequency range from 4.2 to 5.25 GHz and provides good isolation of 29 dB with %bandwidth of 22.3%. The gain observed is above 6.5 dBi at 4.625 GHz and measured value of reflection coefficient is −35 dB at 4.375 GHz [28].

The proposed MIMO DRA design consists of two identical DRs mounted on the metal ground plane. Isolation between the two antenna elements is enhanced by printing a metal strip on the upper surface of each DR. The isolation is improved to 24 dB and gain is 9 dB at 27.9 GHz. The return loss is −20 dB at 28 GHz with %bandwidth of 3.04% [2].

MIMO antenna design based on metamaterial provides high gain, beam steering and wideband. Hence, massive research is done in this area so as to meet the needs of 5G technology in the field of communication. A simple bowtie antenna is integrated with three pairs of metamaterial arrays to design a dual-beam planar antenna suitable for 5G applications. Bowtie antenna with first pair of high refractive index (HRI) MTM array forms antenna with relatively wide beam, with second pair of HRI MTM array the design provides dual beam and finally the third pair of HRI MTM array provides increased gain and directivity. The MIMO antenna is fed by a tapered feed line which provides better impedance matching. The dimensions of the designed antenna is 30 × 30.5 × 0.508 mm^3. The proposed antenna provides a gain of 7.4 dBi at 26 GHz and reflection coefficient of −35 dB at 25.8 GHz. With respect to end-fire direction, a dual-beam planar bowtie antenna is obtained at 30° which is in the E-plane and is symmetric [15].

The comparison Table 3 indicates that metamaterial-based MIMO antenna provides better isolation with improved percent bandwidth. In reference [15] the observed return loss is comparatively better but the gain is reduced by more than 1.5 dBi with respect to reference [2].

Table 3 Comparison table for Multiple input multiple output (MIMO) antennas

S. No.	Antenna type	Platform used	Frequency range (GHz)	Gain (dBi)	Bandwidth (%)	Return loss (dB)	Ref. No.
1	Metamaterial-based 2 element	HFSS	4.2–5.25	6.5	22.3	−35	28
2	DRA	HFSS	27.25–28.59	9	3.04	−20	2
3	Metamaterial-based Bowtie	HFSS	24.25–27.5	7.4	Not given	−35	15

3 Feeding Methods

The review report presented, covers variety of antenna designs which includes single element antennas, antenna arrays and MIMO antennas.

In case of single element antennas, the available feeding methods include both direct and indirect feeds. The feeding techniques used in the reference papers described in Table 1 are briefly summarized as follows. A microstrip feed line along with substrate integrated waveguide is used. An unbalanced electric field is observed in the aperture due to single feed [17] whereas in reference [1], a simple feed line is printed at the bottom of the substrate to feed the antenna. The antenna designed in reference [18] is fed using an indirect feed method that includes aperture coupled method which is difficult to fabricate. A microstrip feed line along with a one quarter wavelength impedance transformer is designed and fabricated to feed the antenna [19]. The antenna is fed using a polarization conversion unit that produces surface waves orthogonal to each other. The polarization conversion unit is excited through the feeding slot by means of electromagnetically coupled waves that are linearly polarized. The feeding method used enhances the axial ratio %bandwidth [20]. On the bottom of the substrate a microstrip line of 50 Ω is printed and each DRA is excited through a slot etched on the ground plane [21].

The antenna array feeding methods include series feed as well as parallel feed. Series feed helps to minimize side lobes and reduces the feed length but suffers from narrow bandwidth. Whereas parallel feed also known as corporate feed provides larger bandwidth and supplies equal power to all the elements but results in high feed losses. The array antenna described provides separate design for linearly polarized array and circularly polarized array. Both the array antennas are fed using a coplanar waveguide (CPW) line characterized by low dispersion. In both the array design, each subarray consisting of 4 identical elements are fed by 1:4 power divider network resulting in a differential output. All the four subarrays are combined together via CPW line. The CPW line feed suffers from radiation loss [24]. A wideband antenna array is fed using a substrate integrated waveguide (SIW) topology consisting of phase shifter, 1:8 power divider network. In SIW the conductor loss is less due to the greater amount of metal that carries the signal, also the structure corresponds to a standard rectangular waveguide structure using PCB technology [16]. A sequentially phased feeding method is used and is integrated with metasurface to obtain wide bandwidth. The patch antenna is excited via microstrip line and is combined with the sequentially phased structure that provides equal current to the 4 array elements by means of four microstrip lines [3]. The antenna array feeding structure is a series feed network that incorporates proximity-coupled method. Further each element in the array is fed using a power divider network [25]. Each element in the array is excited using an inset fed microstrip line which in turn excites the parasitic element printed on the other substrate. The input signals are equal in amplitude but different in phase so as to support beam scanning [8].

In MIMO antennas accurate feeding networks are required to minimize mutual coupling between the antennas. The two antennas are fed using a simple coaxial feed. The outer diameter of the coaxial feed is 4.1 mm whereas the inner diameter is 1.3 mm. The design provides isolation of 29 dB [28]. Each dielectric resonator is fed via a microstrip line represented by a rectangular slot. When one port is excited the other port is terminated to a matched load [2].

4 Conclusion

The literature review is based on various antenna designs that include single element antennas, array antennas followed by MIMO antennas. The antennas discussed in the survey provides high gain and wideband which are the key parameters in 5G technology and plays a vital role in mm-wave wireless communication systems. In MIMO antenna and array antenna design important parameter to be considered is the mutual coupling. A wide range of techniques are available to minimize mutual coupling. The single element antennas discussed provides a maximum gain of 9.3 dBi with a bandwidth of 19.64% [20], further the gain and bandwidth can be enhanced by choosing array antennas. In case of array antennas described in the review papers, reference [8] provides antenna array design with a gain of 19.88 dBi and a bandwidth of 24.4%. Finally, MIMO antennas referred in the literature survey provides a maximum isolation of 29 dB [28] but the gain is merely 6.5 dBi. The gain can be improved further to 9 dBi by using DRA [2]. The review report summarizes the antenna designs mainly in three frequency bands that include 3–8 GHz, 22–36 GHz and above 50 GHz. The parametric study includes comparison of gain, return loss, radiation efficiency, bandwidth and axial ratio.

References

1. An W, Li Y, Fu H, Ma J, Chen W, Feng B (2018) Low-profile and wideband microstrip antenna with stable gain for 5G wireless applications. IEEE Trans Antennas Propag 17(4):621–624
2. Zhang Y, Deng J-Y, Li M-J, Sun D, Guo L-X (2019) A MIMO dielectric resonator antenna with improved isolation for 5G mm-wave applications. IEEE Trans Antennas Propag 18(4):747–751
3. Ta SX, Park I (2017) Compact wideband circularly polarized patch antenna array using metasurface. IEEE Antennas Wirel Propag Lett 16:1932–1936
4. Dixit AS, Kumar S (2020) A survey of performance enhancement techniques of antipodal vivaldi antenna. IEEE Access 8:45774–45796
5. Tutuncu B, Torpi H, Urul B (2018) A comparative study on different types of metamaterials for enhancement of microstrip patch antenna directivity at the Ku-band (12 GHz). Turkish J Electr Eng Comput Sci 26(3):1171–1179
6. Alhawari ARH, Ismail A, Mahdi MA, Abdullah RSAR (2012) Antipodal vivaldi antenna performance booster exploiting snug-in negative index metamaterial. Prog Electromagn Res 27(1):265–279
7. Zhu S, Liu H, Wen P, Du L, Zhou J (2018) A miniaturized and high gain doubleslot Vivaldi antenna using wideband index-near-zero metasurface. IEEE Access 6:72015–72024

8. Khalily M, Tafazolli R, Xiao P, Kishk AA (2018) Broadband mmWave Microstrip Array Antenna With Improved Radiation Characteristics for Different 5G Applications. IEEE Trans Antennas Propag 66(9):4641–4647
9. Wu S, Yi Y, Yu Z, Huang X, Yang H (2016) A zero-index metamaterial for gain and directivity enhancement of tapered slot antenna. J Electromagn Waves Appl 30(15):1993–2002
10. Bhaskar M, Johari E, Akhter Z, Akhtar MJ (2016) Gain enhancement of the Vivaldi antenna with band notch characteristics using zero-index metamaterial. Microw Opt Techn Lett 58(1):233–238
11. Si L, Zhu W, Sun H (2015) A compact, planar, and CPW-fed metamaterial-inspired dual-band antenna. IEEE Antennas Wireless Propag Lett vol 12, pp 305–308
12. Si L et al (2015) A uniplanar triple-band dipole antenna using complementary capacitively loaded loop. IEEE Antennas Wirel Propag Lett 14:743–746
13. Dadgarpour A, Zarghooni B, Virdee BS, Denidni TA (2016) Single end-fire antenna for dual-beam and broad beamwidth operation at 60 GHz by artificially modifying the permittivity of the antenna substrate. IEEE Trans Antennas Propag 64(9):4068–4073
14. Li J, Zeng Q, Liu R, Denidni TA (2017) Beam-tilting antenna with negative refractive index metamaterial loading. IEEE Antennas Wirel Propag Lett 16:2030–2033
15. Jiang H, Si L-M, Hu W, Lv X (2019) A symmetrical dual-beam Bowtie antenna with gain enhancement using metamaterial for 5G MIMO applications. IEEE Photonics J 11(1)
16. Dixit AS, Kumar S (2020) A miniaturized antipodal vivaldi antenna for 5G communication applications. In: 7th international conference on signal processing and integrated networks (SPIN), Noida, India, pp 800–803
17. Zeng J, Luk K-M (2019) Single-layered broadband magnetoelectric dipole antenna for new 5G application. IEEE Trans Antennas Propag 18(5):911–915
18. Asaadi M, Afifi I, Sebak A-R (2018) High gain and wideband high dense dielectric patch antenna using FSS superstrate for millimeter-wave applications. IEEE Trans Antennas Propag 6:38243–38250
19. Sadananda KG, Abegaonkar MP, Koul SK (2019) Gain equalized shared-aperture antenna using dual-polarized ZIM for mmWave 5G base stations. IEEE Antennas Wirel Propag Lett 18(6):1100–1104
20. Jia Y, Liu Y, Gong S, Zhang W, Liao Guisheng (2017) A Low-RCS and high-gain circularly polarized antenna with a low profile. IEEE Trans Antennas Propag 16:2477–2480
21. Li J, Zeng Q, Liu R, Denidni TA (2017) Beam-tilting antenna with negative refractive index metamaterial loading. IEEE Trans Antennas Propag 16(3):2030–2033
22. Dixit AS, Kumar S (2020) The enhanced gain and cost-effective antipodal Vivaldi antenna for 5G communication applications. Microw Opt Technol Lett 62:2365–2374. https://doi.org/10.1002/mop.32335
23. Hao Z-C, He M, Fan K, Luo G (2017) A planar broadband antenna for the E-band gigabyte wireless communication. IEEE Trans Antennas Propag 65(3):1369–1373
24. Li M, Luk K-M (2014) Low-cost wideband microstrip antenna array for 60-GHz Appl. IEEE Antennas Wirel Propag Lett 62(6):3012–3018
25. Diawuo HA, Jung YB (2018) broadband proximity-coupled microstrip planar antenna array for 5G cellular applications. IEEE Antennas Wirel Propag Lett 17(7):1286–1290
26. Syrytsin I, Zhang S, Pedersen GF, Morris AS (2018) Broadband mm-wave microstrip array antenna with improved radiation characteristics for different 5G applications. IEEE Antennas Wirel Propag Lett 66(9):4648–4657
27. Taheri MMS, Abdipour A, Zhang S, Pedersen GF (2019) Integrated millimeter-wave wideband end-fire 5G beam steerable array and low-frequency 4G LTE antenna in mobile terminals. IEEE TransVeh Technol 68(4):4042–4046
28. An W, Li Y, Fu H, Ma J, Chen W, Feng B (2019) Broadband extremely close-spaced 5G MIMO antenna with mutual coupling reduction using metamaterial-inspired superstrate. Opt Express 27(3):3472–3482

Things-to-Cloud (T2C): A Protocol-Based Nine-Layered Architecture

N. D. Patel, B. M. Mehtre, and Rajeev Wankar

Abstract The Internet of Things (IoT) is rapidly becoming a part of our regular activities and environment. According to Cisco, by 2030, it is expected that 500 billion devices will be connected to the IoT. Every passing day, new devices and sensors coming in the market, which use different protocols to connect to the network. With this expected growth of the Internet of Things and the emergence of Fog and Cloud Computing, there is a need for a standard protocol suite to handle different aspects of standardization and cloud protocols in the IoT environment that is not covered by currently existing Three, four, and five-layered IoT models. A T2C architecture is proposed that provides the most prominent standard protocols into nine layers and shows why this extension is essential. The proposed T2C architecture fulfills the prescribed activities for smart sensors and fog networking in the backbone, which brings significant benefits, scalability, and low-power consumption in the IoT architecture. It emphasizes on how IoT device classification can be mapped to our proposed nine-layered T2C Architecture.

Keywords Internet of Things · IoT protocols · IoT layers · Things-to-Cloud (T2C)

N. D. Patel (✉) · B. M. Mehtre
Centre of Excellence in Cyber Security, Institute for Development & Research in Banking Technology (IDRBT), Hyderabad, India
e-mail: ndpatel@idrbt.ac.in

B. M. Mehtre
e-mail: bmmehtre@idrbt.ac.in

N. D. Patel · R. Wankar
School of Computer & Information Sciences (SCIS), University of Hyderabad (UoH), Hyderabad, India
e-mail: wankarcs@uohyd.ac.in

G. Ranganathan et al. (eds.), *Inventive Communication and Computational Technologies*, Lecture Notes in Networks and Systems 145,
https://doi.org/10.1007/978-981-15-7345-3_68

1 Introduction

Things-to-Cloud (T2C) is a bridge between the Internet of Things, Fog Computing, and Cloud Computing and are putting together IoT and Cloud premises, and calling it as Things-to-Cloud (T2C). Internet of Things (IoT) is a system of interconnected electromagnetic tracking devices, computing devices, and smart digital devices. These devices are capable of transferring the data to the cloud via fog computing. The IEEE has defined an IoT architecture, domains of IoT, the definition of IoT, and identification of public IoT domains. Logvinov et al. [1] introduced three-tiered architecture with Sensing layer, Networking & Data Communication layer, and Application layer.

Several attempts were made in the past to provide a layered architecture for IoT based on operations they perform. In the last two-three years, the gap between IoT and Cloud has been bridged due to the introduction of Fog Computing, which communicates with both these components. IoT devices are becoming more powerful and capable of performing various tasks. Due to this advancement, there is a need to re-look-up five-layered architecture [2]. In this paper, nine-layered T2C architecture connect sensor is proposed that enables IoT devices to the cloud and also explained the required layers for each type of IoT devices. Based on operations performed in each layer, the common standard protocols are classified to handle this operation. The nine layers are IoT devices layer, Connectivity layers, Link Protocol layers, Transport & Network layers, Session & Communication layers, Data Aggregation & Processing layers, Data Storage & Retrieval layer, Business Model layer, and Business Application (Apps) layer. At every layer, suggested different protocols to be used. Also, briefly reviewed Things-to-Cloud (T2C) management protocol standards and the current state of security concerns related to specific protocols. A brief comparison between various IoT protocols and how to choose among these protocols is presented.

The rest of the paper is organized as follows: Sect. 2 presents different IoT Architectures available in the literature and present four IoT architecture, in Sect. 3 presented the proposed Things-to-Cloud (T2C) which is a protocol-based nine-layered architecture, in Sect. 4 discussed IoT Security & Privacy trust framework, and Sect. 5 gave a summary of lessons and future scope of IoT.

2 IoT Architectures

The IoT architecture introduced for smart service domains has to reach the main demands of having Intelligence, Scalability, Interoperability, Think of Security Rights, Challenges in Securing, Managing, and Optimizing. Several IoT architectures are proposed in the literature, but they do not adhere to common architecture design. IoT Basic Model is essentially the architecture which allows us to gather data from various sources, where data has never been traditionally looked at with Three-layer IoT Architecture expressing of the Application Layer, Network Layer,

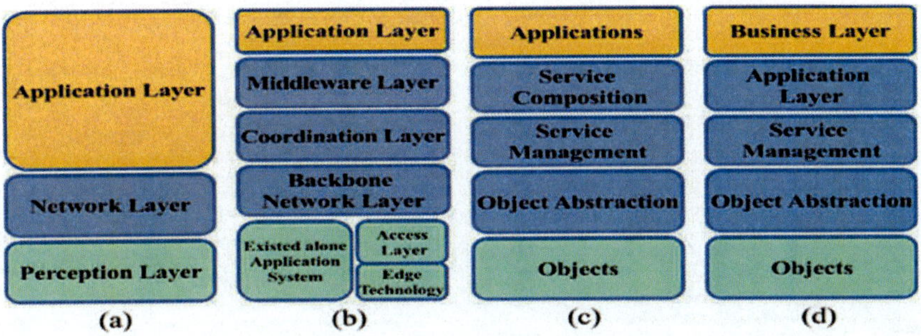

Fig. 1 Layers wise IoT architecture. **a** 3-layer [4]. **b** Middle-ware-based. **c** SOA-based. **d** 5-layer [2]

and Perception Layer has been presented in Fig. 1a [3, 4]. In the recent survey, another architecture has been considered that add more abstraction to the IoT model [2–4].

Also, three-layer architecture has been extended to Middle-Ware-based architecture and address a different type of communication medium Fig. 1b. It has been extended to SOA-based architecture while having a layer like "Service Composition" to run on resource-constrained devices in Fig. 1c. The five layer architecture is quite similar to the three layers of architecture. Khan et al. [5] extended the three layers architecture to five layers in Fig. 1d that helps to visualize a clear view of IoT architecture with Objects Layer, Object Abstraction Layer, Service Management Layer, Application Layer, and Business Layers for the easier operation for the IoT ecosystem.

The main limitations of these IoT architectures, the three-layer architecture does not cover all underlying technologies that transfer data to an IoT platform. The SOA-based architecture takes a big fraction of the time and energy of the device to communicate with other devices and integrate the required services. The five-layer model is rather complex and has substantial computational requirements, and thus it requires compelling devices for its hosting. These models do not cover all standards in the real IoT environment. For example, "Object Abstraction Layer and Service Management Layer" does not cover all Cloud Protocols; neither does it integrate the required services.

3 Protocol-Based Nine-Layered Architecture for Things-to-Cloud (T2C)

T2C Architecture provides an overview of all the IoT protocols that would be helpful for the performances of the IoT Services. In Fig. 2, a layered diagram of the essential standard protocols and technologies is presented. In the T2C Architecture, extended Object Abstraction Layer to Connectivity Layer, Link Protocol Layer, and Transport & Network Layer. The Service Management Layer has been extended into three

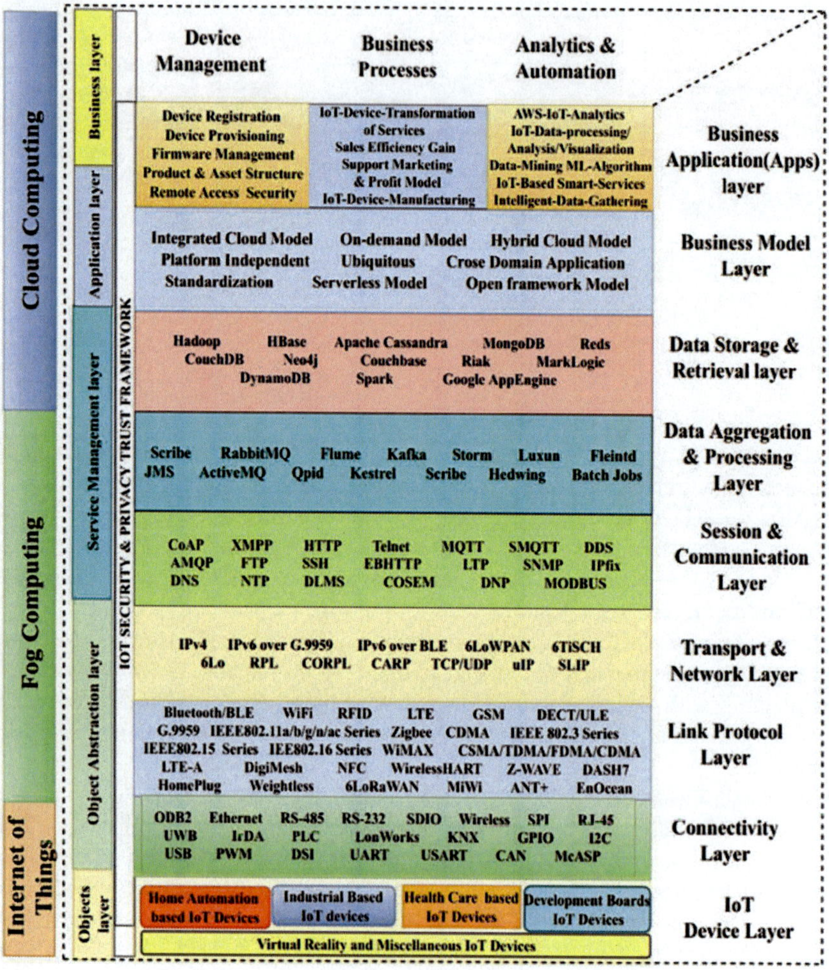

Fig. 2 Protocol-based nine-layered architecture for Things-to-Cloud (T2C)

different layers; Session & Communication layer, Data Aggregation & Processing layer, and Data Storage & Retrieval layer.

The Business Model layer is added to reach the Business Application layer with the help of an open framework. This extension is brought to reduce the maintenance cost, better services & support, smoother logistics, improved experience, and enhanced security with the low-power. The T2C protocol-based architecture defines a protocol-based networking architecture on implementing protocols in nine layers. Different groups (W3C, IETF, IEEE, EPCglobal, ETSI) are the protocols provided in support of IoT ecosystem. The T2C protocol-based nine-layered architecture provides the most prominent Standard protocols into nine broad categories.

IoT Devices Classification and Mapping With T2C Layers:

The IoT device layer is classified into five parts: Home Automation-Based IoT devices, Industrial-Based IoT devices, Health Care-Based IoT devices, Development Board IoT devices, and Virtual Reality & Miscellaneous IoT devices. The detailed classification of IoT devices can be seen in Fig. 3. These devices connect to Link Protocol Layer and receive the data feed over the Transport & Network layer using 2G, 3G, 4G, 5G network technologies. This will not interfere with Transport & Network layer protocols for messages by-passing in the Session & Communication layer. The Session & Communication Layer sends the data feeds to the Aggregation & Retrieval layer to provide a low-latency platform for managing real-time, unified, and high-throughput data feeds. For storage purposes, the Aggregation layer sends the data feeds to the Data Storage & Retrieval layer and pop the data feeds with the help of an open framework to the Business Model layer.

However, not all of these nine layers have to be aggregated together to deliver an IoT device. Moreover, based on the inherent nature of IoT devices, some layers may not be required in them. A mapping for IoT devices classification with T2C Layers in Table 1 is provided.

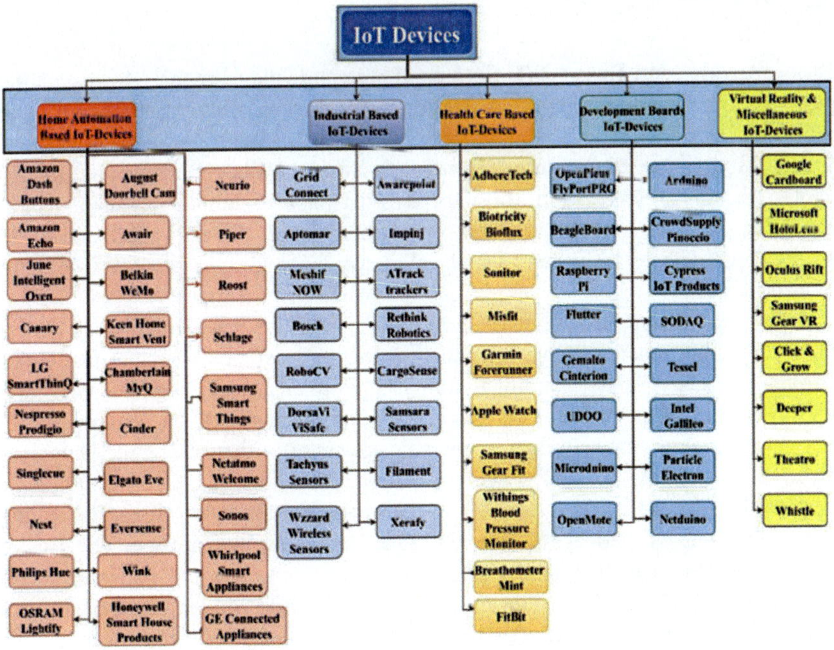

Fig. 3 IoT devices classification

Table 1 IoT devices and required layers

IoT devices/T2C layers	Home automation	Industrial	Health care	Development boards	Virtual reality and miscellaneous
Connectivity		✓		✓	
Link protocol	✓	✓	✓	✓	✓
Transport & network	✓	✓	✓	✓	✓
Session & communication	✓	✓	✓	✓	
Data aggregation & processing		✓		✓	
Data storage & retrieval				✓	
Business model	✓	✓		✓	
Bussiness application	✓	✓	✓	✓	✓

3.1 IoT device Layer

Home Automation-Based IoT devices: These little devices that consist of the perfect combination of the various sensors to prepare our home intelligently and protected i.e., HD Camera, High-Quality Microphone, Smart Temperature, Night-Vision, Motion Detection, Smart Thermostats, Smart Lights, Smart Skybell, Intelligently Tracking, Teleportation, 360 degrees IR blaster, UV, Humidity, Noise detector, Vibration, Monitor Indoor Air, Outdoor Weather, Energy Consumption, and many more. These devices intelligently analyze sensor data and send us alerts in case of unusual activity.

Industrial-Based IoT devices: Industrial-IoT (IIOT) devices are smart assets providing many benefits, including improvements to Productivity Monitoring, Operational Efficiency Safety, Control Liability Performance, and Diagnostics Regulatory. IIoT eliminates workers from many organizations and industries, allowing IoT devices to act autonomously, decreasing risks associated with worker injury or human error system failures.

Health Care-Based IoT devices: In the health care industry, it is challenging to collect samples, patient information, and diagnosing in real-time. The Health Care-based IoT devices help in monitoring and keeping track of patients' health continuously and are responsible for performing specific actions when the critical condition is triggered. It is expected that by 2025, 80% of all health care organizations will have implemented IoT technology. Some companies have also designed other health-related IoT products like Thermometers, Fitness Sensors, Baby Monitors, and more.

Development Board IoT Devices: Development Board IoT devices have a sensor embedded at the IoT Development Board. In development boards, realized how a sensible, smart device acquires, processes, and communicates information to the cloud platform. IoT software and hardware work together to build the IoT Technology stack. In the Development Boards market, there are more than 100 + Single Board Computer (SBC)/Boards available like *Arduino, BeagleBone, Intel, Raspberry Pi families*, etc.

Virtual Reality & Miscellaneous IoT Devices: IoT is also used for the low-Frequency approach to Augmented Reality (AR)/Virtual Reality (VR). There are so many devices designed primarily for enterprise use, *Google Cardboard* for nostalgic toy lovers, *Microsoft HoloLens* augmented reality smart device which can assist with training, design work, communication, and more.

3.2 *Connectivity Layer*

While developing any IoT solution big question which arises is how to connect to the above layer? There are so many connectivity protocols available in the literature. The most popular connectivity protocols are discussed and suggest how to choose the most suitable protocols for an IoT solution. The Connectivity layer protocols provide connectivity to models and interfaces to the set of sensors and will further continue into detailed features regarding the pros and cons of each of them:

OBD-2: Connected Car-The future of automobile, IoT OBD (On-Board Diagnostics) device is the best car tracking and diagnostic devices for someone who wants to track a car's location, experience travel rates, identify the problem of a check engine light, and automatically get emergency help after an accident. *Ethernet*: It is external communication in a local area network. It controls the transmission of the information and avoids simultaneous transmission by various IoT devices.

RS-485: It is an industrial protocol. This kit consists of the features of the Wi-fi connectivity, which establish the real-time overall achievement, authenticity, and accuracy for T2C while transferring information between the Application and the fundamental layer.

RS-232: It is an industrial protocol. This kit includes the features of the LoRaWAN connectivity plus the benefits of having an RS-232 connection to be integrated with industrial devices.

SDIO: SDIO is an evaluation board. It supports 802.11 series protocols and low-cost IoT USB interface.

Wireless: In the IoT, there are a series of factors, all working in tandem, to transmit the system records in real-time. Many purposes fall below the IoT umbrella.

SPI: SPI device, gadgets communicate in full-duplex mode using a master-slave structure with a single grasp. It's faster than asynchronous serial.

RJ-45: RJ-45 is a dual-port connector, including with smart sensors, operator terminals, remote input/output, gateways, and transmitters. It supports industrial protocol stacks.

UWB: Ultrawideband technology could be an undiscovered option for some IoT applications. A UWB was affiliated to a DCU (data collecting unit), and they send their IoT sensor data to the server, wirelessly [6].

IrDA: The Infrared Data Association (IrDA) offers a requirement for an entire set of protocols for Wi-fi infrared communications, and the name "IrDA" also mentions that set of protocols.

PLC: PLC is a Power Line Communication. For the IoT sensor network, Smart Grid, Chaauvenent et al. [7] proposed G3-PLC (Narrowband Powerline Communication) standard. PLC-IoT Application approves agencies of all sizes to upgrade and enhance their legacy systems for the future.

LonWorks: LonWorks is a conversation protocol in the Local operating network (Lon), so it is called LonWorks and open protocol technology. It is a networking platform and individually built address that caters to the needs of applications.

KNX: KNX is a communication protocol developed for home and building automation. KNX-IoT is the state of the art technology. It also supports diverse interactions between KNX and 3rd party equipment driven by relevant use cases from various categories.

GPIO: A GPIO (General Purpose Input/Output) port controls both incoming and outgoing signals. It is a standard interface for electronic devices and connects to microcontrollers. It can be used with System-on-Chip (SoC) Modules, Sensors, Displays, and Diodes. It supports Windows and Linux Operating Systems.

I2C: Inter-Integrated Circuit (I2C), I2C is a serial and synchronous communication protocol, so they enable apps to speak with Inter-Integrated Circuit devices. In I2C devices, EEPROMs, real-time clocks, processors, sensors are used as a control interface. I2C devices have respective data interface (Video Decoders, Digital TV Tuners, Audio processors).

USB: USB stands for Universal Serial Bus. It sends and receives a data feed along with clock pulses. It is swift compared with USART and USB. USB have three generations, i.e., USB1.0, USB2.0, USB 3.0. The USB cable is the only medium that connects Arduino and Computer.

PWM: Use Pulse-Width Modulation (PWM) signals for motor control. PWM controls analog circuits with digital outputs. PWM required less heat dissipation, greater efficiency, and higher power output.

DSI: Digital Serial Interface (DSI) capabilities allow signals communication (high-speed serial interface) between machines and sensors. Machines to be captured and analyzed the data feeds. DSI could help us to connect machines, devices, things, and sensors in various environments. DSI research & development helps to our Military to reduce the expense of display controllers.

UART: Universal Asynchronous Receiver/Transmitter (UART), It's not a communication protocol like I2C, SPI, and more, but it's a stand-alone IC. The main purpose is to transmit, broadcast, and receive serial data.

USART: It is a Universal Synchronous & Asynchronous Data Transmitter & Receiver. USART pushes and pops a block of data with class pulses. It is a duplex communication. It is fast compared with UART.

CAN: Controller Area Network (CAN) is a CSMA-CD/ASM protocol. CAN is an interface for Telit cellular module, as well as a positioning component for the Global Navigation.

McASP: Connected McASP is simple to establish a sensor available to work on your existing device, machinery. You need to install the gateway on the devices or machines which wanted to track and start monitoring them remotely.

3.3 Link Protocol Layer

In the IoT ecosystem, connected with millions of devices in the system. The information and data feeds in the Transport & Network layer is securely transferred. Almost all IoT devices need some wireless or wired connectivity interface that gives them the ability to collect information from IoT devices and transmit the information in The Transport & Network layer with different standard link protocols. These protocols make use of radio waves to transmit the collected information across a network. Link Protocol layer works as a two-way traffic, data or information received from the IoT devices will also pass through the IoT Gateway. Different protocols standards are explained briefly.

Bluetooth/BLE: Bluetooth/BLE designed for connecting devices to the Internet with low energy consumption. It is an ultra-low-power connected devices in short range. Matti et al. [8] investigate Ipv6-based connection over BLE. It is appropriate for future IoT scenarios.

Wi-fi: Any of several standards for short-range wireless data feed transmission. Wi-fi is a wireless fidelity technology. It is a real-time high-speed less communication protocol [9] for cyber-physical control applications.

RFID: RFID tags or smart labels for radio-frequency identification, RFID tags are the small wireless accessory that helps identify people, things, and objects [10]. RFID is a built-in part of our life, which increases convenience and productivity.

LTE: LTE is a high-speed wireless communication based on EDGE/GSM and HSPA/UMTS. LTE will pass large packets of data and streamline the IoT Service. It reduces the latency of information delivery.

GSM: It is established on eGPRS and construct as a high-capacity, long-range, low-power cellular gadget for IoT applications.

DECT/ULE: DECT (Digital Enhanced Cordless Telecommunications) and ULE (Ultra-Low Energy) is an ordinary European widespread for cordless phones. They have specific a low-power air interface technological know-how that can be used for IoT applications.

G.9959: G.9959 performed at low-cost, low-bandwidth, and half-duplex reliable wireless communication. It is planned for real-time functions and required authenticity for low strength consumption [11].

IEEE 802.11 series: 802.11 series a/b/g/n/ac/..etc., also known as Wi-fi for between two wireless clients. It includes a limited range of being connected technology containing wireless LANs (WLANs). IEEE 802.11 wireless communication medium access standard. The basic of 802.11 series MAC layer features include; simultaneous occurrence Frame, Null Data Packet, Short Mac Frame, Efficient Bidirectional Packet Exchange, and Increase Sleep Time.

Zigbee: It is a high-level intelligence communication protocol based on the IEEE 802.15.4 and can develop personal area networks, helps to a medical device for data collection, and home automation with low-power. ZigBee Pro offers greater aspects such as security, the use of the balanced symmetric-key exchange, scalable the usage of random address assignment, and better overall performance the use of environment-friendly many-to-one routing mechanisms.

CDMA: C-Code D-Division M-Multiple A-Access P-Protocol (CDMA). It used 2G and 3G wireless communications. The CDMA-based protocol can support IoT traffic in the upcoming 5G-MNA (Mobile Network Architecture).

Ethernet IEEE 802.3: Ethernet 802.3 is a physical communication in a LAN. Its wired connectivity & networking for computer and data applications.

IEEE 802.15/802.16: IEEE 802.15 is based on direct sequence spread spectrum, DSSS techniques. IEEE 802.16 is a Wireless Communications Standards (WCS) for metropolitan area networks (MANs).

WiMax: Worldwide Interoperability for Microwave Access (WiMax), produce 4G and 5G of wireless Internet. WiMax has average bandwidth in the local LAN connection and offers wireless broadband.

CSMA/TDMA/FDMA/CDMA: There are some methods for Signal. Different users share a common channel using F-Frequency D-Division M-Multiple A-Access (FDMA), T-Time D-Division M-Multiple A-Access (TDMA), C-Carrier S-Sense M-Multiple A-Access (CSMA) or C-Code D-Division M-Multiple A-Access (CDMA). These are secure and authentic MAC protocol. They have advance Encryption-Algorithms to encrypt the messages and calculate the coherence and cohesion.

LTE-A: L-Long T-Term E-Evolution Advanced (LTE-Advanced) is a standard to construct Machine-to-Machine (M2M) communication and IoT devices networks. LTE-A is a flexible, scalable, and low-cost protocol related to other cellular protocols [12]. LTE-Advance is a new evolution to help band-aid to increases data speeds, faster, performance, greater stability in the future.

DigiMesh: It is the high-latency for information transfer, as the network-dictates in the time-frames for data. They increase network range, and every device performs like a router [13].

NFC: Near Field Conversation (NFC) is a set of communication protocols that authorize to electronic devices. It is a flexible system such as a smart-phone and set up authorize communication by bringing them inside four cm of every other [14]. NFC at a receiver apparatus and radio transmitter. The best example is Beacons that come with a built-in NFC tag, and ANT+ provides such strengthens.

WirelessHART: It is a wireless sensor networking science-primarily based on the H-Highway A-Addressable R-Remote T-Transducer (HART) Protocol. It is advanced developed as an inter-operable Wi-fi standard and multi-vendor. WirelessHART was characterized by the necessities of development field system networks. WirelessHART, the first worldwide industrial Wi-fi general (IEC 62591), is built on the pinnacle of the IEEE 802.15.4 widespread [15].

Z-WAVE: A Z-Wave network consists of IoT units, and a foremost controller additionally acknowledged as a smart domestic hub. Z-Wave is one of the oldest and most commercially profitable IoT protocols, while Thread is one of the most current protocols [16]. Z-Wave protocol used for home appliances to monitoring and control.

DASH7: This protocol is a wireless verbal exchange protocol for pushing RFID that operates in a universal accessible I-Industrial S-Scientific M-Medical (ISM) smart band, and it is suitable for all T2C requirements [17].

HomePlug: It is designed for consistently and especially for home automation and smart grid applications. It is specially constructed to scale down the value and energy utilization of smart HomePlug-AV while maintaining its coverage, interoperability, and reliability.

Weightless: Weightless is another Wi-fi WAN, designed with the aid of the Weightless Special Interest Group (SIG). There are two units of standards, like Weightless-N and Weightless-W. There is the first developed aid with low cost and low energy machine-to-machine conversation with the time division multiple access [18].

6LoRaWAN: It is a recently bobbing up wireless technology constructed for low-power WAN networks with bi-directional communication, low cost, security, and mobility for IoT purposes [19].

MiWi: In the MiWi development, there are some different protocols like Miwi protocol, Miwi Mesh, and Miwi P2P. These protocols are used for the short-range area.

ANT+: ANT+ provides such strengthens and features a goal for the IoT. It manages the ANT network and defines how to transmit information across the system in a consistent way. ANT+ is an enabled sensor to help health care services.

EnOcean: EnOcean is an energy harvesting and energetically exploitation of mechanical motion and other potentials from the IoT Ecosystem, such as temperature difference and indoor lights for building Automation and home.

3.4 Transport and Network Layer

The Transport & Network layer is responsible for assigning some addresses and routing of data feeds. At this layer, the data packets are delivered from the Link Protocol layer to the destinations by using the IP address. These layers communicate data to The Session & Communication layer. There are some mechanisms of Network layer encapsulation protocols (IPv4, IPv6 over BLE, IPv6 over G.9959, 6LoWPAN, 6TiSCH,

6Lo, TCIP/UDP, IP, SLIP). IP version 4 (IPv4) is the network layer protocol. IPv4 has five classes, i.e., Class-A IPv4, Class-B IPv4, Class-C IPv4, Class-D IPv4, and Class-E IPv4. Commonly used Class-A, Class-B, and Class-C. The Class-D and

Class-E fixed for multi-casting and some experimental purposes. IPv4 has limited address space (approximately 4.3 billion devices), so needed more address space because there will be more than 50 billion IoT devices alone by the year 2025. IPv6 has a 128-bit address space (approximately 320 billion devices). In IPv6, there are three types of addresses in standard, i.e., Uni-Cast Address, Multi-Cast Address, AnyCast Address [20].

6LoWPAN, 6-IPv6 L-Low P-Power W-Wireless P-Personal A-Area N-Network (6LoWPAN) [21] developed for IoT devices in a WSN. It is a modified version of IPv6. 6TiSCH, allows IPv6 addresses to pass by T-Time S-Slotted C-Channel H-Hopping (TSCH) mode. Resource-Constrained Nodes (6Lo) is IPv6 protocol for data-links that were left excluded by 6TiSCH and 6LoWPAN. These protocols reduce communication by the following three optimizations, i.e., Header Compression, Fragmentation, and Link Forwarding. The consequence of imparting these protocol standards is to spotlight the main challenge of between particular specific MAC requirements and the variety of protocols. So, the IPv6 and the standard protocol stacks based on the future growths of IoT devices. It has sufficient address space for the IoT ecosystem while using IPv6.

These are some Routing Protocols (RPL, CORPL, CARPL) presented in IoT. RPL is a measure vector protocol created by IETF in 2012. RPL: IPv6-Routing Protocol for Lossy Networks and Low-Power. It calculated the de facto routing protocol for IoT. C-Cognitive RPL (CORPL) is an aberrant addition of RPL protocols that are designed for cognitive networks and appropriates the opportunistic sending to deliver packets at each Hop (networking). CA-Channel-Aware R-Routing P-Protocol (CARP) is the only distributed Hop (networking)-based routing protocol that is produced for IoT-devices-based network applications. CARP is used for under the water's surface and submarine communication and sends the information mostly.

3.5 *Session & Communication Layer*

In the Session & Communication layer, communicate (Request/response, Publisher/subscriber) between Server-2-Server (S2S), Device-2-Device (D2D), Device-2-Server (D2S), and Server-2-Device (S2D). These protocols are very lightweight to publish and receive messages. These protocols designed for constrained IoT devices with low-bandwidth. Things-to-Cloud (T2C) has many standard communication protocols, which are briefly highlighted in this section. Those protocols are Application structured-based and software application-specific. For example, if a software application has been constructed with XML, it receives a piece of code in its headers, XMPP may be the considerable option to appoint in the middle of session layer protocols. At the same time, if the Application is energy sensitive, then needed to select MQTT. It would be a predictable, trustworthy option. However, that appears with the other additional implementation. If the request application needs REST serviceability as it will be HTTP-based, then CoAP would be the genuine, authentic alternative if not the single one. The list of alternatives are: CoAP,

XMPP, HTTP, Telnet, MQTT, SMQTT, DDS, AMQP, FTP, SSH, WEB HTTP, LTP, SNMP, DNS, NTP, IPfix, DLMS, DNP, MODBUS, COSEM, etc.

CoAP: The Constrained Application Protocol (CoAP) used with constrained (i.e., lossy, low-power) networks and constrained nodes. CoAP is the most promising protocol for smart IoT devices and M2M applications. CoAP is prescribed in IETF RFC 7252.

XMPP: XMPP is a lightweight Publish or Subscribe resource-based constrained IoT devices [22]. X-eXtensible M-Messaging & P-Presence P-Protocol (XMPP) reduces interoperability between different networks.

HTTP: HTTP is applied for data transfer on the Internet, and it is mostly used for IoT devices that lack to publish a lot of data feeds.

Telnet: Telnet is a Client/Server protocol used in the local area network. It is based on a trust-able connection-oriented transport.

MQTT: MQTT is broadly used in T2C Architecture due to the low power consumption. It stands for M-Message, Q-Queuing, T-Telemetry, T-Transport. MQTT is an M2M/IoT connectivity protocol. It is a dial-up connection with home automation, health care providers, Industrial applications, and small device scenarios. It is very small in size, does the efficient distribution of information, minimized data packets, and has low power usage to one or many receivers.

SMQTT: Singh et al. [23] proposed a secure SMQTT protocol for IoT. SMQTT is an extension of MQTT and used in lightweight attribute-based encryption.

DDS: D-Data, D-Distribution, D-Service (DDS) developed M@M communication by Object Management Group. DDs are the low footstep of our devices to the cloud.

AMQP: A-Advanced M-Message Q-Queuing P-Protocol (AMQP), this is an open-source protocol for asynchronous messaging. AMQP gives better performance compared to other protocols [24].

FTP: F-File T-Transfer P-Protocol (FTP) is a protocol for exchanging files between a client and a server in the network.

SSH: SSH works in the between client and server model. SSH uses public key cryptography.

EBHTTP/LTP: E-Embedded B-Binary HTTP (EB-HTTP) and L-Lean T-Transport P-Protocol (LTP) are lightweight protocols. That protocol works on a low data rate, low computation complexity, and low energy consumption [25].

SNMP: S-Simple N-Network M-Management P-protocol (SNMP) is the most long-range protocol. It supports conventional network equipment like IoT devices, scanners, printers, etc.

IPfix: I-Internet P-Protocol F-Flow I-Information E-Export (IPFIX), it consists of a unidirectional protocol for export and format the data.

DNS: D-Domain N-Name S-Service (DNS), it resolves the host Internet Protocol-Addresses (IP-A) rendering IoT services. There are some Service Discovery protocols, i.e., m-Multicast D-Domain N-Name S-System (mDNS), DNS S-Service D-Discovery (DNS-SD), U-Universal P-Plug n-and P-Play (UPnP), DNS N-Name A-Autoconfiguration (DNSNA), S-Simple D-Discovery S-Service P-Protocol (SDSP), DNS SEC-Security DNSSEC [26].

NTP: N-Network T-Time P-Protocol (NTP), NTP exchange of time-stamped messages in the long-range. NTP-client implements a time-request exchange with IoT devices and data packet exchange protocol, called Synchronization Protocol for IoT (SPoT) [27].

DLMS: D-Device L-Language M-Message S-Specification (DLMS) or D-Distribution L-Line M-Message S-Specification (DLMS) is an advance Utility IoT protocol. DLMS is used for network connectivity that helps smart metering, thermal energy, water, and gas.

COSEM: CO-Companion S-Specification for E-Energy M-Metering (COSEM), COSEM is an interface that is used in the communicating of energy-related equipment.

DNP: D-Distributed N-Network P-Protocol (DNP). It is a set of communication protocols that communicate with a master station and IEDs of RTUs.

MODBUS: Modbus has allowed thousands and thousands of automation units to talk with each other. The embedded operating system starts the communication between the stations, and they talk with another device in cyclic repetition in the Modbus Communication Protocol.

3.6 Data Aggregation & Processing Layer

When IoT devices send data feeds, huge data, required an endpoint to act something with the enormous data. It is a partitioned, distributed, and replicated log service or messaging system. *Storm* help to process in a real-time environment. *Kafka is* getting the data feeds and sending at other destination at large scale. Kafka generates IoT data events and gives a continuous stream of data to sources [28]. Other alternate solutions are Scribe, RabbitMQ, Flume, Storm, Luxun, Flint, JMS, ActiveMQ, Qpid, Kestrel, Scribe, Hedwig, Batch Jobs. They help to monitor the traffic of the data in The Data Aggregation & Processing layer.

3.7 Data Storage & Retrieval Layer

This layer handles the data passes between IoT devices and the business model layer. In the recent past, a lot of attention has been migrated from computing to Data, which can saw in the form of several data processing software products in the market. From SQL to NoSQL, a normal relational database to Big Data all are used to provide backend solutions. Hadoop, HBase, Apache Cassandra, MongoDB, Reds, CouchDB, Neo4j, Couchbase, Riak, MarkLogic, DynamoDB, Spark, Google AppEngine dominate the field. They provide Data-Partitioning, Data-ingest, Data-quality, Data-Governance, Data-Cleansing & Processing, Predictive Capabilities & Normalization in a Distributed Database System (DDS).

This layer is responsible for providing fast services on high-performance distributed databases, highly scalable, and reconfigurable servers to handle huge amounts of data sets. This allows the user to do reliable, scalable, distributed computing, and cost-effective solutions.

3.8 *Business Model Layer*

The basic Business model depends on Cloud Services such as Data as a service, Process as a service, and product as a service. In a new era, industry and people transform their business models and adopt a new business model for social change with the presence of IoT Ecosystem. This business model provides the business value, business purpose, business methods, and business processes by using Integrated Cloud Model, On-demand Model, Hybrid Model, Platform Independent, Ubiquitous, Cross-domain Application, Standardization, Server-less Model, and Open Framework Model, etc.

3.9 *Business Application (Apps) Layer*

Business Application (Apps) layer is divided into three parts, i.e., Device Management, Business Processes, and Analytic & Automation. The function of the Device Management Component is device registration, device provisioning, firmware management, product & asset structure with security, and remote access. The Business Processes Component, on the other hand, is responsible for the transformation of services between the IoT devices, Marketing for IoT devices, and analyzing the impact of IoT device manufacturing on the market. Finally, in the Analytics & Automation Component, provide tools for processing data coming from IoT devices. It also has a component for data visualization, data mining, and intelligent data gathering services. A Machine learning library can also be incorporated.

With this proposed approach, a new nine-layered architecture, which we call it Things-to-Cloud (T2C), is introduced in which the data can travel from sensor to the cloud and come back to the user with a lot of value-added into. It is evident that all layers are not required every time.

4 IoT Security & Privacy Trust Framework

The importance of IoT Security in the proposed nine-layered architecture is essential. In the architecture, few things to take care of: Enable SSL/TLS, Security in cloud VM, and Database, and Bigdata Security. So, the Online Trust Alliance (OTA) framework, and it is an Internet Society Initiative [29]. OTA has released two trust frameworks for IoT, i.eis used, "IoT Security & Privacy Trust Framework v2.0 [29]" and "IoT Security & Privacy Trust Framework v2.5 [29]." This framework includes a set of essential principles required to help secure smart IoT devices and their data feeds. This framework is divided into four areas, i.e., User Access & Credentials, Security Principles, Notifications & Related Best Practices, and Privacy Disclosures & Transparency. This particular framework can be incorporated in the proposed layer, assuming that all the communication between layers is secure.

5 Conclusions and Future Work

A nine-layered protocols based on T2C architecture is proposed. In the protocol environment architecture, different protocols are discussed. These protocols have been developed by W3C, IETF, EPCglobal, ITU, IEEE standards, and products that are developed by different companies using these standards. An attempt has been made to decentralize the functionality into layers and clearly dividing the role, protocol, and software stacks that can be used to realize a real-world scenario. Any application development of T2C can be designed and implemented using these proposed layers. The purpose of this paper is to provide an insightful understanding of this whole ecosystem and the mapping of IoT devices with the respective T2C layers. It is a foundation for IoT practitioners and researchers to understand standard protocol classification and to gain an insight into the IoT protocols and standards to understand overall T2C architecture. In the future, layer-wise attacks are planned and find out protocol-based countermeasures.

References

1. Logvinov O, Kraemer B, Adams C, Heiles J, Stuebing G, Nielsen M, Mancuso B (2016) Standard for an architectural framework for the internet of things (iot) IEEE, Tech. Rep., September 2016, p 2413
2. Al-Fuqaha A, Guizani M, Mohammadi M, Aled-hari M, Ayyash M (2015) Internet of things: a survey on enabling technologies, protocols, and applications. IEEE Communications Surveys & Tutorials 17(4):2347–2376
3. Wu M, Lu T-J, Ling F-Y, Sun J, Du H-Y (2010) Research on the architecture of Internet of things. In: 2010 3rd international conference on Advanced Computer Theory and Engineering (IC-ACTE). IEEE, vol 5, pp V5–484
4. Yang Z, Yue Y, Yang Y, Peng Y, Wang X, Liu W (2011) Study and application on the architecture and key technologies for iot. In: 2011 international conference on multimedia technology. IEEE, pp 747–751
5. Khan R, Khan SU, Zaheer R, Khan S (2012) Future internet: the internet of things architecture, possible applications and key challenges. In 2012 10th international conference on frontiers of information technology, IEEE, pp 257–260
6. Lee Y, Kim J, Lee H, Moon K (2017) Iot-based data transmitting system using a uwb and rfid system in smart warehouse. In: 2017 ninth international conference on ubiquitous and future networks (ICUFN). IEEE, pp 545–547

7. Chauvenet C, Etheve G, Sedjai M (2017) G3-plc based iot sensor networks for smartgrid. In: Power line communications and its applications. IEEE
8. Magno M et al (2019) A bluetooth-low-energy sensor node for acoustic monitoring of small birds. IEEE Sens J 20(1):425–433
9. Wei Y-H, Leng Q, Han S, Mok AK, Zhang, W (2013) Rt-wifi: real-time high-speed communication protocol for wireless cyber-physical control applications. In: 2013 IEEE 34th real-time systems symposium. IEEE
10. Juels A (2006) Rfid security and privacy: a research survey. IEEE J Sel Areas Commun 24(2):381–394
11. Brandt A, Buron J (2015) Transmission of ipv6 packets over ITU-T G. 9959 networks, Tech. Rep., 2015
12. Hasan M, Hossain E, Niyato D (2013) Random access for machine-to-machine communication in LTE-advanced networks: issues and approaches. IEEE Commun Mag 51(6):86–93
13. Medina BE, Manera LT (2017) Retrofit of air conditioning systems through an wireless sensor and actuator network: an iot-based application for smart buildings. In: 2017 IEEE 14th international conference on networking, sensing and control (ICNSC)
14. Tan GW-H, Ooi K-B, Chong TS (2014) Nfc mobile credit card: the next frontier of mobile payment? Telematics Inf 31
15. Chen M, Han NS, Mok AK, Zhu X (2014) Wirelesshart and IEEE 802.15. 4e. In: 2014 IEEE international conference on Industrial technology (ICIT). IEEE, pp 760–765
16. Samuel SSI (2016) A review of connectivity challenges in iot-smart home. In: 2016 3rd MEC international conference on big data and smart city (ICBDSC). IEEE, pp 1–4
17. Cetinkaya O, Akan OB (2015) A dash7-based power metering system. In: 2015 12th annual IEEE consumer communications and networking conference (CCNC), IEEE, pp 406–411
18. Salman T, Raj J (2019) A survey of protocols and standards for internet of things. arXiv preprint arXiv:1903.11549
19. Sinha RS, Wei Y, Hwang SH (2017) A survey on LPWA technology: LoRa and NB-IoT. Ict Express 3(1):14–21
20. Levine MS, Kretchmar JM (2018) Correlating nameserver ipv6 and ipv4 addresses, US Patent App. 15/944,224, August 2018
21. Thubert P, Nordmark E, Chakrabarti S, Perkins C (2018) Registration extensions for ipv6 over low-power wireless personal area network (6lowpan) neighbor discovery, Tech. Rep., 2018
22. Saint-Andre P (2011) Extensible messaging and presence protocol (xmpp): Core, Tech. Rep., 2011
23. Singh M, Rajan M, Shivraj V, Balamuralidhar P (2015) Secure mqtt for internet of things (iot). In: 2015 Fifth international conference on communication systems and network technologies. IEEE, pp 746–751
24. Bhimani P, Panchal G (2018) Message delivery guarantee and status update of clients based on iot-amqp. In: Intelligent communication and computational technologies. Springer, Berlin, pp 15–22
25. Chen X (2014) Constrained application protocol for internet of things. https://www.cse.wustl.edu/~jain/cse574-14/ftp/coap
26. Lee S, Jeong JP, Park J-S (2016) Dnsna: Dns name autoconfiguration for internet of things devices. In: 2016 18th international conference on advanced communication technology (ICACT), IEEE, pp 410–416
27. Mani SK, Durairajan R, Barford P, Sommers J (2018) A system for clock synchronization in an internet of things. arXiv preprint arXiv:1806.02474
28. Dobbelaere P, Esmaili KS (2017) Kafka versus rabbitmq: a comparative study of two industry reference publish/subscribe implementations, ACM
29. Online Trust Alliance (OTA). http://otalliance.org

Role of Internet of Things (IoT) in Protection of Soil and Plant Life from Acid Rain Disasters

S. Ayyasamy, Daniel Felips Jhosiah, K. Prince Wesly, C. Aravind, and V. Swetha

Abstract Agriculture is one of the key sources of income to India. India is the 7th largest agricultural exporter worldwide and is ranked first globally for its net cropped area. The use of IoT to automate the irrigation process in fields, ensure the efficient growth of crops. In this approach, moisture sensors detect changes in soil moisture levels and accordingly turn the sprinklers on/off, to ensure crops are well-watered at all times. This is made possible with the aid of a dedicated mobile application. In addition to detecting changes in the moisture level of the soil, sensors can be used to detect the potential hydrogen levels (pH) changes in soil following a downpour. Pollution is a major problem in developing cities such as Coimbatore. This man-made disaster yields in the formation of acidic rain. Acid rain (pH 3.0 to pH 4.0) causes damages to crops by attacking the leaves and cotyledons (in the case of seedlings). It also kills useful microorganisms and leaches away essential nutrients from the soil by altering its pH level-thereby causing direct damage to crops. The common range of soil pH is anywhere between pH 5.5 and pH 7.0. Hence, it is clear that acid rain damages plants. pH sensors placed in soil sense changes in soil pH when acid rain strikes. Sensors are programmed to alert farmers when soil pH rises or falls from the set level. Sprinklers are then activated by farmers manually. The proposed approach ensures optimal soil pH conditions that are vital for soil life, plant growth and ultimately the ecosystem.

Keywords Acid rain · Crops · Soil · Root · Internet of Things (IoT) · Alkaline · pH levels

S. Ayyasamy (✉) · D. F. Jhosiah · K. Prince Wesly · C. Aravind · V. Swetha
Department of Computer Science and Engineering, Dr.N.G.P. Institute of Technology (Dr.N.G.P. IT), Coimbatore, India
e-mail: ayyasamyphd@gmail.com

D. F. Jhosiah
e-mail: dfelips@gmail.com

G. Ranganathan et al. (eds.), *Inventive Communication and Computational Technologies*, Lecture Notes in Networks and Systems 145,
https://doi.org/10.1007/978-981-15-7345-3_69

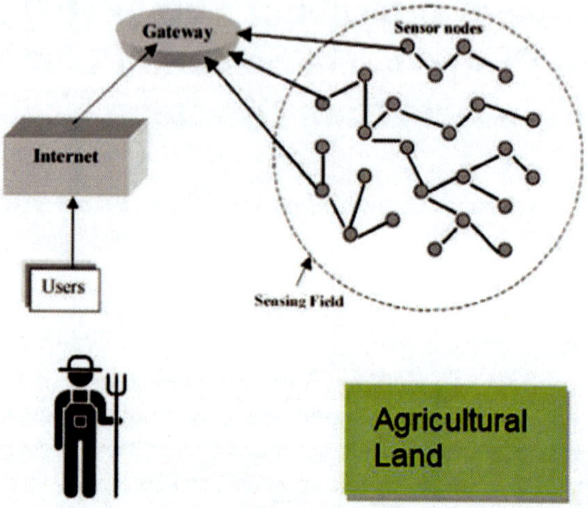

Fig. 1 Wireless sensor networks

1 Introduction

Intelligent farming helps farmers farm effectively with the aid of the Internet of Things (IoT). IoT is a technology that combines hardware, software and sensors. This makes it possible for users to access their devices anytime from anywhere (ubiquitous computing). When it comes to farming, crops and soil are treated as "objects". Sensors track and collect information from these objects. The data is then sent over the cloud (using Wi-Fi or satellite communication) and made available to farmers (even when they are at a considerable distance from the fields).

The Internet of things is composed of wireless sensor network (WSN) nodes. WSN is made up of nodes, gateways and software. Nodes, i.e., sensors (in our case pH sensors) sense data and transmit it wirelessly to the gateway (Arduino UNO board) which is connected to the same network as the farmer's mobile device via a Wi-Fi module (Node MCU). From here the data is collected, analyzed and is used to notify the farmer via a dedicated mobile application.

Figure 1 illustrates the process and Architecture of WSN [14].

2 Related Works

Various methodologies regarding the welfare of crop and soil life using the concept of IoT have been proposed. Here are some of them in brief.

One of the approaches is to detect changes in pH. This idea was proposed by Xiong Zhang, Zhouhu Deng, Shuai Lu, Wenqiong Zhang, Baopeng Lai, Zhiyong Zhang and Junfeng Yan [2]. The pH values sensed are then uploaded to data centers wirelessly. From this, the trend in acid rain pattern can then be predicted. This principle is applied only for the realization (occurrences) of acid rain in large areas and not in the protection of crops from it.

Dr. D. K. Sreekantha, Kavya A. M., proposed that IoT could be used in the implementation of a mobile application or an online platform (in other words a website) to monitor aspects such as the presence of weed among crops, soil moisture and animal intrusion [3]. This monitoring is done virtually by the farmer via a camera connected to the sensing unit wirelessly. However, this approach does not stress the importance of acid rain monitoring.

3 Problem Identification

3.1 Pollutant

Pollution is a major problem faced by developing cities such as Delhi, Madurai and Coimbatore. This is caused by both men-made and natural disaster causes. Acid rain comes from various sources. These include factories, waste incinerators, Sulfur Dioxide (SO_2) from volcanoes and vehicles. According to the report from environmental NGO Greenpeace, India is the world's largest emitter of anthropogenic SO_2 [13]-the main ingredient for acid rain. As shown in Table 1, India ranks first among 25 other emitting countries [4, 11], producing over 4586 kilotonnes per year (kt/year). Figure 2 shows that Chennai ranks 29th in the SO_2 Emission Hotspot list [4, 11].

Countries such as China, USA and Ukraine show a decreasing trend in SO_2 emissions. Iran, Mexico, Saudi Arabia, Turkey and Russia have a constant emission rate. India on the other hand, is the only country to show a steep increase in SO_2 emissions since 2005. Its emission rates in 2016 are double than what it was 11 years ago. The fact can be deduced from Fig. 3. [11].

Acid rain is formed as follows: $SO_2 + H_2O \rightarrow H_2SO_4$ (**Sulfuric Acid**: the major constituent of acid rain). **Nitrogen Dioxide** (NO_2) catalyzes acid rain formation. It too is a major air pollutant produced from high temperature combustion in vehicles and natural disasters such as thunderstorms.

Based on Fig. 4 it can be said that there is a steep increase in SO_2 concentration in Madurai between the years 2012 and 2017. Coimbatore shows a slight increase. Concentrations of 100 $\mu g/m^3$ are enough to trigger an acid rain attack. Due to the rapidly rising population and car ownership, concentrations of these pollutants are bound to increase. If precautions are not taken to control pollution, Coimbatore and Madurai may become susceptible to acid rain in the years to come.

Table 1 Country wise anthropogenic SO_2 emissions in 2018 estimated by NASA from identified point sources—top 25 emitter countries

Country	SO_2 emission from hospitals in 2018 (kt/year)
India	4586
Russia	3633
China	2578
Mexico	1897
Iran	1820
Saudi Arabia	1783
South Africa	1648
Ukraine	979
U.S.	967
Turkey	919
Kazakhstan	832
Australia	696
Cuba	637
United Arab Emirates	406
Qatar	398
Serbia	394
Kuwait	381
Bulgaria	350
Venezuela	340
Peru	305
Indonesia	298
Iraq	258
Bosnia and Herzegovina	242
Morocco	216

3.2 *Effect of Acid Rain on Plant Life*

India produces vegetables such as ginger, okra, onions, cabbages etc. in high numbers. The growth of such crop is hindered by acid rain. Normal rainwater has pH values ranging from (pH 4.0 to pH5.0) [7]. This does not affect crop growth in any way. However, when pH levels fall to pH 3.0, the rain becomes acidic enough to threaten crop growth [7].

Cotyledons (first parts of a plant emerging from seeds) are attacked by this acidic rain [7]. They act as an energy source and are vital to a plant's growth. Exposure to acid rain causes them to bend relative to their initial position affecting the growth posture of plants. Acid rain also attacks leaf surfaces in plants by damaging the epidermal cells. [7] Fig. 5a and b show the effects of acid rain on a cabbage plant [7].

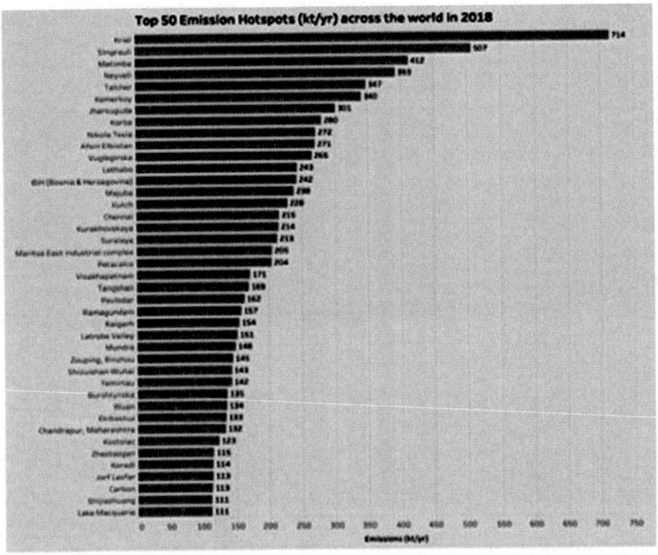

Fig. 2 Top 50 Emission Hotspots (kt/year) across the world in 2018

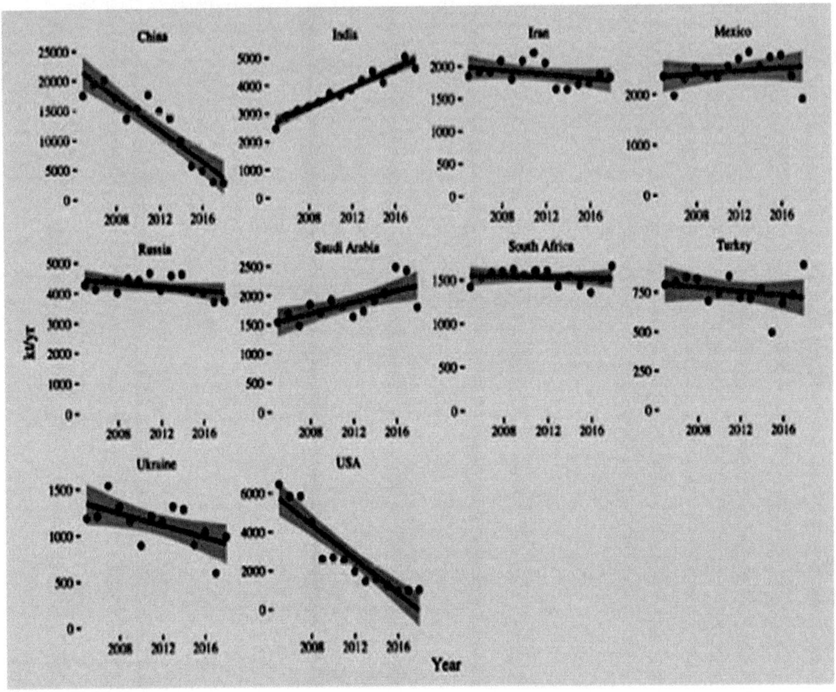

Fig. 3 Trends in anthropogenic SO_2 emissions by country since 2005

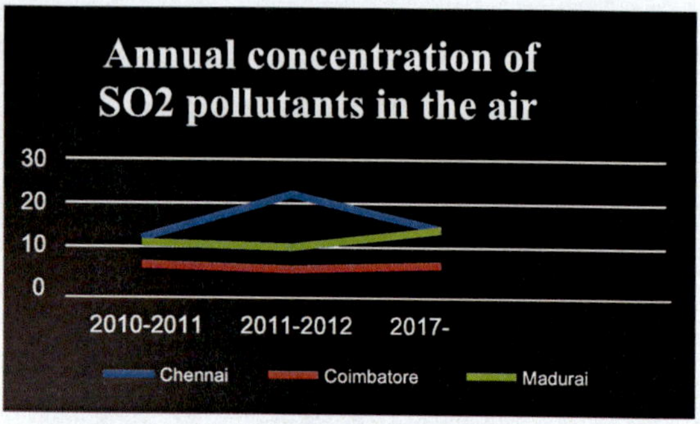

Fig. 4 The graph was plotted using data collected from the National Air Quality Monitoring Programme (NAMP) Annual Average Concentration of Air Pollutants Database (the table given above), for the years 2010–2011, 2011–2012 and 2017–2018 respectively

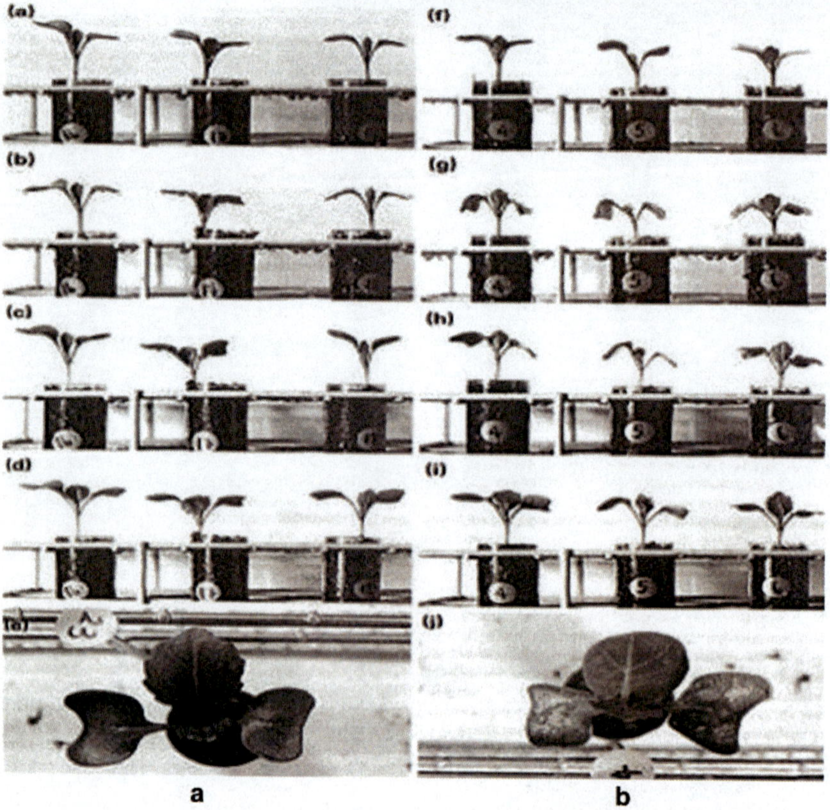

Fig. 5 **a** Exposure to rain with pH values between (pH 5 and pH 6). **b** Exposure to acid rain with values of pH 3

Fig. 6 Silvery to whitish transformation of leaves in presence of acid rain

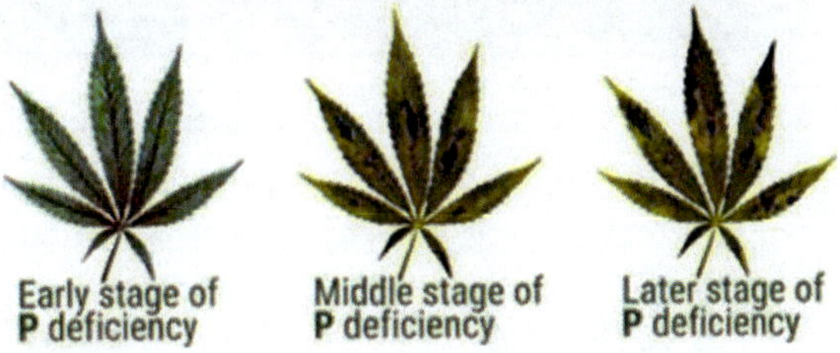

Fig. 7 Phosphate (P) deficiency in plants

The above results were taken before, after, 4–5 h later and 1 day after for each case of rain.

In reference to Fig. 5a it can be observed that both the seedlings and leaves stay healthy when exposed to normal rainwater. Figure 5b indicates the deterioration of seedlings and leaf surfaces upon exposure to acid rain. This is proved by the fact that the plant is withering to time and development of spots on leaf surfaces. Figure 6 provides a detailed look at the effect of acid rain on leaf surfaces [8, 9]—the silvery to the whitish transformation of leaf tone when attacked by acid rain.

The increased acidity of soil due to acid rain releases aluminum in its toxic form. This in effect inhibits the phosphate intake of plants thereby leading to phosphate deficiency [1] as shown in Fig. 7.

3.3 *Effect of Acid Rain on Roots*

Root length, number of roots sprouted and root structure (Fig. 8) are affected by acid rain [10]. Figure 8 [12] shows this result. Figure 9a shows that roots sprouted in acidic environments (pH $\leq 3.0 -$ pH 4.0) are few in number compared to those at pH 5.5 and above. A similar pattern is observed on root length (Fig. 9b.). This reduces the water intake of plants. Therefore acid rain (pH 3.0–pH 4.0) affects root growth.

4 Proposed System

A solution to this issue of acid rain is proposed. Sensor units are placed in the soil at regular intervals to detect the drop in pH levels in the event of acid rain. A mobile

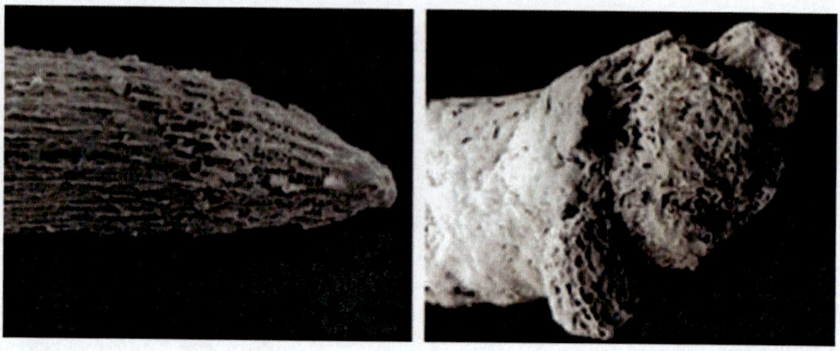

Fig. 8 Damage to root structure (healthy root versus affected root)

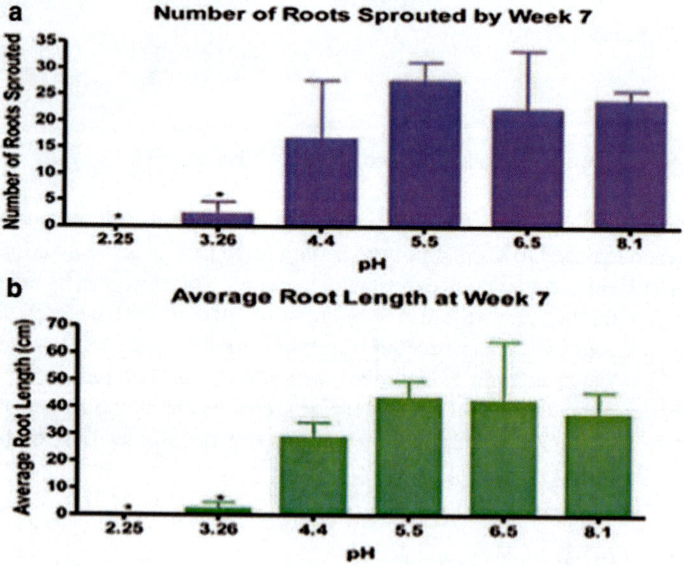

Fig. 9 a, b Effect of acid rain on root population and root length respectively

application connects to these units. When pH levels deviate from the set level, the farmer is alerted.

He/she responds by controlling (open/close) sprinklers containing an alkaline solution (NaOH) to neutralize the acidic rain falling on crops and optimize soil pH values to pH 4–pH 8 range. The process is illustrated in Fig. 10.

In Fig. 11 the sensor unit contains an Arduino Uno board (powered by a suitable source), Node MCU (Microcontroller Unit) for Wi-Fi connection to smart phone, a pH meter board to interface with a pH electrode (via the BNC-Bayonet Neil Concelman connector located on the pH meter board). The entire unit is sealed in a water-proof casing and placed on the soil. An outlet in the casing allows the pH electrode to be fixed into the soil. This is so that changes in soil pH can be sensed.

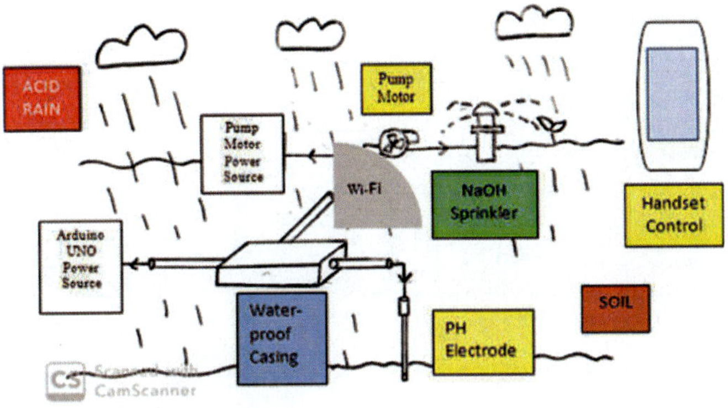

Fig. 10 The proposed system architecture

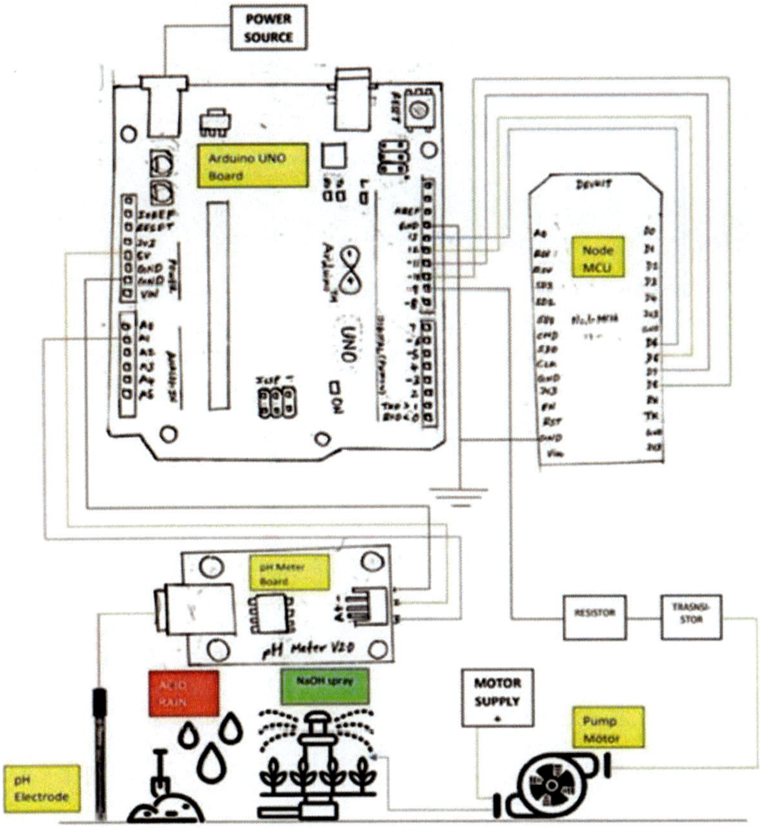

Fig. 11 The proposed hardware architecture

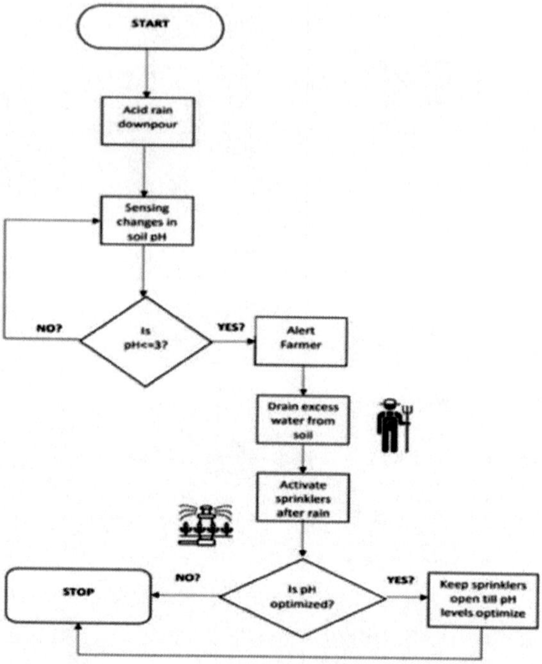

Fig. 12 Flow chart

The pump motor is used to control the operation of sprinklers containing a suitable alkaline solution (NaOH). It is interfaced to the Arduino. Power is supplied via motor supply. A mobile application (app) with a user-friendly interface is set up using IFTT (If This Then That) application and ThingSpeak. It connects and obtains readings from the sensors.

Alerts are sent to the app when pH levels fall from the set level (the default value of soil pH). Excess water is drained from the soil with the help of an effective irrigation mechanism (Fig. 12).

The sprinklers are then activated (after the rain subsides) by the famer to cancel out the acidic effect on the soil. Sodium Hydroxide neutralizes Sulfuric Acid to from Sodium Sulfate and Water:

$$2NaOH + H_2SO_4 \rightarrow Na_2SO_4 + 2H_2O$$

5 Results

It is known that the common range of soil pH is from pH 5.0 to pH 7.0 (can reach values up to pH 9.0). This is shown by point A in Fig. 13.

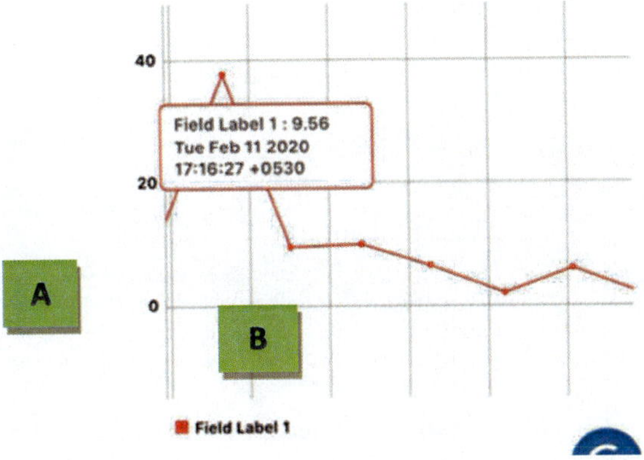

Fig. 13 Acid rain monitoring via thing view

When acid rain strikes, pH values fall below a pH of 4.0. In this case pH levels of value pH, 2.06 were reached. This can be seen from point B in Fig. 13. Under these conditions, the notification: "**Warning: Acid Rain!!**" is triggered via the IFTT application as shown in Fig. 14.

Acid rain won't affect the plants immediately. Hence it is advisable to wait till the rain comes to halt before proceeding with the remedy. The warning from the application alerts the farmer of an acid rain event. When the rain subsides, the farmer drains excess rainwater from the soil.

Pipes containing suitable quantities of alkaline solutions (enough to neutralize the acid) is opened. When the pH optimizes, a second notification "**Close the taps! pH optimized**" (so soil pH doesn't become too alkaline) (Fig. 14). The effect is the rise in pH values to optimum levels as illustrated in Fig. 15.

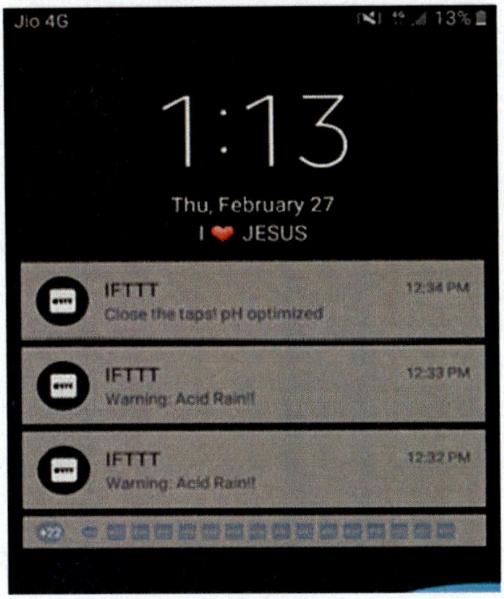

Fig. 14 Warning from IFTT app

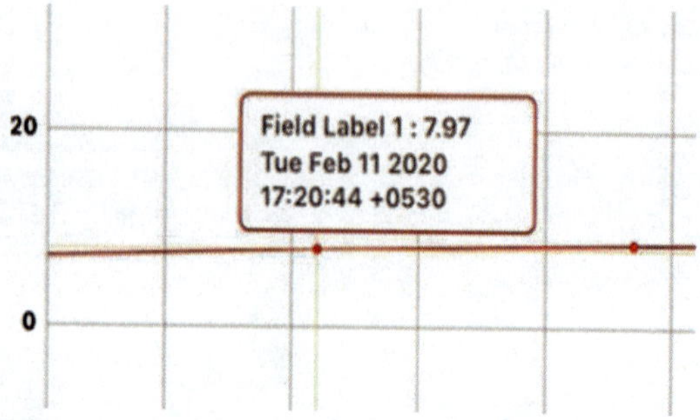

Fig. 15 Neutralizing acidity

Timeline of results obtained is shown graphically in Fig. 16. The *X*-axis represents the time at which readings were taken and *Y*-axis represents the corresponding pH values for each time slot. Readings for soil pH values of pH 6.49 are taken (Fig. 16a). A steep reduction to pH 2.06 (Fig. 16b) occurs when acid rain strikes. Application of suitable alkali optimizes pH levels to pH 8.97 (Fig. 16c) post-acid rain event. These descriptions are proved in Fig. 16a–c respectively.

6 Conclusion

From what has been discussed so far, existing methodologies for crop protection involve the measurements of parameters essential for a crop's growth and reporting to the farmer. A majority of these approaches uses soil pH, temperature, humidity levels, soil moisture etc. and use the data collected to provide fertilizer and crop recommendation to farmers. The approach proposed in this paper differs from the rest when it comes to protecting crops from acid rain. In the event of an acid rain, changes in soil pH are sensed and if any deviation of values from the fixed point occurs, farmers are alerted and sprinklers containing a suitable alkali are opened to optimize soil pH (prevents leaching) and protects crops externally. A farmer-friendly interface for the mobile application puts the farmer in full control of the process. Hence it can be concluded that the proposed approach is well versed when it comes to protecting soil and crop life from acid rain.

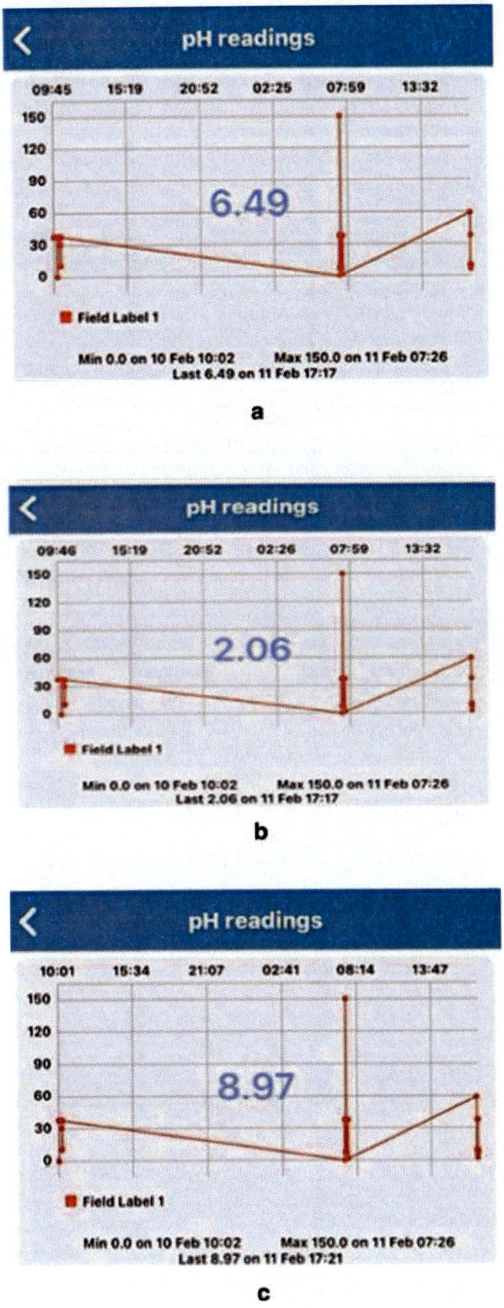

Fig. 16 pH readings of the soil taken before, during and after (neutralization with the alkali) the acid rain attack

References

1. Royal Queen Seeds, Cannabis Blog (2017) Phosphorus deficiency in cannabis plants. In: 11th August 2017. https://www.royalqueenseeds.com/blog-phosphorus-deficiency-in-cannabis-plants-n594
2. Zhang X et al (2015) Research and realization of acid rain measuring method based on Internet of Things technology. In: Sixth international conference on intelligent systems design and engineering applications, pp 608–611
3. Sreekantha DK et al (2017) Agricultural crop monitoring using IoT—a study. In: 11th international conference on intelligent systems and control (ISCO), pp 134–139
4. The News Minute, TNM Staff, August 20th 2019. https://www.thenewsminute.com/article/india-worlds-largest-emitter-so2-chennai-29th-position-report-107542
5. Impacts of Acid Rain on soils. http://www.air-quality.org.uk/16.php
6. Wikipedia, Soil pH. https://en.wikipedia.org/wiki/Soil_pH
7. Simon JM, Capron et. al (1986) The contrasting response to simulated acid rain of leaves and cotyledons of cabbage (Brassica Oleracea l.). New Phytol 103:311–324
8. Francisco B et. al (2006) Effects of simulated acid rain on leaf anatomy and micromorphology of leaf anatomy and micromorphology of Genipa Americana L. (Rubiaceae). Braz Arch Biol Technol 49(2)
9. Aggie Horticulture Network (2009, July) https://nph.aggie-horticulture.tamu.edu/vegetable/problem-solvers/cucurbit-problem-solver/leaf-disorders/air-pollution-injury/. Department of Environment ENVIS Centre Government of Tamil Nadu, Air Pollution Database in Tamil Nadu, 2014, (tnenvis.nic.in)
10. Pasumpon AP (2019) Enhanced edge model for big data in the internet of things based applications. J Trends Comput Sci Smart Technol (TCSST) 1(1):63–73
11. Greenpeace, Greenpeace India, August 19th 2019, https://www.greenpeace.org/india/en/publication/3951/global-so2-emission-hotspots-database-ranking-the-worlds-worst-sources-of-so2-pollution-2/
12. Department of Primary Industries and Regional Development (2018) Effects of soil acidity, 17th September 2018. https://www.agric.wa.gov.au/soil-acidity/effects-soil-acidity
13. India Today, Press Trust of India (2019). https://www.indiatoday.in/education-today/gk-current-affairs/story/india-biggest-sulphur-dioxide-emitter-greenpeace-report-1582382-2019-08-09
14. Wireless Sensor Networks (2008) (cs.usfca.edu)

IoT-Based Wardrobe and Steel Closet Theft Detector

S. Shrinidhi, S. Vinuja, and E. Prabhu

Abstract In recent years, the increasing burglary has been creating a great sense of anxiety, in assuring the safety and security of the essentials placed in the wardrobe. This paper focuses on providing a user-friendly and cost-effective mechanism, to ensure the safety and security, by intimating the user in case of any burglary. The above purpose is achieved by (1) periodically monitoring the inner walls of the wardrobe, (2) monitoring the continuous darkness that prevails in the wardrobe when the doors are closed. The monitoring process involves the presence of an ultrasonic sensor (HC-SR04) on one side of the opening door of the wardrobe, which sends ultrasonic waves to detect the precise distance of the wall of the wardrobe facing the sensor when it is closed. The photoresistor sensor (LDR) senses even the slightest presence of light in the wardrobe, in case of it being opened. The monitored distances and light-sensing data are stored in an authenticated IoT platform of ThingSpeak using ESP-8266. Any inaccuracy in this uploaded data from the previously set data causes the IFTTT application to intimate the user of the wardrobe being opened, through notification and e-mail services.

Keywords IoT · ThingSpeak · Smart wardrobes · Sensors · Theft detector

S. Shrinidhi · S. Vinuja · E. Prabhu (✉)
Department of Electronics and Communication Engineering, Amrita School of Engineering, Coimbatore, Amrita Vishwa Vidyapeetham, Coimbatore, India
e-mail: e_prabhu@cb.amrita.edu

S. Shrinidhi
e-mail: shrin.asian@gmail.com

S. Vinuja
e-mail: vinuja.s2000@gmail.com

G. Ranganathan et al. (eds.), *Inventive Communication and Computational Technologies*, Lecture Notes in Networks and Systems 145,
https://doi.org/10.1007/978-981-15-7345-3_70

1 Introduction

Recently, the alarming rates of burglary have emphasized the need for a more secure smart lock system like biometric devices. The presence of smart wardrobes relieves the users from the stress of having the essentials safe in case of their absence. Regardless of using these special devices, the rates of burglary crimes continue. In the absence of an authenticated user, these devices can be accessed with an identical copy of biometric data, obtained using the tools of forensic sciences. The electronic locks, digital locks and the complex mechanical locks can be accessed using their respective counter technologies.

An ultrasonic sensor (HC-SR04) and a photoresistor sensor (LDR) can continuously monitor the current conditions of the wardrobe. The monitored data is then channeled into an IoT platform of ThingSpeak, wherein the obtained data is compared with the predefined data module. Any inaccuracy or discrepancy in this data, caused due to the opening of the wardrobe, triggers the ThingSpeak *platform.* The details of the user are stored in the cloud and provide a *quick* insinuation to the user of the same, through notification from the IFTTT app and e-mail services.

2 Background

2.1 Sensors and Modules

The wardrobe and steel closet theft detector employs the usage of the ultrasonic sensor (HC-SRO4), a photoresistor and the Wi-Fi module (ESP-8266). The machine is modeled in such a way that it is supposed to be attached to one of the opening doors of the wardrobe. The distance is periodically checked using HC-SR04, which measures the distance of the opposite surface of the door from the sensor. The sensor contains a transducer that sends and receives ultrasonic pulses of high-frequency sound waves. By sensing the echo pattern of the received waves, the object's distance is found. The distance is given by Eq. (1).

$$\text{Range} = \text{Time off light}/\text{Speed of sound} \tag{1}$$

The maximum and the minimum distance which can be detected using HC-SR04 are 400 cm and 3 cm, respectively. If there is no object to be detected then, a high-level signal of 38 ms occurs in the output pin. HC-SR04 gives the precise value since it has high a penetrating power and can sense any material in any adverse condition. The darkness inside the wardrobe is observed using a photoresistor sensor. This sensor is a resistor that has values in megaohms in the absence of light and a very few ohms in the presence of light. There are two types of photoresistors which are intrinsic and extrinsic. Regardless of what type of photoresistor used, both reveal a drop in resistance or rise in conductivity with an increase in the intensity of incident

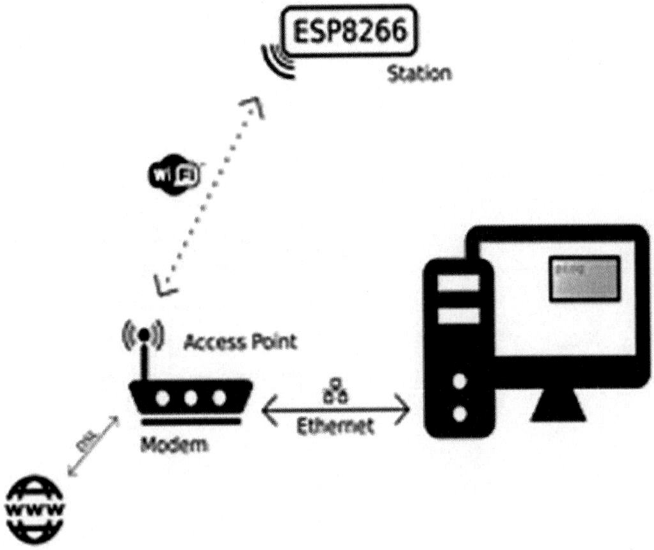

Fig. 1 Working of ESP-8266

light. Normally, the photoresistors take a few tens of milliseconds to acknowledge the light when it is made to strike after total darkness, but when light is removed off, it takes one second or more to reach its final level of resistance. The purpose of using the photoresistor is that it is bi-directional, it offers quick response, and it is highly accurate.

The recorded data from the Arduino to the cloud is updated using an intermediate chip called ESP-8266. ESP-8266 is the first Wi-Fi module at a very low cost. This module is manually connected to the Arduino UNO and is programmed in Arduino IDE using the AT commands to provide a connection with the local Wi-Fi network. The communication of ESP-8266 with the local Wi-Fi network is depicted in the Fig. 1. Figure 2 gives the overall flow of the proposed prototype.

2.2 *Arduino UNO*

To program the sensors and to make them interactive, Arduino UNO is used. All the sensors and modules required for the prototype are connected manually to the Arduino board. It is then programmed in the Arduino integrated development environment (IDE) using the C++ language. The board is first connected to the laptop/computer through a portable USB cable to store the code in the Arduino UNO. Once the code is loaded, the board can be powered using a power bank or any other battery of use.

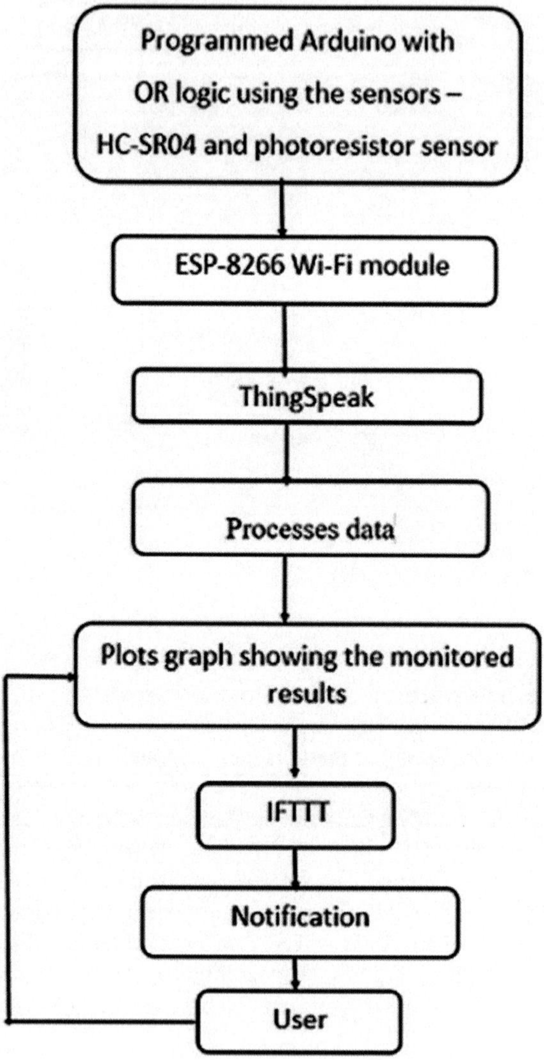

Fig. 2 Architectural framework of the proposed system

2.3 IOT Analytics Platform and Services

ThingSpeak interface: The collected data from the sensors is transmitted to ThingSpeak. The ThingSpeak allows an open visualization of all the details collected as a plot in a graph, for the user to monitor whenever required. The link from Arduino to ThingSpeak is depicted in Fig. 3.

Fig. 3 Communication of Arduino with ThingSpeak

IFTTT: IFTTT is a free Web-based applet, where triggers can be given from other platforms to perform any action. Here, the triggers are configured to send e-mail and notification in the IFTTT app.

3 Related Works

In this generation, all the information which is physically available in this world is transformed to us in the form of figures or data. This is achieved predominantly by the sensors. The sensors used in this project are mainly used for detecting the distance of an object and the presence of light [1, 2]. The data collected from these sensors is made to interface with the Arduino UNO for the working in [3, 4]. Arduino UNO is preferred in this prototype as it is inexpensive and can be supported on Windows, Linux, Mac OSX, iOS and Android [5].

The data thus generated is stored in the cloud through Arduino for further analysis. This is facilitated by using ESP-8266. Since the cloud used here is ThingSpeak, its further connections to the Arduino IDE have been referred from [6–8].

The data stored in this cloud is further integrated with the IFTTT application to send a notification through its app and an e-mail [9]. This is where the IoT comes into picture wherein the two applications are made to interact without any human touch thus making this prototype cost-effective, efficient and accurate [10, 11]. To track stocks available in the kitchen, an inventory system using IoT was developed [12]. The reconfigurable hardware approach for the implementation of the digital logic was discussed in [13].

4 Proposed Methodology

Several case studies reveal that the wardrobe thefts happen, either by breaking the lock of the wardrobe, shattering the doors of the wardrobe, or by drilling holes on the surface of the wardrobe. In many such cases, the thefts happen only in the absence of people in their respective homes. Wardrobe and the steel closet theft detector is a prototype designed to facilitate such people to get information whenever someone is trying to access their wardrobe without their knowledge. Further, they can also monitor their wardrobe periodically using their mobile phones and can check whether their wardrobes have been approached by someone. If any of the initially discussed scenarios take place, the wardrobe is definite to undergo damages and changes. To monitor and record such changes from breakage or opening, sensors such as HC-SR04, photoresistor sensor are used and ESP8266 connected manually to the Arduino UNO. The designed system is to be placed on the inner side of one of the opening doors of the wardrobe. The circuit diagram of the entire working model is given in Fig. 4.

Step 1: Monitoring changes in wardrobe

Whenever a thief breaks open the lock or the doors of the wardrobe, there occurs a change in the distance between the doors and the facing wall of the wardrobe. This distance is monitored and recorded by using an ultrasonic sensor. The monitored distance is then compared with the initially programmed distance obtained when the

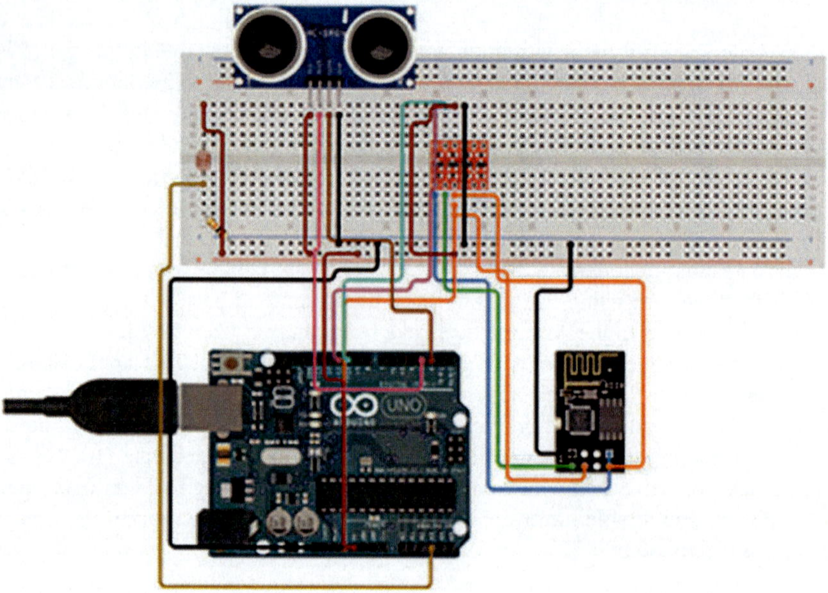

Fig. 4 Circuit diagram of the proposed working model

wardrobe is closed. Any change in the newly measured distance from the initially set distance programs the Arduino IDE to take the value as 1, else the value will be recorded as 0.

When a thief drills holes on the surface of the wardrobe, there can be two cases:

Case 1: DAY Ideally, the inner space of the wardrobe appears dark when it is closed. During the day, whenever the thief tries to access the valuables by drilling a hole on the surface of the wardrobe, a reasonable amount of light penetrates the wardrobe.

Case 2: NIGHT At night, if a thief chooses the path of drilling to access the valuables, he would ideally choose to carry a torch or any other sources of light alongside.

In either of the above two cases, one observes that the light penetrates into the wardrobe. Hence, light is sensed inside the wardrobe. To sense this presence of light due to the opening of the wardrobe, a photoresistor is used. Its values are periodically recorded. Similar to the previously discussed sensor, the value is recorded as 1 in the presence of light and 0 in its absence to denote a safe condition.

Step 2: Assigning solution using OR gate ideology

The Arduino IDE is programmed in such a way to obtain a value of 1 in the cloud when any one of the sensors gives the value 1 or when both the sensors give the value 1. Only when both the sensors observe the value 0, the value 0 is stored in the cloud. It works on the principle of an OR gate, as given in Table 1.

Step 3: Linking the solution with ThingSpeak using ESP8266

ESP8266 Wi-Fi module is manually connected to Arduino UNO, and the Arduino IDE uses the SoftwareSerial library for doing serial communication with ESP8266. This Wi-Fi module requires the respective Wi-Fi address and password to be mentioned. It is further programmed using AT commands for wireless communication such as to pass the values recorded by Arduino UNO to the cloud simultaneously. The cloud that is used here is ThingSpeak where all the data are collected and visualized as a plot graph. For plotting the collected data, a new channel is created in ThingSpeak, and the API key mentioned here must be given in the Arduino IDE program.

The wardrobe owner can periodically check this plot for ensuring whether the wardrobe is secured or not. This graph plot will be updated every 15 s. As already mentioned, the solution from Arduino UNO, which is finally in the form of binary

Table 1 OR gate truth table

A	B	$F = A$ OR B
0	0	0
0	1	1
1	0	1
1	1	1

digits, is recorded in this cloud as a plot. For the prototype, the ideal distance is fixed as 10 cm. The change in the plots according to the four OR gate conditions is given in Figs. 5, 6, 7 and 8.

These plots can be examined only by the user unless he has changed the channel settings to public view. As all the information from the sensors is finally transfigured to binary digits, there is no factual data that can be accessed by any hackers. Further, ThingSpeak is a secured cloud, and the API key generated is unique for different channels. Hence, it cannot be easily accessed by anyone. Even if the third party tries to hack, there is no such appreciable data in this cloud which can help them to retrieve the actual information.

Step 4: Integrating ThingSpeak with IFTTT

Now that the values are recorded in ThingSpeak, it is further integrated with IFTTT which is especially used for alerting the wardrobe owner whenever the thief tries to access the wardrobe without their knowledge. For this purpose, a new ThingHTTP is created in ThingSpeak. In IFTTT, a URL is given which in turn must be specified in the newly created ThingHTTP. Now, a new React is created in ThingSpeak, wherein the channel name, ThingHTTP name and the condition as to when the solution is 1 (when the thief is trying to access and there are changes in wardrobe), a trigger must be created in IFTTT application, are to be mentioned. Thus, whenever the condition is met, the trigger is activated in the IFTTT application and is made to send a notification and an e-mail to the respective wardrobe owner. The IFTTT is also an app, and the user gets notifications from this app when the trigger condition is met. Figure 9 shows how IFTTT alerts through mail and notifications on mobile phones.

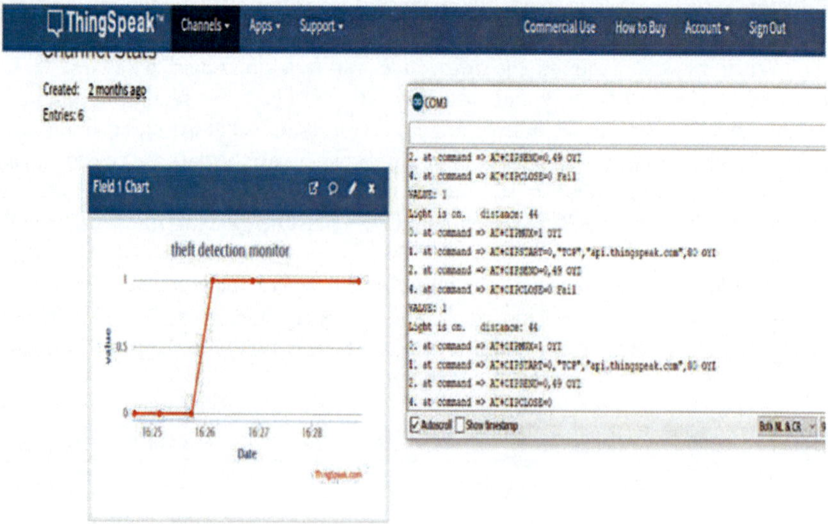

Fig. 5 When light is ON and distance is not 10 cm

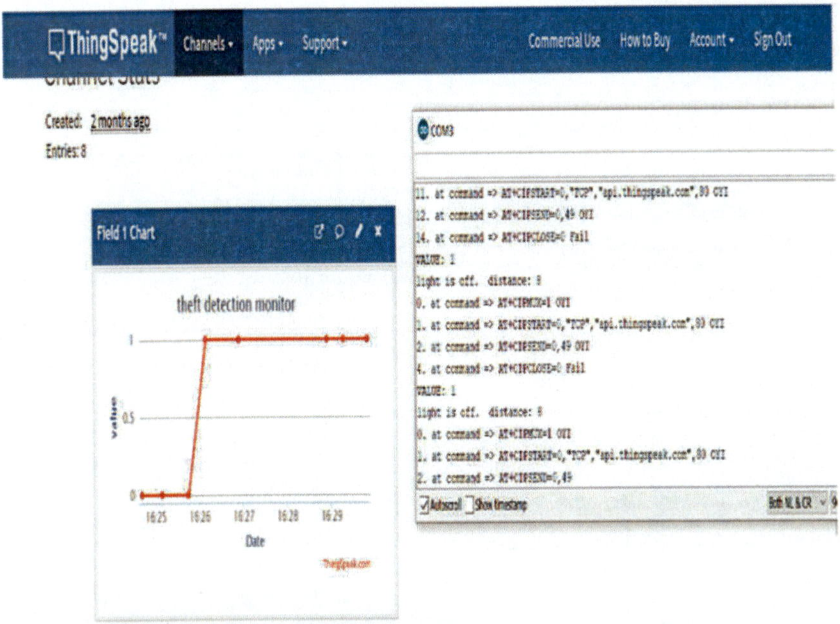

Fig. 6 When light is OFF and distance is not 10 cm

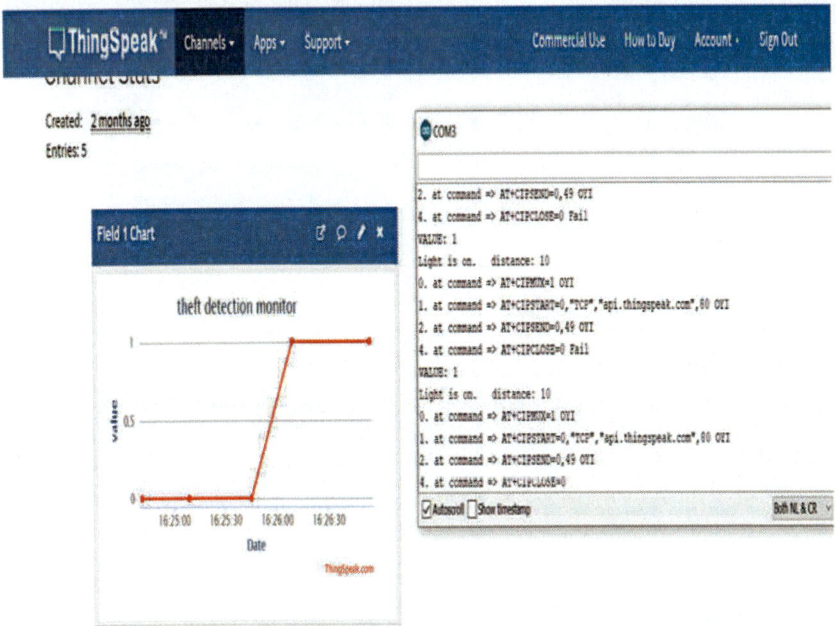

Fig. 7 When light is ON and distance is 10 cm

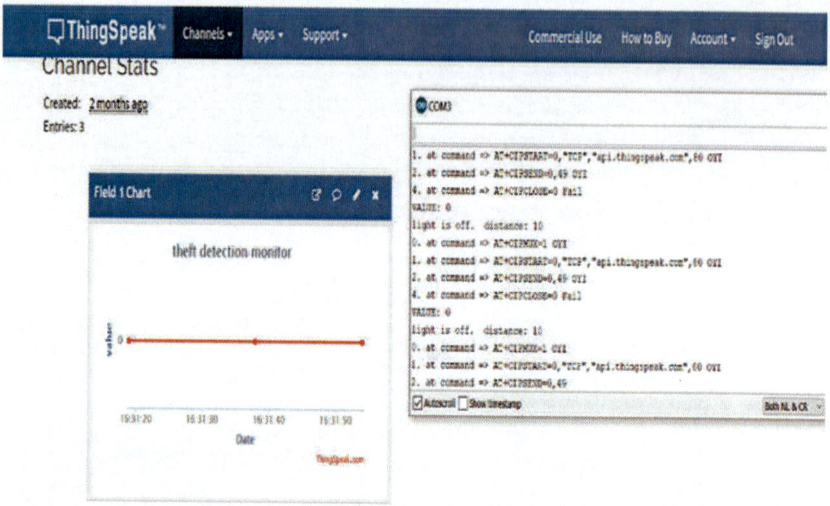

Fig. 8 When light is OFF and distance is 10 cm

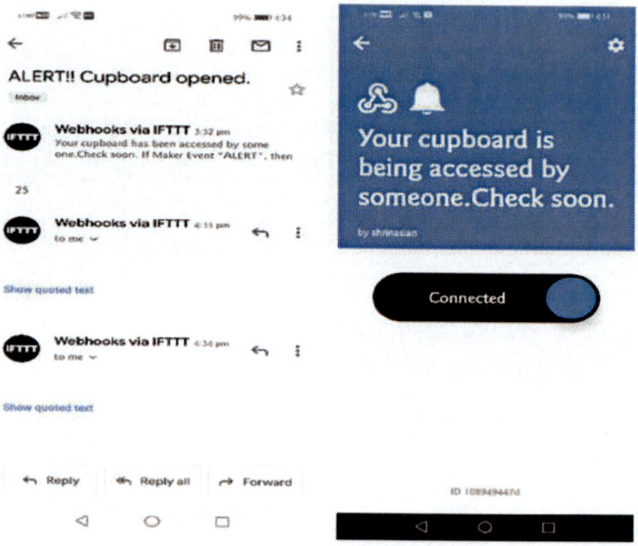

Fig. 9 Notification in user's mobile through mail and app stating wardrobe is being accessed by someone

5 Result and Discussions

The developed prototype uses the IFTTT application instead of the GSM modem for sending notifications, and so, the user may now wonder whether this prototype is accurate enough to alert. To prove the proposed model to be accurate, the system was tested under various conditions to make sure of the efficiency and the accuracy at which the wardrobe owner gets an alert. It is plotted as a graph given in Fig. 10, where the straight line indicates that the user has received the alert within a few seconds of the incident.

If the user has further insecurity about the wardrobe's safety, apart from tracking the ThingSpeak cloud, they can also keep an eye on the IFTTT app, notifications menu which is provided with two options namely view activity and heck now which is shown in Fig. 11. In view activity, the user gets all the records of notifications of when the cupboard has been accessed along with the time specified in it as shown in Fig. 12. The check now option can be opted by the user to check currently whether the wardrobe is safe or not. If there occurs any notification or e-mail from IFTTT, then the wardrobe has possibly been accessed. The check now option is just a backup for the user, in case if the user is very much uncertain about the wardrobe. Otherwise, there occurs alerts regarding the threat.

The practical application of the proposed model has specific pros and cons:

Pros:

1. The model is user-friendly, cost-effective and demands no specific skills from the user for its initial installation.
2. It occupies minimum space and is placed out of sight.
3. It provides precise values under all conditions of the day and night.

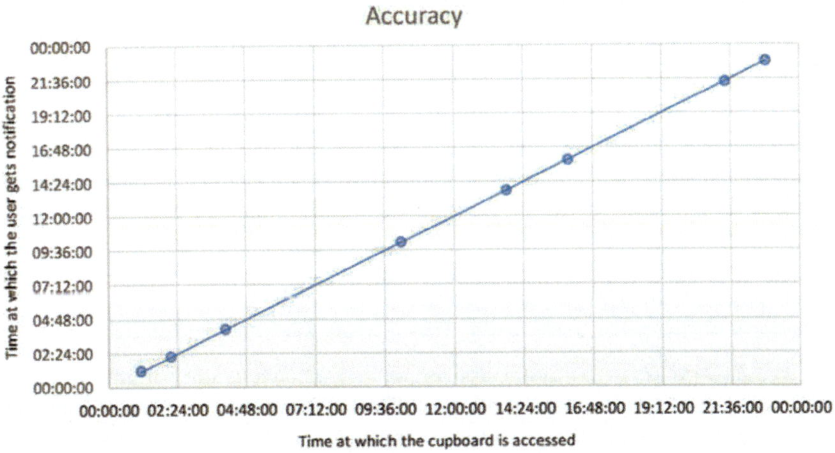

Fig. 10 Time at which wardrobe was accessed versus time at which users received the notification using IFTTT

Fig. 11 Menu options

4. The sensors provide accurate results irrespective of the material under observation.

Cons:

1. ThingSpeak records information every 15 s.
2. The model does not prevent the burglary from happening but rather the intimates the user of the misdeed.

6 Conclusion and Future Scope

There is an increase in the theft cases in the villages, where a row of houses in the same street are being targeted at once. By hiring the proposed system, people can be alerted to safeguard their properties. This system is cost-effective, and hence, it can be acquired by anyone. If one has more than one wardrobe in their house, they can buy as many as required. This product can further be operated using a switch to

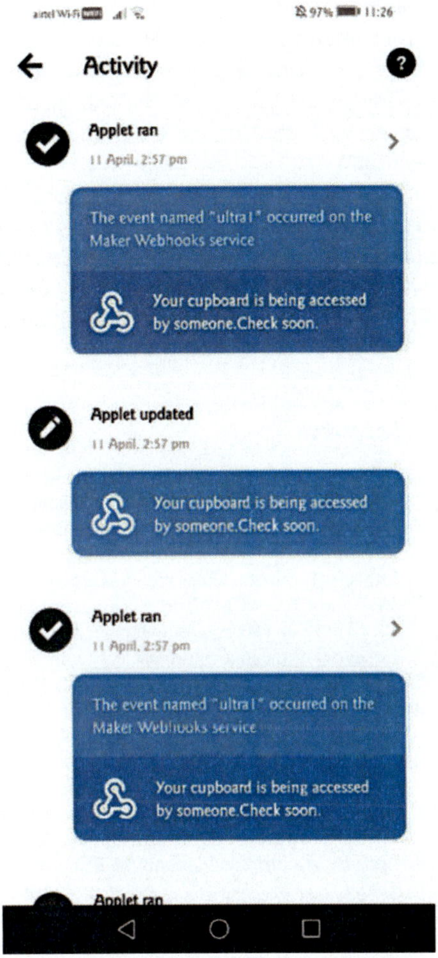

Fig. 12 Record of notifications

have control over power consumption. In many of the case studies, it is seen that the burglars even go to the extent of killing people. Hence, it is recommended to operate this model even when people are alone to get alerts about the theft.

The future work of this proposed system is to convert this prototype model into a full-fledged working product and to make it available in the market for better usage. Further, a suitable vibration sensor can also be added to make the system more sensitive to touch, and the prototype can be linked with image processing using the surveillance camera to capture the image of the thief for easier investigations. The IFTTT application can also be integrated to send notifications to neighbors and the nearby police station to alert them of the theft for the right timely action to be taken. The usage of the ESP8266 Wi-Fi module connected to the Arduino UNO for wireless communication, using AT commands to retrieve the results obtained in the

Arduino and the passage of this information to the cloud platform of ThingSpeak for further analysis is indeed a time-consuming process with a delay of 15 s to provide an alert. In order to reduce the response time of the model, the IoT cloud platform of ThingSpeak can be replaced with an efficient GSM module which can directly be connected to the Arduino UNO, thus enabling a quicker action in sending SMS to the user.

This product can also be hired for alerting and securing motors, vehicles and many other commodities. The proposed model can also be employed to use in the main door of the house. While considering the fact of a full-fledged product, a set of advancements available at that time of market expectations will also be incorporated within the proposed system to make it an effective product to work along.

References

1. Arun Francis G, Arulselvan M, Elangkumaran P, Keerthivarman S, Vijaya Kumar J (2020) Object Detection Using Ultrasonic Sensor, Int J Inno Technol Explor Eng (IJITEE) 8(5):207–209
2. https://www.instructables.com/id/How-to-use-a-photoresistor-or-photocell-Arduino-Tu/
3. Badamasi YA (2014) The working principle of an Arduino. In: 2014 11th ınternational conference on electronics, computer and computation (ICECCO), Abuja, 2014, pp 1–4
4. https://www.theengineeringprojects.com/2018/06/introduction-to-arduino-uno.html
5. https://robotronicspro.blogspot.com/2014/09/why-arduino-is-preferred-over.html
6. https://medium.com/@cgrant/using-the-esp8266-wifi-module-with-arduino-uno-publishing-to-thingspeak-99fc77122e82
7. https://medium.com/@angelinmaryjohn/monitoring-data-from-sensors-using-esp-8266-and-arduino-bb9132d88488
8. Shanmugam M, Singh M (2018) Arduino based IOT platform for remote monitoring of heart attacks and patients falls. J Comput Sci 14(4):574–584
9. https://www.instructables.com/id/ThingSpeak-IFTTT-Temp-and-Humidity-Sensor-and-Goog/
10. Gubbi J, Buyya R, Marusic S, Palaniswamia M (2013) Internet of Things (IoT): a vision architectural elements and future directions. Future Gener Comput Syst 29(7):1645–1660
11. Albishi S, Soh B, Ullah A, Algarni F (2017) Challenges and solutions for applications and technologies in the Internet of Things. Procedia Comput Sci 124:608–614
12. Lakshmi Narayan SP, Kavinkartik E, Prabhu E (2018) IoT Based food ınventory tracking system. In Advances in signal processing and ıntelligent recognition systems (SIRS 2018). Communications in computer and ınformation science, vol 968. Springer, Singapore
13. Paleri PL, Ramesh SR (2015) Early stage FPGA architecture development by exploiting dependence on logic density. Int J Appl Eng Res 10(11):28889–28902

Intrusion Detection and Prevention Systems: A Review

Vaishnavi Ganesh and Manmohan Sharma

Abstract In this paper, the various methods for intrusion detection and prevention are discussed based on the four review papers. In the first paper, the support vector machines and decision trees machine learning techniques are being used to train the data set for detecting DOS attacks in WSN and found that decision trees are more efficient than support vector machines. In the second paper, the routing attacks such as sinkhole and selective forwarding are being detected in the Internet of things for which two detection and prevention algorithms, i.e. key match algorithm (KMA) and cluster-based algorithm (CBA) are used. Same intrusion detection and prevention system which is developed for wired networks cannot be used for wireless networks. In the third paper, a system called wireless intrusion detection prevention and attack system 'WIDPAS' is proposed which attacks the attacker and sends warning to the administrator. In the fourth paper, a method is used which uses filter firewall, honeypot intrusion detection, anomaly intrusion detection and prevention firewall to protect an organization's network.

Keywords WSN · DOS attack · Support vector machines · Decision trees · Sinkhole attack · Selective forwarding

1 Introduction

In the whole world, wherever networking is involved there intrusion comes into the picture. To protect confidential information in many databases is of uttermost importance. Hence, intrusion detection and prevention systems become necessary.

V. Ganesh (✉)
Priyadarshini Indira Gandhi College of Engineering, Nagpur, India
e-mail: vaishnavi.ganesh8@gmail.com

M. Sharma
Lovely Professional University, Phagwara, India
e-mail: manmohan.sharma71@gmail.com

G. Ranganathan et al. (eds.), *Inventive Communication and Computational Technologies*, Lecture Notes in Networks and Systems 145,
https://doi.org/10.1007/978-981-15-7345-3_71

There are several types of attacks like denial-of-service (DoS) attack, man-in-the-middle(MITM) attack, sinkhole attack, selective forwarding, flooding attack, wormhole attack, etc. DoS attack means overcrowding the server in a network by dummy messages to create bottlenecks in the network so that legitimate traffic could not reach the server. This happens, especially in E-commerce field. A site's server can be purposely overcrowded by dummy packets by its counterpart site [1].

Similarly, intrusion detection is necessary for the field of wireless sensor networks. WSNs are in every field like agriculture, industries, military, telecommunications, medical and health fields. In medical fields, it is used to monitor patients' movements, tracking their locations and monitoring elderly patients [2].

Some of the intrusion detection and prevention systems (IDPS) developed in various fields are as follows:

(a) An IDPS has been proposed whose performance is compared with the performance of Snort, a tool for network intrusion detection system(NIDS) available in the market. In the proposed system, prefix and random indexing method are implemented to all Snort rules and primary patterns are created which reduce packet inspection time and also reduce false-positive rate even at high network traffic [3].
(b) A synchrophasor-specific intrusion detection system (SSIDS) tool has been designed for detecting malicious cyberattacks in a synchrophasor system, which involves a heterogeneous whitelist and behaviour-based approach for detecting known and unknown attacks [4, 5].
(c) To protect the control area network(CAN) bus in connected cars, an IDPS has been developed. The control area network bus is a serial car bus network that connects electronic control units and sensors in a system for control applications, which can give real-time information about the car [6].

2 Review of Paper 'Using Machine Learning to Detect DoS Attacks in Wireless Sensor Networks' [2]

Low Energy Aware Cluster Hierarchy (LEACH) is a hierarchically based routing protocol which is used to collect a data set representing WSN features in different attacking scenarios. This data set includes information about nodes in the network, the sender node, the cluster head(CH) and the receiver node. In this paper, four types of DoS attacks blackhole, greyhole, flooding and scheduling of time-division multiple access (TDMA) attacks are considered that could target LEACH protocol [2].

(i) Blackhole attack---In this attack, the malicious node sends false replies to the sender node about routes without looking into the routing table. Also, it refuses to redirect the message from the sender. Even if it agrees to transfer a packet, it will choose the path in which attacker node is involved and compulsorily redirects the packet to the follow this path involving attacker node [7].

(ii) Greyhole attack is a selective packet dropping attack. Here, malicious nodes drop incoming packets in a certain order with a fixed probability.
(iii) Flooding attack means overcrowding the server node by sending unnecessary packets to consume the network's energy, capacity and waste the time of legitimate node [8]. Scheduling attack is similar to flooding attack, i.e. it also consumes the network's energy, space, time and WSN resources but at a scheduled date and time [1, 8].

In this paper, two machine learning techniques, i.e. decision trees and support vector machines are used.

(A) Decision trees

The decision tree is an algorithm for making decisions. It is a graphical representation used to solve a particular problem by arriving at various solutions by following different possibilities, which start from a root node and arrive at the end of the tree called leaves. Leaves are the terminating nodes. So from the root node, one can arrive at different leaves by following different paths [2].

The reason why they have selected these two attacks is that they are the most dangerous attacks. The greyhole attack is dangerous because here the malicious node appears to be a normal node, and it drops any packets randomly. It does not always drop packets like blackhole attack. Hence, it cannot be concluded that the received packet at the destination node is the actual packet sent by the source or an altered one.

(B) Support vector machines

Support vector machine (SVM) is a supervised machine learning algorithm that can be used for both classification or regression tasks. SVM is based on the concept of decision planes that define decision boundaries. Decision plane is the one that separates a group of objects in different classes.

Here following two experiments were conducted.

(a) SVM on a full data set with all four attacks.
(b) SVM on a reduced data set with selected two attacks, i.e. flooding and greyhole attack.

It was found that in SVM classification with reduced data set, i.e. data set with the selected attacks, the time taken is less than in SVM classification of detecting attacks in a full data set with all four attacks. Also, the number of correctly classified instances was more in the former than in the latter.

The overall conclusion of this paper was decision trees achieved better classification results than SVM, i.e. it was observed that higher true-positive detection rate and lower false-positive detection rate were obtained with decision trees than with support vector machines [2].

3 Review of Paper 'Detection and Prevention of Routing Attacks in the Internet of Things' [10]

Due to the increase in the number of dummy packets sent by illegitimate nodes on the Internet, traffic is increased, because of which there is an increase in bottleneck problems and a decrease in the overall efficiency of the network system. The most important attacks which have been taken into consideration here are selective forwarding and sinkhole attack.

It selectively forwards a packet and drops any packets randomly and whole original message is broken down and it is hard to detect that the data packet has been altered. This attack deteriorates the entire routing path.

The overall target of the malicious node is to disturb the integrity and reliability of the system.

They have used two algorithms here in this paper, i.e. key match algorithm (KMA) with hash values and cluster-based algorithm (CBA) with path matrix [10].

(A) Key match detection algorithm (KMA)

Here, the attacker uses the altered node to send wrong information about itself, about routing paths and routing table to its neighbouring nodes and thus disturbs the entire network. It so happens that an incorrect view of the network is created because of these lossy links in the IoT. So, it is necessary to detect the incorrectness and provide the correct routing information.

Here, three algorithms are designed under the KMA approach:

The first algorithm key match algorithm is finding the nodes in the coverage area from the source node to the destination. Here, actually for every point in between source and destination, its x and y coordinates are calculated. Then in the *x*-direction, the distance is calculated between the source and the node under consideration. If it is less than the destination's x-coordinate, then it is seen that if it is also less than 200. If it is so, then it is included in the coverage set of nodes between source and destination. Also, a key match algorithm is used which checks the key of the intermediate nodes in the path. if a mismatch is found in the key of the node, then it is concluded as a malicious node [10].

The second algorithm is for detection of a suspected node in the network. Here, the degree of a node is calculated. If it is greater than 20% of the degree of its neighbour node, then it is considered to be a far away node. If a node having a wrong degree tries to attract the traffic, then it is considered to be a malicious node. Because the degree of a node means incoming and outgoing paths from that node, in which information is present in the routing table also. So if a node has the incoming paths of outgoing paths different from the routing table, then it means its degree is wrong in the routing table, which cannot be true. Hence, only one possibility is left out that this node with a wrong degree is malicious.

In the proposed work, firstly the key is distributed among the various nodes of the network, and the key is crosschecked at every node. Unless and until the whole message does not collaborate and not verified, the data is not transferred forward.

The mislead is being detected using message authentication algorithm in which after crosschecking if there is any mismatch then the message is dumped there itself and not carried forward.

The third algorithm is prevention algorithm using a hash function. The hash function is used for prevention from selective forwarding and sinkhole attack. Here, generatehashfunction is used to generate a key based on random number generation for every node. This key of a node is counterchecked by its neighbouring nodes and if a mismatch is found then that node is blocked as a malicious node and also the data transfer is stopped [10].

(B) Cluster-based detection algorithm (CBA)

If an attacker node is present in the routing path, then it can try to attract maximum traffic by advertising low rank value. The attacker node may try to consume the energy of legitimate nodes. Also, this node may try to perform selective forwarding attack in which it drops packets randomly in any order. Hence, cluster-based algorithms are also necessary along with key match algorithm, in which network is divided into small clusters for enhancing the routing process. Routing attacks are evaluated with the help of path matrix [9] for lossy networks and by dividing cluster heads.

Here, the following two algorithms are designed under CBA approach

In Algorithm 4, we try to cluster the nodes having limitations of their transmission power and figure out their instant neighbours. This algorithm is used to determine the intruders with wrong identities to conduct various kinds of attacks in an IoT environment. Firstly, the transmission limits for every node in the network is calculated and a neighbour table is maintained that consists of the identities of the nodes within their transmission range. This is an effective technique for routing protocol for low power and lossy network (RPL) networks because if an intruder attempts to alter the identity it can be determined from the respective neighbour table. Also, the neighbour table consists of a group of nodes with similar transmission ranges, so they would consist of similar nodes with small differences in rank values [10].

To find out that which node is suspected, vicinity values are calculated that is stored in the node table. Prevention of malicious node is done here using group authentication technique, which is Algorithm 5 [10].

So this paper concludes that above-mentioned algorithms KMA and CBA can detect malicious nodes in any network and the network can be protected from them. The true-positive rate can be increased more if these two detection techniques are combined [10].

4 Review of Paper 'a Proposed Wireless Intrusion Detection Prevention and Attack System' [11]

In this paper, an intrusion detection and prevention system for wireless networks is proposed which is called 'WIDPAS', i.e. wireless intrusion detection prevention and attack system.

In the wireless networks that work within IEEE 802.11 frequency, the users use this frequency within the company. The data is analysed every 20 s to find whether the behaviour exhibited by nodes and packets is normal or suspicious. The proposed system will detect the type of attack if the attacker allows that and also determines the location of the attacker. All things, i.e. user's behaviour, the networks that they are using and the system's updation of its database are always monitored by the system's administrator. The following figure shows types of wireless attacks [11] (Fig. 1).

The supervisory system calls always monitor the network and alert the administrator if some malicious act is detected.

Based on the behaviour of the packets, the system classifies them as:

- Reliable packets
- Suspicious packets [11].

If bad behaviour is detected from the suspicious waves, then the system will do the following things as steps of protection:

- Detect the attack type whether it is DoS, or MITM, spoofing, etc., based on the behaviour of the attacker.
- Registers the attack category and saves the attacker's information, whether the attack falls under anamoly, signature attack or hybrid attack type.
- Collects maximum information about the attacker and the devices that are used.
- Sends warnings to your network administrator.

The main aim here is to attack the attacker if the attack type allows us to do so.

If the attacker tries to manipulate the data packets of legitimate nodes, then the proposed system tries to attack the network of the attacker and prevents it from doing fraud to its employees. Now, the actual working of the proposed system WIDPAS is discussed as follows:

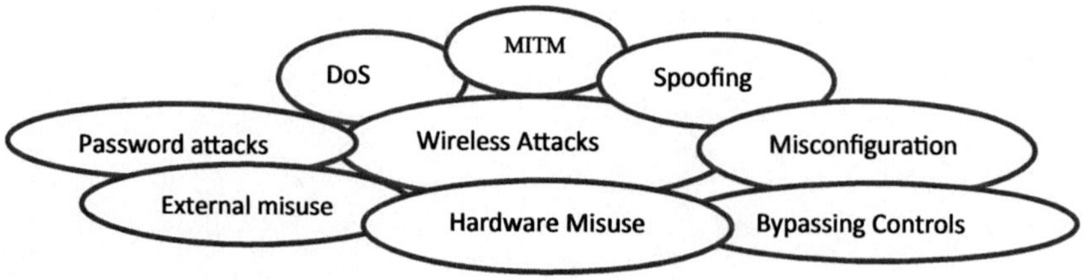

Fig. 1 Types of wireless attacks

- The system collects data through some of the available tools.

This data is then compared with the already stored data about various attacks which shows the system behaviour. Various machine learning algorithms can be used for this comparison purpose.

- After the analysis of data, i.e. after studying the behaviour of the network, the proposed WIDPAS system should conclude that if the system under study is a normal system or some suspicious behaviour is detected in it [11].
- If a RogueAP attack is detected, then to isolate it and to protect the hosts from such an attack, a warning is sent to the system administrator [11].

The overall conclusion of this paper is that a system called WIDPAS is proposed for intrusion detection and prevention in wireless networks. The overall efficiency of the WIDPAS system is improved as it can attack the attacker and protect the wireless network from various attacks by sending warnings to the administrator and thus protects the legitimate nodes in the network [11].

5 Review of Paper 'a Comprehensive Protection Method for Securing the Organization's Network Against Cyberattacks' [12]

In this paper, a comprehensive method for securing an organization's network is proposed. For that, following components are used. Filter firewall(FF), analog detection controller(ADC), honeypot intrusion detection(HID) and prevention firewall(PF). The intrusion is detected in a network when a message passes through the above-stated stages in the above-described order. ADC and HID are the agents which detect intrusion. So finally when HID detects a network's behaviour as normal, it is checked with ADC's detection also and accordingly ADC's false positives and false negatives are altered. Similarly, if HID detects a network's behaviour as a suspicious one then also it is compared with the computed detection results of ADC and accordingly, changes are made in ADC's false positives and false negatives.

For designing an intelligent system which can detect different types of attacks under different situations, a comprehensive mechanism should be designed which can detect attacks at different levels of the system and parallel it can synchronize the agents also. Because of this synchronization, the agent's false positives and false negatives can be corrected. And thus in the network, previously blocked sources may be allowed in a network which was otherwise detected as faulty and vice versa [12].

The method is described below:

Filter firewall(FF):

1. Analyses all incoming traffic including nodes, their packets transmitted, their services offered, etc.

2. If a known attack from the blacklist is found then the proposed system blocks it otherwise passes the entire traffic to ADC.

Anomaly detection controller (ADC) Agent:

3. The packets entering ADC are stamped according to their category and they are forwarded to the honeypot intrusion detection for detection of possible attacks.
4. Compares the incoming traffic against normal behaviour and suspicious behaviour. By doing this, it checks for anomalous attacks [8].
5. If an attack is detected then updation of the PF is done to block the associated traffic.
6. If an attack is not detected, then the PF is informed to allow the traffic.

Honeypot intrusion detection (HID):

7. The effect of incoming traffic is checked on the system. If some critical system states are being altered by the incoming traffic, then malicious behaviour is detected otherwise the traffic is concluded as normal behaviour. Whatever may be the result that is updated to the PF.

Prevention firewall (PF):

8. Now here validation of the results of ADC is done as follows, i.e. synchronization between the results found by both agents that is ADC and HID is done and accordingly, ADC's results are altered.

(I) If an attack is found at HID, it is seen that if the same attack is found at ADC also.

(a) If the same attack is found at ADC then block that incoming traffic.

(b) If this attack is not detected at ADC then update ADC's false negative.

(II) If no attack is detected at HID, then it is seen if this attack is found at ADC or not [12].

(c) If no attack is found at ADC, then it is considered as a normal behaviour and traffic is allowed to pass on.

(d) If the attack is detected at ADC, then update ADC's false positive.

In this manner, a synchronization is done between the agents ADC and HID. And because of this synchronization, earlier blocked sources by ADC can now be considered as legitimate ones and allowed to pass as normal traffic and vice versa [12].

So the overall conclusion of this paper is that traditional methods of protection such as ad IDS, IPS, firewalls and honeypots are not enough to protect a company's network. Rather a combination of these is required to protect an organization's network. Here, the result obtained by honeypot IDs is analysed by preventional firewall and is compared to the results of anomaly Ids, where its false positives and false negatives are modified. Owing to this, previously blocked legitimate traffics could be released now which were once detected as faulty because of the wrong detection of anomaly IDs [12].

6 Conclusion

So the conclusion is to design an intrusion detection and prevention system for protecting an organization's network, similar to WIDPAS in which the legitimate nodes from several attacks by alerting the administrator, using the cluster-based algorithm to form clusters and use decision trees machine learning technique on the data set. Here, routing tables are maintained which will be maintaining all information of a particular node under consideration. A new node entering a network should first register itself with its locality's base station, which will maintain such type of routing table. So here any if any malicious node enters a network, it cannot interfere with other nodes or with their packets, as its entry in the network will be immediately detected with the help of already stored information in the routing table, which will be a known authorized one. So such type of routing tables can be maintained in every cluster. The whole network will be divided into several clusters. Here, decision tree machine learning algorithm can be used to train the data set in detecting any type of entry of malicious nodes in the network. The cluster-based algorithm and decision trees ML techniques are chosen because they gave the best results in the above papers. Also, tried to implement honeypot intrusion detection along with the analysis of prevention firewall whose results would be compared with the results obtained from anomaly intrusion detection for reducing false-positive and false-negative cases.

References

1. Almomani I, Al-Kasasbeh B, AL-Akhras M (2016) WSN-DS: a dataset for intrusion detection systems in wireless sensor networks. J Sens 1–16. http://dx.doi.org/10.1155/2016/4731953
2. Abdulaziz I. Al-issa1, Mousa Al-Akhras1+2, Mohammed S. ALsahli1, Mohammed Alawairdhi1 (2019) Using machine learning to detect DoS attacks in wireless sensor networks. In: IEEE Jordan International Joint Conference on Electrical Engineering and Information Technology
3. Firoz Kabir M, Hartmann S (2018) Cyber security challenges: an efficient ıntrusion detection system design. In: IEEE international young engineers forum
4. Yang Y, McLaughlin K, Sezer S, Littler T, Pranggono B, Brogan P, Wang HF (2013) Intrusion detection system for network security in synchrophasor systems
5. Jokar P, Leung VCM (2016) Intrusion detection and prevention for ZigBee-based home area networks in smart grids. In: IEEE Transaction on Smart Grid
6. Sharma P, Moller DPF (2018) Protecting ECUs and vehicles internal networks. In: IEEE conference
7. Archana M, Binu GS (2017) Vinod G (2017) A survey on defense mechanisms against black hole and gray hole attacks in wireless sensor networks. IJEDR 5(1):177–182
8. Almomani I, Al-Kasasbeh B (2015) Performance analysis of LEACH protocol under denial of service attacks. In: 2015 6th international conference on information and communication systems (ICICS), Amman, pp 292–297. https://doi.org/10.1109/iacs.2015.7103191
9. De Couto DS, Aguayo D, Bicket J, Morris R (2005) A highthroughput path metric for multi-hop wireless routing. Wirel Netw 11(4):419–434
10. Choudhary S, Kesswani N (2018) Detection and prevention of routing attacks in internet of things. In: 2018 17th IEEE International Conference On Trust, Security And Privacy In Computing And Communications

11. Nada JA, Al-Mosa MR (2018) A proposed wireless intrusion detection prevention and attack system. IEEE978-1-7281-0385-3/18/$31.00 ©2018 IEEE
12. Kbar G, Alazab A (2019) A comprehensive protection method for securing the organization's network against cyberattacks. 978-1-7281-2600-5/19/$31.00 ©2019 IEEE https://doi.org/10.1109/ccc.2019.00005

A Novel Task Scheduling Model for Fog Computing

Navjeet Kaur, Ashok Kumar, and Rajesh Kumar

Abstract Fog computing is introduced to improve the performance of cloud computing by deploying fog node(s) near the edge of the network in order to process the data locally. A fog node that acts like a mini cloud with limited resources handles the incoming real-time data, processes it locally and responds back to the edge device. This process has advantage of achieving minimum delay which is considered as a main drawback of cloud computing these days. As fog computing is in its early stage of development, so there are many issues and challenges associated with it, like limited resources of the fog node and processing of real-time task(s) with optimal use of available resources which is also known as task scheduling in fog computing. Task scheduling is one of the important aspects of fog computing that, if carried out efficiently, can largely improve the delay of a service, reduce energy consumption and cut down network traffic. This paper investigates various issues and challenges of fog task scheduling with existing research. Further, the papers also provide solution to various aspects of task scheduling through a novel task scheduling model for fog computing environment.

Keywords Fog computing · Job scheduling · Scheduling algorithm · Task scheduling · Task scheduling model

N. Kaur · A. Kumar (✉)
Chitkara University Institute of Engineering and Technology,
Chitkara University, Patiala, Punjab, India
e-mail: ashok.kr@chitkara.edu.in

N. Kaur
e-mail: navjeet.kaur@chitkara.edu.in

R. Kumar
Department of Computer Science and Engineering,
Thapar Institute of Engineering and Technology, Patiala, India
e-mail: rakumar@thapar.edu

G. Ranganathan et al. (eds.), *Inventive Communication and Computational Technologies*, Lecture Notes in Networks and Systems 145,
https://doi.org/10.1007/978-981-15-7345-3_72

1 Introduction

In this modern era, every human is surrounded by a variety of sensors than before, from wearing a smart watch to driving a smart car having numerous sensors installed. These sensors generate a lot of data every microsecond. According to a forecast by International Data Corporation (IDC) [1], global data generated from various Internet of things (IoT) devices was 33 Zettabytes (ZB) in 2018 and expected to rise to 175 ZB by 2025. Further, IDC also predicted that by 2025, 6 billion customers will interact with data every day because of the billions of IoT devices connected across the globe, which are expected to create over 90 ZB of data by 2025. The processing and mining of this huge data pose numerous challenges to cloud computing. Various such issues and challenges are discussed by [2, 3]. Shi et al. [4] and Hu et al. [5] suggested edge computing to address these issues and gave solution to offload the cloud traffic to lower part of the network or at the edge of the network. Although edge computing does not follow any standard architecture of how and where the intelligence should be deployed. Thus, in order to solve this, fog computing later emerged as a need to address many growing day to day concern of IoT developments like real-time data processing, network traffic and congestion with growing bandwidth requirement. Fog computing acts as a middle layer between IoT sensors and cloud computing with the overall motive of scheduling tasks locally, i.e., near the data production in order to achieve low latency, less bandwidth, reduction of network traffic and scalability. The processing of data near the data production area save huge amount of network bandwidth which generally get consumed a lot in centralized cloud data processing system. In centralized cloud systems, location of the data warehouse might be far away from the source result in consumption of huge network bandwidth in processing information which in parallel increase the cost with time. Further, centralized cloud approach also prone to faults and failures, results in delay of service which might not affordable for real-time applications. Therefore, enabling capabilities near the edge of the network is the ideal solution to these issues by processing requested task in a region close to the user. The notion of processing requested task by allocating it optimal resources for execution is called as task scheduling. The task scheduling in fog computing environment is an important area of research as there are so many issues associated with it such as

- Heterogeneous nature of fog nodes and tasks
- Dynamic nature of the requested tasks
- Identification of authenticated, authorized and priority tasks
- Finding an optimal task-resource pair
- Load balancing among fog nodes
- Deployment of an fast task scheduling algorithm as per the situation.

So, in order to deal with these issues, a task scheduling model for fog computing environment is highly required. Research on tasks scheduling in fog computing environment has not been well established, although many research achievements have been obtained for task scheduling, but researchers are still exploring various aspects of its improvement [6–8].

The motivation behind this research work is to address different issues related to task scheduling in fog computing environment. To the best of our knowledge, no scheduling model or algorithm till now focus on so many different aspects of task scheduling. Although different researches work on different issues of scheduling for fog computing environment like load balancing among fog nodes [9–14], identification of priority tasks [15–18], determining optimal task-resource pair [19, 20], evaluating user requirements dynamically [21] and secure task scheduling [19, 22, 23].

The main contribution of this paper is developing a task scheduling model for fog environment which focuses on different issues like

- The model provides a secure authentication and authority check system for each request placed.
- The model considers priorities of tasks by creating a priority queue based on task(s) having short deadlines.
- The model considers the resource limitations of fog nodes and takes decisions as per their resource capacities.
- The model also considers the optimal task-resource pair by ranking different task(s) and resource(s).

Further, this paper is divided into different sections where Sect. 2 covers the general introduction of task scheduling in fog computing and related background of the task scheduling models developed by various researchers, Sect. 3 covers the problem statement and proposed model. The conclusion is provided in Sect. 4.

2 Related Work

Task scheduling in fog computing is a tedious and challenging task. A fog node also known as fog manager acts as a task scheduler for managing the task scheduling process for fog computing environment [24]. This section focuses on the existing research carried out to deal with various issues pertaining to task scheduling process the paper focused on like security, task priorities, load balancing and optimal task-resource pairing.

2.1 Task Prioritization in Fog Task Scheduling

A lot of research has already been carried on setting priority of task(s) through different ways. Mukherjee et al. [15] proposed a scheduling policy that schedules tasks while meeting deadlines on time. The authors maintained a high priority and a low priority queue which are filled up on the basis of delay-deadline. The proposed algorithm also discussed offloading of task to fog nodes in order to balance the workload.

Oueis et al. [16] proposed work is about finding an optimal fog cluster and allocate resources in multi-user environment. The scheduling algorithm is divided into two major phases where in first phase is local computational resources allocation at small cells (SC) where incoming task is sorted for priority according to metrics like earliest deadline first with priority (EDF-PC), earliest deadline first with latency (EDF-LAT), computation size with latency (CS-LAT). The sorting also defines the scheduling algorithm to be deployed for each task execution.

Liu et al. [17] presented a task scheduling model and algorithm based on an improved classification mining technique Apriori called I-Apriori. The proposed TSFC algorithm also defines the priority of a task to get scheduled according to the minimum completion time out of n-tasks.

Islam et al. [18] proposed an algorithm based on Ford-Fulkerson algorithm and priority queue to perform task scheduling and load balancing in fog computing. The authors presented a fog model having different fog layers like fog data collection layer, fog edge layer and fog intermediate computation layer. They used various notations for fog nodes like fog edge data node, master node and slave node. The scheduling strategy in the model depends upon task finish time (FT). A task within the minimum FT was given higher priority whiling scheduled and a queue with maximum number of tasks was given the priority.

2.2 *Security in Fog Task Scheduling*

There are various researchers who worked on secure fog task scheduling process like Sujana et al. [19] contributed in two main fields of fog computing, i.e., security and task scheduling. They provided a collaborative model called trust-based stochastic scheduling algorithm. The main objective was to find an optimal task-VM pair in order to efficiently schedule the task in fog environment. The algorithm was of type stochastic scheduling and considered stochastic top level (STL) as a factor for assigning priorities to tasks.

Rahbari et al. [22] proposed an algorithm that focused on securing a fog network through authentication, integrity and confidentiality. Further, the algorithm allocates VM to workflow using metaheuristic and data mining techniques.

Auluck et al. [23] introduced a non-preemptive real-time security aware scheduling (RT-SANE) on the network edge. The algorithm considers security and deadline as the two main parameter to schedule tasks. They categorized security levels among tasks and resources as private job send to trusted resource, semi-private send to semi-trusted resource and public job send to non-trusted resource.

2.3 Ranking of Fog Nodes

There are many authors who worked on ranking fog nodes in a fog computing environment using different parameters like Naha et al. [21] proposed resource ranking and provisioning (ReRaP) algorithm which focused on meeting deadline and allocates resources as per changing requirement of user to minimize time, cost and delay within the overall purpose of meeting the deadlines. The authors solved the problem by ranking resources. The ranking of resources was done on the basis of available processing, bandwidth and latency. The proposed algorithm ranked all resources belonging to fog devices, fog servers and cloud.

Bendlidia et al. [25] proposed an algorithm that is based on fuzzy quantifiers in order to rank fog nodes. The ranking of fog nodes was done on two parameters called least satisfactory proportion (LSP) and greatest satisfactory proportion (GSP) in order to rank the fog nodes. The parameters were calculated using fuzzy and linguistic quantifiers which further depend upon different fog node parameters like distance, price, latency, bandwidth and reliability.

Abreu et al. [26] proposed method of ranking fog nodes/cloudlets on the basis of divisions, score and rounds. The method proposed to place the cloudlets in different divisions holding some threshold value. A cloudlet meeting the criterion on threshold value is placed in a division and its score is periodically updated to promote/demote its ranking with respect to other cloudlets. The periodical ranking of a cloudlet is called as a round. The ranking method is used when scheduling occur inside fog node or cloud node rather than just coordinating between cloudlets and cloud.

2.4 Task-Resource Pairing in Fog Task Scheduling

The main objective of a task scheduler in fog computing is to do optimal task-resource pairing and there are researchers like Sujana et al. [19] already discussed under security and Gad-Elrab and Noaman [20] who focused on this issue. Gad-Elrab and Noaman [20] proposed two-tier bipartite graph with fuzzy clustering task allocation approach (2tBiFTA) to solve the multi-objective problem (META) considering minimization of delay, cost and energy. The complete method is divided into five different phases which are: task leveling phase, task fuzzy clustering phase, task entity-tier decision phase, task prioritizing phase and finally task quota-tier allocation phase. This phase calculates the relationship between a set of allocated task nodes and virtual machines on a fog/cloud node using bipartite matching graph. An integer value represents this relationship and higher the value of integer decides the ideal paring of task-VM pair for final execution of the task.

3 Problem Description and Proposed Solution

This section presents the problem statement and the proposed solution model explained through various phases addressing various issues of task scheduling for fog computing.

3.1 Problem Statement

The main aim of this research is finding an optimal task scheduling model for fog environment. The key issue is finding an optimal task-resource pair for task scheduling in fog computing, achieving minimum delay, cost and energy. Finding an optimal task-resource pair is associated with various other issues like decision about task(s) priority, searching for an optimal resource from the available resource pool, supportive system to cope with increasing demand of resources and finally execution of task(s) with optimal use of resources. Prior work in this field is quite limited and there is no single model which covers so many different aspects of tasks scheduling.

3.2 Proposed Task Scheduling Model

The proposed model solves the problems associated with finding an optimal task-resource pair. The fog task scheduling model is divided into four main phases which are

- Request collection
- Authentication/authorization check
- Request priority ranking
- Selection of scheduling node
- Resource ranking
- Task-resource pairing and scheduling.

General assumptions:

- Assumption 1: Every phase is being managed by different managers where authentication/authorization check is being managed by a security manager (SM), resource ranking is being done by resource manager (RM) and request ranking, selection of scheduling node and task-resource pairing with scheduling is handled by a fog manager(FM). A manager is considered as a managing node or machine which is intelligent enough to do this kind of processing.
- Assumption 2: Each FM is managing a number of fog node(s) (FN) which further manages virtual machines (VM). For simplicity, we consider only one VM per FN and taken FN or VM terms consecutively in resource ranking phase.

- Assumption 3: For simplicity, only one FM is being considered for the whole system.

Requests Collection This phase is responsible for collection of data from the IoT sensors and maintained in the form of the queue in first come first serve (FCFS) manner. The maintained queue will be called as request queue (RQ) and tasks T will be maintained as t1, t2, t3, ...so on, where t1 is the first request, t2 is the second request and so on. Each task is containing three properties, maintained in the form of array are deadline, resource demand and cost. The output of this phase is the list of tasks arranged in form of FCFS basis.

Mathematically, it can be represented as
Queue1: [t1: deadline: d1, resource demand: rd1, cost: c1, t2: deadline: d2, resource demand: rd2, cost: c2, t3:deadline: d3, resource demand: rd3, cost: c3 ...].

Authentication/Authorization Check This phase is handled by a security manager (SM) which ensures each task from the RQ is checked for authentication and authorization where authentication checks the identity and authorization grant access to the system. If the request is validated, it is send to the next phase of evaluation, otherwise request is declined.

Request Priority Ranking In this phase, all validated requests are assigned with a unique priority number (UPN) where priority number 1 indicated highest priority. All the tasks are queued in increasing order of their priority number. Note that we calculate priority in terms of task having least deadline, where the first task in the

Table 1 Notations

Symbol	Definition
RR	Resource ranking
SRTF	Shortest remaining time first
D	Deadline
RD	Resource demand
C	Cost
PC	Processing capacity
W	Workload
HC	Hop count
i, j	Variables
x	any random value
TaskList[j].D	Deadline of task j
TaskList[j].RD	Resource demand of task j
TaskList[j].C	Cost paid by task j
ResourceList[j].PC	Processing capacity of resource j
ResourceList[j].W	Workload of resource j
ResourceList[j].HC	Hope count of resource j

queue is the task with the minimum deadline assign with priority 1 then second task with priority 2 and so on. If there is case when two task comes out with same deadline, then we compare resource demand and cost in order to assign UPN. Algorithm 1 represents the request priority ranking (RPR) method, symbols used in the algorithms are described in Table 1. In Algorithm 1, input is the list of task with various attributes value of deadline, resource demand and cost paid, and output is priority task list in increasing order of number as priority 1 is the highest priority task, priority 2 is second highest and so on. Priority is highly depend upon the deadline of the task.

Algorithm 1 RPR, Input: TaskList <instances $< D,RD,C >$ >, Output: PriorityList <instances $< D,RD,C >$ >, $n \rightarrow TaskList.length$

```
for i=1,2... (n-1) do
  for j=1,2... (n-1) do
    if TaskList[j].D > TaskList[j+1].D then
      swap(j,j + 1)
    else
      if TaskList[j].D == TaskList[j+1].D then
        if TaskList[j].RD > TaskList[j+1].RD then
          swap(j,j + 1)
        else
          if TaskList[j].RD == TaskList[j+1].RD then
            if TaskList[j].C > TaskList[j+1].C then
              swap(j,j + 1)
            end if
          end if
        end if
      end if
    end if
  end for
end for
```

Selection of Scheduling Node First task from the queue is chosen for scheduling by checking its resource demand. The resource demand (RD) of the task is checked against the resource availability (RA) among n available fog nodes. If the RD value is less or equal than RA value, then the task is send to fog broker for further processing, otherwise it will send it to the cloud broker for processing at the cloud center.

Resource Ranking The phase considers the fact the resource manager (RM) will result in providing the list of all resource available fog nodes (FN) as per the RD of a task. For example, for a particular priority one task RD, the resource manager results in a list of five available FNs like ResultFN: [FN1, FN2, FN3, FN4, FN5] where each FN from the list, capable of executing the selected task. Further, each FN contains three major properties: processing capacity, workload and hop count. The main responsibility of this phase is to choose the best FN out of the list provided by the RM. All the FN are ranked according to its processing capability, i.e., higher the processing capacity higher the rank of the FN. If there is any conflict arises in processing capability, then minimum workload and hop count are evaluated further to rank a FN. Algorithm 2 explained the resource ranking (RR) algorithm in brief, symbols used in the algorithms are described in Table 1. In Algorithm 2, input is the list of resources with attributes processing capacity, workload and hop count, and output is the resource ranking list which indicates rank of each resource containing optimal attributes according to the task requirements. It is highly depend on processing capacity of the resources.

Algorithm 2 RR, Input: ResourceList <instances $<PC,W,HC>$ >, Output: RRankList <instances $<PC,W,HC>, m \rightarrow ResourceList.length$

```
for i=1,2... (m-1) do
  for j=1,2... (m-i-1) do
    if ResourceList[j].PC < ResourceList[j+1].PC then
      swap(j,j+1)
    else
      if ResourceList[j].PC == ResourceList[j+1].PC then
        if ResourceList[j].HC < q ResourceList[j+1].HC then
          swap(j,j+1)
        else
          if ResourceList[j].HC == ResourceList[j+1].HC then
            if ResourceList[j].W < ResourceList[j+1].W then
              swap(j,j+1)
            end if
          end if
        end if
      end if
    end if
  end for
end for
```

Task-Resource Pairing and Scheduling A task is pair with the highest rank resource, i.e., Rank1FN out of the total list of available resources and update the processing capacity and workload parameters of the FN. The updated information is also passed to the RM for further updation. We are assuming each FN is also maintaining a queue of tasks allocated to it and called as a ready queue. Each task from the ready queue, execute tasks in preemptive shortest remaining time first (SRTF) manner considering the arrival and burst time of the task. SRTF always chooses the task with shortest remaining time of completion with preemptive approach where processor execute the selected task for one time quantum and then look for a task with the shortest time remaining. In this way, the approach achieves the lowest turnaround time as compare to other scheduling approaches [27].

Figure 1 is the detailed workflow of the proposed model where request(s) received from the IoT sensors are arranged in the FCFS queue. Each task from the queue go for a security check before scheduling. Only validated requested are processed further by assigning a UPN number which represents the priority of the task using a ranking method called as RPR. Further, RD of the first high priority task is evaluated against the available pool of resources. If the RD of the task is intensive and there is no FN

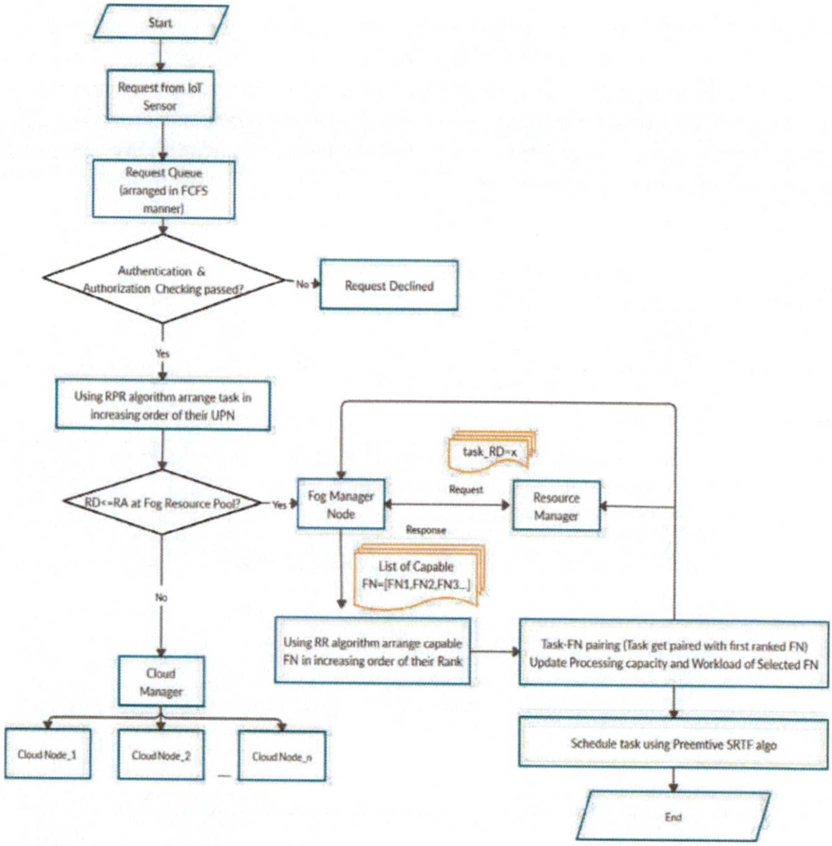

Fig. 1 Proposed task scheduling model workflow

available that task is sent to the cloud manager for further processing, otherwise fog manager will execute it. FM pass the requirement of resource demand to the RM, who manages all the resources of different FNs. RM responds by providing a list of resources which are capable of executing the task. In order to create a optimal task-fog pairing, fog manager rank these resources using RR algorithm. Finally, tasks are executing using preemptive SRTF algorithm.

3.3 Pros and Cons of the Proposed Work

The proposed scheduling model works on various important aspects of task scheduling in fog computing. Although, the model tries to cover all important issues of task scheduling but there are area where it can be improved in order to make it more beneficial as per the increasing demand of IoT environment. Various pros and cons of the system are

Pros:

- The model provides a secure authentication and authority check system.
- The model focuses on time sensitive data by assigning priority to tasks.
- The model takes decisions as per their available resource capacities.
- The model dynamically rank different task(s) and resource(s) rather than a static approach.

Cons:

- Workload balancing or load balancing is the area which can be improved in the model.
- SRTF algorithms usually deal with many context switches in a short period of time which leads to starvation problem. So, an improvement on the algorithm is needed.

4 Conclusion

Task scheduling is the most challenging task for fog computing because of limited resources at the edge of the network. Existing literature considered limited aspects of fog task scheduling where it is found out that optimal task-resource pairing is the overall motto of the scheduler in fog task scheduling. Finding the optimal right resource is a tedious task as the optimal resource is the one having right skill set for a particular task to be executed in minimum period of time. The right pairing of equipped resources as per task requirements, created a optimal pairing of task-resource(s). In this research, optimal task-resource pairing is done using a different ranking strategy on different tasks and resources. Moreover, various challenges associated with fog task scheduling are investigated and we proposed a task scheduling model for fog computing environment which focus on various important aspects of task scheduling like security, task priorities, fog node capacities and optimal task-resource pairing with the motive of achieving minimum delay and energy. The future work will focus on deploying dynamic behavior to the proposed model and implementation on a complex simulation environment. Further, the model can also be improved in terms of better loading balancing and task scheduling algorithm.

References

1. Reinsel D, Gantz J, Rydning J (2018) Data age 2025: the digitization of the world from edge to core
2. Dillon T, Wu C, Chang E (2010) Cloud computing: issues and challenges. In: 2010 24th IEEE international conference on advanced information networking and applications. IEEE, pp 27–33

3. Zhang Q, Cheng L, Boutaba R (2010) Cloud computing: state-of-the-art and research challenges. J Internet Serv Appl 1(1):7–18
4. Shi W, Cao J, Zhang Q, Li Y, Xu L (2016) Edge computing: vision and challenges. IEEE Internet Things J 3(5):637–646
5. Hu YC, Patel M, Sabella D, Sprecher N, Young V (2015) Mobile edge computing—a key technology towards 5G. ETSI White Pap 11:1–16
6. Mahmud R, Kotagiri R, Buyya R (2018) Fog computing: a taxonomy, survey and future directions. Internet of everything. Springer, Singapore, pp 103–130
7. Yi S, Li C, Li Q (2015) A survey of fog computing: concepts, applications and issues. In: Proceedings of the 2015 workshop on mobile big data, pp 37–42
8. Liu Y, Fieldsend JE, Min G (2017) A framework of fog computing: architecture, challenges, and optimization. IEEE Access 5:25445–25454
9. Wu HY, Lee CR (2018) Energy efficient scheduling for heterogeneous fog computing architectures. In: 2018 IEEE 42nd annual computer software and applications conference (COMPSAC), vol 1. IEEE, pp 555–560
10. Wan J, Chen B, Wang S, Xia M, Li D, Liu C (2018) Fog computing for energy-aware load balancing and scheduling in smart factory. IEEE Trans Ind Inform 14(10):4548–4556
11. Nazir S, Shafiq S, Iqbal Z, Zeeshan M, Tariq S, Javaid N (2018) Cuckoo optimization algorithm based job scheduling using cloud and fog computing in smart grid. In: International conference on intelligent networking and collaborative systems. Springer, pp 34–46
12. Wang Y, Guo C, Yu J (2018) Immune scheduling network based method for task scheduling in decentralized fog computing. Wirel Commun Mob Comput 2018
13. Kazemi M, Ghanbari S, Kazemi M (2020) Divisible load framework and close form for scheduling in fog computing systems. In: International conference on soft computing and data mining. Springer, pp 323–333
14. Bhatia M, Sood SK, Kaur S (2020) Quantumized approach of load scheduling in fog computing environment for IoT applications. Computing 102:1097–1115
15. Mukherjee M, Guo M, Lloret J, Iqbal R, Zhang Q (2019) Deadline-aware fair scheduling for offloaded tasks in fog computing with inter-fog dependency. IEEE Commun Lett
16. Oueis J, Strinati EC, Barbarossa S (2015) The fog balancing: load distribution for small cell cloud computing. In: 2015 IEEE 81st vehicular technology conference (VTC spring). IEEE, pp 1–6
17. Liu L, Qi D, Zhou N, Wu Y (2018) A task scheduling algorithm based on classification mining in fog computing environment. Wirel Commun Mob Comput 2018
18. Islam T, Hashem M (2018) Task scheduling for big data management in fog infrastructure. In: 2018 21st international conference of computer and information technology (ICCIT). IEEE, pp 1–6
19. Sujana JAJ, Geethanjali M, Raj RV, Revathi T (2019) Trust model based scheduling of stochastic workflows in cloud and fog computing. In: Cloud computing for geospatial big data analytics. Springer, pp 29–54
20. Gad-Elrab AA, Noaman AY (2020) A two-tier bipartite graph task allocation approach based on fuzzy clustering in cloud-fog environment. Future Gener Comput Syst 103:79–90. https://doi.org/10.1016/j.future.2019.10.003
21. Naha RK, Garg S, Chan A, Battula SK (2020) Deadline-based dynamic resource allocation and provisioning algorithms in fog-cloud environment. Future Gener Comput Syst 104:131–141
22. Rahbari D, Kabirzadeh S, Nickray M (2017) A security aware scheduling in fog computing by hyper heuristic algorithm. In: 2017 3rd Iranian conference on intelligent systems and signal processing (ICSPIS). IEEE, pp 87–92
23. Auluck N, Rana O, Nepal S, Jones A, Singh A (2019) Scheduling real time security aware tasks in fog networks. IEEE Trans Serv Comput
24. Yang Y, Zhao S, Zhang W, Chen Y, Luo X, Wang J (2018) Debts: delay energy balanced task scheduling in homogeneous fog networks. IEEE Internet Things J 5(3):2094–2106
25. Benblidia MA, Brik B, Merghem-Boulahia L, Esseghir M (2019) Ranking fog nodes for tasks scheduling in fog-cloud environments: a fuzzy logic approach. In: 2019 15th international wireless communications & mobile computing conference (IWCMC). IEEE, pp 1451–1457

26. Abreu DP, Velasquez K, Assis MRM, Bittencourt LF, Curado M, Monteiro E, Madeira E (2018) A rank scheduling mechanism for fog environments. In: 2018 IEEE 6th international conference on future internet of things and cloud (FiCloud). IEEE, pp 363–369
27. Vignesh V, Sendhil Kumar K, Jaisankar N (2013) Resource management and scheduling in cloud environment. Int J Sci Res Publ 3(6):1–6

A Bidirectional Power Converter with Shunt Active Filter for Electric Vehicle Grid Integration

Ganesh Anam and M. R. Sindhu

Abstract The tremendous increase in green house gases and fossil fuel depletion rates led to electric vehicles (EV) as alternate options. The impacts of power system network of electric vehicles for charging power converter configurations, control strategies, charging standards are studied in detail by researchers. Major issues faced with EV charging station are large amounts of current and voltage distortions. This paper presents a case study of EV charging station in a scaled down laboratory model. Studies reveal the necessity of installing suitable compensators at EV charging stations, and shunt active filters with suitable controller are implemented in the system. The improvement in system performance with the installation of harmonic compensator is discussed in this paper.

Keywords EV charging stations · Power quality · Shunt active filter · Harmonic compensation

1 Introduction

Recently, a large number of electric vehicles are promoted due to excessive emission rates of IC engine vehicles and also due to depletion of fossil fuels. Major issues with large-scale implementation of electric vehicles are power quality issues, less availability of charging stations, and lack of coordination among EV chargers in charging stations [1–3]. Indian government targets 100% electric vehicles by 2030. Attempts are initiated with the help of National Electric Mobility Mission Plan (NEMMP) [4]

G. Anam (✉)
Department of Electronics and Communication Engineering, Amrita School of Engineering, Coimbatore, Amrita Vishwa Vidyapeetham, Coimbatore, India
e-mail: anamganeshreddy@gmail.com

M. R. Sindhu
Department of Electrical and Electronics Engineering, Amrita School of Engineering, Coimbatore, Amrita Vishwa Vidyapeetham, Coimbatore, India
e-mail: sindhumadassery1@gmail.com

G. Ranganathan et al. (eds.), *Inventive Communication and Computational Technologies*, Lecture Notes in Networks and Systems 145,
https://doi.org/10.1007/978-981-15-7345-3_73

and Faster Adoption and Manufacturing of Hybrid and Electric Vehicles (FAME). Electric vehicle manufacturers have adopted IEEE 519 [5], IEC 1000-3-2 [6], and SAE [7] standards for selection of EV charging infrastructure.

These large-scale implementations of electric vehicles lead to critical issues in power grid. Peak load demand and total harmonic distortion are the commendable drawbacks of EV charging stations. Many EV promotion schemes are introduced by the government, which certainly lead to drastic increment in launch of new electric vehicles. EV charging system includes power electronic devices and their operation generates large amounts of source current harmonics because of which there is a distorted voltage drop across the source impedance that results in voltage distortion at the PCC and other consumers at the PCC will receive distorted supply voltage, which will cause mal-operation of some protection devices, over temperature at PFC capacitors, transformers, motors, and cables. In [8], de Melo et al. proposed bidirectional battery charger with V2G/V2H capabilities. Non-isolated schemes of EV chargers and their effect on power quality are studied in detail in [9]. In [10, 11], battery charger circuits are controlled by means of an aggregator, which allocates power delivered through individual chargers. V2G integrator with harmonic compensation and reactive power control capability is discussed in [12–15]. In [16], Melo et al. discussed results of voltage and current harmonic analysis at EV charging points in Portuguese distribution grid. Studies show that 12% and 20% THDs are introduced by Nissan's Leaf and Mitsubishi i-MiEV, respectively.

If many EV charging points are installed in a charging station, the power quality issues are seen to be extremely high. These current harmonics are propagated throughout the network and degrade quality of voltages. It also causes increased power losses, excessive loading of transformer, reduced efficiency, excessive temperature, and insulation/winding failures. These factors ensure the necessity of detailed case study and analysis of adverse effects of EV chargers. There are many power quality improvement techniques available like high power factor converter, multi-pulse converters, passive filters, PWM inverter, active filters, hybrid shunt active filter, and unified power quality converters. They are the simplest compensators configuration. But it has many drawbacks such as high no load losses, bulky size, and fixed compensation characteristics, whereas shunt active filter has the advantages like lower switching frequency, smaller rated components, and feasibility for future improvements.

This paper presents case study of harmonic effects of EV chargers with the help of a scaled down laboratory model. The author extends work with the installation of suitably controlled shunt active filter and is discussed in detail in the coming sections.

2 Case Study

The main components of the electric vehicle charger are transient suppressor, voltage rectifier, DC/DC converter, and monitors. The simplified block diagram of the EV charger circuit is shown in Fig. 1.

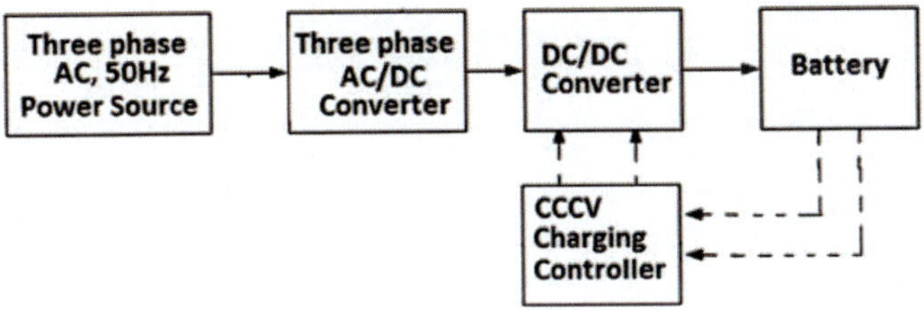

Fig. 1 Block diagram of EV charging station

Small-scale laboratory model of an EV charging station is set up in the laboratory for conducting case studies of harmonic distortions introduced by EV chargers under various scenarios. Laboratory set up is shown in Fig. 2.

These case studies under different charging scenarios show generated current harmonic distortions that are too high as shown in Fig. 3. Harmonic analysis is carried out using Fluke 435 Power Quality Analyzer.

The results shown in Fig. 3 prove that the current distortions with EV charger are too high compared to standard limits. To maintain these recommended IEEE 519 standards, mentioned in Tables 1 and 2, mitigation techniques are certainly required.

Various harmonic mitigation techniques are suggested by researchers such as passive filters and active filters. Passive filters may cause resonances under system specific conditions. In addition, passive filters are to be designed on the basis of

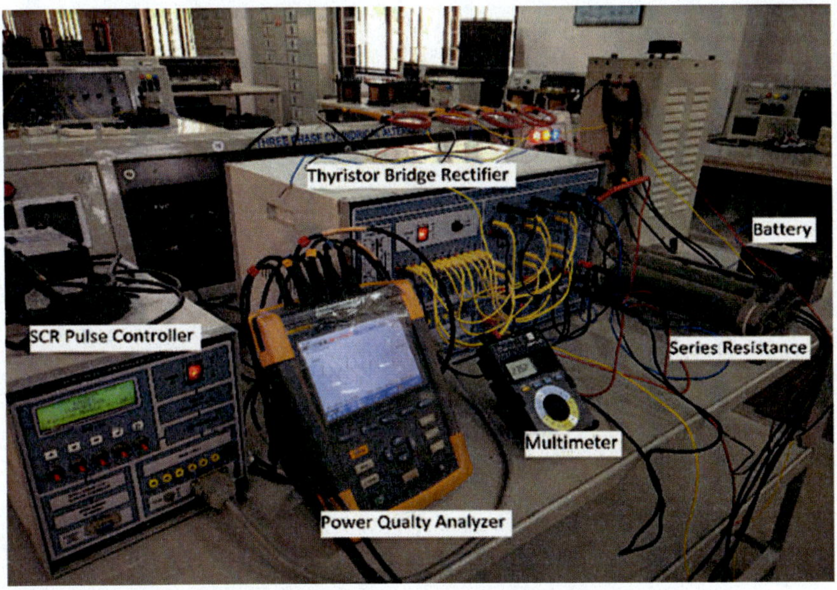

Fig. 2 Scaled down laboratory setup for harmonic analysis of and EV charger

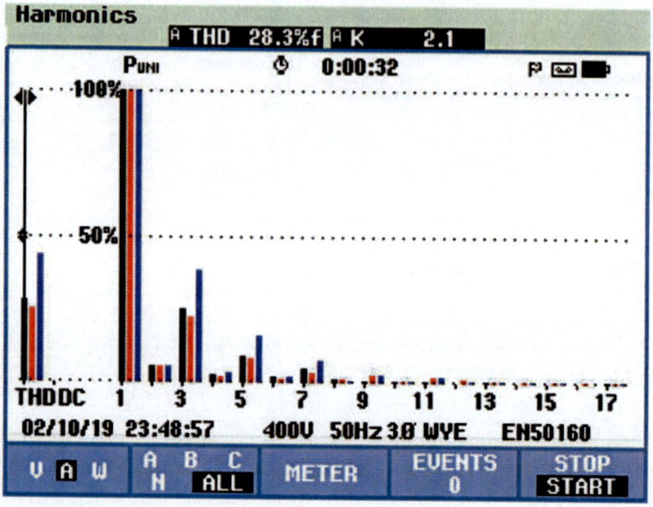

Fig. 3 Harmonic analysis with an EV charger

Table 1 Current distortion limits for systems (120 V to 69 kV) [5]

Maximum harmonic current distortion (in percent of I_L)						
Individual harmonic order (odd harmonics)						
I_{sc}/I_L	$3 \leq h < 11$	$11 \leq h < 17$	$17 \leq h < 23$	$23 \leq h < 35$	$35 \leq h \leq 50$	TDD
<20	4.0	2.0	1.5	0.6	0.3	5.0
20 < 50	7.0	3.5	2.5	1.0	0.5	8.0
50 < 100	10.0	4.5	4.0	1.5	0.7	12.0
100 < 1000	12.0	5.5	5.0	2.0	1.0	15.0
>1000	15.0	7.0	6.0	2.5	1.4	20.0

I_{sc} = maximum short-circuit current at PCC
I_L = maximum demand load current
Even harmonics are limited to 25% of harmonic limits above

Table 2 Voltage distortion limits [5]

Bus voltage at PCC	Individual harmonic (%)	Total harmonic distortion (THD%)
$V \leq 1.0$ kV	5.0	8
1 kV < $V \leq 69$ kV	3.0	5.0
69 kV < $V \leq 161$ kV	1.5	2.5
161 kV > V	1.0	1.5

reactive power requirement and harmonic orders to be filtered. Passive filters were tried with charging stations that are set up earlier and found to be less effective under dynamic charging situations [11]. Hence, active filters are mostly preferred in power grid integration applications. In this paper, authors implement active filter for harmonic cancellation.

3 Shunt Active Filter and Its Control Technique

Shunt active filter injects selected harmonic frequency compensation signals of suitable magnitude. Determination of APF rating is highly significant, while considering its harmonic filtering characteristics. In [11], authors present their research results as the harmonic current appears to be amplified or the rating of active filter becomes excessively high.

The operation of this filter is controlled by effective control algorithms. Many control algorithms are presented by researchers such as instantaneous reactive power theory (IRPT), DC bus voltage control, and synchronous detection algorithm. Author selects instantaneous reactive power theory for the control of active filter. The fundamental real and reactive power components represent DC signal, and it is filtered to generate the source side reference current signals.

According to instantaneous reactive power theory, the instantaneous active and reactive power are computed by sensing three-phase source voltages and load currents as shown in Eq. (1). Three phase source voltage denoted as (V_{sa}, V_{sb}, V_{sc}) are then converted into $\alpha\beta$ orthogonal coordinates (V_α, V_β) as

$$\begin{bmatrix} V_0 \\ V_\alpha \\ V_\beta \end{bmatrix} = \sqrt{\frac{2}{3}} \begin{bmatrix} \frac{1}{\sqrt{2}} & \frac{1}{\sqrt{2}} & \frac{1}{\sqrt{2}} \\ 1 & -\frac{1}{2} & -\frac{1}{2} \\ 0 & \frac{\sqrt{3}}{2} & -\frac{\sqrt{3}}{2} \end{bmatrix} \begin{bmatrix} V_{sa} \\ V_{sb} \\ V_{sc} \end{bmatrix} \tag{1}$$

Similarly, load currents (I_{la}, I_{lb}, I_{lc}) are also converted into $\alpha\beta$ orthogonal coordinates as

$$\begin{bmatrix} I_{L0} \\ I_{L\alpha} \\ I_{L\beta} \end{bmatrix} = \sqrt{\frac{2}{3}} \begin{bmatrix} \frac{1}{\sqrt{2}} & \frac{1}{\sqrt{2}} & \frac{1}{\sqrt{2}} \\ 1 & -\frac{1}{2} & -\frac{1}{2} \\ 0 & \frac{\sqrt{3}}{2} & -\frac{\sqrt{3}}{2} \end{bmatrix} \begin{bmatrix} I_{la} \\ I_{lb} \\ I_{lc} \end{bmatrix} \tag{2}$$

The instantaneous values of active power p_L and reactive power q_L are obtained as shown in (3).

$$\begin{bmatrix} p_L \\ q_L \end{bmatrix} = \begin{bmatrix} V_\alpha & V_\beta \\ V_\beta & -V_\alpha \end{bmatrix} \begin{bmatrix} I_{L\alpha} \\ I_{L\beta} \end{bmatrix} \tag{3}$$

Instantaneous value of active power p_L consists of DC and AC parts, $\bar{p}_L$ and $\tilde{p}_L$ respectively. The instantaneous reactive power q_L also has two components, $\bar{q}_L$ and $\tilde{q}_L$ respectively. The reference three-phase compensating currents I^*_{ca}, I^*_{cb} and I^*_{cc} are calculated as shown in (4).

$$\begin{bmatrix} I^*_{ca} \\ I^*_{cb} \\ I^*_{cc} \end{bmatrix} = \sqrt{\frac{2}{3}} \begin{bmatrix} 1 & 0 \\ -\frac{1}{2} & \frac{\sqrt{3}}{2} \\ -\frac{1}{2} & -\frac{\sqrt{3}}{2} \end{bmatrix} \begin{bmatrix} V_\alpha & V_\beta \\ -V_\beta & V_\alpha \end{bmatrix} \begin{bmatrix} \tilde{p}_L \\ p_L \end{bmatrix} \quad (4)$$

The effectiveness of the shunt active filter with IRPT control algorithm is tested through simulation studies. Simulation system of identical rating as the one selected for experimental studies is considered in this work. Simulation results are explained in detail in the following section.

4 Simulation Results

The system selected for harmonic analysis of EV charger is shown in Fig. 4. During charging mode, the AC-DC interfacing converter acts as three-phase diode bridge rectifier.

4.1 Rectifier Mode of Operation

The bidirectional DC-DC converter is controlled by CCCV controller in which it is designed in a way that it charges the battery in constant current (CC) mode till it reaches 80% SOC and then it switches to constant voltage (CV) mode till the SOC reaches 100%.

According to simulation results shown in Fig. 5, harmonic distortion in source

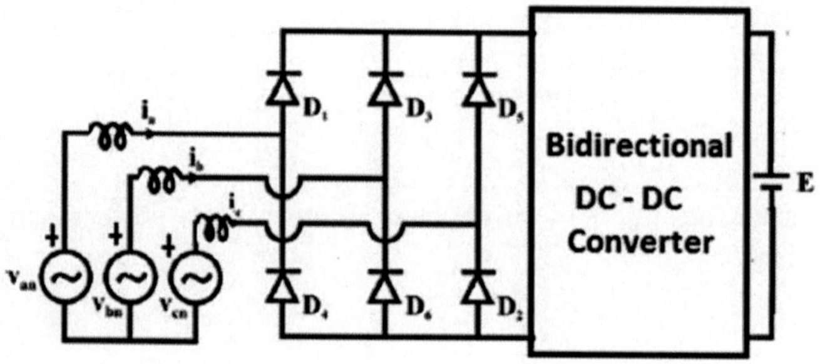

Fig. 4 System for harmonic analysis of EV charger

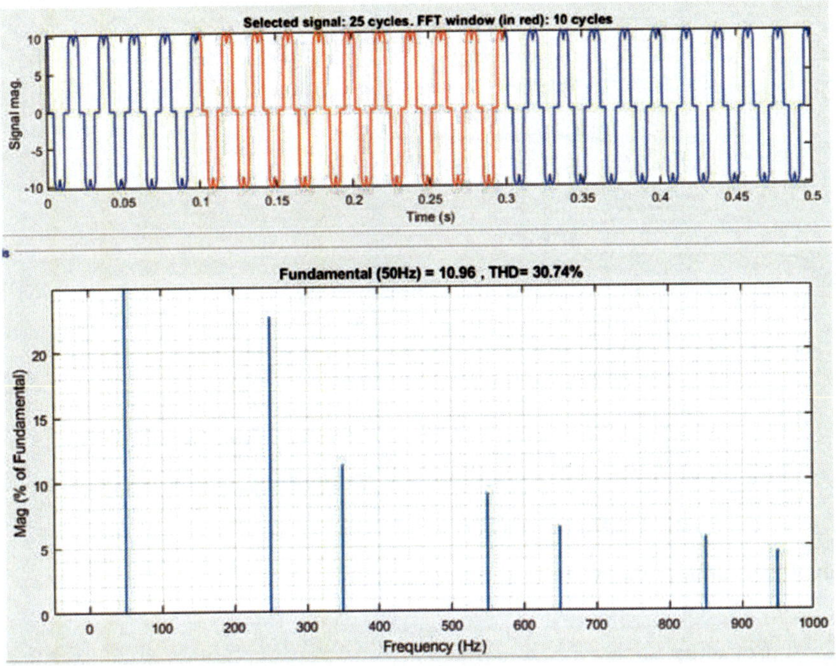

Fig. 5 Source current harmonic analysis—EV charger

current is too high and almost matching experimental results. To keep up with IEEE—519 standard limits, shunt active filter with instantaneous reactive power theory controller is implemented. The schematic diagram of the compensated system is shown in Fig. 6.

The harmonic analysis was carried out in the compensated system and corresponding results are shown in Fig. 7. Comparison of source current THDs of EV charging system without and with compensation is shown in Table 3. These results indicate the effectiveness of the shunt active filter for source current mitigation.

The harmonic analysis was carried out in the compensated system and corresponding results are shown in Figs. 8, 9, 10, 11, 12, and 13. Comparison of source current THDs of EV charging system without and with compensation are shown in Table 3. Rating of active power filter is significant while considering harmonic filtering characteristics. For three-phase active filter, it is computed as $\sqrt{3}$ times product of voltage at the point of common coupling and compensation current. System specifications are shown in Table 4.

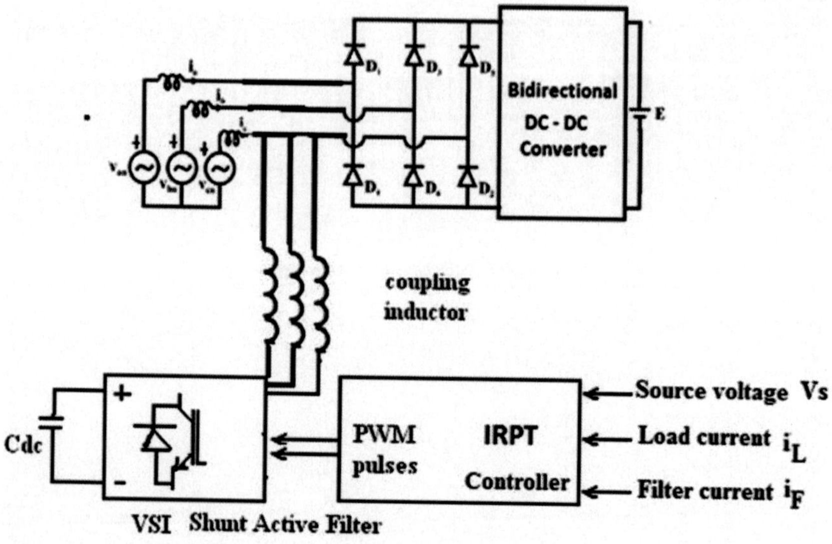

Fig. 6 Schematic diagram of the harmonic compensated system

4.2 *Rectifier Mode of Operation*

The battery discharges to the grid through three-phase inverter. Sinusoidal PWM scheme is employed to reduce source current harmonic injection, and resulting waveforms are shown in Figs. 14, 15 and 16.

5 Conclusion

The fast-growing EV technology utilizes charging infrastructure which includes nonlinear loads such as power electronic converter. As large numbers of electric vehicles are launching, current harmonic distortions in the power grid become increased and their characteristics are dynamically changing. This paper investigates whether properly designed and controlled shunt active filter along with EV charger is a viable option. The studies show the effectiveness of active filter under different states of charging conditions. These results are validated through simulation and experimental studies. The effectiveness and optimized operation of the scheme depend on the prediction of compensation currents. Inclusion of machine learning techniques, which are capable of forecasting presence of EV in parking lot, SoC of EV battery, grid power quality conditions, will certainly boost performance of the compensator.

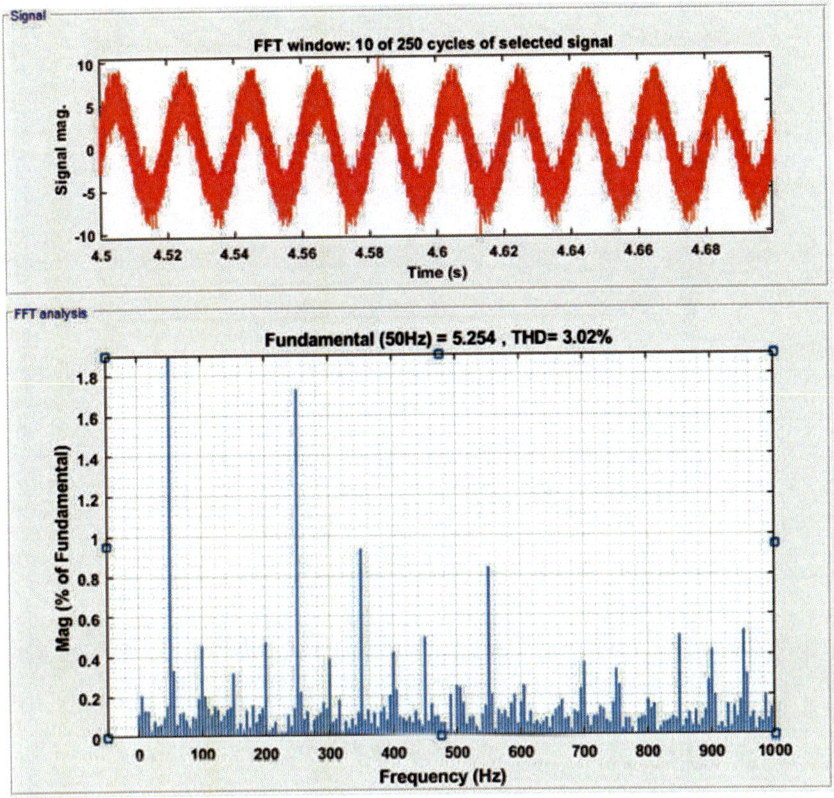

Fig. 7 Source current harmonic analysis during charging—with compensation

Table 3 Comparison of source current THDs—without and with shunt active filter

Source current THD	Source current THD after harmonic compensation
30.74%	3.02%

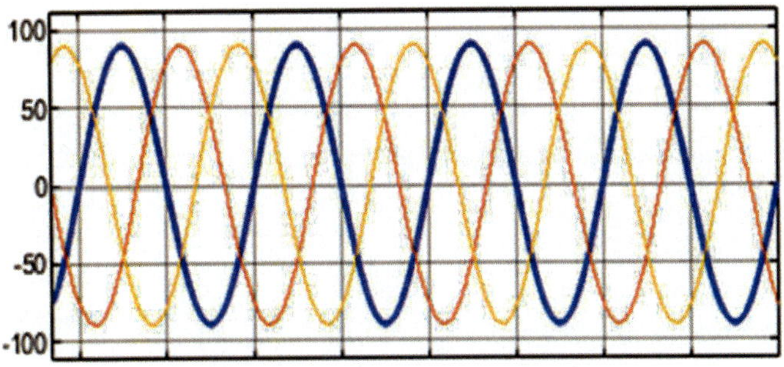

Fig. 8 Source voltage—with compensation filter

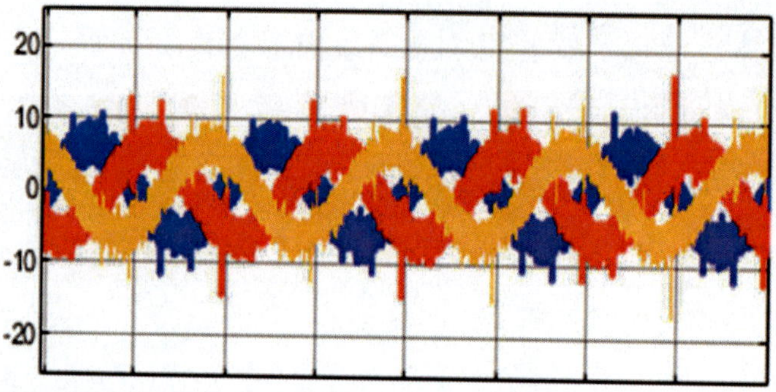

Fig. 9 Source current—with compensation filter

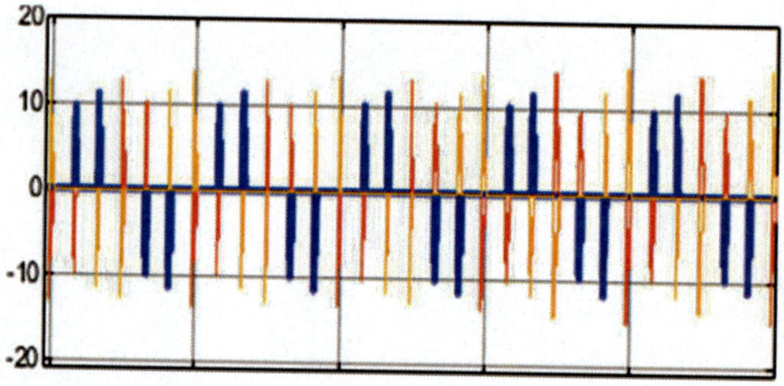

Fig. 10 Load current—with compensation filter

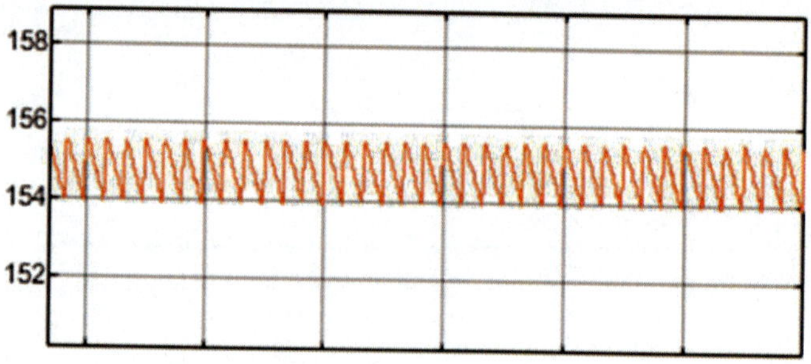

Fig. 11 DC link voltage—with compensation filter

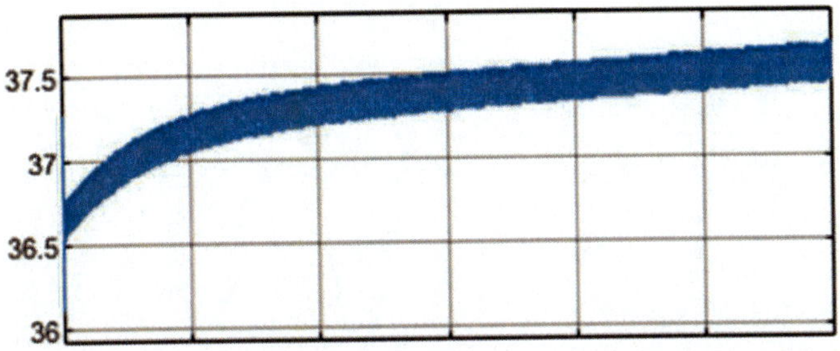

Fig. 12 Battery voltage—with compensation filter

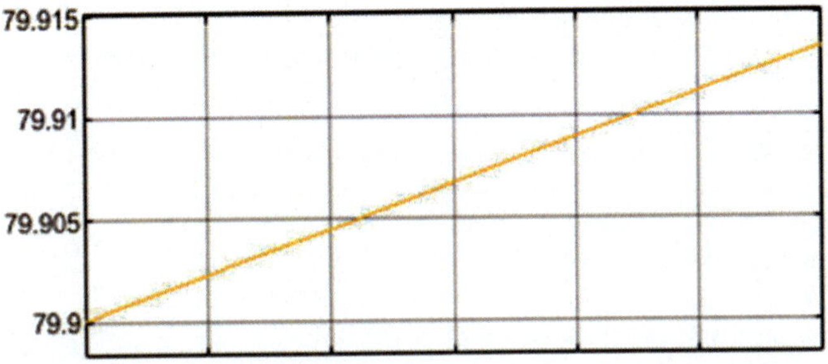

Fig. 13 State of charge—with compensation filter

Table 4 System specifications

Parameter	Value
V_{BAT_MAX}	37.8 V
I_{BAT_MAX}	14 A
C_{BUS}	1 mF
f_{sw}	20 kHz
$C_{CHOPPER}$	1 μF
$L_{CHOPPER}$	1 mH
C_{FILTER}	1 mF
L_{FILTER}	0.5 mH

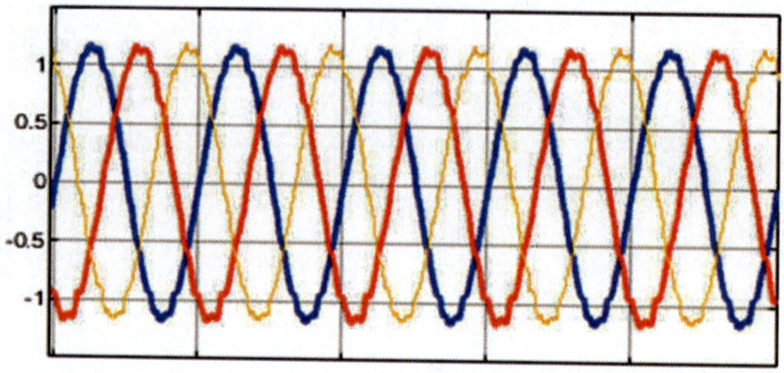

Fig. 14 Source current—inverter mode

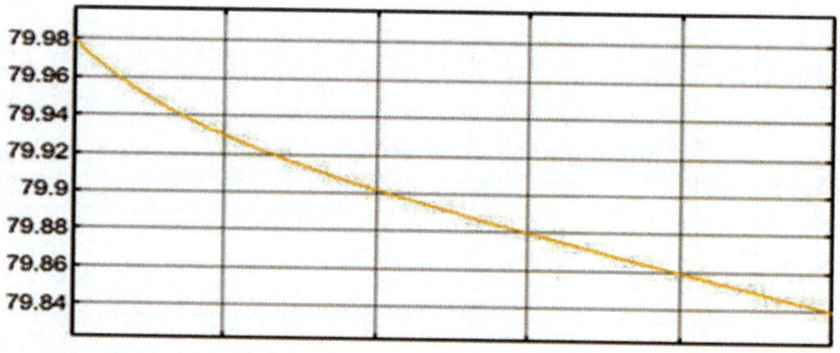

Fig. 15 State of charge—inverter mode

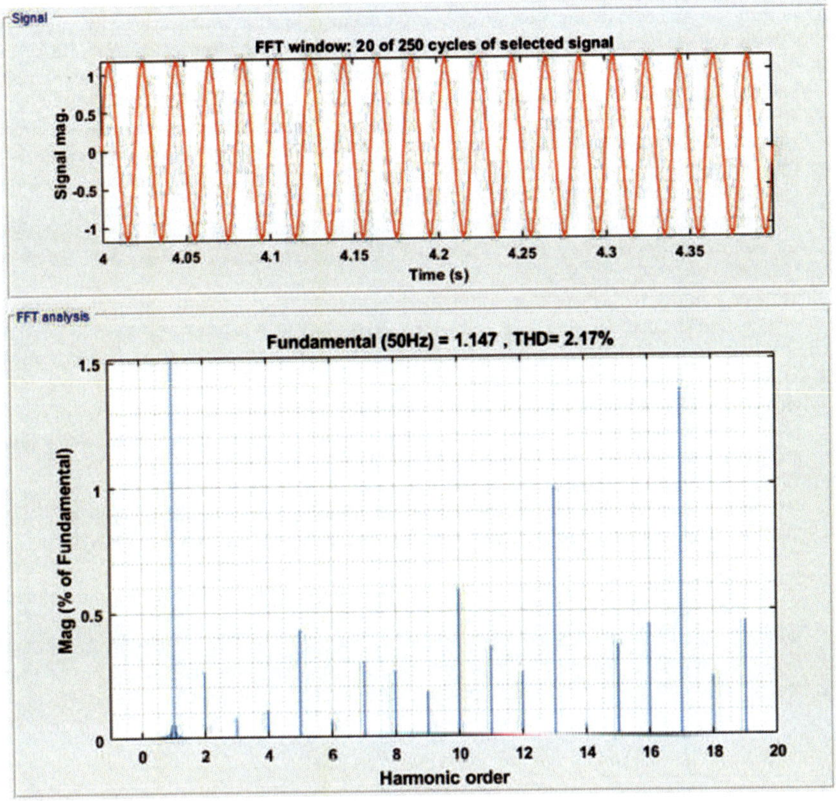

Fig. 16 Harmonic analysis of EV charger (discharging)

References

1. Karmaker AK, Roy S, Ahmed MR (2019) Analysis of the impact of electric vehicle charging station on power quality issues. In: 2019 international conference on electrical, computer and communication engineering (ECCE), Cox's Bazar, Bangladesh, 2019, pp 1–6
2. Kutt L, Saarijarvi E, Lehtonen M, Molder H, Niitsoo J (2013) A review of the harmonic and unbalance effects in electrical distribution networks due to EV charging. In: Proceedings of 2013 IEEE international conference on environment and electrical engineering
3. Pai S, Sindhu MR (2019) Intelligent range predictor for green transport. In: Proceedings of first international conference on material science and manufacturing technology 2019, 12–13 Apr 2019
4. National Electric Mobility Mission Plan 2020, Department of Heavy Industry, Ministry of Heavy Industry and Public Enterprise, Government of India
5. IEEE 519-2014 recommended practices and requirements for harmonic control in electrical power systems
6. International Electrotechnical Commission (2000) Electromagnetic Compatibility (EMC)—Part 3-2: Limits—Limits for Harmonic Current Emissions (Equipment Input Current 16A per phase); IEC Standards 1000—3-2. International Electrotechnical Commission (IEC), Geneva, Switzerland

7. Society of Automotive Engineers (SAE) (2010) Power quality requirements for plug—in electric vehicle chargers. SAE Standard J2894. SAE International, Warrendale, PA, USA
8. Neves de Melo H, Trovao JPF (2018) A controllable bidirectional battery charger for electric vehicles with vehicle—to-grid capability. IEEE Trans Vehic Technol 67(1):114–123
9. Giri M, Isha TB (2017) Comparison of non-isolated schemes for EV charging and their effect on power quality. In: Proceedings of IEEE conference on circuit, power and computing (2017)
10. Chang GW, Shee T-C (2002) A comparative study of active power filter reference compensation approaches. IEEE Trans Power Deliv 02:7803–7519
11. Zhou N, Wang J, Wang Q, Wei N, Lou X (2014) Capacity calculation of shunt active power filters for electric vehicle charging stations based on harmonic parameter estimation and analytical modeling. Energies 7:5425–5443
12. Verma AK, Singh B, Shahani DT (2012) Electric vehicle and grid interface with modified PWM rectifier and DC–DC converter with power decoupling and unity power factor. In: Proceedings of the 5th India international conference power electronics (IICPE), Delhi, India, 6–8 Dec 2012
13. Milanes MI, Montero, Martinez MAG, Gonzalez-Romera E, Romero-Cadaval E, Barrero-Gonzalez F (2016) Active and reactive power control strategies for electric vehicles in smart grids. In: Proceedings of the 10th international conference on compatibility, power electronics and power engineering, Poland, 2016, pp 114–119
14. Vittorias I, Metzger M, Kuz D, Gerlich M, BachmaierG (2014) A bidirectional battery charger for Electric Vehicles with V2G-V2H capability and reactive power control. In: Proceedings of the IEEE Transportation Electrification Conference and Expo, Dearborn, MI, USA, 2014, pp 1–6
15. Renjith R, Sindhu MR (2017) Power quality enhancement in plug-in hybrid electric vehicles. J Adv Res Dyn Control Syst 18(Special issue):1208–1232
16. Melo N, Mira F, de Almeida A, Delgado J (2011) Integration of PEV in Portuguese distribution grid: analysis of harmonic current emissions in charging points. In: Proceedings of 11th international conference on electrical power quality and utilisation (EPQU), 2011, p 16

Formal Verification of IoT Protocol: In Design-Time and Run-Time Perspective

V. Geetha Lekshmy and Jinesh M. Kannimoola

Abstract IoT systems consist of smart devices ranging from a simple surveillance camera to pacemaker and mission-critical rockets. Though these systems are designed and developed systematically, it may malfunction due to hidden bugs that are uncovered only after deployment. Model checking and run-time verification are well-established procedures in formal methods to ensure the correctness of systems. We combine both these methods together to guarantee that IoT systems deployed in critical scenarios are fail-safe. This work aims at creating an end-to-end verification framework for IoT systems. Our system consists of (1) a design-time model for MQTT protocol based on the system specification, (2) a run-time model extracted from the execution trace of MQTT implementation and (3) the essential features of systems described in the temporal logic specification. The correctness of these models are checked against the specification using model checking and run-time verification approaches.

Keywords Model checking · Runtime verification · IoT systems

1 Introduction

Internet of things (IoT) systems with controllers and actuators have pervaded human lives. Seamless interconnection of heterogeneous devices with minimum computing power gained momentum and began to be used in wide range of applications in health care, flight control systems, security systems, etc. Many of these systems are so critical that they should be prone to few errors [9]. We must ensure the correctness of such systems from the initial stage of software engineering to the deployed

V. Geetha Lekshmy (✉) · J. M. Kannimoola
Department of Computer Science and Applications,
Amrita Vishwa Vidyapeetham, Amritapuri, India
e-mail: geethalekshmy@am.amrita.edu

J. M. Kannimoola
e-mail: jinesh@am.amrita.edu

G. Ranganathan et al. (eds.), *Inventive Communication and Computational Technologies*, Lecture Notes in Networks and Systems 145,
https://doi.org/10.1007/978-981-15-7345-3_74

environment. Formal verification and specification are the building blocks to develop fail-proof critical systems. It ensures that essential features of the system specified in the unambiguous formal languages are holding in the entire life cycle of software development. In this work, we investigate how to combine two verification approaches to guarantee the correctness of IoT systems. We illustrate our approach using a widely used telemetry protocol in the IoT ecosystem.

Model checking [6] is an automated technique of verifying logical properties of a system that is modeled as a finite state machine. Model checking aids to uncover the errors at the design stage, thus preventing the errors to be propagated to the deployment stage. The system properties to be verified are expressed in a specification language and model checkers explore all possible paths of the execution to ensure that key properties are satisfied in all stages of the system. The counterexample generated from model checkers serves as a debugging method on models to understand the path which fails to satisfy the system properties and it helps to patch the design-time flaws in the system. However, model checking suffers from state-space explosion. To overcome this limitation symbolic model checking [4] is used, where states are represented symbolically and not explicitly. In [19], kripke structure (finite state machine) of modeled system is represented with a Boolean formula. SPIN [12] is an open-source explicit state model checker and NuSMV [5] is an open-source symbolic model checker. Run-time verification (RV) [10] analyzes program executions against the program specifications to uncover properties that are not satisfied by the system in run-time. Run-time verification also includes techniques for monitoring the execution of systems and detecting and correcting anomalies, preventing system failure. Unlike model checking, run-time verification focuses on a single run rather than all possible execution paths. IoT system comprises of heterogeneous devices working together and in this scenario proposal of strategies ensuring system correctness is challenging [18].

In this work, we are presenting a framework that combines both model checking and run-time verification for proving the correctness of Message Queue Telemetry Transport (MQTT)protocol. MQTT is one of the popular communication protocols in IoT systems designed for constrained environment. The MQTT protocol is modeled in the PROMELA language, and the key properties of protocol are specified in linear temporal logic (LTL). The SPIN model checker performs the symbolic model checking over the PROMELA model and validate the correctness of LTL specification. Both models and specifications are created from the OASIS standard of MQTT protocol [2]. The run-time model is generated from the execution trace of the implemented system. We are using execution logs of Mosquito MQTT broker to produce the state-space representation of MQTT system. We assure that the run-time model is also compliant with LTL properties.

The remainder of this paper is organized as follows. Section 2 discusses the related literature in this field; Section 3 detailed our methodology; Section 4 presents verification of the system; and finally, Section 5 presents conclusions and areas of further work.

2 Related Work

Hinrichs et al. [11] in their positional paper convey the idea of combining model checking and verification as a single technique. They have illustrated examples of systems where algorithm correctness is checked with model checking and data structure correctness is checked at run-time. In this system, there is a clear separation of properties that could be verified using model checking and run-time verification. The work [15] extends a model checker DIVINE to support run-time verification. DIVINE model checker has two modes, run and verify. In run mode, a single execution of the program is explored, where assertions and all behaviors are checked. In verify mode along with the program, an environment that contains a stand-in operating system is given as input to the model checker.

Ankush Desai et al. proposed a framework [7] that combines both model checking and run-time verification for developing robotic systems. The key properties to be satisfied by the robotic software systems are specified in a high-level language P and the system is model checked. Here, model checking is done with assumptions about the correctness of interfaces of the verified software with the physical world and other software components. These assumptions that are to be satisfied by the interfaces are specified in signal temporal logic (STL) and it is monitored at run-time and the result is given as feedback to the software stack of a robotic system for decision making. Samir Ouchani in [22] formally models an IoT system using process algebra. This IoT system model is verified using PRISM (a model checker). Key system properties checked are written in probabilistic computational tree logic (PCTL). Torjusen et al. [24] proposed a technique where run-time verification enablers are included in adaptive security for smart IoT (ASSET) in e-health-based IoT systems. Adaptive security systems learn and adapt dynamically, by changing the parameters/structure of the system. It predicts threats of the system and makes security decisions to overcome the threats.

A monitor for run-time verification of IoT systems using Constraint Application Protocol (CoAP) is described in [14]. This non-intrusive, passive monitor captures CoAP messages in a particular network and generates simple events based on it. These events are articulated with event calculus and then processed by the complex event processing engine to find out abnormal behavior. İnçki and Ari [13] extend this approach where the activity/interaction of nodes in the IoT system is considered as events and modeled using message sequence charts. This message sequence chart is processed and behavior of the system is expressed as formulae of event calculus. From this, complex event patterns are generated as event processing language (EPL) which is given as input to Esper engine (Java-based complex event processing engine) for verification. Leotta et al. in [17] have used UML state machines to specify the expected system behavior. This specification is converted to trace expression manually. This is translated to prolog clauses, which is given as input to a run-time monitor. The monitor observes the system execution and records the events. Then, the trace is compared with the specification for detecting abnormal behavior.

Aktas et al. [1] proposed a system for run-time verification with self-healing capability in the IoT domain. There are two external services, self-healing service and predictive maintenance service, which the system under test interacts. Events in the IoT system are captured along with provenance information about IoT devices and sent to the run-time verification system for detection of faults. On detecting any abnormality, verification system triggers services for corrective actions by the self-healing system. Run-time verification of timed properties is proposed by Pinisetty et al. [23]. Given both expected and observed timed property of the system, a monitor is generated that takes the current execution of the system and predicts whether this execution satisfies or violates the given property in the future. In [16], Caroline Lemieux et al. proposed a mining tool, Texada that mines the traces and extracts properties that satisfy the LTL specification. The system takes logs and LTL formulae and outputs instances of LTL formulae, where the atomic propositions in the LTL formulae are replaced by events in the log. The property instances generated are checked for their validity on all traces.

3 Methodology

The proposed system consists of two components, (1) model checker and (2) run-time verifier. Figure 1 illustrates the overall architecture of our proposed system.

For the illustration, we include the subset of MQTT protocol [3] such as connection establishment, publish and subscribe. In the model checking component, a model is constructed from the specification and correctness properties are specified in linear temporal logic (LTL). In the run-time verification component, an existing stable implementation of MQTT protocol is executed and the events and client-server interactions are recorded in logs. We generate the state space from the log and the same set of correctness properties specified in design-time modeling are verified.

A brief description of tools/languages used to develop this system is given below. **SPIN Model checker**: SPIN [12] is a tool for analyzing the consistency of concurrent systems. Promela (process meta-language) is a modeling language used in the SPIN that facilitates modeling of concurrent processes and its communication. The model described in the Promela consists of proctype definitions that correspond to the behavior of processes in the system. The communication between these processes is realized using channels. Each proctype definition in Promela corresponds with a transition system. The global behavior of system is represented as a single transition system, obtained by interleaving individual transition for each proctype. SPIN verifies the correctness of a concurrent system by executing the process definitions and validating the system against correctness claims specified in linear temporal logic (LTL) and assert statements. SPIN generates counterexample for the failed claims and it helps to improve the model. LTL is a model temporal logic to specify how the behavior of the system changes over time. LTL claims are converted to Buchi automaton in SPIN model checker [8].

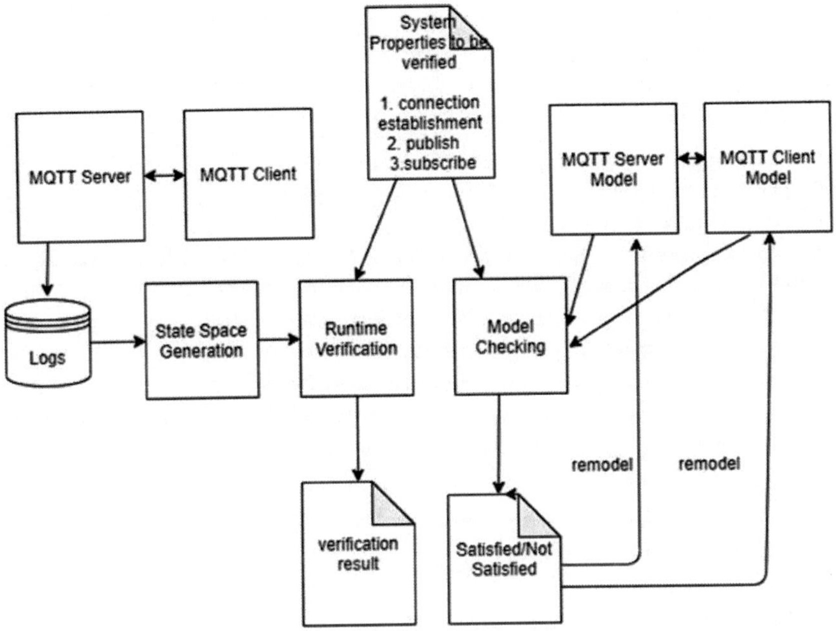

Fig. 1 System architecture

MQTT Protocol: Message Queue Telemetry Transport (MQTT) protocol [3] is a publish-subscribe protocol used for communication between nodes in an IoT system. This protocol requires less memory footprint, thus suitable for resource-constrained devices. MQTT protocol operates in a pre-defined way. Every node (client) communicates with a broker (server) by exchanging control packets. In this work, we deal with the following properties of MQTT.

1. Connection Establishment: This process requires exchange of CONNECT and CONNACK packets between broker and node. For every connect request from the node, the broker should respond with CONNACK with appropriate response code.
2. Subscribe Message: A node can subscribe to a topic by sending SUBSCRIBE packet to the broker, and the broker must respond with a SUBACK with appropriate reason code. The message delivered by the broker depends on the quality of service (QoS) in subscribe and QoS associated with the published message.
3. Publish Message: A node can publish a topic by sending PUBLISH packet containing information including data, QoS, etc. Broker should respond appropriately depending on QoS. This protocol defines three delivery semantics based on quality of service (QoS) expected out of the protocol.

    a. QoS = 0, "at most once" where the message is sent once and there is no guarantee that the receiver will receive it.

b. QoS = 1, “at least once” where the message is guaranteed to be delivered at least once. However, node may receive duplicates of the message.
c. QoS = 2, “exactly once” ensures that the message will be received by the client exactly once, without duplication.

MQTT protocol is lightweight and is suited for systems with component nodes communicating asynchronously. However, it supports only small payload and is not appropriate for systems which require large data sequences.
Mosquitto: Mosquitto [21] is an open-source MQTT broker. A broker acts as an intermediate entity between two communicating nodes. Nodes send data pertaining to a particular topic to the broker. The broker forwards the message to those nodes who have subscribed the topic. The performance of different publicly deployed MQTT brokers is evaluated in [20]. The brokers are tested for their message load handling for 1000 real-time messages subscribed over a topic, which takes least time to deliver messages from server to client at QoS level 0 and 2.

3.1 Design-Time Modeling

We have modeled node as well as broker as proctype in Promela. Communication between these processes occurs through two synchronous Promela channels and defined as:

```
chan ctob=[0] of {int,mtype};
chan btoc=[0] of {int,mtype};
```

The client process sends a connect request (CONNECT) over channel *ctob* along with its identification (clientid). This clientid is used to uniquely identify a client in the server side. The server responds with an acknowledgement (CONNACK) with appropriate response code over channel *btoc*. Similar to CONNECT request, it sends a publish (PUBLISH)/subscribe (SUBSCRIBE) request to the server and receives an acknowledgement (PUBACK/SUBACK) back. The following code snippet gives client proctype implementation.

```
/*Type of messages defined*/
mtype{CONNECT,CONNACK,ERROR,PUBLISH,SUBSCRIBE,
DISCONNECT,SUBACK,PUBACK};
/*Defining channels for communication
from client to server and vice-versa*/
chan ctob=[0] of {int,mtype};
chan btoc=[0] of {int,mtype};
proctype client(int client_id){
 int c_id;
 mtype m1,m2;
```

```
bool con=0; bool cack=0;
bool sub=0; bool pub=0;
bool suback=0; bool puback=0;
do
 ::con==0 ->
 atomic{m1=CONNECT;ctob!client_id,m1; con=1;}
 ::con==1 ->btoc??eval(client_id),m2;
 if
  :: m2==CONNACK ->cack=1; m2=ERROR;
  :: m2==SUBACK -> printf("suback received");suback=1;
  :: m2==PUBACK ->printf("pub ack received");puback=1;
  :: else skip;
 fi;
 ::(con==1)-> ctob!client_id,PUBLISH;pub=1;
 ::(con==1)->  ctob!client_id,SUBSCRIBE;sub=1;
od;
}
```

Following is the model of MQTT broker that responds to the CONNECT, SUBSCRIBE and PUBLISH. Server identifies each client using a clientid and keeps track of its CONNECT request. All clients share a common channel with the server for communication, and a particular client reads a message addressed with its own clientid.

```
connected con_clients[5];
proctype broker(){
 mtype m1,m2;
 int client_id;
 do
  :: ctob?client_id,m1  ->
  if
   :: m1==CONNECT ->
      con_clients[client_id].con_request=1 ;
   m2=CONNACK; btoc!client_id,m2;
   con_clients[client_id].con_ack=1;
  ::else
    printf("process subscribe/publish");
   if
    :: m1==SUBSCRIBE -> m2=SUBACK;btoc!client_id,m2;
    :: m1==PUBLISH -> m2=PUBACK;btoc!client_id,m2;
   fi;
  fi;
 od;
}
```

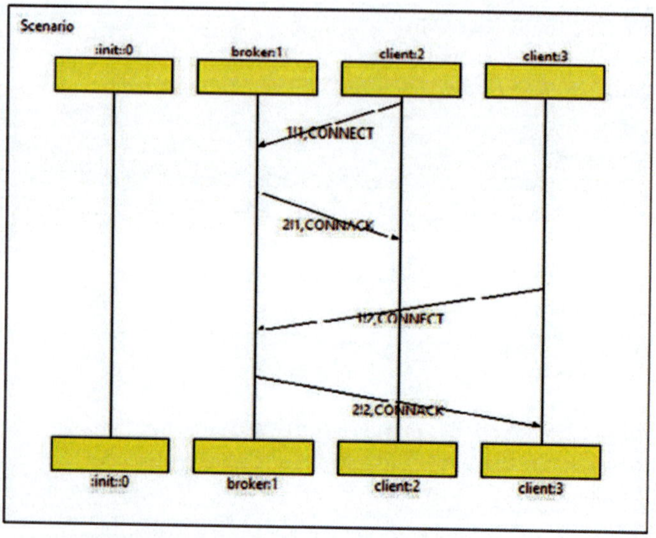

Fig. 2 Sequence of MQTT client-broker interaction in SPIN

This model can accept up to five clients. The following *init* process creates two clients and a broker. The broker waits for the request from the clients and send appropriate response to client which is read by corresponding client.

```
init{
  run broker();
  run client(0);
  run client(1);
}
```

Figure 2 illustrates the working of MQTT client-broker model developed with two clients sending connect requests and receiving appropriate responses in SPIN.

3.2 *Run-Time Modeling*

Run-time data is extracted from the execution trace of the MQTT implementation. As we discussed, we are collecting log of Mosquitto MQTT broker. It includes all the activities like connection establishment, message subscription, etc. Figure 3 shows the example of Mosquitto log. This log is created by running an instance of Mosquitto MQTT broker and instances of MQTT clients. The clients include instance of MQTT spy (an open-source utility to monitor MQTT activity) and MQTT client implemented in Java using Paho Java client (an MQTT client library in Java).

```
1580272791: Sending CONNACK to Geethalekshmi080225260 (0, 0)
1580272793: New connection from 127.0.0.1 on port 1883.
1580272793: New client connected from 127.0.0.1 as Geethalekshmi095217161 (p2, c1, k60).
1580272793: No will message specified.
1580272793: Sending CONNACK to Geethalekshmi095217161 (0, 0)
1580272821:   PUBLISH from Geethalekshmi080225260 (d0, q2, r1, m1, 'data', ... (3 bytes))
1580272821: Sending PUBREC to Geethalekshmi080225260 (m1, rc0)
1580272821: Received PUBREL from Geethalekshmi080225260 (Mid: 1)
1580272822: Sending PUBCOMP to Geethalekshmi080225260 (m1)
1580272839: Received SUBSCRIBE from Geethalekshmi095217161
1580272839:    data (QoS 2)
1580272839: Geethalekshmi095217161 2 data
1580272839: Sending SUBACK to Geethalekshmi095217161
1580272839: Sending PUBLISH to Geethalekshmi095217161 (d0, q2, r1, m1, 'data', ... (3 bytes))
1580272839: Received PUBREC from Geethalekshmi095217161 (Mid: 1)
1580272839: Sending PUBREL to Geethalekshmi095217161 (m1)
1580272839: Received PUBCOMP from Geethalekshmi095217161 (Mid: 1, RC:0)
```

Fig. 3 Snap shot of log file of MQTT server

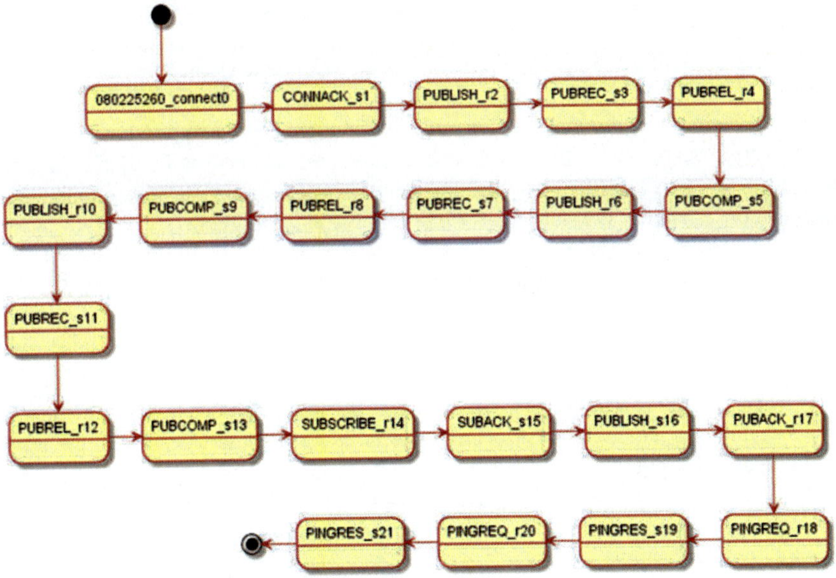

Fig. 4 State diagram of MQTT broker

We have used a Python parser to obtain the run-time model, it is essentially state space over the linear time. Figure 4 is a example of such state space generated from log.

4 Verification

We have verified the following properties in our design-time and run-time model.

1. Connection establishment—A broker that receives a CONNECT control packet should return a CONNACK with appropriate response code. A node should sent only one CONNECT request to the broker after network connection is established.
2. a. A publish message with QoS = 1 should be responded with PUBACK message
   b. A publish message with QoS = 2 should follow a sequence of control packets as given in Fig. 5.
3. A subscribe message received by the broker with QoS ≥ 1 should be responded with SUBACK message with appropriate response code.

We have expressed this properties as the following LTL statement.

$$\Box(\text{con\_request\_client1} \Rightarrow \Diamond\text{con\_response\_client1})$$

$$\Box(\text{con\_request\_client2} \Rightarrow \Diamond\text{con\_response\_client2})$$

This specification ensures that a connection request received by the broker at any time should be followed by a response at any time in the future. For example, con_request_client1 is the variable that keeps track of request from client1 and con_response_client1 is a variable for tracking response to client1.

$$\Box(\text{pub\_client1} \ \& \ \text{pub\_client1\_qos}) \Rightarrow \Diamond\text{puback\_client1})$$

$$\Box(\text{pub\_client2} \ \& \ \text{pub\_client2\_qos}) \Rightarrow \Diamond\text{puback\_client2})$$

In this specification variable, pub_client1 keeps track of publish request from client1 and variable pub_client1_qos stores the quality of service associated with the publish request. If a publish request arrives and its QoS value is 1, then variable puback_client1 should be set to true, where puback_client1 keeps track of PUBACK message that is sent to client1. The similarly we can define the second statement for client2

$$\Box((\text{publish\_r} \Rightarrow \Diamond\text{pubrec\_s}) \Rightarrow \text{pubrel\_r} \Rightarrow \Diamond\text{pubcomp\_s})$$

This specification ensures if a PUBLISH message is received by the broker from a client, it will respond with a PUBREC message, which will in-turn be followed by receiving of PUBREL and sending of PUBCOMP messages.

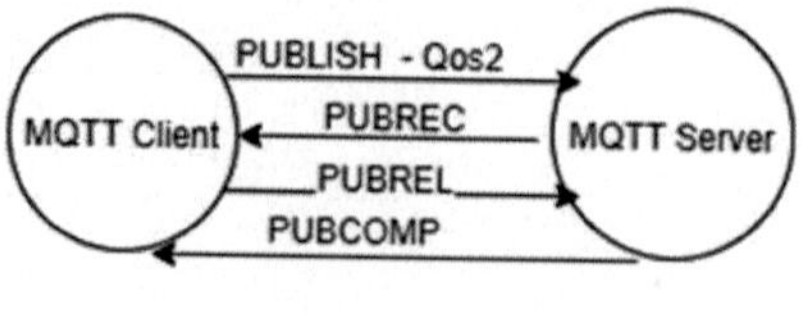

Fig. 5 Publish message with QoS 2

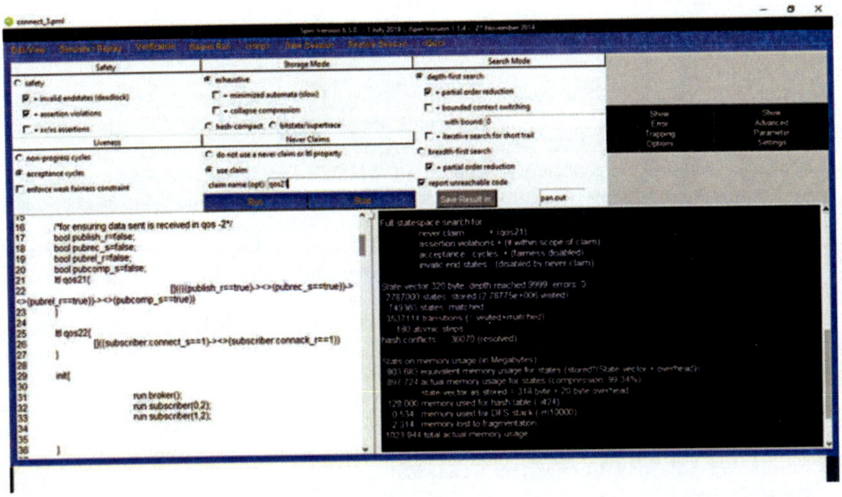

Fig. 6 Verification in iSpin

The verification process ensure that both design-time model and run-time model are satisfied the above-mentioned LTL properties.

Design-time Verification: We verify satisfiability of LTL statement in Promela model. SPIN model checker performs the verification process over the model using symbolic model checking. If the property is not satisfied by model, SPIN generates the counterexample and it gives execution path leads to the failure. Figure 6 gives an example of SPIN verification.

Run-time verification: It checks whether all the essential features designed are actually implemented and it is behaving as expected when the system is deployed. LTL properties are checked for compliance with the state space of a running MQTT implementation. The broker-client implementation we selected satisfies all the specified LTL properties.

5 Conclusion and Future Scope

This work is a step toward the integration of model checking and run-time verification in IoT systems. We have developed design-time and run-time model for MQTT protocol and verify the correctness. We used LTL language to specify the correctness properties and it helps to eliminate the ambiguity in the specification. The challenge involved in this approach is to make the process work seamlessly from model checking to run-time verification. This work can be extended to generate a monitor automatically from LTL specification so that run-time verification will be fully automated.

References

1. Aktas MS, Astekin M (2019) Provenance aware run-time verification of things for self-healing internet of things applications. Concurr Comput: Pract Exp 31(3):e4263
2. Andrew Banks KB, Briggs Ed, Gupta R (2019) MQTT Version 5.0
3. Banks A, Briggs E, Borgendale K, Gupta R (2019) MQTT Version 5.0. OASIS Standard
4. Burch JR, Clarke EM, McMillan KL, Dill DL, Hwang L-J (1992) Symbolic model checking: 10^{20} states and beyond. Inf Comput 98(2):142–170
5. Cimatti A, Clarke E, Giunchiglia E, Giunchiglia F, Pistore M, Roveri M, Sebastiani R, Tacchella A (2002) Nusmv 2: an opensource tool for symbolic model checking. In: International conference on computer aided verification. Springer, pp 359–364
6. Clarke EM (1997) Model checking. In: International conference on foundations of software technology and theoretical computer science. Springer, pp 54–56
7. Desai A, Dreossi T, Seshia SA (2017) Combining model checking and runtime verification for safe robotics. In: 17th international conference on runtime verification (RV), pp 172–189
8. Gastin P, Oddoux D (2001) Fast LTL to Büchi automata translation. In: International conference on computer aided verification. Springer, pp 53–65
9. Hassan WH et al (2019) Current research on internet of things (IoT) security: a survey. Comput Netw 148:283–294
10. Havelund K, Roşu G (2018) Runtime verification-17 years later. In: International conference on runtime verification. Springer, pp 3–17
11. Hinrichs TL, Sistla AP, Zuck LD (2014) Model check what you can, runtime verify the rest. In: HOWARD-60, vol 42, pp 234–244
12. Holzmann GJ (1997) The model checker SPIN. IEEE Trans Softw Eng 23(5):279–295
13. İnçki K, Ari I (2018) A novel runtime verification solution for IoT systems. IEEE Access 6:13501–13512
14. İnçki K, Arı İ, Sözer H (2017) Runtime verification of IoT systems using complex event processing. In: 2017 IEEE 14th international conference on networking, sensing and control (ICNSC). IEEE, pp 625–630
15. Kejstová K, Ročkai P, Barnat J (2017) From model checking to run-time verification and back. In: International conference on runtime verification. Springer, pp 225–240
16. Lemieux C, Park D, Beschastnikh I (2015) General LTL specification mining. In: 2015 30th IEEE/ACM international conference on automated software engineering (ASE). IEEE, pp 81–92
17. Leotta M, Ancona D, Franceschini L, Olianas D, Ribaudo M, Ricca F (2018) Towards a runtime verification approach for internet of things systems. In: International conference on web engineering. Springer, pp 83–96
18. Leotta M, Clerissi D, Franceschini L, Olianas D, Ancona D, Ricca F, Ribaudo M (2019) Comparing testing and runtime verification of IoT systems: a preliminary evaluation based on a case study. In: Proceedings of the 14th international conference on evaluation of novel approaches to software engineering. SCITEPRESS—Science and Technology Publications, Lda, pp 434–441
19. McMillan KL (1993) Symbolic model checking. Springer, Boston, MA, pp 25–60
20. Mishra B (2018) Performance evaluation of MQTT broker servers. In: International conference on computational science and its applications. Springer, pp 599–609
21. Mosquitto E (2018) An open source MQTT broker. Eclipse Mosquitto ™[cit. 2018-04-23]. Dostupné z: Mosquitto.org
22. Ouchani S (2018) Ensuring the functional correctness of IoT through formal modeling and verification. In: International conference on model and data engineering. Springer, pp 401–417
23. Pinisetty S, Jéron T, Tripakis S, Falcone Y, Marchand H, Preoteasa V (2017) Predictive runtime verification of timed properties. J Syst Softw 132:353–365
24. Torjusen AB, Abie H, Paintsil E, Trcek D, Skomedal Å (2014) Towards run-time verification of adaptive security for IoT in eHealth. In: Proceedings of the 2014 European conference on software architecture workshops. ACM, p 4

Automatic Network Scanning System for Monitoring 4G and 5G Network Elements

N. Lakshitha Karthik, Shreya S. Gowda, S. B. RudraSwamy, and B. M. Sagar

Abstract Network monitoring is an empiric approach in which each network feature is tracked, and its performance is continuously evaluated to maintain and optimize its efficiency for the detection of faults in 4G and 5G infrastructure, and their output is simultaneously analyzed to preserve and maximize their availability. Network elements monitoring should be retrospective, tracking connection problems, and bottlenecks proactively help to recognize concerns at the initial stage, to prevent call drops. The automatic system would reduce the work of admin and network element owners, which might assist in monitoring various network elements. This paper proposes an automatic network scanning system for monitoring the network elements and interfaces in 4G and 5G architecture, which provides the network elements statistics, various features for the network elements like monitoring interfaces, tracking network elements (HSS, S-GW, AMF, SMF). The framework offers alert notifications of the network elements to their respective network element proprietors through various mails.

N. Lakshitha Karthik (✉)
Masters in Information Technology, Information Science and Engineering Department, Rashtreeya Vidyalaya College of Engineering (RVCE), Bengaluru, India
e-mail: lakshithakarthik@gmail.com

S. S. Gowda (✉)
Masters in Networking and Internet Engineering, Electronics and Communication Engineering Department, JSS Science and Technology University (SJCE), Mysuru, India
e-mail: shreyasgowdapr1996@gmail.com

S. B. RudraSwamy
Electronics and Communication Engineering Department, JSS Science and Technology University (SJCE), Mysuru, India
e-mail: rudra.Swamy@sjce.ac.in

B. M. Sagar
Information Science and Engineering Department, Rashtreeya Vidyalaya College of Engineering (RVCE), Bengaluru, India
e-mail: sagarbm@rvce.edu.in

G. Ranganathan et al. (eds.), *Inventive Communication and Computational Technologies*, Lecture Notes in Networks and Systems 145,
https://doi.org/10.1007/978-981-15-7345-3_75

Keywords Network elements monitoring · Interface monitoring · Network elements statistics

1 Introduction

Telecommunication is a set of nodes associating with different kinds of communication devices. The control of network communication involves the contact network associated with the operating system. As distributed networks progressed, network management had turned into an annoyance for admin workers. Similarly, it was hard to store all network devices in a single list. A characteristic structure is a network coordinating approach. Network interruption occurs if there is no communication between two devices or a specific interface. Manually controlling the network elements takes too much time and can also produce human errors that could impact the system, it will save the time of multiple network element owners, and human errors can be minimal [1]. It helps to increase productivity and yields to minimize human interference. The automatic scanning system supports functions ranging from checking pingable IP address, error rate, and IP address failure [2]. The program automatically tracks the network elements and their functionality and warns the owners of the network elements if there are any network-related issues. Automatic network element monitoring tool performs monitoring and evaluating the performance of any network elements through a simple and intuitive Web interface, an administrator can review all collected network element data such as pingable network elements, updates, CPU load, disk use, and memory use. The proprietors of network elements [3] will obtain the error mails in the form of alerts that help in identify the problem of their respective network elements and act accordingly. It saves the time of the network element owners and the admin. The proposed system monitors network elements from four locations like Bangalore, Tampere, Hangzhou, and Irving. The software allows control of the elements of the network and the applications running on it. This proposed model is carried out for corporate networks, and virtual private network access is necessary for connecting remote server location.

2 Recent Works

Network elements remote monitoring subsystem addresses challenges in developing an application to control remote servers of a corporate network. The software allows control of the server's physical accessibility and operation, technical requirements (disk space, RAM), and real-time hosting applications. The subsystem immediately reports errors to the device operator/manager, both through a user-friendly GUI and by e-mail [2] as shown in Fig. 1. To provide high reliability and usability in telecommunication networks, dynamic monitoring is implemented. One of the most critical

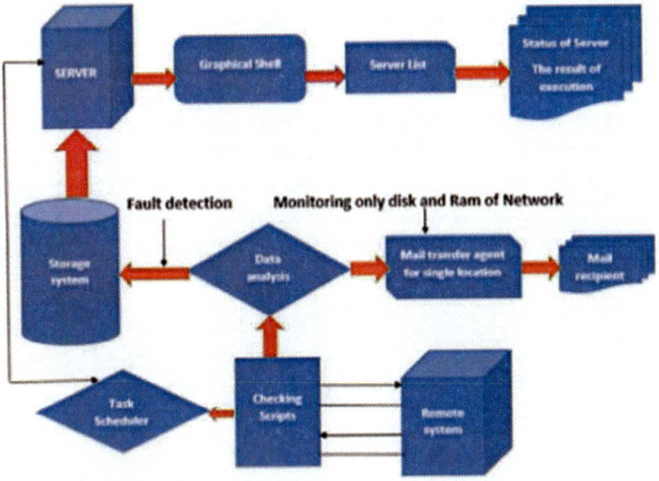

Fig. 1 Structure of remote server monitoring system

problems with advanced surveillance devices is the definition of the network baselines. These baselines provide thresholds to decide if the observed values represent the normal behavior of the network, with a given background, for example, time. The size and complexity of current (and future) network sites make manually determining and setting baselines and adjusting the thresholds to changes in network conditions challenging for each device operator and metric. This results in using default baselines and/or setting baselines just once and never changing the baselines [3].

Increased mobile communication growth has resulted in increased demand for users and traffic through the implementation of Open-Air Interface Control Software. Consumers of existing 3G networks had trouble accessing them. Therefore, the 4G network is established. Many experimental applications and development tools have migrated to 4G networks. OAI is a hardware and software platform for the 4G network that can be used to incorporate various applications and technologies. In this paper, a structure was developed to address the inefficiencies created by incorporating SDN and OAI technologies into 4G networks. This enables setup and control of eNodeBs and UEs [4]. The alternative value network design for industrial 5G is to explore the distribution of roles and value creation between 5G market players by offering alternative useful network configurations. This also provides a comparison table of the value networks and the simulation of them in the scenario matrix to compare and illustrate the future potential of various value networks.

3 Methodology

The overall device architecture is demonstrated in Fig. 2. The software will handle network elements from four sites including Bangalore, Irving, Tampere, and Hangzhou. It manages the network element information of those four sites and transmits updates by using the simple mail transfer protocol (SMTP) from each location to the owners of the network elements. All network item location information is controlled by the device proposed. The framework employs the advanced Spring MVC Java technology.

3.1 Monitoring Network Elements and Interfaces of 4G(LTE) Architecture

The network components of 4G LTE are cell phones, radio network [5], mobility administration (ME), eNodeB, gateway service (S-GW), packet data network gateway (P-GW), home subscriber system (HSS), and regulation mechanism and charging rules (PCRF). User hardware consists of subscriber identity module (SIM) and mobile phones and other devices following LTE protocol.

Radio adapter is a wireless network between the LTE mobile terminals and eNodeB. MME applications include paging, authentication, functioning like visitor location register (VLR) of the 3G network. S-GW handles routing and forwarding of user data packets. P-GW responsible for IP connectivity and linking 4G to 3G or 4G with Wi-Fi and any other different network. HSS performs user identification and

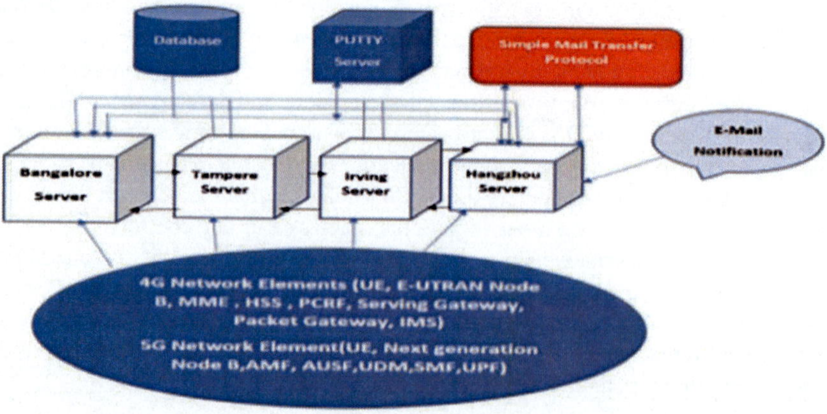

Fig. 2 System architecture of the automatic network scanning for monitoring network elements system

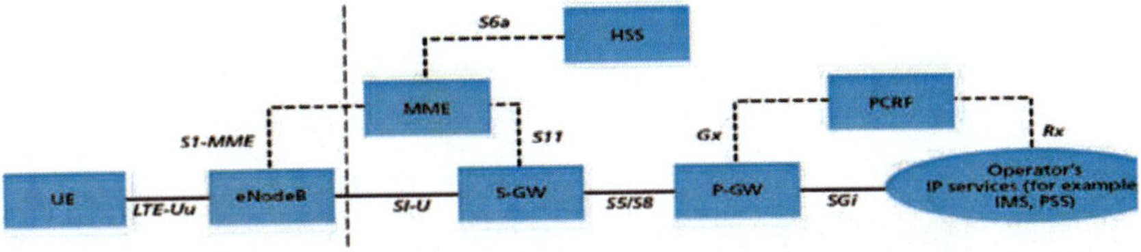

Fig. 3 4G LTE network architecture

addressing, user database and offers user subscription service. The entire 4G (LTE) architecture is depicted in Fig. 3.

Monitoring S1–MME and S6 interface S1-MME is the interface between eNodeB and MME chargeable for evolved packet system (EPS) handles handover signaling procedure, the paging procedure, and transfer signaling messages. It performs transition and device accessibility procedures between eNodeB and MME as users switch between bases. To prevent any network outages or IP address errors, controlling these interfaces is critical. The goal of monitoring the network elements and the interface module is to inspect all the network elements and alert the network element proprietors if network elements have any errors. The s6a interface is between MME and HSS in the LTE network. When an IP address is disabled, it should be tracked and activated mail to network elements owner. Monitoring the unknown and disabled IP address via the putty server using s6ad and s1-mme commands is shown in Fig. 4.

Monitoring Home Subscriber System (HSS) and Serving Gateway (S-GW) in 4G The home subscriber network is the database for mobile communications that stores all information about subscribers in global networks. It also requires the name of all memberships in global Internet subscribers (IMSI). IMSI is used to identify the identification module of the user and to include a key to the registry of a place of residence. It is linked to membership details with subscribers. HSS uses real-time transfer protocol (RTP) to access audio and video files on the Internet. The monitoring of the RTP status is one of the important functions because it handles voice over IP).

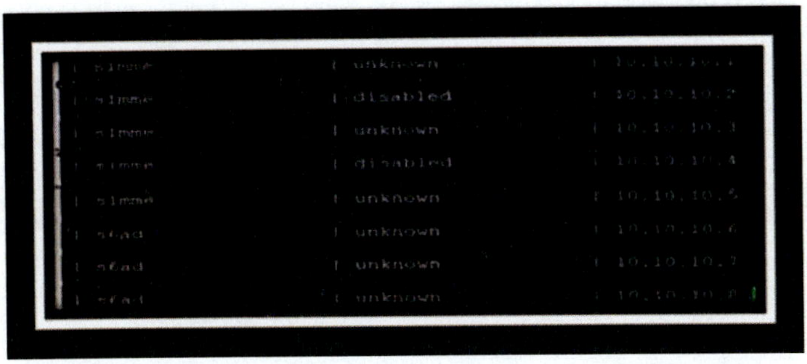

Fig. 4 Monitoring s1-mme, s6a interface in the putty server

Gateway executes functions within a packet's heart. S-GW is operated using an automated JavaScript and triggers e-mail to the owner of the network components when the throughput value of gateway exceeded above the threshold value.

Monitoring UE (User Equipment) and E-UTRAN, GNodeB in 4G, and 5G User equipment included/s mobile finishing, universal integrated circuit card (UICC), and terminal equipment. UE retains all communication functions and ends the flow of data. Next-generation node B (gNodeB) is a base station in 5G that transmits and receives radio resource signals between UE and access mobility function (AMF). Similarly, SIM card saves user and holds user identification, home network, and international network identity and security key data. eNodeB is a baseline that covers femtocells, eNB is linked to the S1 interface and connected to the close base station with the evolved packet core network. eNB and gNb inspect UE in one or more cellular networks. Via automation, the total number of UEs, number of eNBs, number of gNBs are controlled. The UE, eNB, and gNB statistics monitoring takes IP, location, username, password as the input from the user interface and monitored by obtaining the data and processing using SSH commands to acquire the counts of UE, eNBs, and gNBs as shown in Fig. 5. It helps in obtaining the network element name and the corresponding count of UE, eNB, and gNB in day-wise, month-wise, and year-wise data which are categorized according to dates, and it should display the 12 h back data automatically in GUI. These details are picturized using charts.

UE, E-UTRAN Node B and next generation Node B statistics

VNF/VM: 2	**Gateway :1**
Default bearers :2	**Dedicated bearers: 0**
Ipv4 bearers:3	**Ipv6 bearers :0**
Active bearers :0	**Idle bearers :3**
Number of MMEs :3	**Number of PGWs: 2**
Rf Peers: 0	**Number of E-UTRAN Node B: 17**
Paging in progress: 0	**Allocated Paging Buffers: 0**
Total Number of UEs :18	**Total Number of Idle UEs: 13**
Ga DRT Request in queue: 0	**Number of next generation Node B: 16**

Fig. 5 Statistics count of UE and E-UTRAN and GNodeB

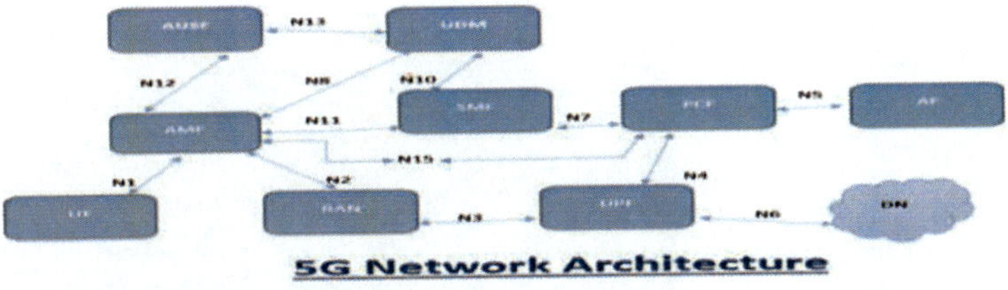

Fig. 6 5G network architecture

3.2 *Monitoring Network Elements and Interfaces of 5G Architecture*

The 5G core network is designed for services that are accessed using a standard programming interface. At the surface, the 5 G architecture is very different from the 4G. The 5G core has evolved from the 4 G packet core into two methods: first, controller and user plane separation of the 4G, and second, reorganization of 4G control and user panel functions into services. Using virtualization in 5G core networks, they can be limited through size. Various core network elements can be run as virtual machines to communicate. The transition of the 5G core network control plane to a cloud provider lowers implementation costs. As shown in Fig. 6, the 5G core is a network of interconnected services. Authentication server feature, unified data management (UDM) [6], access and mobility management (AMF), session management feature, user plane function (UPF), data network (DN) are main physical components.

Monitoring the N2 interface The N2 interface allows communication between the radio access network and the 5G core of the control plane and manages device hardware, packets, and resource management activities. Network radio uses device N2. N2 is responsible for radio network configurations system management, bugs, usability management status. N2 framework manages downlink/uplink radio network configurations for user hardware background management. 5G–N2 interfaces can model gNodeB and access mobility function. N2 is a benchmark between the gNodeB and access mobility feature [7] to facilitate services relevant to user hardware and non-user hardware. It covers activities like hardware upgrades, handling the radio [8], and paging capabilities. The purpose of monitoring the network elements and the interface module is to inspect all the network elements and alert the proprietors of the network elements if there are any errors in the network elements. Disabled states are monitored using the command for N2 [1] interface as shown in Fig. 7.

Monitoring Access and Mobility Management Function (AMF) in 5G The AMF handles the bulk of tasks performed by the mobility functions in a 4 G network. AMF ends the control plane interface (N2) of the radio network that manages network storage signaling, encryption, and integrity protection and authenticates logging and link handling user's applications. Access and mobility management function gathers

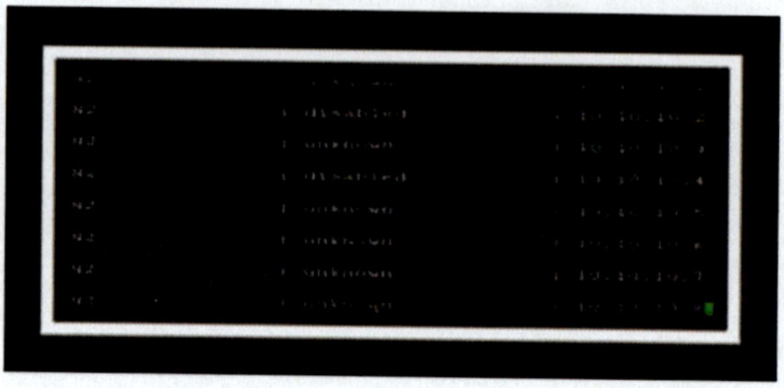

Fig. 7 Monitoring N2 interface in the putty server

all user hardware link and session-related information but only handles link and mobility management tasks. The session management feature passes all messages via the reference interface N11. AMF manages mobility features and handles data sessions between gNodeB within the next generation of radio networks. Code snippet of interfaces from AMF added as an array [9] and monitoring for disabled or unknown states and alert mail sent to respective owners as shown in Fig. 8.

Monitoring Unified Data Management (UDM) and Session Management Function (SMF) in 5G The UDM offers other SBA features, such as the AMF, SMF, and NEF programs. Usually, the UDM is known as a stateful message store, storing information in local memory. However, the UDM can also be stateless, externally storing information inside a unified data repository (UDR). The UDM is similar to the home subscriber server (HSS), which offers passwords for authentication by using the AMF and SMF to collect data and background from subscribers. AUSF and UDM [9] are used as pods that handle traffic. Pods are installed over docker using kubernetes. A pod has one or more shared storage networks, a group of more containers, and specifications on how to run over docker. The container is referred to as a group of the combined microservices.

SMF carries out the session management functions handled by the 4G mobility feature, serving, and packet gateway. Session function assigns IP addresses to user devices, storage signals connected to the network, sends service quality and policy information to a radio network and manages network transmissions. The SMF also communicates with policy control over the N7 interface and the information of its user [10] profile stored within the unified information management (UDM) feature (N10). The SMF takes on the role previously held by the HSS. N8 interface used

```
public void main() throws IOException, JSchException
{
    AMFMonitor amfmonitor = new AMFMonitor();
        List<String> intfTypes = Arrays.asList("N1","N2","N8","N11","N12","N14","N15","N17","N20","N22");
        List<String> NEs = new ArrayList<String>();
        PrintWriter out = null;
        Session session = null;
        Channel channel = null;
        StringBuffer buffer = new StringBuffer();
            if (buffer != null && (buffer.toString().contains("disabled") || buffer.toString().contains("unknown"))) {
                System.out.println(buffer);
                out.println(buffer);
                out.flush();
            }

static void sendMailToUser(String key, String mailBody, List<AttachmentModel> attachmentList, String subject, String from)
        throws SQLException, IOException, ClassNotFoundException, NamingException, MessagingException {
            MailManager mailManager = new MailManager();
            MailModel mailModel = new MailModel();
            mailModel.setTo(key);
            mailModel.setFrom(from);
            mailModel.setSubject(subject);
            mailModel.setMailMessage("Hi,<br><br>Below interfaces are in disabled state, please take action.");
            mailModel.setContent(mailBody);
            mailModel.setAttachmentList(attachmentList);
            mailManager.sendMail(mailModel);
    }
```

Fig. 8 Code snippet of interfaces in AMF

Fig. 9 Monitoring N8, N12 interface in putty server

between AMF, SMF, and http ping status represent success or fail [10] for each IP, and in case of failure, alert e-mails are triggered to respective owners. N8 and N12 interface monitoring as shown in the PuTTY server is shown in Fig. 9.

4 Results and Discussion

In this paper, the framework proposed is categorized into multiple modules according to the features of the network elements. Several features of proposed modules are monitoring the interfaces of 4G and 5G architecture like S1-MME, S6, N2, N8, N12, and other network elements like HSS, eNodeB, gNodeB, UE and Gateways (S-GW), AMF, AUSF, UDM, SMF.

The graphical user interface component in which the network owner can enter detail for the network feature is shown in Fig. 10. It has different fields such as the name of the network element, the network element IP address, types of network

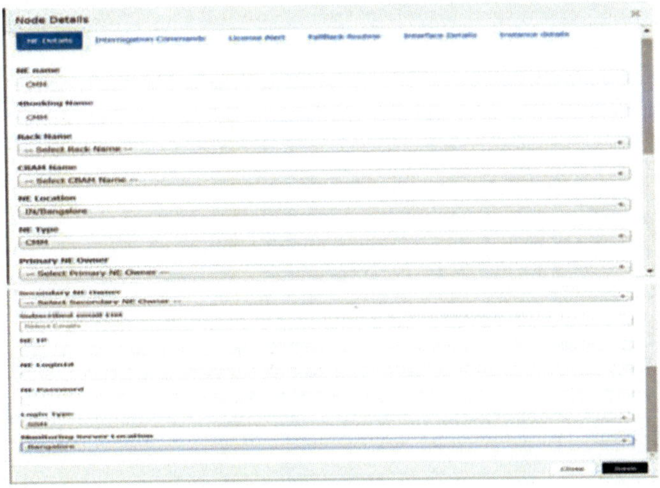

Fig. 10 Adding network elements details to the automated scanning system

elements, login id, password, location, owners of network elements, notifications such as license warnings, fallback notifications, and subscribed e-mail-id to get mail if any of the elements fail.

Display of 4G and 5G network's interface information which will be monitored if any of the IP addresses of these interfaces are down is shown in Fig. 11.

Warning emails will send to proprietors of network elements (NE) is depicted in Fig. 12. Reports of failed IP address and connectivity issues with other elements will appear when network outage takes place. This mail is triggered using JavaScript and storing and details of disabled and unknown IP addresses in the MySQL database.

The real-time transfer protocol (RTP) test in HSS is shown in Fig. 13, if there is an error in HSS according to a different location. When the error is detected, it will check the error status and cause mail to the owner of products. The program can collect and update the user equipment and the eNodeB data every half an hour in the central database. Data are accessible through the graphical user interface to the users of the application. Stored data include information status such as date and

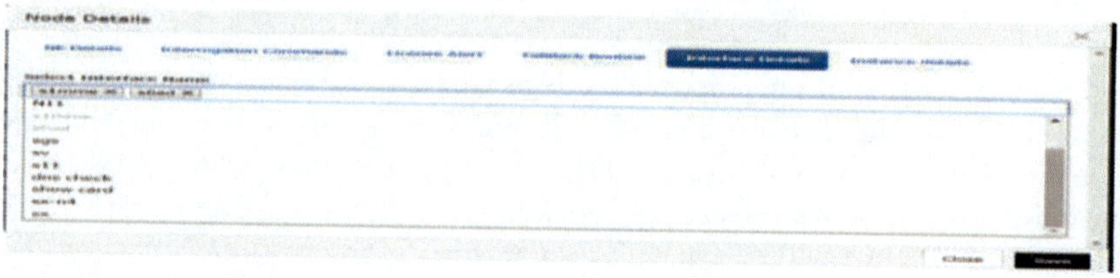

Fig. 11 Interface details of 4G and 5G network (N11, S1-MME, S6)

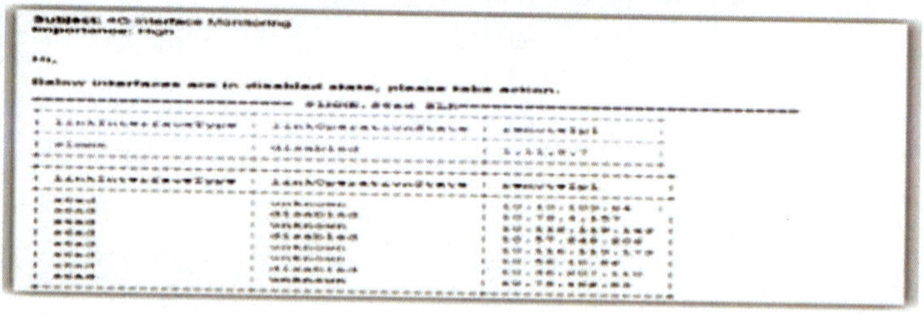

Fig. 12 G interface monitoring mails

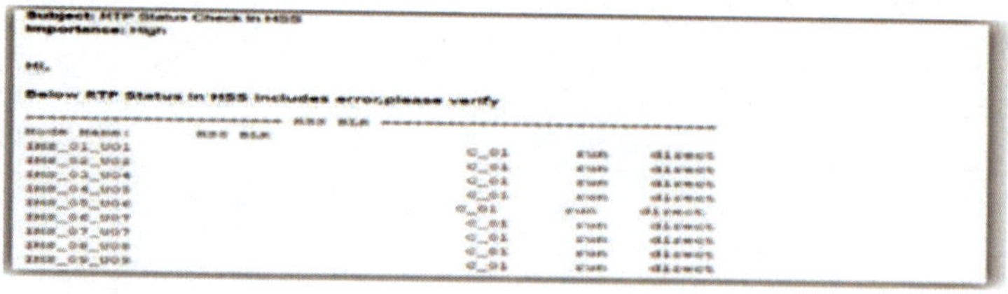

Fig. 13 Monitoring RTP in HSS

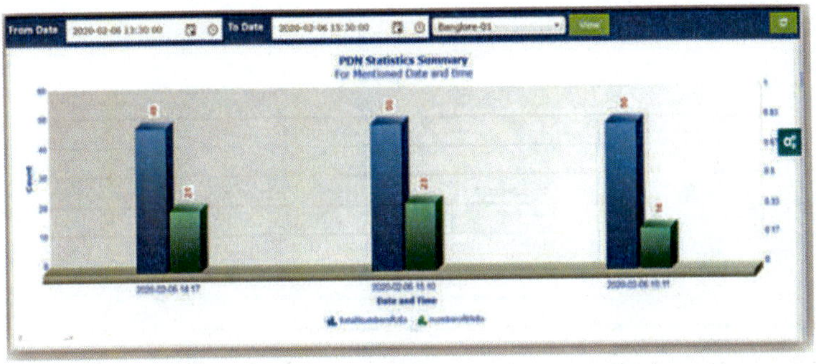

Fig. 14 Month-wise data of user equipment and eNodeB of 4G network in Bangalore location

time, eNodeB number, total user hardware as well as the names of the network unit. The UE and eNodeB counting of each network element is statistically reported on the platform on which the server is built. The data collected every hour are listed in several ways, including one year of usage, one month of usage, one day of usage, and current statistics as shown in Fig. 14. It provides current user hardware statistics and eNodeB network feature information. It gives the number of user devices and eNodeB numbers day to day information.

UE and godeB data are stored and routinely tracked regularly automated scanning systems as shown in Fig. 15 from the Manila server. Once a new network feature is mounted, the data will be fed via the user interface into the application. Information consisting of network element proprietor's login credentials provided as an input to the application in Fig. 10. The application logs into each network entity using these credentials and fetch statistical and network data from the server through packet data network commands and load them into the database.

Data are classified and shown on the UI. Monitoring UE and eNodeB statistics of France location shows in Fig. 16. The Spring MVC architecture of Java is used to achieve so. Through its system view controller architecture, the data transfer from view (front end) to controller (which is made up of logic and resolution) is simplified and eventually reaches the appropriate configuration to store the data in the database.

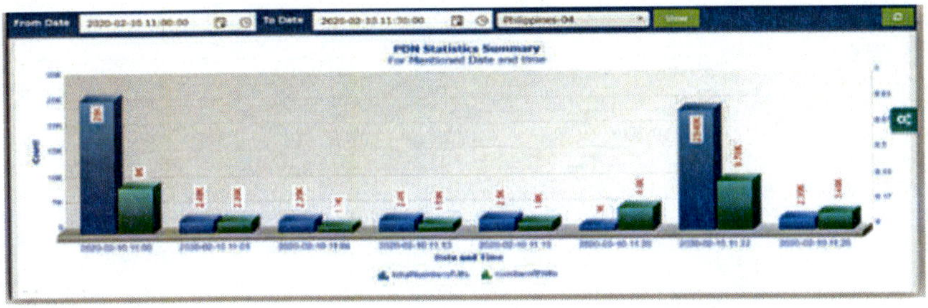

Fig. 15 Year-wise data of user equipment and gNodeB of 5G network in the Philippines location

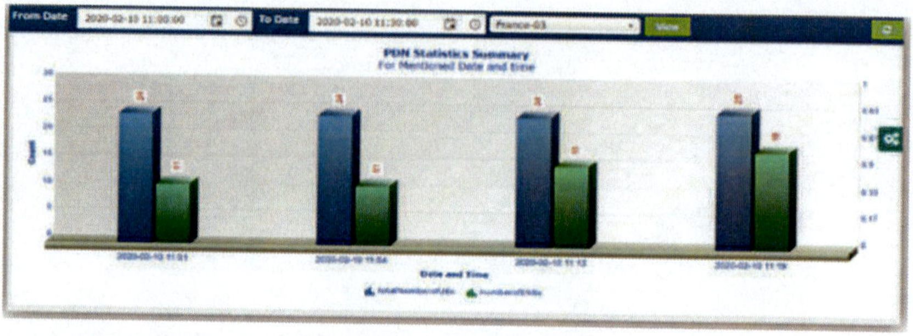

Fig. 16 Monitoring UE and eNodeB count in France's location

Monitoring the threshold value of the serving gateway is shown in Fig. 17, and an e-mail notification warning is sent to the owner of the network element providing whether the network element's throughput value has exceeded the threshold point and given the network element's added IP list.

Figure 18 contains the N2 interface link operation state in the enabled and disabled state sent as a mail to respective NE owners of AMF. Necessary action is carried out to bring the disabled state back to the enabled state.

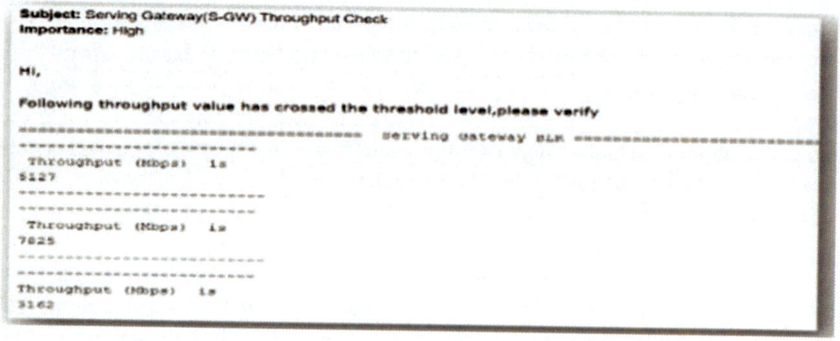

Fig. 17 Monitoring the throughput value of serving gateway (S-GW) of the 4G network in Bangalore location

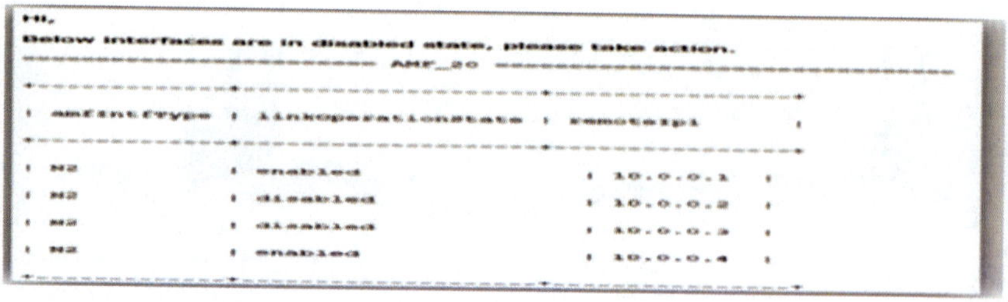

Fig. 18 Monitoring the N2 interface

```
Hi,
Please take action.
----- AUSF_UDM-1--------------------------------------------------------
NAME      READY  STATUS                 RESTARTS    AGE
udmvnf1   0/7    Container Creating         0                 5m29s
udmvnf2   0/7    Pending                    0                 5m29s
```

Fig. 19 Monitoring AUSF and UDM pod names

```
Hi,
Please take action.
+--------------------+-----------------+--------------+-----------------+------------------+-------------------+
| amfHttp2Ping | http2PingTimeOut | numHttp2Ping | http2PingStatus | roundTripTimeSec | roundTripTimeUsec |
+--------------------+-----------------+--------------+-----------------+------------------+-------------------+
| N12~1~10.1.1.1~8080~1 | 5      | 5     | success    | 0      | 464     |
| N12~1~10.1.1.2~8080~2 | 5      | 5     | fail       | 0      | 784     |
| N12~1~10.1.13~8080~3 | 5       | 5     | success    | 0      | 716     |
| N8~1~10.1.1.1~8080~4 | 5       | 5     | success    | 0      | 798     |
| N8~1~10.1.1.2~8080~5 | 5       | 5     | fail   | 0      | 515     |
| N8~1~10.1.1.3~8080~5 | 5       | 5     | success    | 0      | 515     |
```

Fig. 20 Monitoring N8 and N12 interface

Figure 19 shows the pods list that is monitored in ausf_udm in Bangalore location. The footprint of the containers is incredibly low. The container merely requires its framework and a description of all the bins and libraries that it requires to run. Unlike VMs, each contains a full copy of a guest operating system. Compared with VM deployments, scaling is simple.

Figure 20 explains about SMF and AUSF interfaces http2ping status, http2ping time out, round trip time sec. The required action is undertaken upon failure status.

5 Conclusion

An automatic scanning system for monitoring 4G and 5G network elements introduces a functionality in an application that helps reduce the manual effort necessary to monitor the network element in each different location and to reduce network outages. A system designed to manage and configure the network feature via a Web interface and simplifies the entire automated system in just a few clicks. The difficulty of controlling these network elements is influenced by efficient coding in the Java spring framework and offers an easy to usability GUI. The data from all network components are stored in the database and may at any time be accessed by the user. Specific reports and warnings are received before the license expiry date or any device deployment. It improves the data protection that the conventional system does not have. Through the implementation of firewall authentication from each local grid, data security is provided. The proposed system is an e-cloud application that can run only on the VPN cloud servers which makes it inaccessible to run from other places. If the remote server is down, then the whole application does not function which makes the application solely depend on the remote server.

References

1. Singh A, Chawla P, Singh K, Kumar Singh A (2018) Formulating an MVC framework for web development in JAVA. In: 2018 2nd international conference on trends in electronics and informatics (ICOEI), Tirunelveli, pp 926–929
2. Artyukhov OI, Krepkov IM (2018) An integrated network monitoring system for SDN VPN. In: IV international conference on information technologies in engineering education (Inforino)
3. Mijumbi R, Asthana A, Koivunen M (2018) Dynamic baselines for real-time network monitoring. In: 4th IEEE conference on network softwarization and workshops (NetSoft)
4. koh S, Lee S (2017) Implementation of open-air interface control software for 4G network. In: Ninth international conference on ubiquitous and future networks (ICUFN)
5. Abolfazli S, Sanaei Z, Wong SY, Tabassi A, Rosen S (2015) Throughput measurement in 4G wireless data networks: performance evaluation and validation. In: IEEE symposium on computer applications & industrial electronics (ISCAIE
6. Martins RS, Costa (2018) Automatic detection of computer network traffic anomalies based on eccentricity analysis. In: IEEE international conference on fuzzy systems (FUZZ-IEEE)
7. Mwanje S, Decarreau G, Mannweiler C, Naseer-ul-Islam M, Schmelz LC (2016) Network management automation in 5G: challenges and Opportunities. In: IEEE 27th annual international symposium on personal, indoor, and mobile radio communications (PIMRC)
8. Walia JS, Hämmäinen H, Flinck H (2017) Future scenarios and valuenetwork configurations for industrial 5G. In: 2017 8th international conference on the network of the future (NOF)
9. Yigit IO, Ayhan G, Zeydan E, Kalyoncu F, Etemoglu CO (2017) A performance comparison platform of mobile network operators. In: 2017 8th international conference on the network of the future (NOF), pp 144–146
10. Lin Y-H, Yang C-W (2019) An integrated network monitoring system for SDN VPN. In: 20th Asia-Pacific network operations and management symposium (APNOMS)

Comparative Study of Introducing Wavelength Converters in Pre-configured (*P*)-Cycle

Vidhi Gupta and Rachna Asthana

Abstract To protect wavelength division multiplexed (WDM) networks, pre-configured protection cycles (*p*-cycles) found to be very speedy and efficient. To implement *p*-cycles in any network, we may or may not deploy wavelength converters. As the cost of wavelength converters is quite high, fully equipping the networks with wavelength converters makes it very costly. So, instead of making the network 100% wavelength convertible, we are making network convertible in some variation. We compare the spare capacity in terms of route km of fiber length and the number of wavelength converter required for the same. We ensure that making the network wavelength convertible to some extent, i.e., 50% also can perform better without much affecting spare capacity. Also, the hardware cost of converters can be reduced much with lesser use of wavelength converters.

Keywords WDM · *P*-cycles · Wavelength converters · Protection

1 Introduction

The necessity of high data services in today's world is gaining continuous increment in various telecom regions and other business areas. So, protection mechanisms are playing a very vital role in networking. The endpoint devices present in wavelength division multiplexed (WDM) networks communicate to each other via a path called lightpaths. These lightpaths need to be protected, if these get failed due to fiber cuts, etc., then the complete system will be shut down. So, to provide a smooth flow of data services without any disruption, protection becomes essential. The intermediate nodes of the network can switch the incoming lightpath to another node. In this switching mechanism, there are two possibilities, one is that the same wavelength

V. Gupta (✉)
Harcourt Butler Technical University, Kanpur, India
e-mail: er_vidhi02@yahoo.com

R. Asthana
Dr. Ambedkar Institute of Technology for Handicapped, Kanpur, India

G. Ranganathan et al. (eds.), *Inventive Communication and Computational Technologies*, Lecture Notes in Networks and Systems 145,
https://doi.org/10.1007/978-981-15-7345-3_76

switched and the other is that the converted wavelength gets switched. The wavelength is converted by the process called wavelength conversion, and the device is called a wavelength converter. The wavelength converter is a very expensive device. It degrades the signal performance during conversion and also adds complexity to the network [1].

To protect the WDM networks, protection paths are created along with the working paths [2]. Under the normal situation, the traffic is carried by the working path, while in case of breakdown, the same traffic gets switched over to the protection paths. There are several protection mechanisms like ring-based protection, mesh protection, etc. One of the promising protection mechanisms is the ultimate concept of *p*-cycles. It offers enormous advantages of faster restoration speed, i.e., high speed just like ring-based protection as well as high efficiency just like mesh-based protection [3–6].

There are large number of studies based on *p*-cycle protection in which it is assumed that wavelength converter is available at every node providing 100% wavelength conversion [7–9]. 100% wavelength conversion requires every node of the network to be fully convertible. This needs that a wavelength converter should be present at every node. But this will make the network extremely costly due to the higher cost of converters. So, we are investigating the network with a reduced number of converters so that we can cut off the cost of the network to some extent. We have also estimated the effect on spare capacity, in terms of route fiber length required, by reducing the number of converters in the network.

The paper is arranged as follows. Section 2 discusses *p*-cycle protection. Section 3 provides the introduction of wavelength converters. Section 4 elaborates on the previous work. Section 5 introduces our work. Section 6 gives the results followed by the conclusion in Sect. 7.

2 *P*-Cycle Protection

p-cycles stand for pre-configured protection cycles. It is a very effective method of protecting mesh network. It offers the benefits in terms of high capacity efficiency as a mesh protection method and high speed as a ring protection method [4, 5, 10, 11]. There can be various types of *p*-cycles depending on the constraint associated with the pattern.

Two of the types are as follows.

(i) Link *p*-cycle: It protects the working capacity present on a link as shown in Fig. 1.
(ii) Node-encircling *p*-cycle: The failure node gets protection by it thereby forming cycles and keeping the protected nodes enclosed within as shown in Fig. 2.

When we design a WDM network with *p*-cycles, first of all, we route the traffic path following the given demand. It is done to reserve that working capacity for any particular requirement. The remaining available capacity apart from working capacity is called the spare capacity [12, 13]. The *p*-cycles are formed in this spare

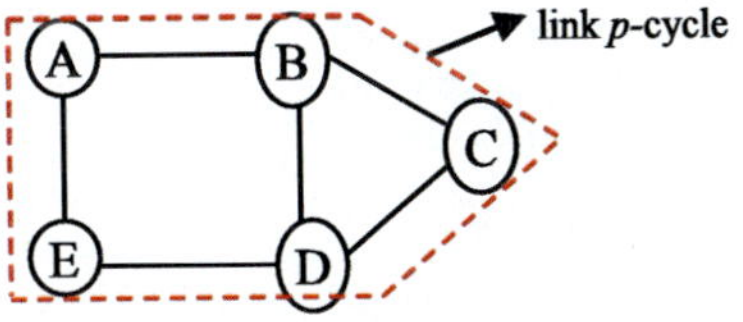

Fig. 1 5 nodes 6 spans network topology with a single link *p*-cycle

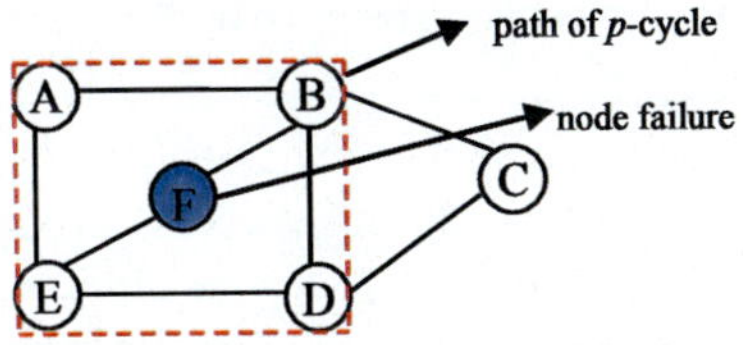

Fig. 2 Node-encircling *p*-cycle

capacity. Out of all cycles, the set of efficient *p*-cycles is selected to protect all working connections, respectively. In another way, it is to find the minimum spare capacity needed by setting up *p*-cycles for protecting the working capacity on all the links. The effectiveness of this strategy is calculated by capacity efficiency which is determined as the ratio of reserved capacity to form *p*-cycles to the total capacity required for working capacities present on all links.

p-cycles can protect the links that are present on the cycle, called on-cycle links and to the links which are not on the cycle but their end nodes are present on the cycle called as straddling links. Together, the on-cycle and straddling protection provided by *p*-cycles consequence in greater efficiency. The protection provided by *p*-cycles is also very fast and simple as only end nodes of the failed link need to perform switching action that results in switching time of the order of 50–150 ms [14]. Thus, the idea of *p*-cycle is found to be very attractive and advantageous in protecting the WDM networks [15].

Let us consider an example with a network topology of 5 nodes and 6 spans. To provide protection, *p*-cycles are formed. Figure 1 shows the network with a single link *p*-cycle drawn with a dashed line. This *p*-cycle can protect the links associated with either on-cycle links or straddling link. Figure 3 shows the failure of the on-cycle link AB; the *p*-cycle protects in the manner as shown by the path AEDCB and is referred to as on-cycle link protection provided by the *p*-cycle.

Figure 4 shows the failure in the straddling link BD; two alternate protection paths are provided by the *p*-cycle as shown by the path BAED and BCD, referred to as straddling link protection provided by the *p*-cycle.

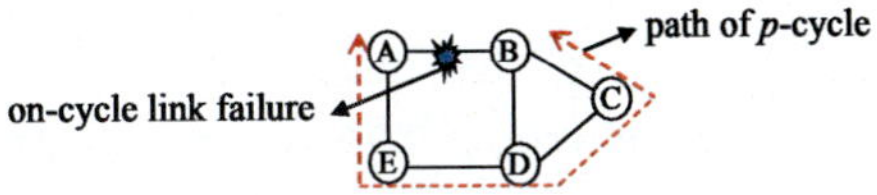

Fig. 3 5 nodes 6 spans network topology with an on-cycle link failure

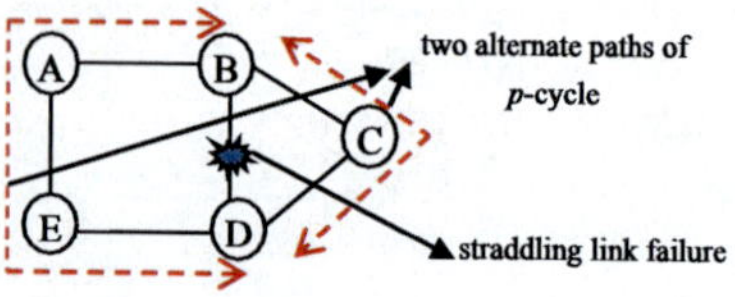

Fig. 4 5 nodes 6 spans network topology with straddling link failure

3 Wavelength Converter

As mentioned above, the WDM network nodes can switch the same wavelength to the corresponding link or can perform conversion to another wavelength. This conversion of wavelength enhances the utility of available wavelengths and called wavelength conversion [16]. It provides the benefit of wavelength reuse in any network. By this wavelength reuse, the same wavelength can be used in any network many times provided that the same wavelength does not get overlapped on any link [17, 18]. It is so because two lightpaths cannot have a similar wavelength on the same link. Another benefit of converting wavelength is the reduction in the blocking probability. Blocking probability can be defined as the ratio of the number of rejected lightpaths to the number of connected lightpaths. If a network possesses the capability of wavelength conversion, then there will be less rejected paths. This will add flexibility to the network thereby providing wavelength change as soon as free channels get available. The wavelength conversion paths called a virtual wavelength path (VWP), whereas a continuous path without wavelength conversion is called a wavelength path (WP) [19, 20]. There are mainly three types of wavelength conversions:

1. Full Wavelength Conversion: It provides full conversion of wavelength from one incoming wavelength to every other wavelength present in the network.
2. Fixed Wavelength Conversion: It provides a fixed conversion from one wavelength to the other wavelength.
3. Limited Wavelength Conversion: It provides conversion of any wavelength into a limited set of other wavelengths.

The wavelength conversion in any network is achieved by the device called as wavelength converters. Besides the various advantages, it offers in converting wavelengths they have certain drawbacks also. These wavelength converters are a very costly device that adds the hardware expense of the network. They increase the complexity of the network. Also, the conversion process degrades the signal performance.

There are various approaches to attain the wavelength conversion. The four main types are as follows:

1. Optoelectronic approach: It allows the input optical signal to be transformed into electronic form. This signal after regeneration gets retransmitted with the help of laser to another wavelength. The various types of optoelectronic regeneration can be (i) 1R (regeneration without reshaping or retiming). (ii) 2R (regeneration with reshaping). (iii) 3R (regeneration with reshaping and retiming).

2. Optical gating: It allows the use of an optical device like a semiconductor optical amplifier (SOA) whose characteristics are varied with input signal intensity. These variations are transferred at other wavelengths to another unmodulated probe signal providing wavelength conversion using the principle of cross-gain modulation (CGM).
3. Interferometric techniques: It is based on the principle of cross-phase modulation (CPM) that uses an interferometer like Mach–Zehender interferometer (MZI) that causes a variation of the phase to provide wavelength conversion.
4. Wave mixing: It is based on the nonlinearity of the transmission medium that causes the phenomenon of four-wave mixing to provide wavelength conversion.

4 Previous Work

The previous work is done so far considered the network topology with working paths as well as *p*-cycles routed with 100% or full wavelength conversion at every node [10, 14]. To meet this criterion, it is necessary to place converters at every node of the network so that it will remain present for all working paths as well as for *p*-cycles. If W is the number of wavelengths to be used and L is the number of links passing through any node, then the $W * L$ number of wavelength converters will be required at that particular node. Say, e.g., if we are using 25 wavelengths and 10 links are passing through any node, then that node will be requiring ($25 * 10 = 250$) 250 wavelength converters as the worst-case scenario. If a single node requires so many converters, then all the nodes comprising a network will be requiring much more wavelength converters. It will be a summation of the number of converters present at all nodes of the network. This gives an upper limit to it.

To protect these networks, *p*-cycles are formed. Converters are required for the *p*-cycles also. As wavelength converters add the cost and complexity, using a large number of converters, i.e., around 250 per node will be undesirable. Also, in addition to cost and complexity, wavelength converters degrade the signal performance during the process of conversion. So, instead of placing converters at every node, we are exploring the WDM network with a lesser number of converters and not with a full conversion.

5 Our Work

We have considered a network topology with a unit traffic matrix with the assignment of paths according to the approximate link length in kilometer (km). It is assumed that one fiber is used for establishing working paths, while the other fiber is used for protection paths on which there is the formation of *p*-cycles. These *p*-cycles are formed on the same wavelengths in the protection fiber which are used for working paths in the working fiber. With the work done in previous approaches, it is assumed that wavelength converters are present at every node. But this adds the network hardware cost of converters. So, instead of placing converters at every node, we are comparing the protection performance in terms of spare capacity required in route km of fiber length, by placing converters with some variation. We are also calculating the number of converters required for the same. We have taken four variations of 25% conversion, 50% conversion, 75% conversion, and 100% conversion. It is observed that making a 50% wavelength convertible can reduce the number of converter requirements to a great extent without much increase in the spare capacity. This will make our network cost-efficient. It is so as instead of placing converters at all the nodes, we are just placing converters at half of the nodes only, i.e., 50%.

The spare capacity is calculated using the following integer linear programming (ILP).

Sets

S set of spans indexed by j
P set of p-cycles indexed by p
V set of wavelengths indexed by v.

Parameters

l_j length metric on span j
$w_{j,v}$ working capacity on span j at wavelength v
$x_{p,j}$ 1 if p-cycle p protects span j as on-cycle, 2 as straddling and 0 otherwise
$\delta_{p,j}$ 1 if p-cycle p passes through span j 0 otherwise.

Variables

$n_{p,v}$ required number of unit capacity copies of p-cycle p at wavelength v
$s_{j,v}$ spare capacity required on span j at wavelength v.

Objective

$$\min \sum_{\forall j \in S} l_j \sum_{\forall v \in V} S_{j,v} \tag{1}$$

Subject to

$$w_{j,v} \le \sum_{\forall p \in P} x_{p,j} n_{p,v} \quad \forall j \in S \quad \forall v \in V \tag{2}$$

$$S_{j,v} = \sum_{\forall p \in P} \delta_{p,j} n_{p,v} \quad \forall j \in S \quad \forall v \in V \tag{3}$$

$$n_{p,v} \ge 0 \quad \forall p \in P \quad \forall v \in V \tag{4}$$

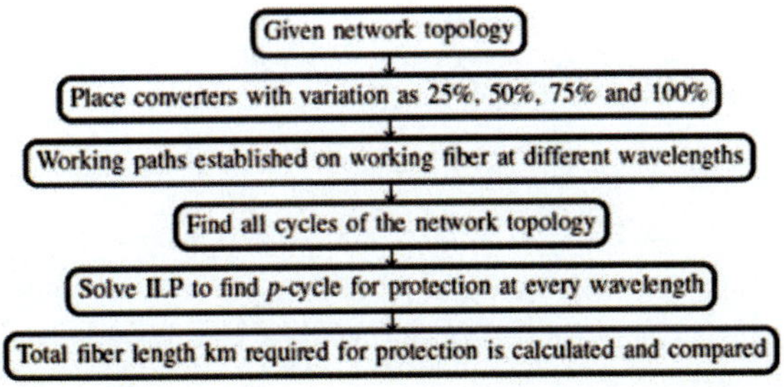

Fig. 5 Flowchart of the proposed work

Equation (1) represents the objective of minimizing the total spare capacity in fiber km length required to form *p*-cycles at different wavelengths. Equation (2) ensures that all the working capacity of every span at each wavelength gets 100% protection for a single failure. Equation (3) gives the required spare capacity on every span at each wavelength to form the *p*-cycles. Equation (4) ensures that the number of unit capacity copies of *p*-cycles formed at every wavelength should be an integer greater than zero.

If wavelength converters are not at all used in the network [21], then all working paths are to be switched over to the separate protection fiber. Without converters, there will be a wavelength blocking effect [22]. It will also unable to reuse the wavelength resulting in the wastage of wavelengths. If wavelength converters are used at every node providing 100% conversion, then it will make the network highly expensive. All these are problems with previous approaches. So, we have explored an optical mesh network along with *p*-cycle-based protection but with variations in the placement of wavelength converters. Figure 5 shows the flowchart of the proposed work.

6 Results

We have considered two test network NET1 with 19 nodes 28 spans and NSFNET with 14 nodes 21 spans as shown in Figs. 6 and 7, respectively, with approximate link length in km as shown. We have considered unit traffic matrix from each node to every other node with the approximate link distance. The shortest path algorithm is used for routing, and then, wavelength assignment is done accordingly [23]. *p*-cycles are formed using the breadth first search (BFS) algorithm to protect every wavelength. This algorithm traverses or searches tree or graph data structures, starting at the root and exploring along with adjacent nodes thereby recursively proceeding. ILPs are solved with ILOG CPLEX 9. This software makes use of decision optimization technology to optimize, build, and set up optimization models using arithmetical and constraint programming. The numbers of wavelengths used for NET1 are 24 and for NSFNET are 16. We have compared spare capacity required in fiber length km for the formation of *p*-cycles by placing converters with a variation of 25% nodes to be wavelength convertible, 50% nodes to be wavelength convertible, 75% nodes

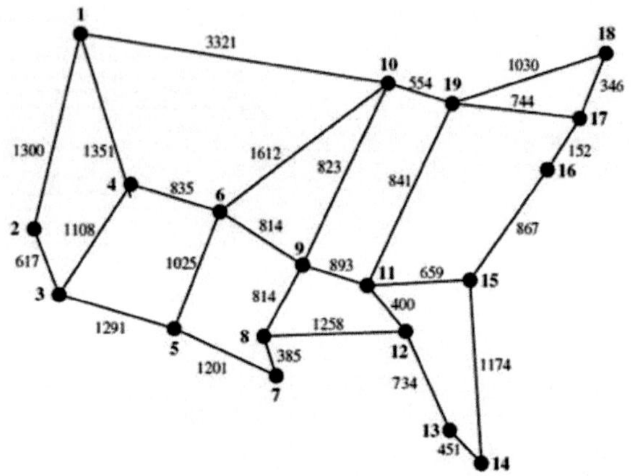

Fig. 6 NET1 with 19 nodes and 28 spans with approximate link length in km

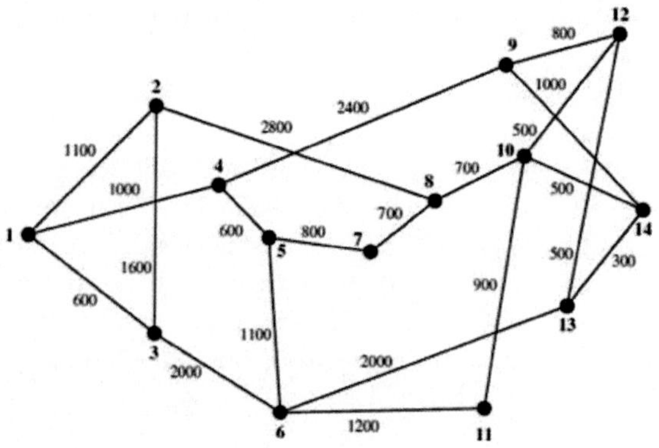

Fig. 7 NSFNET with 14 nodes and 21 spans with approximate link length in km

to be wavelength convertible and 100% nodes to be wavelength convertible as the previous case. We have also estimated the number of wavelength converters required with these variations.

For NET1 of 19 nodes and 28 spans, the spare capacity in fiber length km comes to 334,266 by placing converters at 25% nodes of the network with the number of converters to 336 and the percentage of hardware cost reduction to 75%. It is found to 319,642 by placing converters at 50% node of the network with the number of converters to be 744 and the percentage of hardware cost reduction to 44.64%. It is found to be 306,782 by placing converters at 75% node of the network with the number of converters to be 1080 and the percentage of hardware cost reduction to 19.64%. It is found to be 247,084 by placing converters at 100% node of the network with the number of converters to be 1344.

Table 1 illustrates the spare capacity in fiber link km required and the number of wavelength converters required for the same. Figures 8 and 9 show the graph of spare capacity and number of wavelength converters with nodes variations, respectively, for NET1.

Table 1 Comparison of NET1 in terms of spare capacity (fiber length km) and number of converters required

Cases	Nodes variation with converters (%)	Spare capacity (fiber length km)	Number of converters required
A	25	334,266	336
B	50	319,642	744
C	75	306,782	1080
D	100	247,084	1344

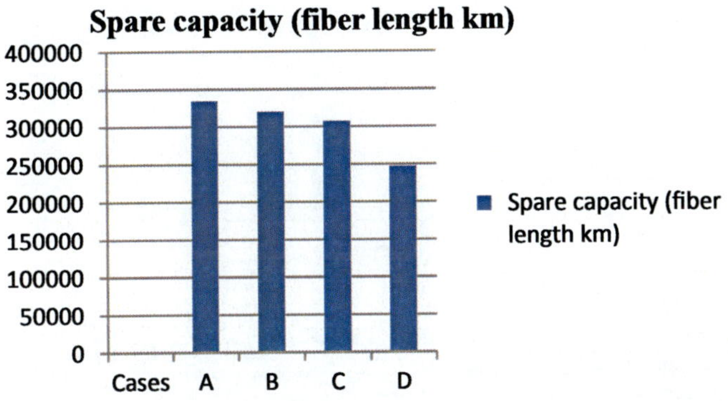

Fig. 8 Bar graph of spare capacity (fiber length km) versus nodes with converters for all the cases for NET1

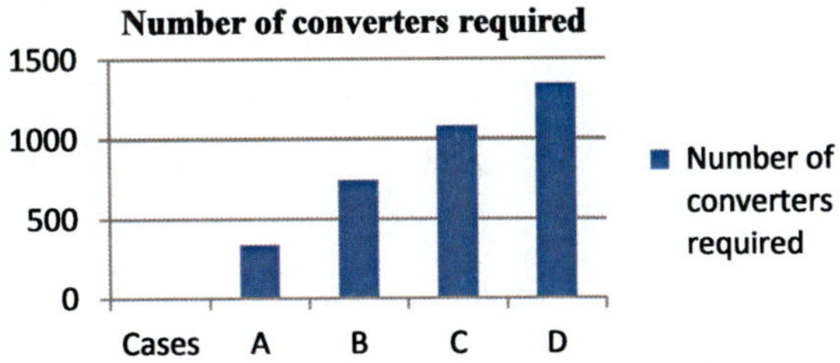

Fig. 9 Bar graph of the number of converters required versus nodes with converters for all the cases for NET1

For NSFNET of 14 nodes 21 spans, the spare capacity in fiber length km comes to 212,700 by placing converters at 25% node of the network with the number of converters to 192 and the percentage of hardware cost reduction to 71.42%. It is found to be 205,300 by placing converters at 50% node of the network with the number of converters to 336 and the percentage of hardware cost reduction to 50%. It is found to be 197,200 by placing converters at 75% node of the network with the number of converters to 528 and the percentage of hardware cost reduction to 21.42%. It is found to be 153,100 by placing converters at 100% node of the network with the number of converters to be 672.

Table 2 Comparison of NSFNET in terms of spare capacity (fiber length km) and number of converters required

Cases	Nodes variation with converters (%)	Spare capacity (fiber length km)	Number of converters required
A	25	212,700	192
B	50	205,300	336
C	75	197,200	528
D	100	153,100	672

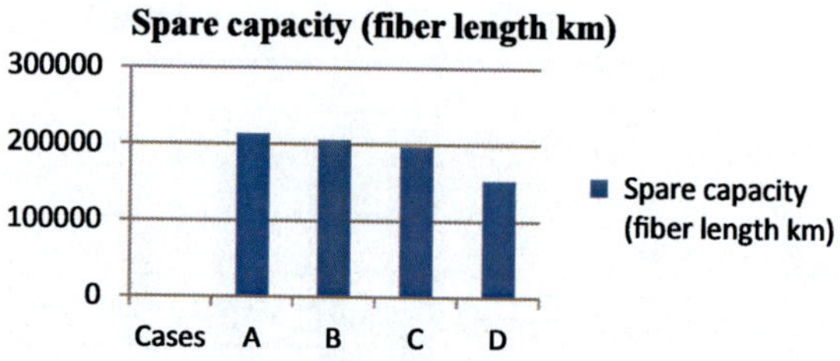

Fig. 10 Bar graph of spare capacity (fiber length km) versus nodes with converters for all the cases for NSFNET

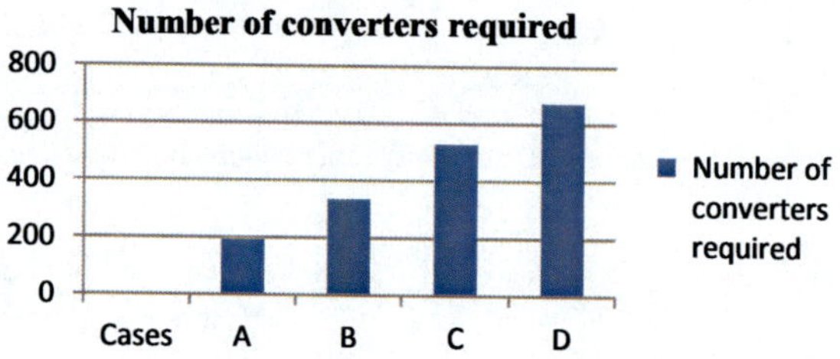

Fig. 11 Bar graph of the number of converters required versus nodes with converters for all the cases for NSFNET

Table 2 illustrates the spare capacity in fiber link km required and the number of wavelength converters required for the same. Figures 10 and 11 show the graph of spare capacity and the number of wavelength converters with nodes, respectively, for NSFNET.

7 Conclusion

In our work, we have compared the spare capacity in fiber length km and the number of converters required by placing the converters with a variation of 25, 50, 75, and 100% wavelength conversion. It is found that we can reduce the number of converters required in any network without much effect in the spare capacity with a 50% wavelength conversion. It has reduced the network expense in terms of hardware cost as we are now not using converter at every node assuming 100% wavelength conversion as done previously. Thus, the cost of converters is deducted as lesser converters can now be used. In the present work, we have considered the situation with the worst-case scenario for wavelength converters, but in the future, the actual number of converters can be calculated.

References

1. Asthana R, Singh YN (2004) Protection and restoration in optical networks. IEEE J Res 50(5):319–329
2. Asthana R, Garg T, Singh YN (2004) Critical span protection with pre-configured cycles. In: Proceedings of the international conference photonics, held at cochin during, 9–11 Dec
3. Schupke DA, Scheffel MC, Grover WD (2003) Configuration of *p*-cycles in WDM networks with partial wavelength conversion. Photonic Netw Commun 5(3):239–252
4. Doucette J, Grover WD (2001) Comparison of mesh protection and restoration schemes and the dependency on graph connectivity. In: Proceedings of 3rd international workshop design of reliable communication networks. Budapest, Hungary, 7–10 Oct, pp 121–28
5. Grover WD, Stamatelakis D (1998) Cycle-oriented distributed preconfiguration: ring-like speed with mesh-like capacity for self-planning network restion. In: Proceedings of ICC
6. Asthana R, Singh YN (2008) Distributed protocol for removal of loop backs and optimum allocation of *p*-cycles to minimize the restored path lengths. IEEE J Lightwave Technol 26(5):616–628
7. Jaiswal DC, Asthana R (2018) Power-efficient *p*-cycle protection with power conscious routing in elastic optical networks. In: International conference on current trends towards converging technologies (ICCTCT), Coimbatore, pp 1–6
8. Grover WD, Doucette J (2002) Advances in optical network design with *p*-cycles: joint optimization and pre-selection of candidate *p*-cycles. In: Proceedings of IEEE LEOS summer topicals. Mont Tremblant, Quebec, 15–17 Jul, pp 49–50
9. Zhang Y, Zhang Y, Shen G (2017) Extending FIPP *p*-cycles to protect straddling paths for dual failure network protection. In: Asia communications and photonics conference (ACP), Guangzhou, pp 1–3
10. Schupke DA, Gruber CG, Autenrieth A (2002) Optimal configuration of *p*-cycles in WDM networks. In: Proceedings of IEEE international conference on communications (ICC), vol 5. New York, April/May, pp 2761–2765
11. Asthana R, Singh YN (2007) Second phase reconfiguration of restored path for removal of loop back in *p*-cycle protection. IEEE Commun Lett 11(2):201–203
12. Asthana R, Singh YN (2006) Removal of loop back in *p*-cycle protection: second phase reconfiguration. In: Proceedings of 10th IEEE international conference on communication systems (IEEE ICCS 2006), held at Singapore during Oct 30 to Nov 1
13. Grover WD (2002) Understanding *p*-cycles, enhanced rings, and oriented cycle covers. In: Proceedings of 1st international conference optics communications and network. Singapore, 11–14 Nov, pp 305–308
14. Grover WD, Stamatelakis D (2000) Bridging the ring-mesh dichotomy with *p*-cycles. In: Proceedings of DRCN workshop
15. Zhang Y, Zhang Y, Shen G (2017) Extending FIPP *p*-cycles to protect straddling paths for dual failure network protection. In: 2017 Asia communications and photonics conference (ACP), Guangzhou, pp 1–3
16. Lee KC, Li K (1993) A wavelength-convertible optical network. IEEE J Lightwave Technol 11:962–970
17. Kaminski PM et al (2019) Characterization and optimization of four-wave-mixing wavelength conversion system. In: J Lightwave Technol 37(21):5628–5636

18. Anjum OF et al (2019) Channel selective wavelength conversion by means of inter modal four wave mixing. In: 2019 optical fiber communications conference and exhibition (OFC), San Diego, CA, pp 1–3
19. Qian J, Yang T, Gao M, Xiang L, Shen G (2016) Seamless all-optical bidirectional wavelength converter. In: 2016 15th international conference on optical communications and networks (ICOCN), Hangzhou, pp 1–3
20. Venkatesh T, Srinivasan SH (2004) Normalized cuts for wavelength converter placement. In: IEEE workshop on high performance switching and routing
21. Gupta V, Asthana R, Singh YN (2020) *p*-cycle protection without wavelength converters. ICICT, 26–28 Feb
22. Gupta V, Asthana R (2020) Study of wavelength converter placement in *p*(pre-configured)-cycle protection. In: Fourth international conference on computing methodologies and communication (ICCMC), Erode, pp 459–464. https://doi.org/10.1109/iccmc48092.2020.iccmc-00086
23. Ramaswami R, Sivarajan KN (1995) Routing and wavelength assignment in all-optical networks. IEEE/A CM Trans Netw 3(5):489–500

A Novel Approach to Reduce False-Negative Alarm Rate in Network-Based Intrusion Detection System Using Linear Discriminant Analysis

Sona Solani and Nilesh Kumar Jadav

Abstract In recent years, Security is remaining as a major concern in various state-of-the-art computers and network systems. Expanding technologies and their relevant framework can also be the focal point, to exploit it. To prevent those frameworks, network systems, servers, files, and systems, enterprises use the intrusion detection system [IDS] and it tries to avoid certain traffic otherwise, it can lead to an intrusion in the system. Nowadays, Machine Learning and Deep Learning techniques used to keep an efficient IDS in almost every area. In this paper, We used a supervised classification algorithm, onto a Network-based IDS dataset—"UNSW-NB15 dataset", and compared it with different algorithms to increase the efficiency of the IDS dataset which can further reduce the false negative alarm rate. We had applied a feature selection algorithm to sort out favorable features that can reduce the false alarm and then use Linear Discriminant Analysis as a classification algorithm that will refine our result analysis. The result will be based on Accuracy, False-Negative alarm rate, and ROC-AUC Score.

Keywords Intrusion detection system · Machine learning · Deep learning · Linear discriminant analysis · False-negative alarm rate

1 Introduction

The Internet has been used as an essential part of human life in the twenty-first century because we can use it in different areas such as healthcare, business, entertainment, education as well as for chatting, audio and video calling also [1]. Technologies such as big data, IoT, cloud computing, AI, etc. help to reduce human power. However, with the usage of these technologies, a large amount of data will be produced at a time [2].

S. Solani (✉) · N. K. Jadav
Department of Computer Engineering, Marwadi University, Rajkot, Gujarat, India
e-mail: ssolani305@gmail.com

N. K. Jadav
e-mail: nileshjadav991@gmail.com

G. Ranganathan et al. (eds.), *Inventive Communication and Computational Technologies*, Lecture Notes in Networks and Systems 145,
https://doi.org/10.1007/978-981-15-7345-3_77

There are so many security systems are available to protect this data. However, some are available in hardware while some are software or both. Cybersecurity provides the security of network, information, and computer system from unauthorized activity. There are mainly five pillars of cybersecurity: confidentiality, integrity, availability, accountability, and assurance. In which, confidentiality, availability, and integrity are used for information security to protect data. Every technology comes up with different threats and vulnerabilities. Attackers are always trying to perform unauthorized activities to steal the data. And this type of activity is known as intrusion and cybercriminals are known as intruders. An intrusion or attack can be characterized as, "an arrangement of activities that endeavor to compromise the security targets".

An IDS is a set of procedures to identify malicious network traffic. It is software, works the same as the firewall Yet, that shields the network as well as a system. We can use it in the middle of a system. IDS can work with both systems such as host-based and network-based which have been explained in [3]. Host-based IDS monitors the OS system and files activities while network-based IDS supervise and prevent the system from unauthorized network traffic. Network-based IDS is easy to identify and detect, rather than host-based because we can define the rules in network-based and analyze the attack and also identify the intention of cybercriminals. Host-based IDS monitors the system in offline mode and at a time so many tasks are going on. So, sometimes a security person can't monitor every task and activity. There are another two types of IDS based on detection techniques: signature-based and anomaly-based. In signature-based, there are predefined patterns, and those patterns are analyzed by tool and it will give the future result based on previous patterns. This technique is easy to use and also fast for execution. While in anomaly-based technique, it is hard to identify because it is generally used for unknown attacks. An Intrusion Prevention System is a control system of IDS that helps to find potential threats and respond to them. Traditional IDS tools are open source and we can download them from the internet very easily. Snort, Suricata, OSSEC, Bro NSM are the most famous tools for NIDS because these tools also give the flexibility to customize them as per our requirement.

Artificial Intelligence is a computer intelligence in which machine is capable to learn from experience and give the prediction for the future. The machine can act and respond like a human. AI can be applying in various fields such as healthcare, business, education, autonomous vehicles, robotics, environment, cybersecurity, space technology, etc. The best examples of AI are self-driving cars, robot (SOPHIA), virtual voice assistant (Google Assistant, Siri, Cortana), chatbots, fitness tracker, etc. Nowadays, AI covers 40% of the jobs globally, and the demand for machine learning, computer vision, and natural language processing are increasing day by day because these technologies help us for a better existence. Of course, there is a tremendous amount of data gathered in the cloud servers and data warehouses so we need to take care of the privacy and security of these confidential data. AI has mainly 4 stages: Reactive machines, Limited memory, Theory of mind, and Self-awareness, which has been explained in [2]. Machine Learning utilizes algorithms to gather information, gain from that information, and make informed decisions based on what it has learned while Deep learning algorithm creates layers to make an "artificial neural system" that can learn and make keen judgments all alone. Deep learning has so many types of neural networks for various types of data such as Recurrent Neural Networks that have the ability to learn temporal sequence by processing arbitrary input sequences using internal memory units [4]. While the other type of neural network is Capsule Neural Network which has been designed to avoid the few limitations of Convolution Neural Network like loss of information, deduction in data dimension for acquiring the spatial invariance, etc. [5].

IDS serves with signatures or patterns and using those patterns numerous suspicious requests filtered out through IDS. However, there are mostly two troubles to be confronted when we are utilizing it. The first challenge is some legitimate websites are significant and trustworthy but filtered out as a suspicious website and second is some requests are unauthorized but filtered out as a genuine website and these are known as False-positive rates and False-negative rate respectively. If alarm did not raise than difficulties will be created such as: unauthorized access, data misuse or fraud issues for large organization. So, now IDS need more configuration so it can ready to recognize suspicious action with lower rate of false negative alarm rate. Traditional IDS tools want constant human monitoring and high-computational power. With the help of AI-based techniques, we can avoid these problems and able to increase the efficiency of the IDS.

The rest of the paper followed as—Sect. 2 provides the literature survey of AI-based IDS methods. Section 3 illustrates the proposed framework. Section 4 explains the proposed methodology. Section 5 gives the result and detailed analysis of result and Sect. 6 summarize the paper.

2 Related Work

Data mining and AI-based methods such as machine learning related methods are using in misuse-based and anomaly-based IDS. However, the data mining technique is used to discover knowledge from data and focused on it while the machine learning algorithm focuses on classification and prediction from a particular domain [6]. To use ML algorithms, we want the IDS dataset, which includes the all-important features use to detect intrusions. In [7], the authors surveyed more than 30 datasets and available ML repositories. The most common IDS datasets are KDD'99, NSL-KDD, Kyoto 2006+, AWID, and UNSW-NB15. UNSW-NB15 is the most prominent dataset for a Network-based IDS and also an updated version of the NSL-KDD dataset. It includes more than 40 features and 9 categories of attacks [8]. And AWID dataset is suitable for the wireless network. Dataset plays a vital role in machine learning techniques because a small dataset has a small amount of data so it can give better accuracy and took a small amount of time for execution but, it is not efficient. In [9, 10], authors have been used the Kyoto 2006+ dataset in which they applied different classification and clustering algorithms. However, Kyoto 2006+ dataset has 24 features in which several features are important. In [11], authors have been applied the combination of a tree-based classifier and implemented Naive Bayes and Random Forest algorithms on the NSL-KDD dataset. Also, in [12], the authors applied the combination of random forest classifiers and decision trees to build an efficient IDS. While in [13], authors have been used only random forest classifier as an ensemble method. In [14], they compared the performance of SVM, Random Forest, and Extreme Learning Machine for intrusion detection.

In [15], they apply the Time-Varying Chaos PSO technique to do two things parameter setting and feature selection for two classifier techniques namely multiple criteria linear programming and SVM. Using the combination of MCLP/SVM and TVCPSO, they gained a high detection ratio and lower false rate. In [16], the prediction is based on a classifier, which uses a policy function to train a novel reinforcement learning model in Neural Network. They have been used two environments simulated and live. Simulate environment used for randomly extracted new samples from training the dataset and also generating rewards for them and the live environment used in reinforcement learning algorithms based on interaction. In [17], the author developed one artificial neural network which uses for deep packet inspection-based intrusion detection systems. They took repetitive tenfold cross-validation to calculate an average accuracy, ROC curve, and FPR. In [18], the authors detected intrusion using combine classification methods based on the Artificial Bee Colony and Artificial Fish Swarm Algorithm. Both techniques are based on swarm optimization. They use Fuzzy C-Means Clustering (FCM) method to divide the training dataset and Correlation-based Feature Selection (CFS) technique applied to remove irrelevant features. In [19], they used neural network-based skip-gram modeling to detect false positive rate. This algorithm has implemented in fully unsupervised learning, which use to predict suspicious requests without any prior knowledge. In [20], authors have been used multiple negative selection algorithm, and it is capable to identify the accurate possible intrusions as true or false without any operator input. In [21], authors have been derived the method to identify an efficient Network-based IDS using the MAPE-K Control Loop. MAPE-K stands for Monitor, Analyze, Plan, Execute, and Knowledge. This framework base on self-taught learning that means the machine can exploit unlabelled data to improve a supervised classification problem. In [22], authors have been proposed the comparison between the ANN and optimizer-based ANN technology. They combine the convolution neural network with SVM and apply it to the KDD-99 dataset. They have been compared the result of CNN-SVM with ANN, optimizer-based ANN techniques such as ANN with Genetic Algorithm and ANN with swarm particle optimization.

3 Proposed Framework

We created one general and simple framework of an efficient IDS using AI-based techniques. For our research, we took one pre-recorded dataset, which has been created by the traditional IDS tools OSSEC or Snort. This dataset has a large amount of raw data, but, we took only 700,000 instances in which we did preprocess and sampling of the data and divided it into training and testing. Here, we create and

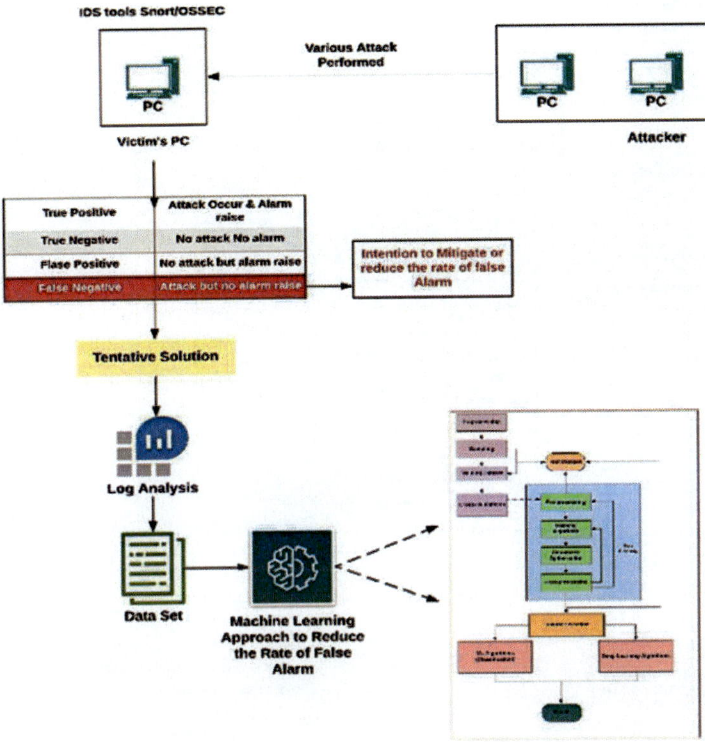

Fig. 1 The architecture of network-based IDS using AI techniques

propose a general framework so, cross-validation and data cleaning are the optional steps and, we do not use them. We use Recursive Feature Elimination (RFE) as a feature selection technique to reduce the computational cost and also avoid the unrequired features in this dataset. After successfully selected appropriate features, we can apply a suitable machine learning algorithm. Here, we implemented Linear Discriminant Analysis as our classification algorithm and compared it with the Random Forest classifier, Gradient Booster classifier, and simple ANN classifier with 100 and 200 epochs. Figures 1 and 2 illustrate the proposed architecture of NIDS using AI techniques and Flowchart of machine learning and deep learning model execution respectively.

4 Proposed Methodology

Most of the AI-based techniques depended on mathematics, data structure, Dynamic Programming, and Greedy Algorithms. In the front-end side, algorithms are very easy in implementation using PYTHON, R, and MATLAB. However, in the backend

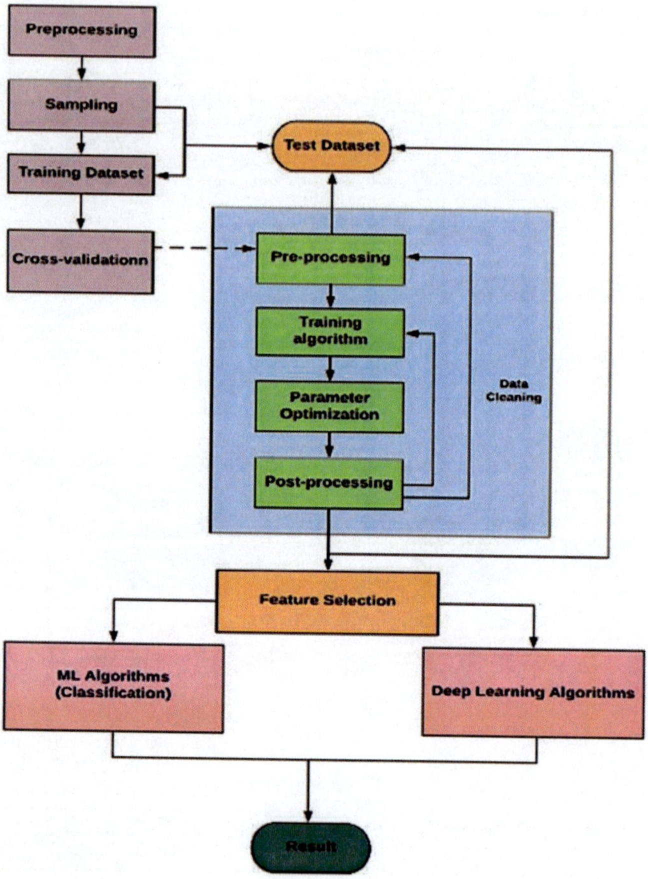

Fig. 2 Flowchart of machine learning and deep learning model execution

side Linear Algebra use for optimization purposes. Also, it requires time to analyze because of the large amount of data. Python programming language is the best suitable language for AI-based techniques because it provides the libraries which include the mathematical functions and required data structure. The proposed methodology first analyzes the incoming request using different parameters such as source and destination's IP address, port no., jitter, network protocol service, etc. And, after log analysis, it will create the labeled dataset and give the output that particular request is coming from the authorized sever or not, and if not it will also define the attack category for a particular unauthorized request. So, after creating the dataset machine learning model will be implemented in it and it will give the prediction that defines categories of the class label and attack label is correctly classified or not.

4.1 Dataset Description

The UNSW-NB 15 dataset was produced by using an IXIA PerfectStorm software to extricate a hybrid of current ordinary and contemporary assault exercises of network traffic. This dataset contains 2,540,044 records which are put away in four CSV files [8]. Here, we have been used only 1 CSV file which includes 700,001 raw data for prediction. In this dataset, there are 9 different types of attacks such as fuzzers, backdoors, exploits, generic, shellcode, worm, analysis, reconnaissance, and DoS. The Argus, Bro-IDS tools are used and twelve algorithms are developed to generate 49 features with the class label.

4.2 Feature Selection Algorithm

Recursive Feature Elimination is a wrapper-based feature selection method that wants one machine learning algorithm and uses its performance as evaluation criteria. RFE algorithm uses recursively eliminating attributes and creating a model on those attributes that left. It uses an accuracy metric to rank the features according to their importance, 1 denoting most prominent. It also gives its support, True being an appropriate feature and False being an inappropriate feature. RFE follows a simple workflow for feature selection. Suppose we have n features in the dataset than in every round 1 irrelevant feature will be eliminated and remaining features will pass for the second round and so on. Here, we have 49 features so RFE will create various subsets of 49 features and applied to learn algorithms on that particular subset and after that derive the performance of that particular subset using feature importance and coefficient. If any particular feature does not adjust with the learning algorithm than RFE will remove or eliminate that feature and at the end of all the calculation, RFE will give the most prominent features of that particular algorithm. For LDA, RFE has been given 22 most prominent features for prediction.

4.3 Linear Discriminant Analysis

LDA uses as a dimensionality reduction technique for preprocessing step into Machine Learning application. The purpose of the LDA is to convert the features from high-dimensional into low-dimension space to reduce the difficulty of dimensional costs. LDA will work as a classifier and reduce the dimensions of data and a neural network will perform the classification task. LDA algorithm includes the statistical properties of data that have been used to calculate for each class and based on that predictions will be measured. Suppose, we have multiple class so we have to calculate within-class scatter matrix S_w (1) which means we have to calculate the covariance matrix $S_1, S_2, S_3 \ldots$ for each class and add them using Eq. (2) and calculate mean μ for each class using (3) and use both the equation to calculate Eq. (1)

$$S_w = \sum_{i=1}^{c} S_i \tag{1}$$

Here, $i = 1, 2, 3, 4, \ldots$ (no. of classes)

$$S_i = \sum_{x \in Di}^{n} (x - \mu_i)(x - \mu_i)^T \tag{2}$$

$$\mu_i = \frac{1}{n_i} \sum_{x \in D_i}^{n} x_k \tag{3}$$

To compute between the class-scatter matrix S_B, we need to first calculate the difference of particular class mean and then multiply it with its transpose (4):

$$S_B = (\mu_1 - \mu_2) - (\mu_1 - \mu_2)^T \tag{4}$$

After calculating the within-class scatter matrix (S_W) and between-class scatter matrix (S_B) Find the best LDA projection vector Similar to PCA we find these using eigenvectors having largest eigenvalues According to Eqs. (1) and (4):

$$S_W^{-1} S_B V = \lambda V \tag{5}$$

And with the use of the highest value of eigenvector (5) Dimension Reduction will be based on projection vector W and no. of input data samples X we can calculate the final value of dimension cost Y:

$$Y = W^T X \tag{6}$$

5 Result and Analysis

We compared the result of LDA with Random Forest classifier, Gradient Booster classifier, and simple ANN classifier with 100 and 200 epochs. The result of all the algorithms comparison will be based on Accuracy, False-Negative Alarm rate (FNAR), and ROC-AUC Score. Here, Table 1 shows that LDA has the least accuracy, ROC-AUC score, and minor difference in false-negative alarm rate than other algorithms. But, why we choose LDA is better than others? The reason is, simple LDA has a closed-form solution hence it has no hyperparameters. Also, LDA has been executed very smoothly and efficiently and it hardly took 5–6 min to run while

Table 1 Result comparison of ML and DL algorithms

Algorithms	Accuracy (%)	False-negative alarm rate	ROC-AUC score
LDA	95.20	0.0323	0.49
RF	96.86	0.0314	0.5
GB	96.79	0.032	0.5
ANN-100	96.82	0.004	0.92
ANN-200	96.79	0.0152	0.51

other algorithms have been taken half an hour to execute. After getting these all the results, we can analyze that in cybersecurity, time management is the most important thing than any other aspect. Because, if the attacker attack and model will take huge time to run that it might be possible that our data will be stolen before alarm raised.

Figure 3 illustrates that 199,912 requests have been correctly classified as legitimate requests and 6679 requests are legitimate requests but classified as suspicious requests. Figure 4 shows the bar chart of all learning algorithms' results using performance matrices. ROC curve is plotted through the true-positive rate on Y-axis and false-positive rate on X-axis. The values of both rates are found for many thresholds from 0 to 1 and plotted on the graph and we will get ROC-AUC-Score. Accuracy is the measurement of overall classification performance and correctly identifies the value.

$$\text{Accuracy} = \frac{\text{TN} + \text{TP}}{\text{TN} + \text{TP} + \text{FP} + \text{FN}}$$

FNAR use to identify suspicious requests which are classified as legitimate.

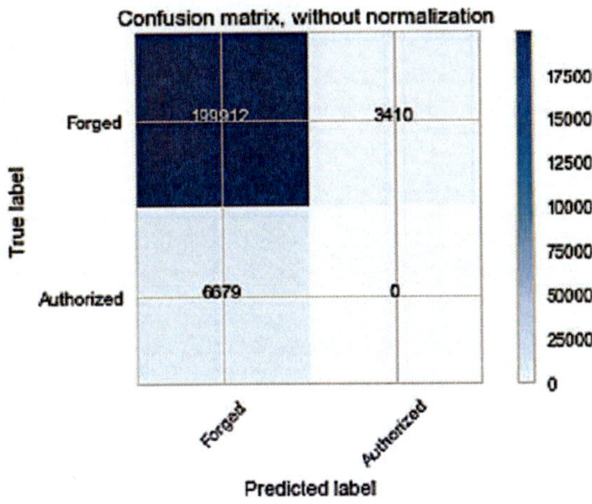

Fig. 3 Confusion matrix of linear discriminant analysis

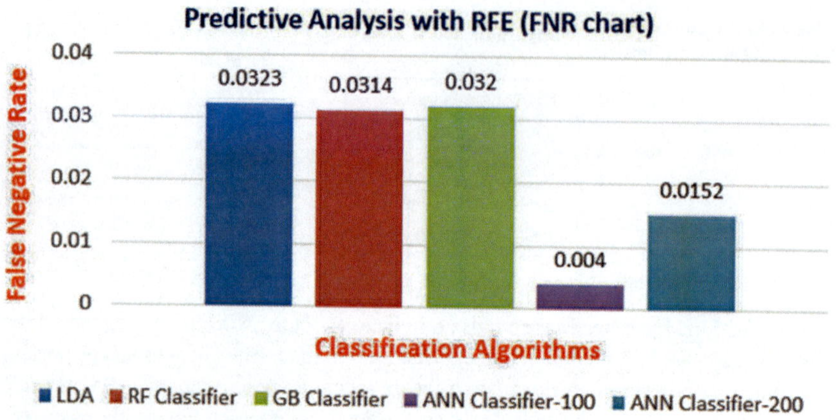

Fig. 4 Experimental result of FNAR

$$\text{False - Negative alarm rate} = \frac{\text{FN}}{\text{FN} + \text{TN}}$$

6 Conclusion

The objective of this research is to reduce the effect of the false-negative alarm rate in a network-based intrusion detection system. IDS is a very sensitive device because it is used to control any type of intrusions or attacks in the computer network system. There are so many confidential information stored in server and if there are no IDS available then it might be possible that attacker can find the vulnerability in the system and will convert that vulnerability into an exploit and can access the system, which we are trying to prevent. Machine Learning and Deep Learning algorithms are used to collect a large amount of data and try to use it to detect unknown or known network traffic and also protect the network and system.

References

1. Kumar G, Kumar K, Sachdeva M (2010) The use of artificial intelligence based techniques for intrusion detection: a review. Artif Intell Rev 34(4):369–387
2. Solani S, Jadav N (2019) A survey on intrusion detection system using artificial intelligence. In: Proceeding of the international conference on computer networks, big data and IoT (ICCBI-2019), pp 67–80
3. Liao H, Richard Lin C, Lin Y, Tung K (2013) Intrusion detection system: a comprehensive review. J Netw Comput Appl 36(1):16–24
4. Raj JS, Ananthi JV (2019) Recurrent neural networks and nonlinear prediction in support vector machines. In J Soft Comput Paradigm 2019(1):33–40

5. Vijayakumar T (2019) Comparative study of capsule neural network in various applications. J Artifi Intell 1(01):19–27
6. Buczak A, Guven E (2016) A survey of data mining and machine learning methods for cyber security intrusion detection. IEEE Commun Surv Tutorials 18(2):1153–1176
7. Ring M, Wunderlich S, Scheuring D, Landes D, Hotho A (2019) A survey of network-based intrusion detection data sets. Comput Secur 86:147–167
8. Moustafa N, Slay J (2015) The significant features of the UNSW-NB15 and the KDD99 data sets for network intrusion detection systems. In: 2015 4th international workshop on building analysis datasets and gathering experience returns for security (BADGERS)
9. Park K, Song Y, Cheong Y (2018) Classification of attack types for intrusion detection systems using a machine learning algorithm. In: 2018 IEEE fourth international conference on big data computing service and applications (BigDataService)
10. Zaman M, Lung C (2018) Evaluation of machine learning techniques for network intrusion detection. In: NOMS 2018–2018 IEEE/IFIP network operations and management symposium
11. Kevric J, Jukic S, Subasi A (2016) An effective combining classifier approach using tree algorithms for network intrusion detection. Neural Comput Appl 28(1):1051–1058
12. Bhavani T, Rao M, Reddy A (2019) Network intrusion detection system using random forest and decision tree machine learning techniques. In: First international conference on sustainable technologies for computational intelligence, pp 637–643
13. Farnaaz N, Jabbar MA (2016) Random forest modeling for network intrusion detection system. Procedia Comput Sci 89:213–217
14. Ahmad I, Basheri M, Iqbal M, Rahim A (2018) Performance comparison of support vector machine, random forest, and extreme learning machine for intrusion detection. IEEE Access 6:33789–33795
15. Bamakan SM, Wang H, Yingjie T, Shi Y (2016) An effective intrusion detection framework based on MCLP/SVM optimized by time-varying chaos particle swarm optimization. Neurocomputing 199:90–102
16. Caminero G, Lopez-Martin M, Carro B (2019) Adversarial environment reinforcement learning algorithm for intrusion detection. Comput Netw 159:96–109
17. Shenfield A, Day D, Ayesh A (2018) Intelligent intrusion detection systems using artificial neural networks. ICT Express 4(2):95–99
18. Hajisalem V, Babaie S (2018) A hybrid intrusion detection system based on ABC-AFS algorithm for misuse and anomaly detection. Comput Netw 136:37–50
19. Carrasco R, Sicilia M (2018) Unsupervised intrusion detection through skip-gram models of network behavior. Comput Secur 78:187–197
20. Pamukov M, Poulkov V (2017) Multiple negative selection algorithm: improving detection error rates in IoT intrusion detection systems. In: 2017 9th IEEE international conference on intelligent data acquisition and advanced computing systems: technology and applications (IDAACS)
21. Papamartzivanos D, Gomez Marmol F, Kambourakis G (2019) Introducing deep learning self-adaptive misuse network intrusion detection systems. IEEE Access 7:13546–13560
22. Nagar P, Menaria H, Tiwari M (2019) Novel approach of intrusion detection classification deeplearning using SVM. In: First international conference on sustainable technologies for computational intelligence, pp 365–381

A Low Cost IoT Enabled Device for Monitoring Agriculture Field and Smart Irrigation System

Sai Surya Kiran Pokala and A. A. Bini

Abstract Agriculture plays a crucial role in the economy of the country. It provides employment opportunities for a large number of people in the country. The traditional techniques used by the farmers are monoculture, pesticide resistance, erosion, water depletion and pest detection etc. The major challenge in agriculture is the usage of excessive water in the crops and monitoring the nutrients in the soil. Usage of large proportions of pesticides in agriculture leads to less productive crops and damages the health of the soil. All these problems in agriculture are degrading the development of the country. Modernisation of agriculture is the smart solution to this problem. The proposed work aims to develop an IoT device for smart crop field monitoring system and automated irrigation system using the wireless sensor networks (WSN) and monitors the temperature, humidity and measures the content of moisture in the soil. The data received from the sensors is stored and analysed in the cloud platform. The system provides a user-friendly web interface for the farmers to access the information about the crop and content of moisture in the soil from the cloud platform.

Keywords Automated irrigation · Wireless sensor networks (WSN) · Crop field monitoring · Internet of things(IoT)

1 Introduction

In India, 70% of the population depends on agriculture for their livelihood and increase the Gross Domestic Product (GDP) of the country. A large proportion of the farmers use traditional farming techniques which makes it difficult to analyse

S. S. K. Pokala (✉) · A. A. Bini
Computer Science and Engineering, Indian Institute of Information Technology, Kottayam, Kerala, India
e-mail: saisuryapokala6@gmail.com

A. A. Bini
e-mail: bini@iiitkottayam.ac.in

G. Ranganathan et al. (eds.), *Inventive Communication and Computational Technologies*, Lecture Notes in Networks and Systems 145,
https://doi.org/10.1007/978-981-15-7345-3_78

the crop conditions. The traditional irrigation methods such as pot irrigation, drip irrigation, strip irrigation and furrow irrigation are used to supply water to the plants [1]. To reduce the wastage of water in agriculture there is a need to modernise the irrigation methods using IoT, Cloud Computing and Machine learning etc. Farmers face difficulties in delivering the water to the plants [2]. The Internet of Things is the smart solution to address this problem. Crop fields can be monitored by using cost-effective wireless sensor networks. The content of moisture in the soil indicates the level of water in the soil [3]. The deployment of the IoT enabled systems in the field to monitor the climatic conditions and supply water to the plants at precise time and quantity. This system uses application software for data analytics in the cloud and a microcontroller to read the data from sensors.

The main objectives of the proposed system are:

- To create an IoT device for monitoring the crop field using sensors(soil moisture, temperature, Humidity).
- To analyse the sensors data and provide information to the farmers.
- To automate the irrigation by comparing the level of soil moisture with the threshold value.

This paper is organised in such a way that Sect. 2 deals with the Literature survey, Sect. 3 describes the experimental setup of the system, Sect. 4 describes the proposed system Sect. 5 depicts the implementation details of the IoT enabled system and Sect. 6 shows the results of real-time data of the sensors which are stored in the cloud platform.

2 Literature Survey

Arduino is cost-effective and open-source extensible software. Users can have the feasibility to extend the model and improve the design. Most of the microcontrollers are limited to windows only whereas the Arduino can run on various operating systems [4]. Balaji et al. [5] This paper proposes an IoT device for monitoring the crop by using the wireless sensor networks. Sensor data related to the crops are collected from the sensors by using Arduino ATMega 328 and Wifi Module to transfer the data to the cloud database. Various sensors are used to update the field condition to the farmers. GSM Technology is used for sending the notification to the farmer. Suma et al. [6] used wireless sensor networks to read the real-time sensor data from the sensors and ZigBee acts as two-way communication between sensors and Arduino. Zigbee lacks a communication range to retrieve the sensor data. Due to some issues, complete automation in irrigation is not reached by the developed system. Rupanagudi et al. [7] This model provides information to the farmers about the condition of the crop and images of the crops are sent to the farmers to detect the pests present in the crops and to take precautionary measures for reducing the crop damage to the farmers. Danita et al. [8] deployed the system in greenhouse farming to monitor the temperature and humidity for making work easier to the farmer. Cooling fans are deployed along with the system to maintain the temperature and humidity. The data received from the sensors is stored and analysed in the cloud database. The system is not cost-effective. Ghanshala et al. [9] A smart crop monitoring system is developed in this paper, in which the system analyses the nutrients present in the soil, temperature and humidity using the wireless sensor networks. The data retrieved from the sensors is stored and analysed using machine learning algorithms to monitor the requirements of nutrients in the soil for better production crops. Sabri et al. [10] This paper proposes a method to increase the yield of the crops by using the efficient automated irrigation to the crops. The soil moisture sensor is deployed in the crops to measure the content of moisture in the soil. An email notification is sent to the farmers about the condition of the crop.

3 Experimental Setup

3.1 Soil Moisture Sensor

The soil moisture sensor is a sensor which is used to measure the content of water inside the soil. The readings of the sensor are constant in salty conditions. The temperature sensor is inbuilt in the soil moisture sensor as it measures the temperature in the soil [11]. The sensor calculates the averages of water content throughout the length of the sensor [12]. The sensor is developed with both analog and digital modes. Soil moisture is measured in bars. Soil moisture sensor contains two probes and these probes pass electricity into the soil and get the resistance back from the soil to measure the moisture content in the soil. This sensor is largely used in agriculture and for irrigation.

Soil moisture consists of 3 pins VCC, GND, V0 in which VCC of soil moisture is connected to the Vin of Arduino UNO, ground pins of both arduino and sensor are connected and V0 of the sensor is connected to D6 of arduino UNO

3.2 Temperature and Humidity Sensor (DHT11)

Temperature and humidity sensor is a low-cost sensor in which the temperature ranges from 0 to 50 °C and humidity ranges from 20 to 90% RH [13]. It contains voltage collector to collector (VCC), DATA, ground (GND) and one pin is not in use. DHT11 is small in size, utilises less power and the range of signal transmission is 20 m.

DHT11 consists of 3 pins VCC, GND, DATA in which VCC of DHT11 is connected to the 5 V power supply of arduino UNO, GNDs of both sensor and arduino UNO are connected and DATA pin is connected to D7 of arduino UNO.

3.3 Water Level Sensor

Water level sensor is used to detect the change in the level of water in the tank. It is an ultra low cost, low power consumption, lightweight and reliable [14].

Water level sensor consists of 3 pins VCC, GND, DATA in which VCC of sensor is connected to the 5 V power supply of arduino UNO, GNDs of both sensor and arduino UNO are connected and DATA pin is connected to A0 of arduino UNO.

3.4 ThingSpeak

ThingSpeak is a cloud platform which is used to store, analyse and visualise the stream data in graphical form. It is an open-source IoT application to retrieve the data using the HTTP protocols over the Internet [15]. Data from the sensors is collected using WiFi module and API key is used for connecting the WiFi module and ThingSpeak.

ESP8266 consists of RX, VCC, GPIO 0, RESET, CH_PD, GPIO 2, TX, GND inn which GNDs are connected in between ESP8266 and arduino UNO, VCC of ESP8266 is connected to 3.3 V of arduino UNO, TX of ESP8266 is connected to D3 of arduino UNO and RX of ESP8266 is connected to D2 of arduino UNO.

3.5 Relay Module

The relay module is a switch which can be controlled by microcontrollers. It is used for remote device switching. They are used to control high voltage signals coming from the hardware devices [16]. It gets control over the irrigation by comparing the levels of other sensors.

VCC of the relay module is connected to 5 V power supply of arduino UNO, GNDs of both relay module and arduino UNO are connected and INPUT1 of relay module is connected to D3 of arduino UNO. A negative terminal of the motor is connected to Normally open port in K1 of relay module, copper wire is connected between a common port in K1 of relay module and one port of AC socket. The positive terminal of the motor is connected to the other port of the AC socket.

3.6 PIR Motion Sensor

PIR sensor is used to detect the movement of animals or humans near the agriculture field. The sensor can cover a distance of 7 m with an angle of 120° [17].

PIR motion sensor consists of 3 pins VCC, GND, OUTPUT in which VCC of DHT11 is connected to the 5 V power supply of arduino UNO, GNDs of both sensor and arduino UNO are connected and OUTPUT pin is connected to D13 of arduino UNO.

4 Proposed System

Overview of the proposed system shown in Fig. 1, the components used in the proposed system are Arduino, WiFi Module, relay module, water pump motor, temperature, humidity, water level sensor and PIR motion sensor to detect the presence of animals near the crop field. Arduino IDE is a software which is used as an interface for executing the programs for Arduino Uno [18]. Thingspeak is an API used in the system to store and analyse the sensor data received from sensors [19]. WiFi module (ESP8266) acts as an interface to transmit the data from the used hardware to the thingspeak. Relay module is a remote switching device in which it switches on if the sensor crosses the threshold value [20]. The sensors are connected to the arduino uno using the jumping wires to read the data from the sensors by executing the arduino code. The received data is stored in thingspeak and the farmers can access the public channel in the software to view the information. The proposed system sends a notification to the farmer about the climatic conditions of the crop and automates the irrigation by using the sensor data. Two users from Fig. 1 represent the farmers who will be able to view the results. A web based user interface is enabled to the farmers to view the results provided by the cloud platform.

Figure 2 describes the workflow for the IoT enabled device for crop monitoring

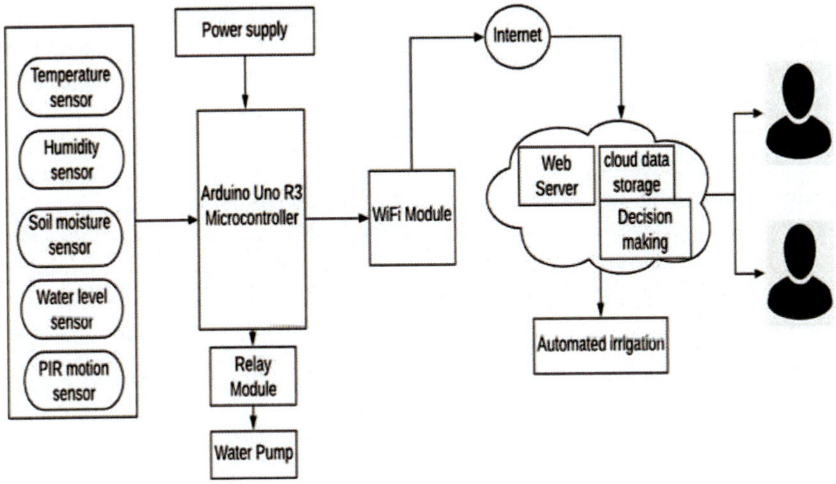

Fig. 1 Block diagram for IoT enabled crop monitoring and irrigation automation

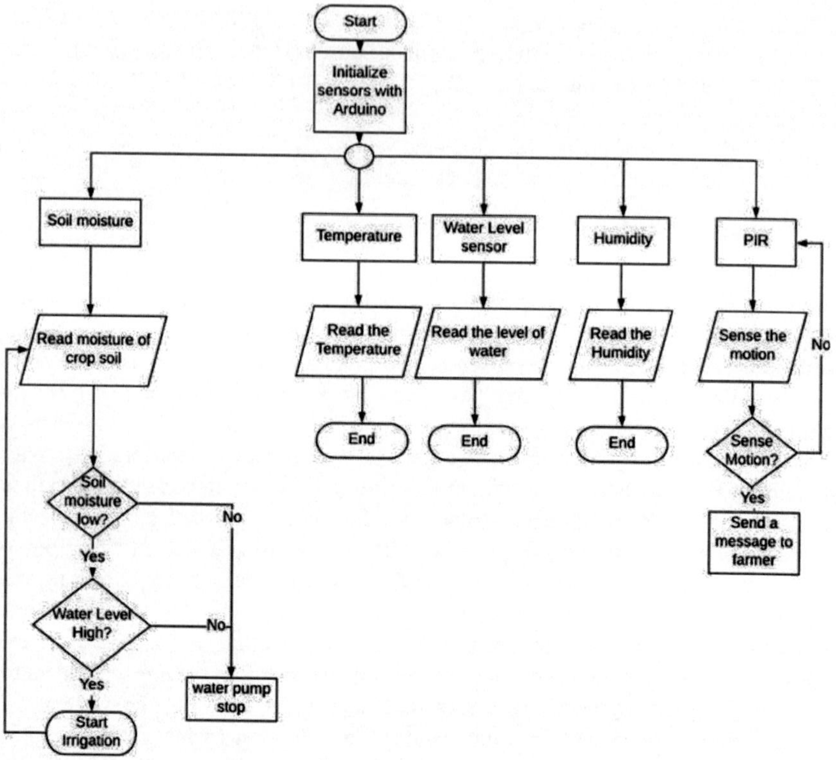

Fig. 2 Work flow of IoT enabled smart crop monitoring and irrigation automation

and automated irrigation. The algorithm shows the process involved in the arduino software to deliver water to the plants automatically. The proposed system involves 4 sensors denoted as S_1, S_2, S_3, S_4 and this sensor data is stored in the cloud database. Using the data provided from the soil moisture sensor and water level sensor the automation of irrigation is processed in the crop. If the value of S_1 is low and S_3 is high then the motor will pump the water from the tank, else in other conditions the motor will not pump the water from the tank. Automated irrigation is triggered by soil moisture sensor (S_1). The information about the climatic conditions of the crop is sent to the farmers using the data from the thingspeak cloud service. Analysis of this sensor data gives a clear cut idea for the farmers to take the correct decision for the crops.

Algorithm: IoT Enabled Device for Monitoring Agriculture Field and Smart Irrigation System **Input**: The values from the 3 sensors $S_1 \leftarrow$ Soil Moisture, $S_2 \leftarrow$ Temperature, $S_3 \leftarrow$ water level sensor, $S_4 \leftarrow$ Humidity

Output: Email notification to the farmer about the condition of the crop.

1. Start

2. Read the values from the sensors S_1, S_2, S_3, S_4.
3. Print Stream Datasheet.
4. IF S3 is High && S1 is Low
    4.1. THEN GO TO 8
    4.2. Else GO TO 4
5. IF S3 is High && S1 is High
    5.1. THEN GO TO 9
    5.2. Else GO TO 4
6. IF S3 is Low && S1 is Low
    6.1. THEN GO TO 9
    6.2. Else GO TO 4
7. IF S3 is Low && S1 is High
    7.1. THEN GO TO 9
    7.2. Else GO TO 4
8. Start Irrigation i.e. Motor ON
9. Stop Irrigation i.e. Motor OFF
10. Send email notification to the farmers
11. IF power is ON
    11.1. THEN GO TO 2. else 12
12. Device is in OFF State
13. Stop

5 Implementation

The Hardware implementation of IoT devices for crop monitoring and irrigation automation is shown in Fig. 3, the sensors such as soil moisture, temperature, humidity, water level and PIR motion are connected to the arduino uno. DHT11 is a sensor for measuring both temperature and humidity. Sensors used in this system provide values for every 17 s. Arduino is directly connected to WiFi module to send the sensor data from arduino to cloud storage by running a software program in arduino IDE. Irrigation is automated when the moisture content in the soil is reduced. The data stored in thingspeak cloud service is processed and analysed to send a message about the condition of the crop.

Fig. 3 Hardware implementation of smart crop monitoring and irrigation automation

6 Results and Discussion

Figure 4 shows the graphical representation of the soil moisture sensor and temperature sensor data. Twilio API is used for sending messages to the farmer [21]. The climatic condition of the crop can be checked by the farmer in any area. Figure 5 shows the thingspeak datasheet for the soil moisture sensor. A friendly web based user interface is created for farmers to visualise the data received from sensors in a graphical form.

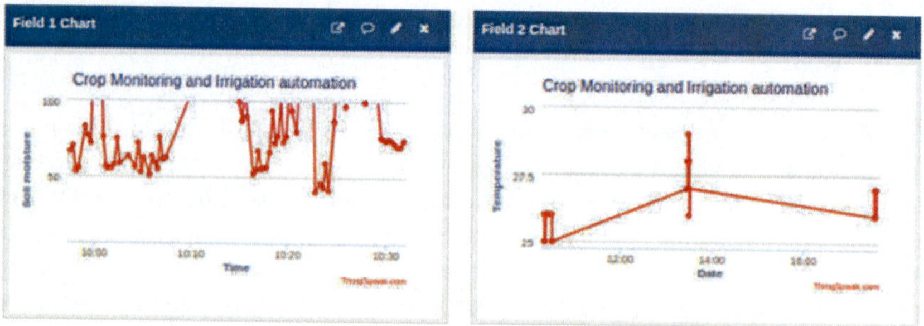

Fig. 4 Real time data analysis of Soil moisture sensor and temperature sensor

Fig. 5 ThingSpeak streamed data sheet for soil moisture sensor

Timestamp	entry_id	moisture
2019-11-30 04:27:30 UTC	167	66
2019-11-30 04:27:47 UTC	168	70
2019-11-30 04:28:04 UTC	169	53
2019-11-30 04:28:21 UTC	170	55
2019-11-30 04:29:01 UTC	171	83
2019-11-30 04:29:18 UTC	172	77
2019-11-30 04:29:38 UTC	173	72
2019-11-30 04:30:09 UTC	174	208
2019-11-30 04:30:26 UTC	175	208
2019-11-30 04:30:55 UTC	176	76
2019-11-30 04:31:12 UTC	177	56
2019-11-30 04:31:29 UTC	178	55
2019-11-30 04:31:46 UTC	179	55
2019-11-30 04:32:03 UTC	180	57
2019-11-30 04:32:20 UTC	181	75
2019-11-30 04:32:37 UTC	182	58
2019-11-30 04:33:28 UTC	183	63
2019-11-30 04:34:13 UTC	184	56
2019-11-30 04:34:31 UTC	185	72
2019-11-30 04:34:48 UTC	186	53
2019-11-30 04:35:05 UTC	187	62
2019-11-30 04:35:40 UTC	188	51
2019-11-30 04:35:57 UTC	189	63
2019-11-30 04:36:14 UTC	190	59
2019-11-30 04:36:31 UTC	191	55
2019-11-30 04:36:48 UTC	192	76
2019-11-30 04:37:05 UTC	193	61
2019-11-30 04:37:22 UTC	194	62

7 Conclusion and Future Scope

In this paper, a smart IoT enabled device is proposed for efficient crop monitoring and irrigation automation for agriculture fields. The experimental tests proved that the created device is cost effective and optimised the use of water resources in agricultural land. The IoT enabled device is successfully implemented and checked in various cultivating areas. By using this approach, farmers can make the right decisions about the crop by considering the information provided by the device. Deployment of this smart device in agricultural fields improves the production of the crops. The automation of irrigation is developed to reduce the utilisation of labour in the agriculture fields. Usage of solar panels for producing electricity is efficient as the electric power supply is expensive. This proposed system is reliable and scalable to any kind of other monitoring devices.

In future, this system can be improved by using data mining algorithms to predict the water requirement to the crops. This prediction helps to deliver an exact amount of water to the plants in the agriculture fields.

References

1. AshifuddinMondal M, Rehena Z (2018) Iot based intelligent agriculture field monitoring system. In: 2018 8th international conference on cloud computing, data science & engineering (Confluence), IEEE, pp 625–629
2. Rao RN, Sridhar B (2018) IoT based smart crop-field monitoring and automation irrigation system. In: 2018 2nd international conference on inventive systems and control (ICISC), IEEE, pp 478–483
3. Agilan VC, Rajasoundaran S (2019) Analysis of IoT based crop field monitoring and leaf nutrients. In: IJIRMPS Int J Innov Res Eng Multidisc Phys Sci 7(3)
4. https://www.arduino.cc/en/guide/introduction
5. Balaji GN, Nandhini V, Mithra S, Priya N, Naveena R (2018) Iot based smart crop monitoring in farm land. Imperial J Interdisc Res (IJIR) 4:88–92
6. Suma N, Samson SR, Saranya S, Shanmugapriya G, Subhashri R (2017) IOT based smart agriculture monitoring system. Int J Recent Innov Trends Comput Commun 5(2):177–181
7. Rupanagudi SR, Ranjani BS, Nagaraj P, Bhat VG, Thippeswamy G (2015) A novel cloud computing based smart farming system for early detection of borer insects in tomatoes. In: 2015 international conference on communication, information & computing technology (ICCICT)
8. Danita M, Mathew B, Shereen N, Sharon N, Paul JJ (2018). IoT based automated greenhouse monitoring system. In: 2018 second international conference on intelligent computing and control systems (ICICCS), IEEE, pp 1933–1937
9. Ghanshala KK, Chauhan R, Joshi RC (2018) A novel framework for smart crop monitoring using internet of things (IOT). In: 2018 first international conference on secure cyber computing and communication (ICSCCC), IEEE, pp 62–67
10. Sabri FN, Hanif NH, Janin Z (2018) Precision crop management for indoor farming. In: 2018 IEEE 5th international conference on smart instrumentation, measurement and application (ICSIMA), IEEE, pp 1–4
11. Mohanraj I, Ashok Kumar K, Naren J (2016) Field monitoring and automation using IOT in agriculture domain. Procedia Comput Sci 93:931–939
12. https://www.baselinesystems.com/mediafiles/pdf/SensorSpec.pdf
13. https://components101.com/dht11-temperature-sensor
14. http://www.powersupplyindia.com/pdf/WLC-AllA.pdf
15. https://www.mathworks.com/help/thingspeak/
16. https://randomnerdtutorials.com/guide-for-relay-module-with-arduino/
17. https://components101.com/hc-sr501-pir-sensor
18. Maidin SN, Azizan A, Abas H, Sam SM, Hamid MJA (2018) Review on implementation of IoT for environmental condition monitoring of crop or farm in agriculture sector. Open Int J Inf (OIJI) 85–95
19. Rahman SSB, Ahmed I, Ahmed F (2020) An IoT based model of a nitrogen detection system for soil samples. In: Proceedings of the international conference on computing advancements, pp 1–5
20. Islam A, Akter K, Nipu NJ, Das A, Rahman MM, Rahman M (2018) IoT based power efficient agro field monitoring and irrigation control system: an empirical implementation in precision agriculture. In: 2018 international conference on innovations in science, engineering and technology (ICISET) (pp. 372–377). IEEE
21. Parameswaran G, Sivaprasath K (2016) Arduino based smart drip irrigation system using Internet of Things. Int J Eng Sci 5518

Deep Network for Network Intrusion with Concept Drift

Shivam Prasad, Osho Agyeya, Prateek Singh, and Shridevi S. Krishnakumar

Abstract A deep learning approach has been proposed for the classification of cluster instances as being intrusive or not intrusive. Mini-batch Adam optimizer was used due to a large number of hidden layers in the model. Massive amounts of data accumulated for training prevented the model from overfitting. After extensive testing of data with various algorithms, it was found that deep learning model with Adam optimizer outperformed others.

Keywords Anomaly detection · Deep learning · Keras · Adam · Machine learning · Concept drift

1 Introduction

Privacy is paramount for people. Institutions handling information have to make sure that the systems and algorithms are dependable and secure. This can be achieved by modern algorithms for machine learning and data mining. By advancements in the digital era, our life is surrounded by electronic devices. All these devices are connected through the Internet and generate massive amounts of data every second. Although remarkable progress has been made in the last decade, and also had a surge in the rise of cyberattacks. As more devices get interconnected, it makes them more vulnerable to being misused and taken advantage of. The problem is that the

S. Prasad (✉) · O. Agyeya · P. Singh · S. S. Krishnakumar
School of Computing Science and Engineering, VIT, Chennai, India
e-mail: shivam13juna@gmail.com

O. Agyeya
e-mail: oshoagyeya123@gmail.com

P. Singh
e-mail: prateeksingh0001@gmail.com

S. S. Krishnakumar
e-mail: shridevi.s@vit.ac.in

G. Ranganathan et al. (eds.), *Inventive Communication and Computational Technologies*, Lecture Notes in Networks and Systems 145,
https://doi.org/10.1007/978-981-15-7345-3_79

rate at which the data has overtaken the improvement in security protocols. Network anomaly detection is one of the most important security protection techniques besides preexisting techniques such as firewalls, anti-virus software, and malware detection tools. Proposed network anomaly detection provides adaptive security that other algorithms can't provide. A strong security protocol is a requisite for our safety and privacy, and need to make sure the model built is secure, works with state-of-the-art methods and generates accurate results quickly.

1.1 Reasons for Using a Machine Learning Approach

Massive amounts of research have been conducted in this field [1]. Machine learning/deep learning is the best methods as of today which can adapt to ever-increasing data regarding intrusion [2]. These algorithms detect anomalous patterns which generic algorithms find hard to locate. Machine learning is scalable, present-day constraints are memory and time.

Machine learning involves building a predictive model for a set of data. Consider a set of training data with x as observation, y as the class label corresponding to each element of training data x. and have a function $y = f(x)$ where f is unknown. It is the function that has to model based on the observed data and class labels. Another subset of machine learning, called deep learning has been implemented for detecting malicious intrusions. Primarily batch method of neural networks has been used. A different set of tests have been performed using the batch method. Each method has its benefit, for instance in a time or data sensitive environment a stochastic gradient descent method is preferred to enable the model to converge to global minimum faster.

2 Related Work

A substantial number of anomaly detection methods have been implemented in Machine Learning in Cyber Trust [3]. Chapter 5 demonstrates various learning algorithms like winnow, stagger, Naive Bayes, ensemble Methods. They have used a supervised machine learning algorithm that builds a predictive model using training data. Another model has been proposed here for predicting unlabeled data. Every learning method can be built and expressed in 3 stages: input shape of the data has to be defined that we plan to train our algorithm on; hyper-parameters are to be tuned for optimal learning; performance analysis of our algorithm for comparative analysis [4]. This decides how our algorithm worked on the unclassified data. More focus has been given to the batch learning method, and its results have been used as a benchmark for further improving accuracy.

2.1 Detecting Malicious Instances

The method used in this paper is STAGGER. It is the ability to deal with concept drift. Instead of storing and feeding on data normally it uses a distributed concept description which contains class nodes that are linked to nodes of attributes values by probability-arcs. In this algorithm, the previous probabilities associated with each arc represent the need and if it is enough, they are again updated on association learning theory. STAGGER doesn't just adjust the probabilities, new nodes can also be added to make sure new concepts can be dealt with. Also, stagger can diminish its probabilities over time, depending on the data it feeds. It is a predictive algorithm that can handle concept drift [5]. The fact of how stagger deals with concept drift have been tried to implement with deep networks.

2.2 Concept Drift

This paper has been built on the concepts obtained from the survey conducted in 2015 [6]. There's been a lot of work to acclimate with something called concept drift [7]. Concept drift is explained further means that the statistical values of a variable that is the target, which any considered model is trying to predict, tend to change in unforeseen ways which make the prediction of the algorithm less accurate. In the proposed implementation of the algorithm, a suitable action has been taken to make sure it keeps changing with data flow, which has been mentioned later. The difference between the concept drift and noise needs to be understood. An algorithm should not adapt to noise too much which might cause irregularities. More often than not algorithms interpret noise as drift and adjusting to changes. A suitable algorithm should not get affected by robustness and sensitivity to concept drift. All kinds of concept drift should be taken care of, which are of mainly two types: sudden and gradual. Quite often changes or drift in data is not because of change in the concept of a target but due to change in the underlying data distribution. The class label may remain the same while only the data distribution changes. The algorithm needs to know which type of drift it is dealing with. STAGGER (Schlimmer and Granger) was the first algorithm capable of handling concept drift. The generic approach so far has been instance selection, instance weighting and ensemble learning (or learning with a various description of concept) [8]. Most common concept drift handling technique is instance selection. It works based on windows. Here, the window size is the size of the current extent of concept drift. The proposed model handles the concept of drift differently.

3 Proposed Solution

The problem being dealt with is intricate and requires high adaption. After trial and error, it was found that deep neural networks perform the best. As neural networks are an artificial copy of human brain cells, they can learn based on training input and with appropriate programs they can be even trained to forget learned parameters that are no longer relevant because of cases like concept drift. Deep learning is more complicated replication of the human brain. Industry grade neural networks can have 100's of layers. Each layer of nodes trains on a unique set of features that is the output of the previous layer. The more layers there are in a neural network, the more intricate features our nodes can recognize, also termed as feature hierarchy. Deep learning is preferred for a hierarchy of increasing complexity which makes such network handle massive dimensional data sets with billions of parameters which again pass through nonlinear functions. Keras framework was utilized to code the neural network which gave substantial results.

Further, the proposed algorithm satisfies the following conditions:

- Flexibility.
- It should be resistant to garbage data or outliers and react to data that changes, instead of noise.
- It should be able to deal with intricate drift in data.
- Scalability without compromise to accuracy.

3.1 Pre-Processing the Data

In data pre-processing, all the other version of the attack form has been replaced with keyword 'attack'. Classifier made has to classify an instance of data as either of two, 'normal' and 'attack'. After that, label encoder has been used for labeling 3 instances that are present in TCP, that is TCP, UDP or it can be ICMP. Then label encoder has been used again in HTTP column for it can be HTTP, STMP, finger and various other types. Another column in which label encoding has been used is to encode Flag (v4) for encoding different kinds of flag RSTOS0, RSTR, RSTO etc. Standard-scaler has been used as pre processing algorithm to standardize features by removing the mean and scaling it to unit variance. Each feature scaling and centering happens independently by calculating relevant statistics on the samples of the training set which we used to train our algorithm on. Also, the standardization of the data set is a common requirement for most of the machine learning algorithms. They perform better if individual features are normally distributed, implying a Gaussian distribution with zero mean and unit variance. A feature whose variance is in much more magnitude than others can retard the performance of a machine learning algorithm. Train test split has been used to shuffle and categorize the data.

3.2 Kernal Initializer and Activation Function

Uniform kernel initializer has been used. It is a way to set the initial random weights of deep neural network layers. There is something known as bias initializer but adding it did not have any positive effect on our result so it was discarded. The activation function is an important feature of the neural networks. It decides whether the information the neuron is receiving is relevant for the given information or should it be ignored. ReLU (Rectified Linear units) has been used for this purpose. It can back-propagate the errors. It doesn't activate all the neurons in a layer at the same time. It activates only a few neurons and uses them for making the network sparse and computation easier. Yes, it has its fair share of disadvantages of having vanishing gradients. In case the neuron dies and gradient is zero, the weights are not updated during backpropagation.

3.3 Designing Layers for Deep Learning

Since a lot of data based on packets and other features was available, several dense hidden layers have been designed. In the first layer (input layer) have 16 nodes with input dimension as 30. Kernel initializer is uniform and activation function is RELU. Now in the hidden layer there are 18 units with activation RELU. Drop out as 0.20. Batch size is set as 100. Dropout [9] prevents overfitting in deep learning. There are a large number of hidden layers which means a large number of parameters to take care of. Large networks are also extremely slow considering they have to compute so many parameters and combine predictions of its layers. The main idea behind dropout is that drop neurons are dropped along with their connections. It prevents units from adapting to each other. Also, at test time it is easy to approximate effect of averaging the predictions of all these "thinned" networks obtained by randomly droppings neurons. Individual node is either dropped out of the net with probability 1-p or kept with any random probability p, so the network that left is thinned. For each hidden layer, 0.25 fraction of nodes will be ignored, as 0.25 is set as dropout parameter. Then in the testing phase after using all activations, the network size will be reduced by 0.25.

Dropout will force our network to learn more intricate features that are more useful concerning different subsets of other neurons. It roughly doubles the number of iterations that are typically required to converge. Since have even lesser number of neurons the training time is less for each epoch. So with H hidden units, each of the units that can be dropped and have an exponential number of models. In testing phase the entire network is usually taken care of and each activation is reduced by a factor of dropout parameter. This significantly reduces over-fitting and gives huge improvement than other contemporary regularization methods. So after placing dropout for 0. Next hidden layer will have 60 units along with same kernel initializer and activation RELU. Followed by couple of hidden layers which has 60 units with

Fig. 1 Architecture of neural network

same parameters as before. Final output layer has 1 unit with same kernel initializer and activation as sigmoid (with a threshold of 0.5), since the output is supposed to be binary, i.e. 0 or 1.

3.4 Optimizer, Loss and Accuracy Metric

The optimizer is one of the two arguments we need to have for compiling a deep learning model. Adam is the optimizer. Adam [10] is a method for stochastic optimization. It was introduced for first-order gradient-based optimization of stochastic objective functions. It is based on adaptive estimates of lower-order moments. It is computationally efficient and has minuscule memory requirements. It is not variant to a diagonal rescaling of the gradients and is well suited for big data sets or a large number of parameters. All the hyperparameters used in this method have intuitive meaning and do not require manipulation often. Binary cross entropy has been explained well in Stochastic Runtime Analysis of the Cross-Entropy Algorithm [11]. As cross entropy deals with probability distribution of two units with same events exposed or the average amount of bits required to identify a particular event drawn from a set, that is if the scheme of coding is fitted for a not normal probability distribution rather than true ones. Loss parameter has been set to binary cross entropy. Accuracy metric's job is to compute subset accuracy, the set of labels predicted by algorithm implemented. Libraries provide many ways to calculate and report with a plethora of standard metrics while training deep neural network.

4 Results

Results have been presented after running the designed deep network over the entire data set. It turned out to be the best with an accuracy of 0.9981. This method didn't only achieve the highest overall accuracy but also performed very well on the examples outside test data set i.e. random data contrived with features of anomalous data. Dr. Bernhard Pfahringer of the Austrian Research Institute for Artificial Intelligence has an accuracy of 0.995 which was highest for all the methods used in research till now. On training the entire network without dropout, accuracy was less than 0.97. After checking how dropout affects the performance of the implemented network, different variations of dropout parameters have been tested, in which 0.25 had maximum value among all others. Some of the methods didn't perform good enough in terms of overall accuracy, primarily because of poor performance in classifying negative instances.

Stagger was a really good example considering its probabilistic nature and accuracy of 0.925 but even with tweaked hyperparameters, its accuracy did not compare to that of a deep network. Deep neural networks outperform others. Table 1 constitutes the performance of models in 47,000 samples. Training models of compared algorithms had different data. They have been trained in 3700 examples.

Table 1 Results obtained from various models

Method	Accuracy
Artificial neural network	0.9981
Dr. Bernhard Pfahringer	0.9951
Kneighbors	0.899
Decision trees	0.911
Naive Bayes	0.961
Quadratic discriminant analysis	0.901
Kernel miner	0.833
Quadratic discriminant analysis	0.901

5 Conclusion

In previous sections, a large number of issues such as accuracy, type of data and the learning rate were discussed. There are certain restrictions on deep networks. Such deep learning algorithms are good only if they're given enough time and a large amount of data [9]. One of the drawbacks of neural networks is that they can't learn nearly fast enough considering parameters of the reasonably short network is close to 20,000. A considerably large amount of time to train all those parameters for the reasonably accurate outcome is needed, but once trained, they outperform most of the other algorithms. On an HP Omen 17, W2490tx running at 3.2 GHz and 16 GB of RAM, training deep network took more than 15 h to train on the data. Besides the other machine learning algorithms took less than a night to train on the model. For one or more new example it will take nearly the same time for processing, but for several examples, the deep network again needs more time than other algorithms to train. All such algorithms have also been tested with ten-fold cross-validation set. Proposed network has been tested with a various number of batch size, with the highest accuracy in a batch size of 100. Further research can be done by training another class of neural networks on the data.

References

1. Peng-Lin L (2013) Research of anomaly detection of laboring statistical data based on DBSCAN cluster algorithm. In: IEEE, Changchun
2. Perlovsky L, Shevchenko O (2014) Cognitive neural network for cyber-security
3. Tsai JJ, Philip SY, Machine learning in cyber trust security, privacy, and reliability
4. Singh S (2018) Using the power of deep learning for cyber security. Available: https://www.analyticsvidhya.com/blog/2018/07/using-power-deep-learningcyber-security/
5. KS D, Ramakrishna BB (2013) An artificial neural network based intrusion detection system and classification of attacks. Available: https://pdfs.semanticscholar.org/79c8/850c8be533c241a61a80883e6ed1d559a229.pdf
6. Agrawal S, Agrawal J (2015) Survey on anomaly detection using data mining techniques. In: 9th international conference on knowledge based and intelligent information and engineering systems
7. Hoens TR, Polikar R, Chawla NV (2012) Learning from streaming data with concept drift and imbalance: an overview, progress in artificial intelligence
8. Norouzian MR, Merati S (2011) Classifying attacks in a network intrusion detection system based on artificial neural networks. In: 13th international conference on advanced communication technology (ICACT 2011). IEE XPLORE
9. Srivastava N, Hinton G, Krizhevsky A, Sutskever I, Salakhutdinov R (2014) Dropout: a simple way to prevent neural networks from overfitting. J Mach Learn Res 15
10. Kingmaand D, Ba J (2015) Adam: A Method for Stochastic Optimization. In: 3rd international conference for learning representations, San Diego
11. Wu Z, Kolonko M, Möhring RH (2017) Stochastic runtime analysis of the cross entropy algorithm. In: IEEE transactions on evolutionary computation

Paillier Homomorphic Encryption with K-Means Clustering Algorithm (PHEKC) for Data Mining Security in Cloud

G. Smilarubavathy, R. Nidhya, N. V. Abiramy, and A. Dinesh Kumar

Abstract In the current scenario, data mining is very significant in the IT and business sector, to realize the business goals. Data mining helps to achieve these goals, which converts the data into knowledge and plan for achieving the goals in all sectors. In the IT sector, the current necessity is cloud. For data processing, sharing the resource, transfer the data and storage, the cloud is essential. But the main threats in the cloud are privacy and security. Hence, a system Paillier homomorphic encryption with K-means clustering algorithm (PHEKC) is proposed for data mining security. In this method, the cloud server did not know about the uploaded data. Only the client can get the result. From this proposed method, the data is secure and confidential.

Keywords Homomorphic encryption · Cloud · Data mining · K-means · Flocks

1 Introduction

In this era, there has been massive information; the data amount is more gigantic, risk and complex. The security for data is unreliable and overloaded. Data warehouses stored the data. To help the project, data mining is an improvement for keeping the

G. Smilarubavathy (✉) · N. V. Abiramy
Computer Science and Engineering, Dr. N.G.P Institute of Technology, Coimbatore, India
e-mail: smilarupa@gmail.com

N. V. Abiramy
e-mail: abigoldenmani@gmail.com

R. Nidhya
Computer Science and Engineering, Madanapalle Institute of Technology & Science, Madanapalle, Andra Pradesh, India
e-mail: nidhuraji88@gmail.com

A. Dinesh Kumar
Computer Science and Engineering, KL Deemed to be University, Vaddeswaram, Andhra Pradesh, India
e-mail: dineshngpit@gmail.com

G. Ranganathan et al. (eds.), *Inventive Communication and Computational Technologies*, Lecture Notes in Networks and Systems 145,
https://doi.org/10.1007/978-981-15-7345-3_80

data tabs in warehouses. Cloud computing is an Internet-based computing; here the shared resources, data and software are offered to the computers, and others are on demand. Cloud computing has more advantages in the economical sides. In this distributing computing situation, the fame of cloud computing is increased more. For the need for data storage and data processing, cloud computing makes a new tendency. By the use of cloud computing, without installation, consumers and business sectors access their files at any computer with the help of Internet. It gives the technological expertise infrastructure to the user. The cloud computing full prospective is used as data is transferred, stored, processed, retrieved by external cloud providers. But the data vendors are unreliable, for processing the outside data in their restricted circle.

Both private and public sectors, data mining used everywhere. The major applications of data mining are banking, health care, insurance, retail, research enhancement, business and IT sectors [1]. Their main issues are integrity, confidentiality, security and data mining methods in the cloud. In public cloud services, the biggest concern is security. The organization regularly placed an antivirus software, firewalls and IDs. Every security component protects the part of the network, but the security may affect when it interrelates with other components.

2 Related Work

The cloud consists of the network combination, computing resources, application of business, management solutions and storage of data. In this new era, it supports customer service and information technology. Cloud computing provides different schemes at least cost. It enables data storage and also offers many information security levels. Google and Microsoft are the cloud providers which consist of the massive amount of data and adopt different kinds of techniques in data analysis for extracting the important information. Depending upon the search history, they used data mining techniques in it [2, 3].

There are three types of services in cloud: They are software as a serve (SaaS), platform as a service (PaaS) and infrastructure as a service (IaaS). The main technology in organizations is given in the cloud facilities to different firms. Software and hardware are the computing resources in the cloud. Through the Internet, these services are distributed. Instead of providing the host substitute on third-party system [4], firms ignore for constructing their infrastructures for information technology.

In [5, 6] from different aspects, the data mining process is used for analyzing the data and converts it into knowledgeable information. This information is used in various types of application like economics and business. It is significant to discover knowledge. For extracting information and forms, which can be recognized by a human, data mining is used. The providers of cloud computing use data mining for giving cost-effective services to users. The ethical issues like privacy and individuality are disrupted, while the client is not known if their data are collected. If the data is misused by the cloud providers, it leads to severe risk to data privacy. The attacker should not have access authorization in the cloud. Likewise, the attacker has no opportunity for mining the cloud data. But the attackers used the cheap and raw power of cloud computing power for data mining and got the necessary information from cloud [7, 8].

3 Contributions

The main objectives are

- To identify different kinds of threats.
- To improve cloud security via data mining techniques, Paillier homomorphic encryption with K-means clustering algorithm is used.
- To give security while retrieving the backup data.

3.1 System Architecture

In Fig. 1, the client uploads the data into the cloud. Then the uploaded data is stored into the database. After that, K-means clustering is applied to it for obtaining the distributed data between various hosts of the cloud. Paillier homomorphic encryption is applied into it for encrypting the plain text which is stored in the database. Only the authorized user can decrypt the data with the help of the private key. Data computation in the encrypted text is occurred by the homomorphic encryption [9].

4 Proposed Methodology

In this paper, Paillier homomorphic encryption with K-means clustering algorithm is used, which maintains the privacy in content from eavesdrop and snooping. The user knows the input, and the final output is not affected by any attacks, which improves the performance compared to the existing system. Data security on the cloud is mainly focused on this method. AES algorithm with K-means clustering

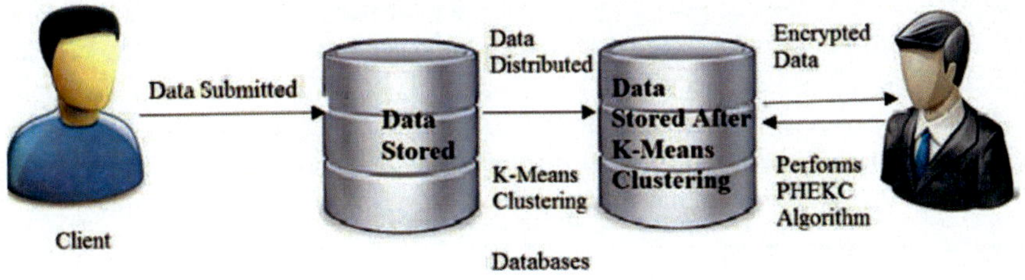

Fig. 1 System architecture

is used to improve the performance. Moreover, Paillier homomorphic encryption is used for extra restriction in security [10]. The private key and public key have to be constructed. This public key encryption is used by the proposed method where PT is the data or plain text. It should be encrypted. It can be classified into three parts as K-means clustering, encryption and decryption.

- The (pri_k, pub_k) considered as a private key and public key, which has to be constructed.
- The encrypted message is considered as $z = Enc(pri_k(d, rv)$ is attained, where $d \in D$ and rv are a random value.
- For getting plain text, decryption $D(pub_k(z) = PT$ is used.

4.1 Data Normalization

Usually, for data submission, a standard XML document is submitted which maintains the data standard, but has to manage a multivariate database in the current era, for example a multi-attribute database, in which the variable value gets through the total number of various attributes [10]. Hence, some of the large value variables probability is high, which can control the whole data. The multi-attribute data is used in the normalization method with the help of private calculation of mean data objects.

- Let Flock X contain $d_x = \sum_{j=1}^{m} d_j^x$ with m number of data entries.
- And Flock Y contains $d_y = \sum_{j=1}^{n} d_j^y$ with n number of data entries.
- Hence, $PT = \frac{d_x + d_y}{x+y}$

Paillier Homomorphic cryptosystem is constructed, which cannot be interrupted by the attacker. This data is locally consistent by using the mean value as $A_i = A - N$ for all objects of data A_i.

4.2 Distance Measurement

After data standardization, a basic K-means cluster is implanted by all cloud host on their private data sets. For every attribute, the cluster center gets initialized, and the data objects get assigned to the neighborhood cluster center with the help of Manhattan or Euclidean distance. It was selected based on the database or any application. Like $U_1^X, U_2^X, \ldots, U_n^X$ for Flock X and $U_1^Y, U_2^Y, \ldots, U_n^Y$ for Flock Y. These cluster centers are locally computed, and the protocol for security is not essential but the cluster update is essential. Find the joint centers and compute it privately.

4.3 Cluster Update

The values of every data object in the ith attribute in jth cluster evaluate sum as $Sum_i^x = Z_{i,j} * m_i$. Here m_i denotes the data object's number for ith cluster $Sum_i^y = Z_{i,j} * n_i$ then the n_i denoted by the data object for jth cluster. The next ith attribute in jth cluster is $z_{i,j} = \frac{Sum_{ix}+Sum_{iy}}{m_i+m_j}$.

To perform the above calculation privately, Paillier Homomorphic cryptosystem is used. The private and public keys pair are randomly constructed by the third party. Using the public key of the third party encrypts the sum value as Flock X and Flock Y. Then public keys and encrypted value are sent to the third party. Combination of centers of local cluster is occurred at the iteration end which forms the center of global cluster. It is needed for subsequent local iteration. Still, the algorithm corrections stand as correct in the distributed environment.

4.4 Criteria for Iteration Stopping

K-mean clustering is naturally iterative. Hence, there should be a condition for stopping the iterations. After fulfilling the need for the output, this condition of iteration is completed. This condition applied in K-means denotes the calculation of consecutive cluster Euclidean distance should be lower than the threshold. i.e., $Dist(Z_j, Z_{i+1}) = Dist(U_i^{X1+1} + U_i^{Y1+1} + U_i^{X11} + U_i^{Y11}) < (U_i^{X11} + U_i^{Y11}) - (U_i^{X1+1} + U_i^{Y1+1}) <$ threshold. For checking, the Flock X calculates $E(U_i^{X1} - U_i^{Y1+1})$ and Flock Y computes $E(U_i^{Y1} - U_i^{Y1+1})$ which is placed locally with the third public key of the third party. The intermediated multiplication encrypted values done by the third party which can be decrypted by the private keys of Flock X and Flock Y is shown as $R = D(E(U_i^{X1} - U_i^{Y1+1}) * E(U_i^{Y1} - U_i^{Y1+1}))$ if $R < threshold$. The outcome is fulfilled, and the iteration value has stopped.

5 Overall Description

Consider the multivariate database $M_D = \{\}$, where m is the attributes number, which contains the data of the user. Horizontally partition the data in the database, which stores into the two locations. i.e., Flock X and Flock Y. Flock $X = \{\}$ and Flock $Y = \{\}$. Using K-means cluster, the data mining is applied for the given data, which maintains the content privacy at both flocks and also prevents from snooping. The flocks can know the input values and output values but no intercept values.

5.1 Formulas for Encryption

For preserving the data privacy for each flock and the midway outcome which are transmitted from one place to another, here encryption is a need for the system. The encrypted data is applied to the specific function; the matching results are created by the pairing key for which the function used to decrypt the plain text. This kind of encryption is considering a homomorphic encryption system. Here, the Paillier cryptography is proposed for achieving the need for this methodology. Here $Enc(x) * Enc(y) = Enc(x + y)$ and $E(x)y = E(x * z)$ in this method where Enc refer as the encryption methodology.

5.2 Assumptions

The proposed approach assumes the attack, in which an attacker can reveal the other flock's data when the database has not secured, while preserves its privacy. In this approach, instead of placing the whole data at one place, first, the client gives the input as the data stored at different places as a chunk. If the data is stored in one place, the attacker can easily attack the whole data. So decentralized method is used to sore the user's data which is horizontally partitioning and stored.

5.3 Distribution of Data

Let us consider $M_D = \{M_{d1}, M_{d2}, \ldots, M_{dn}\}$ is a multivariate relational database in which Flock X has $M_d^X = \{M_{d1}^X, \ldots, M_{dl}^X\}$ and Flock Y has $M_d^Y = \{M_{dl}^Y, \ldots, M_{dl}^Y\}$. Each object of data is considered by a multivariate in the data set, the vector set $M_{da} = a_{j1}, \ldots a_{jn}$ where n is the attribute number. Present, Flock X has a private clustering set $U_{1x}, U_{2x}, \ldots, U_{xK}$ and $U_{1y}, U_{2y}, \ldots, U_{ky}$, and the clustering set is $(CS_1, CS_2, \ldots, CS_K) = \{U_{1x}+U_{1y}, \ldots, U_{Kx}+U_{Ky)}\}$ act as the joint cluster center. The cluster number is denoted by k.

6 Notations

The centers of combined cluster denote as $CS_i = U_x + U_y$ which is the sum of the data in the Flock X and Y, respectively.

Input

(1) The data in the database D_x and D_y depends upon Flock X and Y, which contains m number of data objects

Table 1 Cost comparison with other methodology

	Xiong et al. [11]	Liu et al. [12]	PHEKC
Client	$3T_{mul} + 2T_{hash} + 1T_{\exp}$	$2T_{hash} + 2T_{mul}$	$2T_{mul} + 2T_{hash} + 1T_{sym}$
Server	$1T_{mul} + 2T_{hash} + 1T_{\exp} + 1T_{bm}$	$3T_{mul} + 2T_{hash}$	$3T_{mul} + 3T_{hash} + 2T_{sym}$
Total cost	$4T_{mul} + 4T_{hash} + 1T_{\exp} + 1T_{bm}$	$5T_{mul} + 5T_{hash}$	$5T_{mul} + 5T_{hash} + 3T_{sym}$

(2) K denotes the cluster numbers.

Output

D_x and D_y or D has the cluster combination k.

(1) Local data normalization performed by each party.
(2) Flock X and Flock Y have chosen their cluster as $U_{1x}, U_{2x}, \ldots, U_{xK}$ and $U_{1y}, U_{2y}, \ldots, U_{ky}$ locally and randomly.
(3) Evaluate the K-means of local Flock X and Y.
(4) The cluster centers have been saved as $U_i X_j$, $U_i Y_j$.
(5) Update the cluster securely and resort the object of data to their nearest cluster locally.
(6) Save $U_{1y}, U_{2y}, \ldots, U_{ky}$. In case the existing cluster center and the current cluster center are less than or equal to 1, then iteration has to be stopped.

6.1 Comparison with Other Methodologies

In Table 1, the performance comparison convenience assumes $T_{mul}, T_{sym.}, T_{bm}, T_{hash} \& T_{exp}$ which shows the cost of time for point multiplication, bilinear mapping, symmetric encryption, exponential operation and hash function, respectively. Depending upon the Xiong et al. and Liu.et al. this analysis has been performed. This computation cost of [11, 12] is shown in Table 1. Xiong et al. perform bilinear operations and modular exponentiation, in which computational cost is $4T_{mul} + 4T_{hash} + 1T_{exp} + 1T_{bm}$. This computation cost of Liu.et al. is $5T_{mul} + 5T_{hash}$ which performs the authentication protocol, and it contains the 11.52 ms delay seconds. This proposed method contains a better cost as $5T_{mul} + 5T_{hash} + 3T_{sym}$ when compared to the existing system with the delay of 11.48 ms. Compared to the existing models, the above analysis shows the proposed method is better.

7 Conclusion

The major issue of data mining in the cloud is security and privacy. The cloud services consist of a huge amount of sensitive and confidential data which should be stored in the cloud. It also offers essential information to users. Attackers used sensitive information due to the security issue. These issues are solved by using the proposed PHEKC method. The data of the user is spread into two flocks, and then K-means clustering operation is performed. Then Paillier homomorphic encryption algorithm is used for security purpose, which prevents data from snooping.

References

1. Seifert JW (2007) Data mining and homeland security: an overview. CRS Report, PP 1–1
2. Hao L, Han D (2011) The study and design on secure-cloud storage system. In: Proceedings of the international conference on electrical and control engineering (ICECE'11) Yichang, pp 5126–5129
3. Gupta S, Satapathy SR, Mehta P, Tripathy A (2013) A secure and searchable data storage in cloud computing. In: Proceedings of the 3rd IEEE international advance computing conference (IACC'13), IEEE, Ghaziabad, pp 106–109
4. Dev H, Sen T, Basak M, Eunus Ali M (2012) An approach to protect the privacy of cloud data from data mining basedattacks. In: Proceedings of the 2012 SC companion: high performance computing, network storage and analysis (SCC'12), pp 1106–1115
5. Wang J, Wan J, Liu Z, Wang P (2010) Data mining of massstorage based on cloud computing. In: Proceedings of the 9th international conference on grid and cloud computing (GCC'10). Shanghai, pp 426–431
6. Van Wel L, Royakkers L (2004) Ethical issues in web datamining. Ethics Inf Technol 6(2):129–140
7. Yang Q, Wu X (2006) Challenging problems in data mining research. Int J Inf Technol Decis Mak 5(4):597–604
8. Torgo L (2010) Data Mining with R.: Learning with Case Studies. Chapman & Hall/CRC, New York, pp 345–357
9. Joshi R, Gutal B, Rajkumar G, Suryawanshi M, Wanaskar UH (2015) Data mining using secure homomorphic encryption. Int J Adv Res Comput Commun Eng 4(10):300–302
10. Maral V, Kale S, Balharpure K, Bhakkad S, Hendre P (2016) Homomorphic encryption for secure data mining in cloud. Int J Eng Sci Comput 4533–4536
11. Xiong H, Qin Z (2016) Revocable and scalable certificateless remote authentication protocol with anonymity for wireless body area networks. IEEE Trans Inf Forensics Sec 10:1442–1455
12. Liu J, Zhang L, Sun R (2016) 1-Raap: An efficient 1-round anonymous authentication protocol for wireless body area networks. Sensors 16:728

PortaX Secure Automation System Using Iot—A Survey

Aditya Venkatesh, Aishwarya Alva, Daniya Nausheer, Gagan Deep Shivamadhu, and K. A. Sumithra Devi

Abstract The world is immersed in the idea of automation and it promotes and makes life more accessible. In this day and age, travel has a lot of potentials to be automated, especially airports. Theft/loss of luggage, at airports or somewhere on the route, is one of the main issues to be solved. In this paper, PortaX Secure Automation System using IoT is proposed to handle the theft/loss of the baggage by integrating robots and tracking systems.

Keywords Security · Automation · Object-Following luggage · GPS · GSM · Location-tracking · Arduino

1 Introduction

The PortaX Secure Automation System will integrate robots, which operate in such a way that they obey their owner and also tell the owner that their baggage is protected in the location conveyed. It will use collaborative GPS-GSM systems that are combined using the IoT microcontrollers to provide the above functionalities. The tracking system will primarily be capable of sending the luggage position as it should allow

A. Venkatesh (✉) · A. Alva · D. Nausheer · G. D. Shivamadhu · K. A. Sumithra Devi
Department of Information Science, Dayananda Sagar Academy of Technology and Management, Bangalore, India
e-mail: adityavenkatesh69@gmail.com

A. Alva
e-mail: aishwaryaalva13@gmail.com

D. Nausheer
e-mail: daniyanausheer4576@gmail.com

G. D. Shivamadhu
e-mail: gaganhimamshudeep@gmail.com

K. A. Sumithra Devi
e-mail: deanacademics@dsatm.edu.in

G. Ranganathan et al. (eds.), *Inventive Communication and Computational Technologies*, Lecture Notes in Networks and Systems 145,
https://doi.org/10.1007/978-981-15-7345-3_81

the user, the ability to monitor it. Another approach to do this is to embed a GPS chip integrated onto a GSM or RFID receiver on the luggage, and then enable a GSM device. Additionally, an alarm system could be connected that will let us identify if the bag is lost or out of control, or has been in a breach of security. The Automation-portion of the luggage employing the ideologies of three robots are integrated: Obstacle Avoidance Robot, Line Following Robot and Human Following robot.

The PortaX Secure Automation System will be the combination of two segments i.e. automated portion and the element of the Monitoring system. It can be advantageous at airports, metro stations and typical busy environments, while the system's safety component could be used wherever there is a considerable network connection [1, 2] and will expand the application even to monitor your children and the buses [3, 4]. It is not something that could only be used by a specific part of society. It can be used universally, as everyone can benefit from it for tracking bags and it can have a whole plethora of new applications. The experience will be enhanced by using RFID trackers to make checking in the airport luggage quicker, and can also use it to keep track of attendance in the children's bags. So, this can benefit society in different ways, not just in the field of travel.

2 Methodologies

2.1 Global Positioning System for Object Tracking

Damani, Shah, and Vala proposed the need for an object tracking GPS [5]. The paper examined how the GPS for our predecessors flourished. They specify that the GPS used by US DOD in 1973 had 24 satellites, and so many shortcomings have indeed been overcome to date. Initially developed for military purposes, GPS has now broadened its horizons. They state that GPS is a GNSS, i.e., a Global Satellite Navigation System predominantly used to determine a location or position. The term Geo-Positioning, therefore, used to have different receivers but all cell phones now have an integrated application. The main program for which they used it is Anti-Theft and they use the hardware for arm processors.

They have mounted GSM modules—transmitter and receiver to communicate in the model they have created, and the GPS module calculates the geographic coordinates and sends it from the transmitter to the GSM module in the receiver. GPS-based low-cost vehicle tracking and monitoring device is proposed in [5], it involves transmitting position and vehicle status information and then sending it to the other stationary module; the second part is the receiving module that collects the information transmitted by SMS and processes it in a Google Earth-compatible format to display the location and vehicle status online.

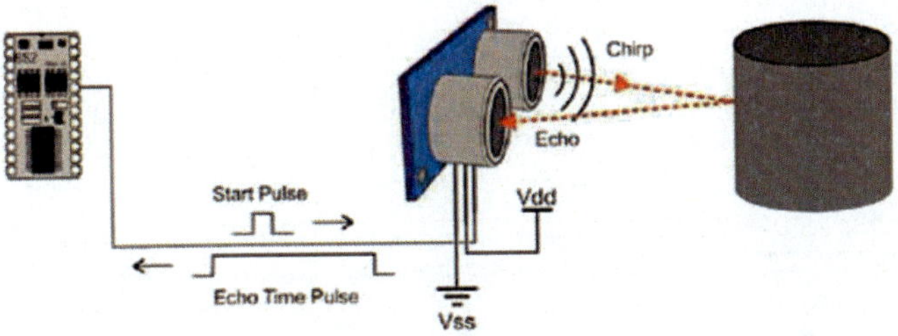

Fig. 1 Working of ultrasonic sensors [6]

2.2 *Obstacle Avoidance Robot*

Bhagat, Deshmukh, Dhonde and Ghag proposed a concept for the Framework for Obstacle Avoidance [6]. Obstacle Avoidance robot stops on its way close to the obstacle, re-routes itself and continues to move. Wall-following, edge detection, line following etc. are a few methods that were discussed in this paper. Methodology following wall is primarily used for robots cleaning the concrete. Blade detection technique can usually be used to avoid obstacles but the main disadvantage is that it has to stop right in front of the target to give the exact measurement. Infrared sensors, ultrasonic sensors, cameras used for machine vision are the sensors that are used here. The Steering Algorithm avoids when the obstacle is detected and takes back the original course (Fig. 1).

The ultrasonic sensor emits the signal at both low and high frequencies. If they detect anything, they reflect the echo signal that was taken via the Echo pin as input to the sensor [4]. They activate Trigger and Echo pin as low first, then push the robot forward (Fig. 2).

When an obstacle is detected Echo pin can give micro-controller input as big. It returns the pulse length in microseconds or it returns 0 if the full pulse was not received within the timeout [7]. This proposed model uses an Obstacle Avoidance Algorithm and a Steering Algorithm in the paper. From the beginning to the end, the Obstacle Avoidance Algorithm's main function is to dodge any kind of obstacles and guide them/redirect them to the endpoint.

2.3 *A New Invention of Alarm Reminder Locking (ARL) Security System*

The prototype proposed in this paper mainly aims at securing the door security which is installed in the doors to enable security at homes, offices, hostel and various other places. This system uses the Arduino controller and a GSM technology which is used to transmit the short message service alert data to the user in case any intruder is trying to gain access to the house or office place. This system is integrated with

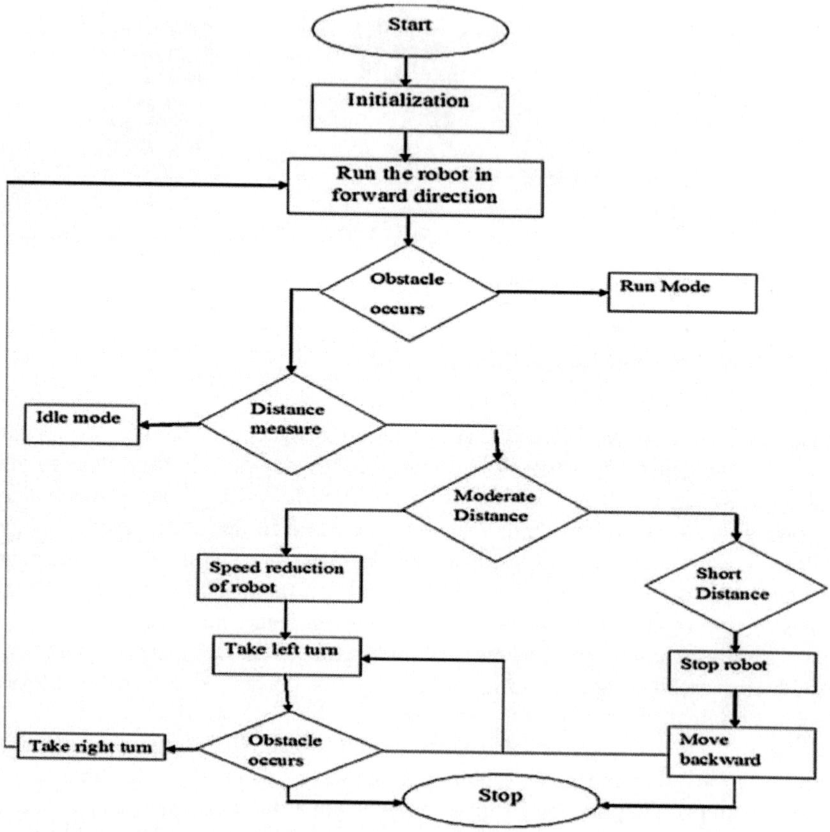

Fig. 2 Obstacle avoidance flowchart [6]

three systems namely, the alarm, the reminder and the lock. This paper has three modes of operation, first is when the user forgets to close the house door, this system will provide the user with the reminder and the buzzer alert. The second being, the automated lock will be activated if the user closes the door but did not lock it manually. The third being, an intruder mode will be activated if any user without authorized access tries to invade the house.

The ARL system is designed to provide a high level of security from unauthorized users from entering the house or office. The GSM module makes use of the SMS technology which informs the user about the activities taking place at the door of the house and also alerts the user if unusual activities take place thus informing the user about the threat caused to the main door of the house. This prototype also helps the user to avoid any causality from occurring and securing the house and the people from any threat. The GSM model used in the SIM900. The door is embedded with magnetic sensors are installed for automated locking and protecting from any threats.

When an intruder tries to breach the security at the entrance, as the magnetic sensor is being tampered, the system will trigger the alarm buzzer to emit the warning and immediately send the alert to the owner about the intrusion and thereby allowing the user to take the precautionary steps to protect the house security. The data taken from the GPS receiver will be sent to the user in the form of SMS using GSM modem. The GPS receiver will retrieve only the $GPRMC data values from the different satellite, then the latitude and longitude data will be sent to the microcontroller AT89C51. AT89C51 will process the data and then deliver to the GSM modem. This integrated system supplied by battery power which GSM required 12 V and GPS and microcontroller requires 5 V with the regulator. In a further development, this system will integrate with the sensor to report the vehicle information to the server to see and save the route and also other information on the computer.

TC35 GSM module will send alarm short message when the wireless sensor network (WSN) node receives unusual data through the Global System for Mobile communication (GSM) network. This module supports standard AT command set and its control by MCU hardware by entering a different AT function command using the serial port. Thus, this prototype helps to enhance the security of homes and office by installing the alarm reminder and locking system to provide more security at homes and also informing the user if there is any intrusion by any unauthorized user through the GSM module via an SMS (Fig. 3).

This model mainly makes use of the GPS and GSM modules to track the progress of the security and also receive updates in the form of messages which are sent to the user. The system sends in a warning when a random intruder who has not been granted access to the house tries to enter the house to rob and threat the environment of the house.

2.4 *Luggage Anti-Lost Wireless Security Card Structure*

In modern times, the security and safety of the user's luggage are one of the most important factors considered for any individual while travelling. This paper aims at providing security to the users while travelling and keeping intact the safety and confidentiality of the user's belongings. The prototype proposed in this paper consists of a wireless card structure which includes a security card which can transmit a wireless signal which also consists of the card number. An external wireless mobile device which is installed with a card number management application is used to receive the wireless signals sent by the security card and to perform logging and managing the user's luggage. This prototype enables the users to track their luggage and also receive a notification if the user's luggage bag is away from the vicinity of the user's environment. Thus, providing the users with a feeling of safety of the user's luggage bag. This prototype enables the users to safeguard their luggage without the threat of losing the luggage bag or being mixed with other fellow travelling passengers. The user can travel worry-free and hassle-free without having to constantly keep an eye on the luggage, if in any case, the luggage is in threat, a notification is sent to the

Fig. 3 the flow process of intrusion situation [12]

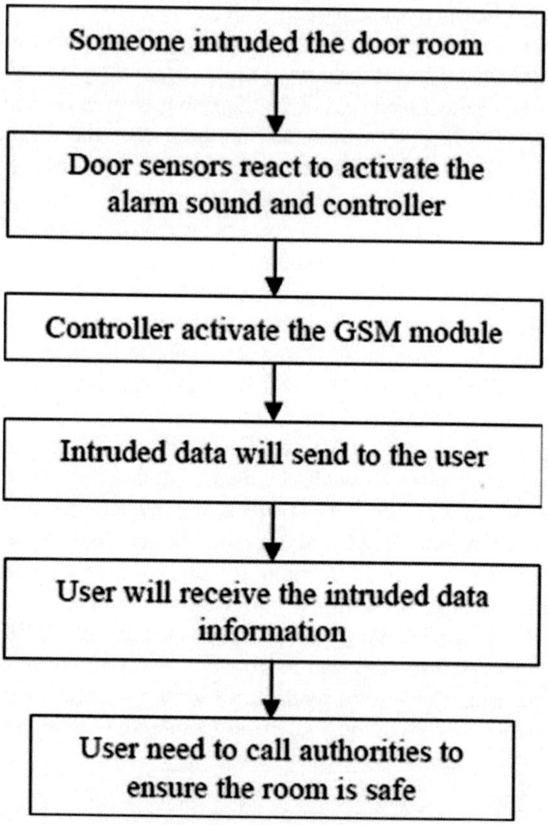

user about his luggage bag. The primary object of the present invention is to provide a luggage anti-lost wireless security card structure. The luggage anti-lost wireless security card structure comprises a security card composed of a thin upper cover and a thin base (Fig. 4).

The security card is provided with a card number wireless transmitter, a power supply unit, an antenna and a switch disposed of in an accommodation space defined between the upper cover and the base. The card number wireless transmitter has at least a built-in card number and is electrically connected to the power supply unit, the antenna and the switch. The card number wireless transmitter can obtain the working power from the power supply unit and can transmit a wireless signal containing the card number from the antenna to the outside under the starting operation of the switch. An external wireless mobile device installed with a card number management APP is used to receive the wireless signal sent by the security card and to perform pairing, login, and management for the card number. When the security card is disposed of in a suitcase or travel bag and the wireless mobile device carried by the owner of the luggage cannot normally receive or read the wireless signal of the security card,

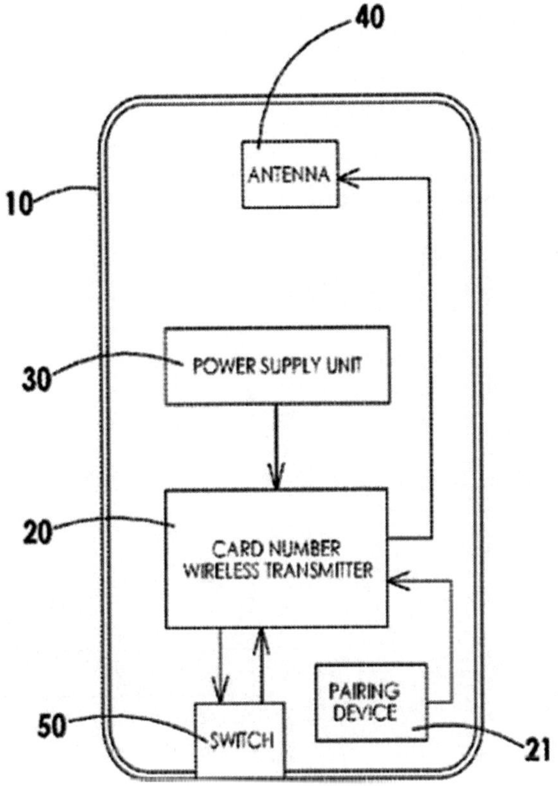

Fig. 4 Implementation of the structure [13]

a warning signal is generated. Thus, the user can travel hassle-free and worry-free without having to be tensed about his luggage bag being stolen or mixed with other fellow passengers or damaged.

2.5 Wearable–Logo Antenna for GPS-GSM Based Tracking Systems

Due to the rise in theft, the anti-theft systems have generated considerable demand and play an important role, particularly in the fashion and automotive industries. One of the most popular and cost-efficient systems is GPS-GSM based tracking devices. Those modules are embedded in items that can always be monitored. The tracking system is incorporated in expensive items such as leather bags, vehicles and other products of pride. It's also permanently embedded on small portable objects like mobile phones and laptops. The GPS-GSM system consists of two main components-the frameworks for data handling and a wearable antenna [8].

The first block comprises of (a) GSM modem (b) GPS receiver (c) Microcontroller. The GSM modem is disabled in the event of theft and alerts the microcontroller. The microcontroller turns the GPS receiver on and sends the GPS coordinates to the GSM modem. The second block is necessary for exchanging data wirelessly. Non-woven conductive fabric is used since it is mostly used alongside leather which is 1.65 mm thick. This material is perfect as it has high mechanical power, no problems with the fraying and low cost. It can also be used in complicated structures/geometry, and it also retains all characteristics of electro textiles. The antenna should be of high F/B ratio and HPBW weight. It should also be able to operate the GPS L1 band and GSM-1800 band respectively.

Two major strategies to reconfigure the antenna's frequency are: (a) Modify the geometry radiating element and (b) Introduce short circuits between radiating element and ground plane (Fig. 5).

As stated in the paper, this tracking system has produced great results and is cost-effective up until now. It can also be used with the operation of the unit without making any and can be fully integrated with leather bags and other small portable devices [7].

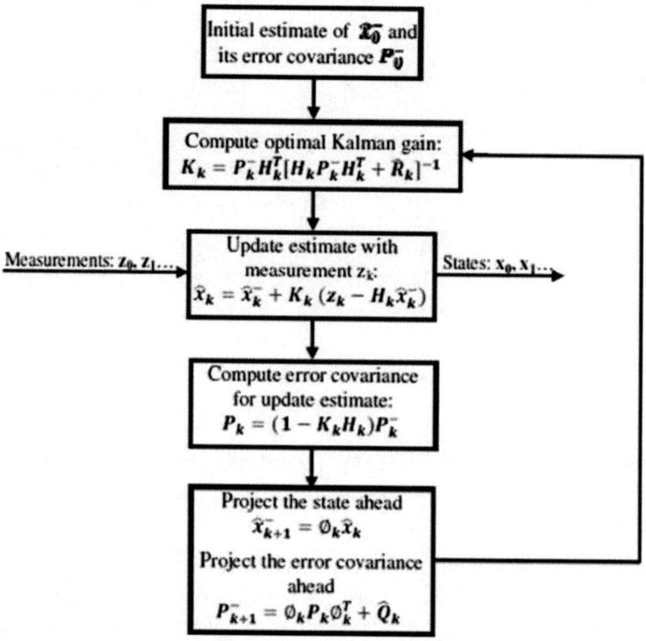

Fig. 5 Kalman filter algorithm flowchart [8]

2.6 A Multipurpose Vehicle Tracking Based on ARM CORTEX-M3, STM32, HMC5883L, MPC-6050, GPS and GSM

They stated in this paper that the ratio of road accidents has increased and that precautions against the susceptible accidents ought to be taken [9]. This project aims to safeguard the vehicle's defense from any vandalism, possible damage or unexpected disaster. Two systems are employed in this plan: Active and Passive. Passive devices are essentially architectured to record a vehicle's location, and then the data will be used for evaluation purposes. The active framework is used to forward the positions of the vehicles to a centralized location in real-time. The active device used in this project is to keep the owner notified about his car, alerts the vehicle owner via messages. Many of the food delivery franchises use this to manage the employees' activities.

This technique involves two parts: a unit installed in the vehicle to be controlled using the GPS and also the framework to provide visual monitoring of the co-ordinates of the vehicles. Often, the coordinates given by the GPS could not be as consistent as in real-time and enhancement was enabled by utilizing a compatible algorithm for this. Vector maps are used to acquire the actual vehicle spot. The data is collected from the pressure sensors and the accelerometer for an emergency alert and then sent as a notification to alert the consumer after interpreting the statistics.

The primary motive of this undertaking is to provide perceived safety for the owner for his car, and also to provide him with the latest updates of the whereabouts and the geographical area of his vehicle. It can also be used to capture the stolen vehicles and help the investigation committee get to the guilty party sooner also. It also informs the consumer that the car has been involved in an accident, if so.

2.7 GPS Based Android Application for Women Security

In this paper, they discuss how useful android applications can be and how it ensures women's safety in the event of any emergencies/attacks [10]. This highlights the application's integrity and how quick action can be taken in an emergency. Women are being mocked and abused in public even in the present time, this android application helps to reduce and works towards a better society. In case of emergencies, the application sends a message along with a Uniform Resource Locator (URL) which is the current location of the user to the emergency contacts registered on those particular devices. A click on the App would send a message to the person's emergency contacts along with the URL. It also triggers a first-person phone call on the list. This app's unique function is that it sends messages continuously every 5 min before you press the Stop button. There is a continuous tracking facility to the victim's position available via SMS.

- SafetyPin—this app lets you get to know where your precise location is, or indeed any place in the world. This keeps a push forward on various harassments so you can be more aware and get to know the place to be better.
- BSafe—this device operates with the slogan 'Never walk alone'. A click of a button sends a message to the selected contacts. It has two modes: (a) danger mode with GPS and real-time tracking, and (b) timer mode with timer activation automatic.
- Sthree Raksha—By a push of a button the nearest police station is alerted and uses the victim's GPS to track the location and also helps catch the culprit thereby reducing the number of crimes. This technology has reportedly saved a lot of lives and primarily helped in crime reduction. Thanks to technology and protection apps like these, women can be comfortable and move about openly without being abused anywhere in society. Few disadvantages are certainly being exploited and do something like this to make the world a crime-free and happier place to live in.

3 Proposed System

This survey is done to find the most favourable, and simpler method to create a secure system: PortaX Secure Automation System using IoT. Now this contains two parts: Human-Following Luggage and the GPS-GSM Enabled Tracker.

In the Human-Following Luggage, Obstacle avoiders that utilize infrared or ultrasonic sensors to make sure there is nothing in front of them and keep traversing a pre-programmed course. However, focusing on a single object and calculating its trajectory is another kind of problem. Attaching a transmitter to the object and a receiver on the follower is the usual solution, but will only use one ultrasonic sensor for a challenge. By having a servo motor sway, the sensor side to side in a thirty-degree arc, measure the distance of an object from both viewable edges. Assuming an object is followed with a flat backside, can know if an object is turning left or right by measuring these values against each other. With this information, our robot can be instructed to turn accordingly and avoid losing the target object. For this project, need a four-wheeled robot with appropriate motor drivers, an Arduino Uno board, a micro servo motor, and an HC-SR04 ultrasonic sensor.

The GPS module is the key element of our baggage monitoring device project and is built on the paper published by Joshi et al. [11]. This device receives satellite coordinates for each interval of time, with time and date. The GPS device transmits information on the tracking position in real-time and sends all these data in the NMEA format. The format of the NMEA consists of several sentences, which need just one sentence. This sentence starts with $GPGGA and contains coordinates, time and other useful information. This GPGGA is referred to as the Global Positioning System Fix Data and can extract coordinates from the $GPGGA string by counting the commas in the string. Assume you take the $GPGGA string and position it in an array, then Latitude can be identified after two commas, and Longitude can be seen after four commas. These latitudes and longitudes could then be mounted in other arrays. In this project, Arduino is used to supporting the entire operation using the GPS Receiver and GSM module. The GPS receiver is used to detect luggage coordinates, the GSM module is used to send SMS, of the coordinates to the user. An available 16 × 2 LCD is usually used to display status messages or coordinates. The GPS module SKG13BL and the GSM module SIM900A is used. The sent message is received by the GSM module which is connected to the system and sends the response message to Arduino. It is read by Arduino and it then extracts the key message from the entire text. It is then compared with Arduino's predefined message. If a match occurs, Arduino reads coordinates, using GSM module, and removes $GPGGA string from the GPS module. This message contains the baggage location coordinates.

4 Conclusion

Luggage safety or security of objects has been one of the most important issues because, in India alone, there are many cases of robbery. Due to India's digitization and campaigns like Digital-India, continues to use technology in a bigger, broader and cleaner way than it could have several decades ago. This venture, would therefore also help to safeguard the belongings and establish a turmoil-free, client travel experience. The proposed project helps assure protection for the consumer's belongings and also includes the latest notifications on the belongings. It also helps to prevent some amount of danger or vandalism that might have been susceptible to belongings.

References

1. AlMashari, R, AlJurbua G, AlHoshan L, Al Saud NS, BinSaeed O, Nasser N (2018) IoT-based smart airport solution. In: 2018 international conference on smart communications and networking (SmartNets), IEEE, pp 1–6
2. Kishan KK, Prashanth KM (2017) Techniques for detecting and tracking of baggages in airports. In: 2017 international conference on recent advances in electronics and communication technology (ICRAECT), IEEE, pp 333–338
3. Uddin MS, Ahmed MM, Alam JB. Islam M (2017) Smart anti-theft vehicle tracking system for Bangladesh based on internet of things. In: 2017 4th international conference on advances in electrical engineering (ICAEE), IEEE, pp 624–628
4. Aziz K, Tarapiah S, Ismail SH, Atalla S (2016) Smart real-time healthcare monitoring and tracking system using GSM/GPS technologies. In: 2016 3rd MEC international conference on big data and smart city (ICBDSC), IEEE, pp 1–7
5. Damani A, Shah H, Shah K, Vala M (2015) Global positioning system for object tracking. Int J Comput Appl 109(8):3977–3984
6. Bhagat K, Deshmukh S, Dhonde S, Ghag S (2016) Obstacle avoidance robot. Int J Sci Eng Technol Res 5:439–442
7. Yassine Z, Darwiche M, Mokhiamar O (2018) GPS tracking system for autonomous vehicles. Alexandria Eng J 57. https://doi.org/10.1016/j.aej.2017.12.00

8. Monti G, Corchia L, De Benedetto E, Tarricone L (2016) Wearable logo-antenna for GPS–GSM based tracking systems. IET Microw Antenn Propag 10(12):1332–1338
9. Abdeen MHU, Khan US, Iqbal J (2016) A multipurpose vehicle tracking system based on ARM CORTEX-M3 STM32, HMC5883L, MPU-6050, GSM and GPS. J Traffic Logist Eng 4(1)
10. Smys S, Raj JS (2019) Performance optimization of wireless adhoc networks with authentication. J Ubiquitous Comput Commun Technol (UCCT) 1(02):64–75
11. Joshi MPR, Patil MVV, Koli MPS, Tade MBS (2017) Device tracking using embedded GPS and zigbee technology. Int J Technol Res Eng 4(8):1175–1180
12. Effendi MSM, Shayfull Z, Saad MS, Nasir SM, Azmi AB (2016) A new invention of alarm reminder locking (ARL) security system. Int J Eng Technol 8(1):465–472
13. Haoxiang W (2019) Trust management of communication architectures of internet of things. J Trends Comput Sci Smart Technol (TCSST) 1(02):121–130

A Review Paper on the Elimination of Low-Order Harmonics in Multilevel Inverters Using Different Modulation Techniques

Kalagotla Chenchireddy and V. Jegathesan

Abstract This paper gives a review on the various modulation techniques that have been used to eliminate lower-order harmonics in multilevel inverters. The output voltage with high quality can be achieved using multilevel inverters. The selection is based on switching losses, power losses, noise, etc. Many authors proposed different modulation techniques such as sine triangular, selective harmonic selection (SHE), model predictive control techniques, space-vector-based modulation techniques. This paper presents a detailed review of existing harmonic elimination methods as well as concludes with the best method.

Keywords Low-order harmonics · Inverter · Space vector modulation

1 Introduction

As of late, the beat width tweak innovations assume the essential job in multilevel inverters. Triangle examination-based PWM and space vector-based PWM are utilized in most of the reasonable applications. Three-stage inverters are utilized for high-power applications. Yong-Chao presented two-changed SVPWM calculations, and these strategies diminished regular mode voltages in three-stage inverter [1]. Irfan Ahmed presented an improved SVPWM strategy. These procedures limit the exchanging frequency in multilevel inverter [2]. Amit Kumar Gupta designed a diminished regular-mode voltage in MLI-utilized SVPWM [3]. Narayanan [4] proposes a novel exchanging successions for a three-level inverter. Yi Deng and so forth proposed [5] a quick summary on SVPWM plot. This plan when compared to other SVPWM techniques comprises quick response, simple count time and

K. Chenchireddy (✉) · V. Jegathesan
Department of EEE, Karunya Institute of Technology and Sciences, Coimbatore, Tamil Nadu, India
e-mail: Chenchireddy.kalagotla@gmail.com

V. Jegathesan
e-mail: jegathesan@karunya.edu

G. Ranganathan et al. (eds.), *Inventive Communication and Computational Technologies*, Lecture Notes in Networks and Systems 145,
https://doi.org/10.1007/978-981-15-7345-3_82

exchanging states. Soumitra Das proposed space-vector-based propelled transport-bracing PWM procedures [6]. Heli Golwala presented four-leg voltage source inverter. This inverter controls three-dimensional space vector beat width adjustment. This adjustment strategy decreased exchanging misfortune in inverter. Four-leg VSI is utilized as shunt compensator in lattice association. This shunt compensator disposed of music, balanced burden and wiped out nonpartisan current [7]. In Ref. [8], lower request music wiped out in multilevel inverter. Space vector beat width tweak strategy killed fifth, seventh, eleventh and thirteenth request music in multilevel inverter. Exchanging misfortunes additionally diminished in MLI. A proto sort 1 kW, 400 V, four-shaft acceptance engine tried and diminished low-request torque swell in the engine. Execution assessed and examined double three-stage AC engines. Six-stage VSI take care of six-stage acceptance engine-controlled DSP [9]. Space vector beat width adjustment-based two-level three-stage inverter presented in Ref. [10]. Reference [11] changed molecule swarm enhancement calculation-limited sounds in three-stage mixture fell multilevel inverter. Srndovic et al. [12] killed first, third and fifth request music in fell H-connect inverter with flight of stairs regulation. The test set-up planned utilized Arduino Due microcontroller. Reference [13] proposed molecule swarm advancement calculation. This calculation totally disposed of fifth, seventh, ninth, eleventh and thirteenth request sounds from the single-stage fell inverter yield voltage. Reference [14] Mohammad Sharifzadeh and others actualized an adjusted particular symphonious relief—beat abundancy balance procedure. This strategy wiped out third, fifth, seventh and eleventh request music in single-stage five-level fell H-connect inverter.

2 Low-Order Harmonic Elimination

In this segment, secured low-request sounds disposal strategies are discussed. Bo guan and shinji doli proposed selective consonant end PWM. This technique is dispensed with third and ninth request music in three-level unbiased point inverter topology. The little force model planned DSP/TMS320C6657 [15]. Reference [16] Qiang Wei, etc., proposed a characteristic inspecting based on space vector tweak. This procedure killed fifth, seventh, eleventh and thirteenth request sounds in three-stage current source inverter. Equipment tried 10 kVA consecutive current source-converter module. This converter intended for high-force low-exchanging recurrence applications. The controller structured DSP TMS320F28335 and FEGA EP4CE10E22C8. Kiadehi et al. [17] proposed close to state PWM procedure. This strategy controlled open-end medium-force enlistment engine drive and eliminated low-request sounds single DC-connected double VSI take care of open-ended engine. Equipment actualized through DSP module RSF562TAADFH. Suresh Lakhimsetty and Somasekhar presented broken decoupled space vector PWM. This method decreased exchanging misfortune in double inverter which take care of open-end winding acceptance engine drive [18]. Reference [19] evolutionary calculations wiped out up to nineteenth request sounds in voltage source inverter which take care of acceptance engine.

In [20] limited and through about low request symphonious torque swells utilizing ideal strategies sine triangle PWM, SHWPWM, recurrence space based PWM and coordinated reference outline techniques. These procedures disposed of torque music sixth and twelfth requests in acceptance engine and furthermore diminished fifth and seventh; eleventh and thirteenth music. Hang yin and Zhiyong Dai diminished lattice flows low-request music in three-stage three-level NPC converter with LCL channel-utilized back-venturing control technique [21]. Reference [22] helper voltage source inverter mitigates both exchanging music and lower request sounds in three-stage voltage source inverters. Matthew Armstrong presented current control parameter randomization procedure for crossing out of low-request music in lattice-associated MLI [23]. K. Gopakumar dispensed with fifth and seventh sounds open-end winding acceptance engine. SVWPM method utilized for end of sounds [24]. M. F. Naguib disposed of fifth and seventh request music current source converter-utilized SV adjustment procedure. Equipment is executed through advanced sign preparing [25]. Reference [26] decreased low-request music in six-switch current source inverter-utilized GTO switches. Dipten decreased low-request consonant in X-associated inverter [27]. Authors in [28] a fake unbiased system is created. This strategy wiped out fifth, seventh, eleventh and thirteenth sounds disposed of in fell staggered inverter. Reference [29] presented Newton-Raphson strategy. This technique disposed of third and fifth request music in seven-level inverter. In [30], Kazem Haghdar and Heidar Ali Shayanfar proposed summed up design search strategy and hereditary calculation. These two techniques killed fifth and seventh request sounds in seven-level inverters. Summed up design search technique shows better outcome contrast and hereditary calculation. In [31], creators presented transient-free internal control circle. This technique disposed of third, fifth, seventh, ninth and eleventh sounds in seven-level inverters. Reference [32] molecule swarm enhancement calculation wiped out fifth, seventh, eleventh and thirteenth request sounds in single-stage fell inverter. Bipolar heartbeat width adjustment wiped out fifth, seventh, eleventh, thirteenth and seventeenth request harmonics [33]. Authors [34] Ricardo Aguilera and others presented particular symphonious end model prescient control method dispensed with fifth, seventh, eleventh and thirteenth request sounds killed in three-stage three-level H-connect inverter. Reference [35] Microprocessor-based heartbeat width adjustment procedure is wiped out up to the eleventh request sounds in inverter which take care of enlistment engine. S. R. Bowes and so forth presented [36–38] space-vector-based sound disposal PWM strategy for control of three-stage acceptance engine. In [39], SHEPWM technique disposed of fifth, seventh, eleventh and thirteenth sounds in single-stage five-level inverter. Chakrapong and Kinnares proposed [40] irregular SVPWM method. This method decreased exchanging misfortune and current waves in VSI-take care of acceptance engine. Reference [41] music examined and planned tedious controller wiped out lower request music. M. H. Etesami and so on [42] proposed provincial serious calculation. This calculation killed fifth, seventh and ninth request sounds in seven-level inverter. Hang Goa and So Forth [43] eliminated fifth, seventh and eleventh request music current source converter. Space-vector-based particular consonant end procedure utilized for disposal of music. Advanced consonant disposal PWM procedure wiped out third, fifth, seventh, ninth, eleventh and thirteenth request sounds in single-stage inverter [44]. Bearer-based heartbeat width regulation strategy dropped music fell staggered inverter [45]. Anping Hu and so forth presented reference direction upgrade-based SVM. This tweak strategy improved symphonious execution in current source converter [46]. In [47], homotopy strategy-based specific symphonious end wiped out lower request sounds in single-stage staggered inverter. Reference [48] proposed honey bee calculation. This calculation dispensed with fifth and seventh request sounds in fell staggered inverter. Particular consonant disposal PWM procedure killed lower request music in arrangement associated with voltage source inverter [49]. Hamid Reza Massrur and so forth presented stochastic procedure. This method limited complete symphonious twisting in staggered inverter [50]. N. V. Nho and M. J. Youn diminished exchanging misfortune in three-stage inverter transporter PWM calculation utilized [51]. Reference [52] diminished exchanging misfortune in three-stage inverter utilized SVWPM strategy. In [53], bearer-based heartbeat width balance wiped out lower request music in square wave inverter. References [54–56] specific symphonious end procedure wiped out low-request sounds in single-stage staggered inverter. Reference [57] unipolar specific consonant disposal PWM wiped out fifth, seventh, eleventh and thirteenth request sounds. Bettayeb and Qidwai [58] evaluated music in power framework utilized hereditary calculation. In [59], sinusoidal regulation procedure restricted music in voltage source inverter. Reference [60] customary-tested PWM strategy disposed of music in inverter drive. Creators yo-han lee and others [61] reduced consonant mutilation factor in staggered inverter-utilized bearer-based SVPWM. In 1990s, a development customized PWM wiped out music in three-stage inverter [62]. Reference [63] adaptable symphonious control method diminished exchanging misfortunes in three-level inverter.

3 Three-Phase Inverter

Figure 1 shows the three-level neutral point clamped (NPC) inverter. NPC inverter

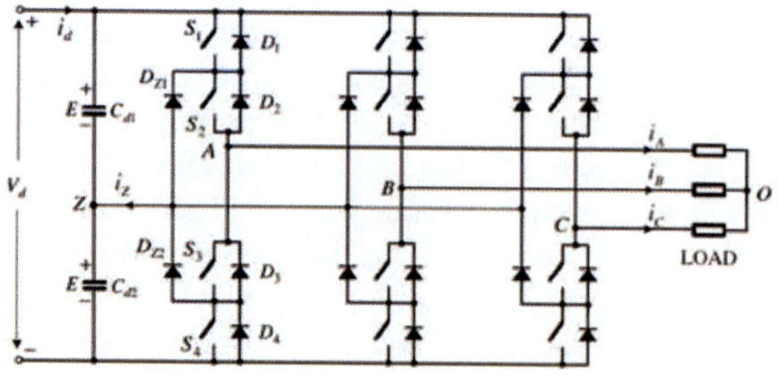

Fig. 1 Three-level NPC inverter

Table 1 Three-phase NPC inverter leg-A switching states

Switching states	Device switching status (phase A)				V_{AZ}
	S_1	S_2	S_3	S_4	
P	On	Off	Off	Off	E
O	Off	On	On	Off	0
N	Off	Off	On	On	−E

phase to neutral only three levels and phase to phase five levels. This multilevel inverter having three legs—Leg-A, Leg-B and Leg-C. Each leg is having four switches S_1, S_2, S_3 and S_4 with anti-parallel diodes. In industries, either insulated gate bipolar transistor or gate commutated thyristor is used as a switching device (Table 1).

4 Space Vector Modulation

Space vector change implies three-stage amount change to two-stage amount. Ordinary space vector tweak gives a higher measure of AC yield voltage same DC transport contrast with sine triangle PWM. SVWPM is one of the picked continuous balance techniques and is usually utilized for advanced control of VSIs. This segment presents the rule and usage of the space vector for three-level inverter. Reference [64] Authors Qamar Muhammad Attique and so forth studied diverse space vector regulation procedures for MLI. Chakrapong and vijit decreased exchanging misfortunes and current waves in voltage source inverter take care of acceptance engine drive. Spasmodic SVWPM strategy is utilized for lessening misfortunes [65]. In [66], Keng-Yuan Chen and Yu-Lin Xie proposed space-vector-based ideal cross-breed beat width tweak. This strategy decreased sounds contortion in five-stage voltage source inverter. In [67], adjusted bearer-based heartbeat width tweak executed disposed of undesirable music AC drives. Fukuda and Kunio Suzuki assessed sounds in three-stage voltage source inverter [68]. In [69], triple sounds dispensed within staggered inverter-utilized bearer-based PWM inverter. Reference [70] beat width regulation decreased sounds in electrical drives. In [71], hereditary calculations are dispensed with sounds in the staggered inverter. Gautam Poddar disposed of music in square wave inverter [53]. In [72] looked at SVPWM and bearer-based PWM procedures. References [73–78] checked on various staggered inverter topologies and distinctive inverter control strategies and consonant disposal techniques. In [79], memetic calculation killed lower request harmonics up to the thirteenth request. Reference [80] offbeat PSO calculation disposed of undesired lower request music. M. H. Etesami and others limited lower request sounds in inverter yield voltage utilized pioneer serious calculation [81]. Equal hereditary calculation limited lower request music in staggered inverter [82]. Reference [83] honey bee calculation disposed of lower request music in the staggered inverter.

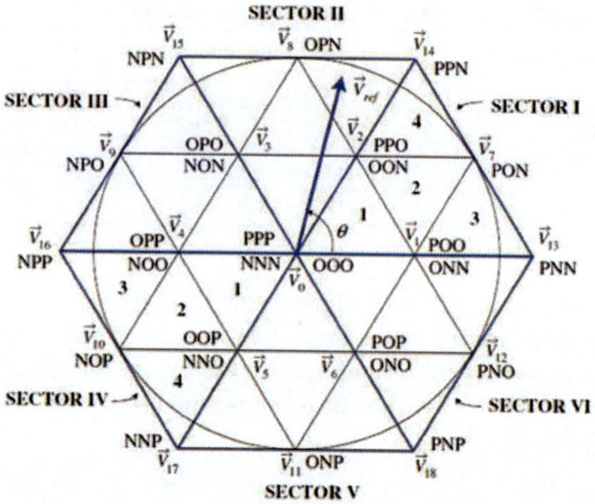

Fig. 2 SVPWM diagram of the NPC inverter

Figure 2 shows space vector modulation for control nonpartisan point inverter. Space vector stage 27 exchanging states and 19 V vectors. SVPWM voltage vectors, four gatherings dependent on voltage size. First vector zero vector voltage extent zero, second vector little vector voltage size Vd/3, third vector medium vector voltage greatness 0.58 Vd and fourth vector 2 Vd/3. Each of the 19 vectors given in Table 2

Reference [84] space vector pulse width modulation reduced current ripples in induction motor drive. In [85], phase-shifted carrier PWM technique controlled cascaded inverter. A variable switching-point predictive current control technique reduced lower-order harmonics in quasi-Z-source inverter [86]. In [87], a novel SVPWM is proposed and controlled three-level NPC inverter. The proportional and integral (PI) control technique real-time calculated for multilevel inverter (MLI) switching angle control. PI control method eliminated third and fifth harmonic contents in H-Bridge seven-level inverter [29]. SHEPWM to eliminate the fifth, seventh and eleventh harmonics in nine-level inverter with neutral point clamped (NPC) structure [88].

The firefly algorithm-based optimization technique the overall THD of the output voltage of 11-Level cascaded multi inverter with equal and non equal dc sources. FFA algorithm is developed to compute the switching angles for minimization of overall voltage THD [89]. Multicarrier pulse width modulation (MCPWM) and optimized harmonic stepped waveform (OHSW) techniques applied solar-fed cascaded 15-level inverter both stand-alone and grid-connected systems which reduce third, fifth and seventh harmonics [90].

Table 2 Voltage vectors and switching states

Space vector		Switching state		Vector classification	Vector magnitude
$\vec{V}_0$		PPP, OOO, NNN		Zero vector	0
$\vec{V}_1$		P-type	N-type	Small vector	1/3 V_d
	$\vec{V}_{1P}$	POO			
	$\vec{V}_{1N}$		ONN		
$\vec{V}_2$	$\vec{V}_{2P}$	PPO			
	$\vec{V}_{2N}$		OON		
$\vec{V}_3$	$\vec{V}_{3P}$	OPO			
	$\vec{V}_{3N}$		NON		
$\vec{V}_4$	$\vec{V}_{4P}$	OPP			
	$\vec{V}_{4N}$		NOO		
$\vec{V}_5$	$\vec{V}_{5P}$	OOP			
	$\vec{V}_{5N}$		NNO		
$\vec{V}_6$	$\vec{V}_{6P}$	POP			
	$\vec{V}_{6N}$		ONO		
$\vec{V}_7$		PON		Medium vector	$\frac{\sqrt{3}}{3} V_d$
$\vec{V}_8$		OPN			
$\vec{V}_9$		NPO			
$\vec{V}_{10}$		NOP			
$\vec{V}_{11}$		ONP			
$\vec{V}_{12}$		PNO			
$\vec{V}_{13}$		PNN		Large vector	2/3 V_d
$\vec{V}_{14}$		PPN			
$\vec{V}_{15}$		NPN			
$\vec{V}_{16}$		NPP			
$\vec{V}_{17}$		NNP			
$\vec{V}_{18}$		PNP			

5 Conclusion

This paper proposes a review of many lower-order harmonic elimination techniques. Selective harmonic elimination, space vector pulse width modulation and carrier-based modulation techniques are discussed in detail. Comparing all these techniques, space vector pulse width modulation (SVPWM) difficult to control sector selection and switching sequence design. But SVWPM technique gives good performance for the elimination of lower-order harmonics.

References

1. Liu YC, Ge X, Tang Q, Gou B (2017) Two modified SVPWM algorithm for common-mode voltage reduction in eight-switch three-phase inverters. Electron Lett 53(10):676–678
2. Chaudhari A, Ahmed I, Borghate VB, Matsa A (2015) Simplified space vector modulation technique for multilevel inverters. IEEE
3. Gupta AK, Khambadkone AM (2007) A space vector modulation scheme to reduce commom mode voltage for cascaded multilevel inverters. IEEE Trans Power Electron 22(5)
4. Das S, Narayanan G (2012) Novel switching sequences for SVM three-level inverter. IEEE Trans Ind Electron 59(3)
5. Deng Y, Teo KH (2013) A fast and generalized space vector modulation scheme for multilevel inverters. IEEE
6. Das S, Narayanan G, Pandey M (2013) Space-vector-based hybrid pulse width modulation for a three-level inverter. IEEE Trans Power Electron
7. Golwala H, Chudanmani M (2015) New space vector based signal generation technique without null vectors and reduce switching losses for a grid-connected four-leg inverter. IEEE
8. Boby M, Arun Rahul S, Gopakumar K, Umanand L (2017) A lower order harmonics elimination scheme for induction motor drives using a multilevel octadecagonal space vector structure with a single dc source. IEEE
9. Hadiouche D, Baghli L, Rezzoug A (2006) Space-vector PWM technique for dual three-phase AC machine: analysis, performance evaluation, and DSP implementation. IEEE Trans Ind Appl 42(4)
10. Huang Y, Xu Y, Li Y, Yang G, Zou J (2017) PWM frequency voltage noise cancellation in three-phase VSI using the novel SVPWM strategy. IEEE Power Electron
11. Routray A, Singh RK, Mahanty R Harmonic minimization in three-phase hybrid cascaded multilevel inverter using modified particle swarm optimization. IEEE Trans Ind Inform
12. Srndovic M, Zhetessov A, Alizadeh T, Familiant YL, Grandi G, Ruderman A Simulations selective harmonic elimination and THD minimization for a single-phase multilevel with staircase modulation. IEEE Trans Ind Appl
13. Taghizadeh H, Hagh MT (2000) Harmonic elimination of cascaded multilevel inverters with nonequal DC sources using Particle swarm optimization. IEEE Trans Ind Electron 57(11)
14. Sharifzadeh M, Vahedi H, Portillo R, Franquelo LG, Al-Haddad K (2018) Selective harmonic mitigation based self-elimination of triplen harmonics for single-phase five-level inverters. IEEE Trans Power Electron
15. Guan B, Doki S (2018) A current harmonic minimum PWM for three level converters aiming at the low frequency fluctuation minimum of neutral point potential. IEEE Trans Ind Electron
16. Wei Q, Wu B, Xu D, Zargari NR (2015) A natural-sampling-based SVM scheme for current source converter with superior low-order harmonics performance. IEEE Trans Power Electron
17. Kiadehi AD, Drissi KE, Pasquier C (2016) Adapted NSPWM for single DC-Link dual-inverter fed open-end motor with negligible low-order harmonics and efficiency enhancement. IEEE Trans Power Electron
18. Lakhimsetty S, Somasekhar VT (2018) Discontinuous decoupled SVPWM schemes for a four-level open-end winding induction motor drive with waveform symmetries. IEEE Trans Power Electron 11(2):280–292
19. Jegathesan V, Jerome J (2011) Elimination of lower order harmonics in voltage source inverter feeding an induction motor drive using evolutionary algorithms. Expert Syst Appl 38(1):692–699
20. Tripathi A, Narayanan G (2015) Evaluation and minimization of low-order harmonics torque in low-switching- frequency inverter fed induction motor drives. IEEE
21. Yin H, Dai Z, Lei X, Lan T (2019) Grid current low-order harmonics suppression of the three-phase grid converter with an LCL filter under the distorted grid voltage. J Eng 2019(7):4675–4680
22. Bai H, Wang X, Blaabjerg F, Loh PC (2017) Harmonics analysis and mitigation of low-frequency switching voltage source inverter with auxiliary VSI. IEEE

23. Armstrong M, Atkinson DJ, Johnson CM, Abeyasekera TD (2005) Low order harmonics cancellation in a grid connected multiple inverter system via current control parameter randomization. IEEE Trans Power Electron 20(4)
24. Pramanick S, Azeez NA, Kaarthik RS, Gopakumar K, Cecati C (2015) Low order harmonics suppression for open-end winding IM with dodecagonal space vector using a single DC-link supply. IEEE Trans Ind Electron
25. Naguib MF, Lopes LA (2009) Minimize low-order harmonics in low-switching-frequency space-vector-modulated current source converters with minimum harmonic tracking technique. IEEE Trans Power Electron 24(4)
26. Lopes LA, Naguib MF (2009) Space vector modulation for low switching frequency current source converters with reduce d low-order non characteristic harmonics. IEEE Trans Power Electron 24(4)
27. Maiti D, Biswas SK (2013) The X-connected inverter: a topology with higher fundamental and reduced low-order harmonic voltages. IEEE Trans Power Electron
28. Tolbert LM, Cao Y, Ozpineci B (2011) Real-time selective harmonic minimization for multi-level inverters connected to solar panels using artificial neural network angle generation. IEEE Trans Ind Appl 47(5)
29. Ahmed M, Sheir A, Orabi M (2017) Real-time solution and implementation of selective harmonic elimination of seven-level multilevel inverter. IEEE J Emerg Sel Top Power Electron 5(4)
30. Haghdar K, Shayanfar HA Selective harmonic elimination with optimal DC sources in multilevel inverters using generalized pattern search. IEEE Trans Ind Inform
31. Zhao H, Jin T, Wang S, Sun L (2015) A real-time selective harmonic elimination based on a transient-free, inner closed-loop control for cascaded multilevel power inverters. IEEE Trans Power Electron
32. Taghizadeh H, Hagh MT (2010) Harmonic eliminated of cascaded multilevel inverters with nonequal DC sources using particle swarm optimization. IEEE Trans Ind Electron 57(11)
33. Agelidis VG, Balouktsis A, Balouktsis I (2004) On applying a mimimization technique to the harmonic elimination PWM control: the bipolar waveform. IEEE Power Electron Lett 2(2)
34. Aguilera RP, Acuna P, Lezana P, Konstantinou G, Wu B, Bernet S, Agelidis VG (2016) Selective harmonic elimination model predictive control for multilevel power converters. IEEE Trans Power Electron
35. Bowes SR, Bullough RI. (1987) Harmonic minimization in microprocessor controlled current fed PWM inverter drives. IEEE Process 134
36. Bowes SR, Grewal S (1999) Novel harmonic elimination PWM control strategies for three-phase PWM inverters using space vector techniques. IEEE Proc Elect Power Appl 146(5)
37. Bowes SR, Grewal S (2000) Novel space-vector-based harmonic elimination inverter control. IEEE Trans Ind Appl 36(2)
38. Bowes SR, Grewal S, Holliday D (2001) Single-phase three-level regular-sampled selective harmonic PWM. IEE Proc Electro Power Appl 148(2)
39. Buccella C, Cecati C, Cimoroni MG (2017) A selective harmonic elimination method for 5-level converters for distributed generation. J Sel Emerg Top Power Electron
40. Charumit C, Kinnares V (2015) Discontinuous SVPWM techniques of three-leg VSI-fed balanced two-phase loads for reduced switching losses and current ripple. IEEE Trans Power Electron 39(4)
41. Chen S, Lai YM, Tan SC, Tse CK (2007) Analysis and deigned of repetitive controller for harmonic elimination in PWM voltage source inverter system. IEEE Trans Power Electron
42. Etesami MH, Farokhnia N, Fathi SH (2015) Colonial competitive algorithm development toward harmonic minimization in multilevel inverters. IEEE Trans Ind Inform 11(2)
43. Gao H, Wu B, Xu D, Aguilera RP, Acuna P Model predictive switching pattern control for current source converters with space-vector-based selective harmonic elimination. IEEE Trans Power Electron
44. Hadji S, Touhami O, Goodman CJ (2007) Vector-optimised harmonic elimination for single-phase pulse-width modulation inverters/converters. IEEE Trans Power Electron

45. Holmes DG, McGrath BP Opportunities for harmonic cancellation with carrier-based PWM for two-level and multilevel caca
46. Hu A, Xu D, Wu B, Wang J, Su J (2014) Reference trajectory optimized SVM for high-power current-source converters to improve harmonic performance and reduced common-mode voltage. IEEE Trans Power Electron
47. Kato T (1999) Sequential homotopy-based computation of multilevel salutations for selected harmonic eliminated in PWM inverters. IEEE Trans Power Circuits Syst Fundam Theory Appl 46(5)
48. Kavousi A, Vahidi B, Salehi R (2012) Applications of the bee algorithm for selective harmonic elimination strategy in multilevel inverter. IEEE Trans Power Electron 27(4)
49. Li L, Czarkowski D, Liu Y, Pillay P (2000) Multilevel selective harmonic elimination PWM technique in series-connected voltage inverters. IEEE Trans Ind Appl 36(1)
50. Massrur HR, Niknam T, Mardaneh M, Rajaei AH (2016) Harmonic elimination in multilevel inverters under unbalanced voltages and switching deviation using a new stochastic strategy. IEEE Trans Ind Inform 12(2)
51. Nho NV, Youn MJ (2005) Carrier PWM algorithm with optimized switching loss for three-phase four-leg multilevel inverters. Electron Lett 41(1)
52. Onederra O, Kortabarria I, de Alegría IM, Andreu J, Gárate JI (2016) Three phase VSI optimal switching loss reduction using variable switching frequency. IEEE Trans Power Electron
53. Poddar G, Sahu MK (2009) Natural harmonic elimination of square-wave inverter for medium-voltage applications. IEEE Trans Power Electron 24(5)
54. Sundareswaran K, Jayant K, Shanavas TN (2005) Inverter harmonic elimination through a colony of continuously exploring ants. IEEE Trans Ind Electron 54(5)
55. Liang TJ, O'Connell RM, Hoft RG (1997) Inverter harmonic reduction using walsh function harmonic elimination method. IEEE Trans Power Electron 12(6)
56. Yang Y, Zhou K, Wang H, Blaabjerg F, Wang D, Zhang B (2014) Frequency adaptive selective harmonic control for grid-connected inverters. IEEE Trans Power Electron
57. Agelidis VG, Balouktsis AI, Cossar C (2008) On attaining the multiple salutations of selective harmonic elimination PWM three-level waveforms through function minimization. IEEE Trans Ind Electron 55(3)
58. Bettayeb M, Qidwai U. (2003) A hybrid least squares-GA-based algorithm for harmonic estimation. IEEE Trans Power Deliv 18(2)
59. Binkowski T (2018) Sinusoidal modulation with higher harmonics limitation. IEEE
60. Bowes SR, Clark PR (1992) Transputer-based harmonic-elimination PWM control of inverter drives. IEEE Trans Ind 28(1)
61. Lee YH, Kim DH, Hyun DS (2000) Carrier based SVPWM method for multi-level system with reduced HDF. IEEE
62. Enjeti P, Shireen W (1990) An advanced programmed PWM modular for inverters which simultaneously eliminates harmonics and rejected DC link voltage ripple. IEEE
63. González FJ and Yin J (2018) Flexible harmonic control for three-level selective harmonic modulation using the exchange market algorithm. IEEE
64. Attique QM, Li Y, Wang K (2017) A survey on space-vector pulse width modulation for multilevel inverter. CPSS Trans Power Electron Appl 2
65. Charumit C, Kinnares V (2015) Discontinuous SVPWM technique of three-leg VSI-fed balance two-phase loads for reduced switching losses and current ripples. IEEE Trans Power Electron 30(4)
66. Chen KY, Xie YL Reducing harmonics distortion in five-phase VSI using space-vector-based optimal hybrid PWM. IEEE Trans Power Electron
67. Gourisetti SN, Patangia H (2013) A summary on modified carrier based real-time selective harmonic elimination technique. IEEE
68. Fukuda S, Suzuki K (1997) Harmonic evaluation of carrier-based PWM methods using harmonic distortion determining factor. IEEE
69. Holmes DG, McGrath BP (2001) Opportunities for harmonic cancellation with carrier-based PWM for two-level and multilevel cascaded inverters. IEEE Trans Ind Appl 37(2)

70. Holtz J, Springob L (1993) Reduced harmonic PWM controlled line-side converter for electrical drives. IEEE Trans Ind Appl 24(4)
71. Ozpineci B, Tolbert LM, Chiasson JN (2005) Harmonic optimization of multilevel converters using genetic algorithms. IEEE Power Electron 3(3)
72. Yao W, Hu H, Lu Z (2008) Comparision of space-vector modulation and carrier-based modulation of multilevel inverter. IEEE Trans Power Electron 23(1)
73. Variath RC, Andersen MA, Nielsen ON, Hyldgård A (2010) A review of module inverter topologies suitable for photovoltaic systems. IEEE
74. Yuan G (2014) Smart PV inverters-DOC sunshot SEGIS-AC program review. IEEE
75. Li W, Gu Y, Luo H, Cui W, He X, Xia C (2015) Topology review and derivation methodology of single-phase transformer less photovoltaic inverters for leakage current suppression. IEEE Trans Ind Electron 62(7)
76. Gupta KK, Ranjan A, Bhatnagar P, Sahu LK, Jain S (2016) Multilevel inverter topology with reduce device count: a review. IEEE Trans Power Electron 31(1)
77. Ellabban O, Abu-Rub H (2016) Z-source inverter topology improvements review. IEEE Ind Electron Mag March 2016
78. Liang X, Andalib-Bin-Karim C (2018) Harmonic and mitigation technique through advanced control in grid-connected renewable energy sources: a review. IEEE Trans Ind Appl
79. Kumle AN, Fathi SH, Jabbarvaziri F, Jamshidi M, Yazdi SS (2015) Application of memetic algorithm for selective harmonic elimination in multilevel inverters. IEEE Trans Power Electron 8(9)
80. Memon MA, Mekhilef S, Mubin M (2017) Selective harmonic elimination in multilevel inverter using hybrid APSO algorithm. IEEE Trans Power Electron
81. Etesami MH, Farokhnia N, Fathi SH (2013) Coconial competitive algorithm developed toward harmonic minimization in multilevel inverters. IEEE Trans Ind Inform 11(2)
82. Roberge V, Tarbouchi M, Okou F (2014) Strategies to accelerate harmonic minimization in multilevel inverters using a parallel genetic algorithm on graphical processing unit. IEEE Trans Power Electron 29(10)
83. Kavousi A, Vahidi B, Farokhnia N (2012) Application of the bee algorithm for selective harmonic elimination strategy in multilevel inverters. IEEE Trans Power Electron 27(4)
84. Narayanan G, Zhao D, Krishnamurthy HK, Ayyanar R, Ranganathan VT (2008) Space vector based hybrid PWM techniques for reduce current ripple. IEEE Trans Ind Electron 54(4)
85. Naderi R, Rahmati A (2008) Phase-shift carrier PWM technique for general cascaded inverters. IEEE Trans Power Electron 23(3)
86. Karamanakos P, Ayad A, Kennel R A variable switching point predictive current control strategy for quasi-z-source inverters. IEEE Trans Ind Appl
87. Wang W, Zhang B, Xie F (2018) A novel SVPWM for three-level NPC inverter based on m-mode controllability. IEEE Trans Ind Electron 65(8)
88. Khoukha I, Hachemi C, El Madjid B. Multilevel selective harmonic elimination PWM technique in the nine level voltage inverter
89. Karthik N, Arul R (2014) Harmonic elimination in cascade multilevel inverters using firefly algorithm. In: 2014 international conference on circuit, power and computing technologies
90. Khan S, Nagar S, Meena M, Singh B (2017) Comparison between SPWM and OHSW technique for harmonic elimination in 15 level multilevel inverter. In: IEEE ICICIC

LNA Architectures for ECG Analog Front End in CMOS Technology

Malti Bansal and Ishita Sagar

Abstract A low noise amplifier, usually abbreviated as LNA, plays a crucial role in the collection of ECG signals for further processing. Through the course of this paper, the numerous topologies of LNA have been thoroughly studied and reviewed. These topologies have been further compared to narrow down the best among them, based on parameters expected from an ideal LNA. According to our analysis, open-loop OTA topology is the best suited LNA topology according to all design parameters taken under consideration for its use in ECG analog front end applications.

Keywords LNA · OTA · ECG · AFE · CMOS

1 Introduction

The rates of diseases and disorders of the human heart are increasing since the past few years, owing to the hectic lifestyles of humans. The heart is the most vital organ for the human body, any disease related to it needs to be identified and cured at the earliest. The first step towards identifying any disease related to the heart is measuring its activity, which is thereby done in the form of electric signals, known as ECG signals. In earlier days, this measurement presented the need for substantial equipment but owing to rapid advancements in the electronics industry, the same equipment has reduced vastly in size. It has also been known to improve in terms of performance. Such considerable achievements, especially in terms of performance, in this field, are a result of the wide use of Low Noise Amplifier (LNA). LNA is one of the many components present in the Analog Front End (AFE) of an ECG acquisition/collection system. The AFE helps in the proper conditioning of a signal once acquired from the external environment. In this paper, different topologies and configurations adopted for LNA in ECG Applications have been reviewed. A

M. Bansal (✉) · I. Sagar
Department of Electronics and Communication Engineering, Delhi Technological University (DTU), Delhi 110042, India
e-mail: maltibansal@gmail.com

G. Ranganathan et al. (eds.), *Inventive Communication and Computational Technologies*, Lecture Notes in Networks and Systems 145,
https://doi.org/10.1007/978-981-15-7345-3_83

comparison between the various topologies has been made leading to a conclusion for the best topology based on certain parameters governing the performance of an LNA for ECG applications.

2 Low Noise Amplifier

Low Noise Amplifier is the most vital component in the process of ECG signal collection. Any ECG signal, acquired using sensors from the human body, is first brought to an LNA. Since LNA is the foremost block of an AFE, many ideal characteristics are expected from an LNA. Owing to the low amplitude of an ECG signal, it needs to be carefully amplified to obtain accurate results for further study. Noise is also a part of the ECG signal when it is measured, which can prove to be very harmful to ECG signals, owing to their miniature amplitude. LNA, belonging to a special class of amplifiers, is widely known and used for amplifying low amplitude signals, such as ECG, without any considerable amplification in the noise present in the signal. The output of LNA is such that the signal becomes suitable for further analog to digital conversions. The characteristics usually considered and studied for a good performance LNA are input-referred noise, gain, CMRR (Common Mode Rejection Ratio), chip area, and power supply. Since all medical equipment is aimed at portability, hence less chip area and low power supply, along with extended battery life become the main aspects for an efficient low noise amplifier. Since noise plays the most important role in ECG signal collection, hence the LNA is highly expected to deliver high values for Noise Efficiency Factor (NEF). Low values for input-referred voltage are aimed at during LNA design for ECG signal collection. Other expected values from a desirable LNA include large values for gain and CMRR. LNA present in the Analog Front End also has connections to other components, which puts up the need for prevention of loading effect [1, 2]. Owing to this requirement, the LNA needs to have high input and low output impedance [1, 2]. Reduction in the overall gain of LNA is a direct consequence of the loading effect. The signals are expected to be transmitted completely without any loss, which requires an efficient matching network [3]. Well defined steps need to be followed in designing LNA with appreciable performance; ranging from biasing, the correct choice of the transistor, stabilization of frequency, etc. [4]. Many different topologies and configurations have been worked upon in the past to achieve the best results.

3 What Is the ECG Signal?

The electrocardiogram (ECG) depicts the plot of the bio-potential generated by the hearts' activity and is used worldwide by doctors for the diagnosis or identification and treatment of numerous cardiovascular diseases. It is collected by electrodes connected to the patients' hearts, thereby generating a graphical representation of

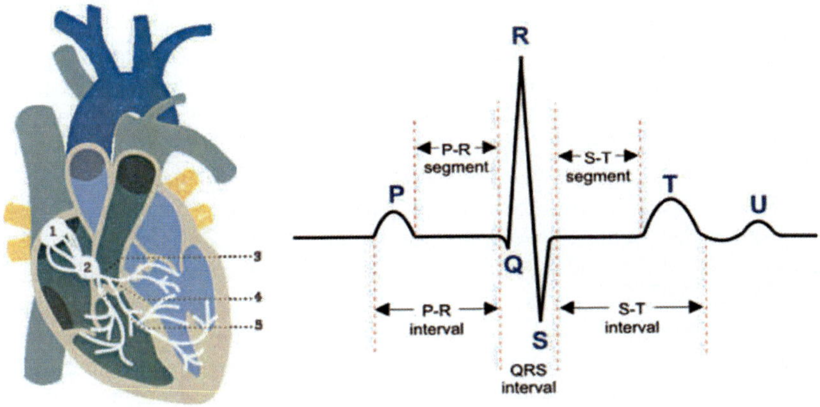

(a) (b)

1: Sinoatrialnode
2: intermediator node between atria
to the ventricles via atrioventricular (AV) node
3: a bundle of His
4: left bundle branches
5: right bundle branches

Fig. 1 **a** The human heart and; **b** it's ECG waveform

the hearts' electrical activity and rhythm. The contractions of the heart create action potentials thereby leading to the formation of electrical current from the heart to the body, which can easily be detected by the electrodes placed on the skin. The amplitude of ECG signals is very miniature ranging from 100 μV to 5 mV; and these signals have a frequency range from 0.1 to 50 Hz.

Figure 1 depicts the human heart and an ideal ECG waveform. The sinoatrial node (1) is the beginning point for the electrical signal in the human heart, which is placed in the right atrium, and moves to the left and right atria, resulting in their contraction and pumping of blood for the ventricles. P on the ECG wave records this activity of the heart. The PR denotes an interval of time, measured in seconds, from the initiation of the P wave till the starting of the complex QRS. Node (2) is the intermediator node between the atria to the ventricles via the atrioventricular (AV) node. The passing of the signal via this node causes it to slow down, hence filling the ventricles with blood. A flat line is observed between P and Q on the ECG monitor resulting from the slowing down of this signal. The electrical conduction via the atria in addition to the electrical impulse's delay suffered in the atrioventricular node is represented by the PR segment. After the departure of the signal from the AV node, it travels beside a pathway known as the bundle of His (3) and moves to the left and right bundle branches (4, 5). Contraction of the heart's ventricles is caused as the signal moves through it, thereby pumping blood into the body and lungs. QRS wave on the ECG waveform depicts this movement. Such waves are viewed as QRS complex as they recur very quickly/rapidly. The ventricles now come back to their normal electrical state, depicted by the wave T. The muscles now relax while stopping to contract, thereby letting the atria to fill with blood. This complete process repeats with every heartbeat. The ST segment provides a connection between the QRS complex and the T wave. It also represents the onset of the ventricles' recovery. The time duration where the ventricles are stimulated, and their recovery post-stimulation, is depicted by the interval QT. A faster heartbeat means the shorter length of QT and vice versa.

The electrodes connected to the skin can be either three leads or twelve leads. A standard form of ECG used includes twelve leads, out of which six leads belong in a plane parallel to the heart and the remaining six are present in the hearts' perpendicular plane. Three electrodes are placed on the heart creating the ECG frontal leads, with a fourth lead as the reference lead. The placement of these leads includes Left Arm (LA), Right Arm (RA), and Left Leg (LL). The reference lead can be put on the Right Leg (RL) [4]. A DC Component ranging from +300 to −300 mV is also a part of the ECG signal resulting from the contact between the electrode and the skin. The electrode and ground potential difference form a common-mode component of about 1.5 V [5]. The range of desirable bandwidth of an ECG signal is from 0.5 to 50 Hz, till 1 kHz in pacemaker detection. 0.05 to 100 Hz is the basic ECG bandwidth for clinical purposes [5]. Many parameters like power-line interference, electrode contact noise, electromagnetic interference, perspiration, and also respiration impact the quality of ECG Signal [6]. Some of these noise sources can be handled well, whereas some might need rigorous attention.

4 Analog Front End for ECG Signals

ECG Signals when collected from a patients' body via electrodes are first presented to the Analog Front End of the receiver system. Weak signals, such as ECG itself, present numerous hindrances to the hardware [7]. The transceivers' input-referred noise must belong to values around 10 μV, owing to the sensitivity requirements of the system [7]. The overall power consumption of the transceiver should be as minimum as possible, with connections to a low voltage supply, if we wish to obtain portability compatibility from our device [7]. Another expected performance parameter from the system is a high value for the common-mode rejection ratio (CMRR). A depiction of an ECG analog Front End is shown in Fig. 2.

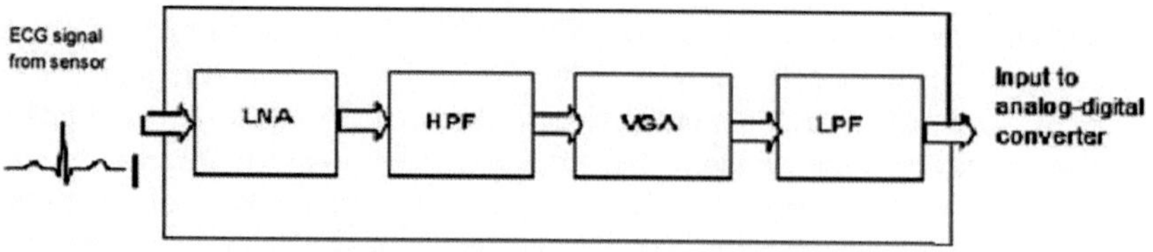

Fig. 2 Analog front end for ECG signals

Electrodes, connected to the patient, used for collecting data for ECG help in the conversion from biological signals to electrical ones, thereby presented as input for the electrical portion of the system. The characteristics of these electrical signals include high DC offset resulting from leakage current. They also offer large impedance values at the interface [8]. So, if we wish to narrow down the hindrances and challenges offered in the design of a front-end low noise amplifier, we need large gain values, minimal noise, small values of form factor accompanied with less overall power consumption [7].

5 Use of LNA for ECG Applications

A wide range of topologies for the study and processing of ECG signals have been reviewed in the past. LNA topologies using Operational Transconductance Amplifier (OTA) have been the focus area of study by many researchers. We review and analyze few of these topologies here and narrow down to the best possible variant, based on parameters such as gain, CMRR, input-referred noise, along with few others.

5.1 *OTA Based LNA with Input and Feedback Capacitors*

Figure 3 presents the design of a low noise amplifier using two-stage OTA in fully differential mode with a couple of input and feedback capacitors.

There are two input capacitors accompanied with feedback capacitors present in the design. The input capacitors help in the provision of high pass cut off frequency [9]. C_{NF1} and C_{NF2} help in the calculation of mid-band gain, accompanied by C_{PF1} and C_{PF2}, helping to calculate the low cut off frequency [9]. The two-stage OTA in a completely differential mode is an integral part of the LNA. Figure 4 depicts this fully differential OTA.

Fully differential OTAs are chosen over single-ended ones due to several reasons, with the most important one being high voltage swing, in addition to protection from environmental and external noise. Such advantages are due to the presence of both differential inputs and outputs [9]. V_{IP} and V_{ON} are applied as the differential inputs for the circuit, thereby creating V_{OP} and V_{ON} as the differential mode outputs. Miller compensation is offered by capacitors C_{C1} and C_{C2}, with the bias voltages being V_{BIAS1}, V_{BIAS2}, and V_{BIAS3} [10]. The first stage of OTA presents a differential gain stage [7]. As NMOS transistors lead to larger flicker noise, hence the input transistors used are PMOS transistors [9]. The common source amplifier using output current load is the second stage present in the OTA [9]. The desired gain of LNA alongside the low pass cut off frequency is a result of the OTA used [9].

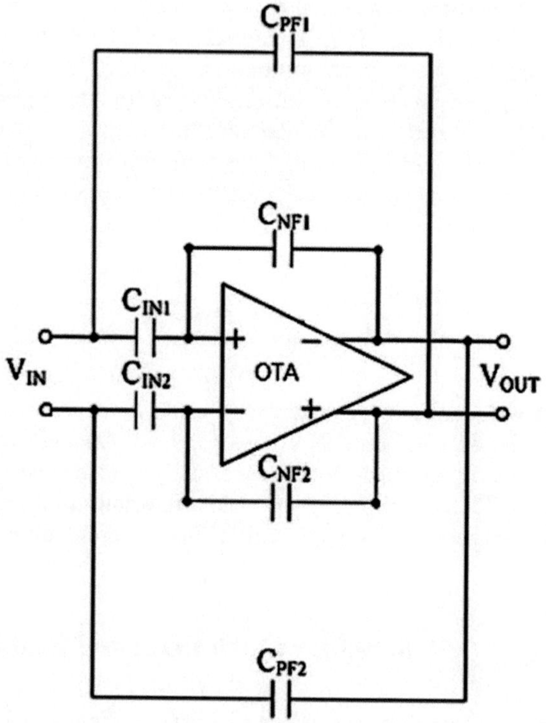

Fig. 3 LNA design using OTA with input and feedback capacitors [7]

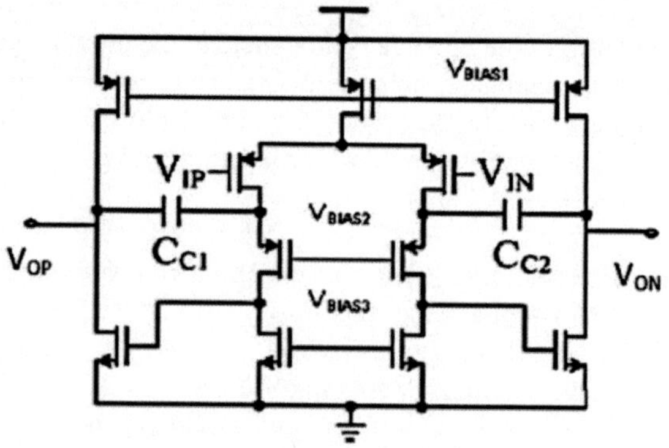

Fig. 4 Fully differential double stage OTA [7]

5.2 *OTA Based LNA with Capacitive Feedback Amplifier*

The presented LNA employs a capacitive feedback amplifier used to amplify incoming ECG signal; and the DC offset due to electrodes is also rejected [7]. The topology is depicted in Fig. 5.

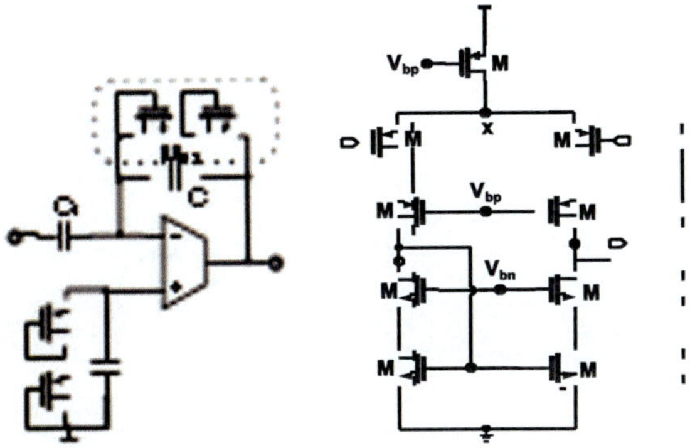

Fig. 5 Closed-loop LNA with telescopic cascode OTA [8]

C_F is the amplifier feedback capacitor and C_{IN} is used as the input capacitor [7]. M_{P1} and M_{P2} are used to create the feedback resistance. LNA input-referred noise results from OTA and feedback resistors [7]. Owing to the low bandwidth, we can neglect the feedback resistors noise with OTA dominant noise source of the LNA [7]. Topologies preferred under the OTA category are majorly folded cascode and telescopic topology because of advantages including large gain and simple dominate pole compensation nature [7]. The number of current mirrors used in the folded cascode is higher; it offers a larger input offset resulting from random device mismatch. The telescopic OTA uses a differential pair alongside a pair of transistors that cause an adjustment in current sources [7]. The gain for telescopic cascode OTA is given by (1)/{(12)/(1) + (34)/(4)} [7]. Increasing the size of the input device aiming at reducing noise is not an effective way to reduce noise. Since the sources of the cascode are in connection with the drain of another transistor, it leads to a reduction in their transconductances hence causing minimal noise contribution [7]. High values of input parasitic capacitances are offered by this circuit. The input signal at receiving end of LNA faces a difference in potential causing less signal to be available for LNA [7]. High values of input capacitor Cp affect the signal to noise ratio of transceiver owing to weak characteristics of incoming ECG signal [7]. This technique aims at reducing OTAs input-referred noise by employing a reduction in transconductance of cascode current source without any increment in power dissipation [7]. It employs a current splitting branch to decrease the current of cascode devices for any value of bias current [7].

5.3 *Open Loop OTA Based Topology*

Open-loop OTAs have also been used to record and analyze weak signals such as ECG itself, with others like neural signals as well. Figure 6 depicts the OTA Topology in open loop mode to directly amplify weak signals [10].

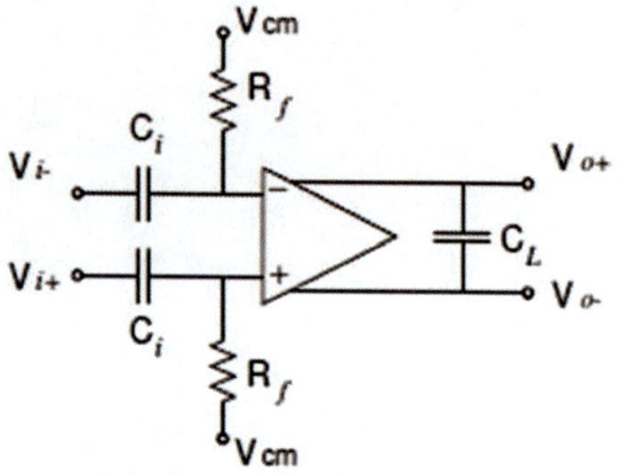

Fig. 6 Open loop topology (OLN) [10]

The input decoupling capacitor C_i in combination with resistor R_f determines the high pass pole frequency, which leads to the setting of input common-mode voltage for OTA [10]. On the other hand, OTA response is responsible for producing the low pass corner frequency [10]. The DC gain of OTA contributes to the mid-band gain causing large variations in it [10]. The input resistor creates noise, which is amplified straight to the output, and the total input-referred noise may include this value as a dominant factor [10]. Therefore the input decoupling capacitance with the mid-band gain affects the Noise Efficiency Factor (NEF) achieved by the system [10]. We can state that to achieve low values of NEF, one might require large values of input decoupling capacitors [10]. It is favorable to have β values near unity to suppress high attenuation of the signal at the input of the system [10]. If one wishes to lower values for Miller multiplication for input capacitance C_{GD}, which would thereby increase the values of parasitic capacitance C_{pi}, other topologies will have to be used.

Another topology under this class can be the capacitive network feedback topology depicted in Fig. 7.

This topology is another simple architecture in which feedback resistor in combination

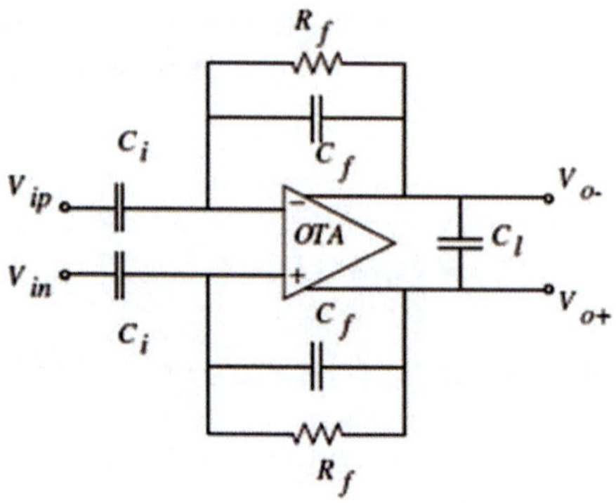

Fig. 7 Capacitive feedback topology (CFN) [9]

with a feedback capacitor are responsible for determination of high pass pole frequency;

and the response of OTA determines the low pass pole frequency [10].

5.4 AC Coupling Capacitor Feedback Operational Amplifier

This is one of the most effective structures to obtain low values of DC offset and low noise at the cost of the smaller area for both portable and implantable biomedical devices.

The design of the LNA is depicted in Fig. 8.

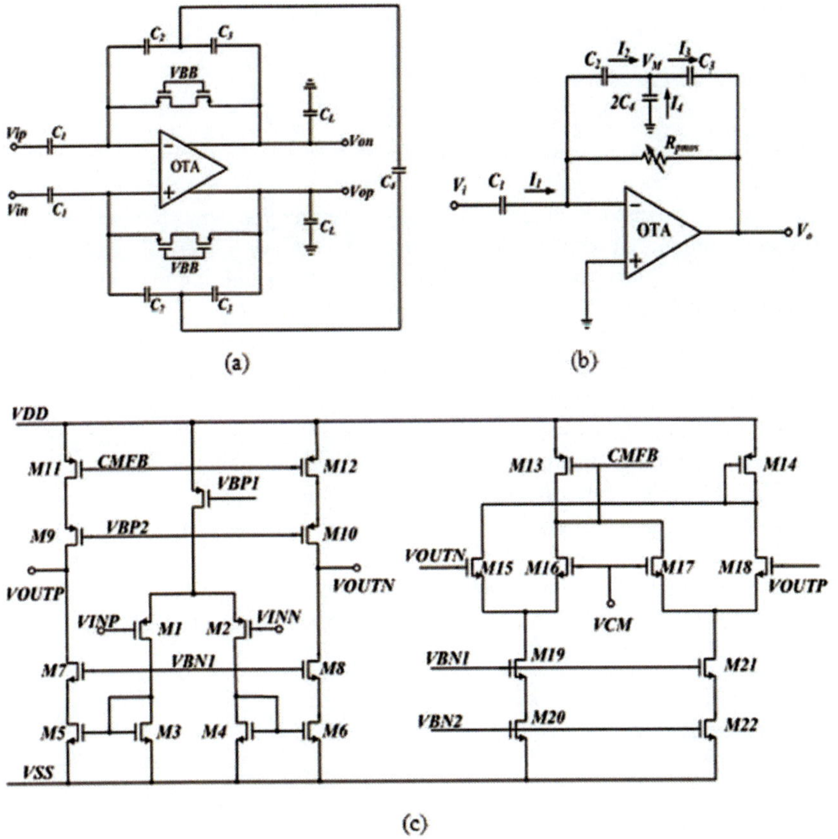

Fig. 8 Structure of LNA [10]. a T feedback preamplifier, b single-ended equivalent model of c schematic of OTA employed in LNA design

In this structure, a pseudo-resistance cell accompanied by a feedback capacitor is used to determine the high pass pole [11]. A MOSFET with its gate biased to a constant voltage (VBB) is used for the realization of the pseudo-resistance cell [11]. A variety of biosignals such as ECG can be gathered by adjustment of the amplifier's low-frequency inflection points via the constant voltage source VBB [11]. As seen in Fig. 8, the capacitor feedback loop is formed using a T-feedback structure [11]. Interconnection between nodes of capacitor C_2 and C_3 is given by capacitor C_4. Such a structure offers smaller values of coupling capacitance at input by an orders' magnitude [11]. A couple of other advantages resulting from this structure include increment in an in-band gain in addition to the decrement in feedback loops' noise [11]. The most vital component of this preamplifier is the OTA which has been shown in Fig. 8c. Also, the MOSFET pseudo-resistance has been considered as a purely passive structure [11].

5.5 *Instrumentation Amplifier Using OTA*

Cascode structures for biomedical signals is not an effective approach, therefore, cascaded structures are used in such cases, for achieving large values of gain. The initial stage of the Instrumentation Amplifier (IA) employed in this section is the low noise, low gain circuit responsible for generating common-mode voltage [12]. The stage followed by the initial stage is a high gain stage in addition to a variable gain differential amplifier [12]. Figure 9 depicts this structure.

Here, A_1, A_2, and A_3 are the OTA employed. A conventional instrumentation amplifier uses resistors which have been replaced with capacitors in this design [12]. The OTA structures used have been depicted in Fig. 10.

As visible from Fig. 10, the input of the OTA structure is constructed using PMOS transistors since they offer up to two magnitude orders of lower flicker noise as compared to NMOS transistors [13]. The sizes of the input PMOS transistors have also been taken up as large values to obtain greater decrement in values of input-referred noise [12]. These transistors also have such biasing as to operate them in the sub-threshold region for reduced values of power consumption [13]. This structure also employs MOS-Capacitors instead of using traditional capacitors to have smaller chip areas [12].

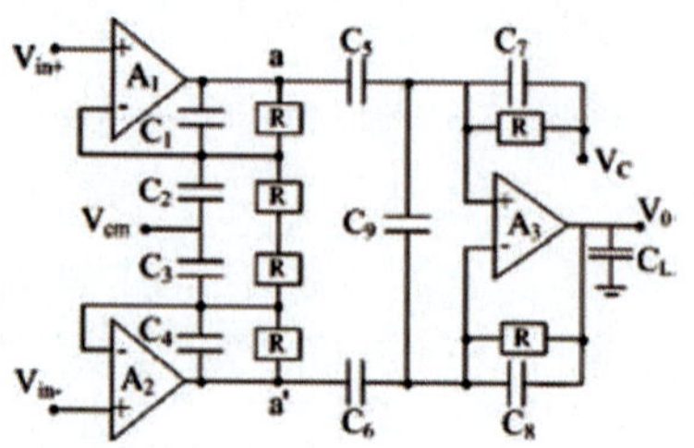

Fig. 9 Instrumentation amplifier circuit [12]

Fig. 10 Structure of OTA [11]

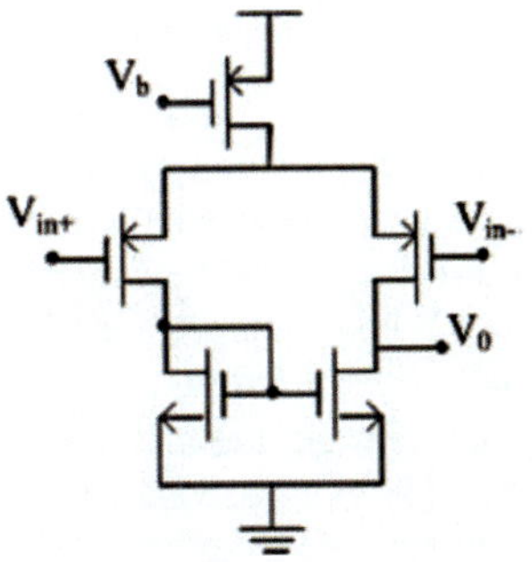

6 Comparison of LNA Architectures for ECG Analog Front End in CMOS Technology

Following the analysis of various architectures discussed so far, a comparison of the architectures is done based on various parameters including gain, power, bandwidth, input referred noise, CMRR, etc. An effort has been put to find out the best architecture suitable for ECG applications. A large gain value is desirable for the processing of ECG signals owing to their weak amplitudes. The gain should be such that the amplitude of the incoming signal can be brought to a minimal level for accurate processing. But, the most efficient architecture could be a factor of the real-time system requirements. A total of seven parameters have been considered for performance analysis of low noise amplifiers for ECG analog front end in CMOS technology and depicted in tabular form (Table 1).

Table 1 Comparison of LNA architectures for ECG analog front end in CMOS technology

Parameter	Ayaz Akram et al. [8]	Nagulapalli et al. [7]	Udupa et al. [9]	Sai Chaitanya et al. [10]	Su et al. [11]
CMOS technology (μm)	0.65	0.18	0.18	0.18	0.18
Input-Referred noise (μVrms)	–	–	–	0.96	7.8
Power (μW)	–	–	–	18.54	0.072
Gain (dB)	30	38.01	60	39.98	46.18
Supply voltage (V)	1.2	0.95	–	1.8	–
High pass cut off frequency (Hz)	–	5.08 m	–	16 k	–
Low pass cut off frequency (Hz)	–	6.99 k	–	0.2-25	–

7 Conclusion and Future Scope

The study of various LNA architectures for ECG analog front end in CMOS technology so far focused on certain parameters between which optimization is needed to achieve the best results in the field. There is a wide variety of architectures that have been developed for LNA for ECG analog front end in CMOS technology, and this discussion on these architectures can be endless. Further improvement in the circuit design and architecture, to achieve better performance, can be analyzed. A good LNA performance for ECG applications should include high gain values with minimal noise. According to our discussion and analysis above, we conclude that the open-loop topology of OTA is the best suited for LNA for ECG analog front end in CMOS technology. It offers the highest gain (60 dB) which is the best-reported value for gain so far.

References

1. Bansal M, Singh D (2019) Design and implementation of low noise amplifier in neural signal analysis. In: Gani A, Das P, Kharb L, Chahal D (eds) Information, communication and computing technology. ICICCT 2019. Communications in computer and information science, vol 1025, Springer, Berlin, pp 12–24
2. Bansal M, Singh D (2017) LNA for neural applications. Int J Comput Math Sci 5(11):74–81
3. Bhalani HV, Prabhakar NM (2015) Rudimentary study and design process of low noise amplifier at Ka band. IJ Publ 3(2):1181–1183
4. Imai Y, Tokumitsu M, Minakawa A (1991) Design and performance of low-current GaAs MMIC's for L-band front-end applications. IEEE Trans Microw Theory Tech 39(2):209–215
5. Jain S, Pathak S, Kumar B (2016) A robust design and analysis of analog front end for portable ECG acquisition system. In: 2016 IEEE region 10 humanitarian technology conference (R10-HTC), Agra, pp 1–5
6. Akram A, Javed R, Ahmad A Android based ECG monitoring system. Int J Sci Res (IJSR) ISSN: 2319-7064
7. Nagulapalli R, Hayatleh K, Barker S, Tammam AA, Yassine N, Yassine B, Ben-Esmael M A low noise amplifier suitable for biomedical recording analog front-end in 65 nm CMOS technology
8. http://www.classes.usc.edu/engr/bme/620/LectureECGNoise.pdf. Last accessed on 20-5-16)
9. Udupa SS, Sushma PS (2017) ECG analog front-end in 180 nm CMOS technology. In: 2017 international conference on intelligent computing, instrumentation and control technologies (ICICICT), Kannur, pp 327–330
10. Sai Chaitanya P, Kalesh B, Karthik N, Shruthi G Design Of Low Noise Amplifier For Seismic Signals
11. Su Y, Liu X (2016) Design of a low noise low power preamplifier used for portable biomedical signal acquisition. In: 2016 9th international congress on image and signal processing, biomedical engineering and informatics (CISP-BMEI), Datong
12. Ghamati M, Maymandi-Nejad M (2013) A low-noise low-power MOSFET only electrocardiogram amplifier. In: 2013 21st iranian conference on electrical engineering (ICEE), Mashhad, pp 1–5
13. Fay L, Misra V, Sarpeshkar R (2009) A micropower electrocardiogram amplifier. IEEE Trans Biomed Circ Syst 3(5)

Community Detection Using Graphical Relationships

Rahul, Prateek Bansal, Priyam Goel, and Purav Nayak

Abstract Community detection in social networks requires a careful effort in capturing relationships between individual nodes and identifying the different communities in the network. Contemporary research in this field has so far not considered nodes to be belonging to multiple communities. In this paper, the Community Detection Problem is solved by converting the social network graph to a low-dimensional space of features is attempted. By adopting flexibility in exploring a node's neighbourhood, overlapping communities are captured. Two kinds of node neighbourhoods based on hop distance and functionality is defined and incorporated both local and global graph representation. Our algorithm is validated by applying it to large social network datasets and visualizing the number of clusters obtained. The clusters depict the various parameters on which the communities have been separated into. The larger the number of clusters, the larger is the complexity of the communities that the dataset has been separated into. On the whole, our work presents a novel approach towards community detection by using overlapping communities in a social network graph.

Keywords Machine learning · Social network · Community detection · Random walk simulation · Feature learning

Rahul · P. Bansal (✉) · P. Goel · P. Nayak
Delhi Technological University, Delhi 110042, India
e-mail: prateekbansal1827@gmail.com

Rahul
e-mail: rahul@dtu.ac.in

P. Goel
e-mail: priyamgoel27071997@gmail.com

P. Nayak
e-mail: puravnayak@gmail.com

G. Ranganathan et al. (eds.), *Inventive Communication and Computational Technologies*, Lecture Notes in Networks and Systems 145,
https://doi.org/10.1007/978-981-15-7345-3_84

1 Introduction

A good application of data usage is to build networks of relationships between individuals. This network can be easily extended to objects. A community can be defined as a highly connected group of individuals or objects present in a network. The focus of community detection is to detect and identify these communities by establishing relationships between individuals and objects. A social network can be viewed as a graphical network with nodes and edges. Community detection in the graph would mean identifying relationships between nodes, in order to identify communities. Community detection has come to use in several domains: in creating recommendation systems on the basis of co-purchasing habits on retailer websites, in identifying influencers on a social network, and many others.

Through our research, the Community Detection Problem is aimed to solve by using the random walk technique, taking into consideration that a node could belong to multiple communities. This helps us in achieving a lot of benefits such as making the exploration and sampling unbiased and flexible and making the whole process time-efficient and space-efficient, which means contributing overall to the scalability of our proposed method. A scalable model is presented with the application of a novel approach: Overlap Node Community.

The Community Detection Problem is solved by implementing our method on the dataset taken from BlogCatalog. Our dataset is a group of bloggers, and the task is to identify the various communities present in this group. The solution to the problem can be identified in two steps:

1. The primary step involves conversion of the dataset, which is present in a graphical format, to a comma-separated value (csv) format, by using the Overlap Node Community algorithm.
2. In the second step, the clustering algorithm is used in order to detect communities.

Conversion of graphical formats to csv will require exploration of the graph. Exploration problems have had two important classical solutions: The breadth-first sampling and exploration strategy (BFS) and the depth-first sampling and exploration strategy (DFS). In most scenarios, the nodes and their neighbourhoods are unable to be completely explored by using BFS or DFS alone. An effective strategy would then be to explore nodes and their neighbourhoods by using a combination of BFS and DFS. Thus, in our research, the Overlap Node Community algorithm for graph exploration is used which employs both BFS and DFS, and alternates between the two strategies with the help of two parameters.

The two parameters s and h are used. The parameter s is known as the 'structural equivalence parameter'. While performing a walk on the graph, suppose the end of a neighbourhood has been reached, then the probability of returning to an already explored node is explained through s. The parameter h is known as the 'homophily parameter' and describes the depth on which the exploration is being performed. The combination of s and h parameters and a search bias value helps decide which between BFS and DFS will be used at that particular stage of exploration. The Overlap Node Community is an ideal trade-off between the two classical strategies.

Our approach using the Overlap Node Community algorithm has been divided into three stages.

I. The first stage describes the preprocessing of transition probabilities. The transition probability for a node describes the probability of making a transition to that node from the current node.
II. The second stage involves simulation of walks. In this stage, the exploration length and the number of times exploration to be performed are defined.
III. The third and final stages optimize these walks by using Stochastic Gradient Descent (SGD) method, so that optimal embeddings are generated using the Overlap Node Community algorithm.

Post the three stages of conversion, the embeddings generated in the first step fed as input to our clustering algorithm, in order to obtain optimal clusters or families. Upon visualizing the results obtained, able to detect and identify communities present in the graph.

2 Related Work

Community Detection Problems have seen a variety of approaches. The initial research in the field of community detection in social networks employed modularity which means maximizing the number of relationships (or edges) inside communities and minimizing edges between communities. Newman introduced the first greedy algorithm based on modularity in 2004 [1].

Another method for the purpose of community detection introduced the concept of random walks. These walks are simulated to identify communities. The premise of this method is that a random walker would most likely stay inside heavily connected areas of the graph. The walktrap algorithm given by Pascal Pons and Matthieu Latapy in 2005 [2] derived its basis from this idea.

What can be considered as a limitation to these algorithms is that they only consider one community per node. This hypothesis could prove wrong for some cases. Several methods have been proposed to take into consideration the idea of overlapping communities where a node can belong to multiple communities. CFinder, a freeware for detecting overlapping communities, has made the clique percolation method popular [3]. This method assumes that a network is composed of cliques, where cliques are subsets of nodes such that every two distinct nodes in a clique will be adjacent to each other. In 2013, Jaewon Yang and Jure Leskovec proposed a community detection algorithm called BigClam [4] in which the authors underscore that community overlaps are denser than the communities themselves, and that the number of communities two nodes share can form the basis for deciding whether they are connected or not.

The more recent literature in community detection focuses on Label Propagation methods which rely on the network structure alone to identify the different communities present in a social network. The Balanced Link Density Label Propagation algorithm given by Ehsan Jokar and Mohammad Mosleh in 2019 [5] tries to improve upon the original Label Propagation methods by substituting random Label Propagation by the use of sensitivity parameter for alternating between different network structures. However, this method does not take into account the overlapping community structures present in a network and the varying relationships that nodes can have with each other based on the structure or functionalities.

Feature extraction and engineering have various applications in machine learning. Conventionally, node features are generated using techniques of feature extraction that generally consist of hand-engineered features on network properties basis [6]. On the contrary, the aim of our research is automation of the entire process by extracting features and modelling it as a representation learning problem without the use of hand-crafted features.

In dimensionality reduction, the properties of different graph matrix representations are used. Some linear, such as PCA, and nonlinear, such as IsoMap, techniques have been proposed for dimensionality reduction [7, 8]. The major drawbacks of these techniques are poor computational as well as statistical performance. For example, eigendecomposition, which is performed on the data matrix, is a fairly computationally expensive operation. Approximations are done to increase efficiency, but they degrade the quality of the solution. Therefore, these methods are not advised due to scalability factor. These methods also optimize the solution in order to satisfy the objectives. This could mean making presumptions about the relationship between the prediction task and the structure of the network. The method of spectral clustering makes a strong and unreasonable assumption that for community detection, graph cuts would be necessary [9].

Latest NLP-based representational learning led to various prospects for feature extraction of discrete articles like words. The Skip-gram model [10] enhances the neighbourhood-preserving likelihood objective to learn the continuous feature word representation. The algorithm examines the words present in a document, and when each time it encounters a new word for the first time, it embeds it in a fashion so as to make it possible to predict surrounding words. These feature representations of words are modelled by optimizing the likelihood objective with negative sampling using SGD [11]. The objective of Skip-gram is on the basis of the distributional hypothesis that asserts that words in similar state of affairs happen to have similar meanings [10].

Some models identify a network as a ‘document’ [12]. A document is an ordered word sequence; similarly node sequences can be sampled from the elemental network, and the network can be transformed into an ordered node sequence. There are various node sampling strategies that result in different representations of learned features. However, there is an absence of a sampling strategy which can be flexible between different sampling strategies and is not tied to a specific one. Our proposed method overcomes this rigidity of approach and uses parameters that let it tune the search space making it flexible.

3 Algorithmic Framework

In this section, formulas are developed for feature learning in a network and model it as a maximum likelihood optimization problem and take a network in the form of a graph $G = (V, E, W)$. Our approach applies to any graph, both directed and undirected, weighted and unweighted. The network is mapped to feature representation using a function as follows—$f: V \rightarrow R_d$—where 'd' is the parameter specifying the number of different dimensions in the representation and f is a matrix with $|V| \times d$ parameters. For every node in the network, we define a network neighbourhood, which is a subset of the set of vertices V.

$$\max_f \sum_{u \in V} \log \Pr(N_d(u)) | f(u)) \tag{1}$$

An assumption is made that the probability of observing one node in the neighbourhood of a source is independent of the probability of observing a different node in the neighbourhood of the same source node.

$$\text{Log} \Pr(N_d(u)) | f(u)) = \prod_{i=1}^{n} \Pr(n_i | f(u) \ (N_i \in N_d(u)) \tag{2}$$

$$\Pr(n_i | f(u)) = \left(e^{(f(N_i) \cdot f(u))}\right) / \left(\sum v \in V e^{(f(v) \cdot f(u))}\right) \tag{3}$$

$$\max_f \sum_{u \in V} \left[-\log Z_u + \sum_{N_i \in N_d(u)} f(N_i) \cdot f(u) \right] \tag{4}$$

Two nodes are symmetric to each other, therefore used a dot product of the pairs as parameters to a softmax unit which formulates the conditional likelihood of nodes being in the neighbourhood of each other. For solving the equation stated, needed to calculate per-node partition given by Z_u is summation of exponential $(f(u) \cdot f(v))$ over $v \in V$. The calculation of this value is costly when done for huge, massive networks with several interconnections and therefore is approximated using the method of negative sampling. Networks are a nonlinear structure in which each node has a different view and keep on changing our sampling strategy, the structure of the neighbourhood and the part of the network under consideration changes. Therefore, a procedure is proposed which is random in nature and samples different neighbourhoods of the source nodes 'u'. It has been arrived at by switching between the classical search strategies described in the following sub-section.

3.1 Classic Search Strategies

The problem comprises generating neighbourhood sets N_s of 'k' nodes and to do so, and have two extreme graph traversal sampling techniques, as described below:

Breadth-first Sampling (BFS)—In this technique, the nodes which are immediate neighbours of a source node 'u' are sampled first, before moving onto further depths from the source node which are at increasing distances from it.

Depth-first Sampling (DFS)—In the DFS sampling approach, the nodes at increasing distance from the source node are traversed, and these nodes consist of the neighbourhood N_s.

Structural Equivalence and Homophily—Two terms are used to define the neighbourhoods of a source node u, namely structural equivalence and homophily. Homophily implies keeping those nodes in the neighbourhood of a source node which are highly interconnected and are part of similar network structures. Structural equivalence states that two nodes which play similar structural roles in networks are known to be structurally equivalent. For example, two nodes which act as hubs or bridges in their respective surroundings would be structurally equivalent to each other. The unique thing about structural equivalence is that unlike homophily, two structurally equivalent nodes can be physically far apart from each other.

The two sampling strategies BFS and DFS are capable of generating neighbourhoods which are closely related to one of structural equivalence or homophily. BFS produces neighbourhoods that closely resemble structural equivalence because of its property of examining the immediate neighbours. By looking at the immediate neighbours, the role played by a node can be determined and structural equivalence can be reached at. Further, a microscopic view is obtained by BFS which reduces the variance of the distribution of nodes which are just a move away from the source node. In DFS, a macro-view of the neighbourhood is obtained, which helps in generating the neighbourhoods based on homophily. However, the sampled nodes can be very far away from the source, and thus, the dependencies that exist between the nodes can be difficult and complex to quantify. Further, there is a problem of high variance in DFS. Both the extreme approaches are flawed in the sense that structural equivalence and homophily are mutually inclusive, and two nodes can be homophilic as well as structurally equivalent to each other simultaneously. Therefore, an approach is presented which employs a combination of both BFS and DFS approaches to sample nodes and generate embeddings.

3.2 Proposed Methodology

Based on the observations made, an exploration strategy is developed which lets us incorporate both structural equivalence and homophily by interpolating between the breadth-first and depth-first approaches and do this by exploring neighbourhoods in a fashion that is a combination of both the aforementioned approaches. A source node, let's say, u is given. For this source, a walk of a fixed length l is run. For $i = 0, 1, 2, 3, \ldots, k$, c_i is the ith node in the walk which starts with $c_0 = u$. The decision of which nodes to traverse in the walk is made through the distribution:

$$P(c_i = x|c_{i-1} = V) = \pi vx/Z \text{ if}(v, x) \in E \qquad 0 \quad \text{Otherwise} \tag{5}$$

Here, π_{vx} is the probability of making a transition from node v to node x (unnormalized). Z is the constant of normalization.

$$a_{5h}(t, X) = 1/s \text{ if } d_{tx=0} \qquad 1 \text{ if } d_{tx=1} \qquad 1/h \text{ if } d_{tx=2} \tag{6}$$

Structural equivalence parameter, '*s*': It determines the probability of immediately re-exploring a node during the walk. If set to a high value (>max(1,*s*)), it is ensured that the probability of sampling an already visited node is very low. It is done only if the node to be sampled in the walk has no other neighbour. This makes the walk moderately exploratory in nature and avoids visiting the same node in 2-hops. If, however, h is kept low (<min(1,*s*)), it leads to backtracking of a step and makes the walk close to the source node.

Homophily parameter, 'h': The homophily parameter allows us to decide whether a local view or global view of the network is wanted. If the parameter is set as greater than 1, then the walk revolves around the local nodes, whereas setting the parameter to a low value less than 1 allows the walk to explore nodes at considerable distance from the source. This makes it resemble DFS and take care of the homophily property (Fig. 1).

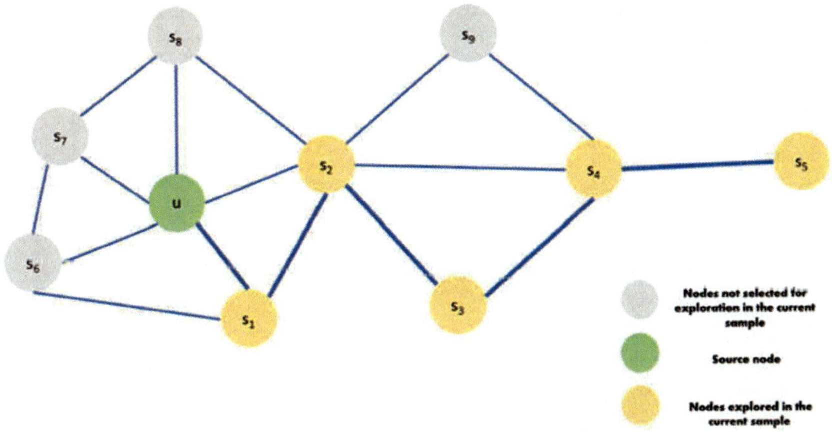

Fig. 1 A particular sampling and exploration of nodes $\{u, s_1, s_2, s_3, s_4, s_5\}$

The random walks simulated during our method have a lot of benefits when compared to the extreme BFS/DFS strategies. The random walks have low space and time complexities. The algorithm takes $O(|E|)$ space for storing immediate neighbours and $O(a^2|V|)$ space for storing the interconnections between neighbours. Here, the parameter 'a' is the average degree of a node in the graph. The space complexity is being $O(|E|)$ for storing immediate neighbours and $O(a^2|V|)$ for storing the interconnections between neighbours where a is the average degree of the graph (usually small for real world networks). The random walk procedure is more efficient because it lets us reuse samples from different source nodes. For example, sampling a walk $\{u, s_1, s_2, s_3, s_4, s_5\}$ of length 6 can result in $N_s(u) = \{s_1, s_2, s_3\}$, $N_s(s_1) = \{s_2, s_3, s_4\}$. Here the nodes s_2 and s_3 are reused for the two distinct source nodes 'u' and s_1.

3.3 *Overlap Node Community Algorithm*

The entire algorithm is divided into three phases:

1. Preprocessing—Computing transition probabilities
2. Simulation of random walks
3. Optimization using Stochastic Gradient Descent (Fig. 2).

For implementing the algorithm depicted in the flowchart, first initialize the hyper-parameters. The number of dimensions has been set to 128, the walk length has been fixed at 80, and 10 walks per source has been taken as the default value. These values have been obtained considering the large-scale nature of the social network graph. Given a graph, 10,312 nodes and 333,983 edges have the average degree of the graph close to 33. By considering a walk length of 80, make sure that the nodes are sampled in relatively fewer number of iterations and the impact on performance is positive. Similarly, for the number of dimensions, if the number of dimensions is beyond 150, the performance tends to saturate. The transition probabilities for each node are calculated. As already explained, it is just the probability of jumping onto a

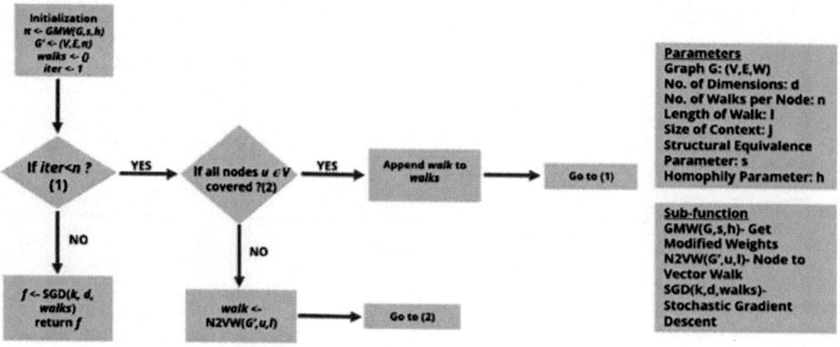

Fig. 2 Learn features function of the overlap node community algorithm

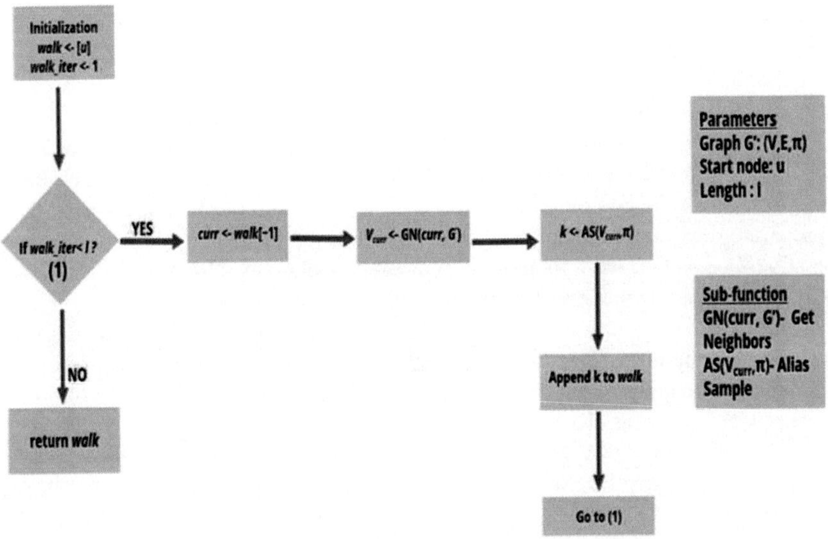

Fig. 3 Overlap node community—simulating walks and generating embeddings

node given that are present at a source node, *u*. At every node that is traversed, transition probabilities are computed and sampling is performed. The precomputation of transition probabilities allows us to achieve sampling during simulation in $O(1)$ time.

Start simulating the walk and keep on checking for the length of walk. If the maximum length has not been achieved and get the neighbours of the node at question, the precomputed transition probabilities, and add the nodes to the current walk. Also added the walk generated to the list of walks. This task is performed till reached the maximum number of iterations (Fig. 3).

The entire process is made unbiased by simulating n walks of length l, each time selecting a different node of the network as a source. Finally, as an optimization step, Stochastic Gradient Descent is applied to the generated walks which finally give us our feature representation as a *d*-dimensional vector.

This feature representation is finally employed for the purpose of community detection, the experimental setup of which is described in the next sub-section along with the results.

4 Experiment

The embeddings generated from our algorithm are applied to a community detection task of identifying the different families present in a dataset. The BlogCatalog dataset is used which consists of 10,312 nodes and 333,983 edges. The different families of bloggers present in the dataset are identified by using the metadata provided by the bloggers. A range of clusters is derived for the dataset and then visualizes the results on a graph. In order to verify whether the number of clusters is optimal or not, a weighted loss function is used which minimizes the variance of the datasets.

The number of clusters is increased and is gradually able to visualize the relationships that the bloggers have with each other. As a part of the experiment, have tried to separate the communities into $n = 2, 3, 4, 6$ and 8 clusters. For $n = 2$, able to separate the blogs into two categories based on the sentiments, i.e. positive and negative. For $n = 3$, incorporated the influence of bloggers along with sentiments as parameters on which the communities of bloggers have been separated. For $n = 4$, move one step further and identify the families of bloggers based on the tags on blogs. As the dimensions are increased, move gradually towards finer levels of granularity by capturing relationships closer to the individual topics of the blogs.

Technical Analysis: This is in alignment with the cluster optimality graph given in Fig. 4a which uses a weighted loss function to calculate the variance of the graph. For $n = 2$ clusters, the loss value is close to 5000, while for $n = 3$ clusters, it reduces to 3500. From $n = 4$ to $n = 6$, the reduction in loss value is of 500 units, from 3000 to 2500, respectively. From $n = 6$ to $n = 8$, this loss reduction is less than 300 units. From $n = 8$ to $n = 10$, there is hardly any reduction in the loss value indicating no further separation of communities is beneficial, and any further separation would mean using trivial and non-essential parameters for separation into communities. Further, our approach takes seconds to run on a large dataset such as the one used, thus confirming the scalability of our approach.

Parameter Sensitivity: Our proposed Overlap Node Community approach involves lots of parameters. The selection of parameters affects the performance of our approach on the BlogCatalog dataset examined. The method achieves optimal performance for low 's' and 'h' values. The low 's' value promoting outward exploration is balanced by a low 'h' value ensuring homophilic tendencies. It makes sure that the walk remains optimal. Further, the number of dimensions 'd', the number of walks and the walk length affect the performance of the model and observe that as the number of dimensions reaches 150, the performance begins to assume a constant value and diminishes if dimensions are increased any further. Also, given the large number of nodes and sampling size, the performance increases as increased the number of walks and walk length of each source node. This increase in performance starts to diminish as go beyond the threshold of 100 walk length.

5 Conclusion

This paper allowed us to combine the benefits of both strategies: BFS and DFS while performing random walks on the graph. BFS is known to explore local neighbourhoods and, hence, can be used when analysing structural equivalence in networks where the local structure of nodes is important. DFS, whereas, is known for free

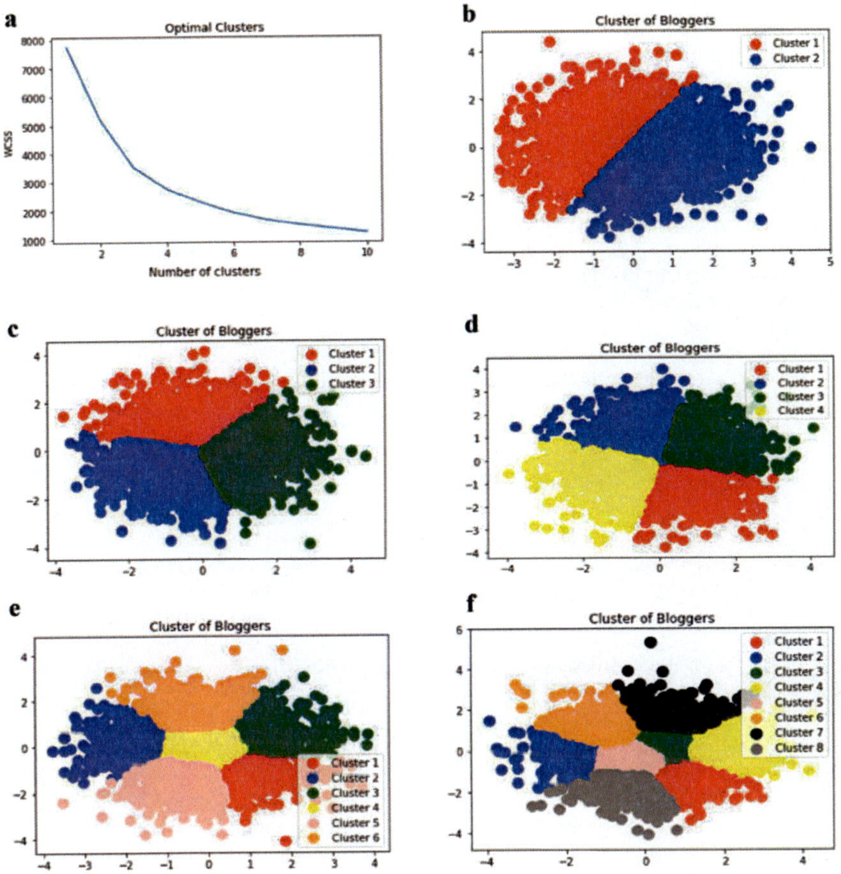

Fig. 4 **a** Optimal number of clusters using overlap node community, **b** visualization using two clusters, **c** visualization using three clusters, **d** visualization using four clusters, **e** visualization using six clusters, **f** visualization using eight clusters

exploration of network neighbourhoods. Thus, DFS finds its application in discovering homophilous communities. A strategy called LINE [13] samples nodes and optimizes the likelihood over hop-1 and hop-2 neighbours. Although it is easy to characterize such an exploration, the technique loses its flexibility when exploring nodes at further depths.

The algorithm in our research is flexible as well as controllable as it uses the structural equivalence parameter '*s*', the homophily parameter '*h*' and a search bias to switch between BFS and DFS. While these parameters can be intuitively interpreted, the best results are obtained when these are learned from available data. Furthermore, our research, with the help of its adaptable clustering algorithm, takes account of the fact that a node could belong to multiple communities. This combination of Overlap Node Community and Clustering is scalable and robust to perturbations.

References

1. Clauset A, Newman MEJ, Moore C (2004) Finding community structure in very large networks. Phys Rev E 70(6)
2. Pons P, Latapy M (2006) Computing communities in large networks using random walks. J Graph Algorithms Appl 10(2):191–218
3. Gao W, Wong KF, Xia Y, Xu R (2006) Clique percolation method for finding naturally cohesive and overlapping document clusters, pp 97–108
4. Yang J, Leskovec J (2015) Defining and evaluating network communities based on ground-truth. Knowl Inf Syst 42:181–213
5. Jokar E, Mosleh M (2018) Community detection in social networks based on improved label propagation algorithm and balanced link density. Phys Lett A 383:718–727
6. Henderson K, Gallagher B, Li L, Akoglu L, Rad TE, Tong H, Faloutsos C (2011) It's who you know: graph mining using recursive structural features, KDD
7. Belkin M, Niyogi P (2001) Laplacian eigenmaps and spectral techniques for embedding and clustering. NIPS
8. Tenenbaum JB, Silva VD, Langford JC (2000) A global geometric framework for nonlinear dimensionality reduction. Science 290:2319–2323
9. Jia H, Ding S, Xu X et al (2014) The latest research progress on spectral clustering. Neural Comput Appl 24:1477–1486
10. Bamler R, Mandt S (2017) Dynamic word embeddings via skip-gram filtering
11. Tschoepe M (2019) Beyond SGD: recent improvements of gradient descent methods
12. Perozzi B, Rfou RA, Skiena S (2014) DeepWalk: online learning of social representations. KDD
13. Tang J, Qu M, Wang M, Zhang M, Yan J, Mei Q (2015) LINE: large-scale information network embedding, pp 1067–1077

Author Index

G. Ranganathan et al. (eds.), *Inventive Communication and Computational Technologies*, Lecture Notes in Networks and Systems 145,
https://doi.org/10.1007/978-981-15-7345-3